RESEARCH AND EVALUATION FOR BUSINESS

MARILYN K. PELOSI, PH.D.
Western New England College
Springfield, MA

THERESA M. SANDIFER, PH.D.
Southern Connecticut State University

UMA SEKARAN, PH.D.
Southern Illinois University at Carbondale

JOHN WILEY & SONS, INC.

ACQUISITION EDITOR	Deborah Berridge
DIRECTOR OF NEW MARKET DEVELOPMENT	Karen Allman
PRODUCTION SERVICES MANAGER	Jeanine Furino
COVER DESIGNER	Karin Gerdes Kincheloe
PRODUCTION MANAGEMENT SERVICES	Susan L. Reiland
TEXT DESIGNER	Nancy Field
COVER ILLUSTRATION	Norm Christiansen

This book was set in New Baskerville by Progressive Information Technologies and printed and bound by Courier Companies. The cover was printed by Lehigh Press.

This book is printed on acid-free paper.

Library of Congress Cataloging-in-Publication Data
Pelosi, Marilyn K.
   Research and evaluation for business / Marilyn K. Pelosi, Theresa M. Sandifer, Uma Sekaran.
     p. cm
   Includes bibliographical references and index.
   ISBN 0-471-39088-7 (cloth/CD-ROM ; alk. paper)
   1. Commercial statistics. 2. Business—Research. I. Sandifer, Theresa M.
   II. Sekaran, Uma. III. Title
   HF1017.P373 2001
   001.4'024'338—dc21   00-036497

ISBN 0-471-39088-7

Printed in the United States of America

10 9 8 7 6 5 4

PREFACE

OUR GOALS IN WRITING THIS BOOK

Welcome to *Research and Evaluation for Business!* This text was created specifically for your course at the University of Phoenix.

The goal of the text is simple—to introduce statistics in the context of business research. We believe that statistics is best understood by doing it and using it. Introducing business statistics in the context of research will allow you a greater understanding of the role of statistics and why one might want to use it.

Thus, the pedagogical focus of *Research and Evaluation for Business* is data driven, emphasizing statistical reasoning, interpretation and decision making. We want to teach the "practice" of statistics; therefore, the emphasis is on comparison and interpretation rather than rote calculation. To support this approach we include discussions of practice as opposed to theory. For example, we discuss when it is OK to be in violation of an assumption and what to watch out for in applying techniques. In order to focus on decision making, the use of the computer is emphasized to keep students from getting bogged down in the details of the calculations.

All the cases found in *Research and Evaluation for Business* are based on real business situations and data that are relevant. Most chapters begin with a motivating example based on real-world data. This example is then threaded throughout the chapter and to a lesser extent throughout the book. The data used in many of the examples and exercises are based on our consulting experiences. In order to protect the confidentiality of the companies involved, the data may be rescaled or altered, which does not hinder the usefulness of the data from a pedagogical perspective.

We have adopted a conversational tone so that students may learn the material using vocabulary that is familiar to them. The explanations have an *informal* flavor in order to allow the student to see that the techniques make logical sense. When necessary, we deviate from "pure" mathematics in the interest of understanding, being sure to note when this occurs. The student should see *why* and *how* the technique works. With this kind of understanding, students will know *what* technique to use and *when* to use it.

This book is intended for use in a two-semester introductory business research statistics course. Each workshop is marked with a title page listing the chapters to be covered in that workshop.

KEY FEATURES OF OUR ACTIVITY-BASED LEARNING APPROACH

In writing this book, we reflected on our many years of teaching experience and examined how people learn new things. The pedagogical features which follow represent the activity-based model we have developed over the years in our own class-rooms. We have found that students . . .

• Learn by Trying

Try It Now! Exercises are checkpoints embedded in the text, with spaces provided for students to do them, making sure that they understand the basics before moving on.

 **TRY IT NOW!**

The Glue Company
Selecting a Simple Random Sample

Select a sample of 5 tubes of glue for the glue company. You can assume that each tube of glue has a 5-digit ID number, which the company uses to track its inventory.

Answers are right there (upside down) so that students will have immediate feedback.

• Learn by Practicing

Learning It! Exercises allow students to practice the basics of computation and include all the techniques covered in the chapter.

2.2.5 Exercises—Learning It!

2.1 The President of the United States wishes to see how popular he/she is after 2 years in office.
  (a) What is the population of interest?
  (b) Identify which of the reasons for taking a sample (listed on page 24) apply in this case. (There are more than one.)
  (c) Identify two variables or characteristics of the members of this population that you may wish to study.

Thinking About It! Exercises emphasize critical thought and ask students to interpret the results of statistical techniques and make decisions based on their analysis.

Thinking About It!

1.1 We have said that quality and statistics go hand in hand. What product or service have you used recently that did not have the quality you thought it should have?
  (a) Describe the product or service and the area in which it was lacking in quality.
  (b) What corrective action would you recommend?

1.2 Describe a situation in which you had a paradigm shift or change.

Doing It! Exercises use large real-world data sets and ask students to perform the statistical analyses and interpret them to solve a real problem. As students progress in their statistical thinking, these exercises will help them to make judgments about "what to do next." The data sets for these exercises are provided on the accompanying CD-ROM in Excel v 4.0 worksheet format.

Doing It!

6.46 The company investigating the golf balls is not satisfied with the limited analysis that it has done. The managers have collected a good deal of data, but they are not sure how to look at them and interpret the output. They decide to hire you to help them understand what the golf balls are doing and how they compare to each other. In addition to measures of the balls' performance, such as the variable *Carry*, the managers know that other factors, both internal (ball-related) and external (environment-related), could affect performance. They tested 36 of each type of ball at three different times using a machine to launch the balls. Data were recorded on 14 different variables. A portion of the data is shown below:

Ball	*Model*	*S1*	*S2*	*S3*	*Wgt*	*Dw*	*Dd*	*Head*	*Temp*	*Carry*	*Tot Dist*	*Date*	*Time*
1	M1	81	81	82	45.3	0.145	0.0110	686	77	257	270	8/20	8:15
2	M1	83	83	84	45.2	0.151	0.0111	688	77	255	267	8/20	8:15
3	M1	81	82	84	45.2	0.145	0.0105	687	77	256	267	8/20	8:15
4	M1	81	81	83	45.3	0.144	0.0117	688	77	255	271	8/20	8:15
5	M1	83	81	82	45.5	0.146	0.0108	687	77	255	268	8/20	8:15

• Learn by Using the Computer

We have chosen Microsoft Excel because it is the leading analytical software used in the business world. There is a greater chance that students will practice statistical thinking if we show them how to do statistics in a software package that is most likely to be on their desktop at home or at work.

6.7.1 Calculating Summary Statistics in Excel

Suppose that we want to calculate a set of summary statistics for the Golf Ball data. Figure 6.9 shows a portion of that data in an Excel worksheet.

The Data Analysis ToolPak has a function that creates a set of summary statistics for a set of data. To access this function, select **Data Analysis** from the **Tools** menu and choose **Descriptive Statistics** from the list of Analysis tools. The dialog box is

	A	B	I	J	K	L	M	N
1	Ball#	Model #	Head	Temp	Carry	Tot Dist	Date	Time
2	1	M1	686	77	257	270	8/20	8:15
3	2	M1	688	77	255	267	8/20	8:15
4	3	M1	687	77	256	267	8/20	8:15
5	4	M1	688	77	255	271	8/20	8:15
6	5	M1	687	77	255	268	8/20	8:15
7	6	M1	687	77	256	267	8/20	8:15
8	7	M1	687	77	255	264	8/20	8:15
9	8	M1	690	78	258	269	8/20	8:15
10	9	M1	686	78	252	257	8/20	8:15
11	10	M1	687	78	256	268	8/20	8:15
12	11	M1	687	78	253	263	8/20	8:15

FIGURE 6.9 The golf ball data

When appropriate, the final section of each chapter describes how to implement the techniques taught in the chapter in Excel. In some instances Excel has built-in features to do the necessary statistical calculations. The sections at the end of each chapter walk students through each step in using these features. While Excel does not perform all of the statistical functions found in the text, we have developed some macros to do these calculations. Macros are included on the accompanying CD-ROM and the text explains how to use them.

If students are not familiar with using Excel, we have included a short introduction to the basics in Appendix B.

• Learn by Discovery

In the process of trying things, we often discover ideas that eventually lead to understanding. Within most of the chapters students will find *Discovery Exercises*. These exercises are written to achieve a high level of student understanding by directing their line of thinking, leading to an "aha" discovery of a key concept. They are designed to help students see that statistical thinking within research is logical and that the formulas make sense when students discover where they came from and see why they work.

Discovery Exercise 2.2
INTRODUCTION TO SAMPLING

Suppose the data shown below represent an entire population. They show the number of people in 50 families living in a small college town in New England. (If you did Discovery Exercise 2.1 then you will recognize this as the same data set.) Now a 2-digit ID number has also been included.

New England families: large amount of variability
Average number of people in 50 families: 4.50

ID: 01	1	ID: 02	4	ID: 03	5	ID: 04	7	ID: 05	8
ID: 06	3	ID: 07	9	ID: 08	8	ID: 09	8	ID: 10	8
ID: 11	4	ID: 12	9	ID: 13	9	ID: 14	1	ID: 15	6
ID: 16	4	ID: 17	1	ID: 18	9	ID: 19	9	ID: 20	7
ID: 21	8	ID: 22	2	ID: 23	3	ID: 24	1	ID: 25	9
ID: 26	1	ID: 27	7	ID: 28	5	ID: 29	1	ID: 30	1
ID: 31	1	ID: 32	6	ID: 33	8	ID: 34	2	ID: 35	9
ID: 36	4	ID: 37	1	ID: 38	1	ID: 39	1	ID: 40	3
ID: 41	4	ID: 42	2	ID: 43	4	ID: 44	9	ID: 45	4
ID: 46	1	ID: 47	3	ID: 48	8	ID: 49	1	ID: 50	1

Step 1: Select a sample of 5 numbers from this population. Use the table of random numbers to do this. Record your sample in the table at the top of page 39.

Several of the *Discovery Exercises* are designed to help students apply the techniques and interpret the results in terms of making a business decision. Most of these exercises require them to *write* a memo or short business report to explain their decisions and recommendations.

Most chapters end with summary tables of key terms and formulas. These features are useful when reviewing the material for quizzes and/or exams. Page references are provided so it is easy to return to the more detailed discussion in the chapter.

Key Terms

Term	Definition	Page reference
Biased sample	A **biased sample** is a sample that does not represent the population.	34
Census	A **census** is a study of the population.	24
Continuous data	**Continuous data** are data that can take on any one of an infinite number of possible values over an interval on the number line.	43
Descriptive statistics	**Tools of descriptive statistics** allow you to summarize the data.	45
Discrete data	**Discrete data** are data that can take on only certain values. These values are often integers or whole numbers.	43
Inference	An **inference** is a deduction or a conclusion.	46

CONTENTS: RES 341

CONTENTS: RES 342

RES 341: RESEARCH & EVALUATION I

WORKSHOP 1

INTRODUCTION: THE ROLE OF STATISTICAL THINKING IN BUSINESS RESEARCH

QUALITY PRODUCTS

If you asked someone to describe the quality of a product made in Japan in the 1950s, what do you think the response would be? Most likely the person would tell you that the Japanese product made in the 1950s was of low quality, not reliable, and probably junk. Now let's roll forward to the 1990s and ask the same question. The person would most likely tell you that the Japanese product is of high quality, reliable, and of great value.

What happened between 1950 and 1990? During this time period Japanese industries started paying increasingly more attention to the quality of their products. At the same time American companies were enjoying the postwar boom and paying little, if any, attention to quality. Interestingly enough, it was an American statistician named W. Edwards Deming who helped the Japanese focus on quality. Deming tried to get American manufacturers to listen to him but they were not interested at the time. Today, American companies, both service industries and manufacturing concerns, are paying close attention to quality in order to compete internationally.

1.1 CHAPTER OBJECTIVES

You may be wondering why a story about product quality is used as an introduction to a book on statistics. Students often feel that taking a course in statistics causes unnecessary pain and that they will never again use the statistical techniques they learn. The objective of this book is to refute both of these points: The course need not be painful and you will certainly be asked to research problems by collecting and analyzing data in order to make informed business decisions.

The objective of this chapter is to start you down this path. Specifically this chapter covers the following topics:

- Dispelling the Myths About Research and Statistics
- What Managers Should Know About Research and Statistics
- Statistical Thinking—A New Paradigm for Management
- Some Situations That Call for Research and Statistical Thinking
- Types of Research
- Steps in Conducting Business Research
- Key Components of Research and Statistical Thinking
- Organization of This Book

1.2 DISPELLING THE MYTHS ABOUT RESEARCH AND STATISTICS

We should first dispel some of the common myths about research and statistics. Listed below are three myths we have run across regularly:

Myth 1: *"If I had one hour left to live, I would choose to live it in statistics class because it would seem to last forever!"*

A student's lament

Myth 2: *"There are three kinds of lies—lies, damned lies, and statistics."*

Benjamin Disraeli

Myth 3: *"If it moves, it's biology; if it changes color, it's chemistry; if it breaks, it's physics; if it puts you to sleep, it's statistics."*

Bob Hogg, University of Iowa

If you are like most people you can relate to one of these myths about the "S-word," *statistics*, or *sadistics* as some people refer to it. Statistics is boring and not useful! This book will lead you to another view of the dreaded S-word—one that sees statistics not as a sleeping pill but as a way in which to view all sorts of exciting, amazing, and valuable things.

 TRY IT NOW!

Myths and Fears
Identifying Some of Your Own

Be honest. Write down your myths and fears about this subject right now. Get them out in the open so you can deal with them directly and put them behind you. To get you started a common student fear is listed at the top of the next page.

- *I am worried about all of the math in this course.*

You have taken the first step. It is easier to combat the myths about research and statistics when we acknowledge them. Good work. Next we need to see why it is important to learn about the tools of research and statistics.

1.3 WHAT MANAGERS SHOULD KNOW ABOUT RESEARCH AND STATISTICS

As a manager you will be making several decisions each day at work. What would help you to make the right decisions? Will it be your experience on the job, your sixth sense or hunch, or will you just trust to good luck? For sure, all these will play a part *after* you have thoroughly investigated or researched the problem, analyzed some data, and generated some alternative solutions to choose from. Whether or not managers realize it, they are constantly engaged in research as they try to find solutions to the day-to-day problems, big and small, that they face at work. Some of the main tools of research are statistical techniques because in attempting to understand a problem you will almost always gather and analyze data. Statistical techniques allow you to see the information in the data. Thus, an understanding of statistics is needed to do the research necessary for you to be a good manager.

The use of statistical techniques has long played an important role in quality control and quality improvement in business and industry. Unfortunately, for too long quality has been relegated to the "quality department" and not integrated into the whole organization. The purchasing department would purchase the raw materials needed at the cheapest price without regard to quality and then would throw the material "over the wall" to the engineers. The engineers would make the product, not worrying about the process, since they knew that the quality department would inspect the product before it went out the door. Businesses in the United States have learned that this way of approaching quality does not work and does not allow them to compete internationally. Quality is everyone's job!

Quality and statistics go hand in hand.

One of the reasons that people prefer to have the quality department handle quality issues is that thinking about quality leads to thinking about data analysis. Data analysis requires the use of statistical techniques, which are often viewed as difficult. But statistical thinking is not difficult to comprehend. Everything we do can be thought of as a series of steps that are connected. Each time we repeat a step it will not be precisely the same. Reducing how much the step changes from time to time is one way to improve things. Suppose your company manufactures copy machines. You would like to know that the raw materials you use to make the copy machine are always the same, and the customer would like to know that the copy machine will perform the same from day to day. Statistical thinking is logical, uses data, recognizes the interdependence of activities, and looks at how things vary.

Four issues must be addressed if statistics is to become an integral part of management:

- Managers must understand *why* they need to possess statistical knowledge.
- Current and future managers (that's you) must *have* this knowledge.
- Steps must be taken to ensure that the knowledge is *used.*
- The *payoff* from the knowledge and its application must be measured.

To achieve this integration of statistics into business and industry we are going to have to change the way we see the world. We need a new pair of glasses.

1.4 STATISTICAL THINKING—A NEW PARADIGM FOR MANAGEMENT

Look at Figure 1.1. What do you see? Ask the person next to you what he or she sees.

FIGURE 1.1   Two faces or a vase?

Some of you will see two faces and some of you will see a vase. You probably can see both of these things, but which one did you see first? There is no right or wrong answer here. What you see first depends on your viewpoint, your lens, your glasses. This is your **paradigm;** this is how you *see* things. Each of us has a paradigm or view of the world that has been developed based on our individual experiences. The hard part is changing our lens. There will be many times when you must change your paradigm in order to see the same picture in a new way. This book is designed to help you shift your paradigm so that you can see the world through the lens of statistical thinking.

A *paradigm* is commonly used today to mean a model, theory, perception, assumption, or frame of reference. It was originally a scientific term.

Did you ever take a class on the great artists of the Italian Renaissance? Even if you did not you might have heard of Michelangelo, Leonardo da Vinci, Raphael, and Donatello. If you study Michelangelo, you would probably study his statue of David, and if you study Leonardo you would certainly learn about his painting of the Mona Lisa. Then you would begin to see David and Mona used in all sorts of advertisements and logos as far away from art as software and tomato sauce! Look at the ones shown in Figure 1.2.

FIGURE 1.2   Advertisements using the statue of David and the Mona Lisa

Are these ads new? No, you are just more aware of their existence after having studied about them. This is similar to what happened when you looked at the two faces/vase illustration in Figure 1.1. Once you become aware of the second way of looking at the picture you see both things each time you look at the picture. Similarly, once you study Michelangelo and Leonardo your view is different because you are now aware of their works and begin to see them everywhere. You have a new paradigm or view of the world. Your paradigm has *shifted*.

This book is about changing your paradigm to one that allows you to *see* things using statistical thinking. Statistical thinking is public property. Everyone owns it and everyone must use it—not just statisticians. Over the short term, it can improve the quality of decisions; over the long term, it can help turn people into leaders.

You will learn to see data everywhere you turn and to see the information hidden in the data as well. This information will help you make informed investment, marketing, and management decisions and will help you develop quality processes and product designs. How is this possible? By learning and understanding the tools of statistics you will be able to see the world differently. The tools of statistics are actually quite logical and simple to use, yet many people do not use them because of lack of exposure to statistics or dislike of math. Hence, these people often do not make well-informed decisions. You will be different because you will learn the tools of statistics as decision-making tools and not as mathematical manipulation. Think about this book as a management text, not a math text. You will soon see what we mean.

1.5 SITUATIONS THAT CALL FOR RESEARCH AND STATISTICAL THINKING

After finishing this course and graduating from college, you may find yourself facing decisions similar to any one of the following scenarios. In each case you must make decisions based on what you see in the data. A short discussion of possible approaches you might use to investigate each situation is provided. This should help convince you that the tools of statistics make logical sense.

EXAMPLE 1.1   Tissue Manufacturer

Complaints Become Opportunities

Most large companies that manufacture consumer goods receive feedback from their customers. A large manufacturer of paper goods keeps track of consumer complaints for its facial tissue product line. The company receives these complaints through a toll-free number that appears on the product. Next time you pick up a box of tissues see if you can find the toll-free number for making complaints.

The data taken consist of a transcription of the actual complaint and a classification of the complaint into a specific category. The complaint categories are dispensing, foreign material, odor, miscounts, and packaging. There are three subcategories of packaging complaints: damaged, misleading, and defective. The 25 packaging complaints for January 1998 are

Damaged	Misleading	Defective	Damaged	Misleading
Defective	Defective	Damaged	Damaged	Defective
Damaged	Misleading	Misleading	Damaged	Damaged
Misleading	Defective	Damaged	Damaged	Damaged
Defective	Damaged	Misleading	Defective	Damaged

You must research the complaints about packaging and make some recommendations to management. What is your decision?

Possible approaches and things to think about in making a decision:

- Make a table summarizing how many of each type of packaging complaint occurred.

- Make a graph from this table.

- Look to see if there are more complaints of one type than others and if this happens in other months. ∎

Let's consider another situation.

EXAMPLE 1.2   Golf Ball Design

New Product Design

All companies are currently facing increasing competition at the national and international level. In the face of this competition, American manufacturers are moving to a focus on quality. Customer feedback described in the previous scenario is one aspect of this movement. This change is not simple, but rather encompasses all aspects of the business and all the employees of the company. Managers are learning to listen to the creative ideas of their employees, breaking down the hierarchical management layers, and are monitoring their production processes. In doing so, more and more data are being collected on every aspect of the business. All parts of the organization must work together as a team to achieve Total Quality Management (TQM).

A large manufacturer of golf balls is incorporating some of these concepts. The company is studying two golf balls of equal cost to determine which ball design performs better. Data on how far the ball carried when hit are given (in yards) for 12 balls of each design:

Design 1						Design 2					
257	259	255	256	260	258	254	252	256	255	255	257
260	259	259	257	255	260	253	255	254	254	256	255

You must make a recommendation on which design to pursue. What is your decision?

Possible approaches and things to think about in making a decision:

- For each design, find the average distance the balls carried.
- Compare these averages.
- Think about whether or not the difference between these averages is large.
- Think about whether the differences that you see in the averages could be caused by something other than the ball design. ∎

Here is one more example.

EXAMPLE 1.3   Dress Down Day

Impact on Employee Morale

Many companies have designated one day a week as dress down day. This idea has been implemented even in long-time conservative companies such as IBM. What reason(s) would management have for adopting such a policy? In April 1992, Levi Strauss and Co. conducted the first national survey on business casual dress issues. This phone survey gathered data and opinions from managers from a wide range of industries.

At the company you work for, your boss has conducted a survey of employees. Thirty employees were asked if they strongly agree (SA), agree (A), disagree (D), or strongly disagree (SD) with each of the following two statements:

Statement 1: Casual dress improves morale.

A	SA	A	A	A	D	SD	SA	A	A	D	D	A	A	SA
A	A	D	D	A	A	SA	SD	A	D	A	A	D	A	SA

Statement 2: I do my best work when dressed casually.

D	SD	A	A	A	A	SA	A	A	A	D	D	A	A	SA
A	A	A	D	A	D	SD	D	A	D	A	A	D	A	SA

Your boss would like your recommendation regarding the likely increase in productivity if a dress down day is adopted by your company. What is your decision?

Possible approaches and things to think about in making a decision:

- Summarize the results.
- Examine what percentage of people strongly agreed, agreed, disagreed, or strongly disagreed with the various statements presented to them.
- Consider how productivity was defined in this survey.
- Look to see if the responses are different in different industries. ∎

Discovery Exercise 1.1
STARTING TO THINK STATISTICALLY

This exercise allows you to begin to think statistically. Although you have not yet learned any statistical techniques, you will be surprised to discover that many of the ideas you suggest are the basis for some of the techniques you will learn about later in this book. Do not try to solve the problem, but rather focus on what you might do with the data to analyze the situation and what additional information you would like to know about the data or the situation. Remember that statistical thinking is logical, uses data, recognizes the interdependence of activities, and looks at how things vary.

Part I. Reduction in Sick Days Used

It has been a widely held belief that a switch to "participative management" would increase employees' "buy-in" to the company. One of the benefits that should be realized is a reduction in the number of sick days that employees use when they feel that they are an important part of the team. A company that has made the switch in some departments wonders if this has been true. The company decides to sample 25 employees from each of two manufacturing departments. The first department has been using a participative management style for almost two years, and the second is still using a traditional management style. The data on the number of sick days used by each employee in the past 12 months are

Participative					Traditional				
1	3	5	5	6	0	5	6	7	9
1	4	5	6	7	3	5	7	7	9
2	4	5	6	8	4	6	7	7	10
2	4	5	6	8	4	6	7	8	11
3	4	5	6	8	5	6	7	8	11

Has the participative management style reduced the number of sick days?

1. What else would you like to know about these data sets before comparing them?

2. How could you compare the two data sets using graphs?

3. What numerical calculations would you perform to compare the two data sets?

Part II. Are We There Yet?

The amount of time it takes to travel is clearly dependent on how far you travel and the type of vehicle you are driving. The data shown below give information about how long it took to travel the distance (in miles) when the vehicle was either a car or a truck. The travel times (in minutes) are listed as the Y values. The distances traveled are listed as the $X2$ values. The vehicles used are listed as $X1$ values with the following coding scheme: $X1 = 0$ if a truck was driven and $X1 = 1$ if a car was driven.

(continued)

Type of vehicle (car = 1, truck = 0)	Distance (miles)	Travel time (min)
X1	X2	Y
0	55	35
0	102	56
1	33	23
0	20	12
1	55	65
0	48	34
0	53	23
0	22	12
1	45	35
1	44	46
1	12	14
0	45	34

Look at the data shown and think about how you would analyze the impact that distance and type of vehicle has on travel time.

1. What else would you like to know about the data?

2. How could you display the relationship between the travel time and distance in a graph?

3. How would you predict the time it would take to travel a distance of 90 miles driving a car?

A picture is worth 1000 words. Did you ever hear that saying? Why is this true? Perhaps it is because your eye can process and understand the millions of tiny dots of paint when they are all put together in a meaningful fashion. In this way you see a beautiful picture. Well, that is exactly what the tools of statistics do for you. They put together millions of bits of data in a meaningful fashion. In this way you can make the best decision based on what you see in the data.

1.6 TYPES OF RESEARCH

The examples in the previous section show that as a manager you will be doing business research. You may need to understand why customers are complaining, you may need to research whether customers would purchase a new product, or you may need to decide whether a new management policy will improve productivity. In each of these cases you need the tools of research to make an informed decision.

What does the term **research** conjure up in your mind? Do you see the mad scientist at work in his or her lab? What does this have to do with being a good manager? The following definitions will help us to get started.

> *Research* is simply the process of finding solutions to a problem after a thorough study and analysis of the situation and data.

Since problems crop up in most settings, research can be used in the science lab, the courtroom, the university, the healthcare environment, the home, and of course in business. The term **business research** simply tells us that the problem to be investigated is in the work setting. Business research comprises a series of steps with the goal of finding answers to the issues that are of concern to the manager.

> *Business research* is the process of finding solutions to a specific problem encountered in the work setting.

 **TRY IT NOW!**

Research in Business

Some Commonly Researched Areas

You have already seen that customer complaints, new product designs, and new management policies are areas of business research. Based on your own work experiences, list three additional areas where it would be helpful to use the tools of business research.

Business research can be undertaken for two different purposes. One is to solve a current problem faced by the manager. For example, a particular product may not be selling well and the manager might want to find the reasons for this in order to take corrective action. Such research is called **applied research.** The results are applicable only to this particular product in this particular company at this particular time.

> *Applied research* is research done with the intention of applying the results of the findings to solve specific problems being experienced in the organization.

ANS. ANSWERS WILL VARY. SOME POSSIBILITIES INCLUDE EMPLOYEE TRAINING, JOB SATISFACTION, MARKET SHARE, TURNOVER, COLLECTION OF ACCOUNTS RECEIVABLE, AND THE ORDERING PROCESS.

In contrast to this, you might be trying to understand how certain problems that occur in organizations can be solved. The results are more widely applicable and add to the general knowledge in some functional area such as management. For instance, you might research how people respond to change in organizations. You are not looking at one specific instance of change but rather at the way people manage change in general. This is called **basic research.**

> *Basic research* is research done mainly to increase the understanding of certain problems that commonly occur in organizational settings.

Most of the examples and work that you will be doing in this course will be applied research. Let's consider some examples.

EXAMPLE 1.4 Oxford Health Plans, Inc.

Computer Problems

Oxford Health Plans, Inc., saw trouble brewing. The company was experiencing computer problems. Turnover among Oxford's programmers was unusually high and processing of claims became a nightmare. Clients started canceling their policies, claims for bypass surgery and such were way up, and premiums paid out relative to clients' medical expenses were close to 85%.

Clearly this company needs to conduct some applied research to investigate its problems. ■

Here's another example of a company that needs to use applied research.

EXAMPLE 1.5 Xerox Corporation

Sales Problems

Xerox still relies on old-fashioned and slow-selling analog copiers for more than half its revenue, and despite its double-digit growth in digital products and services, the company's sales rose just 4%.

Clearly Xerox needs to look at what should be done to increase and promote the company's sales. ■

Now you try one!

TRY IT NOW!

Research in Business
Identifying Applied Research

In the previous Try It Now, you identified three areas that would benefit from research. Which of these would be applied research?

The next example is an instance of basic research.

EXAMPLE 1.6   The Bank

Factors in Job Satisfaction

From her first days as a clerical employee in a bank, Sandra had observed that her colleagues, though extremely knowledgeable about the ins and outs of banking, were doing very little to improve the efficiency of the bank in the area of customer relations and service. They took on a minimum workload, availed themselves of long coffee and lunch breaks, and showed no interest in their dealings with customers or management.

When Sandra left the bank and did her dissertation for her Ph.D., her topic of investigation was job involvement or the ego investment of people in their jobs. The conclusion of her investigation was that the single most important factor is the match between personality characteristics and the characteristics of the job.   ■

In summary, we have seen that business research can be either applied or basic; it is the process of finding solutions to a business problem after a thorough study and analysis of the situation and data. Whenever possible, we would like our business research to be as organized, rigorous, and error free as the type of research we would do in a science lab. Sometimes this is possible but often due to difficulties in measurement and collection of data about people's feelings, attitudes, perceptions, and emotions, it is not as easy to conduct business research in the same manner as you would do scientific research.

1.7 STEPS IN CONDUCTING BUSINESS RESEARCH

As you gain some experience with doing business research, you will find that although the situations and problems are varied, the steps necessary to successfully complete the research can be generalized to apply in any situation. Many authors have summarized these steps and you may find some procedures with 5 steps, some with 6 or 8 steps. These procedures simply differ in how the necessary work is divided up into manageable and logical steps.

In this book we will use the following steps for conducting business research, which are attributed to Dr. Anthony Poet of the University of Phoenix and based on work done at Nova Southeastern University in the mid-1990s. As you continue with the material in this book, keep this framework for business research in mind.

Step 1. Problem Identification

Step 2. Statement of Desired Goal or Outcome

Step 3. Research Evidence and Hard Data

Step 4. Outcomes

Step 5. Identification of Possible Cause(s)

Step 6. Proposed Solutions

Steps for Business Research

At step 1 you identify that a problem exists or there is an opportunity for improvement. The objective of this step is to develop a statement that reflects the current situation. Problems or opportunities for improvement can be identified from customer complaints, employee surveys, focus groups, brainstorming sessions, new regulations, or a variety of other places. For example, the amount of time it takes to get a new product may be too long, there may be too many errors in the billing process, customers may be on hold too long, or there may be too many returns of a particular product. All of these situations represent problems and/or opportunities for improvement. Identification of a problem is an important first step in conducting research. It is the reason for collecting and analyzing data. Otherwise, data analysis occurs in a vacuum and is not connected to business decisions.

EXAMPLE 1.7   Mail-Order Company

Step 1

You work for a mail-order company and you have been receiving an increasing number of complaints from customers about the amount of time they have to wait to talk to a customer service representative. You state the problem as follows: Customers are waiting an average of 5 minutes before being assisted by a customer service representative. ∎

At step 2 you identify the desired goal or outcome of the research. This is generally the exact opposite of the problem.

EXAMPLE 1.8   Mail-Order Company

Step 2

For the mail-order company, the desired goal or outcome might be that the average time a customer is on hold be reduced to 2 minutes or less. ∎

Notice that you are not determining how to solve the problem at this step but you are stating the desired goal. Both the statement of the problem and the statement of the desired goal should be as specific as possible to guide the data collection. For example, a problem statement that customers are on hold too long is not specific enough. Likewise, a desired goal statement that customer wait time be reduced is also not specific enough.

Step 3 involves collecting and analyzing data that are specific to your problem as well as investigating the literature. You do this to learn what you can from your own environment and the work that others have done. At this step the researcher must provide real/hard data to show that the problem exists, is measurable, and is not created out of self-interest. More details on investigating the literature can be found in Chapter 8.

This step will involve using the tools of statistics and much of the remainder of this text is devoted to this material. Remember that statistics is a tool to allow you to see the information hidden in data in order to solve the business problem. Always keep your eye on the goal or outcome of the research. The statistical techniques are presented with this in mind. Many chapters begin with a motivating problem statement (step 1 of this procedure) and explain the statistical techniques in terms of that problem.

EXAMPLE 1.9   Mail-Order Company

Step 3

In examining the research to solve the customer waiting time problem, you might look at other industries that have a comparable problem. For example, perhaps a help desk might have experienced a similar problem or perhaps even calls coming into an insurance company might have some elements in common with your situation. When you examine your own company, you would collect some hard data to support the existence of a problem. For example, you might find that 15 out of 20 customers studied were on hold for more than 5 minutes. ∎

These outcomes should be consistent with the overall goal articulated in step 2. The outcomes should parallel the findings of step 3. These outcomes should be measurable, observable, and realistically achievable. They will be used to evaluate whether or not a successful solution has been found.

EXAMPLE 1.10   Mail-Order Company

Step 4

For the mail-order company, you may have decided to collect data on the time of day the customer calls, the number of operators on duty, the experience level of the operators on duty, the number of items the customer orders, the amount of time it takes for the operator to access the appropriate information in the database, and the name of the manager on duty. You may also have decided to call back a few customers and collect some information in this manner.

You might decide that one reasonable outcome would be that, when tallied at the end of the year, 90% of the calls will have been placed on hold for 2 minutes or less. ■

Once the literature, as it relates to the validation of the problem, has been reviewed and your own data have been collected and analyzed, you can identify some possible causes of the problem. Possible causes were most likely identified at step 4.

EXAMPLE 1.11   Mail-Order Company

Step 5

In doing your literature review, you learned that the experience of the customer service representative is really critical in other industries. This is confirmed by your data analysis using a technique known as hypothesis testing. You also discover that even the most experienced customer service representative cannot go any faster than the time it takes to access the appropriate database with the product availability information. In this case you have identified two causes of the problem: insufficient training for the customer service representatives and a database access time that is too slow. ■

The last step of the research process calls for you to identify solutions and actions to solve the problem. It is possible that one action might be to further investigate a new problem that has been identified as a result of this research. You should always refer back to steps 1 and 2 at this point in the research to be sure that your solutions are addressing the problem and striving to reach the stated goal.

EXAMPLE 1.12   Mail-Order Company

Step 6

For the mail-order company, you might propose a new training program for new customer representatives or you might design a mentoring program whereby new employees work side by side with experienced employees for some period of time. You might also ask the information systems department to improve the database access time. ■

Notice that your solutions will often generate additional research opportunities and the continuous improvement cycle continues. You should always come back to the problem and validate your outcomes for program success. In the mail-order example, you might want to collect data at some future point to see whether the new training you have instituted is in fact helping to reduce customer waiting time. The information systems department might have to begin a research project of its own to determine how to improve database access time.

In conducting any business research, it is important to use a methodical stepwise procedure such as the one explained in this section. A common mistake in attempting to solve problems is to jump from step 1 to step 5 or 6 without going through the other steps. It is easy to think you have the causes and/or a solution without really understanding the problem. It is also easy to jump to erroneous conclusions because of incomplete data or inappropriate data analysis. You must begin to think statistically in order to do business research. The next section explains the major components of statistical thinking which will help to keep the research systematic and thorough.

1.8 KEY COMPONENTS OF STATISTICAL THINKING

It is important to understand the components of this new paradigm that we have called statistical thinking. There are three key components:

- We must use data whenever possible to guide us.
- We must look for connections and relationships.
- We must understand why data values differ from each other.

Now it is time to introduce an official definition of statistics.

> *Statistics* is a branch of mathematics dealing with the analysis and interpretation of masses of data.

With statistical thinking as the new paradigm, your way of thinking about numbers has begun to change. You must begin to discipline yourself to think systematically and to collect data systematically. Then you must learn how to see the information contained in the data you have collected. Finally, you must use the information to make informed business decisions. The rest of this book is designed to increase your inventory of specific research and statistical skills to help you see the information contained in the data and strengthen your decision-making skills. In highly successful organizations, the statistical thinking paradigm will be linked to other important paradigms, such as providing leadership, promoting teamwork, working toward continuous improvement, creating innovative channels of communication, and delighting customers.

1.9 ORGANIZATION OF THIS BOOK

Each chapter of this book contains many of the same elements that you have seen in this introductory chapter. Each element of the chapter is designed to help you change your paradigm so that you may see the world through research and statistical thinking. The common elements are listed here:

- Many chapters begin with a *business problem* based on an actual situation. The business problem will motivate the material and will be carried throughout the chapter as you learn the techniques needed to solve the problem. Some chapters may have more than one motivating problem.

- The *objectives* of the chapter are listed in the first section.

- Worked *examples* are provided for each statistical concept, tool, or other important idea. These are always labeled with the word **EXAMPLE** and are visually separated from the text.

- *Try It Now!* exercises are provided to give you some practice. These exercises are like checkpoints. Do not skip them. Space is provided for you to work the problems right at that point and the answers are given at the end of the exercise so you may check your work immediately. *Try It Now!* exercises are easily found because the open book graphic shown in the margin is always displayed with them.

- One or more *Discovery Exercises* are offered. These exercises are optional and if they are skipped you will not miss any information. They are designed to help you discover some concept or idea. They will help you see that statistical thinking is logical and that the formulas make sense when you discover where they came from and why. They are easily identified by their design. Your instructor will probably use some but not all of these exercises.

- *Side margin notes,* which contain keywords, hints, and tips, are included throughout. These help you to review the material and locate key formulas, definitions, and concepts.

- *Definitions* are presented in shaded boxes for easy location. Any term that has a definition box will appear in the end-of-chapter summary.

- *Formulas* are identified with a label in the margin. All such formulas will appear in the end-of-chapter summary.

- More practice problems are provided at the end of most sections. (Chapter 1 does not have any of these.) These are like the *Try It Now!* exercises. They are labeled *"Exercises—Learning It"* and by doing them you will master the basics.

- Many chapters give detailed instruction on how to use Microsoft Excel to perform most of the statistical analyses described in the chapter. These instructions are found in the last section of each chapter. In many cases, if Excel does not provide a convenient or correct way to do the analysis, we provide an Excel macro, along with directions for using it, that performs the function. These macros are found on the disk with the data files.

- A *Chapter Summary* includes tables of key terms and formulas introduced in the chapter.

- *End-of-Chapter Exercises* are provided. At the end of most chapters there are three types of exercises: *Learning It!* (similar to the end-of-section exercises), *Thinking About It!,* and *Doing It!* The *Thinking About It!* exercises ask you to go beyond the basics. You are asked to think about the business implications of the computational result and your thinking may be directed beyond the immediate details of the problem. Sometimes a set of data is presented in an exercise in one chapter and then used again in an exercise in a later chapter. In these cases the data are always redisplayed with the problem statement to avoid your flipping back and forth in the text. In addition, if an exercise is built on a previously presented exercise, the exercise number corresponding to the related exercise is noted in the side margin. The *Doing It!* exercises are linked to large data sets that are provided on disk. These data sets are designed to let you use the techniques of the chapter in a more realistic setting, one in which you have lots of data to analyze and not a great deal of structure.

CHAPTER 1 SUMMARY

In this chapter you learned how important it is for today's business professional to know how to use the tools of research and statistics. The business environment is changing rapidly and one of the key issues facing most businesses is the issue of quality. This is true regardless of whether the business is a manufacturer, a restaurant, a hospital, or an entertainment industry. To remain competitive, businesses must pay attention to the quality of the product or service they provide. At the foundation of the quality movement lies the field of statistics.

Key Terms

Term	Definition	Page reference
Applied research	**Applied research** is research done with the intention of applying the results of the findings to solve specific problems being experienced in the organization.	11
Basic research	**Basic research** is research done mainly to increase the understanding of certain problems that commonly occur in organizational settings.	12
Business research	**Business research** is the process of finding solutions to a specific problem encountered in the work setting.	11
Paradigm	A **paradigm** is commonly used today to mean a model, theory, perception, assumption, or frame of reference.	5
Research	**Research** is simply the process of finding solutions to a problem after a thorough study and analysis of the situation and data.	11
Statistics	**Statistics** is a branch of mathematics dealing with the analysis and interpretation of masses of data.	16

CHAPTER 1 EXERCISES

Thinking About It!

1.1　We have said that quality and statistics go hand in hand. What product or service have you used recently that did not have the quality you thought it should have?

    (a)　Describe the product or service and the area in which it was lacking in quality.

    (b)　What corrective action would you recommend?

1.2　Describe a situation in which you had a paradigm shift or change.

1.3　There has been an increased use of electronic mail in most organizations. This way of communicating with people changes the dynamics of the exchange of information and how people do business. In fact, you might speculate that e-mail would change the way the organization behaves. It is possible that e-mail is used more by middle management than senior management. It is also possible that e-mail is not used to communicate certain types of decisions. A human resource manager asks 10 senior managers and 10 middle managers how many messages they send each day. The data are

    Senior managers: 10, 15, 2, 21, 14, 35, 19, 12, 18, 19
    Middle managers: 25, 27, 22, 24, 23, 21, 25, 26, 29, 19

    (a)　What might you do with these data to investigate if there is a difference in e-mail usage by level of management?

    (b)　What factors, other than level of management, might influence e-mail usage?

1.4　You have just been hired to be the manager of the service department for a car dealership in your town. You begin by examining the service time for all the cars worked on last Friday. The service time is the number of minutes between the time the car arrived at the dealership and the time the service was completed. The data for last Friday are

    15　　125　　45　　65　　35　　50　　20

    (a)　What additional data would you like to examine to get a more complete picture of the current level of service?

    (b)　How would you collect the data identified in part (a)?

1.5　Describe a situation where research will help you as a manager to make a good decision.

1.6　Given the following situation: (a) discuss, with reasons, whether it will fall into the category of applied or basic research, and (b) who will conduct the research.

To Acquire or Not to Acquire: That Is the Question

Companies are very interested in acquiring other firms even when the latter operate in totally unrelated realms of business. For example, Gencore Industries manufacturing asphalt plants for road construction acquired Ingersoll-Rand in 1996, and later acquired yet another company engaged in the business of food processing. Such acquisitions are claimed to "work miracles." However, given the volatility of the stock market and the slowing down of business, many companies wonder whether such acquisitions are becoming too risky. At the same time, they also wonder if they are missing out on a great business opportunity if they fail to engage in this activity. Some research is needed here!

THE LANGUAGE OF RESEARCH AND STATISTICS

THE GLUE COMPANY

All companies are currently facing increasing competition at the national and international level. In the face of this competition, American manufacturers are moving to a focus on quality. Customer feedback is one aspect of this movement. In fact, companies are now building relationships with their vendors to ensure higher quality.

The U.S. automobile industry was hit very hard by international competition. The industry appears to have recovered by paying attention to quality at all levels, right down to the glue that is used in assembling some of the components. One of the large U.S. automobile manufacturers has been building a relationship with the manufacturer of an adhesive product. In doing so, the car company has required a lot of data to support the claims about the performance of the glue. The car company visits the adhesive manufacturer from time to time to further the relationship and discuss product quality problems and solutions.

This type of interaction between the supplier and the customer is much different from the "old days" when the purchasing department bought from the least expensive supplier, which could easily be several different suppliers over a short period of time. There was no long-term commitment on either end of the deal. This short-term mentality led to shortcomings in quality.

2.1   CHAPTER OBJECTIVES

In the first chapter you learned that quality and statistics go hand in hand. You also saw that to truly integrate quality into the organization we must view our environment with a statistical thinking paradigm. Recall that a statistical thinking paradigm is a systematic way of thinking about and applying statistical techniques to data and business problems. The first step in changing your view of the world to a statistical thinking paradigm is to develop a language that can be used to describe this new view. This chapter develops the basic language that you need. The following material is covered:

- The Difference Between the Population and a Sample of the Population
- The Difference Between a Parameter and a Statistic
- Factors That Influence Sample Size: Some Sampling and Sample Size Considerations
- Selecting the Sample
- Types of Data
- The Difference Between Descriptive Statistics and Inferential Statistics
- Basic Summation Notation

2.2   THE DIFFERENCE BETWEEN THE POPULATION AND A SAMPLE OF THE POPULATION

2.2.1 The Population of Interest

In Section 1.5, we saw several situations that called for statistical thinking. If we step back from the details of these situations we see that regardless of the decision that needs to be made, there is always a group of people or things that needs to be studied and understood to make the necessary decision. For the tissue manufacturer it was all the boxes of tissues that the company makes, for the company considering dress down day it was all the employees of the company, and for the golf ball company it was the performance of all the balls made according to each of the two designs. These are examples of what is called the **population** in the statistical thinking paradigm. Thus, *all the boxes of tissues* is the population of interest to the tissue manufacturer, *all the employees* is the population of interest to the company considering dress down day, and *all the golf balls* is the population of interest to the golf ball manufacturer. In each case, the decision maker must learn something about how the population of interest behaves in order to make a recommendation.

> The *population* is everything you wish to study.

EXAMPLE 2.1   The Glue Company

Identifying the Population

For the glue company to be a supplier for the car company it must study and understand the performance of all the tubes of glue made for the car company. Call the type of glue used by the car company type C. Thus, the population of interest is all the tubes of type C glue manufactured by the glue company. ■

 *TRY IT NOW!*

The In-line Skate Company
Identifying the Population of Interest

There have been a number of failures on the braking device of a new model of roller blades that your company manufactures. What is the population of interest?

ANS. ALL BLADES OF THIS MODEL.

Often, when we study a population, we are really interested in knowing about different characteristics of each member of the population. These characteristics are known as **variables.** For each member of the population we may be interested in knowing about one, two, or even more different variables.

> A *variable* is used to represent a characteristic of each member of the population.

EXAMPLE 2.2   The Glue Company

Possible Variables of Interest

In the case of the glue company, we may wish to know how long the glue will stick but we may also wish to know how the glue smells. In this case we wish to study two variables or characteristics of the members of the population of all the tubes of type C glue manufactured by the glue company. ∎

TRY IT NOW!

The In-line Skate Company
Identifying Possible Variables to Study

There have been a number of failures on the braking device of a new model of roller blades that your company manufactures. Name two variables or characteristics that you might wish to study.

ANS. NUMBER OF HOURS OF USE UNTIL THE BRAKE FAILS, TYPE OF FAILURE THAT OCCURS

2.2.2 The Sample

Now that you understand what a population is, we need to formally define a sample. The statistical definition of the word sample is much like the normal use of the word. If someone tells you to sample the apple pies made by a bakery, then you take a small piece of one and eat it, or you might eat one whole pie out of the many pies made by the bakery. It is the same in statistics. If you take a sample, you take a small piece of the population and look at it or test it. Thus, we have our next definition.

A *sample* is a piece of the population.

If we think about the population as the big oval (all the pies) shown in Figure 2.1, then a sample from this population might be the small oval shown in the figure. This is clearly not the only sample that could be taken from this population; many different samples may be picked. Some of the samples may overlap and some may be bigger or smaller than the one shown.

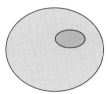

FIGURE 2.1   The population and a sample

 **TRY IT NOW!**

Selecting Different Samples

Draw another oval in Figure 2.1 to represent a different sample that might be selected.

2.2.3 Why Pick a Sample at All?

We have said that the population is everything we wish to study and that the sample is only a piece of the population. A natural question to ask is, Why should we bother to examine a sample when what we really want to know about is the population? Most of the time we cannot study the entire population and must use a sample as a guide.

Reasons why we can't study the entire population

The main reasons are fairly clear when you think about it for a minute:

- It would take too much time to study the entire population.
- It would take too much money to study the entire population.
- It might not be possible to identify all the members of the population.
- If we test the entire population, we might not have anything left to sell.

Let's consider a couple of populations that we have already talked about and see these reasons in action.

EXAMPLE 2.3   The Glue Company

Reason Why the Population Cannot Be Studied

The glue company needs to study the performance of all tubes of type C glue that it sells to the car company. It would clearly take a great deal of time and money to test each and every tube of type C glue that the company manufactures. In addition, each tube that is tested is opened and then cannot be sold. So, if every tube were tested there would be no glue to sell to the car company.   ∎

Here's another example based on one of the situations we looked at in Chapter 1.

EXAMPLE 2.4   Dress Down Day

Reason Why the Population Cannot Be Studied

A company needs to study the productivity of the employees before and after a dress down day is adopted. If there are 5000 employees in this company all doing different jobs, it would be virtually impossible to measure productivity for all of them.   ∎

2.2.4 The Difference Between the Population and the Sample: An Introduction to Sampling Error

It is not necessary to study the entire population.

At this point it might look like we should throw up our hands and go home, as we will never have the time, money, or ability to study an entire population. The good news is we don't need to study the entire population. By studying the behavior of a sample we can get a good idea of the behavior of the population. It will not be a perfect picture of the population, but it will be good enough to guide us in our decision making. It is just like eating a piece of the apple pie. You don't know precisely how all the pie tastes but you have a pretty good idea that it is delicious!

Remember, it is really the population that we wish to understand and study. The sample is a means to this end. By understanding and studying a sample, we gain insight and knowledge about the behavior of the population. The amount of information in the sample is not perfect but adequate.

Imagine that you took a picture of the population but let it develop for only a few seconds. You would be able to get an overall sense of what the population looked like but you would not have all of the details. This is like taking a sample and using the sample to determine how the population behaves. If you let this picture develop a few seconds longer, you would get a little clearer view of the population. This is equivalent to taking a bigger sample. Finally, if you let the picture develop completely, then you would have complete information. This is equivalent to studying the entire population. If you study the entire population, then you have taken what is called a **census.**

A *census* is a study of the entire population.

Since the sample is an imperfect snapshot of the population, you know that there will be differences between the sample and the population. This is a bit disconcerting at first. Unless we study the entire population, which we usually cannot do, we will always have incomplete information about the population. That is, unless we study the entire population, we cannot eliminate what is known as **sampling error.**

> *Sampling error* is the difference between the behavior of the entire population and a sample of that population.

The size of the sampling error is determined by two factors. The first of these factors is the *size of the sample*. Clearly, the larger the sample you take, the more similar the sample will be to the population, thus decreasing the sampling error. The second factor that influences the size of the sampling error is the amount of **variation** that exists in the population. Variation in statistics has the same general meaning that it does in typical language usage.

The size of the sample influences the size of the sampling error.

> The amount of *variation* refers to how different the members of the population are from each other with regard to the variable being studied.

For example, suppose your population of interest is all students taking classes at your university or college. The variable of interest is the age of these students. If all students were exactly the same age then you would say that there was no variation in the age of the members of the population. In this extreme case you would need to take a sample of only one student to have perfect information about the age of members of the population.

The amount of variability in the population influences the size of the sampling error.

However, you will almost never be studying a characteristic that has no variability. Suppose that the ages of the students have a small amount of variability. Let's say the student ages range from 18 to 22 years old. Clearly, you have to take a sample of more than one student to understand how the ages vary in this population. Now suppose the ages of the students range from 17 to 60 years. You have to take a much larger sample to understand how the ages vary in this population. As the amount of variability in the population increases, the sampling error also increases.

2.2.5 Exercises—Learning It!

2.1   The President of the United States wishes to see how popular he/she is after 2 years in office.

(a) What is the population of interest?

(b) Identify which of the reasons for taking a sample (listed on page 24) apply in this case. (There are more than one.)

(c) Identify two variables or characteristics of the members of this population that you may wish to study.

2.2   Tasty Ice Cream Corporation wishes to be sure that all of the half-gallon ice cream cartons do contain one-half gallon of ice cream.

(a) What is the population of interest?

(b) Identify which of the reasons for taking a sample (listed on page 24) apply in this case. (There are more than one.)

(c) Identify two variables or characteristics of the members of this population that you may wish to study.

2.3   A company that manufactures electronic switches wishes to provide the customer with defect-free switches.

   (a) What is the population of interest?

   (b) Identify which of the reasons for taking a sample (listed on page 24) apply in this
       case. (There are more than one.)

   (c) Identify two variables or characteristics of the members of this population that you
       may wish to study.

2.4   A university wishes to know how students view the new athletic center on campus.

   (a) What is the population of interest?

   (b) Identify which of the reasons for taking a sample (listed on page 24) apply in this
       case. (There are more than one.)

   (c) Identify two variables or characteristics of the members of this population that you
       may wish to study.

2.5   The U.S. government wishes to know how many people are unemployed.

   (a) What is the population of interest?

   (b) Identify which of the reasons for taking a sample (listed on page 24) apply in this
       case. (There are more than one.)

   (c) Identify two variables or characteristics of the members of this population that you
       may wish to study.

2.6   The manufacturer of disposable diapers wishes to know how much fluid the diapers can
hold before they leak.

   (a) What is the population of interest?

   (b) Identify which of the reasons for taking a sample (listed on page 24) apply in this
       case. (There are more than one.)

   (c) Identify two variables or characteristics of the members of this population that you
       may wish to study.

2.7   The Coca-Cola Company wishes to know the proportion of people who prefer Coke over
Pepsi.

   (a) What is the population of interest?

   (b) Identify which of the reasons for taking a sample (listed on page 24) apply in this
       case. (There are more than one.)

   (c) Identify two variables or characteristics of the members of this population that you
       may wish to study.

2.8   Before selecting a major, a student decides to study the salaries of individuals working as
accountants.

   (a) What is the population of interest?

   (b) Identify which of the reasons for taking a sample (listed on page 24) apply in this
       case. (There are more than one.)

   (c) Identify two variables or characteristics of the members of this population that you
       may wish to study.

Discovery Exercise 2.1
INTRODUCTION TO SAMPLING AND VARIABILITY

Suppose that each set of data in this exercise represents an entire population. Since we don't yet have
a way to quantify the amount of variability in a population, the data sets are labeled as having a small
amount of variability or a large amount of variability. The first data set shows the number of people in
50 families living in a small college town in New England. The second set of data shows the number of
people in 50 families living in a large city in the South.

New England families: large amount of variability
Average number of people in 50 families: 4.50

1	4	5	7	8
3	9	8	8	8
4	9	9	1	6
4	1	3	9	7
8	2	3	1	9
1	7	5	1	1
1	6	8	2	9
4	1	1	1	3
4	2	4	9	4
1	3	8	1	1

Southern families: small amount of variability
Average number of people in 50 families: 4.36

1	3	4	3	5
3	4	6	6	3
6	4	3	4	6
7	5	4	5	5
4	4	5	5	6
3	4	4	7	4
6	3	5	5	5
5	4	4	4	5
4	5	4	3	4
5	4	2	4	4

Step 1. Select a sample of 5 numbers from each population.

	Sample from NE families	Sample from Southern families
Selection 1		
Selection 2		
Selection 3		
Selection 4		
Selection 5		
Average		

Step 2. Calculate the sample average of the 5 numbers by adding them together and dividing by 5.

Step 3. Calculate how far away the sample average is from the true population average by subtracting the sample average from the population average that is provided for you.

Step 4. Which data set had the greater error—the one with a small amount of variability or the one with a large amount of variability?

(continued)

Repeat steps 1–4 with a sample of 10 numbers.

Step 1. Select a sample of 10 numbers from each population.

	Sample from NE families	Sample from Southern families
Selection 1		
Selection 2		
Selection 3		
Selection 4		
Selection 5		
Selection 6		
Selection 7		
Selection 8		
Selection 9		
Selection 10		
Average		

Step 2. Calculate the sample average of the 10 numbers by adding them together and dividing by 10.

Step 3. Calculate how far away the sample average is from the true population average by subtracting the sample average from the population average that is provided for you.

Step 4. Which data set had the greater error—the one with a small amount of variability or the one with a large amount of variability?

Step 5.   Record the errors in the following table:

	Sample of size 5	Sample of size 10
NE families (large variability)		
Southern families (small variability)		

What happened to the error when you increased the sample size?

2.3   THE DIFFERENCE BETWEEN A PARAMETER AND A STATISTIC

In the last section we saw that by understanding and studying the sample, we can gain insight and knowledge about the behavior of the population. Next we must think about how to describe the behavior of the population to someone else. Suppose you were asked to describe your instructor's behavior in the classroom. You might say that he/she is informative, humorous, organized, and helpful (hopefully)! These are verbal descriptors of your instructor's behavior. If you were asked to provide a visual description of your instructor you might say that she is about 5 feet 2 inches tall, weighs approximately 110 lb, and is probably about 40 years old. These are numerical descriptors of the person. To paint a picture of a population and a sample, we use descriptors that are both verbal and numerical, just like the ones that we used to describe your instructor.

2.3.1 Parameters: Numerical Descriptors of the Population

There are many different ways to describe the behavior of a population. One of them is to use numerical values to paint a picture of the population. For instance, if you are told that all the values in the population fall between 0 and 10, you form a mental image of the population that is quite different from the picture that is conjured up if you learn that all the values in the population fall between 0 and 1000. This is just one example of a numerical way to describe the population. Chapter 6 shows you the traditional numerical measures that are used to describe the population. They are all examples of what are known as **parameters.**

> A *parameter* is a number that describes a characteristic of the population.

The following example gives three different parameters that might be of interest to the glue company.

EXAMPLE 2.5   **The Glue Company**

Parameters of Interest

The glue company may wish to know the average number of days that the glue will stick. In this case the parameter of interest is an average value. The company may also wish to know the longest time the glue will stick. In this case the parameter of interest is the maximum stick time. Likewise, the company might wish to know the earliest time that the glue will fail (come unglued). So the shortest stick time is also a parameter of interest. ∎

The next example shows that the parameter of interest might be a percentage.

EXAMPLE 2.6   Dress Down Day

Parameter of Interest

The company thinking of adopting the dress down day would like to know the percentage of employees that had increased productivity as a result of this policy change. The parameter of interest here is a percentage or a proportion. ∎

Remember that generally we do not know much about the population. If we did we would not need to take a sample and would not need the tools of statistics (and we could all go home now!). We are usually trying to discover information about these parameters. In particular, we are often trying to *estimate the value of these parameters*. This is the job of statistics.

2.3.2 Statistics: Numerical Descriptors of the Sample

Since we most likely will not know the value of the parameters needed to describe the population, we must resort to using the information contained in the sample. It seems logical that a numerical descriptor for the sample might somehow be used to estimate the corresponding measure for the population. This is the right idea.

So we need numbers to describe the behavior of the sample for two reasons: (1) to paint a picture of the sample and (2) to help us estimate the corresponding population parameter. There is nothing difficult here—in fact, it is quite simple. A statistic is nothing more than a number that describes the behavior of the sample. Let's put that into a definition box.

> A *statistic* is a number that describes the behavior of a sample

According to this definition we could dream up any formula that we want to to describe the sample and it would count as a statistic. Surprisingly enough, this is exactly right. There are, however, a few measures that are typically used because they convey some fairly standard type of information. These are the subject of Chapter 6.

Let's look at two examples: the glue company and the company considering the dress down day policy. In each case we will see what statistics might be calculated.

EXAMPLE 2.7   The Glue Company and Dress Down Day

Statistics That Could Be Calculated

The glue company might calculate the average stick time from the sample data. This would be one statistic. The company might also record the shortest and the longest stick times in the sample. These are two more statistics.

Let's assume that the company thinking of adopting the dress down day policy has adopted the policy for one department in the company. The company has taken a sample of employees in this department and measured the productivity of each employee in the sample before and after this policy was adopted. Should the policy be adopted companywide? The statistic to calculate is the percentage of employees in the sample who increased their productivity after the dress down policy was adopted. ∎

Notice that the statistics that might be calculated from the sample data are closely related to the parameters of interest. This is often the case. Remember that when you calculate the average stick time for the sample it is exactly that: a description of the sample. Is the average stick time for the population of all tubes of type C glue the same as the average you found in the sample? Probably not. Is it close? That depends on how well your sample reflects the population. This is the subject of the next two sections.

2.3.3 Exercises—Learning It!

2.9   The President of the United States wishes to see how popular he/she is after 2 years in office.
   (a) What parameter is of interest?
   (b) What statistic might you calculate from the sample?

2.10 Tasty Ice Cream Corporation wishes to be sure that all of the half-gallon ice cream cartons do contain one-half gallon of ice cream.
   (a) What parameter is of interest?
   (b) What statistic might you calculate from the sample?

2.11 A company that manufactures electronic switches wishes to provide the customer with defect-free switches.
   (a) What parameter is of interest?
   (b) What statistic might you calculate from the sample?

2.12 The university wishes to know how students view the new athletic center on campus.
   (a) What parameter is of interest?
   (b) What statistic might you calculate from the sample?

2.13 The U.S. government wishes to know how many people are unemployed.
   (a) What parameter is of interest?
   (b) What statistic might you calculate from the sample?

2.14 The manufacturer of disposable diapers wishes to know how much fluid the diapers can hold before they leak.
   (a) What parameter is of interest?
   (b) What statistic might you calculate from the sample?

2.15 The Coca-Cola Company wishes to know the proportion of people who prefer Coke over Pepsi.
   (a) What parameter is of interest?
   (b) What statistic might you calculate from the sample?

2.16 Before selecting a major, a student decides to study the salaries of individuals working as accountants.
   (a) What parameter is of interest?
   (b) What statistic might you calculate from the sample?

2.4   FACTORS THAT INFLUENCE SAMPLE SIZE: SOME SAMPLING AND SAMPLE SIZE CONSIDERATIONS

The next question is how big does the sample need to be? Do we need to eat half of one pie to know how all the pies in a batch taste? What factors will influence our decision in this matter? We touched on this question when we observed the impact of the sample size and the amount of variability on the size of the sampling error. This section reinforces the connection between the sample size and the two factors we have already studied: amount of variability in the population and the sampling error. In addition, two other factors are introduced.

2.4.1 Size of the Population

The first factor is the **size of the population.** How many tubes of glue are manufactured each day? How many employees are there in the company? How many golf balls are manufactured each hour? How many boxes of tissues are produced in a day? These numbers are the size of the population. Intuitively it should seem that it would take a different sample size to learn about a population of 1000 compared to a population of 100,000. The population size is a factor, but it turns out not to be as important as some of the other factors we have identified.

> The *size of the population* is the number of members of the population. It will be referred to as N.

2.4.2 Extent of Resources Available

The next factor is the amount of time, money and other resources that you have available. If you need a decision in six months rather than next week, the amount of data you could possibly collect and analyze is different. Also, remember that it is expensive to collect and process data, so cost must be a factor. However, we still have not identified the two most important factors.

The amount of resources available impacts sample size.

2.4.3 Amount of Error That Can Be Tolerated

In Section 2.2.4 we learned that whenever you use a sample to draw conclusions about the population you will have some sampling error. And the bigger the sample we pick, the less error we will have in our conclusions. Returning to the ovals in Figure 2.1, you can see that if you take a bigger sample, then you have captured more of the population. Therefore, the conclusions you draw about the population based on the sample will be more accurate. This is shown by the increasing sample sizes in Figure 2.2. The numbers shown in the large oval are the ages of members of the population being studied. You can see that the smallest oval represents a sample size of 5, since there are 5 members of the population in it. The next larger oval represents a sample of size 10. In addition to the 5 members contained in the small oval, it contains 5 more members of the population.

Remember! A sampling error does not imply that you did anything wrong. It results simply because you have an incomplete picture of the population.

If we continue taking increasingly bigger sample sizes, eventually we will end up taking a census, or studying the entire population. This is represented by the largest

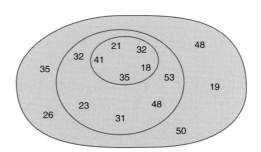

FIGURE 2.2   Bigger and bigger samples

oval. When this happens, the sample size is as big as the population size, N, and there is no sampling error.

It seems, then, that if we can live with a little more error, then we can get by with a smaller sample size. The more costly the error, the larger our sample size will need to be. Thus, the *amount of sampling error* (dictated by the cost of the error) we can live with is a factor in determining sample size. We return to this matter in greater detail in Chapters 9–10.

The amount of error that we can afford is an important factor in determining sample size.

2.4.4 Amount of Variation in the Population

Even if we can tolerate only a small amount of error we still may not need a really large sample size. One other factor is very important: the *amount of variability* that exists in the population. Suppose for a moment that I wish to study all the students in your college. I wish to be very accurate in my conclusions but all of the students feel exactly the same about the issue I am studying. How many students do I need to talk to in order to get very precise information about the population? The correct answer is that only 1 student is needed. A sample size of 1 is adequate despite the need to be very accurate and despite the large size of the population. This is because there is no variation in the population; that is, everyone feels the same way about the issue. Granted, this is not likely to happen, but you can see the impact that this factor has on the sample size needed. The more similar the members of the population are to each other, the less variation there is within the population. So the sample size can be smaller. If the population is highly diversified then you will need to talk to more people to get a sense of how the population feels about the issue.

The amount of variation in the population is a key factor in determining sample size.

A natural question to ask is, How will we know the amount of variability of the population to use in determining the sample size? Considering that we are taking a sample to estimate such parameters as variability in the population, it seems a bit of a problem if we need to know the variability to determine the sample size. We are locked in a loop at this point, a classic "catch 22" situation. We could estimate the amount of variability using information about the amount of variability in a population similar to the one we are studying; we could use previous studies of the same population to give us an idea of the amount of variability; or we could do a pilot study. These are some of the ideas for dealing with this dilemma and they are addressed more completely in Chapter 9.

How will we know the amount of variability in the population?

2.4.5 Summary of Factors Influencing Sample Size

We have identified the following factors that are important in determining the size of the sample needed:

- The amount of variation in the population
- The amount of error that can be tolerated
- The amount of resources available for the project
- The size of the population, N

You will see that the first two factors show up in the formula for determining the size of the sample. These are developed in Chapter 9. For now it is sufficient for you to have a general understanding of the impact that each of these factors has on the sample size determination.

As a final note in this section, we should agree on a label for the sample size. No matter what statistics book you pick up you will always see the **sample size** referred to as **n**. Remember that the size of the population is labeled N.

The *size of the sample* will be referred to as n.

2.5   SELECTING THE SAMPLE

Now that you have a general idea of which factors influence the size of the sample to be selected, we are ready to think about how we could select a sample. In the *Try It Now!* exercise on page 23 you easily drew another oval to represent another sample that could be picked from the population. Let's think for a moment about what qualities we would like our sample to have.

2.5.1 Selecting an Unbiased Sample

Ideally, we would like the sample to be a mini version of the population. Remember that we will be using the sample to understand the population. The sample should thus contain all the key features found in the population. It is easiest to understand what this means by looking at some examples of samples that may *not* be a mini version of the population.

EXAMPLE 2.8   The Glue Company

Selecting a Sample

Suppose the glue company decided to use the first 50 tubes of glue made on a Monday morning as the sample.

Why is this not a good idea? It is possible that the manufacturing process varies as the machine runs for a long period of time. It is also possible that different people operate the machine at different times of the day and on different days of the week. And it is possible that a different batch of raw material might lead to differences in the glue performance. If our sample consists of the first 50 tubes made on Monday morning, we may not see the impact of any of these factors on the glue performance. ∎

EXAMPLE 2.9   Dress Down Day

Selecting a Sample

Suppose the company considering a dress down day decides to use the recently hired female employees as their sample.

Again the problem is that this sample may not truly represent the population of employees of the company. Why not? First of all, the people in this sample are all women. Maybe women in this company are consistently more or less productive than men. Secondly, they are all new hires and are likely to be putting their best effort into the job. At the same time, they are in training and may make more than the average number of errors. For all of these reasons and probably others this group is not a mini version of the employees of this company. ∎

In both of these situations the proposed sample may not let us see all of the variation that exists in the population. Another way to express this is to say that these samples might be **biased.** The word bias in statistics has the same meaning that it has in ordinary usage. It means that the sample is somehow not a fair reflection of the reality we would see in the population. A biased sample would give us an unfair or prejudiced view of the population. This is precisely what we wish to avoid. Let's capture this in a definition.

A *biased sample* is a sample that does not represent the population.

 *TRY IT NOW!*

Stress Relief

Identifying Possible Biases in a Sample

You are studying the methods that students at your school use to relieve stress. You decide to use your statistics class as your sample. Why might this be a biased sample?

2.5.2 Selecting a Simple Random Sample

In selecting any sample, we wish to pick an unbiased or fair sample. We also wish to pick a sample that contains as much information as possible. There are many ways to pick a sample but for the purposes of an introductory course we are going to use what is known as a **simple random sample (srs).** This means that each member of the population has an equal chance of being selected as a member of the sample. This is equivalent to placing the names of all the members of the population in a hat and reaching in and picking out members to be in the sample.

> A *simple random sample* is a sample that has been selected in such a way that all members of the population have an equal chance of being picked.

Simple random sampling is the most obvious way to select a sample. In fact, you might wonder how else you could do it. Suppose you wished to learn about prices charged for advertising in newspapers. Your population would be all companies located in the United States that publish newspapers. If you picked a simple random sample of newspaper companies you might end up with all small companies purely by chance. You might not get any of the "biggies" such as the *New York Times*, the *Boston Globe*, the *Chicago Tribune*, or the *Los Angeles Sun* in the sample. Is this bad? Well, that depends on what variable you are studying. We are interested in prices for advertising in these papers and the larger companies might set the pace for the smaller companies with regard to pricing. In this case you might choose to use something other than simple random sampling in order to guarantee that you had some "biggies" in your sample.

As you learn more about statistics you will see that simple random sampling does not always provide you with the most information for your money. All of the

ANS. THE STUDENTS IN THIS CLASS MAY BE MOSTLY SOPHOMORE BUSINESS STUDENTS.

concepts that you will learn in this book carry over to the other kinds of selection methods, but the formulas might have to change a bit if you use a procedure other than simple random sampling. In Chapter 14 you will learn how to select a sample by designing an experiment.

2.5.3 The Sampling Frame

Now that we have decided that we want to select a simple random sample, we must look at what is needed to pick the sample. Returning to the example of the newspapers from the previous section, we can see that we need a list of all the newspaper publishing companies in the United States. Such a list is known as the **sampling frame.**

> A *sampling frame* is a list of all members of the population.

There are government agencies and private companies, such as Dun & Bradstreet, whose function is to collect data for sampling frames. However, creating a sampling frame may take a fair amount of time, energy, and money. Thus, the list from which you pick your sample may not be accurate or complete. The difficulty in creating these lists is not the focus of this course or this book. However, you should know what a sampling frame is and know enough to watch out for potential problems.

2.5.4 Using a Table of Random Numbers to Select the Sample

Once you have the sampling frame and you have determined your sample size (using the formula developed in Chapter 9), you are ready to select your sample. Although it is easy to understand the idea of placing the name of each member of the population in a hat and selecting the sample by picking names from the hat, this is not a practical way to select the sample. To imitate this process you can use a **table of random numbers.**

> A *table of random numbers* is a table that consists of a list of numbers randomly generated and listed in the order in which they were generated.

Such a table is provided in Table 1 in Appendix A. A portion of this table is shown in Figure 2.3.

Row	Column 1	2	3	4	5	6	7
1	094632795	711501513	537971597	562758635	410398128	182794408	773761503
2	033413186	653475420	289063704	485441982	460744361	328703833	289612212
3	297556368	658953044	738968017	414437050	296126017	075254187	702140315
4	472960570	785645638	574817322	817883255	976076280	843373358	118284363
5	256883707	716249997	378236162	467694224	193707682	380141891	605807481
6	179451522	878902420	602450872	987686989	686677180	242196303	517640224
7	894964682	704841116	241902107	750429362	794778197	693242123	316755091
8	738120861	744470405	873393138	758824215	394004646	496696605	006936567
9	803156944	653387115	716335974	835667154	066959782	908783760	165946696
10	187636922	321421098	638210137	055734541	493193305	566923120	435549770

FIGURE 2.3   Portion of a table of random numbers

A random number table should contain roughly as many zeros as ones as twos and so forth. There is a way to check to see that such a table does in fact contain random numbers, but for our purposes we will assume that the random number table has been correctly generated.

To use this table to identify which members of our population will be selected to be in the sample, we must first assign each member of the population an identification (ID) number. Suppose that you are studying the students at your college or university. Each student probably has a student ID number. Since many schools use the student's social security number as the ID number, let's suppose that the ID number is a 9-digit number. To use a table of random numbers to select a sample of 30 students from this population, you would first select where in the random number table you should start reading. One way to do this is to close your eyes and point to a spot in the table. Suppose that you did this and you selected row 10, column 2 as your starting point. Then you would read the next nine digits from Figure 2.3 as 321421098. This is the ID number of the first member of your sample. The next 9-digit number from the table is 638210137 and this is the ID number of the second member of your sample. You would continue this process until you have selected 30 ID numbers.

The sample obtained in this manner is a simple random sample. However, you would certainly need to read more than 30 ID numbers from the table to get 30 usable ID numbers. Most of the ID numbers you read from the table of random numbers will not correspond to anyone at your school. This would certainly happen, especially if the school is using the student's social security number as the ID number. Although you will eventually find 30 usable ID numbers, you will waste an incredible amount of time.

This method works a little better when you can assign the ID codes to the members of your population in such a way that they are sequential, so that you will not select any nonusable ID numbers. Let's see how this works in the following example.

EXAMPLE 2.10   Student Views

Selecting a Sample Using a Table of Random Numbers

Each student at your college has a mailbox on campus. The mailboxes are numbered from 0000 to 9000. To select a simple random sample of 10 students we can select 10 mailbox numbers at random using the random number table. Suppose we choose to start at row 7, column 3 of the table shown in Figure 2.3. The first student selected has mailbox number 2419, which is a valid number. Continuing to read off 4-digit numbers from this table, the second number selected has mailbox number 0210. The list of all 10 mailbox numbers selected is:

| 2419 | 0210 | 7750 | 4293 | 6279 | 4778 | 1976 | 2123 | 3167 | 5509 |

Since the numbers are organized in 9-digit blocks and we need only 4-digit mailbox numbers, we just kept reading the numbers sequentially, wrapping down to the next row upon reaching the end of column 7. Notice that the selection after 1976 would have been 9324, which is not a valid mailbox number, so it was simply skipped.

The students who have these mailboxes are the students in the sample.   ■

For the most part, this course and this book assume that the sample has been selected and the data have been collected. Your job is to see the information in the data and make some recommendations and/or decisions based on the data. We do not focus on the actual selection of the sample or the job of collecting the data. This is the subject of another course.

TRY IT NOW!

The Glue Company
Selecting a Simple Random Sample

Select a sample of 5 tubes of glue for the glue company. You can assume that each tube of glue has a 5-digit ID number, which the company uses to track its inventory.

Discovery Exercise 2.2
INTRODUCTION TO SAMPLING

Suppose the data shown below represent an entire population. They show the number of people in 50 families living in a small college town in New England. (If you did Discovery Exercise 2.1 then you will recognize this as the same data set.) Now a 2-digit ID number has also been included.

New England families: large amount of variability
Average number of people in 50 families: 4.50

ID: 01	1	ID: 02	4	ID: 03	5	ID: 04	7	ID: 05	8
ID: 06	3	ID: 07	9	ID: 08	8	ID: 09	8	ID: 10	8
ID: 11	4	ID: 12	9	ID: 13	9	ID: 14	1	ID: 15	6
ID: 16	4	ID: 17	1	ID: 18	3	ID: 19	9	ID: 20	7
ID: 21	8	ID: 22	2	ID: 23	3	ID: 24	1	ID: 25	9
ID: 26	1	ID: 27	7	ID: 28	5	ID: 29	1	ID: 30	1
ID: 31	1	ID: 32	6	ID: 33	8	ID: 34	2	ID: 35	9
ID: 36	4	ID: 37	1	ID: 38	1	ID: 39	1	ID: 40	3
ID: 41	4	ID: 42	2	ID: 43	4	ID: 44	9	ID: 45	4
ID: 46	1	ID: 47	3	ID: 48	8	ID: 49	1	ID: 50	1

Step 1:   Select a sample of 5 numbers from this population. Use the table of random numbers to do this. Record your sample in the table at the top of page 39.

	Sample from NE families
Selection 1	
Selection 2	
Selection 3	
Selection 4	
Selection 5	
Average	

Step 2: Calculate the sample average of the 5 numbers by adding them together and dividing by 5.

Step 3: Calculate how far away the sample average is from the true population average by subtracting the sample average from the population average that is provided for you.

Step 4: If you did Discovery Exercise 2.1, compare the sample you selected using the random number table to the sample you selected without the use of the table. Which one had a sample average closer to the population average?

Repeat steps 1–4 with a sample of 10 numbers.

Step 1: Select a sample of 10 numbers from each population. Use the table of random numbers.

	Sample from NE families		Sample from NE families
Selection 1		Selection 7	
Selection 2		Selection 8	
Selection 3		Selection 9	
Selection 4		Selection 10	
Selection 5		**Average**	
Selection 6			

Step 2: Calculate the sample average of the 10 numbers by adding them together and dividing by 10.

(continued)

Step 3:   Calculate how far away the sample average is from the true population average by subtracting the sample average from the population average that is provided for you.

Step 4:   If you did Discovery Exercise 2.1, compare the sample you selected using the random number table to the sample you selected without the use of the table. Which one had a sample average closer to the population average?

2.5.5 Exercises—Learning It!

2.17   The President of the United States wishes to see how popular he/she is after 2 years in office. A sample of 1000 voters is taken from the state of California.
   (a) Why might this be a biased sample?
   (b) How could you get a simple random sample?

2.18   Tasty Ice Cream Corporation wishes to be sure that all of the half-gallon ice cream cartons do contain one-half gallon of ice cream. A sample of 30 cartons are measured. All of the cartons in the sample were filled on Friday.
   (a) Why might this be a biased sample?
   (b) How could you get a simple random sample?

2.19   A company that manufactures electronic switches wishes to provide the customer with defect-free switches. A sample of 5 switches is selected every hour throughout the day. Explain why this is most likely an unbiased sample.

2.20   The university wishes to know how students view the new athletic center on campus. A questionnaire is distributed to people as they enter the building.
   (a) Why might this be a biased sample?
   (b) How could you get a simple random sample?

2.21   The U.S. government wishes to know how many people are unemployed. A sample of 1000 individuals over age 18 is selected from a national listing. Explain why this is probably an unbiased sample.

2.22   The manufacturer of disposable diapers wishes to know how much fluid the diapers can hold before they leak. The diapers are put on children who are playing and fluid is injected every 15 minutes until the diaper leaks. Explain why this is probably an unbiased sample.

2.23   The Coca-Cola Company wishes to know the proportion of people who prefer Coke over Pepsi. A sample of 100 people at a county fair is taken.
   (a) Why might this be a biased sample?
   (b) How could you get a simple random sample?

2.24   Before selecting a major, a student decides to study the salaries of individuals working as accountants. A sample of accountants is selected from the list of alumni of the school that the student is attending.
   (a) Why might this be a biased sample?
   (b) How could you get a simple random sample?

2.6 SOURCES AND TYPES OF DATA

Although we will not study the whole area of data collection, we do need to think about the different sources of data and the kinds of data or variables that we might face.

There are two major sources of data, primary sources and secondary sources. Data that come from a primary source are called **primary data** and data that come from a secondary source are called **secondary data.**

> *Primary data* are data that are obtained and used by the organization or individual that actually collected them.

> *Secondary data* are compiled data that are taken from several primary sources and synthesized or summarized in some way.

Some examples of primary sources of data are the U.S. Bureau of the Census, or an individual or company who distributes a survey on job satisfaction within an organization or industry. Primary data are often referred to as raw data and are most often collected by the person or organization using the data.

Secondary data are data collected from various sources and then either summarized or combined with other data in some way. Some examples of secondary data sources are the *Statistical Abstract of the United States* or the World Bank. Secondary data are data that were not collected by the individual who is using them, but were obtained from another source.

Once we have the data, the statistical analysis that we do depends on the type of data we have. There are two major types of variables, **qualitative** and **quantitative.** In this section, we will concentrate on being able to identify the different types of data. Since there are different statistical techniques for the different types of data, it is important to identify the type of data you have before you analyze them so that you don't use the wrong technique.

2.6.1 Qualitative Data

Qualitative data, also known as **nominal** or **categorical** data, are the simplest form of data. Examples of qualitative data are variables such as gender (male or female) or the expected grade in a course (A, B, C, D, or F). Each item in the sample falls into one of a finite number of possible categories.

> *Qualitative data* describe a particular characteristic of a sample item. They are most often non-numerical in nature.

Suppose you are interested in learning about the length of time that a certain glue, type C, adheres. You collect the data from a sample of glue tubes and after analyzing it you find that there appear to be two groups in the data. Each group has sticking times that cluster around a different value. Why would this occur? When you ask some questions about how the data were collected you learn that some of the tubes came from machine A and some came from machine B. When you return to the sticking times you find that all of the machine A values are clustered at the lower number while those of machine B are clustered at the higher number. The variable *machine* is an example of qualitative data. Knowing which machine produced the tube helps to explain the differences in the glue sticking times that you saw in the data.

 *TRY IT NOW!*

Dress Down Day
Possible Qualitative Variables

The company considering the dress down day tries this policy out with a sample of employees. After the policy has been in effect for some time, the company decides to measure the change in productivity for the employees. It appears that for some employees productivity has increased, while for others it has decreased. What qualitative data might have been collected to help the company understand the differences observed?

The statistical techniques that are used to analyze qualitative data are limited, but are often critical to our understanding of the results of statistical analyses. When you are collecting data it is important to think of qualitative data that may be relevant to the problem to be solved. Qualitative data are easily collected at the time, but almost impossible to reconstruct after the fact. When in doubt, qualitative data should always be collected.

Sometimes numbers are used to classify qualitative data. For example, in surveys that ask for gender we often find that a 0 is used to denote "male" and a 1 is used to denote "female." When this is the case the data are referred to as **nominal** data. The numbers are used simply to represent different categories and have no real meaning as numbers.

> Data that are created by assigning numbers to different categories when the numbers have no real meaning are called *nominal data.*

Nominal data are treated the same way as ordinary qualitative data.

The order in which numbers are assigned to qualitative data may have some meaning. When numbers are used to name ordered categories the data are called **ordinal.** In the case of gender, the numbers 0 and 1 could very easily be reversed and so they have no intrinsic meaning. But suppose you asked a group of people to rank five different versions of a new soft drink. The resulting data might look like this:

Version	3	2	5	1	4
Ranking	1	2	3	4	5

In this case the numbers are not entirely meaningless since they indicate a relative position for each version of the product on a scale. However, you cannot tell from this scale whether the person doing the ranking liked version 3 a lot and really hated the other four or whether versions 3 and 2 were similar and much superior to the remaining three versions. The distances between the assigned numbers are not necessarily equal.

Another example of ordinal data would be when a characteristic of a sample item, such as income, is classified as 1 = low, 2 = medium, and 3 = high. Here, the numbers have a relative ordering, but there is no way to compare 1 to 2 to 3 numerically.

> Data that are created by assigning numbers to categories where the order of assignment has meaning are called *ordinal data.*

Most often, ordinal data are analyzed using the same graphical techniques that apply to other qualitative data. There are some limited statistical techniques that can be used to further analyze ordinal data. These techniques are called nonparametric analyses.

2.6.2 Quantitative Data

Data that are inherently numerical in form are called **quantitative** data. This type of data falls into two different categories: **discrete** and **continuous.**

Discrete data usually result when the data being collected involve *counting* or *enumerating.* The only possible values are positive integers or whole numbers. Examples of discrete data are the number of prior convictions for a person who has been arrested, or the number of defective items in a sample.

> *Discrete data* are data that can take on only certain values. These values are often integers or whole numbers.

Discrete data that result from counting the number of times that something occurs are important. They are the type of data used in analyzing many public opinion polls and other surveys. You will encounter this kind of data many times in the text.

> *Continuous data* are data that can take on any one of an infinite number of possible values over an interval on the number line. These values are most often the result of measurement.

Numerical data that are not discrete are called **continuous.** Continuous data occur most often as the result of measuring and are sometimes referred to as *measurement data.* Continuous data usually consist of real numbers, which, as far as we are concerned, will be in decimal form.

Some researchers make a further distinction with continuous data by classifying the data as interval or ratio data. **Interval data** involve a variable that does not have an absolute zero in its scale. That is, there is no common place that is recognized as the beginning of the scale.

> *Interval data* are data that can be compared only by looking at the difference between two values, or the interval between them.

There are not many examples of interval data. The most common example is temperature. If one temperature measurement is 40° and another is 80°, you can say that one is 40° more than the other, but it does not make sense to say that the 80° measurement is twice as hot (80/40 = 2) as the 40° measurement. Other examples of interval data are variables measured on a logarithmic scale such as pH, sound intensity, and earthquake intensity.

In practice, almost all data are ratio data and are compared either by looking at the difference between the measurements or by taking their ratios.

> *Ratio data* are data that can be compared by looking at either the difference between two values or the ratio of two values.

EXAMPLE 2.11   Study Time

An Example of Continuous Data

Suppose you are interested in learning about the length of time that a typical student spends studying statistics on any given night. Initially, there are an infinite number of possibilities for values, ranging from 0 minutes up. These data are continuous. Once you decide on a measuring device the number of possibilities is limited, but the limitation is caused by the measuring device and not the variable itself. For example, if you measure the time spent studying to the nearest hour, the possible values are 0, 1, 2, ...   If, however, you decide to measure to the nearest one-tenth of an hour (6 minutes), the possible values are 0.0, 0.1. 0.2, . . . , 3.0, 3.1, ... The more precise your measuring device, the more possible values your data can assume.   ■

2.6.3 Exercises—Learning It!

2.25   The President of the United States wishes to see how popular he/she is after 2 years in office.

   (a) What is one variable of interest?

   (b) Is the variable qualitative or quantitative?

   (c) If it is qualitative, is it nominal or ordinal data? If it is quantitative, is it discrete or continuous?

2.26   Tasty Ice Cream Corporation wishes to be sure that all of the half-gallon ice cream cartons do contain one-half gallon of ice cream.

   (a) What is one variable of interest?

   (b) Are the data qualitative or quantitative?

   (c) If the data are qualitative, are they nominal or ordinal? If they are quantitative, are they discrete or continuous?

2.27   A company that manufactures electronic switches wishes to provide the customer with defect-free switches.

   (a) What is one variable of interest?

   (b) Are the data qualitative or quantitative?

   (c) If the data are qualitative, are they nominal or ordinal? If they are quantitative, are they discrete or continuous?

2.28   The university wishes to know how students view the new athletic center on campus.

   (a) What is one variable of interest?

   (b) Are the data qualitative or quantitative?

   (c) If the data are qualitative, are they nominal or ordinal? If they are quantitative, are they discrete or continuous?

2.29   The U.S. government wishes to know how many people are unemployed.

   (a) What is one variable of interest?

   (b) Are the data qualitative or quantitative?

   (c) If the data are qualitative, are they nominal or ordinal? If they are quantitative, are they discrete or continuous?

2.30   The manufacturer of disposable diapers wishes to know how much fluid the diapers can hold before they leak.

(a) What is one variable of interest?

(b) Are the data qualitative or quantitative?

(c) If the data are qualitative, are they nominal or ordinal? If they are quantitative, are they discrete or continuous?

2.31   The Coca-Cola Company wishes to know the proportion of people who prefer Coke over Pepsi.

(a) What is one variable of interest?

(b) Are the data qualitative or quantitative?

(c) If the data are qualitative, are they nominal or ordinal? If they are quantitative, are they discrete or continuous?

2.32   Before selecting a major, a student decides to study the salaries of individuals working as accountants.

(a) What is one variable of interest?

(b) Are the data qualitative or quantitative?

(c) If the data are qualitative, are they nominal or ordinal? If they are quantitative, are they discrete or continuous?

2.7   THE DIFFERENCE BETWEEN DESCRIPTIVE STATISTICS AND INFERENTIAL STATISTICS

So far, we have looked at the concept of population and recognized the usefulness of the sample in gleaning information about the population. We have discussed factors that influence how large the sample needs to be and how we go about picking the sample. And we have looked at the types of data that might be collected from the sample. Now we need an answer to the question, What do I do with the data? The right answer is, It depends. It depends on what type of data you have and on what questions you want answered about the behavior of the population. The rest of this book is devoted to providing tools to answer the question, What do I do with the data? These tools fall into two main categories: the tools of descriptive statistics and the tools of inferential statistics. Although these tools support each other, the jobs they do are quite different.

2.7.1 Descriptive Statistics

The tools of descriptive statistics are usually the first ones encountered in any data analysis. They allow you to *describe* the sample. However, that is all they do. This is generally the first step in any data analysis, once the data have been collected, of course. If you were given the results of a survey on customer satisfaction from 100 customers, the first thing you would want to do is to get a handle on the data by summarizing them. This is precisely what the tools of descriptive statistics do for you: summarize the data.

Tools of descriptive statistics allow you to summarize the data.

There are *graphical or visual descriptive tools,* which generally include bar charts, pie charts, and histograms. These tools are discussed in Chapter 3. Graphical tools help you to see how the data behave and to summarize the data visually. These tools are used all the time. You frequently will see them in newspapers and in reports. Some examples are shown in Figure 2.4 on page 46.

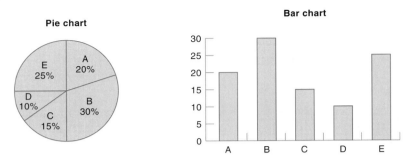

FIGURE 2.4   Pie chart and bar chart

There are also *numerical descriptive tools,* which allow you to summarize the data numerically. Typically, a numerical summary would provide you with such *statistics* as the average, the median, the mode, and the largest and smallest data values. We used the word statistics in the last sentence because that is precisely what these values are. Remember that a statistic is simply a number that describes the behavior of the sample. These numbers are covered in more detail in Chapter 6.

Descriptive tools would be adequate if all we wanted to do was describe the sample. But remember that the sample is a means to an end. At the end of the day it is not the sample that we care about but the population. Somehow we must make the leap from the information contained in the sample to the population behavior. To do this we need the tools of inferential statistics.

2.7.2 Inferential Statistics

Let's start by defining the word **inference.** If your friend told you not to infer that all business students are brilliant, then your friend is telling you not to reach that *conclusion!*

> An *inference* is a deduction or a conclusion.

In our case we wish to draw conclusions about the behavior of the population based on the information in the sample. We have already seen that the sample is only a piece of the population, so we will not have complete information. We have also seen that if the sample is properly selected, then the information contained in the sample will give us reliable information about the population.

How are we going to make this leap from describing the sample data to drawing an inference or conclusion about the behavior of the population? The answer is found in the subject matter of probability. Indeed, without the tools of probability, we could not do anything more than describe the sample data. We could do no inferential statistics.

Think about the population and the sample as the circles shown in Figure 2.5. The larger circle represents the population and the smaller circle represents the sample. This figure is being used to help you see how probability is needed in order to draw inferences about the population from the sample. In reality, the sample circle should sit inside the population circle, because it is a piece of it. But for now let us continue with this picture. The line that moves from the sample to the population is labeled "Inferential statistics." The tools of inferential statistics are used to move from

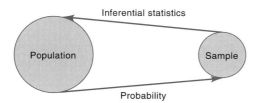

FIGURE 2.5 Relationship between probability and inferential statistics

sample information to population information. Let's save that thought as a definition.

> The ***techniques of inferential statistics*** allow us to draw inferences or conclusions about the population from the sample.

To understand the line that moves from the population to the sample, consider the following situation. You and I have decided to play poker and I have dealt us each 5 cards. You see that you have a full house and are wildly excited. However, your excitement is quickly contained as I reveal to you that I have not one, not two, not three, not four, but *five* aces! Immediately you accuse me of cheating (imagine that!). You quickly conclude that I have not used a standard deck.

Let's take a closer look at how you reached that conclusion. First of all, the deck of cards is our population. You had a sample of 5 cards selected from that population and I had a sample of 5 cards from that population. On the basis of the information in my sample you calculated the probability of observing 5 aces if I was using a standard deck and found it to be zero. There is no chance of getting 5 aces if the population has the behavior of a standard deck. So you quickly rejected the notion that I was using a standard deck. This is precisely what we need to do in general. We need to draw conclusions about a population based on the observed sample and the theory of probability. The conclusions will not always be so obvious as this example. So we have a definition for the line labeled "probability," which leads from the population to the sample.

Reaching a conclusion based on a sample. Did I cheat?

> We will use ***probability*** theory to calculate the likelihood of observing or selecting a particular sample from a population.

Although the branch of mathematics known as probability theory is interesting in its own right, we present only those particular parts of probability that will help us in our ultimate goal: to make inferences about the population from the sample.

Let's summarize Figure 2.5. If we make the trip across the bottom line we would be making some assumptions about the population behavior and determining the likelihood of various samples that might come from this population. If we make the trip across the top line we are taking the information in the sample and drawing conclusions about the population from which it came. The tools of probability ultimately allow us to accomplish what we want, which is to do inferential statistics.

In all of the material that follows this chapter, you will need to be familiar with basic summation notation. If you are familiar with this notation, you may skip the next section. But you may wish to use it as a quick review and refer to it as you proceed through the rest of the book.

2.8   BASIC SUMMATION NOTATION

Before we go much further, it is useful to introduce some shorthand notation for writing statistical formulas. This notation is called the **sigma notation.**

> **Sigma notation** is shorthand notation used to write formulas. It is so named because it uses the Greek capital letter sigma, written as Σ.

2.8.1 Summing the Data

Remember that we agreed to use the letter n *to represent the size of the sample.*

If we have a sample of size n, we can refer to the individual data values as x_1, x_2, x_3, . . ., x_n, where x_1 represents the first data value, x_2 the second data value, and so on. Many of the formulas you encounter will require that you sum the data values. We could write the sum as $x_1 + x_2 + x_3 + \cdots + x_n$ but this will get cumbersome and be awkward to use.

Using the sigma notation, we can write the sum $x_1 + x_2 + x_3 + \cdots + x_n$ as

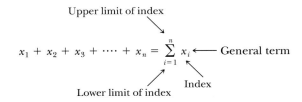

$$x_1 + x_2 + x_3 + \cdots + x_n = \sum_{i=1}^{n} x_i$$

The sigma notation on the right of the equation is read as "the sum of x_i as i goes from 1 to n." The letter under the Σ is called the index and it really doesn't matter what letter you use since you use the same letter in describing the general term. The value of the index indicated on the bottom of the Σ is the starting value for the index and is usually 1, meaning that we are going to start with x_1. The value on the top of the Σ is the ending value for the index and is usually n, the sample size. When the index goes from $i = 1$ to n, this indicates that all of the data values in the sample are being used.

Let's try some examples.

EXAMPLE 2.12   The Glue Company

Using the Summation Notation

The glue company studies the performance (i.e., the stickiness) of 10 tubes of type C glue. The data are recorded in terms of number of days until the glue failed. This is what the company found: 233, 167, 289, 221, 254, 198, 210, 225, 240, 203. Thus, x_1 is 233, the first member of the sample. In a similar fashion we can label the rest of the data values: $x_2 = 167$, $x_3 = 289$, $x_4 = 221$, $x_5 = 254$, $x_6 = 198$, $x_7 = 210$, $x_8 = 225$, $x_9 = 240$, and $x_{10} = 203$. The value of n is 10, the size of the sample.

To find $x_1 + x_2 + x_3 + \cdots + x_{10}$ we will first write it as $\sum_{i=1}^{10} x_i$. This means that we start the value of i at 1 and start with x_1, which we know is 233. Check to see if the value for i is at the upper limit for i, which in this case is 10. Since i is only 1 we continue. Next put a plus sign and change i to be 2 and add in x_2, which we know to be 167. So far we have $233 + 167$. Again we check to see if the value for i is bigger than the upper limit of 10. It is not, so we continue by putting a plus sign and changing i to 3 and adding x_3, which we know is 289. Now we have $233 + 167 + 289$. This process continues until the value of i is 10 and we have written down $233 + 167 + 289 + 221 + 254 + 198 + 210 + 225 + 240 + 203$ and obtained 2240.

 *TRY IT NOW!*

The Mail-Order Company

Using Sigma Notation

A mail-order company wants some information on the daily demand for a product that has been heavily advertised. The company looks at the orders for an 8-day period and obtains the following data:

Demand   31   28   29   32   30   31   29   30

Use the sigma notation to write down the expression which means to add up all the data values.

2.8.2 Summing Differences

When we analyze data we often need to look at how *far away* data values are from some number. Clearly, we could state how far away any one of the data values was from a number by simply subtracting the two numbers. For the glue company data shown in Example 2.12 we could say that x_1 is 9 days away from a number, for instance 224. Remember that x_1 is 233. In addition to being interested in how far away x_1 is from 224, we often have to state how far away the whole data set is from 224. To do this we have to subtract 224 from each value in the data set, each x_i, and then add up all these differences.

EXAMPLE 2.13   The Glue Company

Use of Sigma Notation to Calculate Differences

Suppose we wish to quantify how far away the whole data set is from 224 for the glue company. The calculation we want is

$$(233 - 224) + (167 - 224) + (289 - 224) + (221 - 224) + (254 - 224) + (198 - 224)$$
$$+ (210 - 224) + (225 - 224) + (240 - 224) + (203 - 224)$$
$$= (9) + (-57) + (65) + (-3) + (30) + (-26) + (-14) + (1) + (16) + (-21)$$
$$= 0$$

Using the notation x_1 to represent the first data value and x_2 to represent the second data value, we can write this expression more generally as

$$(x_1 - 224) + (x_2 - 224) + (x_3 - 224) + (x_4 - 224) + (x_5 - 224)$$
$$+ (x_6 - 224) + (x_7 - 224) + (x_8 - 224) + (x_9 - 224) + (x_{10} - 224)$$

Taking this one step further, we can use the sigma notation to make this expression less cumbersome:

$$\sum_{i=1}^{10} (x_i - 224) \qquad\qquad \blacksquare$$

This shorthand way of writing the long summation will help us when we get to Chapter 4. Now you try one of these.

 *TRY IT NOW!*

The Mail-Order Company
Using Sigma Notation to Sum Differences

A mail-order company wants some information on the daily demand for a product that has been heavily advertised. The company looks at the orders for an 8-day period and obtains the following data:

Demand   31   28   29   32   30   31   29   30

Use the sigma notation to write down the expression that means to add up all differences between the data values and the number 30.

2.8.3 Exercises—Learning It!

2.33 Tasty Ice Cream Corporation wishes to be sure that all of the half-gallon ice cream cartons do contain one-half gallon of ice cream. A sample of 30 cartons are inspected. All of the cartons in the sample were filled on Friday. The data (in gallons) are

0.51 0.50 0.49 0.48 0.51 0.50 0.49 0.50 0.50 0.48 0.53 0.50 0.49 0.49 0.51
0.50 0.50 0.51 0.52 0.50 0.48 0.49 0.50 0.47 0.49 0.51 0.50 0.50 0.50 0.49

(a) Write the sigma notation to represent the sum of the 30 data values.
(b) Tasty would like to know how far away from 0.5 (one-half gallon) the data are on the whole. Write the sigma expression to find the sum of the difference between each data value and 0.5.
(c) Evaluate the expression you constructed in part (b).

2.34 The manufacturer of disposable diapers wishes to know how much fluid the diapers can hold before they leak. The diapers are put on 10 children who are playing and fluid is injected every 15 minutes until the diaper leaks. The weight of the diaper (in grams) when it fails is recorded. The data are

503   513   489   499   520   511   525   494   498   501

Write the sigma notation to represent the sum of these data values.

2.35 Before selecting a major, a student decides to study the salaries of individuals working as accountants. A sample of 15 accountants is selected from the list of alumni of the school that the student is attending. The data are

$25,100 $33,000 $27,500 $29,000 $35,000 $26,400 $29,100
$31,050 $32,100 $29,100 $40,000 $21,000 $25,000 $26,500 $33,000

ANS. $\sum_{i=1}^{8} (x_i - 30)$

(a) Write the sigma notation to represent the sum of all the salaries in the sample.

(b) Write the sigma notation to represent the sum of the first 5 salaries in the sample.

(c) Write the sigma notation to represent the sum of the last 5 salaries in the sample.

2.9 SELECTING A SAMPLE IN EXCEL

There are two ways to use Excel to select a simple random sample from a population. We will look at the method that assumes you have the population values already in a worksheet. Suppose that you have a population with 50 two-digit ID numbers, a portion of which is shown in Figure 2.6.

	A	B
1	ID Codes	
2	1	
3	2	
4	3	
5	4	
6	5	
7	6	
8	7	
9	8	
10	9	
11	10	
12	11	
13	12	
14	13	
15	14	
16	15	
17	16	
18	17	
19	18	

FIGURE 2.6 Population of 50 ID codes

2.9.1 Selecting a Random Sample from a Population

You can use the **Data Analysis** tools in Excel to select a random sample. From the **Tools** menu, select **Data Analysis;** the dialog box shown in Figure 2.7 opens.

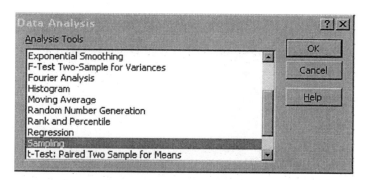

Note: Make sure the **Tools** menu contains the **Data Analysis** option. If not, you may have to run the **MS Office Setup** program to include the **Analysis ToolPak** and then execute the **Tools** Add-**Ins** command.

FIGURE 2.7 Data Analysis dialog box

These steps will allow you to select a random sample from the population:

1. Scroll down the **Analysis Tools** list and select **Sampling.** The dialog box shown in Figure 2.8 opens. You must tell Excel three things to obtain the sample: (1) the location of the population, (2) the type of sampling method and the number of samples, and (3) where you want the sample placed.

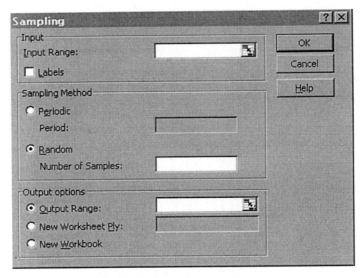

FIGURE 2.8 Sampling dialog box

2. Position the cursor in the box labeled **Input Range** and then highlight the range in the worksheet that contains the data, in this case A1:A51. Since the first row of the population is a title, ID Codes, check the box marked **Labels.**

3. In the section labeled **Sampling Method,** click the radio button for the **Random** sampling method and type "5" in the text box for **Number of Samples:.**

4. In the section labeled Output Options, you can specify that the sample be located in a section of the current worksheet, a new worksheet in the same workbook, or a new workbook. In this example, click the radio button for **Output Range:** and position the cursor in the textbox. Then highlight the cell in the worksheet where you want the output to start. The finished dialog box should look like the one in Figure 2.9.

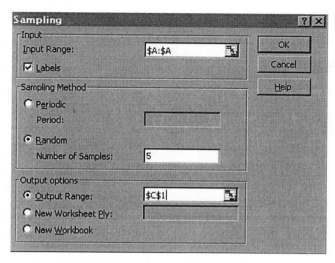

FIGURE 2.9 Completed sampling dialog box

5. Click **OK;** the random sample of 5 ID codes will appear in the location you specified, in this case cells C1:C5, as shown in Figure 2.10. This tells you to sample the items with ID codes 42, 11, 10, 40, and 46.

	A	B	C
1	**ID Codes**		42
2	1		11
3	2		10
4	3		40
5	4		46
6	5		
7	6		
8	7		

FIGURE 2.10 Random sample of 5 ID codes

CHAPTER 2 SUMMARY

In this chapter you learned the basic language of statistics. You learned that the complete group that you wish to study is called the population. Typically there are several characteristics of each member of the population that you wish to know about. These are called variables. There are two types of variables: quantitative and qualitative. Ideally you would like to study the entire population, or take what is known as a census, but most of the time a census is too expensive and too time-consuming. For these reasons and others you select a sample from a list of the members of the population. In doing so you end up with sampling error, because you are not studying every member of the population. The only way to eliminate sampling error is to study the entire population.

We always want our sample to be a fair representation of the population. This is known as an unbiased sample. There are many ways to select the sample but we will use simple random samples in this book. You learned how to select a simple random sample using a table of random numbers or a software tool such as Excel. You were also introduced to some shorthand notation, known as sigma notation, which is used to write equations throughout the remainder of this book.

Key Terms

Term	Definition	Page reference
Biased sample	A **biased sample** is a sample that does not represent the population.	34
Census	A **census** is a study of the entire population.	24
Continuous data	**Continuous data** are data that can take on any one of an infinite number of possible values over an interval on the number line.	43
Descriptive statistics	**Tools of descriptive statistics** allow you to summarize the data.	45
Discrete data	**Discrete data** are data that can take on only certain values. These values are often integers or whole numbers.	43
Inference	An **inference** is a deduction or a conclusion.	46
Inferential statistics	The **techniques of inferential statistics** allow us to draw inferences or conclusions about the population from the sample.	47

(continued)

Term	Definition	Page reference
Interval data	**Interval data** are data that can be compared only by looking at the difference between two values, or the interval between them.	43
Nominal data	Data that are created by assigning numbers to different categories when the numbers have no real meaning are called **nominal data.**	39
Ordinal data	Data that are created by assigning numbers to categories where the order of assignment has meaning are called **ordinal data.**	43
Parameter	A **parameter** is a number that describes a characteristic of the population.	29
Population	The **population** is everything you wish to study.	21
Population size	The **size of the population** is the number of members of the population. It will be referred to as N.	32
Primary data	**Primary data** are data that are obtained and used by the organization or individual that actually collected them.	41
Probability	We use **probability** theory to calculate the likelihood of observing or selecting a particular sample from a population.	47
Qualitative data	**Qualitative data** describe a particular characteristic of a sample item. They are most often non-numerical in nature.	41
Quantitative data	Data that are inherently numerical in form are called **quantitative** data.	43
Ratio data	**Ratio data** are data that can be compared by looking at either the difference between two values or the ratio of two values.	44
Sample	A **sample** is a piece of the population.	23
Sample size	The **size of the sample** will be referred to as n.	33
Sampling error	**Sampling error** is the difference between the behavior of the entire population and a sample of that population.	25
Sampling frame	A **sampling frame** is a list of all members of the population.	36
Secondary data	**Secondary data** are compiled data that are taken from several primary sources and synthesized or summarized in some way.	41
Sigma notation	**Sigma notation** is shorthand notation used to write formulas. It is so named because it uses the Greek capital letter sigma, written as Σ.	52
Simple random sample	A **simple random sample** is a sample that has been selected in such a way that all members of the population have an equal chance of being picked.	35
Statistic	A **statistic** is a number that describes the behavior of a sample.	30
Table of random numbers	A **table of random numbers** is a table that consists of a list of numbers randomly generated and listed in the order they were generated.	36
Variable	A **variable** is a characteristic of each member of the population.	22
Variation	The amount of **variation** refers to how different the members of the population are from each other with regard to the variable being studied.	25

CHAPTER 2 EXERCISES

Learning It!

2.36 A golf ball manufacturer wishes to know how far its newly designed super fly ball will go.
  (a) What is the population of interest?
  (b) Identify which of the reasons for taking a sample (listed on page 24) apply in this case. (There are more than one.)
  (c) What parameter is of interest?
  (d) What statistic might you calculate from the sample?
  (e) What type of data would you collect?

2.37 A mall vendor wishes to know how people like his new chocolate chip cookies.
  (a) What is the population of interest?
  (b) Identify which of the reasons for taking a sample (listed on page 24) apply in this case.
  (c) What parameter is of interest?
  (d) What statistic might you calculate from the sample?
  (e) What type of data would you collect?

2.38 A golf ball manufacturer wishes to know how far its newly designed super fly ball will go. A sample of 50 golf balls is hit and the distance is measured.
  (a) Why might this be a biased sample?
  (b) How could you get a simple random sample?

2.39 A mall vendor wishes to know how people like his new chocolate chip cookies. Fifty shoppers are stopped and given free samples of the new cookies. Do you think this sample is representative of the population of mall shoppers? Explain why or why not.

2.40 A golf ball manufacturer wishes to know how far its newly designed super fly ball will go. A sample of 50 golf balls is hit and the distance is measured in yards. The data are

301	299	307	310	291
300	297	286	297	302
296	290	291	306	302
301	296	287	311	309
309	290	292	287	276
299	302	304	300	309
290	281	308	306	307
317	277	301	298	285
289	288	306	311	293
297	291	285	302	300

  (a) Write the sigma notation to represent the sum of the data values.
  (b) The golf ball manufacturer wishes to know how different the values are from 300. Write the sigma notation to subtract 300 from each data value and sum the differences.

Thinking About It!

2.41 Would your information be accurate if you studied every member of the population? Why or why not?

2.42 In trying to understand the population, what is the only way to eliminate the possibility of reaching the wrong conclusion?

2.43 Is it likely that the value of the statistic calculated from the sample is exactly the same as the value of the population parameter?

2.44　Remembering that the percentage of voters who support the President of the United States is an unknown number, what might you suggest using as a best guess for this percentage?

2.45　There is a classic story about how the wrong inferences can be reached if the sample is biased. Just prior to the presidential election between Truman and Dewey the *Chicago Tribune* selected a national sample of voters from the telephone book. These voters overwhelmingly supported Dewey in the presidential race and so the newspaper confidently printed the headline "Dewey Wins" before the results were finally tallied. As we know, the headline was wrong: Truman became the 33rd President in 1945. What mistake did the *Chicago Tribune* make?

Doing It!

2.46　The data in this problem have been collected over the past 11 years of futures trading of Frozen Concentrate Orange Juice (FCOJ). The FCOJ is traded on the New York Cotton Exchange. The unit of trading is 15,000 pounds (one contract = 15,000 lb). The price quoted is in cents per pound. The data show the average price for the month.

　　Data file: FCOJ.XXX

(a)　Consider all the months of data to be the population and find the population average.

(b)　Select 50 simple random samples of 30 months and find the sample average for each of the samples. Which sample average is closest to the population average? Which sample average is the farthest from the population average?

(c)　Select 50 random samples of 40 months and find the sample average for each of the samples. Which sample average is closest to the population average? Which sample average is the farthest from the population average?

(d)　Select 50 random samples of 50 months and find the sample average for each of the samples. Which sample average is closest to the population average? Which sample average is the farthest from the population average?

(e)　What happens to the accuracy of the sample average as you increased your sample size from 30 to 40 to 50?

GRAPHICAL DISPLAYS OF DATA

ALUACHA BALACLAVA COLLEGE

Aluacha Balaclava College (ABC) is having trouble with faculty compensation. The faculty claim that they have not received pay raises for several years and that new faculty are being hired in at much higher salaries than existing faculty members are getting. The former Provost has just resigned under pressure and the President has hired a new Provost to design a new compensation package that is fair and reasonable. The new Provost knows that, prior to designing anything new, she must first understand the current system and some of the history of the system. She decides that her first step must be to collect some data on faculty salaries.

After some thought, she decides that faculty salaries depend on other factors, so she has the Management Information Systems (MIS) department gather data for each faculty member on Current Rank (Professor, Associate Professor, Assistant Professor, and Instructor), number of years of service to the college, and current salary. The data, which they provide to her in the form of a list, are not very enlightening:

Rank	Years of service	1996–97 salary
ASST	22	53316
PROF	11	64375
ASSO	7	63501
ASSO	6	59426
ASSO	20	49058
PROF	4	94969
ASST	21	54762
ASSO	9	55516

To gain any information from the data she will need to summarize them, probably with some graphical displays.

3.1   CHAPTER OBJECTIVES

In Chapter 2 you learned about the different types of data and about how data can and should be collected. You learned that people collect data to provide information about something they are interested in understanding, such as why brakes on rollerblades fail or why customers complain about tissues. Usually there is some problem that needs to be solved, like a new brake design, when this understanding is gained.

In this chapter we look at ways to display different types of data. The chapter covers the following material:

- Graphical Methods for Qualitative Data: frequency tables, bar charts, and pie charts
- Graphical Methods for Quantitative Data: frequency tables and histograms
- Other Graphical Methods: dotplots and Pareto diagrams

3.2   ORGANIZING DATA

If one objective of statistics is to obtain information about a set of data, then we need to organize the data in some way. When data are collected the initial result is usually a *list* of the observations for each variable. This is referred to as the *raw data*. Raw data, such as those given to the Provost at the college, provide almost no information. Statistics provides some tools or techniques for turning raw data into information.

3.2.1 The Frequency Distribution

One way to organize data is to consolidate them by determining how many times each value in the data occurs and making a table that summarizes this information. This is called a **frequency table** or **frequency distribution.**

> A *frequency table* or *frequency distribution* is a table containing each category, value, or class of values that a variable might have and the number of times that each one occurs in the data. The frequency of the ith class is denoted f_i.

Frequency Tables for Qualitative Data

Creating a frequency table for qualitative data is not difficult. A basic frequency table has two columns. In the first column of the table each row lists one of the values for the variable of interest. The second column lists the corresponding number of times that the value occurred in the set of data. Later on we will add some additional columns to the table, but right now we will work with only two. Figure 3.1 shows the setup for a typical frequency table:

Category	Frequency
Category 1	f_1
Category 2	f_2
Total	**n**

FIGURE 3.1   A frequency table

EXAMPLE 3.1   **ABC Faculty Salaries**

Creating a Frequency Table

The Provost at ABC knows that salary will differ for each faculty member, depending on several factors. One such factor is the current rank of the faculty member. She decides to look at the faculty by rank to see how the faculty is made up as a whole. She puts the data in the form of a frequency table as shown below:

Rank	Frequency
Professor	55
Associate Professor	67
Assistant Professor	77
Instructor	8
Total	**207**

From the frequency table she sees that while Assistant Professor has the highest frequency, the three highest ranks are not all that different. There are obviously not many faculty at the rank of Instructor.   ■

The order for the categories in the frequency table is not important. If there is a logical order, or if you know what the classifications will be before you fill out the table, then you might use a particular order. If you are creating the categories as you make the table, you might list them in the order in which they first appear in the raw data.

EXAMPLE 3.2   **Student Distribution**

Creating a Frequency Table

The faculty in the School of Business at a university are concerned that many students are delaying taking the Introductory Statistics course required by all majors. The course can be taken as early as the freshman year, but traditionally is taken by sophomores. To determine if the concerns are justified, a section of Introductory Statistics is selected at random and each student is classified as a freshman (F), sophomore (S), junior (J) or senior (Sr) according to the number of credits completed to date. The raw data for the 28 students are shown below:

Sr	Sr	S	S	S	Sr	Sr
S	S	S	S	S	F	F
Sr	S	Sr	Sr	Sr	F	F
J	S	J	S	S	Sr	Sr

To use the data to answer their questions, the faculty organize the data in the form of a frequency table:

Classification	Frequency
Freshman	4
Sophomore	12
Junior	2
Senior	10
Total	**28**

From the table they see that very few students taking the course are freshmen or juniors, and that the rest are about evenly divided between sophomores and seniors.   ■

While the information obtained from the frequency table is certainly better than that from the raw data, it is still not quite as informative as it might be. The values in the frequency table are influenced by the sample size. For large samples the individual frequencies will be much larger numbers than for small samples. This makes comparisons of different samples difficult. To solve this problem we must find a way to express the frequency so that the sample size does not matter. One way to do this is to use the **relative frequency.**

> The *relative frequency* of a classification is the number of times an observation falls into that classification represented as a portion of the total number of observations. It can be expressed as a *fraction, decimal,* or *percentage.*

Thus, to find the relative frequency for the ith classification, rf_i, we use

$$\text{rf}_i = \frac{\text{Frequency of } i\text{th classification}}{\text{Sample size}} = \frac{f_i}{n}$$

EXAMPLE 3.3   ABC Faculty Salaries

Finding Relative Frequency

To get a better idea of how the faculty were distributed among the ranks, the provost decided to calculate the relative frequency for each rank. The calculations and results are shown in the table below:

Rank	Frequency	Relative frequency	Relative frequency (%)
Professor	55	55/207	26.6
Associate Professor	67	67/207	32.4
Assistant Professor	77	77/207	37.2
Instructor	8	8/207	3.9
Total	**207**	**207/207**	**100.0**

Note: If you add the percentages for relative frequency, you will not get 100% here, because the percentages have been rounded off.

It appears from the relative frequencies that the three largest classes are not quite as different as she had thought, but no single class constitutes a majority.   ∎

Note: Use the actual variable name for the classification to provide the most information.

It does not matter whether you use fraction, decimal, or percentage to calculate the relative frequency. Percentage is easiest for most people to understand. When using percentage you need to be careful about rounding. When you report relative frequency to the nearest percent or even one decimal place, the values may not sum to exactly 100%. This is not a problem.

EXAMPLE 3.4   Student Distribution

Calculating the Relative Frequency

To get a better picture of the students in the Introductory Statistics course, the person studying the data decides to look at the relative frequency for each classification.

Freshmen:

$$\text{rf}_1 = \tfrac{4}{28} = 0.143 \qquad \text{or, as a percentage} = 0.143 \times 100 = 14.3\%$$

Sophomores:

$$\text{rf}_2 = \tfrac{12}{28} = 0.429 \qquad \text{or, as a percentage} = 0.429 \times 100 = 42.9\%$$

If these calculations are not readily apparent to you, then you should get out a calculator and practice them until they are!

Juniors:

$$\text{rf}_3 = \tfrac{2}{28} = 0.071 \qquad \text{or, as a percentage} = 0.071 \times 100 = 7.1\%$$

Seniors:

$$\text{rf}_4 = \tfrac{10}{28} = 0.357 \qquad \text{or, as a percentage} = 0.357 \times 100 = 35.7\%$$

Adding a column to the frequency table gives

Year in school	Frequency	Relative frequency (%)
Freshman	4	14.3
Sophomore	12	42.9
Junior	2	7.1
Senior	10	35.7
Total	**28**	**100.0**

From this table the School of Business can see that 35.7% of the students in the class are seniors, and 35.7% + 7.1% = 42.8% of the students delayed the course beyond the sophomore year. ∎

TRY IT NOW!

Student Grades

Creating a Frequency Table

A professor in an Introductory Statistics course knows that while students dread taking the course, they also have unusually high expectations for their performance. She surveys (anonymously, of course) her students and asks them what grade they expect to get in the course. The raw data are

A	C	B	A	A	B	B
B	A	B	A	A	C	B
B	A	F	C	B	D	C
B	B	D	B	A	B	A

Make a frequency table for the data. Include both frequencies and relative frequencies.

Frequency Tables for Integer Data

Remember from Chapter 2 that one type of quantitative data you might collect results from *counting* the number of times that something occurs in a set of data or from *ranking* or *rating* objects. This type of data takes on integer values.

Creating a frequency table for this type of data is exactly the same as creating one for qualitative data. Each value of the variable represents one category or classification in the table. The only real difference is that since the numbers for these data have some meaning, the categories must be in numerical order. For qualitative data the order in which you list the categories is not rigid.

EXAMPLE 3.5   Student Attitudes

Frequency Table for Integer Data

The statistics professor who has been trying to learn about her class is also interested in her students' attitudes toward the study of statistics. She is aware that most of the people are enrolled in the course because it is required, and she wonders if this is reflected in their attitude toward the subject. As part of the last anonymous survey on grades, she asks the class to rank "the importance of statistics in my life" on a scale of 0 to 10, where 0 = of no importance whatsoever and 10 = the most important thing I will ever study. The raw data from the study are

0	10	8	0	0	8	6
5	1	6	1	4	10	8
7	0	10	9	4	10	10
7	6	10	9	1	5	1

Since this is not very informative she decides to create a frequency table for the data:

Rating	Frequency	Relative frequency
0	4	4/28 = 14.3%
1	4	4/28 = 14.3%
2	0	0/28 = 0.0%
3	0	0/28 = 0.0%
4	2	2/28 = 7.1%
5	2	2/28 = 7.1%
6	3	3/28 = 10.7%
7	2	2/28 = 7.1%
8	3	3/28 = 10.7%
9	2	2/28 = 7.1%
10	6	6/28 = 21.4%
Total	**28**	**28/28 = 100%**

From the frequency table she sees that most people seemed to have attitudes on either end of the scale, with fewer in the center, neutral, zone.   ■

Most data that are truly integer in nature occur from counting, rankings, or ratings. Sometimes data that are really measurements are recorded as integers. In this case it is important to think about where the data came from and the number of possible values before you pick the classes for your frequency table. Since the data are numbers, you cannot leave out a class (value) if there are no observations in it. If the number of possible values is much larger than 15 or 20 you probably do not want to treat the data as integer data. Instead you should use the methods for continuous data shown later in this section.

EXAMPLE 3.6   ABC Faculty Salaries

Looking at Integer Data

The provost looking at faculty salaries also knows that salary depends on the length of time that the faculty member has been at the College. She decides to summarize these data using a frequency table, and notices that the data are integer data.

She also notices that the shortest length of time involved is 0 years and the longest is 41 years. She realizes that time is a measurement and that it is integer data because it was recorded to the nearest whole year. She sees that treating these data as integer data will not be appropriate.   ∎

Since quantitative data are *ordered* by nature, it is also interesting to ask questions like "What percentage of the class rates statistics on the bottom half of the scale? neutral? the top half?" You can answer the questions by summing the relative frequencies, or by using the **cumulative relative frequency.**

> The *cumulative relative frequency* of a class is the sum of the relative frequencies of all classes at or below that class represented as a portion of the total number of observations. It can be expressed as a fraction, decimal, or percentage.

EXAMPLE 3.7   Student Attitudes

Cumulative Frequencies

For the data from the Student Attitudes example, the professor calculated the cumulative frequency for each class in the table:

Rating	Frequency	Cumulative relative frequency
0	4	4/28 = 14.3%
1	4	8/28 = 28.6%
2	0	8/28 = 28.6%
3	0	8/28 = 28.6%
4	2	10/28 = 35.7%
5	2	12/28 = 42.9%
6	3	15/28 = 53.6%
7	2	17/28 = 60.7%
8	3	20/28 = 71.4%
9	2	22/28 = 78.6%
10	6	28/28 = 100.00%
Total	**28**	

From the table, the professor can easily see that 35.7% of the students had negative attitudes toward statistics. To find the percentage of students that have neutral attitudes, she must do some arithmetic first. From the table she can see that 42.9% of the students were in the classes from 5 down. Since she knows that 35.7% were in the classes from 4 down, it seems that 42.9% − 35.7% = 7.2% have a neutral attitude toward statistics. This agrees with the relative frequency from Example 3.5.

To find the percentage with positive attitudes, again a little arithmetic is needed. Since she knows that 42.9% are in the classes from 5 down, the remainder must be in the classes from 6 up. She finds that 100% − 42.9% = 57.1% have positive attitudes toward statistics. Teaching this class may be an uphill battle!   ∎

At this point you must be thinking that it would be simpler to use the relative frequencies and just add up the ones you are interested in, like you did for the qualitative data. Sometimes this is true, but when there are a lot of classes the procedure becomes very tedious. There are two other reasons for learning how to deal with cumulative relative frequencies: (1) When you are using statistics or spreadsheet software to analyze data, the cumulative frequencies are part of the output, which makes it a much easier job; and (2) certain statistical tables such as probability tables are cumulative and there is just no other way to use them.

TRY IT NOW!

New Product Survey
Cumulative Relative Frequencies

A marketing research firm conducted a survey of consumers who invariably use a particular brand of bath soap. The consumers were given a competitor's version of the same product with nonallergenic enhancements and asked whether they would consider buying the new product. Their answers were given on a scale of 1 to 5 where 1 = would not ever buy this product and 5 = will buy this product immediately. The raw data from the survey are given below:

5	4	1	4	2	2
3	1	4	4	2	4
3	1	3	3	5	4
4	4	2	3	2	5
4	3	4	4	1	5

Create a frequency table for the data that includes both frequency and cumulative relative frequency.

Which ratings indicate a negative attitude toward the new product? What percentage of the people surveyed had a negative attitude?

Which rating indicates a neutral attitude? What percentage of the people surveyed had a neutral attitude?

Which ratings indicate a positive attitude? What percentage had a positive attitude?

Frequency Tables for Continuous Data

Another type of quantitative data that is frequently encountered in statistics is *measurement* data. This type of data can assume many different values in a sample. For this reason, we cannot use each value in the sample as a category for the frequency table. It is possible that each value will occur only once in the sample! This would result in entirely too many classes for the frequency table and no real information. Thus, before we can create a frequency table for measurement data we must figure out how to define the categories or classes for the data.

Two questions must be answered to create the frequency table for continuous data:

Question 1: How many classes should there be in the table?

Question 2: How large (wide) should each class be?

While there are no absolute rules for creating the frequency classes for continuous data, there are some guidelines that you can apply to answer the questions.

Question 1: There are two basic rules for determining the number of classes for the frequency table.

Rule 1: The number of classes in the frequency table should be approximately equal to the square root of the sample size, n. That is,

$$\text{Number of classes} = \sqrt{n}$$

You can see already that this is just a guideline, since the result of the calculation may not be a whole number. When this happens we usually just take the integer part of the answer. For example, if our sample size were $n = 30$, then $\sqrt{30} = 5.477$ so we would plan on 5 classes.

Rule 2: The number of classes should not be less than 5 or more than 20. The reasons for this rule have to do with the kind of information that we are looking for in the frequency table. The reasons for it will be more obvious when we talk about the graphical methods for displaying data.

Once you have decided on the number of classes for the frequency table you can answer the second question.

Question 2: To determine how large each class should be we need to think about what the table must accomplish. Once the classes have been determined, each piece of data in the sample must be able to be counted in *one and only one* of the classes. This tells you that

1. The classes must span the entire data set.

2. The classes cannot overlap.

Think about the number of classes as the number of steps you can take to get from the smallest data value (minimum) to the largest data value (maximum). Then the size of each class should be

$$\frac{\text{Maximum} - \text{Minimum}}{\text{Number of classes}}$$

Figure 3.2 gives an illustration of the rules.

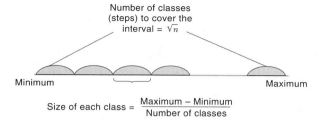

FIGURE 3.2 Number of classes and class interval for continuous data

Once the number of classes and size of each class are determined you can define each class precisely. That is, you can specify the smallest and largest data values that will fall in each class. At this point it is probably time for an example.

EXAMPLE 3.8 ABC Faculty Salaries

Setting up the Class Intervals for Continuous Data

The provost of ABC decides to create frequency tables for the years of service data. Although she is going to use a computer package to actually create the table, she knows that some software asks the user for input on the number of classes to use as well as the widths or starting points.

She knows that she has 207 faculty members in the sample, so she calculates $\sqrt{207} = 14.39 \approx 14$ and decides that she will use about 14 classes. Since the highest data point is 41 years and the lowest is 0, she calculates that the width of each class should be

$$\text{Width} = \frac{41 - 0}{14} = 2.93 \text{ years}$$

∎

EXAMPLE 3.9 On-Time Rates

Creating Class Intervals for Continuous Data

A major commuter railroad has been looking at the on-time rates for service to a particular city. The railroad collects data for 30 randomly selected days and finds the percent of trains on time each day:

60.4	92.9	67.0	91.6	93.7
71.9	66.4	48.0	66.6	94.6
83.2	68.0	74.3	78.6	40.9
76.4	62.9	67.1	68.5	29.5
69.2	45.8	49.1	34.7	65.1
72.1	40.6	35.0	60.8	54.3

To make some sense of the data the railroad decides to create a frequency table. The first task is to decide how many classes the table should have. The railroad first calculates

$$\sqrt{30} = 5.48$$

and so decides to use 5 classes.

Next the railroad must decide how large each class should be. The data show that the maximum value is 94.6% and the minimum value is 29.5%. The total distance that needs to be covered by the classes is

$$94.6 - 29.5 = 65.1$$

and the size of each class must be

$$\frac{65.1}{5} = 13.02$$

∎

To determine the starting and ending points (boundaries) for each class we must remember that the classes cannot overlap and that there cannot be any gaps in the interval. There are several ways to accomplish this, but we will define a class as containing all observations *more than* the lower boundary, *up to and including* the upper boundary:

Lower boundary $< x \leq$ Upper boundary

where x represents the value of the variable being studied.

Different texts will use different methods for defining intervals, but they will all yield similar results. The interval definition used here is consistent with the major spreadsheet software such as Excel and Lotus 1-2-3. Most statistical packages, such as Minitab and Data Desk, define the classes to *include* the lower boundary and to go up to but *not include* the upper boundary. As we will see later in the chapter, this causes some differences in the final output, but it is not a problem.

Remember that the intervals have been defined so that they contain the values *above* the lower boundary. To make sure that the classes you define include the minimum, you need to start just *below* the minimum value in the data set and add the class width. The lower boundary for the next class is the upper boundary for the previous class.

EXAMPLE 3.10 ABC Faculty Salaries

Creating the Class Intervals

The provost decides that for the data on the number of years of service she will use a computer software package. She will let the software create the frequency distribution for her and see what it comes up with:

Class	Frequency
$-2.928571 < x \leq 0.000000$	15
$0.000000 < x \leq 2.928571$	19
$2.928571 < x \leq 5.857143$	19
$5.857143 < x \leq 8.785714$	34
$8.785714 < x \leq 11.714286$	16
$11.714286 < x \leq 14.642857$	14
$14.642857 < x \leq 17.571429$	13
$17.571429 < x \leq 20.500000$	17
$20.500000 < x \leq 23.428571$	33
$23.428571 < x \leq 26.357143$	20
$26.357143 < x \leq 29.285714$	4
$29.285714 < x \leq 32.214286$	2
$32.214286 < x \leq 35.142857$	0
$35.142857 < x \leq 38.071429$	0
$38.071429 < x \leq 41.000000$	1

The results are not quite what she expected, but she sees that there are 15 classes and that the width of each class is 2.92857 years. ∎

In Example 3.10 we saw that the frequency table generated by a software package used intervals that had six decimal places! While this is not bad, it might be more appealing to use intervals that look more like the original data, with the same precision.

EXAMPLE 3.11 On-Time Rates

Creating Class Intervals for Continuous Data

In the On-Time Rates example we found that the width of each class should be 13.02. The data are measured only to the nearest tenth, so we will use increments of 13.0. Since the smallest value in the data is 29.5, we will start just below that value, at 29.4, and add 13.0. That is, the first class interval contains all values, x, of the variable such that

$$29.4 < x \le 42.4$$

The next class is formed by starting at 42.4 and again adding 13.0:

$$42.4 < x \le 55.4$$

The remainder of the classes are

$$55.4 < x \le 68.4$$
$$68.4 < x \le 81.4$$
$$81.4 < x \le 94.4$$
$$94.4 < x \le 107.4$$

It is okay to wind up with one more class than you planned on. Remember the rules are only guidelines.

We need this last class for the maximum observation, 94.6%.

To create the frequency table for the data we count the number of data values that fall in each interval and complete the table.

On-time rates	Frequency	Relative frequency (%)	Cumulative relative frequency (%)
$29.4 < x \le 42.4$	5	16.7	16.7
$42.4 < x \le 55.4$	4	13.3	30.0
$55.4 < x \le 68.4$	9	30.0	60.0
$68.4 < x \le 81.4$	7	23.3	83.3
$81.4 < x \le 94.4$	4	13.3	96.6
$94.4 < x \le 107.4$	1	3.3	100.0

From the frequency table the railroad management can obtain information such as, "On 83.3% of the days the on-time rate was no greater than 81.4%" or "The on-time rate was greater than 94.4% on only 3.3% of the days." This allows the railroad to judge its performance and take appropriate action. ■

You may be wondering why you need to know how to do this, when you can use a computer to generate the frequency distribution for you. There are two reasons that this knowledge is necessary. First, some software packages require input from the user about how the frequency table is set up. Knowledge of some basic rules will help to answer those questions. The second reason is that even if there is a default setup for the frequency table, the result may not be acceptable. Remember the frequency table for the faculty salaries!

What are we looking for in a frequency table? How can we tell if the one we have created is good? We already know that we need the classes to be the same size and that they must cover all of the data values. What else is important?

Many times you may want the intervals to be defined in nice, round, numbers that are integer data or that end in 0.5 or some other convention. Certainly, it is more appealing to look at classes with numbers like 25, 35, 45, . . . , but software packages follow rules very similar to the ones outlined here and the results are very often not what you want.

Figure 3.3 shows the frequency table produced by Microsoft Excel for the **On-Time Rate** data.

Bin	Frequency	Cumulative %
29.50	1	3.33
42.52	4	16.67
55.54	4	30.00
68.56	10	63.33
81.58	6	83.33
More	5	100.00

In Excel, the number in the Bin column is the highest number that is included in the interval, so the first bin is all numbers less than and including 29.50, and the last bin is all numbers greater than 81.58.

FIGURE 3.3   Microsoft Excel frequency table

It is possible to vary both the starting point of the first class and the interval width, within reason. Remember that the decisions for the number of classes and interval widths are guidelines based on the sample size and the range of the data. As long as your final results do not vary too much from the guidelines you can adjust them.

EXAMPLE 3.12   ABC Faculty Salaries

Adjusting Class Boundaries

After looking at the results given by the statistical software, the provost feels that the class intervals do not use numbers with which people will identify. There is no reason to use so many decimal places when the data are measured to the nearest year. Perhaps she should adjust the class intervals a little bit so that they use nice, round numbers. After looking at the data she decides to start just below 0, at -1 rather than at the -3 suggested by the software. She is less sure about the width of each class. Using a class width of 3 years would not be too different from the 2.93 calculated value, but she feels that a width of 5 years might be more appealing. She decides to try both ways and obtains the following frequency tables.

Class	Frequency	Class	Frequency
$-1 < x \leq 2$	34	$-5 < x \leq 0$	15
$2 < x \leq 5$	19	$0 < x \leq 5$	38
$5 < x \leq 8$	34	$5 < x \leq 10$	42
$8 < x \leq 11$	16	$10 < x \leq 15$	23
$11 < x \leq 14$	14	$15 < x \leq 20$	29
$14 < x \leq 17$	13	$20 < x \leq 25$	51
$17 < x \leq 20$	17	$25 < x \leq 30$	8
$20 < x \leq 23$	33	$30 < x \leq 35$	0
$23 < x \leq 26$	20	$35 < x \leq 40$	0
$26 < x \leq 29$	4	$40 < x \leq 45$	1
$29 < x \leq 32$	2		
$32 < x \leq 35$	0		
$35 < x \leq 38$	0		
$38 < x \leq 41$	1		

The first set of changes does not alter the frequency distribution very much. There are 14 classes instead of 15, and the end classes are fairly empty. The second set of changes reduces the number of classes to 10 and there are fewer empty or nearly empty classes. She is not sure if this matters, but she will keep it in mind.   ■

EXAMPLE 3.13   On-Time Rates

Adjusting Class Boundaries

After examining the frequency table for the On-Time Rates problem, the railroad managers decide that they would prefer to see the data classified with rounder, more understandable numbers. They look at the data and decide to start the first interval at just above 25.0% and use a width of 10%. The new frequency table would look like this:

On-time rates	Frequency	Relative frequency (%)	Cumulative relative frequency (%)
$25.0 < x \leq 35.0$	3	10.0	10.0
$35.0 < x \leq 45.0$	2	6.7	16.7
$45.0 < x \leq 55.0$	4	13.3	30.0
$55.0 < x \leq 65.0$	3	10.0	40.0
$65.0 < x \leq 75.0$	11	36.7	76.7
$75.0 < x \leq 85.0$	3	10.0	86.7
$85.0 < x \leq 95.0$	4	13.3	100.0
Totals	**30**	**100.0**	**100.0**

The changes result in seven classes rather than six because the first interval starts at a lower number and the interval widths are not as large.

Another possibility for class intervals would have been to start at just above 25% and use an interval width of 15%. The frequency table for that arrangement would be

On-time rates	Frequency	Relative frequency (%)	Cumulative relative frequency (%)
$25.0 < x \leq 40.0$	3	10.0	10.0
$40.0 < x \leq 55.0$	6	20.0	30.0
$55.0 < x \leq 70.0$	11	36.7	66.7
$70.0 < x \leq 85.0$	6	20.0	86.7
$85.0 < x \leq 100.0$	4	13.3	100.0
Totals	**30**	**100.0**	**100.0**

This time the frequency table has only five classes because we used an interval width of more than 13.0%. ∎

 **TRY IT NOW!**

Assignment Times

Creating a Frequency Table for Continuous Data

The instructor for an introductory statistics class wonders about the complaints that she is hearing about the time it takes to complete a computer assignment. The assignments are designed to be done in about 25 minutes. She asks the members of the class to time how long it takes to do the next assignment and to hand the data in with the assignment. The data, in minutes, she obtains are

22.8	27.0	27.9	30.4	33.4
24.8	27.2	27.9	31.1	33.9
24.8	27.4	28.2	31.4	35.3
26.0	27.4	29.4	32.4	35.7
26.0	27.4	29.6	33.1	36.3
26.1	27.6	29.8	33.2	40.4

Approximately how many classes should the frequency table have?

What should the class width be?

Create a frequency table for the data.

3.2.2 Exercises—Learn It!

3.1   The administrators of a local university are trying to determine whether the amount of parking that they have is adequate. They take a random sample of 25 entering freshmen and ask them what mode of transportation they will use to travel to classes. The responses are

Car	Car	Car	Car	Car
Car	Walk	Car	Car	Car
Public	Car	Walk	Car	Car
Other	Car	Bicycle	Car	Car
Public	Car	Car	Public	Walk

(a) Create a frequency table for the responses.

(b) What mode of transportation was the most popular? the least popular?

(c) Does any class in the frequency table constitute a majority? If so, which one?

3.2   After the incredible advertising blitz that accompanied the release of Microsoft's Windows 95 operating system, a local computer society wondered what operating platform its members were using. The society surveyed 30 members and obtained the following data:

Macintosh	Macintosh	Macintosh	W 3.1*	W 3.1
W 3.1	W 3.1	W 3.1	OS /2	W 3.1
W 3.1	W 3.1	W 3.1	W 3.1	W NT
W 3.1	W 3.1	W 3.1	W NT	W 3.1
W 3.1	W 95	Macintosh	W 95	W 3.1
OS /2	DOS	W 3.1	W 3.1	W 3.1

* The W 3.1 category includes Workgroups for Windows 3.11 users.

(a) Create a frequency table for the data on operating systems.

(b) What percentage of the people surveyed used Windows 95?

(c) What percentage of the people surveyed did not use a Microsoft Windows operating system?

(d) Was any operating system used by a majority of those surveyed?

3.3 A manufacturing company has been receiving complaints from its warehouse operators about the condition of the product cases that are being stored. The complaints indicate that most of the cases are defective in some way and that it is creating logistics problems for the warehouse employees. The managers of the company decide to take a random sample of the cases produced and examine them for various defects. The data they obtain are

Dented	No defect	No defect	Dented	No defect
No defect	Torn	Crushed	No defect	No defect
Unsealed	No defect	No defect	No defect	No defect
No defect	No defect	No defect	No defect	No defect
No defect	No defect	No defect	Dented	No defect
No defect	Unsealed	No defect	No defect	No defect

(a) Create a frequency table for the defect data.

(b) Based on the data, does the warehouse claim appear justified? Why or why not?

3.4 A survey in a computer newsletter for employees at a large corporation asked readers how many times per week they logged onto the Internet using any one of the available services at the company. The responses from 50 employees are shown below:

```
1   7   3   4   1
0   7   0   1   8
3   3   4   2   4          50
5   0   3   3   5
3   3   7   4   1
7   5   3   3   4
4   3   3   3   7
6   8   0   7   2
7   0   3   9   3
4   4   4   4   3
```

(a) Create a frequency table for the data.

(b) What percentage of those surveyed logged onto the Internet exactly 5 times per week? more than 5 times per week?

(c) What percentage of those surveyed logged on 4, 5, or 6 times per week? less than twice per week?

3.5 As part of a cost/benefit analysis on service contracts for photocopiers, a large company surveyed all departments with copy machines and asked how many times per week their machines needed to be serviced. The data obtained are

```
3   2   3   0   3
0   2   2   2   1
3   2   2   1   0
2   3   2   1   2
3   4   2   7   1
2   2   1   3   1
0   3   1   1   2
```

(a) Create a frequency table for the data. Include relative frequency and cumulative relative frequency.

(b) What percentage of the copiers need repairs more than twice per week?

(c) What percentage need repairs at least four times per week?

(d) Do a majority of the copiers require repairs more than three times per week?

3.6 To determine whether it was reasonable to request that students use computers to write reports, a school surveyed a typical fifth-grade class and asked the students how many computers they had at home. The responses are

1	0	1	1	2
2	2	0	1	0
0	0	3	3	0
0	1	1	2	1
1	2	1	1	0

(a) Create a frequency table for the data. Include relative frequency and cumulative relative frequency.

(b) What percentage of the students did not have access to computers at home?

(c) What percentage of the students had easy access (2 or more) to computers at home?

(d) Based on the data collected, do you think that asking students to use computers for reports is reasonable? Why or why not?

3.7   As part of a study to decide whether to renew the contract of a food service company at a local university, the administration surveyed a group of students who used the service on a regular basis. The students were asked to rate the food quality on a scale of 1 to 5, where 1 was extremely bad and 5 was extremely good. The results are

2	4	3	4	1
3	2	3	1	3
1	1	4	1	4
1	3	3	4	1
4	2	3	4	3
3	3	3	1	3

(a) Create a frequency table for the data.

(b) What rating had the highest frequency? The lowest?

(c) Based on the data, do you think that the students like the food provided by the company? Why or why not?

3.8   A local computer group surveyed high school students to find out how many hours per week the students spent logged on to various on-line services. The data are

0.2	4.1	11.9	15.5	17.8
0.8	4.2	13.0	15.5	17.9
1.0	6.9	13.4	15.5	18.4
1.4	8.0	13.5	15.5	18.7
2.5	8.2	14.9	16.7	19.5
2.9	8.6	15.1	16.8	19.7

Create a frequency table for the amount of time spent on-line by students. Include relative frequency and cumulative relative frequency.

3.9   A large corporation that relies on a secretarial service group to do much of its work is trying to decide whether it needs more workers in the group. One of the criteria that will be used to make the decision is the turn-around time experienced by users of the service. The corporation surveys a number of employees who use the service on a regular basis and asks them for the turnaround time (in hours) for the last job they submitted. The following data are obtained:

14	19	21	25	26	29
15	19	22	25	26	29
16	20	22	25	26	30
16	20	23	25	26	31
18	20	23	25	27	31
18	20	23	25	28	35
18	21	24	26	29	40

(a) Create a frequency table for the data. Include relative frequency and cumulative relative frequency.

(b) What percentage of the people who use the service wait more than one day (24 hours) for their jobs to be done?

(c) Modify the class intervals for your frequency table to allow you to easily answer the questions, What percentage of those surveyed wait 20 hours or less? more than 30 hours?

(d) *Approximately* what percentage waited between 20 and 30 hours?

3.10 The Bureau of Weights and Measures conducts random checks of products in various supermarkets. The Bureau decides to check half-gallon containers of a particular brand of orange juice. The contents of each of the containers tested (measured to the nearest 0.1 oz) are

64.8	65.2	65.6	65.7	65.9
64.8	65.3	65.7	65.7	66.0
64.9	65.3	65.7	65.7	66.1
65.1	65.4	65.7	65.8	66.1
65.2	65.4	65.7	65.8	66.3

(a) Create a frequency table for the data.

(b) If the containers are subject to bursting when filled with more than 66 oz. of fluid, what percentage of the containers are in danger of bursting?

3.11 The local refuse company in charge of recycling in a town is receiving complaints from the people at the end of the pickup route. The complaints indicate that their recycling is not being picked up at the end of the route because the drivers say that the trucks are too full. In an effort to determine how much capacity is really needed for the route, the refuse company randomly samples the newspaper recycling that is put out and measures each pile to the nearest 0.1 inch. The data obtained are

8.3	10.2	12.2	13.1	13.9
8.5	10.4	12.3	13.3	14.1
8.8	11.0	12.3	13.3	14.3
9.1	11.0	12.3	13.4	14.3
9.7	11.2	12.3	13.4	14.4
9.8	11.4	12.5	13.4	14.4
9.9	12.0	12.7	13.6	14.4
9.9	12.0	12.8	13.6	14.8
10.1	12.2	12.9	13.8	15.2

(a) Create a frequency table for the data. Include relative frequency and cumulative relative frequency.

(b) Modify the frequency table you created in part (a) so that you can easily use it to answer the following questions:

(c) What percentage of the newspaper recycling piles are more than a foot in height?

(d) What percentage of the recycling piles are 10 inches or less?

(e) *Approximately* what percentage of the recycling piles are between 10 and 14 inches?

3.3   GRAPHICAL DISPLAYS OF DATA

While the frequency distribution does a good job of organizing and summarizing a set of data, it does not have much of a visual impact. To create a more immediate impression from a data set we need to create a pictorial (graphical) representation of the information in the frequency distribution. Most of the work is already done in creating the frequency table. The displays have the same basic structures with some changes to accommodate the type of data being displayed.

3.3.1 Graphical Displays for Qualitative Data

There are two methods that you can use to display qualitative data, a **bar chart** and a pie chart.

A *bar chart* represents the frequency or relative frequency from the table in the form of a rectangle or bar.

Creating a Bar Chart for Qualitative Data

In a bar chart, one of the axes is used to represent the categories from the frequency table and the other axis is used to represent the frequency or relative frequency for the categories. For qualitative data, assignment of the axes is a matter of preference. We will use the *x* axis for the categories and the *y* axis for the frequencies and relative frequencies to be consistent with the graphs for the other types of data.

Steps for creating a bar chart

Step 1: Draw a pair of axes, *x* and *y*.

Step 2: At evenly spaced intervals on the *x* axis put tick marks and label them with the categories from the frequency table.

Step 3: Scale the *y* axis so that the category with the highest frequency or relative frequency can be graphed. Choose the scale so that you can distinguish different frequencies or relative frequencies from each other.

Step 4: At each category on the *x* axis, draw a rectangle (bar) whose height is equal to the frequency or relative frequency for the category. The bases of the rectangles must be the same width and the bars should not touch each other.

Step 5: Label the axes and give the graph an appropriate title.

Often there is no predetermined order for the categories for qualitative data, but you might want to think about the order in which you place them, since the order may affect the impact of the graph. Some common choices are by descending or ascending frequency or alphabetical order.

EXAMPLE 3.14   ABC Faculty Salaries

Creating a Bar Chart

The Provost at the college decides to create a bar chart for the data she has collected on faculty rank. The frequency table for the data was

Rank	Frequency	Relative frequency (%)
Professor	55	26.6
Associate Professor	67	32.4
Assistant Professor	77	37.2
Instructor	8	3.9
Total	**207**	**100.0**

She decides to use increasing rank order for the bars on the chart and the resulting bar chart looks like the one below:

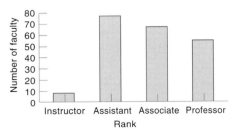

Distribution of Faculty by Rank

You may wonder why you would want to know how to create bar charts by hand when it is very likely that you will use a computer to make the charts. Most computer software requires the user to make decisions about how the chart should be set up. Knowing the basics of how the charts are made helps you to make these decisions.

EXAMPLE 3.15 **Student Distribution**

Creating a Bar Chart

The School of Business would like to create a bar chart of the data that it collected on the year of students who are taking the Introductory Statistics course. The frequency table for the data is

Year in school	Frequency	Relative frequency (%)
Freshman	4	14.3
Sophomore	12	42.9
Junior	2	7.1
Senior	10	35.7
Total	**28**	**100.0**

Since in this case there is a natural order for the categories we will use it when labeling the *x* axis. The *y* axis needs to accommodate a frequency as high as 12, so the scale for that axis will go from 0 to 12 by twos. The completed bar chart is shown in the figure.

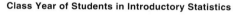
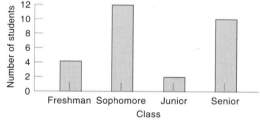

One thing to remember when creating a bar chart is that there are no absolute rules. The idea is to create a chart that conveys information to the viewer. In general, it is desirable to keep the number of categories in a bar chart to no more than ten. When there are too many categories, the axis becomes crowded and difficult to read and the viewer is distracted. To accomplish this, categories with very low frequencies are often consolidated into a single category labeled "Other." Even though the "Other" category might have a higher frequency than some of the categories, it must appear last in the bar chart.

 **TRY IT NOW!**

Student Grades

Creating a Bar Chart

The instructor who surveyed her students about expected grades wants to create a bar chart from the data. The frequency table for the data is

Grade	Frequency	Relative frequency (%)
A	9	32.1
B	12	42.9
C	4	14.3
D	2	7.1
F	1	3.6
Total	**28**	**100.0**

Create a bar chart for the data using relative frequency on the *y* axis. Be sure to label the axes and include an appropriate title.

Pareto Analysis

An important application of bar charts in the area of management is in Pareto analysis. Very often, managers are faced with numerous situations and problems that are not of equal importance. This principle, known as the principle of "the vital few and the trivial many" or the "80–20 rule," was named the Pareto principle in the 1940s by J. M. Juran, one of the major contributors in the field of Total Quality Management (TQM).

The idea behind the principle is that, in most cases, 80% of the phenomena being observed fall into only 20% of the categories. For example, in quality control, most customer complaints can be attributed to only a few of the defect categories; in marketing, a small number of customers account for most of the sales; and in human services and employee relations, only a few of the employees account for most of the personnel problems. If managers can identify the key few categories, they can improve quality, increase sales, and improve employer/employee relations.

A tool that helps accomplish this is known as a **Pareto diagram.**

> A *Pareto diagram* is a bar chart in which the categories are plotted in order of decreasing relative frequency. In addition to the bars, the cumulative relative frequency of the categories is plotted on the same graph.

EXAMPLE 3.16   Tissue Defects

Pareto Diagram

A large consumer products company that manufactures facial tissues has been keeping track of customer complaints and classifying them according to the type of complaint. The company has created a frequency table for the complaints for the previous month. The data, in order of decreasing frequency, are

Complaint category	Frequency	Relative frequency (%)	Cumulative relative frequency (%)
Dispensing	127	58.0	58.0
Packaging	44	20.1	78.1
Miscounts	32	14.6	92.7
Softness	12	5.5	98.2
Odor	3	1.4	99.6
Pricing	1	0.5	100.0
Total	**219**	**100.0**	

To create a Pareto diagram for these data, first construct a bar chart using the relative frequency (see figure).

To make the bar chart a Pareto diagram we must add a new, different scale to the right side of the graph. This second y-axis scale must allow us to graph the cumulative relative frequency, so it must go from 0 to 100%. At the location of each bar, a point is plotted that represents the cumulative relative frequency for that bar. At the end, the points are connected to create a line. The Pareto diagram for the tissue complaint data is shown below:

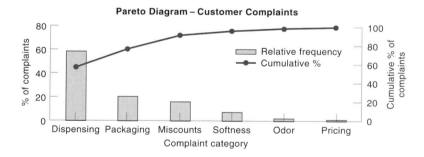

From the Pareto diagram the company can see that almost 80% (78.1%) of the customer complaints that they received in the last month were about dispensing or packaging. If they concentrate on fixing the defects that generate those complaints they can go a long way toward reducing customer complaints. ■

Other Uses for Bar Charts

Bar charts are not limited to situations where the data are frequencies. Very often bar charts are used to display data for different categories where the data are some kind of quantitative measure for each category. For example, a bar chart could be used to display and compare sales revenues for a sample of different software products. In this case, the *y* axis of the chart would represent the value of the variable being studied.

EXAMPLE 3.17   Food Sales

Bar Charts for Nonfrequency Data

A company did a study of sales for a group of foods that are advertised as having "added nutrients." The company looked at supermarket sales for a 52-week period and found the following:

Product	Manufacturer	Sales ($ million)
Hawaiian Punch	Procter & Gamble	125.7
Yoo Hoo Chocolate Flavored Drink	Austin Nichols	32.9
Life Savers Flavor Pops	Agway	9.4
Wonder Calcium Enriched White Bread	Ralston-Continental Baking	1.0
Vicks Vitamin C Lemon Drops	Procter & Gamble	0.5

The company created a bar chart for the data using the product for the *x* axis and the sales (in millions of dollars) for the *y* axis. The chart obtained is shown below:

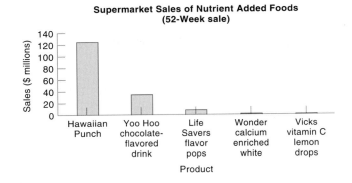

Supermarket Sales of Nutrient Added Foods
(52-Week sale)

Creating a Pie Chart for Qualitative Data

Qualitative data can also be displayed using a **pie chart.** This type of chart is often used when the categories of the data represent some part or portion of a whole.

> A *pie chart* represents data in the form of slices or sections of a circle. Each slice represents a category and the size of the slice is proportional to the relative frequency of the category.

Drawing a pie chart by hand requires the use of a protractor to accurately measure the sizes of the slices and is fairly tedious. Since the number of degrees in a circle is 360, you can use the relative frequency to determine how many degrees should be allocated to each slice. Because of this it is usually best to use computer software to create one.

EXAMPLE 3.18   Class Year

Creating a Pie Chart

The School of Business wants to use the data on class year for a presentation to the faculty. They think that a pie chart might have the best impact. From the relative frequency table they determine the following:

Class year	Relative frequency (%)
Freshman	14.3
Sophomore	42.9
Junior	7.1
Senior	35.7
Totals	**100.0**

They use Microsoft Excel to create a pie chart from the relative frequency table:

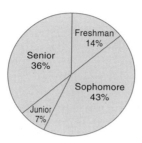

3.3.2 Graphical Displays for Quantitative Data

The tool used to display quantitative data is called a **histogram.** A histogram is very similar to a bar chart, but since numbers are naturally ordered, the x axis of the graph must be scaled to reflect this. There are slight differences for dealing with integer and continuous data.

Histograms for Integer Data

As in a bar chart, you use a rectangle to represent each possible data value, with the height of the bar corresponding to the frequency or relative frequency for that value. Remember that integers are numbers and they have a definite ordering. The x axis must accommodate all of the possible values, whether or not there were any observations of the value. The rectangles are centered on the data values as in a bar chart, but the bars are contiguous; that is, they touch each other.

EXAMPLE 3.19   Student Attitudes

Creating a Histogram for Integer Data

The instructor who is interested in student attitudes toward statistics would like to create a graphical display of the data. She feels that this might provide another view of the situation. The frequency table for the data is

Rating	Frequency	Relative frequency (%)
0	4	14.3
1	4	14.3
2	0	0.0
3	0	0.0
4	2	7.1
5	2	7.1
6	3	10.7
7	2	7.1
8	3	10.7
9	2	7.1
10	6	21.4
Total	**28**	**100.0**

To create a relative frequency histogram for the data the *x* axis will be scaled so that the numbers 0 through 10 appear at the center of the bars. The *y* axis will be scaled to accommodate the relative frequencies, the highest of which is 21.4%. The histogram is shown below:

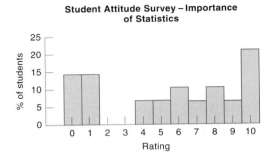

From the histogram the instructor sees that the really negative attitudes are separate from the rest of the group and that the most typical response was a 10. The remainder of the responses were evenly spread out over the 4 through 9 ratings. Although there is no new information in the graph, the visual display does add another dimension to the information.   ■

 **TRY IT NOW!**

New Product Survey

Cumulative Relative Frequencies

The marketing research firm that is conducting a survey about customer attitudes toward a new brand of soap would like to look at its data graphically. The frequency table for the data is shown below:

Rating	Frequency	Relative frequency (%)
1	4	13.3
2	5	16.7
3	6	20.0
4	11	36.7
5	4	13.3
Total	**30**	**100.0**

Create a relative frequency histogram for the data.

Histograms for Continuous Data

A histogram for continuous data differs from the one for integer data in that each rectangle represents a class interval, which is a *range of values*. For this reason, the rectangles are not centered on values, but begin and end at each of the class boundaries.

EXAMPLE 3.20   ABC Faculty Salaries

Creating a Histogram for Continuous Data

The provost at ABC wants to make a histogram of the data on number of years of service. She decides to use the frequency distribution that uses intervals of 5 years:

Class	Frequency
$-5 < x \le 0$	15
$0 < x \le 5$	38
$5 < x \le 10$	42
$10 < x \le 15$	23
$15 < x \le 20$	29
$20 < x \le 25$	51
$25 < x \le 30$	8
$30 < x \le 35$	0
$35 < x \le 40$	0
$40 < x \le 45$	1

The histogram is shown below:

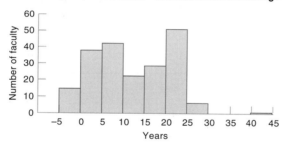

Faculty Years of Service – Aluacha Balaclava College

She sees at first glance that the data are certainly heavily concentrated at the bottom half of the class intervals and that the upper half of the intervals are very sparse.   ∎

Sometimes you will see histograms that do not use the endpoints of the intervals to label the *x* axis. In this case, the bars of the histogram are centered on the axis tick mark and the *midpoint* of the interval is used instead.

EXAMPLE 3.21   On-Time Rates

Creating a Histogram for Continuous Data

The railroad company interested in on-time rates would like to look at a graphical display of the data that have been collected. The frequency table for the data is shown at the top of the next page.

On-time rates	Frequency	Relative frequency (%)
$25.0 < x \leq 35.0$	3	10.0
$35.0 < x \leq 45.0$	2	6.7
$45.0 < x \leq 55.0$	4	13.3
$55.0 < x \leq 65.0$	3	10.0
$65.0 < x \leq 75.0$	11	36.7
$75.0 < x \leq 85.0$	3	10.0
$85.0 < x \leq 95.0$	4	13.3
Totals	**30**	**100.0**

The railroad decides to create a relative frequency histogram. To do this the y axis must be scaled so that it can accommodate percentages from 0 to 36.7%. So the railroad decides to go from 0 to 40% in increments of 5%. The tick marks on the x axis will be the values at the beginning of each of the class intervals. The last tick mark will be at the end of the last interval. The histogram is shown below:

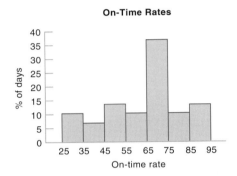

From the histogram the railroad company sees that the most typical on-time rates were in the 65–75% range, and that they were evenly distributed among the other classes. ■

3.3.3 Displaying Small Data Sets

The rules for creating histograms are really not suitable for data sets with less than 25 observations. This is because the number of classes should not be less than 5 and we determine the number of classes by taking the square root of the sample size. Often when we collect data we do not have more than 25 observations. Is there a way to display these types of data sets graphically? The answer is a graphical method called a **dotplot.**

> In a *dotplot,* each observation is plotted as a point on a single, horizontal axis. The axis is scaled so that each of the data points can be located uniquely on the axis. When there is more than one observation with the same value the points are "stacked" on top of each other.

A dotplot can show many of the same features of the data as a histogram.

EXAMPLE 3.22   Bank Customers

Creating a Dotplot for a Data Set

A bank is interested in understanding the number of customers that arrive hourly on Fridays preceding holiday weekends. It takes data for eight hours on each of two different Fridays and obtains the following:

| 10 | 14 | 15 | 14 | 19 | 12 | 11 | 14 |
| 15 | 14 | 20 | 19 | 11 | 12 | 17 | 17 |

After looking at the data the bank decides to use a dotplot to display them because there are not very many observations. An axis is created that is scaled from 10 (the smallest value) to 20 (the largest value) in increments of one, since the data are integer data:

The dotplot shows that the number of customers per hour varies widely over the interval, although in 25% (4/16) of the hours there were 14 customers, and in 69% (11/16) there were at least 14 (14 or more) customers. The bank decides to use these data to plan the number of tellers that should be working on these Fridays. ■

3.3.4 Using the Computer to Create Graphical Displays

You may be wondering why it is necessary to learn how to create bar charts, pie charts, and histograms by hand, when you will almost certainly be using a computer to create them. The graphs produced by computer software are superior to those produced by hand, but a good deal of critical thought goes into creating a graph that conveys *information* to the viewer. It is not so important that you *create* the charts by hand as it is that you know *how* they are created.

Most computer software packages ask the user for input when creating graphs. Knowing how the charts are created will help you make decisions about how the charts look. In addition, not all software allows the user to make the same decisions. The default graphs produced by computer software offer great starting points for the final graphical display, but they are not usually the end product that you are looking for.

When you create a bar chart by hand you have the freedom to put the bars in any order you want and to combine categories as you wish. Some software will allow you to choose ascending or descending order of frequency for the bars, others will require you to sort the frequency table first, either by frequency or alphabetical order of categories. The same is true about combining categories and using frequency or relative frequency. Some computer software packages such as Excel will not create frequency distributions for categorical data and will make bar and pie charts only from already existing frequency tables. If you did not know what the chart should contain you would not know how to process the raw data that you collect.

Creating histograms differs among software packages. There are differences in the way you go about defining and creating the chart. Almost all software packages will create a histogram using a default set of class intervals, but they differ widely in what they will allow the user to specify. You saw in the previous section that when you are trying to make graphical displays to compare different samples, you need to be able to control the *x* axis of the graph. Microsoft Excel will allow you to specify the ending values for each of the classes, but when it creates the histogram from the data, it puts the endpoints in the *center* of the bar. Minitab and Data Desk will allow you to specify endpoint values or midpoint values, or the number of classes you want to use, but not both.

There are also differences in the way the final chart looks. Remember that Minitab and Data Desk define their intervals to include the lower value and go up to

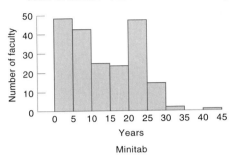
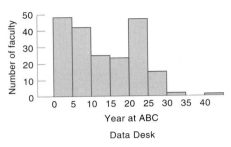

FIGURE 3. 4   Histograms from Minitab and Data Desk for ABC years of service

but not include the upper value. The differences in processing causes the histograms from the different software packages to look a little different. Figure 3.4 shows the histograms produced by each package for the data on faculty years of service for Aluacha Balaclava College.

If you compare these histograms to the one shown on page 83, you see that the major difference is in the first class. In the original histogram, this class contains only those people with 0 years of service ($-5 < x \le 0$), while in Minitab and Data Desk it contains all people with 0, 1, 2, 3, or 4 years of service ($0 \le x < 5$). This difference does not really change the general appearance of the histograms. Knowing the basics of creating histograms enables you to make smarter, more informed decisions, no matter what software you might use.

3.3.5 Exercises—Learning It!

Requires Exercise 3.2

3.12   The data that the computer club collected on personal computer operating systems is given again here:

Macintosh	Macintosh	Macintosh	W 3.1*	W 3.1
W 3.1	W 3.1	W 3.1	OS /2	W 3.1
W 3.1	W 3.1	W 3.1	W 3.1	W NT
W 3.1	W 3.1	W 3.1	W NT	W 3.1
W 3.1	W 95	Macintosh	W 95	W 3.1
OS /2	DOS	W 3.1	W 3.1	W 3.1

* The W 3.1 category includes WFW 3.11 users.

(a) Suppose you were asked to create a bar chart for the data. What order would you pick for the categories? Why?

(b) Make the bar chart for the data.

(c) Does the bar chart indicate that the majority of the people surveyed had switched to Windows 95? Were a majority of those surveyed users of some Windows operating system?

3.13   An annual survey by J. D. Power and Associates reported in *The Connecticut Post*, 27 May 1994, rated vehicles on the number of problems reported by owners per 100 vehicles. The data for the top 11 cars are shown below:

Car	Reported Problems
Lexus LS 400	32
Lexus GS 300	48
Lexus SC 300/400	52
Geo Prizm	56
Acura Legend	57
Infiniti J-30	61
Mercury Grand Marquis	61
Toyota Camry	63

(continued)

Car	Reported Problems
Toyota Tercel	67
Lexus ES 300	68
Lincoln Town Car	68

(a) Create a bar chart for the number of reported problems per 100 vehicles.

(b) Comment on any interesting features of the data.

3.14 Columbia University conducted a survey of 637 members of the Authors Guild and the Dramatists Guild, all of whom have published at least one book, one play, or three magazine articles. As part of the survey, the members were asked to classify the amount of money that they made from writing. The frequency table below gives the results of the survey:

Revenue from Writing	Number of Members
More than $50,000	57
More than $20,000 but less than $50,000	127
More than $1000 but less than $20,000	192
More than $0 but less than $1000	102
No money at all	159
Total	**637**

(a) Create a relative frequency bar chart for the data.

(b) From the data would you conclude that a majority of those surveyed can or cannot earn a living solely from writing?

(c) Although the data are numerical, why did you have to make a bar chart rather than a histogram?

(d) What problem do you see with the way that the classes were defined?

3.15 The company that collected data on the number of weekly repairs needed by copy machines decided that it wanted a graphical display of the data to include in a budget report. The data are repeated below: *Requires Exercise 3.5*

```
3   2   3   0   3
0   2   2   2   1
3   2   2   1   0
2   3   2   1   2
3   4   2   7   1
2   2   1   3   1
0   3   1   1   2
```

Create a histogram for the data.

3.16 After collecting the data on packaging defects, the company decided to ask you how it could make a graph of the data to use in a brainstorming session with the employees in the department. The data are repeated here: *Requires Exercise 3.3*

Dented	No defect	No defect	Dented	No defect
No defect	Torn	Crushed	No defect	No defect
Unsealed	No defect	No defect	No defect	No defect
No defect	No defect	No defect	No defect	No defect
No defect	No defect	No defect	Dented	No defect
No defect	Unsealed	No defect	No defect	No defect

(a) What type of graph would be appropriate for these data?

(b) Create a graphical display of the data.

(c) What order did you select for the categories? Why?

3.17 A company has recently become concerned about the number of young engineers that seem to be leaving the company and decides to collect some data to try to understand why the engineers are leaving. During the exit interviews with employees the human resources department asks several questions about the reasons the employee has for leaving the company. A table showing the reasons for leaving and the appropriate codes is shown at the top of the next page:

Dislike engineering	DE
Relocation	L
Raise in pay	P
More responsibilities	R
Relations with manager	M
Colleagues	C
Type of work	B

The human resources department looks at the results from the exit interview process for 40 engineers that have left the company in the last two years and find the following:

DE	P	B	R	R	R	R	DE
P	L	M	DE	M	M	R	P
R	P	L	P	B	P	DE	L
R	R	P	P	P	R	C	C
M	P	R	P	P	M	P	L

(a) Create a Pareto diagram for the data.

(b) What category has the highest relative frequency?

(c) Does it appear that the 80–20 rule is true for these data? Why or why not?

3.18   After a change in the channels provided by the cable television service in a town, subscribers were asked to rate the new service on a scale of 1 = worst service I have ever had to 10 = best service I have ever had. The responses from 40 random subscribers are

1	4	5	7	8
2	4	5	7	9
2	4	5	7	9
3	4	6	7	9
3	4	6	7	9
3	4	6	7	9
3	4	6	8	9
3	5	7	8	10

(a) Create a relative frequency histogram of the ratings.

(b) Are the majority of the ratings favorable, unfavorable, or neither?

3.19   The company that is worried about losing its engineering staff also looked at the length of time, in months, that the engineers in the sample worked for the company:

16.7	26.0	28.0	29.6	30.6	32.6	34.0	35.8
20.9	26.6	28.1	29.7	30.8	32.6	34.6	37.4
25.2	26.6	28.5	30.2	31.1	33.2	34.8	37.8
25.6	27.4	28.8	30.2	31.7	33.5	34.9	38.3
25.8	27.7	28.8	30.3	32.1	33.6	35.2	44.6

Create a relative frequency histogram for the data.

Requires Exercise 3.11   3.20   The refuse company decides to use the data it collected in a presentation to the town. The data are shown again here:

8.3	10.2	12.2	13.1	13.9
8.5	10.4	12.3	13.3	14.1
8.8	11.0	12.3	13.3	14.3
9.1	11.0	12.3	13.4	14.3
9.7	11.2	12.3	13.4	14.4
9.8	11.4	12.5	13.4	14.4
9.9	12.0	12.7	13.6	14.4
9.9	12.0	12.8	13.6	14.8
10.1	12.2	12.9	13.8	15.2

(a) Create a relative frequency histogram for the data.

(b) From the histogram, does it appear that a majority of the newspaper piles put out for recycling exceed 12 inches?

3.21   The Bureau of Weight and Measures that looked at orange juice containers from super-   *Requires Exercise 3.10*
       markets wants to create a graphical display of the data it obtained.

64.8	65.2	65.6	65.7	65.9
64.8	65.3	65.7	65.7	66.0
64.9	65.3	65.7	65.7	66.1
65.1	65.4	65.7	65.7	66.1
65.2	65.4	65.7	65.8	66.3

Create an appropriate graphical display of the data.

3.4   DESCRIBING AND COMPARING DATA

Since the reason for displaying data is to gain understanding, it is important to know what we can learn from the graph of a data set. For quantitative data we usually want to know what a typical observation might be, and how the actual observations differ from the typical values. We would also like to be able to compare different data sets and make some decisions based on the comparisons.

3.4.1 Describing Quantitative Data

In statistics, the features of interest for a set of numerical data can be classified as **center, shape,** and **variability.**

> The *center* of a set of data describes where, numerically, the data are centered or concentrated.

> The *shape* of a set of data describes how the data are spread out around the center with respect to the symmetry or skewness of the data.

> The *variability* of a set of data describes how the data are spread out around the center with respect to the smoothness and magnitude of the variation.

Together these three features describe the distribution of the data. To describe the data it is useful to picture the distribution of the data as being represented by a smooth curve that captures the "shape" of the histogram. One way to achieve this curve is to plot a point at the top of each bar of the histogram and then connect the dots as shown in Figure 3.5.

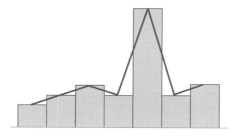

FIGURE 3.5   Histogram with line showing the shape of the distribution

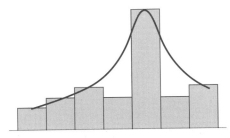

FIGURE 3.6   Histogram with smoothed curve representing the distribution

If you smooth out the plotted line you get a curve that looks like the one shown in Figure 3.6. From the curve you see that the distribution is centered at the bump or high point of the curve. Since the bump in the distribution is the location of the class with the highest frequency, it locates the most typical data points.

The shape of a distribution concerns how the data are spread out on either side of the center, that is, whether they are **symmetric** or **skewed.**

> When data are evenly spread out on both sides of the center, we describe the distribution of the data as *symmetric.*

A typical symmetric distribution is shown in Figure 3.7.

FIGURE 3.7   Typical symmetric distribution

> When the data are not evenly spread out on either side of the center then we refer to the distribution as being *skewed.*

Skewness has a direction associated with it, either left (negative) or right (positive). The direction of the skew describes the side on which the distribution of the data covers a larger distance, that is, the direction in which the distribution "tails off" more slowly. Figure 3.8 shows both right and left skewed distributions.

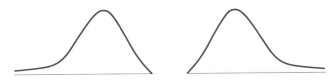

FIGURE 3.8   Typical skewed distributions

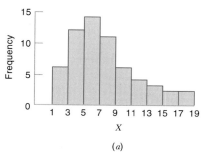

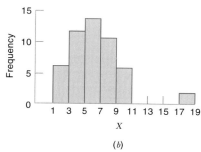

(a)   (b)

FIGURE 3.9 Histograms with (a) skewed data and (b) extreme values

When data are skewed, either left or right, the tailing off of the data is continuous and gradual as shown in Figure 3.9a. When the tailing off involves a gap in the data—a place where classes in the frequency histogram have no observations—as shown in Figure 3.9b, the data are not really skewed. More likely the data you have contain some extreme or unusual observations. We talk about these extreme values more in Chapter 6 when we discuss outliers.

In addition to center and shape we would like to describe how much the data differ from the center or typical values. This is not easy to do without some way to measure the differences. We examine measures of variability in Chapter 6. At this point we can describe the variability of the distribution in two ways: in terms of (1) the "smoothness" of the curve and (2) the total spread of the data.

When data are not very variable, the frequency of observations decreases steadily as you move away from the center. Sometimes when data are highly variable, the distribution will be jagged. That is, the frequency of data values will not decrease steadily as you move away from the center. Another way to describe variability is to describe the distance from one end of the data to the other. Figures 3.10 and 3.11 show histograms with different degrees of variability.

Dotplots can also be used to describe data distributions, although because they represent small data sets it is often more difficult to describe the data.

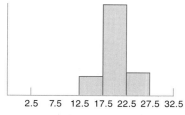

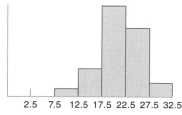

2.5   7.5 12.5 17.5 22.5 27.5 32.5   2.5   7.5 12.5 17.5 22.5 27.5 32.5

FIGURE 3.10 Histograms with different variability showing spread of data

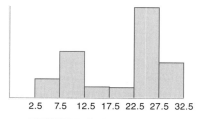

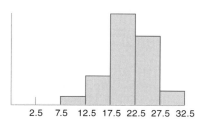

2.5   7.5 12.5 17.5 22.5 27.5 32.5   2.5   7.5 12.5 17.5 22.5 27.5 32.5

FIGURE 3.11 Histograms with different variability showing smoothness

EXAMPLE 3.23   ABC Faculty Salaries

Describing Data Distributions

The ABC provost looked at the distribution of years of faculty service to see if she could determine how the data are distributed.

Years of Service – Aluacha Balaclava College

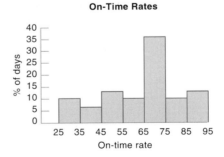

She sees that the data are highly variable and there is no smooth shape to the distribution. In addition, there does not seem to be a unique center or shape. She wonders why this might be. Perhaps another variable is affecting the picture. It might make more sense to look at years of service for each rank separately. ∎

Sometimes when you collect data and make a histogram there is no cohesive picture to be seen. This often happens when there are other, underlying variables that have an effect on the variable you are studying. When this happens it is useful to separate the data into groups on the basis of another variable and look at them again.

EXAMPLE 3.24   On-Time Rates

Describing Data Distributions

The railroad company that is looking at on-time rates obtained the histogram below:

On-Time Rates

From the histogram we see that the data have a distinct center and that, typically, the on-time rate was between 65% and 75%. The frequency of the data seems to drop off rapidly from the center, and the data are rather uniformly distributed among the other classes. The data seem a little variable. The distribution is skewed to the left, which means that the data are more spread out on the low side (four classes) than on the high side (two classes). ∎

The choice of the number of class intervals for a histogram is very important. If your histogram has too many intervals for the number of data values, then the data might appear to have a lot of variability when, in fact, classes are empty or have few observations because of the lack of data. If your histogram has too few intervals, then the distribution will look like a lump and important features of the data might be hidden. It is often important to look at the data in several different ways to make sure that you are getting a true and consistent picture. We see later on that it is possible to distort graphical displays and bias an analysis by playing with the scale.

TRY IT NOW!

Assignment Times

Creating a Frequency Table for Continuous Data

The instructor for the introductory statistics class wants to see, in graphical form, the data she has collected on the amount of time it took the students to do the assignment. The frequency distribution for the data is

Time	Frequency	Relative frequency (%)
$22.70 < x \le 26.22$	6	20.0
$26.22 < x \le 29.74$	11	36.7
$29.74 < x \le 33.26$	7	23.3
$33.26 < x \le 36.78$	5	16.7
$36.78 < x \le 40.30$	0	0.0
$40.30 < x \le 43.82$	1	3.3

Create a relative frequency histogram for the data.

Use the histogram to describe the distribution of the times the students took to complete the assignment.

ANS.   **Time to Complete Statistics Assignment**   TYPICALLY BETWEEN 26.22 AND 29.74 MINUTES, SKEWED RIGHT, NOT VERY VARIABLE.

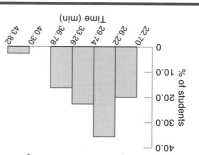

3.4.2 Comparing Data Distributions

One of the major reasons for doing statistical analyses of data is to obtain facts for making *informed* decisions. As a result, we often need to make comparisons between or among samples taken from different populations. When comparing different data sets, we make those comparisons based on the qualities of the data you just learned: center, shape, and variability.

To make valid comparisons it is critical that the data be displayed in the same way. Since the graphical displays have a *visual* impact, it is important that the graphs used are comparable in scale so that the viewers are not misled.

To make visual comparisons about center, shape, and variability, the class intervals for the different graphs should be the same. When this is not the case it is difficult if not impossible to compare these qualities in a meaningful way. The following example shows how using different scales can mask the visual impact of the display.

EXAMPLE 3.25   Loan Application Times

Comparing Data Distributions

A large bank with many branch offices has developed two different procedures for filling out mortgage loan applications. Each procedure is being used in a different branch of the bank in comparable locations. After the procedures are used for several months, some data are collected on the amount of time (in minutes) it took for loan candidates to complete the application. The data for each branch are

Branch 1						Branch 2			
37.0	39.3	42.1	44.3	47.8	40.4	42.8	43.4	44.3	46.1
37.2	39.9	42.4	44.5	49.3	40.8	42.8	43.5	44.4	46.4
37.6	41.2	42.6	45.3	49.5	41.5	42.8	43.5	44.7	46.5
38.1	41.5	42.7	45.7	50.3	41.5	43.1	43.8	44.7	47.1
38.8	41.6	43.0	46.1	52.1	41.6	43.2	44.0	45.1	47.9
39.2	41.9	43.4	46.2	54.7	42.3	43.3	44.2	45.4	48.6

To better understand the data and to compare the loan application times for the two branches a computer software package is used to create histograms for each of the data sets. The histograms show that the application times for Branch 1 are typically between 37.5 and 42.5 minutes and that the data are slightly skewed to the right. The loan application times for Branch 2 are typically between 43 and 45 minutes and the data are more symmetric than for Branch 1. On first glance it appears that the first bank is using the better method. Or is it?

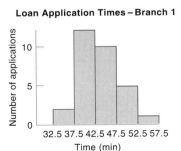

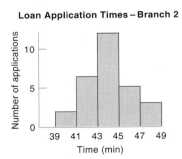

The bank decides to re-create the histograms, this time using a common scale for the *x* axis. The results are shown at the top of page 95. From the second set of histograms it appears that the times for Branch 1 are much more variable than those for

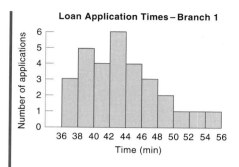

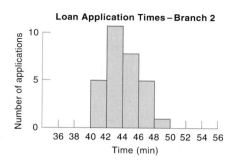

Branch 2. It also seems that the typical times for Branch 1 are really not that much different than those for Branch 2. They can see from the second set of graphs that the method used for Branch 2 is more consistent than that for Branch 1 and it has comparable application times. ■

In the previous example it does not really matter that the y axis of the histograms is frequency rather than relative frequency, because the number of observations is the same for each sample. If this is not the case it is necessary to use relative frequency in order to make valid and meaningful comparisons.

Discovery Exercise 3.1
THINKING ABOUT VARIABILITY

A manufacturer of compact discs uses two different suppliers for the jewel boxes used to hold the discs. There have been problems with these boxes in the past. The inside width of the jewel box has critical specifications of 119.0 ± 0.2 mm. If the case is too narrow the disc will not fit in it and if it is too wide, the front label insert slips around. Because it is time to renew the purchasing contracts for the jewel boxes, the CD manufacturer decides to take a look at a sample of the boxes from each supplier. The data (in mm) for each source are

Supplier A					Supplier B				
118.7	118.9	119.0	119.1	119.2	118.8	118.8	118.9	118.9	118.9
118.8	119.0	119.0	119.1	119.2	118.8	118.8	118.9	118.9	119.0
118.8	119.0	119.0	119.1	119.2	118.8	118.8	118.9	118.9	119.0
118.9	119.0	119.1	119.2	119.2	118.8	118.8	118.9	118.9	119.0
118.9	119.0	119.1	119.2	119.3	118.8	118.9	118.9	118.9	119.1

Make a relative frequency histogram of the data for each supplier.

(continued)

Describe the distribution of jewel box widths for each supplier and compare them.

Your company has decided to single source its supply of jewel boxes. The purchasing agent in charge of the accounts argues that supplier B should not get a renewed contract since he observes that the jewel boxes from that source are not centered at the target specification of 119.0 mm while the jewel boxes from supplier A are right on target. Can you explain to him why, although his observation is true, his decision to use supplier A is not necessarily correct? What factor has he failed to consider?

Which supplier would you recommend that your company use? Write a short memo to the manager with your recommendation and your supporting reasons.

3.4.3 Exercises—Learning It!

3.22 A university collected some data on the amount of money ($) that students spend on textbooks in a typical semester:

214	241	248	258	269
233	244	249	260	274
234	245	250	262	276
236	247	253	262	277
239	248	254	263	277
241	248	254	265	281

(a) Create a histogram of the data.

(b) Where is the center of the distribution located?

(c) Are the data symmetric or skewed? If they are skewed, are they left or right skewed?

(d) Describe the variability of the data.

3.23 The company that is looking at the length of time that engineers stayed at the company wants to summarize its data, shown below:

16.7	26.0	28.0	29.6	30.6	32.6	34.0	35.8
20.9	26.6	28.1	29.7	30.8	32.6	34.6	37.4
25.2	26.6	28.5	30.2	31.1	33.2	34.8	37.8
25.6	27.4	28.8	30.2	31.7	33.5	34.9	38.3
25.8	27.7	28.8	30.3	32.1	33.6	35.2	44.6

Requires Exercise 3.19

(a) Describe the center of the data.

(b) Describe the shape of the data.

(c) Describe the variability of the data.

3.24 The refuse company decides to use the data it collected in a presentation to the town. The data are shown again:

8.3	10.2	12.2	13.1	13.9
8.5	10.4	12.3	13.3	14.1
8.8	11.0	12.3	13.3	14.3
9.1	11.0	12.3	13.4	14.3
9.7	11.2	12.3	13.4	14.4
9.8	11.4	12.5	13.4	14.4
9.9	12.0	12.7	13.6	14.4
9.9	12.0	12.8	13.6	14.8
10.1	12.2	12.9	13.8	15.2

Requires Exercises 3.11, 3.20

Describe the distribution of the data. Remember to include center, shape and variability.

3.25 The company that is looking at turnaround time from the Office Services department wants to summarize its data:

Requires Exercise 3.9

14	19	21	25	26	29
15	19	22	25	26	29
16	20	22	25	26	30
16	20	23	25	26	31
18	20	23	25	27	31
18	20	23	25	28	35
18	21	24	26	29	40

(a) How long is a typical turnaround time for a job?

(b) What shape is the distribution of turnaround times?

(c) How would you describe the variability in turnaround times?

3.5 CREATING GRAPHICAL DISPLAYS USING EXCEL 97

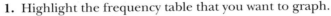

The graphical displays that you learned about in this chapter can be created easily using Microsoft Excel 97. The power of using a computer package such as Excel to create the graphs is that you can modify them easily. This leads to better graphical displays because it eliminates much of the drudgery. This section will cover the basics of creating a chart using Microsoft Excel, as well as the specific methods for creating bar charts, pie charts, and histograms.

3.5.1 The Basics of Creating a Chart in Excel

Chart Wizard belongs to the Standard Toolbar.

In most cases, graphs and charts in Excel are easily created by using a special assistant called the *Chart Wizard.* The steps for creating a chart or graph using the Chart Wizard are:

1. Highlight the frequency table that you want to graph.
2. Invoke the Chart Wizard by clicking on the icon on the toolbar.
3. Follow the directions and hints from the Chart Wizard.
4. Edit the graph to include any other features or changes that you want.

3.5.2 Creating a Bar Chart in Excel

A Bar Chart

A Column Chart

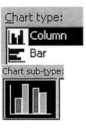

Excel differentiates between *bar* and *column* charts. The only real difference is the direction (horizontal vs. vertical) of the bars. What we call a bar chart in statistics is called a column chart in Excel. Using the *Chart Wizard* to create a bar or column chart in Excel assumes that the data have already been tabulated in the form of a frequency table. Once this is done, it is relatively easy to create the chart.

Suppose we want to create a bar chart for the faculty rank data for Aluacha Balaclava College. Figure 3.12 shows the frequency table as part of an Excel worksheet.

To use the *Chart Wizard* to create a bar or column chart, select the range that contains the frequency table, in this case **A2:B6,** and click the *Chart Wizard* icon on the Standard Toolbar.

There are four steps to creating the chart:

1. In the first step you select the *Chart Type.* Select **Column Chart** as the *type* and **Clustered Chart** as the *sub-type,* and then click the **Next>** button.
2. The second step of the process tells Excel where the frequency table is located and how the data are arranged, in columns or in rows. The default setting in this box is usually correct. You can see what your chart will look like. If it does

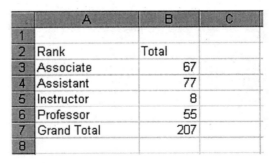

	A	B	C
2	Rank	Total	
3	Associate	67	
4	Assistant	77	
5	Instructor	8	
6	Professor	55	
7	Grand Total	207	

FIGURE 3.12   Excel worksheet with frequency table

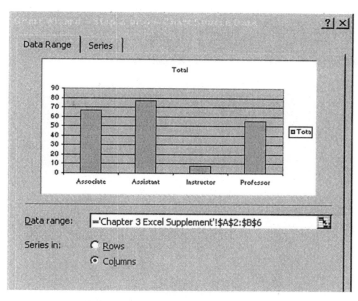

FIGURE 3.13   The Chart Source Data dialog box

not appear the way you expect, experiment with the settings. Figure 3.13 shows what you should see at this point in the process.

3. Once you have indicated where the data are located, you can click on the Next button and go to the third step of the process. In this step you will be making decisions about the format of the chart. The dialog box for this step has six tabs — Titles, Axes, Gridlines, Legend, Data Labels, and Data Table. In the **Titles** section, click the **Chart title** box and type **"Aluacha Balaclava College — Faculty Rank".** Next, press TAB and type **"Rank".** Finally, press TAB and type **Number of Faculty".** Figure 3.14 shows a portion of the dialog box at this point. Click the **Gridlines** tab and clear the **Major gridlines** check box. Then Click the **Legend** tab and clear the **Show legend** check box. Finally, click the Data Labels tab and click the **Show value** check box. Notice that while you are making these choices, you can see what your chart will look like. When you are finished with this step, click the Next button and begin the fourth and last step.

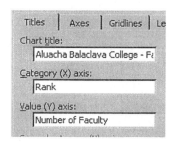

FIGURE 3.14   Adding titles to the chart

4. The fourth step of the process lets you select where you want the chart to appear. You can either put the chart on a new worksheet or locate the chart on an existing sheet. To have the chart located on the current sheet, check the **As object in** radio button and click **Finish.** Excel will place the chart on the current worksheet. Figure 3.15 (page 100) shows the chart.

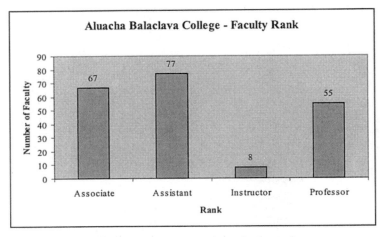

FIGURE 3.15   Finished Excel column chart

3.5.3 Creating a Pie Chart in Excel

Creating a pie chart in Excel uses the same procedure as that for creating a bar chart. The major difference occurs in the first dialog box of the Chart Wizard, where you select a different type of chart: In the first dialog box of the Chart Wizard, select Pie as the type of chart you want. The second step defines the location of the data, just as it did for the bar chart. The real differences are in step 3, where you determine the format of the chart. The dialog box for a pie chart has three tabs—Title, Legend, and Data Labels. The Title and Legend boxes are the same as those for the bar chart. The Data Labels tab allows you to select how the pie slices are labeled. Figure 3.16 shows the dialog box with the selections to label each slice with the category and the percent. The chart to the right updates to show how your selections will look. Note that if you are going to label the slices of the pie, it is not necessary to have a legend show as well.

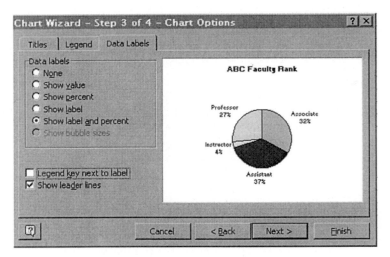

FIGURE 3.16   Formatting the pie chart

3.5.4 Creating a Pareto Diagram in Excel

As you learned in this chapter, a Pareto diagram is a special kind of bar chart used to identify trouble spots in a process. It is really a bar chart with the bars sorted by decreasing frequency. It also has a line that plots the cumulative frequency for the categories.

To create a Pareto diagram in Excel, you must first sort your frequency table by using the Sort command. You can do this in two ways. The quick way is to highlight the range that you want to sort—in this case, it is the data in the frequency table for Total—and click on the *Sort Descending* button on the Standard Toolbar. This will sort the range you have selected and will also sort the adjacent data accordingly. You can also highlight the entire table you want to sort and select **Sort** from the **Data** menu. In the dialog box, select the field you want to sort by and indicate whether you want to sort in ascending or descending order.

Sort Descending

To create a Pareto diagram, you will also need a column for cumulative relative frequency. You can do this using the SUM function to keep a running total similar to the way you would update a checkbook register. The steps are:

1. Label the cell adjacent to Total in the frequency table Cumulative and copy the frequency of the first category to the adjacent cell.

2. In the cell directly below this, you will enter a formula using the SUM function. You tell Excel that you are entering a formula or function by typing an equal sign "=" and entering the formula or function name. In this case, the function you want to use is SUM, so you type this directly after the equal sign. The SUM function has two parts here, the location of the current subtotal and the location of the data you want to add to that subtotal. For example, looking at the data table in Figure 3.17, you see that the current subtotal is in cell C3 and you want to add the data in cell B4. You can see the formula in Cell C4. When you hit **Enter**, the formula will calculate the value you have indicated.

	A	B	C
1			
2	Rank	Total	Cumulative
3	Associate	67	67
4	Assistant	77	=SUM(C3,B4)
5	Instructor	8	
6	Professor	55	

FIGURE 3.17   Formula for Cumulative Frequency

3. Copy this formula to the rest of the cells in the column labeled Cumulative and you will see the cumulative frequencies calculated.

4. Now you need to calculate the cumulative relative frequencies. Using the next column (you can label it Cumulative Relative Frequency) in the worksheet, you will enter the formula—that is, the cumulative frequency divided by the total. In cell D4 type "=C4/207" and press enter. The cumulative relative frequency for the first category is calculated. Copy the formula to the remainder of the cells in this column.

Creating the Pareto diagram is easy using one of the custom charts in the Chart Wizard. After starting the Chart Wizard, the steps are as follows:

1. In the first dialog box for chart type, click on the tab labeled Custom Types and scroll down until you see a chart called Line–Column on 2 Axes as shown in Figure 3.18 (page 102). This will plot the first data series in your table as columns and the second as a line. The left axis corresponds to the bars and the right to the line. Click **Next▷** to proceed.

2. The second step lets you specify the location of the data and see that your chart is set up correctly. When this is done, click **Next▷** to continue.

Note: If the ranges are not adjacent, hold down the CTRL button while you highlight the ranges.

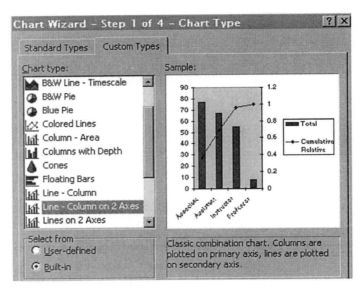

FIGURE 3.18   Custom Chart Type menu

3. The third step allows you to format the chart by adding labels. You might want to use the Legend tab to relocate the legend to the bottom of the chart so that it does not interfere with the right axis. When the graph is the way you want it, click **Next>** to continue.

4. In the fourth step you can specify the location of the chart and click **Finish** to see your result. The finished chart is shown in Figure 3.19.

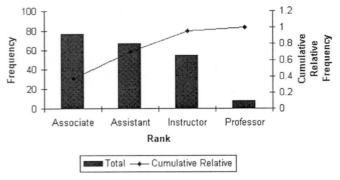

FIGURE 3.19   Pareto diagram

3.5.5 Creating a Frequency Table in Excel

In the previous section, you learned how to use the Chart Wizard to create a bar chart in Excel when you have the frequency table available. If you have the raw data instead of the frequency table, then you need to use **Pivot Table** to create it.

A **Pivot Table** in Excel is a tool that you will use often to manipulate raw data that are qualitative. You can create a pivot table by using the Pivot Table Wizard in Excel. The steps for creating the frequency table are:

1. Highlight the data for which you want to create the frequency table.

2. Invoke the Pivot Table Wizard in Excel by using the **Data > Pivot Table Report** menu.

	A	B	C
1	Rank	Service (Yr)	Salary
2	ASST	22	53316
3	PROF	11	64375
4	ASSO	7	63501
5	ASSO	6	59426
6	ASSO	20	49058
7	PROF	4	94969
8	ASST	21	54762
9	ASSO	9	55516
10	ASSO	18	45932

FIGURE 3.20   Raw data for faculty rank

3. Follow the directions that the Pivot Table Wizard gives you.

4. Copy or edit the pivot table to include any changes you want.

Remember that the Provost at Aluacha Balaclava College received the data in raw form; that is, the data had not been summarized. A portion of the data for Faculty Rank is shown in Figure 3.20.

The frequency table for these data will have a column for each rank that appears and a column for the frequency, or number of times that it appears. You can also add columns for the relative frequency and cumulative relative frequency. Highlight the range that contains the data you want to summarize—in this case, the data are in the range **A1:A208**—and follow these steps:

1. From the **Data** menu select **Pivot Table Report.** The Pivot Table Wizard will start by asking what type of data you want to work with. Since you are using data that are already in an Excel worksheet, simply leave the default radio button checked and click on **Next>** to continue.

2. Since you have already highlighted the range that contains your data, you can proceed by clicking **Next>.** If you have not indicated where your data are located, you can do that now by highlighting the range in the worksheet. The next step of the process will define the table you want to create. The dialog box is shown in Figure 3.21.

3. At the right of the dialog box, there are small rectangles called field buttons, which represent all of the variables you have highlighted. In this case, Rank is

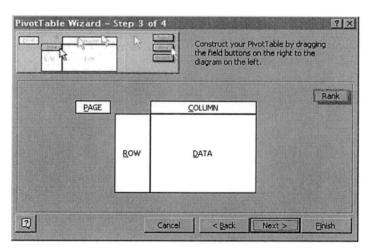

FIGURE 3.21   Creating the table

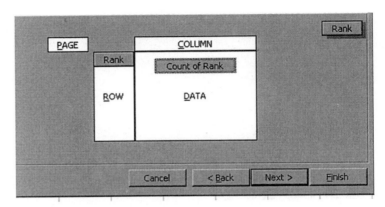

FIGURE 3.22   Defining the table

the only one. You create the table by dragging the variable names to the places you want them to appear. Since we want the ranks to appear as rows in the table, click on the Rank box and drag it to the area marked **ROW,** as indicated at the top of the dialog box. To get the frequencies, you must tell the Pivot Wizard how to process the data. To do this, drag the field button for Rank into the area of the dialog box marked **DATA.** The dialog box should now look like the one in Figure 3.22.

The default operation in a Pivot Table is to provide a count (frequency) of each value. If you want to change this, or if you want to provide additional information in the table, you can double click on the field in the DATA area and select a different data operation such as Sum or Average. When you have finished this step, click **Next>** to continue.

4. The last step of the process is to tell Excel where you want the Pivot Table to appear. You can have it appear in the current worksheet or on a new worksheet. To have it appear in the current sheet, select the **Existing worksheet** radio button and then click on the cell in the worksheet where you want the left corner of the table to appear. Click **Next>** and the pivot table shown in Figure 3.23 will appear in the worksheet location you indicated.

If you want to change the pivot table so that it looks better, you can click on the cell and type in exactly what you want to appear. For example, if you want the ranks to appear as Associate Professor, Assistant Professor, etc., simply edit those cells in the table.

Count of Rank	
Rank	Total
ASSO	67
ASST	77
INST	8
PROF	55
Grand Total	207

FIGURE 3.23   Finished pivot table

Note: Make sure the **Tools** *menu contains the* **Data Analysis** *option. If not, you may have to run the* **MS Office Setup** *program to include the* **Analysis ToolPak** *and then execute the* **Tools\Add-Ins** *command.*

It is possible to change other features of the table by double clicking on the cell in the table with the field label and clicking on the Advanced button. This will allow you to change the order of the categories and several other options.

Once you have the frequency table, you can create a bar chart, pie chart, or Pareto diagram as explained previously.

3.5.6 Creating Histograms in Excel

Quantitative data are easier to process using Excel because there are built-in functions for many of the statistical tools. Selecting **Data Analysis** from the **Tools** menu accesses these functions, as shown in Figure 3.24. Macros for other types of analyses are available on the disk that comes with this book.

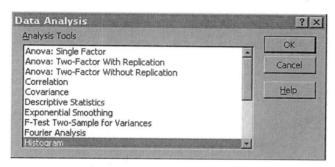

FIGURE 3.24   Data Analysis dialog box

The basics of creating a histogram in Excel are quite simple. Suppose you want to create a histogram for the data on Years of Service for Aluacha Balaclava College. Figure 3.20 shows the data as part of an Excel file.

To produce a basic histogram, simply follow these steps:

1. From the Data Analysis dialog box, select Histogram and click **OK.** The **Histogram** dialog box will open.

2. Position the cursor in the **Input Range** text box and highlight the range that contains the data, **B1:B208.**

3. Click the checkbox for **Labels** since the range contains the variable name.

4. Excel calls the class intervals for continuous data Bins. Leave the Bin text box empty this time and let Excel pick the class intervals.

5. Click the radio button for **Output Range,** position the cursor in the text box, and click on the cell where you want the top left corner of the output to appear.

6. Finally, click the checkbox for **Chart Output.** The completed dialog box appears in Figure 3.25.

7. Click **OK** and the output will appear as shown in Figure 3.26 (page 106).

FIGURE 3.25   Completed histogram dialog box

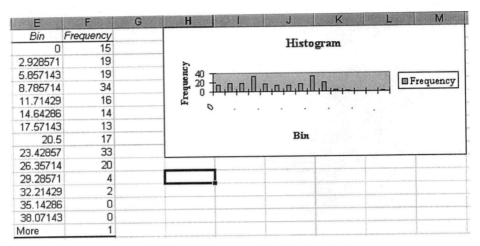

Bin	Frequency
0	15
2.928571	19
5.857143	19
8.785714	34
11.71429	16
14.64286	14
17.57143	13
20.5	17
23.42857	33
26.35714	20
29.28571	4
32.21429	2
35.14286	0
38.07143	0
More	1

FIGURE 3.26   Output from histogram tool

Clearly this is not the output you hoped for, but it is a starting point. There are two basic problems with this output. First, Excel used rules very similar to the ones you learned in this chapter for creating the class intervals or Bins. As a result, the numbers are not particularly friendly. The number that shows in each bin is the *upper* boundary for that class interval. A bin represents all of the data values that are greater than the previous bin number, up to and including the current bin number. That is, the number 2.928571 represents all of the data greater than 0 and less than or equal to 2.928571 ($0 < x \leq 2.928571$). The first class is all data values that are less than or equal to zero, and the last is all data values greater than 38.07143.

The second problem is with the format of the graph. No labels can be seen on the axis, the graph is squashed, and the titles and axis labels are not very useful. This can be fixed by clicking on the parts of the chart that you do not like and editing them. A set of steps for editing the chart in Figure 3.26 is given below:

- Click on the chart to select it. Handles appear on the corners and sides.
- Click on the bottom handle and drag it down to increase the height of the chart, as shown in Figure 3.27. You see that the axis labels become visible.

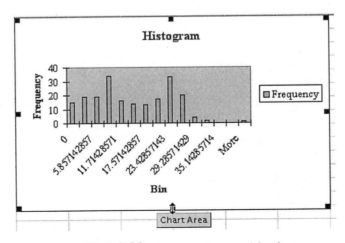

FIGURE 3.27   Changing the size of the chart

To change any portion of the chart, you simply double click on it and the appropriate dialog box opens. There are many changes that can be made to the chart. We will focus on a few that will produce an acceptable finished product.

The major problems with the chart are the way the x axis is labeled, the titles, and the gaps between the bars. We will fix some of these easily by making some changes to the bins in the frequency table. The frequency table and the chart are *linked* so that making changes to the table will update the chart.

- To get a little more room for the chart, click on the legend box and press the Delete key to remove it.

- Since the original data are integers, it really does not make sense to have the classes use six decimal places. Highlight the range in the worksheet that contains the frequency table and select **Cells** from the **Format** menu. The Format Cells dialog box opens. The box has six tabs: Number, Alignment, Font, Border, Patterns, and Protection. Click on the tab for Number and select Number from the list under **Category.** In the textbox for **Decimal places:** type 0, because you want integers. The dialog box is shown in Figure 3.28. Click **OK** to continue. The bins will change to integer values and you will see the changes on the chart.

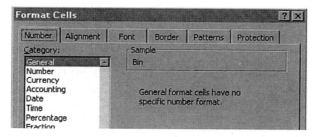

FIGURE 3.28   Format Cells dialog box

- Before we proceed, we really need to change the last bin label. Excel automatically labels the bin More, but we know it should be a number. From the updated frequency table, you can see that the class width is 3, so the last number in the bin column should be $38 + 3 = 41$. Click on the cell that contains the word "More" and change it to 41. The worksheet should look like the one in Figure 3.29.

    Notice that this changed the orientation of the text on the axis. That happened because the word "More" was too long to fit horizontally. If this change

When you have an extreme outlier in the data, Excel will include it in that last interval regardless of the width of the other intervals.

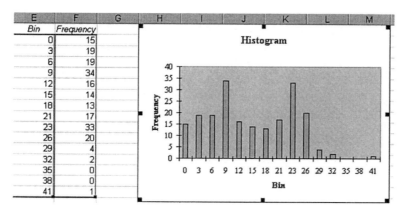

FIGURE 3.29   Edited bins and chart

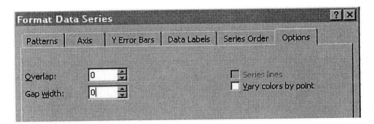

FIGURE 3.30   Changing the gap width

had not occurred, you could have changed the orientation of the text as follows:

Note: Clicking and drag-
ging the red diamond
will change the orien-
tation of the text to any
angle you want.

- Move the cursor to the chart and click on the *x* axis labels. The ToolTips box
will say "Category Axis" when your cursor is in the right location. The Format
Axis dialog box will open. The box has five tabs: Patterns, Scale, Font, Number,
and Alignment. To change the orientation of the text on the axis, click on the
Alignment tab. Click on the red diamond to orient the text horizontally.

We still need to change the title and axis labels and to remove the gaps between
the bars.

- To change any title or label, simply click on the title and type the new text.
- To remove the gaps, double click on any of the histogram bars and the **Format
Data Series** dialog box will open. This box allows you to change the way the bars
in the histogram look. For example, from the Patterns box you can change the
color of the bars or from the Data Labels box you can have the frequency for
each bar print above the box. To change the spacing between the bars, click on
the **Option** tab, and in the text box for **Gap width,** type 0 as shown in Figure
3.30.

Click **OK,** and the chart should look like the one in Figure 3.31.
Even though Excel defines the bin numbers as the endpoints of the class intervals, it prints them in the center of the histogram bars. There is no way to change
this. It is one of the problems that arise because Excel is not a statistical package.
There are many other changes that you can make to a chart in Excel. Experiment with them and see what improvements you can make to the histogram.

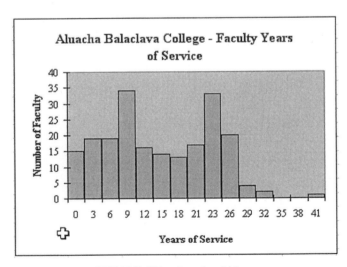

FIGURE 3.31   Completed histogram

CHAPTER 3 SUMMARY

Raw data are nothing more than a list of words or numbers. The purpose of descriptive statistics is to turn *data* into *information*. One way to do this is by summarizing the data with a frequency distribution and then using a graphical display that is appropriate for the type of data. Qualitative data can be displayed using *bar charts* or *pie charts*, while quantitative data are usually displayed using a *histogram* or *dotplot*.

Graphical displays of data can be used to describe data in terms of *center, shape,* and *variability*. They can also be used to compare different data sets.

Although it is possible to create graphs by hand, using a computer software package such as Minitab or Excel greatly enhances the quality of the finished product. It is important, however, to know how each graphical technique works so that you can use the software intelligently and make useful (and not just flashy) graphs.

Key Terms

Term	Definition	Page reference
Bar chart	A **bar chart** is a graph that represents the frequency or relative frequency from a frequency table in the form of a rectangle or bar.	76
Center	The **center** of a distribution tells where on the number line the data are centered or concentrated.	89
Cumulative relative frequency	The **cumulative relative frequency** of a class is the sum of the relative frequencies of all classes at or below that class represented as a portion of the total number of observations. It can be expressed as a fraction, decimal, or percentage.	64
Dotplot	A **dotplot** is a graph used for small data sets, in which each observation is plotted as a point on a single, horizontal axis.	84
Frequency table	A **frequency table** or frequency distribution is a table containing each category, value, or class of values that a variable might have and the number of times that each one occurs in the data.	59
Histogram	A **histogram** is very similar to a bar chart, but since numbers are naturally ordered, the *x* axis of the graph must be scaled to reflect this.	81

(continued)

Term	Definition	Page reference
Pareto diagram	A **Pareto diagram** is a bar chart in which the categories are plotted in order of decreasing relative frequency. The cumulative relative frequency of the categories is plotted on the same graph.	78
Pie chart	A **pie chart** is a graph in which a circle is used to represent the whole, and each "slice" is used to represent one of the categories. The size of the slice is proportional to the relative frequency of the category.	80
Relative frequency	The **relative frequency** of a classification is the number of times an observation falls into that classification represented as a portion of the total number of observations. It can be expressed as a *fraction, decimal,* or *percentage.*	61
Shape	The **shape** of a distribution describes how the data are spread out around the center with respect to symmetry or skewness.	89
Skewed	When the data are not evenly spread out on either side of the center then we refer to the distribution as being **skewed.**	90
Symmetric	When data are evenly spread out on both sides of the center, we describe the distribution of the data as **symmetric.**	90
Variability	The **variability** of a distribution describes how the data are spread out around the center with respect to the magnitude of the variation.	89

CHAPTER 3 EXERCISES

Learning It!

3.26 In a study on housing costs in major international cities, data were collected on the construction cost ($ per square meter) for housing in 23 capital cities. Here are the data:

Jakarta	65	Kingston	157	London	560
Dar Es Salaam	67	Bogota	171	Toronto	608
Karachi	87	Johannesburg	192	Seoul	617
Beijing	90	Rio de Janeiro	214	Hong Kong	641
New Delhi	94	Melbourne	383	Singapore	749
Istanbul	110	Algiers	500	Paris	990
Manila	148	Washington, D.C.	500	Tokyo	2,604
Bangkok	156	Madrid	510		

(a) What kind of graphical display is appropriate for these data? Why?

(b) Create a dotplot of the data.

(c) Use the data to describe housing construction costs for these cities.

(d) Do any of the data values appear to be unusual? If so, which one(s)?

3.27 An article in *Dartmouth Life* (15 September 1994) reported on the number of students from different classes who enrolled in various majors as freshmen. The data for the class of 1986 are:

Major	1986
Anthropology	10
Art History	28
Asia	11
Biology	31
Chemistry	28

(continued)

Major	1986
Classics	2
Comparative Literature	5
Drama	13
Earth Science	8
Economics	90
English	156
French	23
Geography	23
Government	147
History	102
Mathematics	35
Music	10
Philosophy	21
Physics	13
Psychology	70
Religion	22
Sociology	17

(a) Create a bar chart for these data.

(b) What percentage of the students majored in the sciences?

(c) What order did you pick for the categories on the x axis? Why did you choose this order? Can you think of any other orders that might also be appropriate? If so, what are they?

3.28 The *Chronicle of Higher Education* (25 May 1994) reported on the distribution of college and university faculty by rank. The data are summarized below:

Academic rank	Relative frequency (%)
Professor	36.3
Associate Professor	27.8
Assistant Professor	26.5
Instructor	6.1
Lecturer	2.3
No Rank	1.0

(a) Create a bar chart for these data.

(b) What do the data tell you about the distribution of faculty over the various ranks?

3.29 A report by the Bureau of Labor Statistics says that in 1992, the last year for which such data are available, 1004 homicides occurred while people were on the job. These homicides were classified by the type of job that the victim had:

Job	People killed on job
Managerial and professional	177
Sales	335
Service industry	225
Drivers/factory workers	202
Other	65

SOURCE: *Connecticut Post,* 11 June 1994.

(a) Create a bar chart for these data.

(b) Describe any interesting features of the data.

3.30 The publishing industry has long felt that the field of food magazines is in a recession. They looked at paid circulations for all of the magazines in the field and obtained the following data:

Magazine	Paid circulations
Bon Appetit	1,294,945
Cooking Light	1,119,811
Gourmet	906,299
Food & Wine	734,831

(*continued*)

Eating Well	648,697
*Cook's Illustrated**	160,000
*Fine Cooking**	100,000
*Saveur**	100,000

* Publishers' current estimates.

(a) Create a bar chart for the data.

(b) Does one of the magazines account for a majority of the paid circulations in food magazines? If so, which one?

(c) Do you think that *Bon Appetit, Food & Wine,* and *Gourmet* deserve their reputation as the Big Three? Why or why not?

3.31 As part of a survey conducted by the Bureau of Transportation in 1992, drivers were asked questions about the age of the vehicle that they normally drive. The survey recorded the year that the vehicle was manufactured. Data for 50 drivers surveyed are:

1955	1978	1983	1986	1988
1966	1979	1983	1986	1988
1970	1979	1984	1986	1988
1973	1980	1984	1987	1988
1974	1980	1984	1987	1988
1975	1981	1984	1987	1988
1976	1982	1984	1987	1988
1977	1982	1985	1987	1989
1977	1982	1985	1987	1989
1978	1982	1986	1987	1990

(a) Create a histogram for these data.

(b) Use the histogram to describe the distribution of the age of vehicles driven by the people in the sample.

(c) These data are really integer data. Why would a histogram using each year as a separate class not work for these data?

3.32 In addition to information about vehicles, the Personal Transportation Survey conducted by the Bureau of Transportation also asked questions about households, in particular, the number of people who lived in the household. The data for 45 random respondents are:

1	2	2	3	4
1	2	2	3	4
1	2	2	3	4
1	2	2	3	4
1	2	2	3	4
1	2	3	4	4
2	2	3	4	4
2	2	3	4	5
2	2	3	4	5

(a) Create a relative frequency histogram for the data.

(b) What percentage of the households surveyed had 3 or more people in them?

(c) What percentage of the households had fewer than 4 people?

3.33 A manufacturer of breakfast cereals packs the cereal in 15-ounce boxes. Periodic quality checks must be made to ensure that the boxes are being filled adequately. A random sample of 35 boxes of the cereal are selected and weighed. The data are

14.91	15.23	15.34	15.40	15.48	15.59	15.62
14.96	15.24	15.34	15.42	15.49	15.60	15.62
15.11	15.28	15.36	15.45	15.50	15.60	15.63
15.21	15.30	15.38	15.46	15.52	15.61	15.64
15.22	15.30	15.39	15.48	15.58	15.61	15.67

(a) Create a relative frequency histogram for the cereal box weights.

(b) Use the histogram to describe the distribution of cereal weights.

3.34 A large company buys computer storage media (diskettes) from one supplier. The company wants to know something about how long the diskettes last before they show signs of wear, so it collects data (in hours) for 20 diskettes from different boxes:

486	494	502	508
490	496	504	510
491	498	505	514
491	498	506	515
494	498	507	517

(a) Create a dotplot for the data.

(b) Use the dotplot to describe the distribution of lifetimes for the 20 diskettes.

3.35 A survey of 40 households asked the head of the household how much was spent on clothing ($) during the previous three months. The data are

0.00	0.00	0.00	35.00	210.00
0.00	0.00	0.00	36.75	221.00
0.00	0.00	0.00	60.89	224.00
0.00	0.00	0.00	64.66	303.33
0.00	0.00	10.62	96.00	310.55
0.00	0.00	15.00	114.61	365.95
0.00	0.00	17.21	149.27	365.95
0.00	0.00	21.33	165.37	702.60

(a) Create a frequency table for the data.

(b) Use the frequency table to make a histogram of the data.

(c) Describe the distribution of the amount of money that a household spends on clothing in a three-month period.

Thinking About It!

3.36 The problem of piracy has long been a concern for the computer software industry. Each year, the amount of revenue lost from piracy is staggering. Since the Internet has become more developed, the problems have become even more severe. In 1994 piracy losses were $15.2 billion. A study showed that the losses were distributed over the world as follows:

Region	% of total
Europe	39
Asia	29
Africa/Middle East	2
Latin America	9
North America	21

Source: *The Economist*, 27 July 1996.

(a) Create a bar chart for these data.

(b) Create a pie chart for these data.

(c) Which chart do you think does a better job of displaying the data and what they mean? Why?

3.37 Consumers are increasingly complaining that it is not worth the effort to clip coupons any more. One of the reasons they cite for this is that the life of the coupons is decreasing and it takes more time to sort and discard expired coupons. In an effort to substantiate these claims a consumer group looked at samples of 500 coupons from both 1995 and 1996. The data are summarized as follows:

	Number of Coupons	
Coupon life	**1995**	**1996**
Less than 1 month	28	123
1–4 months	88	140
5–8 months	204	122
9–12 months	141	88
More than 1 year	39	27

(a) Create bar charts for 1995 and 1996.

(b) Compare and describe the data for the two years. Do you think that there has been a change in coupon life? Why or why not?

3.38 The problem of workplace violence is growing. A survey of 600 full-time American workers who were the victims of violence while working reports the following:

Workplace violence survey

Type of Violence	% Who Reported
Harassment	19
Threat of physical harm	7
Physical attack	3

Major effect on worker	Type of violence (%)		
	Harassment	Threat	Physical
Psychologically	49	53	49
Disrupted work life	34	25	25
Physically injured or sick	13	9	17
No negative effect	4	13	9

Create graphical displays of these data. Use any techniques you feel are appropriate to convey the information in the data.

Requires Exercise 3.27 **3.39** The article that looked at majors of students at Dartmouth College also reported data for the classes of 1991 and 1996:

Major	1991	1996
Anthropology	19	17
Art History	20	15
Asia	15	22
Biology	44	108
Chemistry	15	55
Classics	1	8
Comparative Literature	6	6
Drama	8	5
Earth Science	15	7
Economics	66	93
English	143	92
French	15	19
Geography	28	28
Government	172	142
History	125	115
Mathematics	23	14
Music	3	11
Philosophy	28	25
Physics	10	16
Psychology	74	87
Religion	31	8
Sociology	26	15

(a) Create bar charts for the 1991 and 1996 data. Design the charts so that you will be able to compare the two classes and also compare them to the data for the class of 1986.

(b) Did the fact that you needed to make comparisons change the way you might have arranged the categories? Why or why not?

(c) Did the fact that you needed to make comparisons affect the way you scaled the y axis?

(d) Do you think that there is any basis for the claim that there has been an increase in the number of students who are interested in the sciences? Why or why not?

(e) Make comparative bar charts for only those majors that are sciences. Do you think that these charts make trends in the sciences easier to see? Why or why not?

(f) Compare the distribution of majors in general for each of the classes. Do you see any other changes or trends?

3.40 A study in 1992 by the National Center for Health Statistics looked at the death rates for a group of diseases. The data were grouped by sex and race:

Disease	Deaths (per 100,000 people)			
	White Males	**Black Males**	**White Females**	**Black Females**
Heart disease	190.3	264.1	98.1	162.4
Prostate cancer	15.7	35.8	NA	NA
Breast cancer	NA	NA	21.7	27.0
Diabetes	11.6	24.2	9.6	25.8
HIV infection	18.1	61.8	1.6	14.3
Chronic liver disease	11.1	17.2	4.6	6.9

(a) Create a bar chart for the total deaths for each disease. What information do you get from this chart?

(b) Create separate bar charts for each of the four groups.

(c) Compare the deaths for the groups to each other and to the chart for the total deaths. What interesting features do you see in the data?

3.41 The company looking at the life of its diskettes, decides to buy some diskettes from a different vendor and do a comparative test. The company selects 45 diskettes from each supplier and distributes them to a random sample of employees. The employees are asked to use the diskettes as they normally would, but to check them regularly for signs of wear using a disk-checking utility. They are to record how long (in hours) their disk lasts before it exhibits signs of wear. The data, for each supplier, are:

Supplier A						Supplier B				
474	492	498	504	511		487	492	495	497	499
486	492	500	505	511		488	492	495	497	499
489	494	501	506	512		489	492	495	497	499
490	494	501	507	513		489	492	495	497	501
490	494	501	508	514		489	493	495	498	502
490	495	502	508	515		491	493	496	498	503
491	496	502	509	517		491	494	496	498	503
491	498	504	509	519		491	494	496	498	505
491	498	504	510	528		492	494	496	499	506

(a) Create histograms for the lifetimes of diskettes from each supplier. Be sure to choose your scales so that you can make meaningful comparisons.

(b) Describe the distribution of diskette lifetimes for each supplier.

(c) Compare the lifetimes for the diskettes from supplier B, to those from the company's current supplier, supplier A. Which diskettes would you recommend the company use? Why?

3.42 *The New York Times* (28 October 1994) published a state by state report on the number of people incarcerated per 100,000 people in the population. The data are shown in the table below:

District of Columbia	1578	Mississippi	385	Tennessee	278	Wisconsin	172
		California	382	Colorado	272	Hawaii	170
Texas	545	Virginia	374	Alaska	256	Oregon	169
Louisiana	514	United States (Federal)	373	Indiana	256	New Hampshire	167
South Carolina	504			Idaho	253	Massachusetts	165
Oklahoma	501	Ohio	369	Wyoming	247	Utah	154
Nevada	456	New York	361	Kansas	239	Nebraska	148
Arizona	448	Arkansas	355	South Dakota	227	Vermont	138
Alabama	439	Connecticut	331	Pennsylvania	224	Maine	113
Michigan	423	Missouri	321	New Mexico	216	West Virginia	106
Georgia	417	North Carolina	314	Washington	198	Minnesota	100
Florida	404	New Jersey	307	Montana	192	North Dakota	75
Maryland	392	Illinois	302	Rhode Island	185		
Delaware	391	Kentucky	281	Iowa	180		

(a) Create a histogram for the data.

(b) Describe the number of people incarcerated (per 100,000) in the United States.

(c) Do you think that the observation from the District of Columbia has an effect on the way your histogram turned out? If so, what is it?

(d) Recreate the histogram without the data from the District of Columbia.

(e) Use the new histogram to describe the number of people incarcerated (per 100,000) in the United States.

(f) Has your description changed from that of part (b). If so, what is different?

Requires Exercise 3.18   **3.43**   The cable company that is looking at its ratings after changing channel selection wonders if there has been a change in customer feelings about the service. The company looks at the responses from a survey done six months prior to the change and finds the following data:

$$
\begin{array}{ccccc}
2 & 3 & 3 & 4 & 5 \\
2 & 3 & 3 & 4 & 6 \\
2 & 3 & 3 & 4 & 6 \\
2 & 3 & 3 & 4 & 6 \\
2 & 3 & 4 & 5 & 6 \\
2 & 3 & 4 & 5 & 7 \\
3 & 3 & 4 & 5 & 8 \\
\end{array}
$$

(a) Create a graphical display for the ratings data before the service changes were made.

(b) Do you think that the company can conclude that the ratings have improved? Why or why not?

3.44   It has been a widely held belief that the switch to participative management would increase employee "buy-in" to the company. One of the benefits that should be realized is a reduction in the number of sick days that employees use. A company that has made the switch in some departments wonders if this has been true. The company decides to sample 25 employees from each of two manufacturing departments. The first has been using participative management for almost two years and the second is still using a traditional management style. The data on the number of sick days used by each employee in the past 12 months are

Participative Management					Traditional				
1	3	5	5	6	0	5	6	7	9
1	4	5	6	7	3	5	7	7	9
2	4	5	6	8	4	6	7	7	10
2	4	5	6	8	4	6	7	8	11
3	4	5	6	8	5	6	7	8	11

(a) Create a graphical display for each of the samples.

(b) Use the graphical display to describe the distribution of sick days for each department.

(c) Based on the data, do you think there is a difference in the number of sick days taken by each group? Why or why not?

(d) Do you think, on the basis of these data, that the company can conclude that participative management reduces the number of sick days taken by employees? Why or why not?

Doing It!

Datafile:
COMPLAIN.XXX   **3.45**   A large manufacturer of paper goods keeps track of consumer complaints for its facial tissue product line. The company receives these complaints through a toll-free number that appears on the product. The data taken consist of a transcription of the actual complaint language and a classification of the complaint into a specific category. When a complaint is received, the company generally asks for additional information about the specific package. The information is then used to relate the complaints to manufactur-

ing data so that in the future similar problems can be prevented. The categories and subcategories used for classifying complaints are given below:

Category	Subcategory
Dispensing	Sheets tear on removal
	Reach in/fallback
Foreign material	Lint/dust
	Other
Odor	
Miscounts	
Packaging	Defective
	Misleading
	Damaged
	Other

For example, if a consumer calls the company and says, "The box of tissues I bought was not full," the customer service representative classifies this as a miscount. Sometimes complaints can be tricky to classify. If a customer calls, for example, after a product change and says, "I eat tissues and I prefer the taste of the old product," into what category should this complaint be placed?

The company is looking at data for 3 different years. It compiles data for five different variables:

Variable name	Description
Month	Month in which complaint occurred
Year	Year of the complaint
Category	Category of complaint
Subcategory	Subcategory of complaint
Number	Number of complaints of the particular category and subcategory in the time period

A portion of the data file is shown below:

Month	Year	Category	Subcategory	Number
JANUARY	1989	DISPENSING	FALLBACK	80
JANUARY	1989	DISPENSING	SHEETS_TEAR	118
JANUARY	1989	FOREIGN MATL.	LINT/DUST	53
JANUARY	1989	FOREIGN MATL.	OTHER	1
JANUARY	1989	MISCOUNTS	MISCOUNTS	37
JANUARY	1989	ODOR	ODOR	5
JANUARY	1989	PACKAGING	ADVERTISING	1

(a) Create a bar chart for customer complaints by subcategory for January 1989. What subcategory has the largest number of complaints? Does any subcategory constitute a majority?

(b) Create similar charts for January 1990 and January 1991. Compare the three months. Do you notice any similarities or differences?

(c) Create bar charts for the complaint categories for the same three months. Describe and compare the bar charts.

(d) Look at the data for all of 1989. Create a bar chart that displays the total number of complaints by month. Do you notice any trends or interesting features? Does any month or season have considerably more or less complaints than others? Can you offer any explanation for this?

(e) Create similar bar charts of total complaints for 1990 and 1991. Compare these years to 1989 and to each other. Are they similar? Would you expect them to be? Why or why not?

(f) Take the bar chart that you made for January 1989 and modify it so that it is a Pareto chart. Do the same thing for January 1990 and January 1991. As manager of the company, where would you concentrate your resources to eliminate complaints? Why?

(g) Look at the data for July 1989 and make bar charts for both complaint category and complaint subcategory. Compare July 1989 to January 1989. How are they similar? How are they different?

(h) Look at the same data as in part (h), but this time create pie charts. Which chart does a better job at conveying information about customer complaints? Why?

(i) Look at the data by category for each year and make pie charts comparable to the bar charts from parts (d) and (e). Have there been any changes over the three-year period in the type and number of complaints? If so, what are they?

(j) The management of this company asks for your assessment of the complaint data. In particular, the managers would like to know where their biggest problems are and what the trends seem to be over the past three years. Create any additional graphs that will help you answer their questions and prepare a report for them offering your analysis and your suggestions.

Datafile:
FACULTY.XXX

3.46 The provost at Aluacha Balaclava College would like to look at the data she has collected in a different way. She has collected data on salary, rank, school, gender, and tenure. A sample of the data is shown below:

Salary	Years of Service	Rank	School	Gender M/F	Tenure Y/N
48,000	22	ASST	BUSINESS	F	Y
57,956	16	PROF	BUSINESS	M	Y
57,170	7	ASSO	BUSINESS	M	Y
53,500	11	ASSO	BUSINESS	M	N
44,166	20	ASSO	BUSINESS	M	Y
85,500	23	PROF	BUSINESS	M	N
49,302	21	ASST	BUSINESS	M	Y
49,981	10	ASSO	BUSINESS	M	Y

(a) Create histograms of salary for each rank separately.

(b) Compare the distribution of salaries for the four ranks.

(c) Create histograms of salary for each school separately.

(d) Compare the distributions of salary for the different schools.

(e) Create separate histograms of salary by gender and for tenured/untenured faculty and compare them.

(f) Write a report to the provost that summarizes your observations.

WORKSHOP 2

MEASUREMENT: SCALING, RELIABILITY, VALIDITY

4.1 CHAPTER OBJECTIVES

After completing this chapter you should be able to:

- Know how and when to use the different forms of rating scales and ranking scales.
- Explain stability and consistency, and how they are established.
- Be conversant with the different forms of validity.
- Discuss what "goodness" of measures means, and why it is necessary to establish it in research.

4.2 RATING SCALES FREQUENTLY USED

Now that we know the four different types of scales that can be used to measure the operationally defined dimensions and elements of a variable, it is necessary to examine the methods of scaling (i.e., assigning numbers or symbols) to elicit the attitudinal responses of subjects toward objects, events, or persons. There are two main categories of attitudinal scales (not to be confused with the four different types of scales) — the **rating scale** and the **ranking scale.**

> *Rating scales* have several response categories and are used to elicit responses with regard to the object, event, or person studied. *Ranking scales,* on the other hand, make comparisons between or among objects, events, or persons and elicit the preferred choices and ranking among them.

Both the scales are discussed below.

4.2.1 Rating Scales

The following rating scales are often used in organizational research:

Dichotomous scale
Category scale
Likert scale
Semantic differential scale
Numerical scales

Itemized rating scale

Fixed or constant sum rating scale

Stapel scale

Graphic rating scale

Consensus scale

Other scales, such as the Thurstone Equal Appearing Interval scale and the Multidimensional Scale, are less frequently used. We will briefly describe each of the preceding attitudinal scales.

Dichotomous Scale

The dichotomous scale is used to elicit a *yes* or *no* answer, as in the example below. Note that a nominal scale is used to elicit the response.

EXAMPLE 4.1 **Rating Scales**

Dichotomous Scale

Do you own a car?   Yes   No   ■

Category Scale

The category scale uses multiple items to elicit a single response as shown in the following example. This also uses the nominal scale.

EXAMPLE 4.2 **Rating Scales**

Category Scale

Where in northern California do you reside?   ___North Bay
                                                  ___South Bay
                                                  ___East Bay
                                                  ___Peninsula
                                                  ___Other   ■

Likert Scale

The Likert scale is designed to examine how strongly subjects agree or disagree with statements on a 5-point scale with the following anchors:

Strongly Disagree	Disagree	Neither Agree Nor Disagree	Agree	Strongly Agree
1	2	3	4	5

The responses over a number of items tapping a particular concept or variable (as in the following example) are then summated for every respondent. The interval scale is used here, the differences in the responses between any two points on the scale remaining the same.

EXAMPLE 4.3   Rating Scales

Likert Scale

Using the preceding Likert scale, state the extent to which you agree with each of the following statements:

My work is very interesting	1	2	3	4	5
I am not engrossed in my work all day	1	2	3	4	5
Life without my work will be dull	1	2	3	4	5

■

Semantic Differential Scale

Several bipolar attributes are identified at the extremes of the scale, and respondents are asked to indicate their attitudes, on what may be called a semantic space, toward a particular individual, object, or event on each of the attributes. The bipolar adjectives used, for instance, would denote such terms as: Good–Bad; Strong–Weak; Hot–Cold. The semantic differential scale is used to assess respondents' attitudes toward a particular brand, advertisement, object, or individual. The responses can be plotted to obtain a good idea of the perceptions of the respondents. This is treated as an interval scale. An example of the semantic differential scale follows.

EXAMPLE 4.4   Rating Scales

Semantic Differential Scale

Responsive	__	__	__	__	__	__	__	Unresponsive
Beautiful	__	__	__	__	__	__	__	Ugly
Courageous	__	__	__	__	__	__	__	Timid

■

Numerical Scale

The numerical scale is similar to the semantic differential scale, with the difference that numbers on a 5-point or 7-point scale are provided, with bipolar adjectives at both ends, as illustrated below. This is also an interval scale.

EXAMPLE 4.5   Rating Scales

Numerical Scale

How pleased are you with your new real estate agent?

Extremely Pleased	7	6	5	4	3	2	1	Extremely Displeased

■

Itemized Rating Scale

A 5-point or 7-point scale with anchors, as needed, is provided for each item and the respondent states the appropriate number on the side of each item, or circles the relevant number against each item, as indicated in the examples that follow. The responses to the items are then summated. This uses an interval scale.

EXAMPLE 4.6 (i)   Rating Scales

Itemized Rating Scale

Respond to each item using the scale below, and indicate your response number on the line by each item.

1	**2**	**3**	**4**	**5**
Very Unlikely	**Unlikely**	**Neither Unlikely Nor Likely**	**Likely**	**Very Likely**

1. I will be changing my job within the next 12 months.   ___
2. I will take on new assignments in the near future.   ___
3. It is possible that I will be out of this organization within the next 12 months.   ___

Note that the above is a *balanced rating scale* with a *neutral* point.   ∎

EXAMPLE 4.6 (ii)   Rating Scales

Itemized Rating Scale

Circle the number that is closest to how you feel for each item below.

Not at All Interested **1**	**Somewhat Interested** **2**	**Moderately Interested** **3**	**Very Much Interested** **4**				
How would you rate your interest in changing current organizational policies?				1	2	3	4

This is an *unbalanced rating scale* that does *not* have a neutral point.   ∎

The itemized rating scale provides the flexibility to use as many points in the scale as considered necessary (4, 5, 7, 9, or whatever), and it is also possible to use different anchors (e.g., Very Unimportant to Very Important; Extremely Low to Extremely High, etc.). When a neutral point is provided, it is a balanced rating scale, and when it is not, it is an unbalanced rating scale.

Research indicates that a 5-point scale is just as good as any, and that an increase from 5 to 7 or 9 points on a rating scale does not improve the reliability of the ratings.

The itemized rating scale is frequently used in business research, since it lends itself to adaptation as to the number of points used, as well as the nomenclature of the anchors, as is considered necessary to fit the needs of the researcher for tapping the variable.

Fixed or Constant Sum Scale

Here the respondents are asked to distribute a given number of points across various items as shown in Example 4.7. This is more in the nature of an ordinal scale.

EXAMPLE 4.7   Rating Scales

Constant Sum Scale

In choosing a toilet soap, indicate the importance you attach to each of the following five aspects by distributing a total of 100 points among them.

Fragrance	___
Color	___
Shape	___
Size	___
Texture of lather	___
Total points	**100**

∎

Stapel Scale

This scale simultaneously measures both the direction and intensity of the attitude toward the items under study. The characteristic of interest to the study is placed at the center and a numerical scale ranging, say, from $+3$ to -3, placed on each side of the item as illustrated below. This gives an idea of how close or distant the individual response to the stimulus is, as in the example. Since this does not have an absolute zero point, this is an interval scale.

EXAMPLE 4.8   Rating Scales

Stapel Scale

State how you would rate your supervisor's abilities with respect to each of the characteristics mentioned below, by circling the appropriate number.

$+3$	$+3$	$+3$
$+2$	$+2$	$+2$
$+1$	$+1$	$+1$
Adopting Modern Technology	Product Innovation	Interpersonal Skills
-1	-1	-1
-2	-2	-2
-3	-3	-3

Graphic Rating Scale

Here, a graphical representation helps the respondents to indicate their answers to a particular question by placing a mark at the appropriate point on the line, as in the following example. This is an ordinal scale, though the following example might seem like an interval scale.

EXAMPLE 4.9   Rating Scales

Graphic Rating Scale

On a scale of 1 to 10, how would you rate your supervisor?

    ┌10 Excellent
    │
    ┤ 5 All right
    │
    └ 1 Very bad

This scale is easy to respond to. The brief descriptions on the scale points are meant to serve as a guide in locating the rating rather than to represent discrete categories. The **Faces scale,** which depicts faces ranging from smiling to sad (illustrated in Chapter 5), is also a graphic rating scale used to obtain responses regarding people's feelings with respect to some aspect (e.g., how they feel about their jobs).

Consensus Scale

Scales are also developed by consensus, where a panel of judges select certain items they feel measure the concept desired to be measured. The items are chosen particularly based on their relevance to the concept. Such a consensus scale is developed after the selected items are examined and tested for their validity and reliability. One such consensus scale is the **Thurstone Equal Appearing Interval scale,** where a concept is measured by a complex process followed by a panel of judges. Using a pile of cards on which are statements describing the concept, a panel of judges offer their inputs to indicate how close the statements are, or are not, to the concept under study. The scale is then developed based on the consensus reached. However, this scale is rarely used for measuring organizational concepts because of the time involved in developing it.

Other Scales

There are also some advanced scaling methods such as **multidimensional scaling,** where one can visually scale objects, or people, or both, and a conjoint analysis is performed. This provides a visual image of the relationships in space among the dimensions of a construct.

It is to be noted that usually the Likert scale or some form of numerical scale is most frequently used to measure attitudes and behaviours in organizational research.

4.2.2 Ranking Scales

As already mentioned, **ranking scales** are used to tap preferences between two or more objects or items (ordinal in nature). However, such ranking may not give definitive clues to some of the answers sought. For instance, let us say there are four product lines and the manager desires to obtain information that would be helpful in deciding which product line should get the most attention. Let us also say that 35% of the respondents choose the first product, 25% the second, and 20% choose each of products three and four as of importance to them. The manager cannot then decide that the first product is the most preferred since 65% of the respondents did not choose that product! Alternative methods used are the *paired comparisons, forced choice,* and the *comparative scale,* which are discussed below.

Paired Comparison

The *paired comparison scale* is used when among a small number of objects, respondents are asked to choose between two objects at a time. This helps to assess preferences. If, for instance, in the previous example, during the paired comparisons, respondents consistently show a preference for product one over products two, three, and four, that would provide the manager a good indication as to which product line is worth pursuing with vigor. However, as the number of objects to be compared increases, the number of paired comparisons also increases. The number of paired choices for n objects will be $[(n)(n-1)/2]$. The greater the number of objects or stimuli, the greater the number of paired comparisons presented to the respondents, and the greater the respondent fatigue. Hence paired comparison is a good method when the number of stimuli presented is small.

Forced Choice

The *forced choice* offers respondents the opportunity to rank objects relative to one another, among the alternatives provided. This is easier for the respondents to do, particularly if the number of choices to be ranked is limited in number.

EXAMPLE 4.10   Ranking Scales

Forced Choice

Rank your preferences among the following magazines, which you would like to subscribe to, 1 being the most preferred choice and 5 being the least preferred.

Fortune	___
Playboy	___
Time	___
People	___
Prevention	___

■

Comparative Scale

The *comparative scale* provides a benchmark or a point of reference to assess attitudes toward the current object, event, or situation under study. An example of the use of the comparative scale follows.

EXAMPLE 4.11   Ranking Scales

Comparative Scale

In a volatile financial environment, compared to stocks, how wise or useful is it to invest in Treasury bonds? Please circle the appropriate response.

More useful		About the Same		Less Useful
1	2	3	4	5

■

In sum, nominal data lend themselves to dichotomous or category scale; ordinal data lend themselves to any one of the ranking scales—paired comparison, forced choice, or comparative scales; and interval or interval-like data lend themselves to the other rating scales, as seen from the various examples above. The semantic differential and the numerical scales are, strictly speaking, not interval scales, though they are often treated as such in data analysis.

Rating scales are used to measure most behavioral concepts. The ranking scales are used to make comparisons or rank the variables that have been tapped on a nominal scale.

4.3   GOODNESS OF MEASURES

Now that we have seen how to operationally define variables and apply different scaling techniques, it is important to make sure that the instrument we develop to measure a particular concept is indeed *accurately* measuring the variable, and that, in fact, we are *actually* measuring the concept that we set out to measure. The latter ensures that in operationally defining perceptual and attitudinal variables, we have not overlooked some important dimensions and elements nor included some irrelevant ones. The scales developed could often be imperfect, and of course, there are always errors in measurement of attitudinal variables. The use of better instruments will ensure more accurate results, which in turn will enhance the scientific quality of the research. Hence, in some way we need to assess the "goodness" of the measures developed. That is, we need to be reasonably sure that the instruments we use in our research do indeed measure the variables they are supposed to, and that they measure them accurately.

Let us now examine how we can ensure that the measures developed are reasonably good. First an item analysis of the responses to the questions tapping the variable is done, and then the reliability and validity of the measures are established as described below.

4.4 ITEM ANALYSIS

Item analysis is done to see whether the items in the instrument belong there. Each item is examined to see if it is able to discriminate between those subjects whose total scores are high, and those whose scores are low. In item analysis, the means between the high-score group and the low-score group are tested to detect significant differences through the *t*-values (see Chapter 13 for explanation of *t*-tests). The items with a high *t*-value (test which is able to identify the highly discriminating items in the instrument) are then included in the instrument. Thereafter, tests for the reliability of the instrument are done. The validity of the measure is then established.

Briefly, reliability tests *how consistently* a measuring instrument measures whatever concept it is measuring. Validity tests how well an instrument that is developed measures the *particular concept* it is supposed to measure. In other words, validity is concerned with whether we measure the right concept, and reliability is concerned with stability and consistency in measurement. Validity and reliability of the measure attest to the scientific rigor applied to the research study. These two criteria will now be discussed. The various forms of reliability and validity are depicted in Figure 4.1.

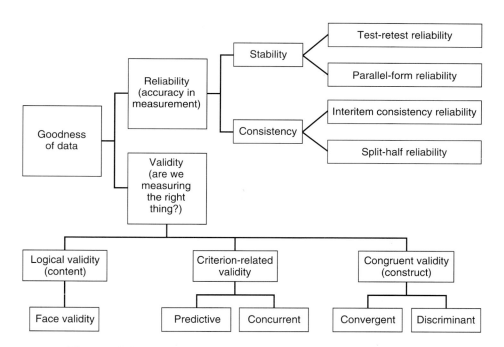

Figure 4.1 Testing goodness of measures: Forms of reliability and validity

4.5 RELIABILITY

The ***reliability*** of a measure indicates the extent to which the measure is without bias (error free) and hence offers consistent measurement across time and across the various items in the instrument.

In other words, the reliability of a measure indicates the stability and consistency with which the instrument measures the concept and helps to assess the "goodness" of a measure.

4.5.1 Stability of Measures

The ability of a measure to maintain stability over time, despite uncontrollable testing conditions or the state of the respondents themselves, is indicative of its stability and low vulnerability to changes in the situation. This attests to the "goodness" of the measure, inasmuch as it stably measures the concept, no matter when it is measured. Two tests of stability are test–retest reliability and parallel-form reliability.

Test–Retest Reliability

> The reliability coefficient obtained with a repetition of the same measure on a second occasion is called *test–retest reliability.*

That is, when a questionnaire containing some items that are supposed to measure a concept is administered to a set of respondents now, and again to the same respondents, say several weeks to six months later, then the correlation between the scores obtained at the two different times from the same set of respondents is called the test–retest coefficient. The higher it is, the better the test–retest reliability, and hence the stability of the measure across time.

Parallel-Form Reliability

> When responses on two comparable sets of measures tapping the same construct are highly correlated, we have *parallel-form reliability.*

Both forms have similar items and the same response format with only the wordings and the ordering of questions changed. What we try to establish here is the error variability resulting from wording and ordering of the questions. If two such comparable forms are highly correlated, we can be fairly certain that the measures are reasonably reliable, with minimal error variance caused by wording, ordering, or other factors.

4.5.2 Internal Consistency of Measures

The internal consistency of measures is indicative of the homogeneity of the items in the measure that tap the construct. In other words, the items should "hang together as a set," and be capable of independently measuring the same concept such that the respondents attach the same overall meaning to each of the items. This can be seen by examining whether the items and the subsets of items in the measuring instrument are highly correlated. Consistency can be examined through the interitem consistency reliability and split-half reliability tests.

Interitem Consistency Reliability

This is a test of the consistency of respondents' answers to all the items in a measure. To the degree that items are independent measures of the same concept, they will be correlated with one another. The most popular test of interitem consistency reliability is the Cronbach's coefficient alpha [Cronbach, L. J., "Response sets and test validating," *Educational and Psychological Measurement* 6 (1946): 475–494], which is used for multipoint-scaled items, and the Kuder-Richardson formulas [Kuder, G. F. & Richardson, M. W., "The theory of the estimation of test reliability," *Psychometrika* 2 (1937): 151–160], used for dichotomous items. The higher the coefficients, the better the measuring instrument.

Split-Half Reliability

Split-half reliability reflects the correlations between two halves of an instrument. Split-half reliability estimates would vary depending on how the items in the measure are split into two halves. Split-half reliabilities could be higher than Cronbach's alpha only in the circumstance of there being more than one underlying response dimension tapped by the measure and when certain other conditions are met as well. Hence, in almost all cases, Cronbach's alpha can be considered a perfectly adequate index of the interitem consistency reliability.

It should be noted that the consistency of the judgment of several raters on how they see a phenomenon or interpret some responses is termed **interrater reliability,** and should not be confused with the reliability of a measuring instrument. As we had noted earlier, interrater reliability is especially relevant when the data are obtained through observations, projective tests, or unstructured interviews, all of which are liable to be subjectively interpreted.

It is important to note that reliability is a necessary but not sufficient condition of the test of goodness of a measure. For example, one could very reliably measure a concept establishing high stability and consistency, but it may not be the concept that one set out to measure.

Validity ensures the ability of a scale to measure the intended concept.

We will now discuss the concept of validity.

4.6   VALIDITY

We are going to examine the validity of the measuring instrument. That is, when we ask a set of questions (i.e., develop a measuring instrument) in hopes that we are tapping the concept, how can we be reasonably sure that we are measuring the concept we set out to measure and not something else? This can be determined by applying certain validity tests.

Several types of validity tests are used to test the goodness of measures. Writers use different terms to denote these validity tests. For the sake of clarity, we can group validity tests under three broad headings: **content validity, criterion-related validity,** and **construct validity.**

4.6.1 Content Validity

Content validity ensures that the measure includes an adequate and representative set of items that tap the concept.

The more the scale items represent the domain or universe of the concept being measured, the greater the content validity. In other words, content validity is a function of how well the dimensions and elements of a concept have been delineated.

A panel of judges can attest to the content validity of the instrument. Kidder and Judd (Kidder, L. H. & Judd, C. H., *Research methods in social relations.* New York: Holt, Rinehart and Winston, 1986) cite the example where a test designed to measure degrees of speech impairment can be considered as having validity if it is so evaluated by a group of expert judges (i.e., professional speech therapists).

Face validity is considered by some as a basic and a very minimum index of content validity. Face validity indicates that the items that are supposed to measure a concept, do on the face of it look like they measure the concept. Some researchers do not see fit to treat face validity as a valid component of content validity.

4.6.2 Criterion-Related Validity

> *Criterion-related validity* is established when the measure differentiates individuals on a criterion it is expected to predict.

This can be done by establishing **concurrent validity** or **predictive validity,** as explained below.

Concurrent validity is established when the scale discriminates individuals who are known to be different; that is, they should score differently on the instrument as in the example that follows.

EXAMPLE 4.12

If a measure of work ethic is developed and administered to a group of welfare recipients, the scale should differentiate those who are enthusiastic about accepting a job and being off welfare from those who do not want to work even when offered a job. Obviously, those with high work ethic values would not want to be on welfare and would be eager to accept a job as soon as possible. Those who are low on work ethic values, on the other hand, might exploit the opportunity to survive on welfare for as long as possible without having to work. If both types of individuals have the same score on the work ethic scale, then the test would not be a measure of work ethic, but of something else.   ■

Predictive validity indicates the ability of the measuring instrument to differentiate among individuals as to a future criterion. For example, if an aptitude or ability test administered to employees at the time of recruitment is expected to differentiate individuals on their future job performance, then those who score low on the test should be poor performers and those who score high should be good performers.

Construct Validity

> *Construct validity* testifies to how well the results obtained from the use of the measure fit the theories around which the test is designed.

This is assessed through **convergent** and **discriminant validity,** which are explained below.

Convergent validity is established when the scores obtained by two different instruments measuring the same concept are highly correlated.

Discriminant validity is established when, based on theory, two variables are predicted to be uncorrelated, and the scores obtained by measuring them are indeed empirically found to be so.

Validity can thus be established in different ways. Published measures for various concepts usually report the kinds of validity that have been established for the instrument, so that the user or reader can judge the "goodness" of the measure. Table 4.1 summarizes the kinds of validity discussed here.

Ways to establish validity   Some of the ways in which the above forms of validity can be established are through (i) *correlational analysis* (as in the case of establishing concurrent and predictive validity or convergent and discriminant validity), (ii) *factor analysis,* a multivariate technique which would confirm the dimensions of the concept that have been operationally defined, as well as indicate which of the items are most appropriate for each dimension (establishing construct validity), and (iii) the *multi-trait, multi-method matrix* of correlations derived from measuring concepts by different forms and different methods, additionally establishing the robustness of the measure.

TABLE 4.1 Types of Validity

Validity	Description
Content validity	Does the measure adequately measure the concept?
Face validity	Do "experts" validate that the instrument measures what its name suggests it measures?
Criterion-related validity	Does the measure differentiate in a manner that helps to predict a criterion variable?
Concurrent validity	Does the measure differentiate in a manner that helps to predict a criterion variable currently?
Predictive validity	Does the measure differentiate individuals in a manner as to help predict a future criterion?
Construct validity	Does the instrument tap the concept as theorized?
Convergent validity	Do two instruments measuring the concept correlate highly?
Discriminant validity	Does the measure have a low correlation with a variable that is supposed to be unrelated to this variable?

In sum, the **goodness of measures** is established through the different kinds of validity and reliability depicted in Figure 4.1. The results of any research can only be as good as the measures that tap the concepts in the theoretical framework. We need to use well-validated and reliable measures to ensure that we engage in scientific research. Fortunately, measures have been developed for many important concepts in organizational research, and their psychometric properties (i.e., the reliability and validity) established by the developers. Thus, researchers can use the instruments already reputed to be "good," rather than laboriously develop their own measures. When using these measures, however, researchers should cite the source (i.e., the author and reference) so that the reader can seek more information, if necessary.

It is not unusual that two or more equally good measures are developed for the same concept. For example, there are several different instruments for measuring the concept of job satisfaction. One of the most frequently used scales to measure job satisfaction, however, is the Job Descriptive Index (JDI) developed by Smith, Kendall, and Hulin (Smith, P. C., Kendall, L., & Hulin, C., *The measurement of satisfaction in work and retirement*. Chicago: Rand McNally, 1969, 79–84). When more than one scale exists for any variable, it is preferable to use the measure that has better reliability and validity and is also more frequently used.

At times, we may also have to adapt an established measure to suit the setting. For example, a scale that is used to measure job performance, job characteristics, or job satisfaction in the manufacturing industry may have to be modified slightly to suit a utility company or a healthcare organization. The work environment in each case is different and the wordings in the instrument may have to be adapted. However, in doing this, we are tampering with an established scale, so it would be advisable to test for the adequacy of the validity and reliability again.

A sample of a few measures used to tap some frequently researched concepts in the management and marketing areas is provided in the appendix to this chapter.

CHAPTER 4 SUMMARY

In this chapter, we saw what kinds of attitude rating scales and ranking scales can be used in developing instruments, after a concept has been operationally defined. We also discussed how the goodness of measures is established by means of item analysis, and reliability and validity tests. We also noted that the Likert scale and other types of interval-type scales such as the numerical scale are extensively used in organizational research, since they lend themselves to more sophisticated data analysis. Finally, we discussed the goodness of measures in terms of reliability and validity and the various ways in which these can be established.

Knowledge of the different scales and scaling techniques helps managers to administer short surveys by designing questions that use ranking or rating scales, as appropriate. Awareness of the fact that measures are already available for many organizational concepts further facilitates mini exploratory surveys by managers.

In the next chapter, we will see the different methods by which data can be collected.

Key Terms

Term	Definition	Page reference
Category scale	A scale that uses multiple items to seek a single response.	121
Comparative scale	A scale that provides a benchmark or point of reference to assess attitudes, opinions, and the like.	126
Concurrent validity	Relates to criterion-related validity, which is established at the same time the test is administered.	130
Consensus scale	A scale developed through consensus or the unanimous agreement of a panel of judges, as to the items that measure a concept.	124
Constant sum rating scale	A scale where the respondents distribute a fixed number of points across several items.	123
Construct validity	Testifies to how well the results obtained from the use of the measure fit the theories around which the test was designed.	130
Content validity	Establishes the representative sampling of a whole set of items that measures a concept, and reflects how well the dimensions and elements thereof are delineated.	129
Convergent validity	That which is established when the scores obtained by two different instruments measuring the same concept, or by measuring the concept by two different methods, are highly correlated.	130

Term	Definition	Page reference
Criterion-related validity	That which is established when the measure differentiates individuals on a criterion that it is expected to predict.	130
Dichotomous scale	Scale used to elicit a Yes/No response, or an answer to two different aspects of a concept.	121
Discriminant validity	That which is established when two variables are theorized to be uncorrelated, and the scores obtained by measuring them are indeed empirically found to be so.	130
Face validity	An aspect of validity examining whether the item on the scale, on the face of it, reads as if it indeed measures what it is supposed to measure.	129
Faces scale	A particular representation of the graphic scale, depicting faces with expressions that range from smiling to sad.	124
Fixed rating scale	*See* Constant sum rating scale.	123
Forced choice	Elicits the ranking of objects relative to one another.	125
Goodness of measures	Attests to the reliability and validity of measures.	126
Graphic rating scale	A scale that graphically illustrates the responses that can be provided, rather than specifying any discrete response categories.	124
Interitem consistency reliability	A test of the consistency of responses to all the items in a measure to establish that they hang together as a set.	128
Interrater reliability	The consistency of the judgment of several raters on how they see a phenomenon or interpret the activities in a situation.	129
Itemized rating scale	A scale that offers several categories of responses, out of which the respondent picks the one most relevant for answering the question.	122
Likert scale	An interval scale that specifically uses the five anchors of Strongly Disagree, Disagree, Neither Disagree nor Agree, Agree, and Strongly Agree.	121
Numerical scale	A scale with bipolar attributes with five points or seven points indicated on the scale.	122
Paired comparison scale	Respondents choose between two objects at a time, with the process repeated with a small number of objects.	125
Parallel-form reliability	That form of reliability which is established when responses to two comparable sets of measures tapping the same construct are highly correlated.	128
Predictive validity	The ability of the measure to differentiate among individuals as to a criterion predicted for the future.	130
Ranking scale	Scale used to tap preferences between two or among more objects or items.	120
Rating scale	Scale with several response categories that evaluate an object on a scale.	120
Reliability	Attests to the consistency and stability of the measuring instrument.	127
Semantic differential scale	Usually a seven-point scale with bipolar attributes indicated at its extremes.	122
Split-half reliability	The correlation coefficient between one half of the items measuring a concept and the other half.	129
Stapel scale	A scale that measures both the direction and intensity of the attributes of a concept.	124
Test-retest reliability	A way of establishing the stability of the measuring instrument by correlating the scores obtained through its administration to the same set of respondents at two different points in time.	128

(continued)

Term	Definition	Page reference
Validity	Evidence that the instrument, technique, or process used to measure a concept does indeed measure the intended concept.	129

CHAPTER 4 EXERCISES

Thinking About It!

4.1 Briefly describe the difference between attitude rating scales and ranking scales and indicate when the two are used.

4.2 Why is it important to establish the "goodness" of measures and how is this done?

4.3 Construct a semantic differential scale to assess the properties of a particular brand of coffee or tea.

4.4 "Whenever possible, it is advisable to use instruments that have already been developed and repeatedly used in published studies, rather than develop our own instruments for our studies." Do you agree? Discuss the reasons for your answer.

4.5 "A valid instrument is always reliable, but a reliable instrument may not always be valid." Comment on this statement.

Doing It!

4.6 Develop and name the type of measuring instrument you would use to tap the following:

   (a) Which brands of beer are consumed by how many individuals?

   (b) Among the three types of exams—multiple choice, essay type, and a mix of both—which is the one preferred most by students?

   (c) To what extent do individuals agree with your definition of accounting principles?

   (d) How much people like an existing organizational policy.

APPENDIX

EXAMPLES OF SOME MEASURES

Some of the measures used in behavioral research can be had from the *Handbook of Organizational Measurement* by J. L. Price (Lexington, MA: D. C. Heath, 1972) and from the *Michigan Organizational Assessment Package* published by the Institute of Survey Research in Ann Arbor, Michigan. Several measures can also be seen in *Psychological Measurement Yearbooks* and in other published books. A sample of measures from the management and marketing areas is provided in this appendix.

MEASURES FROM MANAGEMENT RESEARCH

This section gives a sample of five scales used to measure five variables related to management research.

I. Job Involvement

	Strongly Disagree	Disagree	Neither Agree nor Disagree	Agree	Strongly Agree
1. My job means a lot more to me than just money.	1	2	3	4	5
2. The major satisfaction in my life comes from my job.	1	2	3	4	5
3. I am really interested in my work.	1	2	3	4	5
4. I would probably keep working even if I didn't need the money.	1	2	3	4	5
5. The most important things that happen to me involve my work.	1	2	3	4	5
6. I will stay overtime to finish a job, even if I am not paid for it.	1	2	3	4	5
7. For me, the first few hours at work really fly by.	1	2	3	4	5
8. How much do you actually enjoy performing the daily activities that make up your job?	1	2	3	4	5
9. How much do you look forward to coming to work each day?	1	2	3	4	5

SOURCE: J. K. White and R. R. Ruh (1973). Effects of personal values on the relationship between participation and job attitudes. *Administrative Science Quarterly*, 18, 4, p. 509. Reproduced with permission.

II. Participation in Decision Making

	Strongly Disagree	Disagree	Neither Agree nor Disagree	Agree	Strongly Agree
1. In general, how much say or influence do you have on how you perform your job?	1	2	3	4	5
2. To what extent are you able to decide how to do your job?	1	2	3	4	5
3. In general, how much say or influence do you have on what goes on in your work group?	1	2	3	4	5
4. In general, how much say or influence do you have on decisions that affect your job?	1	2	3	4	5
5. My superiors are receptive and listen to my ideas and suggestions.	1	2	3	4	5

SOURCE: J. K. White and R. R. Ruh (1973). Effects of personal values on the relationship between participation and job attitudes. *Administrative Science Quarterly*, 18, 4, p. 509. Reproduced with permission.

III. Role Conflict

	Very False						Very True
1. I have to do things that should be done differently.	1	2	3	4	5	6	7
2. I work under incompatible policies and guidelines.	1	2	3	4	5	6	7
3. I receive an assignment without the manpower to complete it.	1	2	3	4	5	6	7
4. I have to buck a rule or policy in order to carry out an assignment.	1	2	3	4	5	6	7
5. I work with two or more groups who operate quite differently.	1	2	3	4	5	6	7
6. I receive incompatible requests from two or more people.	1	2	3	4	5	6	7
7. I do things that are apt to be accepted by one person and not accepted by others.	1	2	3	4	5	6	7
8. I receive an assignment without adequate resources and materials to execute it.	1	2	3	4	5	6	7
9. I work on unnecessary things.	1	2	3	4	5	6	7

SOURCE: J. R. Rizzo, R. J. House, and S. I. Lirtzman (1970). Role conflict and role ambiguity in complex organizations. *Administrative Science Quarterly*, 15, p. 156. Reproduced with permission.

IV. Career Salience

Strongly Disagree 1	Disagree 2	Slightly Disagree 3	Neutral 4	Slightly Agree 5	Agree 6	Strongly Agree 7

1. My career choice is a good occupational decision for me. ____
2. My career enables me to make significant contributions to society. ____
3. The career I am in fits me and reflects my personality. ____
4. My education and training are not tailored for this career. ____
5. I don't intend changing careers. ____
6. All the planning and thought I gave for pursuing this career are a waste. ____
7. My career is an integral part of my life. ____

SOURCE: U. Sekaran. (1986) *Dual-Career Families: Contemporary Organizational and Counseling Issues.* San Francisco: Jossey Bass. Reproduced with permission.

V. LEAST PREFERRED COWORKER SCALE (to assess whether employees are primarily people-oriented or task-oriented)

Look at the words at both ends of the line before you put in your "X." Please remember that there are no right or wrong answers. *Work rapidly; your first answer is likely to be the best. Please do not omit any items, and mark each item only once.*

LPC

Think of the person with whom you can work least well. *He may be someone you work with now, or he may be someone you knew in the past.*

He does not have to be the person you like least well, but should be the person with whom you had the most difficulty in getting a job done. Describe this person as he appears to you.

Pleasant	8	7	6	5	4	3	2	1	Unpleasant
Friendly	8	7	6	5	4	3	2	1	Unfriendly
Rejecting	1	2	3	4	5	6	7	8	Accepting
Helpful	8	7	6	5	4	3	2	1	Frustrating
Unenthusiastic	1	2	3	4	5	6	7	8	Enthusiastic
Tense	1	2	3	4	5	6	7	8	Relaxed
Distant	1	2	3	4	5	6	7	8	Close
Cold	1	2	3	4	5	6	7	8	Warm
Cooperative	8	7	6	5	4	3	2	1	Uncooperative
Supportive	8	7	6	5	4	3	2	1	Hostile
Boring	1	2	3	4	5	6	7	8	Interesting
Quarrelsome	1	2	3	4	5	6	7	8	Harmonious
Self-assured	8	7	6	5	4	3	2	1	Hesitant
Efficient	8	7	6	5	4	3	2	1	Inefficient
Cheerful	8	7	6	5	4	3	2	1	Gloomy
Open	8	7	6	5	4	3	2	1	Guarded

SOURCE: Fred E. Fiedler (1967). *A Theory of Leadership Effectiveness.* New York: McGraw-Hill. Reproduced with permission.

MEASURES FROM MARKETING RESEARCH

Below is a sample of some scales used to measure commonly researched concepts in marketing. Bruner and Hensel have done extensive work since 1992 in documenting and detailing several scores of scales in marketing research. For each scale examined, they have provided the following information:

1. Scale description
2. Scale origin
3. Samples in which the scale was used
4. Reliability of the scale
5. Validity of the scale
6. How the scale was administered
7. Major findings of the studies using the scale

The interested student should refer to the two volumes of *Marketing Scales Handbook* by G. C. Bruner and P. J. Hensel, published by the American Marketing Association. The first volume covers scales used in articles published in the 1980s, and volume two covers scales used in articles published from 1990 to 1993. The third volume will cover the period from 1994 to 1997. Also refer to the web site: http://www.siu.edu:80/departments/coba/marketing/osr

I. Index of Consumer Sentiment Toward Marketing

1. Listed below are seven statements pertaining to each of the four marketing areas. There is also a fifth section labeled "Marketing in General." It contains four statements.

For each statement, please "X" the box which best describes how strongly you agree or disagree with each statement. For example, if you strongly agree that the quality of most products today is as good as can be expected, then "X" the Agree Strongly box.

On the other hand, if you strongly disagree that the quality of most products today is as good as can be expected, then "X" the Disagree Strongly box. Remember to "X" one box for each statement.

Product quality	Strongly Disagree	Somewhat Disagree	Neither Agree nor Disagree	Somewhat Agree	Strongly Agree
The quality of most products I buy today is as good as can be expected.	☐1	☐2	☐3	☐4	☐5
I am satisfied with most of the products I buy.	☐1	☐2	☐3	☐4	☐5
Most products I buy wear out too quickly.	☐1	☐2	☐3	☐4	☐5
Products are not made as well as they used to be.	☐1	☐2	☐3	☐4	☐5
Too many of the products I buy are defective in some way.	☐1	☐2	☐3	☐4	☐5
The companies that make products I buy don't care enough about how well they perform.	☐1	☐2	☐3	☐4	☐5
The quality of products I buy has consistently improved over the years.	☐1	☐2	☐3	☐4	☐5

Price of products	Strongly Disagree	Somewhat Disagree	Neither Agree nor Disagree	Somewhat Agree	Strongly Agree
Most products I buy are overpriced.	☐1	☐2	☐3	☐4	☐5
Businesses could charge lower prices and still be profitable.	☐1	☐2	☐3	☐4	☐5
Most prices are reasonable considering the high cost of doing business.	☐1	☐2	☐3	☐4	☐5
Competition between companies keeps prices reasonable.	☐1	☐2	☐3	☐4	☐5
Companies are unjustified in charging the prices they charge.	☐1	☐2	☐3	☐4	☐5
Most prices are fair.	☐1	☐2	☐3	☐4	☐5
In general, I am satisfied with the prices I pay.	☐1	☐2	☐3	☐4	☐5

Advertising for products	Strongly Disagree	Somewhat Disagree	Neither Agree nor Disagree	Somewhat Agree	Strongly Agree
Most advertising provides consumers with essential information.	☐1	☐2	☐3	☐4	☐5
Most advertising is very annoying.	☐1	☐2	☐3	☐4	☐5
Most advertising makes false claims.	☐1	☐2	☐3	☐4	☐5
If most advertising was eliminated, consumers could be better off.	☐1	☐2	☐3	☐4	☐5
I enjoy most ads.	☐1	☐2	☐3	☐4	☐5
Advertising should be more closely regulated.	☐1	☐2	☐3	☐4	☐5
Most advertising is intended to deceive rather than to inform consumers.	☐1	☐2	☐3	☐4	☐5

Retailing or selling	Strongly Disagree	Somewhat Disagree	Neither Agree nor Disagree	Somewhat Agree	Strongly Agree
Most retail stores serve their customers well.	□1	□2	□3	□4	□5
Because of the way retailers treat me, most of my shopping is unpleasant.	□1	□2	□3	□4	□5
I find most retail salespeople to be very helpful.	□1	□2	□3	□4	□5
Most retail stores provide an adequate selection of merchandise.	□1	□2	□3	□4	□5
In general, most middlemen make excessive profits.	□1	□2	□3	□4	□5
When I need assistance in a store, I am usually not able to get it.	□1	□2	□3	□4	□5
Most retailers provide adequate service.	□1	□2	□3	□4	□5

Marketing in general	Strongly Disagree	Somewhat Disagree	Neither Agree nor Disagree	Somewhat Agree	Strongly Agree
Most businesses operate on the philosophy that the consumer is always right.	□1	□2	□3	□4	□5
Despite what is frequently said, "let the buyer beware" is the guiding philosophy of most businesses.	□1	□2	□3	□4	□5
Most businesses seldom shirk their responsibility to the consumer.	□1	□2	□3	□4	□5
Most businesses are more interested in making profits than in serving consumers.	□1	□2	□3	□4	□5

2. Now, I'd like to know how satisfied you are, in general, with each of these four marketing areas. Please "X" the one box that best describes your overall satisfaction with each marketing area.

	Very Satisfied	Somewhat Satisfied	Neither Satisfied nor Dissatisfied	Somewhat Dissatisfied	Very Dissatisfied
The *quality* of most of the products available to buy.	□1	□2	□3	□4	□5
The *prices* of most products.	□1	□2	□3	□4	□5
Most of the *advertising* you read, see, and hear.	□1	□2	□3	□4	□5
The *selling conditions* at most of the stores at which you buy products.	□1	□2	□3	□4	□5

3. Listed below are four questions which ask about how often you have had problems with the products you buy, the prices you pay, the advertising you read, see, and hear, and the stores at which you shop.

After each statement, there are five numbers from 1 to 5. The higher the number means you have experienced the problem more often. The lower the number means you have experienced the problem less often.

For each question, please "X" the box which comes closest to how often the problem occurs. Remember to "X" one box for each question.

	Very Seldom				Very Often
How *often* do you have problems with or complaints about the *products* you buy?	☐1	☐2	☐3	☐4	☐5
How *often* do you have problems with or complaints about the *prices* you pay?	☐1	☐2	☐3	☐4	☐5
How *often* do you have problems with or complaints about *advertising?*	☐1	☐2	☐3	☐4	☐5
How *often* do you have problems with or complaints about the *stores* at which you buy products?	☐1	☐2	☐3	☐4	☐5

SOURCE: J. F. Gaski and M. J. Etzel. (1986). The index of consumer sentiment toward marketing. *Journal of Marketing*, 50, 71–81. Reproduced with permission of American Marketing Association.

II. SERVQUAL-P Battery (to assess the quality of service rendered)

Reliability

1. Provides the service as promised.
2. Is dependable in handling customers' service problems.
3. Performs the service right the first time.
4. All _____'s employees are well-trained and knowledgeable.

Responsiveness

5. Employees of _____ give you prompt service.
6. Employees of _____ are always willing to help you.
7. Employees of _____ are always ready to respond to your requests.
8. _____ gives customers individual attention.

Personalization

9. Everyone at _____ is polite and courteous.
10. The _____ employees display personal warmth in their behavior.
11. All the persons working at _____ are friendly and pleasant.
12. The _____ employees take the time to know you personally.

Tangibles

13. _____ has modern-looking equipment.
14. _____'s physical facilities are visually appealing.
15. _____'s employees have neat and professional appearance.
16. Materials associated with the service (such as pamphlets or statements) are visually appealing at _____.

SOURCE: B. Mittal and W. M. Lassar. (1996). The role of personalization in service encounters. *Journal of Retailing*, 72, 95–109. Reproduced with permission of Jai Press, Inc.

III. Role Ambiguity (Salesperson)

Very False						Very True
1	2	3	4	5	6	7

1. I feel certain about how much authority I have in my selling position. ____
2. I have clearly planned goals for my selling job. ____
3. I am sure I divide my time properly while performing my selling tasks. ____
4. I know my responsibilities in my selling position. ____
5. I know exactly what is expected of me in my selling position. ____
6. I receive lucid explanations of what I have to do in my sales job. ____

A modified version of Rizzo, House, and Lirtzman's (1970) Role ambiguity in complex organizations scale published in *Administrative Science Quarterly*, 15, p. 156

DATA COLLECTION METHODS

5.1 CHAPTER OBJECTIVES

After completing this chapter you should:

- Be conversant with the various data collection methods.
- Know the advantages and disadvantages of each method.
- Make logical decisions as to the appropriate data collection method(s) for specific studies.
- Demonstrate your skills in interviewing others to collect data.
- Design questionnaires to tap different variables.
- Evaluate questionnaires, distinguishing the "good" and "bad" questions therein.
- Identify and minimize the biases in various data collection methods.
- Discuss the advantages of multisources and multimethods of data collection.
- Apply what you have learned to class assignments and projects.

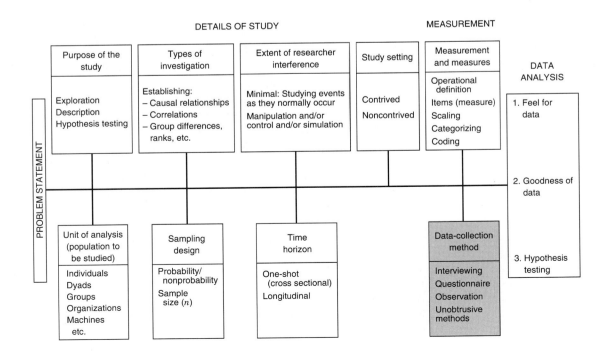

Having examined how variables are measured, we will now discuss how data can be gathered for the purpose of analysis, testing of hypotheses, and answering the research questions. The manner in which data are collected could make a big difference to the rigor and effectiveness of the research project.

Data collection methods are an integral part of research design as shown in the shaded portion in the figure. There are several data collection methods, each with its own advantages and disadvantages. Problems researched with the use of appropriate methods greatly enhance the value of the research, as discussed in this chapter.

5.2 DATA COLLECTION METHODS, SETTINGS, AND SOURCES OF DATA

Data Collection Methods

Data can be collected in a variety of ways, in different settings, and from different sources. Data collection methods include *interviews*—face-to-face interviews, telephone interviews, computer-assisted interviews, and through the electronic media; *questionnaires* that are either personally administered, sent through the mail, or electronically administered; *observation* of individuals and events with or without videotaping or audio recording; and a variety of other *motivational techniques* such as projective tests.

Setting

Data can be collected in any one of the aforementioned ways in the natural environment of the workplace. Data may also be collected in experimental lab settings where variables are controlled and manipulated, or gathered in the homes of the respondents, on the street, in malls, or in a setting where a LAN (Local Area Network) system is available. It is not unusual to find marketers conducting what are known as *intercept interviews* in malls and fairs, to obtain a considerable amount of marketing information.

Sources

Data sources can be **primary** or **secondary.** Individuals, focus groups, and a panel of respondents specifically set up by the researcher whose opinions may be sought on specific issues from time to time are examples of primary data sources. Data can also be obtained from secondary sources, as for example, company records or archives, government publications, industry analysis offered by the media, web sites, the Internet, and so on. In some cases, the environment or particular settings and events may themselves be sources of data, as for example, studying the layout of a plant.

Recall the definitions of primary and secondary data from Chapter 2.

A rich source of data is the aforementioned focus groups. Free-flowing intensive discussions for one to three hours, among small groups of people—say, six to 10 persons—with the researcher as the moderator, help to develop or refine concepts, explore organizational ideas for conducting further research, and so on. On-line focus groups and use of videoconferencing as a means of conducting discussions in focus groups are also gaining popularity.

Data can also be culled from case studies and any of the many sources of secondary data for analysis and application to solve specific problems.

Interviewing, administering questionnaires, and observing people and phenomena are the three main data collection methods in survey research. Projective tests and other motivational techniques are also sometimes used to tap variables. In such cases, respondents are usually asked to write a story, complete a sentence, or offer their reactions to ambiguous cues such as inkblots or unlabeled pictures. It is assumed that the respondents project into the responses their own thoughts, feelings, attitudes, and expectations, all of which can be interpreted by trained psychologists.

Although interviewing has the advantage of flexibility in terms of adapting, adopting, and changing the questions as the researcher proceeds with the interviews, questionnaires have the advantage of obtaining data more efficiently in terms of researcher time, energy, and costs. Unobtrusive methods of data collection such as extracting data from company records have the advantage of accuracy. For instance, attendance records will probably give a truer and more reliable picture of the absenteeism of employees than eliciting the information directly from the respondents. Projective tests are usually administered by researchers who have had training in administering them and interpreting the results. Though some management research has been done using projective techniques, they are more frequently used in marketing research, as we will see later.

Modern technology is increasingly playing a key role in shaping data collection methods. Computer-assisted surveys, which help both interviewing as well as preparing and administering questionnaires electronically, are on the increase. Computer-assisted telephone interviewing (CATI), interactive electronic telephonic surveys, as well as administering questionnaires through the electronic mail (e-mail), are now being used to facilitate data gathering.

The choice of data collection methods depends on the facilities available, the degree of accuracy required, the expertise of the researcher, the time span of the study, and other costs and resources associated with and available for data gathering.

We will now examine the various data collection methods.

5.2.1 Interviewing

One method of collecting data is to interview respondents to obtain information on the issues of interest. Interviews could be unstructured or structured, and conducted either face to face or by telephone or on-line.

The unstructured and structured interviews are discussed first. Some important factors to be borne in mind while interviewing will then be detailed; thereafter, the advantages and disadvantages of face-to-face interviewing and telephone interviews will be enumerated, and finally, computer-assisted interviews described.

Unstructured and Structured Interviews

Unstructured Interviews.   Unstructured interviews are so labeled because the interviewer does not enter the interview setting with a planned sequence of questions that will be asked of the respondent. The objective of the unstructured interview is to cause some preliminary issues to surface so that the researcher can decide what variables need further in-depth investigation. In some situations, a manager might have a vague idea of certain changes taking place in the situation without knowing what exactly they are. Such situations call for unstructured interviews with the people concerned. In order to understand the situation in its totality, the researcher will interview employees at several levels. At the initial stages, only broad, open-ended questions would be asked, and the replies to them would give the researcher an indication of the individuals' perceptions. The type and nature of the questions asked of the individuals might vary according to the job level and type of work done by them. For instance, managers at top and middle levels might be asked more direct questions about their perceptions of the problems and the situation. Employees at lower levels may have to be approached differently.

Clerical and other employees at lower hierarchical levels may be asked broad, open-ended questions about their jobs and the work environment during the unstructured interviews. Supervisors may be asked broad questions relating to their department, their employees, and the organization. The following request, for instance, may be made during the unstructured interview stage:

> "Tell me something about your unit and department, and perhaps even the organization as a whole, in terms of work, employees, and whatever else you think is important."

Such a request might elicit an elaborate response from some people, whereas others may just say that everything is fine. Following the leads from the more vocal persons is easy, especially when the interviewer listens carefully to the important messages that they might convey in a very casual manner while responding to a general, global question. As managers and researchers, we should train ourselves to develop these listening skills and identify the critical topics that are touched on. However, when some respondents give a one-word, crisp reply that is not informative, the interviewer will have to ask questions that cannot be answered in one or two words. Such questions might be phrased as this:

> "I would like to know something about your job. Please describe to me in detail the things you do on your job on a typical day, from eight in the morning to four in the afternoon."

Several questions might then be asked as a follow-up to the answer. Some examples of such follow-up questions include:

> "Compared to other units in this organization, what are the strengths and weaknesses of your unit?"

> "If you would like to have a problem solved in your unit, or eliminate a bottleneck, or attend to something that blocks your effectiveness, what would that be?"

If the respondent answers that everything is fine and she has no problems, the interviewer could say:

> "That is great! Tell me what contributes to this effectiveness of your unit, because most other organizations usually experience several difficulties."

Such a questioning technique usually lowers the respondent's defenses and makes the individual more willing to share information. Typical of the revised responses to the original question is something like, "Well, it is not that we never have a problem, sometimes, we are late in getting the jobs done, crash jobs have some defective items . . . " Encouraging the respondent to talk about both the good things and the not-so-good things in the unit can elicit a lot of information. Whereas some respondents do not need much encouragement to speak, other do; and they have to be questioned broadly. Some respondents may be reluctant to be interviewed, and may subtly or overtly refuse to cooperate. The wishes of such people must be respected and the interviewer should pleasantly terminate such interviews.

Employees at the shop-floor level, and other nonmanagerial and nonsupervisory employees, might be asked very broad questions relating to their jobs, work environment, satisfactions and dissatisfactions at the workplace, and the like—for example:

> What do you like about working here?

> If you were to tell me what aspects of your job you like and what you do not like, what would they be?

> Tell me something about the reward systems in this place.

If you were offered a similar job elsewhere, how willing would you be to take it and why?

If I were to seek employment here and ask you to describe your unit to me as a newcomer, what would you say?

After conducting a sufficient number of such unstructured interviews with employees at several levels and studying the data obtained, the researcher would have a grip on the variables that need greater focus and where more in-depth information has to be obtained.

This sets the stage for the interviewer to conduct further structured interviews, for which purpose the variables would have been identified.

Structured Interviews.   Structured interviews are those conducted when it is known at the outset what information is needed. The interviewer has a list of predetermined questions to be posed to the respondents either personally, or through the telephone, or through the medium of a PC. The questions are likely to focus on factors that had surfaced during the unstructured interviews and are considered relevant to the problem. As the respondents express their views, the researcher will note them down. The same questions will be asked of everybody in the same manner. Sometimes, however, based on the exigencies of the situation, the experienced researcher might take a lead from a respondent's answer and ask other relevant questions not on the interview protocol. Through this process, new factors might be identified and a deeper understanding might result. However, to be able to recognize a meaningful response when it is given, the interviewer must comprehend the purpose and goal of each question. This is particularly important when a team of trained interviewers conducts the survey.

Visual aids such as pictures, line drawings, cards, and other materials are also sometimes used in conducting interviews. The appropriate visuals are shown to the interviewees, who then indicate their responses to the questions posed. Marketing research, for example, benefits from such techniques in order to capture the likes and dislikes of customers to different types of packaging, forms of advertising, and so on. Visual aids, including painting and drawing, are particularly useful when children are the focus of marketing research. Visual aids also come in handy when eliciting certain thoughts and ideas that are difficult to express or awkward to articulate.

When a sufficient number of structured interviews has been conducted and adequate information has been obtained to understand and describe the important factors operating in the situation, the researcher would stop the interviews. The information would then be tabulated and the data analyzed. This would help the researcher to accomplish the task set out to be done, as for example, to describe the phenomena, or quantify them, or identify the specific problem and evolve a theory of the factors that influence the problem or find answers to the research question. Much qualitative research is done in this manner.

Training Interviewers

When several long interviews are to be conducted, it is often not feasible for one individual to conduct all the interviews. A team of trained interviewers then becomes necessary. Interviewers have to be thoroughly briefed about the research and trained in how to start an interview, how to proceed with the questions, how to motivate respondents to answer, what to look for in the answers, and how to close an interview. They also need to be instructed on taking notes and coding the interview responses. The tips for interviewing, discussed later, should become a part of their repertoire for interviewing.

Good planning, proper training offering clear guidelines to interviewers, and supervising their work—all help in profitably utilizing the interviewing technique as a viable data collection mechanism. Personal interviews offer rich data that are spontaneously provided by the respondents, in the sense that their answers do not typically fall within a constricted range of responses, as in a questionnaire. However, personal interviews are expensive in terms of time, training costs, and resource consumption.

Review of Unstructured and Structured Interviews

The main purpose of the unstructured interview is to explore and probe into the several factors in the situation that might be central to the broad problem area. During this process it might become evident that the problem, as identified by the client, is only the symptom of a more serious and deep-rooted problem. Conducting unstructured interviews with many people in the organization could result in the identification of several critical factors in the situation. These factors would then be pursued further during the structured interviews for eliciting more in-depth information on them. This will help identify the critical problem as well as solve it. In applied research, a tentative theory of the factors influencing the problem is often conceptualized on the basis of the information obtained from the unstructured and structured interviews.

Some Tips to Follow in Interviewing

The information obtained during the interviews should be as free of bias as possible. Bias refers to errors or inaccuracies in the data collected. Biases could be introduced by the interviewer, the interviewee, or the situation.

It is important to minimize bias, which has several potential sources.

- The **interviewer** could bias the data if proper trust and rapport are not established with the interviewee, or when the responses are either misinterpreted or distorted, or when the interviewer unintentionally encourages or discourages certain types of responses through gestures and facial expressions.

    Attentively listening to the interviewee, evincing interest in what the respondent has to say, exercising tact in questioning, repeating and/or clarifying the questions posed, and paraphrasing some of the answers to ensure a thorough understanding of the responses, go a long way in maintaining the interest of the respondent throughout the interview. Accurately recording the responses is equally important.

- **Interviewees** can bias the data when they do not express their true opinions but provide information that they think the interviewer expects of them or would like to hear. Also, if the respondents do not understand the questions, they may hesitate to seek clarification. They may then answer questions without knowing what exactly the questions mean, and thus introduce biases.

    Some interviewees may be turned off because of personal likes and dislikes, or the dress of the interviewer, or the manner in which the questions are posed. They may, therefore, not provide truthful answers, but instead deliberately offer incorrect responses. Some respondents may also answer questions in a socially acceptable manner rather than indicate their true sentiments.

- Biases could be **situational** as well, in terms of (1) nonparticipants, (2) trust levels and rapport established, and (3) the physical setting of the interview. *Nonparticipation*, either because of unwillingness or the inability of the interviewee to participate in the study, can bias data inasmuch as the responses of the participants may be different from those who did not participate (which implies that a biased, rather than a representative set of responses is likely to be gathered). Bias also occurs when different interviewers establish *different levels of trust and rapport* with their interviewees, thus eliciting answers of varying degrees of openness. The actual *setting* itself in which the interview is conducted might sometimes introduce biases. Some individuals, for instance, may feel uncomfortable when interviewed at the workplace and not respond frankly and honestly.

In door-to-door or telephone interviews, when the respondent cannot be reached due to unavailability at that time, callbacks and further contacts should be attempted so that the sample does not become biased. The interviewer can also reduce bias by being consistent with the questioning mode as each person is interviewed, by not distorting or falsifying the information received, and by not influencing the responses of the subjects in any manner.

The preceding biases can be minimized in several ways. The following strategies will be useful for the purpose.

Establishing Credibility and Rapport, and Motivating Individuals to Respond

The projection of professionalism, enthusiasm, and confidence is important for the interviewer. A manager hiring outside researchers would be interested in assessing their abilities and personality predispositions. Researchers must establish rapport with and gain the confidence and approval of the hiring client before they can even start their work in the organization. Knowledge, skills, ability, confidence, articulateness, and enthusiasm are therefore qualities a researcher must demonstrate in order to establish credibility with the hiring organization and its members.

To obtain honest information from the respondents, the researcher/interviewer should be able to establish rapport and trust with the interviewees. In other words, the researcher should be able to make the respondent comfortable enough to give informative and truthful answers without fear of adverse consequences. To this end, the researcher should state the purpose of the interview and assure complete confidentiality about the source of the responses. Establishing rapport with the respondents may not be easy, especially when interviewing employees at lower levels. Employees are likely to be suspicious of the intentions of the researchers; they may believe that the researchers are on the management's "side" and are therefore likely to propose reduction of the labor force, increase in the workload, and so on. Thus, it is important to ensure that everyone concerned is aware of the researchers' purpose as being one of simply understanding the true state of affairs in the organization. The respondents must be tactfully made to understand that the researchers are not on any particular side; they are not there to harm the staff, and they will provide the results of research to the organization only in aggregates, without disclosing the identity of the individuals. This would help the respondents feel secure about responding.

The researcher can establish rapport by being pleasant, sincere, sensitive, and nonevaluative. Evincing a genuine interest in the responses and allaying any anxieties, fears, suspicions, and tensions sensed in the situation will help respondents to feel more comfortable with the researchers. If the respondent is told the purpose of the study and how the individual was chosen to be one of those interviewed, there would be better communication flow between the parties. Researchers can motivate respondents to give honest and truthful answers by explaining to them that their contribution would indeed help, and that they themselves may stand to gain from such a survey, in the sense that the quality of life at work for most of them could improve significantly.

Certain other strategies in how questions are posed also help participants to offer less biased responses. These are discussed below.

The Questioning Technique

Funneling.   In the beginning of an unstructured interview, it is advisable to ask open-ended questions to get a general idea and form some impressions about the situation. For example, a question that could be asked might be:

"What are some of your feelings about working for this organization?"

From the responses to this broad question, further questions that are progressively more focused may be asked as the researcher processes the interviewees' responses and determines some possible key issues relevant to the situation. This transition from broad to narrow themes is called the funneling technique.

Unbiased Questions.   It is important to ask questions in a way that would ensure the least bias in the response. For example, "Tell me how you experience your job" is a better question than, "Boy, the work you do must be really boring; let me hear how you experience it." The latter question is "loaded" in terms of the interviewer's own perceptions of the job. A loaded question might influence the types of answers received from the respondent. Bias could be also introduced by emphasizing certain words, by tone and voice inflections, and through inappropriate suggestions.

Clarifying Issues.   To make sure that the researcher understands issues as the respondent means to represent them, it is advisable to restate or rephrase important information given by the respondent. For instance, if the interviewee says, "There is an unfair promotion policy in this organization; seniority does not count at all. It is the juniors who always get promoted," the researcher might interject, "So you are saying that juniors always get promoted over the heads of even capable seniors." Rephrasing in this way clarifies the issue of whether or not the respondent considers ability important. If certain things that are being said are not clear, the researcher should seek clarification. For example, if the respondent happened to say, "The facilities here are really poor; we often have to continue working even when we are dying of thirst," the researcher might ask if there is no water fountain or drinking water available in the building. The respondent's reply to this might well indicate that there is a water fountain across the hall, but the respondent would have liked one on his side of the work area.

Helping the Respondent to Think Through Issues.   If the respondent is not able to verbalize her perceptions, or replies, "I don't know," the researcher should ask the question in a simpler way or rephrase it. For instance, if a respondent is unable to specify what aspects of the job he dislikes, the researcher might ask the question in a simpler way. For example, the respondent might be asked which task he would prefer to do: serve a customer or do some filing work. If the answer is "serve the customer," the researcher might use another aspect of the respondent's job and ask the paired-choice question again. In this way, the respondent can sort out which aspects of the job he likes better than others.

Taking Notes.   When conducting interviews, it is important that the researcher makes written notes as the interviews are taking place, or as soon as the interview is terminated. The interviewer should not rely on memory, because information recalled from memory is imprecise and often likely to be incorrect. Furthermore, if more than one interview is scheduled for the day, the amount of information received increases, as do possible sources of error in recalling from memory as to who said what. Information based solely on recall introduces bias into the research. It is possible to record the interviews on tape if the respondent has no objection. However, taped interviews might bias the respondents' answers because they know that their voices are being recorded, and their anonymity is not completely preserved. Hence, even if the respondents do not object to being taped, there could be some bias in their responses. Before recording or videotaping interviews, one should be reasonably certain that such a method of obtaining data is not likely to bias the information received. Any taping or videotaping should always be done only after obtaining the respondent's permission.

Review of Tips to Follow in Interviewing

Establishing credibility as able researchers with the client system and the organizational members is important for the success of the research project. Researchers need to establish rapport with the respondents and motivate them to give responses relatively free from bias by allaying whatever suspicions, fears, anxieties, and concerns they may have about the research and its consequences. This can be accomplished by being sincere, pleasant, and nonevaluative. While interviewing, the researcher has to ask broad questions initially and then narrow the questions to specific areas, ask questions in an unbiased way, offer clarifications when needed, and help respondents to think through difficult issues. The responses should be transcribed immediately and not be trusted to memory and later recall.

Having looked at unstructured and structured interviews and learned something about how to conduct the interviews, we can now discuss face-to-face and telephone interviews.

Face-to-Face and Telephone Interviews

Interviews can be conducted either face to face or over the telephone. They could also be computer-assisted. Although most unstructured interviews in organizational research are conducted face to face, structured interviews could be either face to face or through the medium of the telephone, depending on the level of complexity of the issues involved, the likely duration of the interview, the convenience of both parties, and the geographical area covered by the survey. Telephone interviews are best suited when quick information from a large number of respondents spread over a wide geographic area is to be obtained, and the likely duration of each interview is, say, 10 minutes or less. Many market surveys, for instance, are conducted through structured telephone interviews. In addition, Computer Assisted Telephone Interviews (CATI) are also possible, and easy to manage.

Face-to-face interviews and telephone interviews have their other advantages and disadvantages. These will now be briefly discussed.

Face-to-Face Interviews

- *Advantages.*   The main advantage of face-to-face or direct interviews is that the researcher can adapt the questions as necessary, clarify doubts, and ensure that the responses are properly understood, by repeating or rephrasing the questions. The researcher can also pick up nonverbal cues from the respondent. Any discomfort, stress, or problems that the respondent experiences can be detected through frowns, nervous tapping, and other body language unconsciously exhibited by the respondent. This would obviously be impossible to detect in a telephone interview.
- *Disadvantages.*   The main disadvantages of face-to-face interviews are the geographical limitations they may impose on the surveys and the vast resources needed if such surveys need to be done nationally or internationally. The costs of training interviewers to minimize interviewer biases (e.g., differences in ques-

tioning methods, interpretation of responses) are also high. Another drawback is that respondents might feel uneasy about the anonymity of their responses when they interact face to face with the interviewer.

Telephone Interviews

- *Advantages.* The main advantage of telephone interviewing, from the researcher's point of view, is that a number of different people can be reached (across the country or even internationally) in a relatively short period of time. From the respondents' standpoint it would eliminate any discomfort that some of them might feel in facing the interviewer. It is also possible that most of them would feel less uncomfortable disclosing personal information over the phone than face to face.

- *Disadvantages.* A main disadvantage of telephone interviewing is that the respondent could unilaterally terminate the interview without warning or explanation, by hanging up the phone. Caller ID might further aggravate the situation. This is understandable given the numerous telemarketing calls people are bombarded with on a daily basis. To minimize this type of a nonresponse problem, it would be advisable to call the interviewee ahead of time to request participation in the survey, giving an approximate idea of how long the interview would last, and setting up a mutually convenient time. Interviewees usually appreciate this courtesy and are more likely to cooperate. It is a good policy not to prolong the interview beyond the time originally stated. As mentioned earlier, another disadvantage of the telephone interview is that the researcher will not be able to see the respondent to read the nonverbal communication.

Interviewing is a useful data collection method, especially during the exploratory stages of research. Where a large number of interviews is conducted with a number of different interviewers, it is important to train the interviewers with care in order to minimize interviewer biases manifested in such ways as voice inflections, different wordings, and differences in interpretation. Good training decreases interviewer biases.

Additional Sources of Bias in Interview Data

We have already discussed several sources of bias in data collection. Biased data will be obtained when respondents are interviewed while they are extremely busy or are in a bad mood. Responses to issues such as strikes, layoffs, or the like could also be biased. The personality of the interviewer, the introductory sentence, inflection of the voice, and such other aspects could introduce additional biases. Sensitivity to the many sources of bias will enable interviewers to obtain relatively valid information.

Sampling biases, which include inability to contact persons whose telephone numbers have changed, could also affect the quality of the data obtained. Similarly, people with unlisted numbers who are not contacted could also bias the sample and, hence, the data obtained. With the introduction of caller identity, it is possible for telephone interviews to be ridden with complexity.

Computer-Assisted Interviewing

With computer-assisted interviews (CAI), thanks to modern technology, questions are flashed onto the computer screen and interviewers can enter the respondents' answers directly into the computer. The accuracy of data collection is considerably enhanced since the software can be programmed to flag the "offbase" or "out-of-range" responses. CAI software also prevents interviewers from asking the wrong questions or in the wrong sequence since the questions are automatically flashed to the respondent in an ordered sequence. This would, to some extent, eliminate interviewer-induced biases.

CATI and CAPI.   There are two types of computer-assisted interview programs: CATI (computer-assisted telephone interviewing) and CAPI (computer-assisted personal interviewing).

CATI, used in research organizations, is useful inasmuch as responses to surveys can be obtained from people all over the world, since the PC is networked into the telephone system. The PC monitor prompts the questions with the help of software and the respondent provides the answers. The computer selects the telephone number, dials, and places the responses in a file. The data are analyzed later. Computerized, voice-activated telephone interviews are also possible for short surveys. Data can also be gathered during field surveys through hand-held computers which record and analyze responses.

CAPI involves big investments in hardware and software. CAPI has an advantage in that it can be self-administered; that is, respondents can use their own computers to run the program by themselves once they receive the software and enter their responses, thereby reducing errors in recording. However, not everyone is comfortable using a personal computer.

The **voice recording system** assists CATI programs by recording interviewees' responses. Courtesy, ethics, as well as legal requirements would require that the respondent's permission to record be obtained before the **voice capture system** (VCS) is activated. The VCS allows the computer to capture the respondents' answers, which are recorded in a digital mode and stored in a data file. They can be played back later, for example, to listen to customers by region, industry, or any combination of different sets of factors.

In sum, the advantages of computer-assisted interviews can be stated simply as quick and more accurate information gathering, plus faster and easier analysis of data. The field costs are low and automatic tabulation of results is possible. It is more efficient in terms of costs and time, once the initial heavy investment in equipment and software is made. However, to be really cost-effective, the surveys should be large and done frequently enough to warrant the heavy front-end investment and programming costs.

Computer-Aided Survey Services.   Several research organizations offer their services to companies that engage in occasional data gathering. For instance, the National Computer Network provides computer survey services for conducting marketing studies. *Advantages of using computer survey services* Some of the advantages of using these services are that (1) the researcher can start analyzing the data even as the field survey is progressing, since results can be transmitted to clients through modem in raw or tabulated form; (2) data can be automatically "cleaned up" and errors, if any, fixed even as they are being collected; (3) biases due to ordering questions in a particular way (known as the ordering effects) can be eliminated since meaningful random start patterns can be incorporated into the questioning process; (4) skip patterns (e.g., if the answer to this question is NO, skip to question #19) can be programmed into the process; and (5) questions can be customized to incorporate the respondents' terminology of concepts into subsequent questions.

Computer surveys can be conducted either by mailing the disks to respondents or through on-line surveys, with the respondents' personal computers being hooked up to computer networks. Survey System provided by Creative Research Systems and

Interview System provided by Compaq Co. are two of the several computer survey systems available in the market.

Advantages of Computer Packages

Field notes taken by interviewers as they collect data have generally to be transcribed, hand-coded, hand-tabulated, and so on—all of which are tedious and time consuming. Computers vastly facilitate the interviewer's job with regard to these activities. Automatic indexing of the data can be done with special programs. The two modes in operation are (1) **indexing** such that specific responses are coded in a particular way, and (2) **retrieval** of data with a fast search speed—going through 10,000 pages in less than five seconds. Text-oriented database management retrieval program allows the user to go through the text, inserting marks that link related units of text. The associated links formed are analytical categories specified by the researcher. Once the links are created, the program allows the user to activate them by opening multiple windows on the screen.

We thus see that computers have a large impact on data collection. With greater technological advancement and a lowering of hardware and software costs, computer-assisted interviews will become a primary data collection mechanism in the future.

Review of Interviewing

Interviews are one method of obtaining data; they can be either unstructured or structured, and can be conducted face to face, over the telephone, or through the medium of the PC. Unstructured interviews are usually conducted to obtain definite ideas about what is and is not important and relevant to particular problem situations. Structured interviews give more in-depth information about specific variables of interest. To minimize bias in responses, the interviewer must establish rapport with the respondents and ask unbiased questions. The face-to-face interview and that conducted over the telephone have their advantages and disadvantages, and both are useful under different circumstances. Computer-assisted interviewing, which entails a heavy initial investment, has an impact on interviewing and in the analyses of qualitative, spontaneous responses. We can expect computer interactive interviews to become an increasingly important mode of data collection in the future. We will now see how data can be gathered through questionnaires.

5.2.2 Questionnaires

> A *questionnaire* is a preformulated written set of questions to which respondents record their answers, usually within rather closely defined alternatives.

Questionnaires are an efficient data collection mechanism when the researcher knows exactly what is required and how to measure the variables of interest. Questionnaires can be administered personally or mailed to the respondents, or electronically distributed.

Personally Administered Questionnaires

When the survey is confined to a local area, and the organization is willing and able to assemble groups of employees to respond to the questionnaires at the workplace, personally administering the questionnaires is a good way to collect data. The main advantage of this is that the researcher or a member of the research team can collect all the completed responses within a short period of time. Any doubts that the respondents might have regarding any question could be clarified on the spot. The researcher also has the opportunity to introduce the research topic and motivate the respondents to give frank answers.

Administering questionnaires to large numbers of individuals simultaneously is less expensive and less time-consuming than interviewing; it also does not require as much skill to administer the questionnaire as to conduct interviews. Wherever possible, it is advantageous to administer questionnaires personally to groups of people because of these advantages. However, organizations are often unable or unwilling to allow use of work hours for data collection, and other ways of getting the questionnaires completed and returned may have to be found. In such cases, employees may be given blank questionnaires which can be collected from them personally after a few days, or the respondents can be provided with self-addressed, stamped envelopes with a request to complete and mail them back by a certain date. Scanner sheets (the answer sheets that are usually provided for answering multiple-choice questions in exams) are usually sent with the questionnaire so that respondents can circle their answers to each question on the sheet, which can then be directly entered into the computer as data without someone having to code and then manually enter them in the computer. Disks containing the questions can also be sent to such respondents who have and can use personal computers.

Mail Questionnaires

Advantages and disadvantages of mail questionnaires

The main advantage of mail questionnaires is that a wide geographical area can be covered in the survey. They are mailed to the respondents, who can complete them at their own convenience, in their homes, and at their own pace. However, the return rates of mail questionnaires are typically low. A 30% response rate is considered acceptable. Another disadvantage of the mail questionnaire is that any doubts the respondents might have cannot be clarified. Also, with very low return rates it is difficult to establish the representativeness of the sample, because those responding to the survey may be totally different from the population they are supposed to represent. However, some effective techniques exist for improving the rates of response to mail questionnaires. Sending follow-up letters, enclosing some small monetary incentives with the questionnaire, providing the respondent with self-addressed, stamped return envelopes, and keeping the questionnaire brief are helpful.

Mail questionnaires are also expected to elicit a better response rate when respondents are notified in advance about the forthcoming survey, and a reputed research organization administers the questionnaire with its own introductory cover letter.

The choice of using the questionnaire as a data gathering method might be restricted if the researcher has to reach subjects with very little education. Adding pictures to the questionnaires, if feasible, might be helpful in such cases. For most organizational research, however, after the variables for the research have been identified and the measures therefor found or developed, the questionnaire is a convenient data collection mechanism. Field studies, comparative surveys, and experimental designs often use questionnaires to measure the variables of interest. Because questionnaires are in common use in surveys, it is necessary to know how to design them effectively. A set of guidelines for questionnaire construction follows.

5.2.3 Guidelines for Questionnaire Design

Sound questionnaire design principles should focus on three areas. The first relates to the wording of the questions. The second refers to planning of issues of how the variables will be categorized, scaled, and coded after receipt of the responses. The third pertains to the general appearance of the questionnaire. All three are important issues in questionnaire design because they can minimize biases in research. These issues are discussed below. The important aspects are schematically depicted in Figure 5.1.

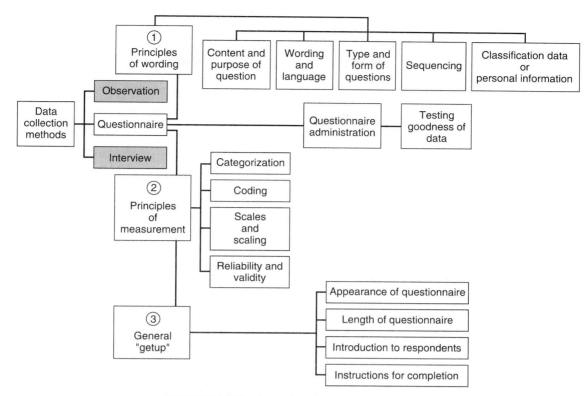

FIGURE 5.1 Principles of questionnaire design

Principles of Wording

The principles of wording refer to such factors as (1) the appropriateness of the content of the questions, (2) how questions are worded and the level of sophistication of the language used, (3) the type and form of questions asked, (4) the sequencing of the questions, and (5) the personal data sought from the respondents. Each of these is explained below.

Content and Purpose of the Question. The nature of the variable tapped—subjective feelings or objective facts—will determine what kinds of questions will be asked. If the variables tapped are of a subjective nature (e.g., satisfaction, involvement), where respondents' beliefs, perceptions, and attitudes are to be measured, the questions should tap the dimensions and elements of the concept. Where objective variables such as age and educational levels of respondents are tapped, a single direct question—preferably one that has an ordinal scaled set of categories—would be appropriate. Thus, the purpose of each question should be carefully considered so that the variables are adequately measured and yet no superfluous questions are asked.

Language and Wording of the Questionnaire. The language of the questionnaire should approximate the level of understanding of the respondents. The choice of words would depend on their educational level, the usage of terms and idioms in the culture, and the frames of reference of the respondents. For instance, even when English is the spoken or official language in two cultures, certain words may be alien to one culture. Terms such as "working here is a *drag*," and "she is a *compulsive* worker," may not be interpreted the same way in different cultures. Because some blue-collar workers may not understand terminology such as "organizational structure," it is essential to word questions in such a way that they are understood by the respondent. If some questions are either not understood or are interpreted differently by the respondent, the researcher will be obtaining the wrong answers to the questions, and responses will thus be biased. Hence, the questions asked, the language used, and the wording should be appropriate to tap respondents' attitudes, perceptions, and feelings.

Type and Form of Questions. Type of question refers to whether the question will be open-ended or closed. Form refers to positively and negatively worded questions.

Open-Ended versus Closed Questions.

> ***Open-ended questions*** allow respondents to answer in any way they choose.

An example of an open-ended question is asking the respondent to state five things that are interesting and challenging in the job. Another example is asking what the respondents like about their supervisors or their work environment. A third example is to invite their comments on the investment portfolio of the firm.

> A ***closed question*** asks the respondents to make choices among a set of alternatives given by the researcher.

For instance, instead of asking the respondent to state any *five* aspects of the job that are interesting and challenging, the researcher might list 10 or 15 characteristics that might seem interesting or challenging in jobs and ask the respondent to rank the first five among these. All items in a questionnaire using a nominal, ordinal, or Likert or ratio scale are considered closed.

Closed questions help the respondents to make quick decisions to choose among the several alternatives before them. They also help the researcher to code the information easily for subsequent analysis. Care has to be taken to ensure that the alternatives are mutually exclusive and collectively exhaustive. If there are overlapping categories, or if all possible alternatives are not given (i.e., the categories are not exhaustive), the respondents might get confused and the advantage of their being able to make a quick decision thus lost.

Some respondents may find even well-delineated categories in a closed question rather confining, and might like the opportunity to make additional comments. This is the reason that many questionnaires end with open-ended questions that invite respondents to comment on topics that might not have been fully or adequately covered. The responses to such open-ended questions have to be edited and categorized for subsequent data analysis.

Positively and Negatively Worded Questions. Instead of phrasing all questions positively, it is advisable to include some negatively worded questions as well, so the tendency of respondents to mechanically circle the points toward one end of the scale is minimized. For example, let us say that a set of six questions are used to tap the variable "perceived success" on a five-point scale, with 1 being "very low" and 5 being "very high" on the scale. A respondent who is not particularly interested in completing the questionnaire is more likely to stay involved and remain alert while answering the questions when positively and negatively worded questions are interspersed in it. For instance, if the respondent had circled 5 for a positively worded question such as, *"I feel I have been able to accomplish a number of different things in my job,"* he cannot circle number 5 again to the negatively worded question, *"I do not feel I am very effective in my job."* The respondent is now shaken out of any likely tendency to mechanically respond to one end of the scale. In case this does still happen, the researcher has an opportunity to detect such biases. A good questionnaire should therefore include both positively and negatively worded questions. The use of double negatives and excessive use of the words *not* and *only* should be avoided in the negatively worded questions because they tend to confuse respondents. For instance, it is better to say, *"Coming to work is no great fun"* than to say *"Not coming to work is greater fun than coming to work."* Similarly, it is better to say *"The rich need no help"* than to say *"Only the rich do not need help."*

Biases in Questions

Double-Barreled Questions.   A question that lends itself to different possible responses to its subparts is called a double-barreled question. Such questions should be avoided and two or more separate questions asked instead. For example, the question

"Do you think there is a good market for the product and that it will sell well?"

could bring a "yes" response to the first part (i.e., there is a good market for the product) and a "no" response to the latter part (i.e., it will not sell well—for various other reasons). In this case, it would be better to ask two questions: (1) "Do you think there is a good market for the product?" and (2) "Do you think the product will sell well?" The answers might be "yes" to both, "no" to both, "yes" to the first and "no" to the second, or "yes" to the second and "no" to the first. If we combined the two questions and asked a double-barreled question, we would confuse the respondents and obtain ambiguous responses. Hence, double-barreled questions should be avoided.

Ambiguous Questions.   Even questions that are not double-barreled might be ambiguously worded and the respondent may not be sure what exactly they mean. An example of such a question is

"To what extent would you say you are happy?"

Respondents might be unsure whether the question refers to their state of feelings at the workplace, or at home, or in general. Because it is an organizational survey, she might presume that the question relates to the workplace. Yet the researcher might have intended to inquire about the general, overall degree of satisfaction that the individual experiences in everyday life—a very global feeling not specific to the workplace alone. Thus, responses to ambiguous questions have built-in bias inasmuch as different respondents might interpret such items in the questionnaire differently. The result would be a mixed bag of ambiguous responses that do not accurately reflect the correct answer to the question.

Recall-Dependent Questions.   Some questions might require respondents to recall experiences from the past that are hazy in their memory. Answers to such questions might have bias. For instance, if an employee who has had 30 years' service in the organization is asked to state when he first started working in a particular department and for how long, he may be unable to give the correct answers and may be way off in his responses. A better source for obtaining that information would be his personnel records.

Leading Questions.   Questions should not be phrased in such a way that they lead the respondents to give the responses that the researcher would like or want them to give. An example of such a question is:

"Don't you think that in these days of escalating costs of living, employees should be given good pay raises?"

By asking such a question, we are signaling and pressuring respondents to say "yes." Tagging the question to rising living costs makes it difficult for most respondents (unless they are the top managers in charge of budget and finances) to say, "No; not unless their productivity increases too!" Another way of asking the question about pay raises to elicit less biased responses would be:

"To what extent do you agree that employees should be given higher pay raises?"

If respondents think that the employees do not deserve a higher pay raise, their response would be "Strongly Disagree"; if they think that respondents should be definitely given a high pay raise, they would respond to the "Strongly Agree" end of the scale; and the in-between points would be chosen depending on the strength of their agreement or disagreement. In this case, the question is not framed in a suggestive manner as in the previous instance.

Loaded Questions.   Another type of bias in questions occurs when they are phrased in an emotionally charged manner. An example of such a loaded question is asking employees:

"To what extent do you think management is likely to be vindictive if the union decides to go on strike?"

The words *strike* and *vindictive* are emotionally charged terms polarizing management and unions. Hence, asking a question such as the above would elicit strongly emotional and highly biased responses. If the purpose of the question is twofold—that is, to learn (1) the extent to which employees are in favor of a strike and (2) the extent to which they fear adverse reactions if they do go on strike—then these are the two specific questions that need to be asked. It may turn out that the employees do not strongly favor a strike and they also do not believe that management would retaliate if they did go on strike!

Social Desirability.   Questions should not be worded such that they elicit socially desirable responses. For instance, a question such as

"Do you think that older people should be laid off?"

would elicit a "no" response, mainly because society would frown on a person who would say that elderly people should be fired even if they are capable of performing their jobs. Hence, irrespective of the true feelings of the respondent, a socially desirable answer would be provided. If the purpose of the question is to gauge the extent to which organizations are seen as obligated to retain those above 65 years of age, a differently worded question with less pressure toward social desirability would be:

"There are advantages and disadvantages to retaining senior citizens in the workforce. To what extent do you think companies should continue to keep the elderly on their payroll?"

Sometimes certain items that tap social desirability are deliberately introduced at various points in the questionnaire and an index of each individual's social desirability tendency is calculated therefrom. This index is then applied to all other responses given by the individual in order to adjust for social desirability biases.

Length of Questions.   Finally, simple, short questions are preferable to long ones. As a rule of thumb, a question or a statement in the questionnaire should not exceed 20 words, or exceed one full line in print.

Sequencing of Questions.   The sequence of questions in the questionnaire should be such that the respondent is led from questions of a general nature to those that are more specific, and from questions that are relatively easy to answer to those that are progressively more difficult. This *funnel* approach, as it is called, facilitates the easy and smooth progress of the respondent through the items in the questionnaire. The progression from general to specific questions might mean that the respondent is first asked questions of a global nature that pertain to the organization, and then is asked more incisive questions regarding the specific job, department, and the like. Easy questions might relate to issues that do not involve much thinking; the more difficult ones might call for more thought, judgment, and decision making in providing the answers.

In determining the sequence of questions, it is advisable not to place contiguously a positively worded and a negatively worded question tapping the same element or dimension of a concept. For instance, placing two questions such as the following, one immediately after the other, is not only awkward but might also seem insulting to the respondent.

1. *I have an opportunity to interact with my colleagues during work hours.*
2. *I have few opportunities to interact with my colleagues during work hours.*

First, there is no need to ask exactly the same question in a positive and a negative way. Second, if for some reason this is deemed necessary (e.g., to check the consistency of the responses), the two questions should be placed in different parts of the questionnaire, as far apart as possible.

The way questions are sequenced could also introduce certain biases, frequently referred to as the ordering effects. Though randomly placing the questions in the questionnaire would reduce any systematic biases in the response, it is very rarely done because of subsequent confusion while categorizing, coding, and analyzing the responses.

In sum, the language and wording of the questionnaire focus on such issues as the type and form of questions asked (i.e., open-ended and closed questions, and positively and negatively worded questions), as well as avoiding double-barreled questions, ambiguous questions, leading questions, loaded questions, questions prone to tapping socially desirable answers, and those soliciting distant recall. Questions should also not be unduly long. Using the funnel approach helps respondents to move through the questionnaire easily and comfortably.

Classification Data or Personal Information.   Classification data, also known as personal information or demographic questions, consist of such information as age, educational level, marital status, and income. Unless absolutely necessary, it is best not to ask for the name of the respondent. If, however, the questionnaire has to be identified with the respondents for any reason, then the questionnaire could be numbered and connected by the researcher to the respondent's name, in a separately maintained, private document. This procedure should be clearly explained to the respondent. The reason for using the numerical system in questionnaires is to ensure the anonymity of the respondent even if the questionnaires should fall into someone else's hands.

Whether questions seeking personal information should appear in the beginning or at the end of the questionnaire is a matter of choice for the researcher. Some people advocate asking for personal data at the end rather than the beginning of the questionnaire. Their reasoning may be that by the time the respondent reaches the end of the questionnaire the individual would have been convinced of the genuineness of the questions posed by the researcher, and hence be more open to sharing personal information. Researchers who prefer to elicit most of the personal information at the very beginning may feel that once respondents have said something about themselves at the very beginning, they may have psychologically identified themselves with the questionnaire, and feel more committed to respond. Thus, whether one asks this information in the beginning or at the end of the questionnaire is a matter of individual choice. However, questions regarding details of income, or other highly sensitive information—if such information is absolutely necessary—are best placed at the very end of the questionnaire.

It is also a wise policy to ask for information regarding age, income, and other sensitive personal questions by providing a range of response options, rather than seeking actual numbers. For example, the variables can be tapped as shown below:

EXAMPLE 5.1   Questionnaire Design

Personal Information

Age (years)	Annual Income
☐ Under 20	☐ Less than $20,000
☐ 20–30	☐ $20,000–30,000
☐ 31–40	☐ $30,001–40,000
☐ 41–50	☐ $40,001–50,000
☐ 51–60	☐ $50,001–70,000
☐ Over 60	☐ $70,001–90,000
	☐ Over $90,0000 ■

In organizational surveys, it is advisable to obtain certain demographic data such as age, sex, educational level, job level, department, and number of years in the organization, even if the theoretical framework does not necessitate or include these variables. Such data will help to describe the sample characteristics while writing the report after data analysis. However, when there are only a few respondents in a department, then asking information that might reveal their identity might be threatening to employees. For instance, if there is only one female in a department, she might not respond to the question on gender, because it might reveal the source of the data, and this apprehension is understandable.

To sum up, certain principles of wording need to be followed while designing a questionnaire. The questions asked must be appropriate for tapping the variable. The language and wording of the questionnaire should be at a level that is meaningful to the employees. The form and type of questions should be geared to minimizing respondent biases. The sequencing of the questions should facilitate the smooth progression of the responses through the questionnaire. The personal data should be gathered with sensitivity to the respondents' feelings, and with respect for privacy.

5.2.4 Cross-Cultural Research

Hitherto, we have discussed instrument development for eliciting responses from subjects within a country. With the globalization of business operations, managers often want to compare the business effectiveness of their subsidiaries in different countries. Researchers engaged in cross-cultural research also endeavor to trace the similarities and differences in the behavioral and attitudinal responses of employees at various levels in different cultures. Such data collected through questionnaires and occasionally through interviews should pay attention, in addition, to cultural differences in the use of certain terms, and to the measuring instruments that should also be tailored to the different cultures as discussed below.

Special Issues in Instrumentation for Cross-Cultural Research

Certain special issues need to be addressed while designing instruments for collecting data from different countries. Since different languages are spoken in different countries, it is important to ensure that the translation of the instrument to the local language is equivalent to the original language in which the instrument was developed. For this purpose, the instrument should be first translated by a local expert. If a comparative survey is to be done between Japan and the United States, and the researcher is a U.S. national, then the instrument must first be translated from English to Japanese. Then, another bilinguist should translate it back to English. This *back translation,* as it is called, ensures *vocabulary equivalence* (i.e., that the words used have the same meaning). *Idiomatic equivalence* could also become an issue where some idioms unique to one language do not lend themselves for translation to another language. *Conceptual equivalence,* where the meanings of certain words could differ in different cultures, is yet another important issue. As an example, the meaning of the concept "love" may differ in different cultures. All these can be taken care of through good back translation by persons who are proficient with the relevant languages and are also knowledgeable about the concerned cultures.

Issues in Data Collection

At least three issues are important for cross-cultural data collection—response equivalence, timing of data collection, and the status of the individual collecting the data. *Response equivalence* is ensured by adopting uniform data collection procedures in the different cultures. Identical methods of introducing the study, the researcher, task instructions, and closing remarks, in personally administered questionnaires, would provide equivalence in motivation, goal orientation, and response attitudes. *Timing of data collected* across cultures is also critical for cross-cultural comparison. Data collection should be completed within acceptable time frames in the different countries—say within three to four months. If too much time elapses in collecting data in the different countries, much might change during the time lag in either country or all the countries.

As pointed out as early as 1969 by Mitchell, in *interview surveys,* the egalitarian-oriented interviewing style used in the West may be inappropriate in societies that have well-defined status and authority structures. Also, when a foreigner comes to collect data, the responses might be biased for fear of portraying the country to a "foreigner" in an "adverse light" [Sekaran, U., "Methodological and theoretical issues and advancements in cross-cultural research," *Journal of International Business* (1983, Fall): 61–73]. The researcher has to be sensitive to these cultural nuances while doing cross-cultural research.

There are other concerns that need to be addressed in cross-cultural research that relate to scaling and sampling. We discussed in Chapter 4 scaling, which becomes a sensitive area in instrument development in cross-cultural research. For instance, a 5-point or a 7-point scale may make no difference in the United States, but could in the responses of subjects in other countries. [See Sekaran, U. & Martin, H. J., "An examination of the psychometric properties of some commonly researched individual differences, job and organizational variables in two cultures," *Journal of International Business Studies* (1982, Spring/Summer): 551–566; and Sekaran, U. & Trafton, R. S., "The dimensionality of jobs: Back to square one," *Twenty-fourth Midwest Academy of Management Proceedings* (1978): 249–262.] H. Barry ["Cross-cultural research with matched pairs of societies," *Journal of Social Psychology* 79 (1969): 25–33], for instance, found that in some countries, a 7-point scale is more sensitive than a 4-point scale in eliciting unbiased responses.

5.2.5 Principles of Measurement

Just as there are guidelines that have to be followed to ensure that the wording of the questionnaire is appropriate to minimize bias, so also are there some principles of measurement that are to be followed to ensure that the data collected are appropriate to test our hypotheses. These principles of measurement refer to the scales and scaling techniques used in measuring concepts, as well as the assessment of reliability and validity of the measures used, which were all discussed in Chapter 4.

As we have seen, appropriate scales have to be used depending on the type of data that need to be obtained. The different scaling mechanisms, which help us to anchor our scales appropriately, should be properly used. Whenever possible, the interval and ratio scales should be used in preference to nominal or ordinal scales. Once data are obtained, the "goodness of data" is assessed through tests of validity and reliability. Validity establishes how well a technique, instrument, or process measures a particular concept, and reliability indicates how stably and consistently the instrument taps the variable. Finally, the data have to be obtained in a manner that makes for easy categorization and coding, both of which are discussed later.

5.2.6 General Appearance or "Getup" of the Questionnaire

Not only is it important to address issues of wording and measurement in questionnaire design, it is also necessary to pay attention to how the questionnaire looks. An attractive and neat questionnaire with appropriate introduction, instructions, and well-arrayed set of questions and response alternatives will make it easier for the respondents to answer them. A good introduction, well-organized instructions, and neat alignment of the questions are all important. These elements are briefly discussed with examples.

Important Aspects of Questionnaires

A Good Introduction. A proper introduction that clearly discloses the identity of the researcher and conveys the purpose of the survey is absolutely necessary. It is also essential to establish some rapport with the respondents and motivate them to respond to the questionnaire willingly and enthusiastically. Assuring confidentiality of the information provided by them will ensure less biased answers. The introduction section should end on a courteous note thanking the respondent for taking the time to respond to the survey. The following is an example of an appropriate introduction.

EXAMPLE 5.2 General Appearance of Questionnaire

Introduction

Department of Management
Southern Illinois University at Carbondale
Carbondale, Illinois 62901

Date

Dear Participant,

This questionnaire is designed to study aspects of life at work. The information you provide will help us better understand the quality of our work life. Because *you* are the one who can give us a correct picture of how *you* experience your work life, I request you to respond to the questions frankly and honestly.

Your response will be kept *strictly confidential.* Only members of the research team will have access to the information you give. In order to ensure the utmost privacy, we have provided an identification number for each participant. This number will be used by us only for follow-up procedures. The numbers, names, or the completed questionnaires will not be made available to anyone other than the research team. A summary of the results will be mailed to you after the data are analyzed.

Thank you very much for your time and cooperation. I greatly appreciate your organization's and your help in furthering this research endeavor.

Cordially,
(Sd)
Anita Sigler, Ph.D.
Professor ■

Organizing Questions, Giving Instructions and Guidance, and Good Alignment. Organizing the questions logically and neatly in appropriate sections and providing instructions on how to complete the items in each section will help the respondents to answer them without difficulty. Questions should also be neatly and conveniently organized in such a way that the respondent can read and answer the questionnaire without eyestrain, and with a minimum amount of time and effort.

A specimen of the portion of a questionnaire incorporating the above points follows.

EXAMPLE 5.3   **General Appearance of Questionnaire**

Organizing and Giving Instructions

SECTION TWO: ABOUT WORK LIFE

The questions below ask about how you experience your work life. Think in terms of your everyday experiences and accomplishments on the job and put the most appropriate response number for you on the side of each item, using the scale below.

Strongly Disagree 1	Disagree 2	Slightly Disagree 3	Neutral 4	Slightly Agree 5	Agree 6	Strongly Agree 7

1. I do my work best when my job assignments are fairly difficult.
2. When I have a choice, I try to work in a group instead of by myself.
3. In my work assignments, I try to be my own boss.
4. I seek an active role in the leadership of a group.
5. I try very hard to improve on my past performance at work.
6. I pay a good deal of attention to the feelings of others at work.
7. I go my own way at work, regardless of the opinions of others.
8. I avoid trying to influence those around me to see things my way.
9. I take moderate risks, sticking my neck out to get ahead at work.
10. I prefer to do my own work, letting others do theirs.
11. I disregard rules and regulations that hamper my personal freedom.   ■

Personal Data. Demographic or personal data could be organized as in the example that follows. Note the ordinal scaling of the age variable.

EXAMPLE 5.4   General Appearance of Questionnaire

Organizing Personal Data

SECTION ONE: ABOUT YOURSELF

Please *circle* the numbers representing the most appropriate responses for you in respect of the following items.

1. Your Age (years)	2. Your Highest Completed Level of Education	3. Your Gender
1 Under 20	1 Elementary school	1 Female
2 20–35	2 High school	2 Male
3 36–50	3 College degree	
4 51–65	4 Graduate degree	
5 Over 65	5 Other (specify)	

4. Your Marital Status	5. Number of Preschool Children (under 5 Years of Age)	6. Age of the Eldest Child in Your Care (years)
1 Married	1 None	1 Under 5
2 Single	2 One	2 5–12
3 Widowed	3 Two	3 13–19
4 Divorced or separated	4 Three or more	4 Over 19
5 Other (specify)		5 Not applicable

7. Number of Years Worked in the Organizations	8. Number of Other Organizations Worked for Before Joining This Organization	9. Present Work Shift
1 Less than 1	1 None	1 First
2 1–2	2 One	2 Second
3 3–5	3 Two	3 Third
4 6–10	4 Three	
5 Over 10	5 Four or more	

10. Job Status

1 Top management
2 Middle management
3 First-level supervisor
4 Nonmanagerial

■

Information on Income and Other Sensitive Personal Data. Though demographic information can be asked either at the beginning or at the end of the questionnaire, information of a very private and personal nature such as income, state of health, and such, if considered absolutely necessary for the survey, should be asked at the end of the questionnaire rather than at the beginning. Also, such questions should be justified by explaining why this information might contribute to knowledge and problem solving, so that respondents do not perceive the questions to be of an intrusive or prying nature (see the next example). Shifting such questions to the end would help reduce respondent bias in case the individual gets irritated by the personal nature of the question.

EXAMPLE 5.5 General Appearance of Questionnaire

Obtaining Sensitive Personal Data

Because many people believe that income is a significant factor in explaining the type of career decisions individuals make, the following two questions are very important for this research. Like all other items in this questionnaire, the responses to these two questions will be kept confidential. Please circle the most appropriate number that describes your position.

Roughly, *my* total yearly income before taxes and other deductions is:	Roughly, the total yearly income before taxes and other deductions of my immediate *family*—including my own job income, income from other sources, and the income of my spouse—is:
1 Less than $36,000	1 Less than $36,000
2 $36,001–50,000	2 $36,001–50,000
3 $50,001–70,000	3 $50,001–70,000
4 $70,001–90,000	4 $70,001–90,000
5 Over $90,000	5 $90,001–120,000
	6 $120,001–150,000
	7 Over $150,000

■

Open-Ended Question at the End. The questionnaire could include at the end an open-ended question allowing respondents to comment on any aspect they choose. The questionnaire would end with sincere thanks to respondents. The last part of the questionnaire could look like the following.

EXAMPLE 5.6 General Appearance of Questionnaire

Asking Open-Ended Questions

The questions in the survey may not have allowed you to report some things you may want to say about your job, organization, or yourself. Please make any additional comments needed, in the space provided.

How did you feel about completing this questionnaire? Check the face in the following diagram that reflects your feelings.

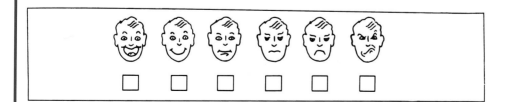

■

Concluding the Questionnaire. The questionnaire ends on a courteous note, reminding the respondent to check that all the items have been completed, as shown in the following example.

EXAMPLE 5.7   General Appearance of Questionnaire

Concluding the Questionnaire

> *I sincerely appreciate your time and cooperation. Please check to make sure that you have not skipped any questions, and then drop the questionnaire in the locked box, clearly marked for the purpose, at the entrance of your department.*   ■

Review of Questionnaire Design

We have devoted a lot of attention to questionnaire design because questionnaires are one of the most common methods of collecting data. The principles of questionnaire design relate to how the questions are worded and measured, and how the entire questionnaire is organized. To minimize respondent biases and measurement errors, all the principles discussed have to be followed carefully.

Questionnaires are most useful as a data collection method, especially when large numbers of people are to be reached in different geographical regions. Questionnaires are a popular method of collecting data because researchers can obtain data fairly easily, and the questionnaire responses are easily coded. When well-validated instruments are used, the findings of the study benefit the scientific community since the results can be replicated and additions to the theory base made.

There are several ways of administering questionnaires. Questionnaires can be personally administered to respondents, can be inserted in magazines, periodicals, or newspapers, mailed to respondents, or electronically distributed through the e-mail—via both the Internet and intranet. Software is also available to ask questions adapting to the subject's response to the preceding question. Companies' web sites can also elicit survey responses, as for example reactions to customer service, product utility, and the like. Global research is now vastly facilitated by the electronic system.

Pretesting of Structured Questions

Whether it is a structured interview where the questions are posed to the respondent in a predetermined order, or a questionnaire that is used in a survey, it is important to *pretest* the instrument to ensure that the questions posed are understood by the respondents (i.e., there is no ambiguity in the questions) and that there are no problems with the wording or measurement. Pretesting involves the use of a small number of respondents to test the appropriateness of the questions and their understandability. Such pretesting helps to rectify any inadequacies, in time, before administering the instrument orally or through a questionnaire to a large number of respondents, and thus reduce biases.

It would be wise to debrief the results of the pretest and obtain additional information from the small group of participants (who would serve the role of a focus group) on their general reactions to the questionnaire and how they felt about completing the instrument.

5.2.7 Electronic Questionnaire Design and Surveys

On-line questionnaire surveys are easily designed and administered when microcomputers are hooked up to computer networks. Data disks can also be mailed to respondents, who can use their own personal computers for responding to the questions. These can, of course, be helpful only when the respondents know how to use the computer and feel comfortable responding in this manner.

Several programs have been developed to administer questionnaires electronically. Disks are inexpensive, so mailing them across the country is not a problem. The PC medium nonresponse rates may not be any higher than the mail questionnaire re-

TABLE 5.1 Advantages and Disadvantages of Interviews and Questionnaires

Mode of Data Collection	Advantages	Disadvantages
Personal or Face-to-Face Interviews	Can establish rapport and motivate respondents. Can clarify the questions, clear doubts, add new questions. Can read nonverbal cues. Can use visual aids to clarify points. Rich data can be obtained. CAPI can be used and responses entered in a portable computer.	Takes personal time. Costs more when a wide geographic region is covered. Respondents may be concerned about confidentiality of information given. Interviewers need to be trained. Can introduce interviewer biases. Respondents can terminate the interview at any time.
Telephone Interviews	Less costly and speedier than personal interviews. Can reach a wide geographic area. Greater anonymity than personal interviews. Can be done using CATI.	Nonverbal cues cannot be read. Interviews will have to be kept short. Obsolete telephone numbers could be contacted, and unlisted ones omitted from the sample.
Personally Administered Questionnaire	Can establish rapport and motivate respondent. Doubts can be clarified. Less expensive when administered to groups of respondents. Almost 100% response rate assured. Anonymity of respondent is high.	Organizations may be reluctant to give up company time for the survey with groups of employees assembled for the purpose.
Mail Questionnaires	Anonymity is high. Wide geographic regions can be reached. Token gifts can be enclosed to seek compliance. Respondent can take more time to respond at convenience. Can be administered electronically, if desired.	Response rate is almost always low. A 30% rate is quite acceptable. Cannot clarify questions. Follow-up procedures for nonresponses are necessary.
Electronic Questionnaires	Easy to administer. Can reach globally. Very inexpensive. Fast delivery. Respondents can answer at their convenience like the mail questionnaire.	Computer literacy is a must. Respondents must have access to the facility. Respondent must be willing to complete the survey.

sponse rates. With increasing computer literacy, we can expect electronic questionnaire administration to play an increasing part in the future.

The advantages and disadvantages of personal or face-to-face interviews, telephone interviews, personally administered questionnaires, mail questionnaires, and questionnaires distributed through the electronic system are tabulated in Table 5.1.

It should be pointed out that information obtained from respondents either through interviews or questionnaires, being self-report data, could be biased. That is the reason why data should be collected from different sources and by different methods as discussed later.

5.2.8 Other Methods of Data Collection

Observational Surveys

Whereas interviews and questionnaires elicit responses from the subjects, it is possible to gather data without asking questions of respondents. People can be observed in their natural work environment or in the lab setting, and their activities and behaviors or other items of interest can be recorded.

Apart from the activities performed by the individuals under study, their movements, work habits, the statements made and meetings conducted by them, their facial expressions of joy, anger, and other emotions, and body language can be observed. Other environmental factors such as layout, work-flow patterns, how close the seating arrangement is, and the like, can also be noted. Children can be observed as to their interests and attention span with various stimuli, such as their involvement with different toys. Such observation would help toy manufacturers, child educators, day-care administrators, and others responsible for children's development, to design and model ideas based on children's interests, which are more easily observed than traced in any other manner.

The researcher can play one of two roles while gathering field observational data—that of a nonparticipant-observer or participant-observer.

Nonparticipant-Observer.   The researcher can collect the needed data in the role of a researcher without trying to become an integral part of the organizational system. For example, the researcher might sit in the corner of an office and observe and record how the manager spends her time. Observation of all the activities of managers, over a period of several days, will allow the researcher to make some generalizations on how managers typically spend their time. By merely observing the activities, recording them in a systematic way, and tabulating them, the researcher comes up with some findings. For this purpose, observers must be physically present at the workplace for extended periods of time, thus making observational studies time-consuming.

Participant-Observer.   The researcher can also play the role of the participant-observer. Here, the researcher enters the organization or the research setting, actually becoming a part of the work team. For instance, if a researcher wants to study group dynamics in work organizations, then she may enter the organization in the role of an employee and observe the dynamics in groups while being a part of the work organization and work groups. Much anthropological research is conducted in this manner, where researchers become a part of the alien culture, which they are interested in studying in depth.

Structured versus Unstructured Observational Studies

Structured Observational Studies.   As we have seen, observational studies could be of either the nonparticipant-observer or the participant-observer type. Both of these, again, could be either structured or unstructured. Where the observer has a predetermined set of categories of activities or phenomena planned to be studied, it is a *structured observational study*. Formats for recording the observations can be specifically designed and tailored to each study to suit the goal of the research.

Usually, such matters that pertain to the feature of interest, like the duration and frequency of the event, and also certain activities that precede and follow the feature of interest, are recorded. Environmental conditions, and any changes in setting, are also noted, if they are considered relevant. Task-relevant behaviors of the actors, their perceived emotions, verbal and nonverbal communication and such, are recorded. Observations that are recorded in work sheets or field notes are then systematically analyzed, with minimal personal inferences made by the investigator. Categories can then be developed for further analysis.

Unstructured Observational Studies.   At the beginning of a study it is possible that the observer has no definite ideas of the particular aspects that need focus. Observing events that are happening may also be a part of the plan as in many qualitative studies. In such cases, the observer will record almost everything that is observed. Such a study will be an *unstructured observational study.*

Unstructured observational studies are claimed to be the hallmark of qualitative research. The investigator might entertain a set of tentative hypotheses that might serve as a guide as to who, when, where, and how the individual will observe. Once the information needed is observed and recorded over a period of time, patterns can be traced, and inductive discovery can then pave the way for subsequent theory building and hypotheses testing.

Advantages and Disadvantages of Observational Studies

There are some specific advantages and disadvantages to gathering data through observation as listed below.

Advantages of Observational Studies. The following are among the advantages of observational studies:

1. The data obtained through observation of events as they normally occur are generally more reliable and free from respondent bias.

2. In observational studies, it is easier to note the effects of environmental influences on specific outcomes. For example, the weather (hot, cold, rainy), the day of the week (midweek as opposed to Monday or Friday), and such factors that might have a bearing on, say, the sales of a product, traffic patterns, absenteeism, and the like, can be noted and meaningful patterns might emerge from these types of data.

3. It is easier to observe certain groups of individuals (e.g., very young children and extremely busy executives) from whom it may be otherwise difficult to obtain information.

Advantages of observational studies

The preceding three advantages are perhaps unique to observational studies.

Disadvantages of Observational Studies. The following drawbacks of observational studies have also to be noted.

1. It is necessary for the observer to be *physically* present (unless a camera or another mechanical system can capture the events of interest), often for prolonged periods of time.

2. This method of collecting data is not only slow, but also tedious and expensive.

3. Because of the long periods for which subjects are observed, observer fatigue could easily set in, which might bias the recorded data.

4. Although moods, feelings, and attitudes can be guessed by observing facial expressions and other nonverbal behaviors, the cognitive thought processes of individuals cannot be captured.

5. Observers have to be trained in what and how to observe, and ways to avoid observer bias.

Disadvantages of observational studies

Biases in Observational Studies

Data observed from the researcher's point of view are likely to be prone to observer biases. There could be recording errors, memory lapses, and errors in interpreting activities, behaviors, events, and nonverbal cues. Moreover, where several observers are involved, interobserver reliability has to be established before the data can be accepted. Observation of the happenings day in and day out, over extended periods of time, could fatigue or bore the observers and introduce biases in the recording of the observations. To minimize observer bias, observers are usually trained on how to observe and what to record. Good observational studies would also establish interobserver reliability. This could also be established during the training of the observers when videotaped stimuli could be used to determine interobserver reliability. A simple formula can be used for the purpose—dividing the number of agreements among the trainees by the number of agreements and disagreements—thus establishing the reliability coefficient.

Respondent bias could also be a threat to the validity of the results of observational studies, because those who are observed may behave differently during the period the study is conducted, especially if the observations are done for a short period of time. However, in studies of longer duration, as the study progresses, the employees become more relaxed and tend to behave normally. For these reasons, researchers doing observational studies discount the data recorded in the initial few days if they seem to be quite different from what is observed later.

Summary of Observational Studies

Observational studies have a formulated research purpose and are systematically planned. Such studies can be structured or unstructured, with the investigator being a participant or a nonparticipant in the study setting. All phenomena of interest are systematically recorded, and quality control can be exercised by eliminating biases. Observational studies can provide rich data and insights into the nature of the phenomena observed. Such studies have offered much understanding of interpersonal and group dynamics. Interestingly, observational data can be quantified.

Data Collection Through Mechanical Observation

There are situations where machines can provide data by recording the events of interest as they occur, without a researcher being physically present. Nielsen ratings are an oft-cited example in this regard. Other examples include collection of details of products sold by types or brands tracked through optical scanners and bar codes at the checkout stand, and tracking systems keeping a record of how many individuals utilize a facility or visit a web site. Films and electronic recording devices such as video cameras can also be used to record data. Such mechanically observed data are error-free.

Projective Methods

Certain ideas and thoughts that cannot be easily verbalized or that remain at unconscious levels in the respondents' minds can usually be brought to the surface through motivational research. This is typically done by trained professionals who apply different probing techniques in order to uncover deep-rooted ideas and thoughts in the respondents. Familiar techniques for gathering such data are word associations, sentence completion, thematic apperception tests (TAT), inkblot tests, and the like.

Word-association techniques, such as asking the respondent to quickly associate a word—say, *work*—with the first thing that comes to mind, are often used to uncover the true attitudes and feelings. The reply would be an indication of what work means to the individual. Similarly, **sentence completion** would have the respondent quickly complete a sentence, such as "Work is —." One respondent might say, "Work is a lot of fun," whereas another might say, "Work is drudgery." These responses may provide some insights into individuals' feelings and attitudes toward work.

Thematic Apperception Tests (TAT) ask the respondent to develop a story around a picture that is shown. Several need patterns and personality characteristics in employees could be traced through these tests. **Inkblot tests,** another form of motivational research, use colored inkblots that are interpreted by the respondents, who explain what they see in the various patterns and colors.

Although these types of projective tests are useful for tapping attitudes and feelings that are difficult to obtain otherwise, they cannot be engaged in by researchers who are not trained to conduct motivational research.

Consumer preferences, buying attitudes and behaviors, product development, and other marketing research strategies make substantial use of in-depth probing. TAT and inkblot tests are on their way out in marketing research since advertisers and others are now utilizing the sentence completion tests and word association tests more frequently. Sketch drawings, collages from magazine pictures, filling in the balloon captions of cartoon characters, and other strategies are also being followed to see how individuals associate different products, brands, advertisements, and so on, in their minds. Agencies frequently ask subjects to sketch "typical" users of various brands and narrate stories about them. The messages conveyed through the unsophisticated drawings are said to be very powerful, helping development of different marketing strategies.

The idea behind motivational research is that "emotionality" ("I identify with it" feeling) rather than "rationality" ("it is good for me" thought)—which is what keeps a product or practice alive—is captured. Emotionality is a powerful motivator of actions and knowledge of what motivates individuals to act is very useful. The failure of attempts to trade in the "New Coke" for "Classic Coke" is an oft-cited example of the emotional aspect. Emotionality is clearly at the nonrational, subconscious level, lending itself to capture by projective techniques alone.

Secondary Data

Secondary data are indispensable for much of organizational research. As discussed in Chapter 2, secondary data refer to information gathered by someone other than the researcher conducting the current study. Such data can be internal or external to the organization and can be accessed through the computer or by going through recorded or published information.

There are several sources of external data, including books and periodicals, government publications of economic indicators, census data, statistical abstracts, databases available, the media, annual reports of companies, and so on. Much internal data could be proprietary and not accessible to all.

Secondary data can be used for such things as forecasting sales by constructing models based on past sales figures, and through extrapolation. Financial databases are also available for research. The Compustat Database contains information on thousands of companies organized by industry, and information on global companies is also available through Compustat.

The advantage of secondary data is savings in time and costs of acquiring information. However, secondary data as the sole source of information has the drawbacks of becoming obsolete and not meeting the specific needs of the particular situation or setting.

5.3 SOME SPECIAL DATA SOURCES

Focus groups, secondary data sources including case studies, and other archival sources such as diaries, desk calendars, schedules, speeches, and the like are good sources of data. Focus groups, which are very useful and frequently used, are now discussed.

5.3.1 Focus Groups

Focus groups are relatively inexpensive and can provide fairly dependable data within a short time frame. They typically consist of 8 to 10 members randomly chosen, with a moderator leading the discussions on a particular topic, item, or product for about two hours. The focus sessions are aimed at obtaining respondents' impressions, interpretations, and opinions as the members talk about the product or service. The moderator plays a vital role in steering the discussions and keeping them on-track.

Focus groups are used for exploratory studies, making generalizations based on the information generated by them, or for purposes of sample survey. Focus groups have been credited with illuminating investigators as to why certain products are not doing well, why certain advertising strategies are effective, why specific management techniques do not work, and so on.

5.3.2 Other Secondary Sources of Information

We have discussed secondary sources of data in some detail earlier. Case studies and other archival records can provide a lot of information for research and problem solving. Such data are mostly qualitative in nature. Included in the secondary sources are the schedules maintained for or by key personnel, the desk calendar of executives, and the speeches delivered by them.

Trace measures or **unobtrusive measures** as they are also called, come from a source that does not involve people. One such is the wear and tear of journals in a university library which offers a good indication of their popularity, their use, or both. The number of different brands of soft drink cans found in trashbags can also provide a measure of their consumption levels. Signatures on checks exposed to ultraviolet rays could indicate the extent of forgery and frauds; actuarial records are good sources for collecting data on the births, marriages, and deaths in a community; and company records disclose a lot of personal information about employees, the level of company efficiency, and other data as well. Thus, these unobtrusive sources of data and their use are also important in research.

5.3.3 Panels

Panels, like focus groups, are another source of information for research purposes. Whereas focus groups meet for a one-time group session, panels (of members) meet more than once. In cases where the effects of certain interventions or changes are to be studied over a period of time, panel studies are very useful. Several individuals are randomly chosen to serve as panel members for a research study. For instance, if the effects of a proposed advertisement for a certain brand of coffee are to be assessed quickly, the panel members can be exposed to the advertisement and their intentions of purchasing that brand assessed. This can be taken as the response that could be expected of consumers if, in fact, they were exposed to the advertisement. A few months later, the product manager might think of changing the flavor of the same product and might explore its effects on this panel. Thus, a continuing set of "experts" serve as the sample base or the sounding board for assessing the effects of change. Such members are called a panel, and research that uses this panel is called a panel study.

The Nielsen Television Index is based on the television viewing patterns of a panel. The index is designed to provide estimates of the size and nature of the audience for individual television programs. The data are gathered through Audimeter instruments attached to the television sets in approximately 1200 cooperating households. The Audimeters are connected to a central computer, which records when the set is turned on, and what channel is tuned. From these data, Nielsen develops estimates of the number and percentage of all TV households viewing a given TV show.

Other panels used in marketing research include the National Purchase Diary Panel, the National Family Opinion Panel, and the Consumer Mail Panel.

Static and Dynamic Panels

Panels can be either *static* (i.e., the same members serve on the panel over extended periods of time) or *dynamic* (i.e., the panel members change from time to time as various phases of the study are in progress). The main advantage of the static panel is that it offers a good and sensitive measurement of the changes that take place between two points in time—a much better alternative than using two different groups at two different times. The disadvantage, however, is that the panel members could become so sensitive to the changes as a result of the continuous interviews that their opinions might no longer be representative of what the others in the population might hold. Members could also drop out of the panel from time to time for various reasons, thus raising issues of bias due to mortality. The advantages and disadvantages of the dynamic panel are the reverse of the ones discussed for the static panel.

We have thus far discussed different methods of collecting data from various sources. Each method has its advantages and disadvantages, as we have seen. We will now see how we can try to obtain data "scientifically."

5.4 MULTIMETHODS OF DATA COLLECTION

Because almost all data collection methods have some biases associated with them, collecting data through multimethods and from multiple sources lends rigor to research. For instance, if the responses collected through interviews, questionnaires, and observation are strongly correlated with one another, then we will have more confidence about the goodness of the collected data. If there are discrepancies in how respondents answer the same question when interviewed, as opposed to how they answer the question in a questionnaire, then we would become suspect and be inclined to discard both data as being biased.

Similarly, if data obtained from several sources are highly similar, we would have more conviction in the goodness of the data. For example, if an employee rates his performance as 4 on a 5-point scale, and his supervisor rates him the same way, we may be inclined to think that he is perhaps a better than average worker. In contrast, if he gives himself a 5 on the 5-point scale and his supervisor gives him a rating of 2, then we will not know to what extent there is a bias and from which source. Therefore, high correlations among data obtained on the same variable from different sources and through different data collection methods lend more credibility to the research instrument and to the data obtained through these instruments. Good research entails collection of data from multiple sources and through multiple data collection methods. Such research, though, would be costly and time-consuming.

5.4.1 Review of the Advantages and Disadvantages of Different Data Collection Methods and When to Use Each

Having discussed the various data collection methods, we will now briefly recount the advantages and disadvantages of the three most commonly used data collection methods—interviews, questionnaires, and observation—and see when each method can be most profitably used.

- **Face-to-face interviews** provide rich data, offer the opportunity to establish rapport with the interviewees, and help to explore and understand complex issues.

Many ideas that are ordinarily difficult to articulate can also be uncovered and discussed during such interviews. On the negative side, face-to-face interviews have the potential for introducing interviewer bias and can be expensive if a large number of subjects are to be personally interviewed. Where several interviewers are involved, adequate training becomes a necessary first step.

Face-to-face interviews are best used at the exploratory stages of research when the researcher tries to get a handle on concepts or the situational factors.

- **Telephone interviews** help to contact subjects dispersed over various geographic regions and obtain responses from them immediately on contact. This is an efficient way of collecting data when one has specific questions to ask, needs quick responses, and has the sample spread over a wide geographic area. On the negative side, the interviewer cannot observe the nonverbal responses of the respondents, and the interviewee can block a call.

  Telephone interviews are best suited for asking structured questions where responses need to be obtained quickly from a sample that is geographically spread.

- **Personally administering questionnaires** to groups of individuals helps to: (a) establish rapport with the respondents while introducing the survey, (b) provide clarifications sought by the respondents on the spot, and (c) collect the questionnaires immediately after they are completed. In that sense, there is a 100% response rate. On the negative side, administering questionnaires personally is expensive, especially if the sample is geographically dispersed.

  Personally administered questionnaires are best suited when data are collected from organizations that are located in close proximity to one another and groups of respondents can be conveniently assembled in the company's conference (or other) rooms.

- **Mail questionnaires** are advantageous when responses to many questions have to be obtained from a sample that is geographically dispersed, or it is difficult or not possible to conduct telephone interviews to obtain the same data without much expense. On the negative side, mailed questionnaires usually have a low response rate and one cannot be sure if the data obtained are biased because the nonrespondents may be different from those who did respond.

  The mailed questionnaire survey is best suited (and perhaps the only alternative open to the researcher) when a substantial amount of information is to be obtained through structured questions, at a reasonable cost, from a sample that is widely dispersed geographically.

- **Observational studies** help to comprehend complex issues through direct observation (either as a participant- or a nonparticipant-observer) and then, if possible, asking questions to seek clarifications on certain issues. The data obtained are rich and uncontaminated by self-report biases. On the negative side, they are expensive, since long periods of observation (usually encompassing several weeks or even months) are required, and observer bias may well be present in the data.

  Because of the costs involved, very few observational studies are done in business. Observational studies are best suited for research requiring non-self-report descriptive data—that is, when behaviors are to be understood without directly asking the respondents themselves. Observational studies can also capture "in-the-stores buying behaviors."

5.5 MANAGERIAL ADVANTAGE

As a manager you will perhaps engage consultants to do research and may not be collecting data yourself through interviews, questionnaires, or observation. However, during those instances, when you will perforce have to obtain work-related information through interviews with clients, employees, or others, you will know how to phrase unbiased questions to elicit the right types of useful responses. Moreover, you, as the sponsor of research, will be able to decide at what level of sophistication you want data to be collected, based on the complexity and gravity of the situation. Moreover, as a constant participant observer of all that goes on around you at the workplace, you will be able to understand the dynamics operating in the situation. Also, as a manager, you will be able to differentiate between good and bad questions used in surveys.

CHAPTER 5 SUMMARY

In this chapter, we examined various data collection methods and different primary sources of data. We discussed the advantages and disadvantages as well as the biases inherent in each data collection method. We also examined the impact of personal computers in data collection. Because of the inherent biases in each of the data collection methods, the collection of data from multiple sources and through multiple methods was recommended. The final decision would, of course, be governed by considerations of cost, and the degree of rigor that the given research goal would require.

Key Terms

Term	Definition	Page reference
Bias	Any error that creeps into the data. Biases can be introduced by the researcher, the respondent, the measuring instrument, the sample, and such.	147
Closed questions	Questions with a clearly delineated set of alternatives that confine the respondents' choice to one of them.	156
Computer-assisted telephone interviews (CATI)	Interviews in which questions are prompted onto a PC monitor that is networked into the telephone system, to which respondents provide their answers.	152
Cross-cultural research	Studies done across two or more cultures to understand, describe, analyze, or predict phenomena.	160
Double-barreled questions	Refers to the improper framing of a question that should be posed as two or more separate questions, so that the respondent can give clear and unambiguous answers.	157
Dynamic panel	Consists of a changing composition of members in a group who serve as the sample subjects for a research study conducted over an extended period of time.	173
Electronic questionnaire	On-line questionnaire administered when the microcomputer is hooked up to computer networks.	166
Face-to-face interview	Information gathering when both the interviewer and interviewee meet in person.	150
Focus group	A group consisting of 8 to 10 randomly chosen members who discuss a product or any given topic for about two hours with a moderator present, so that their opinions can serve as the basis for further research.	172
Funneling technique	The questioning technique that consists of initially asking general and broad questions, and gradually narrowing the focus thereafter on more specific themes.	149
Inkblot tests	A motivational research technique that uses colored patterns of inkblots to be interpreted by the subjects.	171

Term	Definition	Page reference
Interviewing	A data collection method in which the researcher asks for information verbally from the respondents.	144
Leading questions	Questions phrased in such a manner as to lead the respondent to give the answers that the researcher would like to obtain.	157
Loaded questions	Questions that would elicit highly biased emotional responses from subjects.	158
Nonparticipant-observer	A researcher who collects observational data without becoming an integral part of the system.	168
Observational survey	Collection of data by observing people or events in the work environment and recording the information.	167
Open-ended questions	Questions that the respondent can answer in a free-flowing format without restricting the range of choices to a set of specific alternatives suggested by the researcher.	156
Panel studies	Studies conducted over a period of time to determine the effects of certain changes made in a situation, using a panel or group of subjects as the sample base.	172
Participant-observer	A researcher who collects data by becoming a member of the system from which data are collected.	168
Pretesting survey questions	Test of the understandability and appropriateness of the questions planned to be included in a regular survey, using a small number of respondents.	166
Questionnaire	A preformulated written set of questions to which the respondent records the answers, usually within rather closely delineated alternatives.	153
Recall-dependent questions	Questions that elicit from the respondents information that involves recall of experiences from the past that may be hazy in their memory.	157
Social desirability	The respondents' need to give socially or culturally acceptable responses to the questions posed by the researcher even if they are not true.	158
Static panel	A panel that consists of the same group of people serving as subjects over an extended period of time for a research study.	173
Structured interviews	Interviews conducted by the researcher with a predetermined list of questions to be asked of the interviewee.	146
Structured observational studies	Studies in which the researcher observes and notes down specific activities and behavior that have been clearly delineated as important factors for observation, before the commencement of the study.	168
Telephone interview	The information-gathering method by which the interviewer asks the interviewee *over the telephone,* rather than face to face, for information needed for the research.	151
Thematic Apperception Test (TAT)	A projective test that requires the respondent to develop a story around a picture.	171
Unbiased questions	Questions posed in accordance with the principles of wording and measurement, and the right questioning technique, so as to elicit the least biased responses.	149
Unobtrusive measures	Measurement of variables, through data gathered from sources other than people, such as examining birth and death records or counting the number of cigarette butts in an ashtray.	172
Unstructured interviews	Interviews conducted with the primary purpose of identifying some important issues relevant to the problem situation, without prior preparation of a planned or predetermined sequence of questions.	144
Word association	A projective method of identifying respondents' attitudes and feelings by asking them to associate a specified word with the first thing that comes to mind.	170

CHAPTER 5 EXERCISES

Thinking About It!

5.1 As a manager, you have invited a research team to come in, study, and offer suggestions on how to improve the performance of your staff. What steps would you take to allay their apprehensions even before the research team sets foot in your department?

5.2 What is bias, and how can it be reduced during interviews?

5.3 Explain the principles of wording, stating how these are important in questionnaire design, citing examples not in the book.

5.4 What are projective techniques and how can they be profitably used?

5.5 Describe the different data sources, explaining their usefulness and disadvantages.

5.6 How are multiple methods of data collection and from multiple sources related to the reliability and validity of the measures?

5.7 "Every data collection method has its own built-in biases. Therefore, resorting to multi-methods of data collection is only going to compound the biases." How would you critique this statement?

5.8 "One way to deal with discrepancies found in the data obtained from multiple sources is to average the figures and take the mean as the value of the variable." What is your reaction to this?

5.9 How has the advancement in technology helped data gathering?

5.10 How will you use the data from observational study to reach scientific conclusions?

5.11 "The fewer the biases in measurement and in the data collection procedures, the more scientific the research." Comment on this statement.

Doing It!

5.12 A production manager wants to assess the reactions of the blue-collar workers in his department (including foremen) to the introduction of computer-integrated manufacturing (CIM) systems. He is particularly interested in knowing how they would perceive the effects of CIM on:

(a) their future jobs
(b) additional training that they will have to receive
(c) future job advancement

Design a questionnaire for the production manager.

5.13 Seek permission from a professor to sit in on two sessions of his or her class, and do an unstructured, nonparticipant-observer study. Give your conclusions on the data, and include in the short report your observation sheets and tabulations.

5.14 First conduct an unstructured, and later a structured interview, with any professor not known to you, to learn about his or her values and strategy in teaching courses. Write up the results, and include the formats you used for both stages of the research.

5.15 The president of Serakan Co. suspects that most of the 500 male and female employees of the organization are somewhat alienated from work. He is also of the view that those who are more involved (less alienated) are also the ones who experience greater satisfaction with their work lives.

Design a questionnaire the president could use to test his hypothesis.

NUMERICAL DESCRIPTORS
OF DATA

THE GOLF BALL COMPANY

A company that manufactures golf balls is preparing for an advertising campaign. The company wants to compare the ball that it manufactures to a competitor's product to see how they differ. In preparation for the study, sample data are collected on a number of different variables for the two types of balls and you are asked to summarize the data and report on what you find. A portion of the data is shown below:

Ball number	Model number	S1	S2	S3	Weight (g)	Dimple width (mm)	Dimple depth (mm)	Head	Temper- ature (°F)	Carry (yd)	Total distance (yd)	Date	Time
1	M1	81	81	82	45.3	0.1450	0.0110	686	77	257	270	8/20	8:15
2	M1	83	83	84	45.2	0.1510	0.0111	688	77	255	267	8/20	8:15
3	M1	81	82	84	45.2	0.1450	0.0105	687	77	256	267	8/20	8:15
4	M1	81	81	83	45.3	0.1440	0.0117	688	77	255	271	8/20	8:15
5	M1	83	81	82	45.5	0.1460	0.0108	687	77	255	268	8/20	8:15
6	M1	83	83	82	45.3	0.1560	0.0111	687	77	256	267	8/20	8:15
7	M1	81	81	82	45.2	0.1495	0.0111	687	77	255	264	8/20	8:15
8	M1	83	81	82	45.1	0.1505	0.0110	690	78	258	269	8/20	8:15

The company wants to know things like how far most of the balls go when they are hit, how much variation there is in the distance, and what percentage of the balls go beyond a certain distance. You need to figure out which numerical measures will provide the company with the most information.

6.1   CHAPTER OBJECTIVES

Remember that when you looked at different graphical techniques for displaying numerical data you were interested in three characteristics: center, spread or dispersion, and shape. In this chapter we look at different numerical measures that can be used to describe the same features of the data. The chapter covers the following material:

- Numerical measures of center: the mean, the median, and the mode
- Numerical measures of variability: the range and the standard deviation
- Describing a set of data: the empirical rule and boxplots
- Descriptive statistics for grouped data
- Measures of relative standing: percentiles and percentile rank
- Identifying outliers: z-scores and boxplots

6.2   DESCRIBING DATA NUMERICALLY

Although we may say that a picture is worth a thousand words, it is useful to find numerical quantities to describe the data as well. You may wonder why you need to do this, when the graphs let you *see* the data. There are two reasons. First, while you can certainly see the data using histograms and bar charts, it is difficult to *talk* or *write* about pictures. We often need other references to describe the data. A second reason for using numerical descriptors is that we may want to make inferences based on the sample data. To make statistical inferences you need to use *numerical measures*.

When you collect data you may have either a *population* or a *sample* from the population. Numerical measures calculated from the data are known as either **statistics** or **parameters.**

> A *statistic* is a numerical descriptor that is calculated from sample data and is used to describe the sample. Statistics are usually represented by Roman letters.

> A *parameter* is a numerical descriptor that is used to describe a population. Parameters are usually represented by Greek letters.

Most of the time in statistics you will be working with sample data, but you may sometimes have the entire population available for study. We usually use Greek letters to denote parameters and Roman letters to describe statistics.

When you look at numerical descriptors for a set of data, you want to describe the same properties of the data that you described from the graphical displays. You will find, however, that there are several different statistics that you can use to describe each property and that the choice of the statistic is dependent on the problem you are trying to solve.

6.3   MEASURES OF CENTRAL TENDENCY

The golf ball company would like to know about values that represent a "typical" golf ball. The company would like to measure the *center* of its data. We will look at three different statistics that measure central tendency: the *sample mean,* the *median,* and the *mode.*

6.3.1 The Arithmetic Mean

You are probably already familiar with the most common measure of center, the **sample mean.** The mean, or average, as it is commonly known, is calculated by adding all of the data values in the sample and then dividing by the number of values. The symbol for the sample mean is $\overline{X}$ (this is read as "X bar").

$$\overline{X} = \frac{\text{Sum of all the values in the sample}}{\text{Total number of observations}}$$

> The **sample mean** is the center of balance of a set of data, and is found by adding up all of the data values and dividing by the number of observations.

The population parameter that corresponds to the sample mean is the **population mean,** μ (mu).

> The **population mean** is represented by the Greek letter μ (mu).

Using the Σ notation that you saw in Chapter 2, we can write the formula for the sample mean as:

$$\overline{X} = \frac{\sum_{i=1}^{n} x_i}{n} \quad \text{or} \quad \frac{\Sigma x}{n}$$

SAMPLE MEAN

When we talk about a variable or a statistic in general we use *capital* letters, such as X or $\overline{X}$. When we talk about a *specific value* of a variable we use *lowercase* letters such as *x*. For example, we write, "The sample mean, $\overline{X}$, is calculated by . . . ," or "The third value in the sample is $x = 27.2$."

Most of the time in statistics, it is understood that the sum is over the entire sample and so we can leave out the index on the summation sign.

EXAMPLE 6.1   The Mail-Order Company

Calculating the Sample Mean

A mail-order company wants some information about the daily demand for a product that has been heavily advertised. The company wants a measure of what it might typically expect the demand to be. The company looks at the number of orders for a 10-day period and obtains the following data:

Demand	29	28	29	31	30	31	27	29	30	32

Since it is interested in a *typical* value for the demand, the company decides to calculate the sample mean, $\overline{X}$:

$$\overline{X} = \frac{29 + 28 + 29 + 31 + 30 + 31 + 27 + 29 + 30 + 32}{10} = \frac{296}{10} = 29.6 \text{ orders} \quad \blacksquare$$

What Does the Sample Mean Really Measure?

You can think of the sample mean as the *balance point* of the data. The value of $\overline{X}$ balances the higher values against the lower ones. It is easiest to see this when you look at the mean on a dotplot of the data. In Figure 6.1 (page 182) you can see that the mean, 29.6, is in the center of the data and that the data are fairly evenly spread out on both sides of the mean.

Suppose that the data on the right (high) side are more spread out than those on the left. What will happen to the value of the sample mean? Remember, we said

You want to use a dotplot for these data because the sample size is so small.

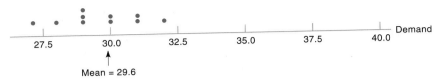

FIGURE 6.1   Dotplot of the data for the mail order company

that the sample mean is the balance point of the data values. When there are a few data points on one side that are far from the bulk of the data (the bump), the sample mean moves toward them in order to maintain balance with the data on the other side. The next example shows what happens.

EXAMPLE 6.2   The Mail-Order Company

The Sample Mean as a Balance Point

Suppose that when the mail-order company looks at its data it finds the following:

Demand	40	28	29	31	30	31	27	29	39	36

Most of the data values are still around 30, but three of them are quite a bit higher. If you calculate the sample mean from the data,

$$\overline{X} = \frac{40 + 28 + 29 + 31 + 30 + 31 + 27 + 29 + 39 + 36}{10} = \frac{320}{10} = 32.0 \text{ items}$$

you see that the sample mean has changed from 29.6 to 32.0. This is not really where the bulk of the data are located. You can see that although the bulk of the data are still located around 30, the value of the sample mean has changed to 32.0 to balance the three data values at the high end.

Mean = 32.0

■

📖 **TRY IT NOW!**

Restaurant Table Times

Calculating the Sample Mean

A restaurant is trying to decide whether it has an adequate number of tables available. The restaurant owner would like some information on the amount of time a table is occupied by a customer. She collects data on the length of time a customer occupies a table for a random sample of 10 customers:

Customer	1	2	3	4	5	6	7	8	9	10
Time (min)	59.3	58.6	62.7	65.4	59.0	67.3	62.8	68.1	59.4	63.7

Calculate the sample mean for the length of time a table is occupied.

EXAMPLE 6.3   The Golf Ball Problem

The Sample Mean

The company that manufactures golf balls is interested in describing the way the two different golf balls behave so that an advertising campaign can be planned. One way to describe the distance that the balls travel is to use the variable *Carry,* which measures the distance (yd) from the point where the ball was hit to the point where it hit the ground. Since the company is interested in *comparing* the two different designs, it will want to look at the designs separately. Using a computer package, you can calculate a set of descriptive statistics for the two different ball designs. From the output you find the following information:

	Type M1	Type M2
Sample size	36	36
Sum of data	9267	9244
Sample mean	257.4	256.8

It would appear from these values that there is not much difference in the way a "typical" ball behaves, but at this point that is just conjecture.

To better understand what the numbers really mean you can locate the values obtained for the mean on a histogram of the data. Perhaps the numbers and graphs together will provide more information.

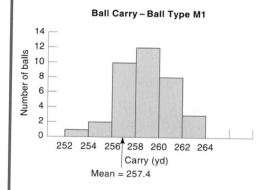

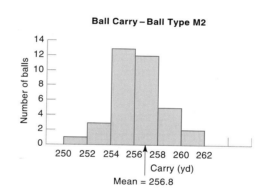

When you locate the mean *Carry* for each ball type on the appropriate histogram you see that the mean appears to be a good measure of the center of the data and that it does not appear to be influenced by extreme values. However, you notice that the sample mean does not provide any information about the number of golf balls that went more than that number of yards or less than that number of yards. This could be useful information, and so it seems that you need another measure of center. ∎

6.3.2 The Sample Median

While the sample mean does measure the center of the data, its value might be influenced by unusually high or low values in the sample and might not present a true picture of the sample data. For this reason we often look at other measures of center in addition to the sample mean so that we will be able to see a better picture of the data.

Another measure of central tendency that is often used is the **sample median.** While the sample mean is a measure of the center of balance of the data and is sensitive to the actual values, the sample median is a measure of the middle of the data after it is sorted from lowest to highest.

> The *sample median* is the value of the middle observation in an ordered set of data.

Finding the sample median requires sorting the data set first. Once this is done, the *sample median* is the value of the observation that is in the middle of the data. The exact location of the middle will depend on whether the number of observations in the sample is even or odd.

STEPS FOR LOCATING THE MEDIAN

Step 1: If the number of observations in the sample, n, is odd, then the median is the value of the observation in the $(n + 1)/2$ position.

Step 2: If n is even, then the median is the average of the values in the $n/2$ and $n/2 + 1$ positions.

EXAMPLE 6.4 Mortgage Waiting Times

Calculating the Sample Median

A bank has been receiving complaints from real estate agents that their customers have been waiting too long for mortgage confirmations. The bank prides itself on its mortgage application processing and decides to investigate the claims. The bank takes a random sample of 15 customers whose mortgage applications have been processed in the last six months and finds the following data:

Wait (days) 8 10 12 12 6 10 6 15 8 7 13 9 6 12 14

What is the median number of days required for mortgage confirmation?

Step 1. Put the data in numerical order from lowest to highest.

Position	1	2	3	4	5	6	7	8	9	10	11	12	13	14	15
Wait (days)	6	6	6	7	8	8	9	10	10	12	12	12	13	14	15

Step 2. Locate the middle observation.

Since $n = 15$ is odd, the middle position is the $(15 + 1)/2 = 16/2 = 8$th observation. The median of the sample is the *value* of the eighth observation, 10 days.

Since the median represents the middle of the data set, it tells the company that about half of the customers waited less than 10 days for a mortgage confirmation and about half of the customers in the sample waited more than 10 days. ∎

EXAMPLE 6.5 The Mail-Order Company

Calculating the Sample Median

Look at the data from the mail-order company in Example 6.1. If we sort the data in numerical order from highest to lowest we obtain:

Position	1	2	3	4	5	6	7	8	9	10
Demand	27	28	29	29	29	30	30	31	31	32

Since $n = 10$ is even, the sample median is the average of the observations in positions $10/2 = 5$ and $10/2 + 1 = 6$. So we find that the median $= (29 + 30)/2 = 29.5$ orders. ∎

TRY IT NOW!

Town Hall Traffic
Calculating the Sample Median

In the past few years the town council of a small town has received complaints that it has become increasingly difficult to cross the main street in town near the library. The council decides to look at traffic flow on the street. It selects a site directly in front of the library where most people try to cross the road and records the number of cars that pass the point in a two-minute period. This is done for 10 two-minute periods at 3:00 P.M. over several weeks and the following data are obtained.

Number of cars   20   27   29   28   37   23   21   28   29   28

Find the median number of cars that pass the site in two minutes.

Remember to SORT the data before you locate the median!

Why Use Two Different Measures?

At this point, you still may be wondering why we need to have two different measures of center. For most sets of sample data, the mean and the median will be very close to each other in value. If you need to decide on a single measure, then the choice will depend on the problem you are trying to solve and what information you need. The median tells you that half of the observations in the sample are above that value and half of the observations are below it. Because it is a measure of *location* it ignores the actual values of the observations and may not fully reflect the sample data. The mean uses all of the data values in its calculation and measures the center of balance of the data. While it can be shifted by extreme values, it does reflect all of the data values equally.

6.3.3 Comparing the Mean and the Median

We know that the mean and the median provide different information about the center of a set of sample data, but can anything additional be gained from knowing both of them?

If we start out with a symmetric, mound-shaped distribution, then the mean and the median are both located at the center of the distribution, at the bump. This is illustrated in Figure 6.2. We know that when the data are more spread out in one di-

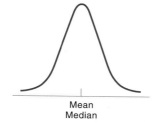

Mean
Median

FIGURE 6.2   Mean and median for a symmetric distribution

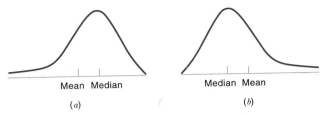

Mean Median

Median Mean

(a)

(b)

FIGURE 6.3 Mean and median for skewed distributions

rection (that is, when the data are skewed), then the mean is pulled toward these values, in the direction of the skew. This is illustrated in Figure 6.3. If we compare the mean and the median then we can learn about the shape of the distribution. In particular,

If	Distribution is	Illustration
Mean = median	Symmetric	Figure 6.2
Mean < median	Skewed left	Figure 6.3a
Mean > median	Skewed right	Figure 6.3b

At this point you might be saying, "Hey, hold on here. The mean and the median will almost never be exactly equal. How different do they have to be to say that the data are skewed?" This is a really good question and it is one that is often ignored. There is no exact answer, but there are several rules of thumb that can be used. One rule is that the mean and the median should differ by at least the width of a class in a histogram of the data. For small data sets or data values that do not have large magnitudes another rule is that they should differ by at least 10% of either measure. Because these are rules of thumb they will not work in all situations. Knowledge of the data, intuition, and experience are always important in interpreting statistics.

EXAMPLE 6.6   **College Graduate Salaries**

Comparing the Mean and the Median

A college wonders about the numbers reported by its placement office on graduates' salaries. The college thinks that the mean seems inflated and wonders whether it gives a good picture of what is really happening. It decides to collect some data on the salaries earned by students graduating from the School of Business and to calculate both the mean and the median. The college takes a random sample of 100 graduates from the past year and obtains the following data:

$25,000	$25,400	$25,600	$25,800	$25,900	$26,200	$26,400	$26,600	$27,000	$28,000
25,100	25,400	25,600	25,800	25,900	26,200	26,400	26,600	27,100	28,200
25,100	25,400	25,600	25,800	26,000	26,200	26,400	26,600	27,100	28,300
25,200	25,500	25,700	25,800	26,000	26,200	26,400	26,600	27,100	28,300
25,200	25,500	25,700	25,800	26,000	26,200	26,400	26,600	27,200	28,400
25,200	25,500	25,700	25,800	26,100	26,200	26,400	26,700	27,300	28,500
25,200	25,500	25,700	25,900	26,100	26,200	26,500	26,700	27,400	28,600
25,300	25,600	25,700	25,900	26,100	26,300	26,500	26,800	27,600	29,400
25,300	25,600	25,700	25,900	26,100	26,300	26,500	26,900	27,700	30,700
25,300	25,600	25,800	25,900	26,100	26,300	26,500	26,900	27,700	30,800

The mean of the salaries is

$$\overline{X} = \frac{2,638,500}{100} = \$26,385$$

The median is the average of the 50th and 51st observations in the data set so

$$\text{Median} = \frac{26{,}100 + 26{,}200}{2} = \$26{,}150$$

In this case, the actual difference between the mean and the median is

$$\$26{,}385 - \$26{,}150 = \$235$$

Clearly the mean is larger than the median, but are they different enough to indicate that the data are skewed? If you apply the 10% rule, you find

$$10\% \text{ of } \$26{,}385 = \$2638.50$$

The actual difference of $235 is much less than $2638.50, and, so, using the 10% rule, the salaries would not appear to be skewed.

*Remember! No **one** statistical tool will provide all of the information in a sample.*

The administrators decide to use a statistical software package to make a histogram to see if this conclusion is consistent with what they see. From the histogram we see that, in fact, the data are skewed right. This is not what the administrators expected.

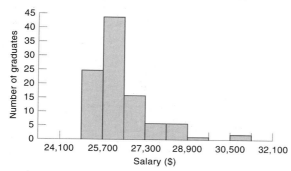

They see that the width of the classes in the histogram is $800 and that the difference between the mean and median, $235, is not more than this either. They understand that this is why it is important to look at data in more than one way.

Knowing that the data are skewed, the college officials decide that they should look at other variables to see why this might be true. This way the placement office will be able to prepare a report that better represents the true situation. ■

 **TRY IT NOW!**

Airline Cancellations

Comparing the Mean and the Median

An airline company is wondering about the number of cancellations that it receives for a particular business commuter flight. The airline takes a random sample of 15 days from the first quarter of the year and obtains the following data:

Number of cancellations 4 9 9 12 12 13 14 14 15 15 16 16 17 17 24

Find the mean and median for the number of cancellations for the commuter flight.

Note: The data have been sorted for you.

(continued)

When compared, do the data appear symmetric or skewed?

Make a dotplot of the data.

From the dotplot, do the data appear symmetric or skewed?

EXAMPLE 6.7   The Golf Ball Problem

Comparing the Mean and the Median

The golf ball company needs another number to describe a "typical" golf ball. Although the mean is a good statistic, you know you can give the company more information if you look at the median, too. You know that the value of the sample median will represent the distance that had half of the golf balls above it and half below it. This will tell the company a little more about how the golf balls behaved.

From the same computer output that gave us the mean, we learn that the median for the M1 balls is 257.5 yd and the median for the M2 balls is 257.0 yd. This information tells the company that 50% of the M1 balls in the sample went farther than 257.5 yards and that 50% of the M2 balls in the sample went farther than 257 yards.

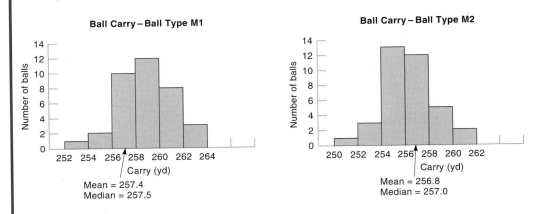

The medians are very close to the sample means. ∎

Discovery Exercise 6.1
THE TRIMMED MEAN

Part I. Investigating the Data

In a report to the administration of a large university, the Psychology Department states that the average class size is greater than the 35 students per class allowed by the university charter. The report indicates that the mean class size is 39.4.

No data are appended to the report, but you can obtain the current enrollments easily. The data you find are

3	14	22	26	42
3	15	23	27	45
5	15	24	28	45
9	17	24	28	190
11	21	25	36	193
13	22	26	38	193

(a) Do you think that the mean is a good measure of center for these data? Why or why not?

(b) By simply studying the data, what do you think a typical class size for the Psychology Department is?

(c) What is the median of the data? Is this close to what you thought?

(continued)

(d) Compare the mean and the median. What does the comparison lead you to believe about the data?

(e) Display the data graphically. Do you still think the same thing?

Part II. Solving the Problem

In the first part of this exercise you saw that neither the mean nor the median give a very good measure of a typical class size. In addition, a comparison of the mean and the median leads you to believe that the data are skewed right! From the histogram you can see that this is not the case.

How can we measure the center of the data when we have extreme values that influence the statistics we usually rely on?

You really can't just discard the extremes without careful investigation of the causes. One way to do this is to use a measure other than the median that removes the effect of the extremes. The *trimmed mean* is a statistic that does this. The trimmed mean allows you to drop a specified percentage of the observations in the data set from *each end.* By doing this you are not simply dropping outliers, but are looking at the center of the data, which may prove to be more reliable. Typically the trimmed mean is used with a percentage of 10%, but the percentage can be varied to suit the specific circumstances.

Computing the Trimmed Mean

To compute a 10% trimmed mean you must first determine how many values will be dropped from each end of the data set.

(a) Find 10% of the sample size.

(b) Do you think that dropping this many values from each end will be effective? Why or why not? If not, how many values do you think would be effective?

(c) Drop the top and bottom 3 observations from the data and recalculate the mean.

(d) Compare the trimmed mean and the median. What do you think about the data? Do you think this is a more accurate representation?

By using the trimmed mean together with the mean and the median you find that the mean was not influenced by the *skewness* of the data, but by extreme values on the high end.

(e) As an administrator at this university, do you think that the Psychology Department can claim that it exceeds the class size specification of 35 students per section? Use all of the information you have gathered to write a memo explaining your decision.

6.3.4 The Sample Mode

There is another measure of center that is used in statistics to measure the center of the data. This measure is the **mode.** In a bar chart the mode is analogous to the bar with the highest frequency.

> The sample *mode* is the data value that has the highest frequency of occurrence in the sample.

It would appear that the mode would be a very good measure of a typical value, but there are some obvious reasons why it will not always provide useful information. One is that, depending on the size of the sample and the number of possible data values, there may not be any repeated values in the sample. *That is, for some samples, the mode may not exist.* For continuous data, where there are many different possible values, we do not usually talk about the mode because of the problems just mentioned. In these cases, we often refer to the **modal class** in a frequency distribution or histogram.

> The *modal class* is the class interval in a frequency distribution or histogram that has the highest frequency.

Another problem with the mode is that there may appear to be more than one mode for a sample. This frequently happens with small samples. When this occurs it is not necessarily a case of two or three values that may occur much more frequently than any others. Rather, it is that most of the values happen to occur more than once.

For these reasons the mode is considered by many people to be an unreliable measure of central tendency. However, sometimes a sample does have more than one distinct mode. This indicates a *bimodal* (two modes) or *multimodal* (many modes) sample and should raise a number of questions in your mind.

1. Is it likely that these data could have two or more distinct centers? What would cause such a phenomenon?
2. Is it possible that the sample represents two or more different populations that were not understood when the data were taken?

EXAMPLE 6.8   **The Clothing Store**

Calculating the Sample Mode

A large retailer of women's clothing is trying to obtain some information that will help it formulate an ordering policy for clothing sizes. The retailer decides to look at a single line of apparel and collect data on the sizes of the items sold in a two-week period. The data after sorting in size order are

6	10	10	12	12	14
8	10	10	12	12	14
8	10	10	12	12	14
10	10	10	12	12	14
10	10	10	12	12	16

A frequency table of the data is shown below:

Size	Frequency
6	1
8	2
10	12
12	10
14	4
16	1

From the frequency table you can see that the mode of the sample is 10. That is, size 10 was sold the most often during the period of study.

What would the sample mean and median be for these data?

Sample mean: $\overline{X} = \dfrac{334}{30} = 11.1$

Sample median: Average of the $\dfrac{30}{2} = 15$th and $\dfrac{30}{2} + 1 = 16$th observations

so

$$\text{Median} = \frac{10 + 12}{2} = 11$$

You can see that neither the mean nor the median is a real size. The mean measures the balance point of the sizes and the median simply tells you the size that had half the sales below it and half above. Neither supplies information that helps to develop an ordering policy. ■

You might wonder why, in the previous example, we did not say that the data are bimodal, since size 12 has a frequency that is very close to the frequency for size 10. Usually when we refer to bimodal or multimodal data we are talking about data with very distinct, different centers. In a histogram this would mean that the two modal classes are not adjacent classes. Figure 6.4 shows a histogram that is bimodal.

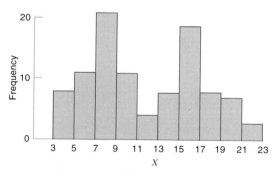

FIGURE 6.4 Histogram of bimodal data

6.3.5 Exercises—Learn It!

6.1   The management of Disney World would like to know something about the amount of time that visitors spend waiting for the monorail at one of the hotels at the resort. They take some sample data of times (minutes) and find the following:

   5.5   9.6   5.1   13.6   6.5   8.6   9.3   9.1   9.5   15.0   9.7   14.1

   (a) What is the mean time spent waiting for a monorail?

   (b) What is the median time spent waiting for a monorail?

In many exercises, the data are already sorted for you.

6.2   A company that sells mail-order computer systems needs to know something about typical weekly sales in order to plan inventory and staffing. The company selects 15 weeks at random from the past year and obtains the data (thousands of dollars) shown below:

Weekly sales 191 222 222 223 223 225 227 228 229 232 234 234 236 244 253

(a) What are the mean and median weekly sales for the company?

(b) What is the mode for the data? Do you think the mode is a good measure of typical weekly sales? Why or why not?

6.3   In manufacturing, knowing the amount of material that is wasted or lost during a process is very important. A company looks at the percent material lost per day for a new manufacturing process and finds the following data:

Daily loss   10   12   12   13   14   14   18   19   19   20

(a) What is the mean percent material loss for the process?   *15.1*

(b) What is the median percent material loss for the process?   *14*

(c) Do you think that either one of these two statistics is preferable for this data set? Why or why not?   *14. looks like losses are fewer*

6.4   To revise an existing inventory system, a company needs to know something about the lead time for orders it places for a critical part. The company looks at the last 20 orders placed and finds that the lead times (days) are:

1	4	6	7
4	5	6	8
4	5	7	8
4	6	7	10
4	6	7	11

(a) What is the mean lead time for an order?

(b) What is the median lead time for an order?

(c) What is the mode lead time for an order?

(d) Which statistic do you think best represents a typical lead time? Why?

6.5   In an effort to understand the cost overruns (the amount by which actual spending exceeds the estimated or budgeted cost) by a particular department in a company, data were collected on the amount of the overrun ($ thousand) for 24 different days:

87.3	93.7	96.8	98.4	100.9	107.8
89.9	94.9	97.0	99.6	101.3	109.7
91.5	96.5	97.1	100.3	105.7	111.5
93.6	96.7	97.3	100.4	107.7	114.2

(a) Find the mean and median daily overrun by the department.

(b) Compare the two statistics. Do you think the data are skewed or symmetric? Why?

6.6   To assess the effectiveness of the Total Quality Management (TQM) a company that manufactures paper products looks at daily production. Data on the number of items per minute that are produced are taken hourly. The sorted data are:

312	323	327
313	325	331
315	325	332
317	326	332
321	326	335

(a) What is the mean number of items produced during the 15-minute period?

(b) What is the median number of items produced during the 15-minute period?

(c) Compare the mean and the median and decide if the data are skewed or symmetric.

6.7   The transportation department of a small city has received numerous complaints about the length of the light cycle at a busy intersection. Drivers complain that there are so many cars backed up at the light that they cannot all get through when it turns green. In an effort to determine the validity of the complaints the transportation engineer decides to collect some data on the number of cars that are waiting at the light when it turns green. The data for 10 randomly selected light cycles are

4   4   5   6   7   8   9   10   11   12

   (a) Find the mean and median of the number of cars waiting at the light.

   (b) Are the data skewed or symmetric?

6.8   A study on the effects of television on behavior in adolescents, uses, as part of the data, the number of hours per day that the television set is turned on in a household. Twenty-six households are randomly selected and the data are

| 3.6 | 3.7 | 3.7 | 3.8 | 3.9 | 3.9 | 3.9 | 3.9 | 3.9 | 3.9 | 4.2 | 4.3 | 4.6 |
| 5.0 | 5.3 | 5.6 | 5.7 | 5.8 | 6.0 | 6.0 | 6.0 | 6.0 | 6.0 | 6.0 | 6.3 | 6.9 |

   (a) Find the mean and median for the number of hours that the television is turned on in a household.

   (b) Do you think that these statistics are good measures for a typical household?

   (c) What is the mode or modal class for the data? Do you think that this statistic provides any information that is not provided by the mean or median? Why or why not?

Discovery Exercise 6.2
INVESTIGATING VARIABILITY

The table contains air-quality data collected by the Environmental Protection Agency. The data show the number of unhealthy days for 14 major U.S. cities in 1989.

City	Number of unhealthy days
Atlanta	3
Boston	1
Chicago	2
Dallas	3
Denver	10
Houston	12
Kansas City	2
Los Angeles	206
New York	9
Philadelphia	18
Pittsburgh	11
San Francisco	0
Seattle	4
Washington, DC	7

Part I

(a) Display these data using a dotplot.

(continued)

(b) Find the typical number of unhealthy days by calculating the average value.

(c) Can you expect every observation to be typical? Why not?

Part II

(a) What you noticed in Part I is that although we know how to measure a typical value we don't have any way to describe the differences from this typical value that we see in the data. This is called *variability*. Using the dotplot, decide if you think this data set has a lot of variability, a little variability, or something in between a lot and a little.

(b) How can we measure this variability?

(c) You might have tried subtracting the smallest value from the largest value in an attempt to measure this variability. If you did so, you calculated what is known as the *range*. If you did not do this already, calculate the range.

(d) In this case the range is a large number, which presumably tells you that there is a large amount of variability in the data. Do you agree? Based on this, do you think that the range is a good measure of variability?

Part III

(a) Hopefully you concluded that the range can, in fact, give a very misleading picture of the amount of variability in the data. The range is quite large for this data set. Why?

(b) What about if we tried measuring how far away each data point is from the middle (i.e., the average) of the data? Let's do this by filling in the table below. Remember that the average is 20.57. The first measurement has been done for you.

City	Number of unhealthy days	Distance from middle
Atlanta	3	$3 - 20.57 = -17.57$
Boston	1	
Chicago	2	
Dallas	3	
Denver	10	
Houston	12	
Kansas City	2	
Los Angeles	206	
New York	9	

(continued)

City	Number of unhealthy days	Distance from middle
Philadelphia	18	
Pittsburgh	11	
San Francisco	0	
Seattle	4	
Washington, DC	7	
Average or typical	**20.57**	

(c) This is still not informative because we want a single number that will tell us what the typical deviation from the middle is. What is one way to measure "typical"?

Now, calculate the typical deviation.

(d) Does this value give us a good idea of how much variation there is in the data? Why not?

(e) What caused the typical variation just calculated to be such a small number? How can we fix this?

Part IV

(a) There are two ways to handle this. One way is to convert all the numbers to positive values by squaring them. The other way is to take the absolute value of the numbers. Fill in the table below. Then calculate the typical or average of the values in column 4 and the average of the values in column 5.

City	Number of unhealthy days	Distance from middle	Absolute value of distance from the middle	Distance from the middle squared
Atlanta	3	$(3 - 20.57) = -17.57$	17.57	$(-17.57)^2 = 308.7$
Boston	1			
Chicago	2			
Dallas	3			
Denver	10			
Houston	12			
Kansas City	2			
Los Angeles	206			
New York	9			
Philadelphia	18			
Pittsburgh	11			
San Francisco	0			
Seattle	4			
Washington, DC	7			
Average	20.57			

(b) Why are these two averages not close in magnitude?

(c) The average of the absolute distances (column 4) is called the *mean absolute deviation* (MAD) and the average of the squared distances (column 5) is called the *variance*. What are the units for the MAD and what are the units for the variance?

(continued)

(d) The MAD is actually easier to interpret but is not often used because absolute values do not "behave well." What can we do to the variance to get the magnitude to be about the same as the MAD? Do it.

What you have calculated is called the standard deviation.

6.4 MEASURES OF DISPERSION OR SPREAD

In Chapter 3 you saw that simply describing the center of the data or a typical data value does not provide complete information about the data set. In addition to knowing what a typical value for the sample is, it is important to know how diverse the values in the sample can be. That is, we need to know how *spread out* or *dispersed* the data values are relative to the typical values. Understanding the *variation* in a set of data is of critical importance in statistics. When people use statistics to make decisions, it is important to understand not only a typical outcome, but all possibilities as well. We will look at two different measures of dispersion, the *sample range* and the *sample standard deviation*.

6.4.1 The Sample Range

The simplest measure of dispersion, the **sample range,** involves looking at the two extreme values in the sample: the highest (maximum) and the lowest (minimum) values.

> The *sample range, R,* is the difference between the maximum and minimum observations in the sample.

The sample range is very easy to calculate and understand. It gives information about the distance from one end of an ordered data set to the other. If the sample data are symmetric, then it also gives information about the spread of the data relative to the measures of central tendency.

EXAMPLE 6.9   The Mail-Order Company

Calculating the Sample Range

Look at the data on demand from the mail-order company. In Example 6.1 we had the sample data:

Position	1	2	3	4	5	6	7	8	9	10
Demand	27	28	29	29	29	30	30	31	31	32

The sample range for this data is

$$R = 32 - 27 = 5$$

This tells the company that the demand for the product has a range or spread of 5 units around its center. In this case the range is a reliable measure of how spread out the demands are. ∎

Information on the range *along with a measure of central tendency* gives you a mental image of the data. If the data are symmetric, then the company would expect that a typical demand for the product is 30 units, and that the demand is evenly spread out on either side of the center. The actual demand might be as low as 27.1 (29.6 − 2.5) or as high as 32.1 (29.6 + 2.5). In Example 6.9 these bounds agree very well with the actual data.

It would seem that the range is a good statistic because it gives a clear picture of the spread and is easy to calculate and understand. The next example illustrates why this is not always the case.

EXAMPLE 6.10   Hold Times

Understanding the Sample Range

A company is wondering whether complaints about the amount of time customers spend on hold for technical service are justified or whether the complaints are a result of the "squeaky wheel" phenomenon. The company takes a sample of 15 customer calls to the technical service phone line and records the amount of time each customer spends on hold.

Customer	1	2	3	4	5	6	7	8	9	10	11	12	13	14	15
Wait (min)	5.6	10.2	6.6	6.9	9.4	6.7	0.6	9.2	7.6	10.7	9.6	2.9	6.0	8.6	4.6

To find the range, find the minimum and maximum observations and subtract them:

$$R = 10.7 - 0.6 = 10.1 \text{ min}$$

What does the range tell the company?

At first glance, the company would think that the length of time that a customer spends on hold has a spread of 10.1 minutes. If the company considers that the largest value in the sample is 10.7 minutes, it could conclude that the data are quite variable. Or are they?

A dotplot of the data shows that the value of 0.6 minute is really quite a distance from the next nearest value, while there are quite a few observations in the

Waiting time on hold (min)

9- to 10-minute range. The image that the company receives from the data is actually quite distorted. If it calculates the average hold time

$$\overline{X} = \frac{105.2}{15} = 7.0 \text{ min}$$

then the company would think that a typical caller is on hold for 7.0 minutes and that the time a customer spends on hold might be as low as 1.95 (7.0 − 5.05) minutes or as high as 12.05 (7.0 + 5.05) minutes. This is certainly not an accurate picture of the data since the mean is biased downward by the few extremely low values and the range was biased upward by the same values! ∎

 **TRY IT NOW!**

Restaurant Table Time
Calculating the Sample Range

The restaurant looking at the turnaround time for its tables, wonders how variable the occupation time for a table really is. The data the restaurant had collected are

Time (min)	59.3	58.6	62.7	65.4	59.0	67.3	62.8	68.1	59.4	63.7

What is the range of turnaround times?

Previously (page 182) you calculated the mean turnaround time to be 62.6 minutes. Using this information and the value for the range, what would the restaurant expect as its lowest turnaround time? its highest turnaround time?

The range is heavily influenced by unusual or extreme values in the sample. We saw that the mean is also influenced by extremes, but the amount of bias introduced is much less, since when you calculate the sample mean you use all of the values in the sample. When you calculate the range you use only the extreme values, so when these are unusual they have a large impact on the statistic. In fact, the sample range is one of the only statistics that gets more *unreliable* as the sample size gets larger. As a rule, when the sample size is more than 25, the sample range should not be used as a measure of variability.

ANS. RANGE = 9.5 MIN; LOWEST = 57.9 MIN; HIGHEST = 67.4 MIN

EXAMPLE 6.11 The Golf Ball Company

Looking at Variation

The company investigating the behavior of the two golf ball designs will need to know about the variation in the characteristics as well as the measure of typical distance. Unless the distance that its golf balls carry is pretty consistent, the company will have customer complaints about any claims it makes in advertising. From the histogram, you saw that the data did not appear to have any unusual values and so you decide to use the sample range as a measure of variation. You look at the computer printout and find the following information:

	Ball type	
	M1	**M2**
Minimum	252	251
Maximum	262	262
Range	10	11
Average	257.4	256.8

From the data it appears that the range is about the same for each of the ball types.

Since we also know that the data are symmetric, by using the sample ranges and sample means together, we can provide the company with a better picture of how the golf balls behave. The M1 balls would appear to carry between a low of 252.4 yards and a high of 262.4 yards. The M2 balls have a low carry of 251.3 yards and a high of 262.3 yards. The fact that these numbers agree with the observed minimums and maximums for the data support the idea that the range is a good measure for this data.

Still, the range does not answer all of the questions. Although you know that an M2 golf ball *could* travel between 251.3 and 262.3 yards, you do not have a measure of how the actual values are spread out within that range. What you would like to know, really, is what the typical variation in travel is. ∎

6.4.2 The Sample Standard Deviation

We have seen that the sample range proves unreliable in the presence of extreme data values. As the size of the sample increases, the chance that the sample will contain an extreme value increases. So for large sample sizes another measure of dispersion or spread must often be used.

If you think about what we are really trying to describe when we measure variability, it seems logical to try to define a measure of how far away from the center of the data a value might be. In fact, what we would probably like to know is, on the average, or typically, how far the values *vary* (differ) from the center. This measure is the **sample standard deviation.** The standard deviation is most often defined relative to another measure of dispersion called the **sample variance.** In practice, the measure that is used is the standard deviation because its units and order of magnitude are the same as those of the actual data.

> The *sample variance, s^2,* is the average of the squared deviations of the data values from the sample mean.

> The *sample standard deviation, s,* is the positive square root of the sample variance.

By their definition alone the sample variance and standard deviation seem overly complicated, especially when compared to the sample range. To calculate the sample standard deviation, you calculate the sample variance first, using the formula

$$s^2 = \frac{\sum_{i=1}^{n}(x_i - \bar{x})^2}{n - 1}$$

To obtain the sample standard deviation, *s*, you take the positive square root of the sample variance to obtain

$$s = \sqrt{s^2}$$

The population variance and standard deviation are represented by the Greek letter σ (sigma), where σ^2 is the *population variance* and σ is the *population standard deviation.*

EXAMPLE 6.12   Hold Times

Calculating the Sample Standard Deviation Using the Definition

The company looking at customer hold times knows that a measure of a typical hold time will not provide enough information about what its customers might encounter. The managers of the company need to know how much the time of a call might vary from the center. They decide to calculate the sample standard deviation for their data on customer hold times:

Customer	1	2	3	4	5	6	7	8	9	10	11	12	13	14	15
Wait (min)	5.6	10.2	6.6	6.9	9.4	6.7	0.6	9.2	7.6	10.7	9.6	2.9	6.0	8.6	4.6

To calculate the sample standard deviation the managers first need to calculate the sample mean, since the standard deviation measures the average of the squared distances from the sample mean. For this set of data we found previously that the mean hold time is 7.0 min.

The next part of the calculation for the sample variance finds the distance of each data value, X, from the sample mean, $\bar{X}$, squares each of them, and then adds them. The table below gives the details of each part of the calculation. Each column of the table below represents one part of the calculation. The last row of the table is the sum of all of the previous rows.

Customer	Wait	$(X - \bar{X})$	$(X - \bar{X})^2$
1	5.6	− 1.4	1.96
2	10.2	3.2	10.24
3	6.6	− 0.4	0.16
4	6.9	− 0.1	0.01
5	9.4	2.4	5.76
6	6.7	− 0.3	0.09
7	0.6	− 6.4	40.96
8	9.2	2.2	4.84
9	7.6	0.6	0.36
10	10.7	3.7	13.69
11	9.6	2.6	6.76
12	2.9	− 4.1	16.81
13	6.0	− 1.0	1.00
14	8.6	1.6	2.56
15	4.6	− 2.4	5.76
Sum	**105.2**	**0.2**	**110.96**

The last part of the calculation divides the sum of the squared deviations by $n - 1$, in this case, 14. So we have the sample variance

$$s^2 = \frac{110.96}{14} = 7.93 \text{ min}^2$$

and the sample standard deviation

$$s = \sqrt{7.93} = 2.82 \text{ min}$$

This tells the company that while an average call lasts 7.0 minutes, the actual call times will vary from that value. Typically, the variations or differences from the average will be about 2.82 minutes. ■

By now you must be convinced that there has to be a really good reason to use the standard deviation instead of the range! In fact, almost nobody actually *calculates* the sample variance and standard deviation using the definition. There is a shortcut formula that reduces the number of calculations you have to perform considerably:

$$s^2 = \frac{n \sum x^2 - (\sum x)^2}{n(n - 1)}$$

SHORTCUT FORMULA FOR SAMPLE VARIANCE

To calculate the sample variance and sample standard deviation this way you need to sum all of the data values and also to square each value and sum the squares. The next example recalculates both the sample variance and standard deviation from the previous example using the new, shorter method.

EXAMPLE 6.13 **Hold Times**

Calculating s Using the Shortcut Formula

Using the same data, you would need to sum both the values and the values squared. The table below gives the details of the calculations:

Customer	Wait (X)	X^2
1	5.6	31.36
2	10.2	104.04
3	6.6	43.56
4	6.9	47.61
5	9.4	88.36
6	6.7	44.89
7	0.6	0.36
8	9.2	84.64
9	7.6	57.76
10	10.7	114.49
11	9.6	92.16
12	2.9	8.41
13	6.0	36.00
14	8.6	73.96
15	4.6	21.16
Sum	**105.2**	**848.76**

Substituting the values in the last row of the table into the formula you get

$$s^2 = \frac{(15)(848.76) - (105.2)^2}{(15)(14)} = \frac{1664.36}{210} = 7.93 \text{ min}^2$$

and

$$s = \sqrt{7.93} = 2.82 \text{ min}$$

which is the same answer we got using the original formula.   ■

*CAUTION! Calculators and spreadsheets can calculate more than one type of standard deviation. Be sure that you are finding the **sample** standard deviation.*

It is probably hard to convince yourself that this is actually any better than the first set of calculations. Actually it is, but unless you do it several times it is hard to see. In fact, there is really no reason to calculate the sample variance and sample standard deviation by hand more than once or twice in your lifetime. Many calculators available today include statistical calculations like the sample mean and sample standard deviation, and these functions are also built into every spreadsheet package. What is important is understanding what the standard deviation measures and how it can be used to interpret sample data.

 **TRY IT NOW!**

Town Hall Traffic Flow
Calculating the Sample Variance and Standard Deviation

The town council looking at the traffic flow problem has seen reports that use the standard deviation, and wants to use it to describe the variability of traffic flow. The data are

| **Number of Cars** | 20 | 27 | 29 | 28 | 37 | 23 | 21 | 28 | 29 | 28 |

What is the sample standard deviation of the traffic flow?

Use whatever method you feel most comfortable with. If you have a statistical calculator learn how to use it NOW!

6.4.3 Interpreting the Standard Deviation— The Empirical Rule

Admittedly, the standard deviation is not as intuitive or appealing as the sample range. From the sample range you get an immediate (although sometimes false) picture of how far the data spread out around the center. The sample standard deviation does not give you the same intuitive response.

One way to understand what information the standard deviation gives is to use the **empirical rule.**

> The *empirical rule* says that for a mound-shaped, symmetric distribution
>
> - about 68% of all observations are within one standard deviation of the mean
> - about 95% of all observations are within two standard deviations of the mean
> - almost all (more than 99%) of the observations are within three standard deviations of the mean.

ANS. MEAN = 27 CARS; s = 4.85 CARS

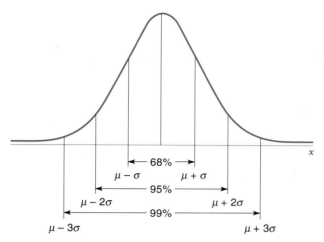

FIGURE 6.5   The empirical rule

The empirical rule (Figure 6.5) is defined for large data sets and distributions that are symmetric and mound-shaped, often called bell-shaped or normal curves. For distributions that are only slightly skewed the empirical rule is surprisingly accurate as well. It provides sets of bounds for data values from a given population.

In general, if we are talking about sample data we will not know the population mean and standard deviation, so it is necessary to substitute $\overline{X}$ and s for μ and σ. To really understand the empirical rule an example is necessary.

EXAMPLE 6.14   Hold Times

The Empirical Rule

Suppose that the company looking at hold times had collected a total of 50 observations on customer hold times. The sorted data are shown below:

0.6	4.6	5.6	6.3	6.8	7.5	7.8	8.3	8.9	9.6
2.9	4.7	6.0	6.3	6.9	7.5	7.9	8.4	9.2	10.1
3.4	5.2	6.0	6.6	6.9	7.6	8.0	8.4	9.2	10.2
3.8	5.5	6.1	6.6	7.0	7.6	8.1	8.6	9.4	10.7
4.5	5.5	6.1	6.7	7.2	7.8	8.2	8.6	9.4	11.1

For a data set this large you should use your calculator or a computer software package to find $\overline{X}$ and s.

To investigate the empirical rule we need to find the sample mean and sample standard deviation of the data. For this data set $\overline{x} = 7.12$ minutes and $s = 2.08$ minutes. We should also verify that the empirical rule applies—that is, that the data we have follow a normal curve. To do this we can make a histogram of the data:

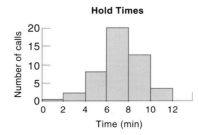

The histogram shows that the data are reasonably symmetric and bell-shaped, so we can use the empirical rule:

$$\bar{x} + s = 7.12 + 2.08 = 9.20 \text{ min}$$
$$\bar{x} - s = 7.12 - 2.08 = 5.04 \text{ min}$$

About 68% of the data values should be between 5.04 and 9.20 min.

$$\bar{x} + 2s = 7.12 + 2(2.08) = 11.28 \text{ min}$$
$$\bar{x} - 2s = 7.12 - 2(2.08) = 2.96 \text{ min}$$

About 95% of the data values should be between 2.96 and 11.28 min.

$$\bar{x} + 3s = 7.12 + 3(2.08) = 13.36 \text{ min}$$
$$\bar{x} - 3s = 7.12 - 3(2.08) = 0.88 \text{ min}$$

Almost all of the data (more than 99%) should be between 0.88 and 13.36 min.

To see how well the data we collected agrees with this rule we can calculate the actual percentage of data values that fall within each interval. The table below contains the number of data values and the percentage that fall within each of the three intervals.

Interval	Number of data values	Percentage of data values	Empirical rule (%)
5.04 to 9.20	36	72	68
2.96 to 11.28	48	96	95
0.88 to 13.36	49	98	99

Looking at the table we see that the actual percentages are slightly different from what the empirical rule predicted. Remember that variation in data is to be expected. In statistics it is important to understand when the variation is within the limits of what is expected or typical. When the variation from what is expected is too large, something may be happening to cause it.

In this case actual percentages are fairly close to those predicted by the empirical rule. The largest difference occurs in the first interval. If the percentages are very different from what is expected then the cause is that the data are probably from a distribution that is not symmetric. The more skewed a data set is, the more it will deviate from the empirical rule in the first two intervals. ■

TRY IT NOW!

Loan Processing
The Empirical Rule

Errors in filling out loan applications can lead to delays in having the loans approved. Bank employees must contact the applicants to correct the errors. This sometimes requires multiple contacts. To understand the extent to which the errors affect the application process a bank collected data on the number of follow-up contacts required before a loan could be processed. The bank looked at 25 different applications and found

```
0   1   2   3   4
0   2   2   4   4
1   2   3   4   5
1   2   3   4   5
1   2   3   4   7
```

Make a dotplot of the data.

From the dotplot, do you think that the assumption that the data have a symmetric, bell-shaped distribution is a reasonable one?

Find the mean and standard deviation of the data.

According to the empirical rule, between what two values should 68% of the observations fall?

Between what two values should 95% of the observations fall?

Between what two values should more than 99% of the observations fall?

6.4.4 z-Scores

In the previous section we saw that the standard deviation can be used to measure how likely it is for a data value to occur. Since, for a symmetric, bell-shaped distribution, 68% of the data values fall within one standard deviation of the mean, it is reasonable to assume that data values in that interval will make up about 68% of the sample values. Certainly, most (more than 99%) of the data values in the sample should be within three standard deviations of the mean! We could use this rule as a measure of how "usual" that data value is. This measure is called the **z-score.**

> A *z-score* measures the number of standard deviations that a data value is from the mean.

To calculate the z-score of a data value we first find the *distance* that the data value is from the mean and then divide by the standard deviation:

$$z = \frac{\text{Distance between the data value and the mean}}{\text{Standard deviation}} = \frac{X - \mu}{\sigma}$$

Remember! When we substitute $\overline{X}$ and s for μ and σ we are relying on large sample sizes to ensure that they are good estimates.

As in the empirical rule, for sample data we substitute $\overline{X}$ and s for μ and σ. A positive z-score indicates that the data value is *above* the mean, while a negative z-score indicates that the data value is *below* the mean.

If you think about the empirical rule together with the z-score you can begin to make some inferences on how data values compare to what is expected from a random sample. Table 6.1 gives you an idea of how this can be used.

TABLE 6.1   Using the Empirical Rule and z-Scores Together

If the z-score is . . .	The empirical rule says it will occur . . .	You can conclude that . . .
Less than −2 or more than 2	About 5% of the time	It is unusual and possibly an outlier
Less than −3 or more than 3	Less than 1% of the time	It is very unusual and probably an outlier

Just because a data value is identified as an outlier does not mean you can discard it from the sample!

It is possible to use the z-score of a data value to identify *outliers*. The problem with doing this is that everyone's definition of unusual is not the same and so inconsistencies in data analysis can result. z-scores are discussed much more extensively in Chapters 7 and 10.

EXAMPLE 6.15   The Golf Ball Company

z-scores

From the histograms, the company investigating the two designs of golf balls did not think that any of the data are potential outliers. Still, it is better to be sure, so the company asks you to calculate z-scores. Rather than calculate the z-score for every data value you know, it will be easier to just look at the maximum and minimum values first. If neither of these have unusual z-scores, then you know that none of the other values will either. From the tables of summary statistics you find the information you need and compute the z-scores shown in the table. The z-scores are indeed in the 2–3 range, but they do not indicate that the extreme sample values are outliers.

Ball type	Carry (yd)		Carry (yd)		z-score
M1	Max	262	$\overline{x} =$	257.4	1.92
	Min	252	$s =$	2.4	−2.25
M2	Max	262	$\overline{x} =$	256.8	2.26
	Min	251	$s =$	2.3	−2.52

 *TRY IT NOW!*

Town Hall Traffic
Calculating z-Scores

The town that was looking at traffic flow in front of the town hall wonders if the observation of 37 cars is unusual. Although the town officials know that their sample size of 10 cars is not large enough to ensure accuracy, they want to use z-scores to look at the data:

Number of cars	20	27	29	28	37	23	21	28	29	28

What is the z-score for the observation of 37 cars?

Comparing the z-score to the empirical rule, do you think that the value is unusual?

6.4.5 Exercises—Learn It!

6.9 In addition to knowing the typical time visitors to Disney World wait for a monorail, management wants to know about the variation in waiting times. The data for the waiting times (min) are *Requires Exercise 6.1*

5.5 9.6 5.1 13.6 6.5 8.6 9.3 9.1 9.5 15.0 9.7 14.1

(a) What is the range of the waiting times?

(b) What are the variance and the standard deviation of the waiting times?

6.10 To understand its customer base, a mail-order book club needs to know something about the frequency with which customers make subsequent purchases. They collect data on the number of purchases that customers make during their initial three-year membership. From a sample of 15 randomly selected customers they obtain the following data:

12 12 14 15 14 11 13 10 14 14 13 14 14 11 15

(a) Find the range of the number of purchases in a three-year period.   *10/15*

(b) Find the variance and the standard deviation of the number of purchases.

(c) Do you think that the range or the standard deviation is a better measure of variation for these data? Why?

Range: gives an easier value to look at & obtain Info

6.11 A professor in an introductory statistics course is interested in the number of hours that students spend doing homework during the week. A random selection of 12 students yields the following:

4.1   2.8   6.1   4.9   4.2   5.5   3.2   5.9   2.7   5.4   6.9   3.7

(a) Find the range and the standard deviation of the number of hours spent doing homework.

(b) According to the empirical rule, between what two amounts of time will 68% of the students spend on homework? 95%? more than 99%?

(c) How do the actual data compare to the predictions from the empirical rule?

6.12 A manager at XYZ Corporation thinks that the number of travel miles claimed in expense reports of its sales personnel has increased in recent months. To substantiate his idea he collects historical data from expense reports filed in the previous year:

2171   1709   2062   2075   1758   1733   1716   1963   1655   1558   1617   1908

(a) Find the range of the number of miles claimed.

(b) Find the standard deviation of the number of miles claimed.

(c) What is the z-score of the smallest data value? Do you think this value is unusual? Why or why not?

(d) What is the z-score of the largest data value? Do you think this value is unusual? Why or why not?

Requires Exercise 6.2   **6.13** The company that sells mail-order computer systems wants to further analyze its typical weekly sales. The data (in thousands of dollars) for the 15 weeks are

Weekly sales 191   222   222   223   223   225   227   228   229   232   234   234   236   244   253

(a) What are the range and standard deviation of the weekly sales?

(b) According to the empirical rule, between what two values will 68% of the weekly sales fall? 95%? more than 99%?

(c) Do you think that the empirical rule is appropriate for these data? Why or why not?

(d) What is the z-score for the week that had $228,000 in sales? What does this tell you about that observation?

6.14 To make a decision about replacing the cars in its current fleet, a company looks at the amount of money spent on repairs in the past 12 months, for a random sample of ten cars:

472   472   603   459   538   601   449   588   539   521

(a) Find the range, variance, and standard deviation of the amount spent on repairs.

(b) Use the empirical rule to describe the distribution of the amount spent on repairs.

(c) Do you think that the use of the empirical rule was appropriate here? Why or why not?

(d) Find the z-score for each data value. Are any of the values unusual? If so, why?

Requires Exercise 6.3   **6.15** Look at the data on percent material lost per day in manufacturing:

Daily loss   10   12   12   13   14   14   18   19   19   20

(a) What is the range of percent material loss for the process?

(b) What is the standard deviation of the percent material loss for the process?

(c) Find the z-score for each data value. Do you think that any of the values are unusual? Why or why not?

6.16 A company that buys blank VHS tapes for video recording is concerned about the actual amount of time that the tapes are able to record. The tapes are rated at 120 minutes, but the company knows that there is variation in the actual recording time. The company collects data on 25 randomly selected tapes and finds the actual recording times (min) listed below:

116	118	119	119	120
117	118	119	119	121
117	118	119	120	121
117	119	119	120	121
117	119	119	120	121

   (a) What is the range of actual recording times for the tapes? What are the variance and standard deviation of the recording times of the tapes?

   (b) Make a dotplot or a histogram of the data.

   (c) Use the empirical rule to find the intervals $\overline{X} \pm 1s$, $\overline{X} \pm 2s$, $\overline{X} \pm 3s$ and mark them on the graphical display.

   (d) Find the actual number of data points that fall in each interval. Does this agree with the predictions of the empirical rule? Why or why not?

6.17 A large company that sells software has had complaints from customers lately that the disks provided by the company fail to work properly after extended use. The company decides to investigate the complaint and looks at the disks from its current supplier. The disks are rated with a life of 500 hours of use before failure. After testing 20 disks until they fail, the company obtains the following data on disk life (hours):

486	494	502	508
490	496	504	510
491	498	505	514
491	498	506	515
494	498	507	527

   (a) Find the range and standard deviation of the time to failure of the disks.

   (b) Make a dotplot of the data. Do any of the data values appear unusual?

   (c) Do you think that the range or the standard deviation is more reliable for this set of data? Why?

   (d) Find the z-score of the two largest and the two smallest disk lives. What does this tell you about the observations?

6.18 The company that is looking at cost overruns by a particular department also wants to look at how the overruns vary: *Requires Exercise 6.5*

87.3	93.7	96.8	98.4	100.9	107.8
89.9	94.9	97.0	99.6	101.3	109.7
91.5	96.5	97.1	100.3	105.7	111.5
93.6	96.7	97.3	100.4	107.7	114.2

   (a) Find the range and standard deviation of the daily overrun by the department.

   (b) Use the empirical rule to give the company a picture of how the overruns vary.

   (c) Find the z-scores of the maximum and minimum overruns. Is either of these two values unusual? Why or why not?

6.5 DESCRIPTIVE STATISTICS FOR GROUPED DATA

So far you have learned how to measure the center and the dispersion of a set of data. All of the examples and exercises that you have looked at have involved calculating these numerical descriptors when you have the individual data observations to work with. In other words, if there were 30 observations of weekly sales then you had all 30 numbers available to you. When you are trying to solve a problem by analyzing data, this is the best situation to be in. You have what is known as **raw** or **ungrouped data.**

> Individual observations are known as *raw* or *ungrouped data.*

   As the name suggests, the data have not been grouped or summarized in any way. You can calculate the measures of the center and dispersion because you have all the observations.

   However, sometimes you do not have access to the individual observations. This may occur for confidentiality reasons or sometimes you have not collected the data yourself. This is clearly the case if you are using secondary data, and much of the data published on the Web are unavailable as raw data. Thus, often the only thing available to you is what is known as **grouped data.**

For example, suppose you wished to compare the salaries of managers in your organization to national values. The human resource manager may not wish to share individual salary values with you but might give you information in the following form:

Salary	Frequency
$0 < x \le \$30,000$	1
$30,000 < x \le \ 60,000$	8
$60,000 < x \le \ 90,000$	3

You should recognize this as a frequency distribution, which was discussed in Chapter 3. You do not know the individual salary values for the 12 people but you have the data in grouped form. This gives us a definition for grouped data.

> *Grouped data* are data that are available only as a frequency distribution. The individual observations are not accessible.

It is clearly better to use grouped data than no data at all in making your business decision. So we need to be able to calculate descriptive statistics for such grouped data. First we look at measures of the center for grouped data.

6.5.1 Measures of the Center for Grouped Data

Note: You will only be able to estimate the mean. You cannot calculate it exactly without the raw data.

There are three measures of the center: the mean, the median, and the mode. First, consider how to estimate the mean of the data set when you have grouped data. Consider the amount of time, in minutes, people occupy a table in a particular restaurant. The manager is interested in the center or the "typical" length of time that the table is occupied. She has only the following frequency table from 32 observations:

Time	Frequency
$25.0 < x \le 35.0$	5
$35.0 < x \le 45.0$	2
$45.0 < x \le 55.0$	4
$55.0 < x \le 65.0$	3
$65.0 < x \le 75.0$	11
$75.0 < x \le 85.0$	3
$85.0 < x \le 95.0$	4

This assumes that the data that fall in each class are evenly spread out on each side of the midpoint.

Remember that to calculate the mean you sum all the data and divide by the sample size. But for grouped data you can't sum the actual data because you don't have them. So, you have to estimate what the values might sum to for each interval. Consider the 5 observations that fall in the first interval between 25 and 35 minutes. We need a way to estimate the sum of those 5 values to begin our estimation of the mean. It seems reasonable to use the middle of the interval as our best "guess" of the actual values in the class. So, you must first find the midpoint of each class. In this dataset, the 5 values for table times that fall between 25 and 35 min are assumed to be spread evenly throughout the interval so that the middle value of 30 minutes is a good representation of the data in that interval. Since there are 5 of them, you multiply the midpoint of 30 by the frequency of 5 to get the contribution to the sum for that interval. This is like adding 5 values of 30 together.

This process is repeated for each interval and then the interval sums are added together and divided by the sample size. The details are shown in the next example.

EXAMPLE 6.16   Restaurant Data

Estimating the Sample Mean for Grouped Data

Time	Frequency, f	Midpoint, m	Frequency $\times$ Midpoint, fm
$25.0 < x \le 35.0$	5	30	150
$35.0 < x \le 45.0$	2	40	80
$45.0 < x \le 55.0$	4	50	200
$55.0 < x \le 65.0$	3	60	180
$65.0 < x \le 75.0$	11	70	770
$75.0 < x \le 85.0$	3	80	240
$85.0 < x \le 95.0$	4	90	360
Total	32		1980

The mean is estimated to be $1980/32 = 61.875$ minutes.

■

This procedure is summarized in the steps below. It gives you a good estimate of the mean when the data are in fact evenly spread out throughout the interval.

Step 1.   Find the midpoint of each class. Call it m_j.

Step 2.   Multiply the midpoint by the class frequency, f_j, to yield $f_j m_j$.

Step 3.   Add up all the interval sums found in step 2.

Step 4.   Divide the sum from step 3 by the sample size, n. Note that the sample size is the sum of all the frequencies.

STEPS FOR ESTIMATING THE MEAN FROM GROUPED DATA

The formula for estimating the mean from grouped data is thus

$$\overline{X} = \frac{\sum\limits_{j} f_j m_j}{n}.$$

ESTIMATE OF SAMPLE MEAN FOR GROUPED DATA

Now consider an example that we have looked at previously. Suppose we had only grouped data for the weekly sales of the mail-order company discussed in Exercise 6.2.

EXAMPLE 6.17   Mail-Order Company Sales

Estimating the Sample Mean for Grouped Data

Weekly sales (thousands of $)	Frequency	Midpoint	Frequency $\times$ Midpoint
$190 < x \le 200$	1	195	195
$200 < x \le 210$	0	205	0
$210 < x \le 220$	0	215	0
$220 < x \le 230$	8	225	1800
$230 < x \le 240$	4	235	940
$240 < x \le 250$	1	245	245
$250 < x \le 260$	1	255	255
Total	15		3435

The mean is estimated to be $3435/15 = 229$ (thousands of dollars).

■

Now you try one.

 *TRY IT NOW!*

Internet User Demographics
Finding the Mean

At the following Web site you can find Survey-Net:

http://www.survey.net/hon1r.html

Survey-Net is the source for user demographics on the Internet. The age distribution of the people who responded to the most recent survey at this site is shown below. (Note that the given age categories do not satisfy our convention regarding the endpoints of the intervals, but we will use the grouped data as provided.)

Age	Frequency	Midpoint	Frequency × Midpoint
0–15	380		
15–17	859		
18–21	1668		
22–30	2300		
31–40	1380		
41–50	891		
51–60	364		
61–70	74		

Estimate the average age of the survey respondents.

You know from Section 6.3 that there are two other commonly used measures of the middle: the median and the mode. If you have only grouped data you can estimate the median and the mode from the frequency table.

Consider the median. Recall that the median is the data value of the middle observation in an ordered set of data; thus it is the value at or below which half (50%) of the data values fall. So to find the median for grouped data we need to find the midpoint of the interval that contains the data value whose cumulative relative frequency is 0.50. Reconsider the restaurant data on the number of minutes customers stay at the table.

EXAMPLE 6.18 Restaurant Data

Estimating the Sample Median for Grouped Data

Time	Frequency	Relative Frequency	Cumulative Relative Frequency
$25.0 < x \le 35.0$	5	0.16	0.16
$35.0 < x \le 45.0$	2	0.06	0.22
$45.0 < x \le 55.0$	4	0.13	0.35
$55.0 < x \le 65.0$	3	0.09	0.44
$65.0 < x \le 75.0$	11	0.34	0.78
$75.0 < x \le 85.0$	3	0.09	0.87
$85.0 < x \le 95.0$	4	0.13	1.00

Remember that the relative frequency is found by dividing the frequency by the total sample size of 32. The cumulative relative frequency is found by accumulating the relative frequency column. A look at the column labeled Cumulative Relative Frequency tells us that the interval from 65 to 75 minutes includes the median, the value whose cumulative relative frequency is 0.50.

Therefore the estimate of the sample median is the midpoint of the interval from 65 to 75 minutes or 70 minutes. ■

The steps for estimating the sample median for grouped data are summarized below:

STEPS FOR ESTI-MATING THE SAMPLE MEDIAN FOR GROUPED DATA

Step 1. Find the relative frequency for each class by dividing the frequency by the sample size.

Step 2. Find the cumulative relative frequencies.

Step 3. Identify the class that contains the median. This will be the first class where the cumulative relative frequency is greater than 0.50.

Step 4. The estimate of the sample median is the midpoint of the class you identified in step 3.

Now you try one.

 TRY IT NOW!

Internet User Demographics
Finding the Median

At the following Web site you can find Survey-Net:

http://www.survey.net/hon1r.html

Survey-Net is the source for user demographics on the Internet. The age distribution of the people who responded to the most recent survey at this site is shown below:

Age	Frequency	Relative Frequency	Cumulative Relative Frequency
0–15	380		
15–17	859		
18–21	1668		
22–30	2300		
31–40	1380		
41–50	891		
51–60	364		
61–70	74		

Estimate the sample median age of the people who responded to this survey.

The third measure of the middle that was discussed for raw data was the mode. Recall that the mode is the data value that has the highest frequency of occurrence in the sample. Using this definition, it is easy to see that the modal class is the class interval in the frequency distribution that has the highest frequency. The estimate of the mode is then the midpoint of the modal class.

EXAMPLE 6.19   Restaurant Data

Estimating the Sample Mode for Grouped Data

Time	Frequency
$25.0 < x \le 35.0$	5
$35.0 < x \le 45.0$	2
$45.0 < x \le 55.0$	4
$55.0 < x \le 65.0$	3
$65.0 < x \le 75.0$	11
$75.0 < x \le 85.0$	3
$85.0 < x \le 95.0$	4

The modal class interval is the time interval from 65 to 75 minutes and the estimate of the mode is therefore 70 minutes.   ∎

6.5.2 Measures of Dispersion for Grouped Data

The three measures of dispersion or spread that we looked at for raw data were the sample range, the sample variance, and the sample standard deviation. Clearly with grouped data the sample range can be estimated by taking the difference between the upper value of the last class and the lower value of the first class. For example, for the sales data in Example 6.17, the sample range would be estimated as $260 - 190 = 70$ (thousands of dollars).

In order to adapt the formula for the sample variance for use with grouped data, we need to take the same approach that we used for estimating the sample mean for grouped data. In particular, we need to adapt the formula for the sample variance shown below to accommodate the fact that we no longer have the individual data values represented by x_i in the formula

SAMPLE VARIANCE

$$s^2 = \frac{\sum_{i=1}^{n} (x_i - \bar{x})^2}{n - 1}$$

Following the same argument that we used for estimating the sample mean, we can use the midpoint of the class as the "typical" value for that class and remember to multiply by the frequency of the class to get the contribution of that class to the sum in the numerator of the formula. Doing this gives the following formula and steps for estimating the sample variance for grouped data.

ESTIMATE OF SAMPLE VARIANCE FOR GROUPED DATA

$$s^2 = \frac{\sum (m_j - \bar{x})^2 f_j}{n - 1}$$

Step 1.   Find the midpoint of each class. Call it m_j.

Step 2.   Subtract the estimate of the sample mean, $\bar{x}$, from each class midpoint. Square the difference.

Step 3.   Multiply the result of step 2 by the class frequency.

Step 4.   Add up the results of step 3 for all classes.

Step 5.   Divide the sum from step 4 by one less than the sample size, $n - 1$. Note that the sample size is the sum of all the frequencies.

Let's look at an example. Reconsider the sales data from Example 6.17.

EXAMPLE 6.20 Mail-Order Company Sales

Estimating the Sample Variance for Grouped Data

Recall that the estimate of the sample mean was found to be 229.

Weekly sales (thousands of $)	Frequency	Midpoint	(Midpoint − Sample Mean)² × Frequency
$190 < x \le 200$	1	195	$(195 - 229)^2(1) = (-34)^2(1) = 1156$
$200 < x \le 210$	0	205	$(205 - 229)^2(0) = 0$
$210 < x \le 220$	0	215	$(215 - 229)^2(0) = 0$
$220 < x \le 230$	8	225	$(225 - 229)^2(8) = (-4)^2(8) = 128$
$230 < x \le 240$	4	235	$(235 - 229)^2(4) = (6)^2(4) = 144$
$240 < x \le 250$	1	245	$(245 - 229)^2(1) = (16)^2(1) = 256$
$250 < x \le 260$	1	255	$(255 - 229)^2(1) = (26)^2(1) = 676$
Total	15		2360

The estimate of the sample variance is $2360/(15 - 1) = 2360/14 = 168.57$ (thousands of dollars)².

Finally, the estimate of the sample standard deviation is found by taking the square root of the sample variance. So for the sales example above, the estimate of the sample standard deviation would be $\sqrt{168.57} = 13.0$ (thousands of dollars).

Now you try one.

 **TRY IT NOW!**

Internet User Demographics

Finding the Sample Variance and Standard Deviation

At the following Web site you can find Survey-Net:

http://www.survey.net/hon1r.html

Survey-Net is the source for user demographics on the Internet. The age distribution of the people who responded to the most recent survey at this site is shown below:

Age	Frequency	Midpoint	(Midpoint − Sample Mean)² × Frequency
0–15	380		
15–17	859		
18–21	1668		
22–30	2300		
31–40	1380		
41–50	891		
51–60	364		
61–70	74		

Estimate the sample variance and the sample standard deviation of the ages of the people who responded to this survey.

6.5.3 Exercises—Learn It!

6.19 A company is concerned about the turnover in its information systems (IS) department. You have been asked to research the situation and you find some information about the length of time IS professionals stay with a company. The data are available only as a frequency distribution and it is shown below:

Time with Company (months)	Frequency
$21 < x \le 25.5$	5
$25.5 < x \le 30$	3
$30 < x \le 34.5$	35
$34.5 < x \le 39$	35
$39 < x \le 43.5$	20
$43.5 < x \le 48$	0
$48 < x \le 52.5$	2

(a) Estimate the mean number of months that an information systems professional remains at a company.

(b) Estimate the median number of months that an IS professional remains at a company.

(c) Find the modal class.

(d) Estimate the standard deviation of the number of months that an IS professional remains at a company.

(e) Your company has experienced an average of 30 months for IS professionals. Based on your research, is your company experiencing more or less turnover than the industry as a whole? Support your answer with calculations.

6.20 Your town is thinking of supplying the residents with containers to be used to recycle newspapers. To decide how deep to make the container, you decide to research the height of newspaper recycle piles in other towns. You find the following grouped data:

Height of Newspaper Pile (inches)	Relative Frequency
$8 < x \le 9$	0.05
$9 < x \le 10$	0.12
$10 < x \le 11$	0.11
$11 < x \le 12$	0.08
$12 < x \le 13$	0.22
$13 < x \le 14$	0.25
$14 < x \le 15$	0.15
$15 < x \le 16$	0.02

(a) Estimate the mean number of inches of newspaper piles.

(b) Estimate the median number of inches of newspaper piles.

(c) Find the modal class.

(d) Estimate the standard deviation of the number of inches of newspaper piles.

(e) Based on your calculations, what would you recommend to the town for the height of the recycle bin?

6.21 You think that your university bookstore is overcharging students. In researching this, you find the following data on the amount of money students spend on textbooks at other schools.

Money spent on textbooks in one semester (\$)	Frequency
$200 < x \le 250$	100
$250 < x \le 300$	250
$300 < x \le 350$	75
$350 < x \le 400$	25

(a) Estimate the mean amount spent on textbooks in one semester.

(b) Estimate the median amount spent on textbooks in one semester.

(c) Find the modal class.

(d) Estimate the standard deviation of the amount spent on textbooks in one semester.

(e) You learn that the average spent on textbooks at your university is \$310. Using this information and your analysis, what can you conclude?

6.22 You think that your statistics instructor is assigning too much homework so you decide to find out how much time students in other sections spend on their statistics homework. You find the following data:

Time (in minutes)	Frequency
$20 < x \le 25$	6
$25 < x \le 30$	11
$30 < x \le 35$	7
$35 < x \le 40$	0
$40 < x \le 45$	2

(a) Estimate the mean amount of time spent doing statistics homework.

(b) Estimate the median amount of time spent doing statistics homework.

(c) Find the modal class.

(d) Estimate the standard deviation of the amount of time spent doing statistics homework.

(e) You have been spending 28 minutes on your statistics homework. Are you interested in changing sections? Why or why not?

6.6 MEASURES OF RELATIVE STANDING

Measures of center and dispersion are certainly important, but they are not the only numerical measures that can be used to obtain information about a set of data. Other measures, called measures of relative standing, or *order statistics,* give information about the position of an observation in the sample. We looked at one measure of relative standing, the median, when we looked at measures of center. Now we look at some additional measures.

6.6.1 Percentiles

It is useful in some real situations to know what data value in a sample has a certain percentage of the sample above or below it. This measure is known as the **percentile** of the data.

The methods used for calculating percentiles vary slightly among software packages, and so the values that you get might differ.

> The *p*th ***percentile*** of a data set is the value that has *p*% of the data at or below it.

Two questions can be asked involving percentiles: What value has *p*% of the data at or below it? and What is the percentile rank of a particular data value? The first question involves finding either a particular percentile or set of percentiles, such as the *deciles* (10%, 20%, . . . , 90%). It can be tedious to do by hand, but is easily done using most statistical software packages. It is of particular use when the observations in the data set are to be compared to each other or some other norm.

EXAMPLE 6.21   The Golf Ball Company

Finding Percentiles

The statistics you have provided so far tell the management of the golf ball company a lot about how its golf balls behave when they are hit, but these statistics will not be useful for making statements like "Ninety Percent of Brand X balls go XXX yards when they are hit!" No number that the company has looked at so far seems to provide that information, although the median comes close. The company wonders if there is a more general statistical measure similar to the median.

Remember! The median gives the data value that has 50% of the values above or below it.

A statistician is consulted who says that the measures the company wants are called percentiles. These percentiles are easily obtained from a statistical software package:

Data vector: GOLFDATA.Carry M1			Data vector: GOLFDATA.Carry M2		
Percentages		Percentiles	Percentages		Percentiles
10	=	255	10	=	254
20	=	255	20	=	255
30	=	256	30	=	255
40	=	257	40	=	256
50	=	257.5	50	=	257
60	=	258	60	=	257
70	=	259	70	=	258
80	=	260	80	=	258
90	=	260	90	=	260

From this information the golf ball company sees, for example, that 20% of the M1 balls went further than 260 yards, compared to only 10% of the M2 balls. This is really the first difference that the company has seen in the way the two balls behaved, but it might just be part of the natural variation. ■

The second question involves finding the **percentile rank** of a particular value in a data set.

> The ***percentile rank*** of a value is the percentage of the data in the sample that are at or below the value of interest.

This measure allows you to determine the *relative standing* of an observation in a set of data. To find the percentile rank of an observation, the data must be put in numerical order. The percentile rank, *P*, is then found by

PERCENTILE RANK

$$P = \frac{b + \frac{1}{2}e}{n}$$

where

b = the number of data values *below* the value of interest

e = the number of data values *equal* to the value of interest

n = the sample size

EXAMPLE 6.22 **Starting Salaries**

Finding the Percentile Rank

Suppose that in the example where we looked at starting salaries, a person in the group who had a starting salary of $26,200 wanted to know how she ranked relative to her peers. Looking at the data you see that

$b = 50$ (there are 50 salaries below $26,200)

$e = 7$ (there are 7 salaries equal to $26,200)

$n = 100$

so

$$P = \frac{50 + \frac{1}{2}(7)}{100} = 0.535 = 53.5\%$$

This tells her that 53.5% of the starting salaries were at or below hers. ∎

 ***TRY IT NOW!***

Aptitude Test Scores

Calculating the Percentile Rank

A group of employees at a manufacturing facility take a test to determine their aptitude for training. The tests are scored on a 400-point scale and are shown here in increasing order:

185	227	241	257	281	299	314	329
195	228	243	261	283	304	318	333
196	234	248	269	283	307	319	335
199	238	250	271	291	309	322	349
223	241	253	272	297	310	328	353

One of the employees who scored 283 wants to know how he stands relative to the other employees who took the exam. What is the percentile rank for the employee's score?

(continued)

What is the percentile rank of the employee that scored 319?

6.6.2 Quartiles

Although calculating percentiles in general can be tedious, there are certain percentiles that are used frequently. These percentiles are the 25th percentile and the 75th percentile, also known as the **first and third quartiles.**

> The *first quartile*, Q_1, is the value in the sample that has 25% of the data at or below it.

> The *third quartile*, Q_3, is the value in the sample that has 75% of the data at or below it.

You may be wondering what happened to the second quartile, but we have already seen it! If you think about the definition of the quartile, it is not hard to see that the second quartile, Q_2, must be the median.

Just like finding percentiles, finding the quartiles for a set of data can be tedious, depending on how you choose to do it. Actually, several methods can be used to find the quartiles of a set of data. As with the percentiles, different software packages use different methods and so it is not unusual to get two slightly different answers when using two different packages on the same data set. The method that we use estimates the quartiles and is not really exact. It gives values that are quite close to the more exact methods and it is much simpler.

If you think about the quartiles you see that they divide the data set in fourths. We know that the median divides the data set in half, and that if you take half of one-half you get a quarter. So, it would appear that if we find the median of each of the two halves of the data we will have our quartiles! It may sound complicated, but it really isn't.

Since percentiles and quartiles are order statistics, finding them requires that the data set be sorted from lowest to highest.

STEPS FOR FINDING THE QUARTILES

Step 1:   Put the data set in order and find the median of the data.

Step 2:   Take the lower half of the data (all of the values that are below the median) and find the median of the lower half of the data. This value will be the first quartile, Q_1.

Step 3:   Take the upper half of the data (all of the values that are above the median) and find the median of the upper half of the data. This value will be the third quartile, Q_3.

When the median is actually one of the data values (when n is odd) do not include the median in either half of the data.

ANS. 55%, 81.25%

EXAMPLE 6.23 Starting Salaries

Finding the Quartiles

The university that was looking at starting salaries of their graduates wants to know among the graduates sampled, what salary has 25% of the students earning less than that salary and what salary has 25% of the students earning more than that salary. In other words, the university wants to know the quartiles of the data. To make this easier the data set is repeated here:

$25,000	$25,400	$25,600	$25,800	$25,900	$26,200	$26,400	$26,600	$27,000	$28,000
25,100	25,400	25,600	25,800	25,900	26,200	26,400	26,600	27,100	28,200
25,100	25,400	25,600	25,800	26,000	26,200	26,400	26,600	27,100	28,300
25,200	25,500	25,700	25,800	26,000	26,200	26,400	26,600	27,100	28,300
25,200	25,500	25,700	25,800	26,000	26,200	26,400	26,600	27,200	28,400
25,200	25,500	25,700	25,800	26,100	26,200	26,400	26,700	27,300	28,500
25,200	25,500	25,700	25,900	26,100	26,200	26,500	26,700	27,400	28,600
25,300	25,600	25,700	25,900	26,100	26,300	26,500	26,800	27,600	29,400
25,300	25,600	25,700	25,900	26,100	26,300	26,500	26,900	27,700	30,700
25,300	25,600	25,800	25,900	26,100	26,300	26,500	26,900	27,700	30,800

1. Since there are 100 observations in the sample, the median must be the average of the 50th and 51st data values:

$$Q_2 = \frac{26,100 + 26,200}{2} = \$26,150$$

Remember! Since 100 is even you average the two middle values to find the median.

The lower half of the data set, everything *below* the median, consists of the first 50 values, and the upper half of the data, everything above the median, consists of the second 50 values.

2. To find the first quartile, find the median of 50 values, which will be the average of the 25th and 26th observations:

$$Q_1 = \frac{25,700 + 25,700}{2} = \$25,700$$

3. To find the median of the 51st through 100th values you will need to average the 75th and 76th observations:

$$Q_3 = \frac{26,600 + 26,700}{2} = \$26,650$$

So, the university finds that 25% of the graduates earn less than $25,700 and 25% of the graduates earn more than $26,650. ■

TRY IT NOW!

Training Aptitude
Finding the Quartiles

The company looking at training aptitude wants to give employees who scored in the top 25% on the test the opportunity to attend a seminar on training. The test scores are

185	227	241	257	281	299	314	329
195	228	243	261	283	304	318	333
196	234	248	269	283	307	319	335
199	238	250	271	291	309	322	349
223	241	253	272	297	310	328	353

(continued)

In the sample, what is the cutoff score for those people who will be able to attend the seminar?

Hint: The value that defines the top 25% is the same as the value that defines the bottom 75%.

Suppose that the company decides that the employees who scored in the bottom 25% need some additional classes on team building. What is the cutoff score for those employees who need the classes on team building?

6.6.3 Displaying the Data Using Boxplots

In Chapter 3 we looked at several different methods for displaying quantitative data. These methods allowed us to look at the data and describe central tendency, shape, variability, and outliers. Both methods, histograms and dotplots, provide displays of the actual data in the data set.

There is another method for displaying a set of data which uses not the individual data values, but rather a set of summary statistics taken from the data. The plot is called a **boxplot** or a **box and whisker diagram.**

> A *boxplot* or *box and whisker diagram* is a graphical display that uses summary statistics to display the distribution of a set of data.

A boxplot summarizes a sample using the quartiles and the median. You know that the median is a measure of center, but how can these statistics be used to show shape and variability? How can they identify outliers?

If you look at the first and third quartiles of a sample, Q_1 and Q_3, you see that 50% of the data in the sample fall between these two values. The distance between these two values is called the **interquartile range (IQR).**

> The *interquartile range* (**IQR**) is the difference between the third and first quartiles, $Q_3 - Q_1$.

We found that one way of describing the variability of a set of data is to use the empirical rule. Remember, the empirical rule says that approximately 68% of the data in the sample should be within one standard deviation of the mean (in the interval $\mu - 1\sigma, \mu + 1\sigma$). The interval from Q_1 to Q_3 is similar to the first interval of the empirical rule.

The main part of the boxplot consists of a rectangle (box) constructed from the median and the quartiles, as shown in Figure 6.6. This provides a partial picture of

Note: In drawing the box the height of the vertical lines is arbitrary. The scale on the x axis should span the values of the data similar to a histogram or dotplot.

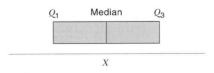

FIGURE 6.6   Box portion of boxplot

ANS. $Q_3 = 312, Q_1 = 241$

the data set. To complete the description with the empirical rule we used two additional intervals, $\mu \pm 2\sigma$ and $\mu \pm 3\sigma$. The boxplot uses multiples of the IQR instead of the standard deviation. The second interval in a boxplot is described by the **inner fences.** This is similar to the 2σ interval of the empirical rule. The last interval is described by the **outer fences.**

Note: The parts of a boxplot are defined from the quartiles, not the median.

> The *inner fences* of a boxplot are located at $Q_1 - 1.5(IQR)$ *and* $Q_3 + 1.5(IQR)$.

> The *outer fences* of a boxplot are located at $Q_1 - 3(IQR)$ *and* $Q_3 + 3(IQR)$.

The inner and outer fences are not drawn on the boxplot but are used to define the "whisker" portion of the plot. The whiskers are constructed by drawing horizontal lines from the quartiles to the smallest (and largest) observations in the sample that are *within* the inner fences. This is illustrated in Figure 6.7.

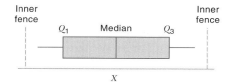

FIGURE 6.7   Boxplot with whiskers

It is easy to see that the median gives us a picture of the center of the data, but you may be wondering how the boxplot shows the shape of the distribution. You know that when data are symmetric the mean and the median are both located at the bump in the distribution and that the distribution tails off at the same rate in both directions. In a boxplot, if the data are symmetric, then you would expect that the median is located halfway between the two quartiles and that the whiskers are the same length.

If data are skewed the median stays located near the bump in the data and one side of the distribution tails at a slower rate than the other. In a boxplot, the median will be closer to one of the quartiles than the other and/or the whisker on the tail side of the distribution will be longer than the other. Figure 6.8 shows boxplots for data that are skewed.

While all this may seem quite complicated, it really is not. It is actually *easier* to make a boxplot than a histogram because the rules are standard and there is no judgment involved in setting it up. Making a boxplot can be separated into two parts, the calculations and the construction.

FIGURE 6.8   Boxplots for skewed data

CREATING A BOXPLOT   **Calculations**

Step 1: Find the median and the first and third quartiles for the data.

Step 2: Calculate the interquartile range (IQR) by finding $Q_3 - Q_1$.

Step 3: Find the values for locating the inner and outer fences:

Lower inner fence (LIF): $Q_1 - 1.5(\text{IQR})$
Upper inner fence (UIF): $Q_3 + 1.5(\text{IQR})$

Lower outer fence (LOF): $Q_1 - 3(\text{IQR})$
Upper outer fence (UOF): $Q_3 + 3(\text{IQR})$

Construction

Step 1: Draw the x axis and select a scale that will allow you to locate the largest and smallest values in the sample.

Step 2: Draw three vertical lines, all of the same height, one at Q_1, one at Q_3, and one at the median.

Step 3: Construct a box using the quartiles as the vertical edges and enclosing the median.

Step 4: Locate the inner fences on the axis. Draw a horizontal line from Q_1 to the smallest value in the sample that is within (larger than) the lower inner fence. Draw a similar horizontal line from Q_3 to the largest sample value that is within (less than) the upper inner fence.

It is probably time for an example!

EXAMPLE 6.24   Starting Salaries

Constructing a Boxplot

Consider the data about starting salaries. We have already summarized the data by constructing a histogram and by looking at sample statistics. What can we learn from a boxplot?

The first step in the calculation portion is done, since in Example 6.23 we found the median and the quartiles for the data to be

Median:	$26,150
Q_1:	$25,700
Q_3:	$26,650

The interquartile range is

$$Q_3 - Q_1 = \$26{,}650 - \$25{,}700 = \$950$$

For the inner fences we find that

$$\text{Lower inner fence} = 25{,}700 - 1.5(950) = \$24{,}275$$

$$\text{Upper inner fence} = 26{,}650 + 1.5(950) = \$28{,}075$$

and for the outer fences we find that

$$\text{Lower outer fence} = 25{,}700 - 3(950) = \$22{,}850$$

$$\text{Upper outer fence} = 26{,}650 + 3(950) = \$29{,}500$$

We now have all of the information needed to draw the boxplot.

Select the scale for the x axis so that it will accommodate the smallest and largest values in the sample, in this case $25,000 and $30,800. We will let the axis go from $24,000 to $31,000 in $1000 increments.

Locating the box portion of the plot is not difficult. To determine how far to extend the whiskers, compare the minimum and maximum values in the data set to the lower and upper inner fences:

Minimum: $25,000	Lower inner fence: $24,275
Maximum: $30,800	Upper inner fence: $28,075

You see that the lower whisker can extend all the way to the minimum data value, but the upper whisker cannot go to the maximum because it is beyond the inner fence. Looking back at the data you see that the largest data value that is less than $28,075 is $28,000. The boxplot will look like the one shown below:

Starting salary ($)

From the boxplot it would appear that the data are only slightly skewed to the right, since the median is located only a little closer to the lower quartile and the right whisker is only slightly longer than the left. ■

6.6.4 Using a Boxplot to Identify Outliers

At this point you may have some questions such as Why did we find the outer fences if we never use them? and What happens to the values from the sample that are beyond the inner fences? Both of these questions are quite reasonable and we are ready to answer them.

When we started talking about boxplots we said that they could be used to identify outliers. The intervals defined by the fences are similar to the 2σ and 3σ intervals from the empirical rule. Sample data that fall between the inner and outer fences are called *possible outliers,* while data values that fall beyond the outer fences are called *probable outliers.* If you are having trouble figuring out the difference between *probable* and *possible,* think about the difference in your reaction when your instructor tells you, "It is *possible* that you will pass this course" vs. "It is *probable* that you will pass this course."

The observations that fall between the inner and outer fences, the possible outliers, correspond to those that fall between two and three standard deviations from the mean using the empirical rule. Remember that less than 5% of the data should fall in this region, so data values that fall there are fairly unlikely events.

The observations that fall beyond the outer fences, the probable outliers, correspond to those that are more than three standard deviations from the mean in the empirical rule. Less than 1% of the data should fall in this region, so data values that fall there are very unlikely events.

The possible and probable outliers are plotted individually on the boxplot using special symbols, such as an open dot for possible outliers and a closed dot for probable outliers.

EXAMPLE 6.25   Starting Salaries

Locating Outliers on a Boxplot

Let's finish up the boxplot for the starting salaries data by identifying the possible and probable outliers. To do this we need to compare the data values that were not included in the whiskers to the inner and outer fences. You remember that nine observations were beyond the inner fence on the upper side of the boxplot. Comparing these values to the values of the inner and outer fences, we find

Upper inner fence:	$28,075
Possible outliers:	$28,200; 28,300; 28,300; 28,400; 28,500; 28,600; 29,400
Upper outer fence:	$29,500
Probable outliers:	$30,700; 30,800

The completed boxplot is shown below:

When the outliers are added to the boxplot the extreme skewness shown in the histogram is evident. Further, now we know that it is not just a case of skewed data but that in fact at least two of the data points are very different from the rest of the data. ■

Now that we know that some of the data might be outliers, you might wonder what we will do about it. You remember that when we sample from populations we want our sample to represent the population of interest. An outlier is a data value that has a large variation from the center. The first thing to do when we have identified outliers is to search for the *cause* of the variation. If there is some identifiable, assignable cause that makes the data value *not* representative of the data, then it is reasonable to drop the observation from the sample and redo the analysis. You cannot drop the observation without cause, but you can note in your report or analysis that an observation might have influenced the results.

Why do we need *another* method for displaying sample data and for identifying outliers? One important reason for using the boxplot is that it relies on statistics that are *invariant* (do not change) to the outliers themselves. When we use z-scores to identify outliers we use the mean and the standard deviation, both of which are sensitive to extreme values. If we have cause to actually delete the extreme observations the mean and standard deviation will change and *so will the definition of an outlier!* In a way, using z-scores to identify outliers is like trying to hit a moving target. The median and quartiles, on the other hand, remain relatively stable. Removing outliers does not really affect the locations of the inner and outer fences and so the definition of an outlier stays the same.

Another reason for using boxplots to display data is that they really do give an excellent picture of the distribution and they make it easy to compare samples from the same or different populations. Multiple boxplots lend themselves to being put on the same axis and make comparisons easier than multiple histograms, each of which requires a separate graph.

EXAMPLE 6.26   The Golf Ball Company

Comparing Data Using Boxplots

As the statistical consultant, you decide to give the management of the golf ball company a final picture by looking at boxplots for each sample. A statistical software package is used to create the plots and the results are shown below:

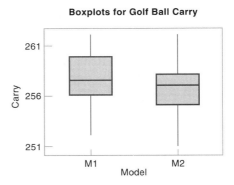

Boxplots for Golf Ball Carry

This is the first time that the two ball types have actually been compared side by side and the boxplots provide new information. From the plots it appears that while the middle 50% of the M1 balls carry between 256 and 259 yards, the middle 50% of the M2 balls only carry between 255 and 258 yards. In addition, based on the size of the plot, it appears that the M2 balls are slightly more variable, but, again, this may be due entirely to the samples selected. ■

**TRY IT NOW!**

Training Aptitude
Finding the Quartiles

The company that administered the training aptitude test to its employees would like a better picture of how the employees performed on the test. The data are

185	227	241	257	281	299	314	329
195	228	243	261	283	304	318	333
196	234	248	269	283	307	319	335
199	238	250	271	291	309	322	349
223	241	253	272	297	310	328	353

Previously (pages 225–226), you found the first and third quartiles of the data set. Use these values to complete the calculations needed for a boxplot.

Draw a complete boxplot of the data.

Were there any outliers? If so, which data values were they?

6.6.5 Exercises—Learn It!

6.23 The management of Disney World thinks that between 5 and 10 minutes is about the right amount of time for someone to wait for a monorail. Looking at the sample data, we find

5.1   5.5   6.5   8.6   9.1   9.3   9.5   9.6   9.7   13.6   14.1   15.0

   (a) What is the percentile rank of the person who waited 9.7 minutes?

   (b) What is the percentile rank of the person who waited 5.1 minutes?

   (c) Approximately what percentage of the people in the sample waited between 5 and 10 minutes?

Requires Exercise 6.7   **6.24** The transportation department of the small city wants to know what percentage of time there were 10 cars backed up at the light. Use the data below:

4   4   5   6   7   8   9   10   11   12

   (a) What is the percentile rank of the observation of 10 cars waiting at the light?

   (b) What is the percentile rank of the observation of 7 cars waiting at the light?

   (c) What information does this give the transportation department?

Requires Exercise 6.10   **6.25** The mail-order book club wants to know about the percentage of its customers that make subsequent purchases. Use the sample of 15 randomly selected customers given below:

10   11   11   12   12   13   13   14   14   14   14   14   14   15   15

   (a) What is the percentile rank of a person who makes 13 subsequent purchases?

   (b) What are the first and third quartiles of the data?

   (c) What information do the quartiles give the book club?

Requires Exercise 6.12   **6.26** The manager looking at the travel expenses of sales representatives would like to look at the data in terms of percentiles:

1558   1617   1655   1709   1716   1733   1758   1908   1963   2062   2075   2171

   (a) Find the first and third quartiles of the data and explain what they mean.

   (b) What is the interquartile range? How would you explain what that tells the manager?

   (c) Make a boxplot of the data.

   (d) Use the boxplot to describe the distribution of travel miles claimed.

   (e) Are any of the observations unusual? Why or why not?

Requires Exercise 6.2   **6.27** The company that sells mail-order computer systems wants to look at typical weekly sales in terms of quartiles. The data (in thousands of dollars) for the 15 weeks are given below:

191   222   222   223   223   225   227   228   229   232   234   234   236   244   253

   (a) Find the first and third quartiles of the data and explain what they mean.

   (b) What is the interquartile range? How would you explain what it means?

   (c) Make a boxplot of the data.

   (d) Use the boxplot to describe the distribution of weekly sales.

   (e) Are any of the observations unusual? Why or why not?

6.28 *The New York Times* (28 October 1994) published a state-by-state report on the number of people incarcerated per 100,000 people in the population. The data are shown in the table below:

District of Columbia	1578	Mississippi	385	Tennessee	278	Hawaii	170
Texas	545	California	382	Colorado	272	Oregon	169
Louisiana	514	Virginia	374	Alaska	256	New Hampshire	167
South Carolina	504	United States (Federal)	373	Indiana	256	Massachusetts	165
Oklahoma	501	Ohio	369	Idaho	253	Utah	154
Nevada	456	New York	361	Wyoming	247	Nebraska	148
Arizona	448	Arkansas	355	Kansas	239	Vermont	138
Alabama	439	Connecticut	331	South Dakota	227	Maine	113
Michigan	423	Missouri	321	Pennsylvania	224	West Virginia	106
Georgia	417	North Carolina	314	New Mexico	216	Minnesota	100
Florida	404			Washington	198	North Dakota	75
Maryland	392	New Jersey	307	Montana	192		
Delaware	391	Illinois	302	Rhode Island	185		
		Kentucky	281	Iowa	180		
				Wisconsin	172		

(a) Make a boxplot of the data.

(b) Use the boxplot to describe the number of people incarcerated in the United States.

(c) Are any of the data values unusual? Why or why not?

6.7 NUMERICAL DESCRIPTORS IN EXCEL

This section covers the basics of using Excel to calculate summary statistics and to create boxplots.

6.7.1 Calculating Summary Statistics in Excel

Suppose that we want to calculate a set of summary statistics for the Golf Ball data. Figure 6.9 shows a portion of that data in an Excel worksheet.

The Data Analysis ToolPak has a function that creates a set of summary statistics for a set of data. To access this function, select **Data Analysis** from the **Tools** menu and choose **Descriptive Statistics** from the list of Analysis tools. The dialog box is

	A	B	I	J	K	L	M	N
1	Ball#	Model #	Head	Temp	Carry	Tot Dist	Date	Time
2	1	M1	686	77	257	270	8/20	8:15
3	2	M1	688	77	255	267	8/20	8:15
4	3	M1	687	77	256	267	8/20	8:15
5	4	M1	688	77	255	271	8/20	8:15
6	5	M1	687	77	255	268	8/20	8:15
7	6	M1	687	77	256	267	8/20	8:15
8	7	M1	687	77	255	264	8/20	8:15
9	8	M1	690	78	258	269	8/20	8:15
10	9	M1	686	78	252	257	8/20	8:15
11	10	M1	687	78	256	268	8/20	8:15
12	11	M1	687	78	253	263	8/20	8:15

FIGURE 6.9 The golf ball data

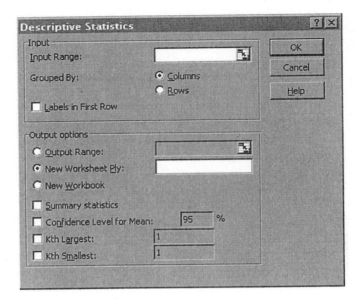

FIGURE 6.10   The Descriptive Statistics dialog box

shown in Figure 6.10. You will see that it is similar to the one you used to create a histogram.

The steps for creating the summary statistics are:

1. Position the cursor in the textbox labeled **Input Range** and highlight the range of data for which you want to calculate summary statistics. If your first row contains the variable name, be sure to check the box labeled **Labels in First Row.**

2. You must specify a location for the output, either a section of the current worksheet, or a new worksheet or workbook. Click on the radio button for your choice. If you select **Output Range,** you must specify a location on the worksheet. Position the cursor in the textbox for Output Range and click on the cell where you want the upper left corner of the results to appear. If you want to put the results in a new worksheet, you have the option of giving the sheet a name in the textbox or just letting Excel create a new, numbered sheet.

3. The last thing you must do is specify what kind of descriptive statistics you want. Click on the box labeled **Summary statistics** and finally click on **OK.** The output will appear in the location you specified and should look like Figure 6.11. The output includes most of the summary statistics that you learned in this chapter, as well as some that you have not yet encountered. The output does not include the quartiles. When a statistic (such as the mode) cannot be computed, the output will read N/A (Not Available).

AE	AF
Carry	
Mean	257.0972
Standard Error	0.274902
Median	257
Mode	258
Standard Deviation	2.332621
Sample Variance	5.441119
Kurtosis	-0.18199
Skewness	-0.04536
Range	11
Minimum	251
Maximum	262
Sum	18511
Count	72

FIGURE 6.11   Output from **Tools> Data Analysis> Descriptive Statistics**

Note: You will have to adjust the column widths to be able to read the first column of the output.

6.7.2 Making a Boxplot in Excel

Excel does not include boxplots as part of the graphs it can create. However, it is possible to create them if you know how Excel creates graphs. We have provided a macro for creating boxplots on the data disk that comes with this book.

To access the macros, you must open the file MacDoIt.xls either from your hard drive or from the floppy disk. This makes the macros available to you during your Excel session. Once you have the file open, you can use the macros in any workbook or worksheet. To use the boxplot macro, follow these instructions:

Note: The MacDoIt workbook must be open to make the macros available.

1. From the **Tools** menu select **Macro>Macros,** as shown in Figure 6.12. The macro dialog box shown in Figure 6.13 opens and you are presented with a list of macros for different statistical tools.

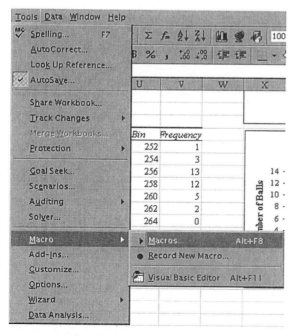

FIGURE 6.12   Menu selection to run a macro

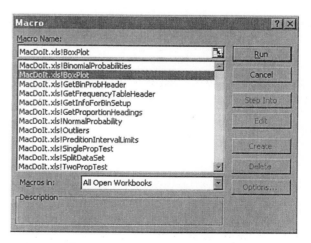

FIGURE 6.13   The Macro dialog box

2. From the list of available macros, highlight **MacDoIt.xls!BoxPlot** and click on **Run.** A small dialog box will open, prompting you to enter the location of your data. You can either enter the name of a range (if you have done this) or just use the cursor to highlight the data range in the worksheet. Click **OK** and the boxplot will be created in a new worksheet. In addition to the boxplot, the macro prints the Min and Max values of the data, and Q_1, Q_3, and the median.

The boxplot and summary statistics for the distance that the golf ball traveled when hit (Carry) for the M1 ball are shown in Figure 6.14.

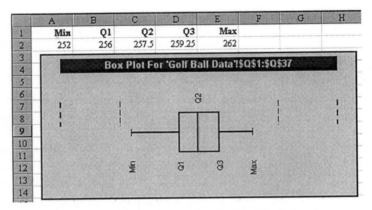

FIGURE 6.14   Finished boxplot for golf ball data

CHAPTER 6 SUMMARY

There are many ways to describe a set of data using sample statistics. No single number will do the job, nor is there any standard way to proceed. The measures that you choose must reflect the characteristics of the data itself. Most of the time the best descriptions come from the use of multiple measures and the conclusions that can be reached by comparing them.

Rather than summarize a sample with a list of numbers it is often useful to create images of the data using combinations of different statistics. An example of this is the *empirical rule,* which gives a picture of the distribution of the data. Another example of using summary statistics together to get a picture of the distribution is the *boxplot.*

Data analysis is not a static tool. You will need to look at a set of data in every way possible to obtain all of the information that it contains. Sometimes different methods will all lead you to the same conclusions and sometimes one method will yield an insight that is hidden in every other method.

Key Terms

Term	Definition	Page reference
Boxplot	A graphical tool that summarizes a sample using the quartiles and the median.	226
Empirical rule	For a symmetric distribution • about 68% of all observations are within one standard deviation of the mean. • about 95% of all observations are within two standard deviations of the mean. • almost all (more than 99%) of the observations are within three standard deviations of the mean.	206
First quartile, Q_1	The value in the sample that has 25% of the data at or below it.	224

(continued)

Term	Definition	Page reference
Grouped data	**Grouped data** are data that are available only as a frequency distribution. The individual observations are not accessible.	214
Interquartile range, IQR	The difference between the third and first quartiles, $Q_3 - Q_1$.	226
Median	The value of the middle observation in an ordered set of data.	124
Modal class	The class interval in a frequency distribution or histogram that has the highest frequency.	192
Mode	The data value that has the highest frequency of occurrence in the sample.	192
Parameter	A numerical descriptor that is used to describe a population.	180
pth percentile	The value in the data that has $p\%$ of the data at or below it.	222
Population mean	μ (mu).	181
Possible outlier	Observations that fall between the inner and outer fences.	229
Probable outlier	Observations that fall beyond the outer fences.	229
Raw data	Individual observations are known as **raw** or **ungrouped data.**	213
Sample mean	The center of balance of a set of data, and is found by adding up all of the data values and dividing by the number of observations.	181
Sample range	The difference between the maximum and minimum observations in the sample.	200
Sample standard deviation, s	The positive square root of the sample variance.	203
Sample variance, s^2	The average of the squared deviations of the data values from the sample mean.	203
Statistic	A numerical descriptor that is calculated from sample data and is used to describe the sample.	180
Third quartile, Q_3	The value in the sample that has 75% of the data at or below it.	224
z-score	A **z-score** measures the number of standard deviations that a data value is from the mean.	210

Key Formulas

Term	Formula	Page reference
Inner fences	$Q_1 - 1.5(\text{IQR})$ $Q_3 + 1.5(\text{IQR})$	227
IQR	$Q_3 - Q_1$	226
Outer fences	$Q_1 - 3(\text{IQR})$ $Q_3 + 3(\text{IQR})$	227
Percentile rank	$P = \dfrac{b + \frac{1}{2}e}{n}$	222
Sample mean, $\overline{X}$	$\overline{X} = \dfrac{\sum_{i=1}^{n} x_i}{n}$ or $\dfrac{\sum x}{n}$	181

Term	Formula	Page reference
Sample mean for grouped data (estimate)	$$\overline{X} = \frac{\sum\limits_j f_j m_j}{n}$$	215
Sample range, R	Max − Min	200
Sample standard deviation, s	$s = \sqrt{s^2}$	204
Sample variance, s^2	$$s^2 = \frac{\sum_{i=1}^{n}(x_i - \overline{x})^2}{n-1}$$	204
Sample variance for grouped data (estimate)	$$s^2 = \frac{\sum\limits_j (m_j - \overline{x})^2 f_j}{n-1}$$	218
z-score	$$\frac{X - \mu}{\sigma}$$	210

CHAPTER 6 EXERCISES

Learning It!

6.29 A manufacturer of pain relievers is interested in studying the amount of time it takes a person to be relieved of headache pain after taking the medication. The manufacturer selects a random sample of 12 people and conducts a study. The data (minutes) are

13.0 12.9 13.2 12.7 13.1 13.0 13.1 13.0 12.6 13.1 13.0 13.1

(a) Find the mean, median, and mode of the relief time.
(b) Find the range and standard deviation of the relief time.
(c) What is the percentile rank of the person who took 12.9 minutes to be relieved of their headache pain?
(d) Find the quartiles for the data set and interpret them.

6.30 A group of elderly people in a town filed a grievance with the town council, saying that the length of the walk signals in the town was inadequate for many elderly people to cross safely. In an attempt to investigate the claim, the town collected data on street crossing times for 12 elderly persons. The times, to the nearest tenth of a second, are

21.4 15.1 13.6 16.0 15.0 19.1 21.0 14.2 15.6 20.1 21.1 22.2

(a) Find the mean, median, and mode for the crossing times.
(b) Find the range and standard deviation of the crossing times.
(c) Compare the mean and the median. Do you think the data are skewed?
(d) What is the percentile rank of the person who took 20.1 seconds to cross the street?
(e) Find the quartiles and interpret them.
(f) Make a boxplot of the data.
(g) Describe the distribution of the crossing times.

6.31 A manufacturer of compact discs decides to take a random sample of the discs and measure the diameter. The diameter of the discs is a critical measurement, since if the discs are too large they will not fit in the players and if they are too small then they will not be read properly. The sample of 10 discs yields the following data (in.):

4.74 4.72 4.76 4.72 4.73 4.72 4.76 4.74 4.75 4.75

(a) Find the mean diameter of the compact discs.
(b) Find the median diameter of the compact discs.
(c) Compare the median and the mean. Do you think the distribution of the diameters is symmetric? Why or why not?

(d) Would it make sense to use the mode as a measure of center for this sample? Why or why not?

(e) Find the range of the diameters of the compact discs.

(f) Find the standard deviation of the diameters of the compact discs.

6.32 The vice president of marketing at a large corporation is wondering how many of the company's employees arrive at work before he does. He decides to collect some data by counting the number of cars that are in the parking lot when he arrives at 6:30 A.M.

$$23 \quad 24 \quad 24 \quad 25 \quad 25 \quad 25 \quad 25 \quad 26 \quad 26 \quad 37$$

(a) Find the sample mean of the data.

(b) Find the median of the data.

(c) Find the range of the data.

(d) Find the standard deviation of the data.

6.33 A company that sells personal computers via mail order would like to know something about the amount of time that a customer spends on the phone with the Technical Support department during a call. The company collects information on the length of a call in minutes for 20 different customers:

$$
\begin{array}{ccccc}
15 & 23 & 34 & 38 & 41 \\
16 & 25 & 37 & 40 & 43 \\
19 & 25 & 38 & 40 & 43 \\
20 & 28 & 38 & 41 & 44 \\
\end{array}
$$

(a) What is the average length of a call to Technical Support? What is the median length of a call?

(b) What is the range of call length? What is the standard deviation of the call length?

(c) Compare the mean and the median. Do you think the distribution of call times is symmetric or skewed? If it is skewed, which way is it skewed?

(d) Use one of the graphical methods you learned in Chapter 3 to display the data. Does the display agree with your answer to part (c)? If not, why not?

(e) Create a frequency distribution for these data and use it to estimate the mean and standard deviation of the call length.

(f) Compare the actual mean and standard deviation to the estimates. How close are they?

6.34 A manufacturer of breakfast cereals packs the cereal in 15-ounce boxes. Periodic quality checks are made to ensure that the boxes are being filled adequately. A random sample of 35 boxes of the cereal are selected and weighed. The data are

$$
\begin{array}{ccccccc}
14.91 & 15.23 & 15.34 & 15.40 & 15.48 & 15.59 & 15.62 \\
14.96 & 15.24 & 15.34 & 15.42 & 15.49 & 15.60 & 15.62 \\
15.11 & 15.28 & 15.36 & 15.45 & 15.50 & 15.60 & 15.63 \\
15.21 & 15.30 & 15.38 & 15.46 & 15.52 & 15.61 & 15.64 \\
15.22 & 15.30 & 15.39 & 15.48 & 15.58 & 15.61 & 15.67 \\
\end{array}
$$

(a) Find the mean, median, and mode for the weight of the cereal boxes.

(b) Find the range and the standard deviation of the cereal box weights.

(c) Describe the distribution of the data using the information from parts (a) and (b).

(d) Find the quartiles of the data and interpret them.

(e) Make a boxplot of the data.

(f) Create a frequency distribution for these data and use it to estimate the mean and standard deviation of the cereal box weights.

(g) Compare the actual mean and standard deviation to the estimates. How close are they?

6.35 A university collected some data on the amount of money ($) that students spend on textbooks in a typical semester:

214	241	248	258	269
233	244	249	260	274
234	245	250	262	276
236	247	253	262	277
239	248	254	263	277
241	248	254	265	281

(a) Find the mean and the standard deviation of the amount spent on textbooks.

(b) Find the intervals $\overline{X} \pm 1s$, $\overline{X} \pm 2s$, and $\overline{X} \pm 3s$.

(c) Make a histogram of the data and mark the intervals on the histogram.

(d) Calculate the actual percentage of the data that fall within each interval.

(e) Compare the actual percentages with those predicted by the empirical rule.

6.36 To begin a study comparing the monthly salaries of the auditors for a particular company to accountants' salaries in the industry in general, a random sample of twenty auditors was taken and their monthly salaries recorded. The data are

2832	3050	3231	3286
2843	3087	3233	3302
2978	3096	3237	3308
3016	3193	3239	3396
3032	3223	3252	3429

(a) Find the mean and the standard deviation of the monthly salaries.

(b) Find the intervals $\overline{X} \pm 1s$, $\overline{X} \pm 2s$, and $\overline{X} \pm 3s$.

(c) Calculate the actual percentage of the data that fall within each interval.

(d) Compare the actual percentages with those predicted by the empirical rule.

(e) Create a boxplot of the monthly salaries.

(f) Use the boxplot to describe the monthly salaries of auditors in the company.

6.37 It has been a widely held belief that the switch to participative management would increase employees' "buy-in" to the company. One of the benefits that should be realized is a reduction in the number of sick days that employees use. The company decides to sample 25 employees from each of two manufacturing departments. The first department has been participative for almost two years and the second is still using a traditional management style. The data on the number of sick days used by each employee in the past 12 months are

Participative Management					**Traditional**				
1	3	5	5	6	0	5	6	7	9
1	4	5	6	7	3	5	7	7	9
2	4	5	6	8	4	6	7	7	10
2	4	5	6	8	4	6	7	8	11
3	4	5	6	8	5	6	7	8	11

(a) Find the mean, median, and mode for the number of sick days used for each group of employees.

(b) Find the range and standard deviation for the two sets of data.

(c) Make a boxplot for each set of data.

Thinking About It!

6.38 Consider the data about the number of cars that are in the parking lot when the vice president arrives: *Requires Exercise 6.32*

23   24   24   25   25   25   25   26   26   37

(a) If the vice president wanted an estimate of the typical number of employees who arrive at work before he does, would the mean or the median be a better statistic to use? Why?

(b) Is the standard deviation or the range more reliable for these data? Why?

(c) Find the z-score of the data value of 37 cars. Does it confirm your suspicion that the data value is unusual?

6.39   Grading homework is a real problem. It takes an enormous amount of time and many students do not do a very good job or copy answers from other students or the back of the book. A teacher of Elementary Statistics decided to conduct a study to determine what effect grading homework had on her students' exam scores. She taught three sections of Elementary Statistics and randomly assigned each class one of three conditions: (1) no homework given, (2) homework given but not collected, and (3) homework given, collected, and graded. After the first exam, she collected the data (exam scores) and made histograms of the data and calculated some numerical measures. The histograms and summary data for each group are:

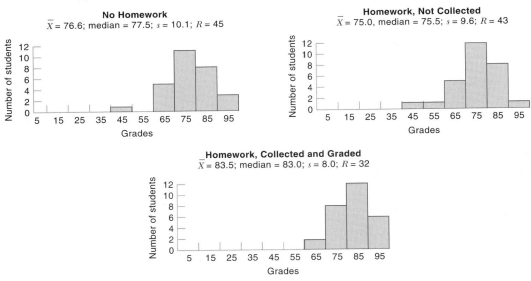

(a) Describe the distribution of the data for each of the groups and use the information to compare the three data sets and come to any conclusions you can about the effects of homework on exam grades.

(b) Do you think that this was a valid way for the instructor to examine the effects of homework on exam scores? Why or why not? What other factors might need to be considered? Based on the data, if you were the instructor, what would you do about assigning homework?

Requires Exercise 6.33   6.40   The mail-order computer firm that collected data on the length of a call to Technical Support is unsure about which sample statistics should be used to summarize the data and what they really mean. The firm calls on you to be its expert. What can you tell the firm about the length of phone calls to its Technical Support lines?

6.41   A company has been recording data on the number of defective items found in the daily production lots after 100% inspection. The data recorded for the last five days had an average of 20 defectives. One of the people on the inspection team thought this seemed strange and realized that on one of the days the number of defectives was really 10 and had been miscopied. When the correct value was written down the average number of defectives changed from 20 to 15. What value had been written down instead of the 10?

Requires Exercises 6.32 and 6.38   6.42   Create a boxplot for the data on the number of cars in the parking lot when the vice president arrives:

   23   24   24   25   25   25   25   26   26   37

(a) What does the boxplot indicate about outliers?

(b) Does this agree with the information obtained from the z-score? Why or why not?

(c) Which do you think is a more reliable indicator in this case?

Requires Exercise 6.17   6.43   The software company that is looking at the time to failure of the diskettes it uses decides to look at an alternative supplier of the product. The data for its current supplier and for the new supplier are

Current Supplier				Alternative Supplier			
486	494	502	508	489	492	495	498
490	496	504	510	489	492	496	499
491	498	505	514	491	493	497	502
491	498	506	515	492	493	497	503
494	498	507	527	492	494	497	505

(a) Find the mean, median, range, and standard deviation for each supplier.

(b) Make a boxplot of the data for each supplier.

(c) Based on these statistics, describe the diskettes for each supplier and make a recommendation to the company on which supplier to choose. Be sure to support your recommendation with references to the data.

6.44 The university that collected the data on the amount that students spend on textbooks wants to include the information in their catalog on typical semester expenses. The current catalog estimates that students spend $225 on books in a semester. *Requires Exercise 6.35*

(a) Do you think the catalog should be changed?

(b) If you think the catalog should be changed, what amount should be used? Justify your choice. If you think the catalog does not need to be changed, explain why.

(c) Suppose that changing catalog copy is costly considering the number of catalogs that are currently on hand. Does this change your answer to part (b)? If so, why? If not, why not?

6.45 The company that wants to compare its auditors' salaries to the industry in general finds that the average monthly salary of accountants in the industry is $3400 with a standard deviation of $50. The company asks you to interpret the data and to show how its auditors' salaries compare to those of the industry in general. *Requires Exercise 6.36*

6.46 As part of an annual program to calibrate their quality inspectors, each inspector in a company is asked to do a 100% inspection of a lot of 500 items. The data on the number of defectives found by each of the 35 inspectors are

8	10	13	14	15	16	17
9	11	13	15	15	16	18
9	12	13	15	15	17	18
9	12	14	15	15	17	19
10	12	14	15	15	17	20

(a) If you did not know the actual number of defectives in the lot, would you use the sample mean, median, or mode as your estimate of the number of defectives in the lot? Justify your decision.

(b) Suppose that you were the training coordinator for the quality inspectors. How would you interpret the data? What would you do as a result?

6.47 The company that is looking at employee sick days asks you to examine the summary statistics and compare the number of sick days taken by the two groups. *Requires Exercise 6.37*

(a) Based on the data, what would you tell the company?

(b) Are the data sufficient for you to conclude that participative management reduces the number of sick days taken by employees?

(c) If your answer to part (b) is yes, why can you reach this conclusion? If your answer to part (b) is no, why do you think this conclusion is not valid?

(d) As a consultant, would you like the company to collect any additional information for you? If so, what information might be useful?

6.48 A study was recently completed by an insurance company concerning a particular surgical procedure. The study looked at the hospital records of 40 patients at two hospitals and compared the length of patient stay. The data were analyzed using Minitab. Output showing the descriptive statistics for the data and boxplots of the two samples is shown on page 244.

Descriptive Statistics Output for Patient Stay (Minitab)

```
Descriptive Statistics

Variable          N      Mean    Median   Tr Mean    StDev    SE Mean
Hospital 1       40     7.725    7.500     7.667     2.562     0.405
Hospital 2       40    10.350   10.000    10.222     3.340     0.528


Variable        Min      Max       Q1        Q3
Hospital 1     2.000   14.000    6.000     9.000
Hospital 2     5.000   18.000    8.000    13.500
```

Boxplots for Patient Stay Data (Minitab)

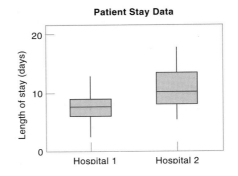

(a) Using the computer output, describe each of the data sets. Be sure to include information about central tendency, variability, shape, and outliers.

(b) Compare the length of patient stay for the two hospitals.

(c) Proponents of hospital 1 say that this hospital is better than hospital 2 at the surgical procedure because the patients from hospital 1 recover faster than the patients from hospital 2. Can you conclude, on the basis of these data, that hospital 1 is better at the surgical procedure than hospital 2?

(d) If you think that the conclusion described in part (c) is reasonable, justify it. If you think this conclusion is not possible, what can you conclude? What additional information would you, as the insurance company president, find useful?

Requires Exercise 6.30   **6.49** The town that is studying crossing times consults you about what the data mean. Currently, the walk signals are set for 15 seconds. Write a report telling the town council what you have found.

6.50 Explain why you should not use the steps for estimating the mean of grouped data for the following data on the length of coupon lives:

Coupon Life	Number of coupons	
	1995	1996
Less than 1 month	28	123
1–4 months	88	140
5–8 months	204	122
9–12 months	141	88
More than 1 year	39	27

6.51 Do the intervals have to be the same width to use the steps to estimate the mean of grouped data? Explain why or why not.

6.52 Sometimes the estimates of the mean and standard deviation that you obtain from grouped data are close to the actual values and sometimes they are off by quite a bit. What are the characteristics of the data that would lead to poor estimates of the mean and standard deviation if you had only grouped data to work with?

Doing It!

*Datafile:
GOLFBALL.XXX*

6.53 The company investigating the golf balls is not satisfied with the limited analysis that it has done. The managers have collected a good deal of data, but they are not sure how to look at them and interpret the output. They decide to hire you to help them understand what the golf balls are doing and how they compare to each other. In addition to measures of the balls' performance, such as the variable *Carry*, the managers know that other factors, both internal (ball-related) and external (environment-related), could affect performance. They tested 36 of each type of ball at three different times using a machine to launch the balls. Data were recorded on 14 different variables. A portion of the data is shown below:

Ball	Model	S1	S2	S3	Wgt	Dw	Dd	Head	Temp	Carry	Tot Dist	Date	Time
1	M1	81	81	82	45.3	0.145	0.0110	686	77	257	270	8/20	8:15
2	M1	83	83	84	45.2	0.151	0.0111	688	77	255	267	8/20	8:15
3	M1	81	82	84	45.2	0.145	0.0105	687	77	256	267	8/20	8:15
4	M1	81	81	83	45.3	0.144	0.0117	688	77	255	271	8/20	8:15
5	M1	83	81	82	45.5	0.146	0.0108	687	77	255	268	8/20	8:15

Variable	Description
Ball	Keeps track of the observation number and goes from 1 to 72
Model	Identifies ball design, M1 and M2
S1, S2, S3	Measurement of circumference at three different points
Wgt	Weight of the ball
Dw, Dd	Dimple width and depth
Head	Head speed of the ball when it is hit
Temp	Environmental temperature in degrees Fahrenheit
Carry	Distance from the point the ball was hit to the point where it first hit the ground, measured in yards
TotDist	Total distance the ball travels from the point where it was hit to its final position
Date	Date of the testing
Time	Time of day that the observation took place. There were 3 different time periods.

(a) For each model number, calculate the mean, median, mode, range, and standard deviation of the variable *TotDist*.

(b) Describe the distribution of *TotDist* for each ball. How does *TotDist* compare with *Carry*? Which measure do you think is better from a statistical point of view?

(c) What can you tell the company about the effects of temperature during the trials?

(d) What can you say about the variable *Head* over the entire trial? Is it consistent for both ball models? Do you think that head speed needs to be considered as a variable in this trial or is it controlled well enough?

(e) Would you expect there to be much difference in performance from one time period to the next? Examine the effects of *Time* for each model separately and describe what you see. Make sure you look at both *Carry* and *TotDist*.

(f) Look at the variables *S1, S2,* and *S3*. The closer these measures are to being equal, the closer to spherical the ball is. Do you think that the balls are spherical? Why or why not?

(g) Look at *Dw* and *Dd*. How do these compare for each ball type?

(h) Think about the problem and analyze the data in any other way that you think might be interesting or important. Prepare a report for the company describing the behavior of each ball type and comparing the two balls. Include any graphs that you think might be useful in illustrating your findings.

Datafile: TISSUES.XXX

6.54 Remember the complaint data from the tissue manufacturer that we looked at in Chapter 3? Recall that one of the largest categories of customer complaints involved sheets tearing on removal. One of the product variables that can cause tissues to tear is tensile strength. The manufacturer has decided to look at the manufacturing process and to investigate the tensile strength of the tissues. Two hundred samples are taken from tissues produced on a single-tissue machine. The samples are taken over three different days. A portion of the data file is shown below:

Day	MDStrength	CDStrength
1	1006	422
1	994	440
1	1032	423
1	875	435
1	1043	445

Day keeps track of the day on which the sample was taken and goes from 1 to 3. *MDStrength* measures the machine-directional strength and is measured in lb/ream. *CDStrength* measures the cross-directional strength and is measured in lb/ream.

(a) Calculate a set of summary statistics for *MDStrength* for the entire three-day period.

(b) Use the summary statistics to describe the *MDStrength* of the tissues. Be sure to comment on central tendency, variability, shape, and outliers.

(c) Display the data for *MDStrength* for the three-day period graphically. How does the graph agree with or disagree with your answer to part (b)?

(d) Calculate a set of summary statistics for each day separately. How does the distribution of *MDStrength* for each of the three days compare to the overall distribution?

(e) Suppose that the *MDStrength* for the product is supposed to have a mean of 1050 and a standard deviation of 50. Do you think that the manufacturing product is meeting these specifications?

(f) Repeat parts (a)–(e) for the variable *CDStrength*. The specifications for *CDStrength* are a mean of 450 and a standard deviation of 25. In addition to a target mean and standard deviation, each variable has defect limits. For example, *MDStrength* should be between 850 and 1075. If the strength is too low the tissue snaps and causes the machine to shut down. If it is too high the tissues are stiff. The *CDStrength* should be between 390 and 480. If the strength is too low then the tissues will tear when you pull on them to remove them from the box. If it is too high, they will be stiff.

(g) Prepare a report for the management of the tissue company telling them how the actual process compares to the specifications. Include any appropriate graphs or tables of data with the report.

WORKSHOP 3

Chapter 7 Probability

PROBABILITY

CREDIT PROBLEMS

One of the problems facing a small business is that of borrowing money. As a result of problems with loan defaults, banks and other lending institutions have imposed more stringent conditions for obtaining loans. As a result, many small businesses find themselves forced to close because they cannot meet the stiffer requirements.

The problems of small businesses affect the cities and towns in which the businesses are located. As a result, the towns would like to know what is going on and how they can help. A medium-sized New England city has been trying to understand the problems facing the small businesses in their town by administering a questionnaire every six months. The questionnaire collects information on the size of the business in terms of annual sales and employees, the type of business, and whether the business is experiencing recession-related problems. A sample of the data is shown below:

Number	Size	Employee	Nature	Problem	Understand	Concerned	Call	Loan	Collateral	Access
1	2	2	1	1	2	1	2	2	0	2
2	1	2	3	1	2	2	0	2	2	1
3	4	3	1	2	1	2	2	2	2	0
4	1	1	1	2	1	2	2	2	2	2
5	1	2	1	2	0	0	0	0	0	0
6	3	1	5	2	0	2	2	2	2	1

The town has employed an analyst to help them interpret the results of the study and to determine how the small businesses in the city are being affected by credit problems.

7.1 CHAPTER OBJECTIVES

In Chapters 3 and 6 you learned different ways to summarize and describe data that were collected. The data were usually random samples from different populations, but the techniques used really described only the sample data.

In this chapter we look at how *probability theory* can be used to measure and predict what is *likely* to happen when data are collected from different populations. This chapter covers the following topics:

- Basic Probability Rules
- Random Variables and Probability Distributions
- The Binomial Probability Distribution
- The Normal Probability Distribution

7.2 BASIC RULES OF PROBABILITY

Probability is a concept that you are most likely familiar with, although perhaps not in a formal sense. Whenever you talk about whether some event is likely to occur, such as whether it will rain tomorrow, you are using the concepts of **probability.**

> *Probability* is a measure of how likely it is that something will occur.

To talk about probability using the language of probability it is necessary to define some terms. In probability and statistics we often speak of **experiments.** These are not experiments in the laboratory sense, but rather are the actions we perform to collect data. As an experiment we might count the number of students who miss statistics class each day or record the color of the car parked next to us in the parking lot.

> An *experiment* is any action whose outcomes are recordable data.

When we perform an experiment, or collect data, we must think about what outcomes might occur so that we know what form our data will take. This is called the **sample space** of the experiment. You remember from the previous chapters that data can be qualitative or quantitative and that we use different techniques when we analyze them.

> The *sample space, S,* is the set of all possible outcomes of an experiment.

As a very simple experiment, consider rolling a single, six-sided die and recording the number of spots on the top facing side. You know that there are six possible outcomes of the experiment, so the sample space is

The brackets { } are used to indicate a sample space.

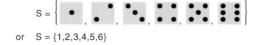

or   S = {1,2,3,4,5,6}

Another example is the sample space for the experiment of tossing a coin. In this case the sample space is

$$S = \left\{ \text{\includegraphics{nickel heads}} , \text{\includegraphics{nickel tails}} \right\}$$

or   S = {H, T}, where H represents a head and T represents a tail

Depending on the experiment and the type of data collected, a sample space might have a *finite* number of elements such as the ones we just described. The word *finite* means that you can count the number of elements. This will most often happen when the experiment results in qualitative data, or in quantitative data that are integer.

When the data from the experiment are quantitative and continuous, then the sample space contains an *infinite* number of possible values and must be described mathematically or in words. For example, suppose that as our experiment we decide to take college students at random and measure their heights. This is an example of *continuous* qualitative data. The number of possible values that we get will be *bounded* (there are probably not any college students shorter than one foot or taller than eight feet) but *infinite* (since height is a measurement and there are an infinite number of measurements between 1 foot and 8 feet). To describe this sample space we might write

S = {all numbers greater than or equal to 1 and less than or equal to 8}

or using mathematical notation

$$S = \{x : 1 \leq x \leq 8\}$$

Figuring out the sample space for an experiment is important in data collection, too. It helps the experimenter think about all of the possible results they might obtain and to plan for them.

EXAMPLE 7.1   Flipping Two Coins

Writing Out a Sample Space

An experiment consists of flipping two standard coins and writing down what is on the side facing up for each of the coins. If we let H represent a head and T represent a tail, then the sample space for this experiment is

$$S = \{HH, HT, TH, TT\}$$

HT and TH are different even if the coins are both quarters, but this is another way to try and think about it.

If you are thinking that HT and TH are the same outcome, then try to think of tossing the coins one after another or think of the problem as if the coins are different, such as a penny and a quarter. In writing out sample spaces it is best not to combine outcomes, but to write each one separately. ■

📖✍ TRY IT NOW!

The Spinner Problem
Writing out the Sample Space

An experiment consists of spinning the two different spinners pictured below:

(continued)

Write down the sample space for this experiment.

You may be wondering what rolling dice and spinning spinners has to do with collecting data and you are right to wonder. In fact, they have little to do with the types of problems you will encounter in statistics, but they do form an easy basis for discussing and understanding the general rules of probability.

7.2.1 Probability of an Event

Several of the sample spaces that you looked at have something in common that make calculating probabilities easy. This feature is that each of the outcomes in the sample space is equally likely to occur on any given trial of the experiment. For example, when you roll a single die, each of the six possible outcomes in **S** is equally likely to occur. The same is true about flipping coins, both one coin and two coins.

Often we are interested in knowing how likely is that that a certain outcome or outcomes of the experiment will occur. We call these outcomes the **events** of interest.

An *event, A,* is an outcome or a set of outcomes that are of interest to the experimenter.

The likelihood that the event A will occur is called the **probability of A** and is written P(A).

The *probability of an event A,* P(A), is a measure of the likelihood that an event A will occur.

When each of the outcomes in a sample space *is equally likely,* then the probability of A can be calculated using the following formula:

$$P(A) = \frac{\text{Number of ways that A can occur}}{\text{Total number of possible outcomes}}$$

Formula for P(A)

or

$$P(A) = \frac{n_A}{N}$$

where n_A is the number of outcomes that correspond to the event, A, and N is the total number of outcomes in the sample space, S.

If you look at the simple experiment of flipping a coin described earlier, you can see that if S = {H, T} and we look at the event A = coin turns up heads, then P(A) = $\frac{1}{2}$, since there is only one way that the event A can occur and there are two possible outcomes of the experiment. This should not be surprising, since it is one of the ways you know probability intuitively.

Although you have been taught to reduce fractions and use decimals, when we are calculating probabilities here we will not do this so that you can see the actual number of outcomes of interest.

Similarly if we look at the experiment of rolling a die where S = {1, 2, 3, 4, 5, 6} and we let A be the event that the number that comes up is even, then

$$P(A) = \tfrac{3}{6} = \tfrac{1}{2}$$

since there are 3 even numbers possible out of the 6 possible outcomes of the experiment. Again, this is not surprising since it is somewhat intuitive.

There are some facts about probabilities that must be true:

*These facts may seem trivial, but try to keep them in mind when you calculate probabilities. If you get an answer that is not a number between 0 and 1 then **it cannot be correct!***

1. $0 \leq P(A) \leq 1$. This says that the probability of an event must be a number between 0 and 1 inclusive. Since you know that probabilities are formed by taking the ratio of the number of ways that A can happen to the total number of outcomes, the numerator is a subset of (smaller than or equal to) the denominator.
2. $P(S) = 1$. This says that the sum of the probabilities for the entire sample space must be equal to 1, or that essentially, when you perform an experiment, something must happen!
3. If an event A MUST happen, then $P(A) = 1$, and if the event cannot happen, then $P(A) = 0$.

EXAMPLE 7.2   Flipping Two Coins

Finding Probabilities

Joe and Tom decide to flip coins to determine who will pay for dinner. They agree that if the coins match Joe will pay and if they do not match Tom will pay. What is the probability that Tom pays for dinner?

Remember from the previous example that the sample space for this experiment is given by S = {HH, HT, TH, TT}. If we let A be the event that Tom pays, then we can write

A = the two coins do not match

It was important to list HT and TH as different outcomes so that the outcomes in S are all equally likely.

Looking at S, we see that the outcomes in the sample space are equally likely to occur and that two outcomes correspond to the event A: HT and TH. Thus, the probability that Tom pays for dinner is

$$P(A) = \tfrac{2}{4}$$

Using the same approach, if we let B be the event that Joe pays for dinner, or

B = the two coins match

then there are two outcomes in S that correspond to B, HH and TT, so

$$P(B) = \tfrac{2}{4}$$

From this we see that the arrangement is fair to both people.   ■

Finding probabilities when the events are equally likely is not hard if you can write down the sample space, S, without much difficulty. These types of problems are known as "classical" probability problems. Although the situations where the ideas of classical probability apply are limited, they do make it easy to illustrate the rules of probability.

 *TRY IT NOW!*

The Spinner Problem
Classical Definition of Probability

In the previous exercise you found the sample space for the spinner example to be

$$S = \{1A, 1B, 1C, 2A, 2B, 2C, 3A, 3B, 3C\}$$

Let A be the event that the first spinner lands on an odd number. Find P(A).

Let B be the event that the second spinner is a vowel. Find P(B).

When the number of possible outcomes of an experiment is large, calculating probabilities can be tedious. At times it is easier to solve the opposite or **complement** of the problem you are interested in! You may wonder how this can be possible. It is possible because it makes use of one of the three rules of probabilities, the one that says that the probabilities must sum to one.

The complement of an event A is often referred to as the event, not A.

The *complement* of an event A, denoted **A′**, is the set of all outcomes in the sample space, S, that do not correspond to the event A.

Since the event A is a set of outcomes in S and the complement of the event A′ is the set of all outcomes in S that do NOT correspond to A, we can see that

$$P(A) + P(A') = 1$$

Formula for the probability of the complement of A

and that we can find P(A) by calculating P(A′) and subtracting it from 1. That is,

$$P(A) = 1 - P(A')$$

For the experiment of rolling a single die, the sample space was

$$S = \{1, 2, 3, 4, 5, 6\}$$

If we define the event A to be that the number that comes up is greater than 2 we can calculate $P(A) = \frac{4}{6}$ directly, since there are 4 outcomes that are larger than 2 out of the 6 possible outcomes.

We notice that the complement of A, A′, is defined as all of the numbers that are NOT larger than 2. Since two outcomes correspond to A′ we know that $P(A') = \frac{2}{6}$ and we can find $P(A) = 1 - P(A') = 1 - \frac{2}{6} = \frac{4}{6}$.

Although this seems to complicate something that was not very complicated, there are times when using the complement of an event to find a probability is much easier. Usually this is when the number of possible outcomes that correspond to the event of interest is a large part of the sample space, S.

Despite what it may look like, the idea of equally likely outcomes is not limited to dice, coins, and spinners! In fact, the ideas of classical probability apply to random sampling as well. Whenever an experiment consists of selecting an item at random from a group of items, the sample space will consist of all of the possible items. If the sample is truly random, then each item in the group is just as likely to be selected as any other item.

EXAMPLE 7.3   Credit Problems

Probabilities and Random Sampling

The Chamber of Commerce in the city looking at credit problems of small businesses administered the questionnaire to 166 random businesses out of 1536 existing small businesses. The Chamber of Commerce was worried that if the surveys were administered by mail the sample might be biased toward companies that were having problems. These companies might be more likely to respond to the survey because it would give them an opportunity to report their problems to a group who might be able to help them. To ensure that the sample was truly representative, the surveys were administered by phone. One of the questions asked was about the size of the company in annual sales. The questionnaire allowed the following responses:

1	Under $1 million
2	$1–5 million
3	$6–10 million
4	$11–20 million
5	Over $20 million

If the business did not respond to the question then the response was coded as a zero (0). The results of the questionnaire are

Size	Number of responses
No response	1
Under $1 million	60
$1–5 million	21
$6–10 million	42
$11–20 million	21
Over $20 million	21
Total	166

The Chamber of Commerce decided to select a business at random and interview it along with the bank with which it does business. What is the probability that the company selected will have over $20 million in sales?

This is an experiment in which the sample space is the set of 166 companies that respond. That is, S = {C1, C2, C3, . . . , C166} and each element of S is equally likely to be chosen. Since we are interested in the event that the company has over $20 million in sales (call that event A) we see that 21 elements (companies) in the sample space correspond to A out the total of 166, so

$$P(A) = \tfrac{21}{165}$$

■

EXAMPLE 7.4 **Product Preference**

Random Sampling

A market research firm has conducted a product preference survey at a large manufacturing company. A group of 250 production workers were asked to use three different pairs of safety glasses and to select the one that they preferred. The results of the survey showed that 120 preferred product A, 85 preferred product B, and 45 preferred product C. The marketing research firm decides to select a worker at random and interview him more thoroughly about his choice. What is the probability that the worker interviewed prefers product B?

If you think about the problem you will see that the experiment consists of selecting a worker at random. Thus, the sample space of the experiment will consist of the 250 workers,

$$S = \{W1, W2, W3, \ldots, W248, W249, W250\}$$

Let B be the event that the worker chosen preferred product B. Then $P(B) = \frac{85}{250}$, since 85 outcomes (workers) in S corresponded to the event B, and there were 250 possible outcomes. ∎

7.2.2 Combinations of Events—OR and AND

Calculating the probability of a single event is not difficult. Much of the time, however, we are interested in looking at more than one event. For example, if an experiment consisted of selecting employees at random from a company, we might be interested in the event A = the employee is an hourly worker, or the event B = the employee participates in the company's stock purchase program. From what we have learned, if we had the data on the number of employees in the company and the number that fell into each category, we could calculate P(A) and P(B) without much trouble. But suppose that the company is interested in how these two events behave *together?* We must look at the events **A OR B** and **A AND B.**

> The event **A OR B** describes the event when either A happens or B happens or they both happen.

> The event **A AND B** is the event that A and B both occur.

Let's look at the problem in terms of one of the smaller problems we studied earlier. When you roll a single die, the sample space is S = {1, 2, 3, 4, 5, 6}. If you let the event A = the number that comes up is even, and the event B = the number that comes up is a 3, then you can easily see that $P(A) = \frac{3}{6}$ and $P(B) = \frac{1}{6}$. What if you were interested in the event that the number that comes up is even OR a 3? The same rules that we used to calculate simple probabilities will apply here.

If we look at the sample space S we can count how many of the possible outcomes correspond to the event A OR B. We see that 2, 4, and 6 correspond to A and that 3 corresponds to B. Thus, using the rules for probability when events are equally likely, we find that $P(A \text{ OR } B) = \frac{4}{6}$, since 4 outcomes out of the 6 possible outcomes correspond to what we are interested in.

You may have noticed two things about this problem. First, the answer, $\frac{4}{6}$, is simply the sum of the two individual probabilities, $\frac{3}{6}$ and $\frac{1}{6}$. This is not a coincidence. Second, the events A and B have no outcomes in common. In probability we refer to these kinds of events as **mutually exclusive.**

> Two events, A and B, are said to be ***mutually exclusive*** if they have no outcomes in common.

When two events are mutually exclusive, then the probability that A occurs or B occurs, **P(A OR B),** is the sum of the individual probabilities. This is known as the **simple addition rule** and is found by

Formula for simple addition rule: P(A OR B)

$$P(A \text{ OR } B) = P(A) + P(B)$$

The simple addition rule easily extends to any number of mutually exclusive events. For example, if A, B C, and D are four mutually exclusive events, then

$$P(A \text{ OR } B \text{ OR } C \text{ OR } D) = P(A) + P(B) + P(C) + P(D)$$

EXAMPLE 7.5 Flipping Two Coins

Finding **P(A OR B)**

Remember that Joe and Tom were matching coin flips to see who would pay for dinner. The sample space for the experiment was S = {HH, HT, TH, TT}.

If A is the event that the two coins both come up tails, and B is the event that the two coins do not match, find P(A OR B).

You can see that one outcome, TT, corresponds to the event A and that two outcomes, HT and TH, correspond to the event B, so that

$$P(A \text{ OR } B) = \tfrac{3}{4}$$

You can also see that A and B have no outcomes in common. That is, A and B are mutually exclusive, so that you should be able to find P(A OR B) by adding P(A) and P(B). Since $P(A) = \tfrac{1}{4}$ and $P(B) = \tfrac{2}{4}$ (these should be easy for you by now),

$$P(A \text{ OR } B) = \tfrac{1}{4} + \tfrac{2}{4} = \tfrac{3}{4} \qquad ■$$

In probability, as in any kind of mathematics, we must be very precise in our use of words. The word OR in probability is *inclusive.* That means when we talk about the event A OR B occurring we mean that A occurs or B occurs or BOTH occur. This was not a problem when we were talking about rolling a die, since there is no way that a number can be both even and a 3!

EXAMPLE 7.6 Credit Problems

Simple Addition Rule

In the example with the Chamber of Commerce and the small businesses, the Chamber of Commerce wants to know the probability that the company selected to be interviewed in depth is one of the largest (sales over $20 million) or one of the smallest (sales under $1 million).

In the last example we defined A to be the event that the company has sales of over $20 million and we found that P(A) was $\tfrac{21}{166}$. We can define event B to be the event that the company chosen has sales under $1 million. What we are trying to determine here is P(A OR B).

It is easy to see that the events A and B are mutually exclusive, that is, that they cannot have any company in common. This is the same as saying that there is no company that has sales under $1 million and over $20 million. We can use the simple addition rule here since the sample space is much too large to work with comfortably:

$$P(A\ OR\ B) = P(A) + P(B)$$

We need to find P(B) first. The results of the survey showed that 60 companies had sales under \$1 million, so $P(B) = \frac{60}{166}$. Thus,

$$P(A\ OR\ B) = \tfrac{21}{166} + \tfrac{60}{166} = \tfrac{81}{166}$$ ■

EXAMPLE 7.7 Product Preference

Finding the Probability of **A OR B**

In the product preference example, a random sample of 250 production workers were asked to use three different pairs of safety glasses and to select the one that they preferred. The results of the survey found that 120 preferred product A, 85 preferred product B, and 45 preferred product C.

 We defined an experiment where a worker was selected at random for further interviewing. We let the event B = the worker preferred product B and we found that $P(B) = \frac{85}{250}$. Let the event C = the worker preferred product C.

 Find the probability that the worker selected for further interviewing preferred product B or C.

 Since the sample space for this experiment does not help us, we must use the simple addition rule. First we need to find P(C). Since 45 outcomes (workers) in the sample space preferred product C, we know that

$$P(C) = \tfrac{45}{250}$$

To find P(B OR C) we use the simple addition rule and find that

$$P(B\ OR\ C) = \tfrac{85}{250} + \tfrac{45}{250} = \tfrac{130}{250}$$ ■

 Up to this point we have been expressing probabilities as fractions without reducing them. This is only to illustrate what numbers we use to calculate the probabilities. In fact, probabilities are often expressed as decimals or percentages. In the previous example you could also say that the probability that the worker selected preferred product B or C is 0.52 or 52%. Looking at the probability as a percentage is really quite illuminating in this case. We see that while product A had the largest number of people preferring it, a majority of the workers prefer something else.

 ***TRY IT NOW!***

The Spinner Problem

Calculating the Probability of **A OR B**

The sample space for the experiment of spinning the two spinners is

$$S = \{1A, 1B, 1C, 2A, 2B, 2C, 3A, 3B, 3C\}$$

Let A be the event that the first spinner comes up a 1 and let B be the event that it comes up a 3.

 Find the probability that A OR B occurs using the sample space.

Now find the same probability using the simple addition rule.

Why are the two answers the same?

The simple addition rule for probability that you just learned applies only when the events of interest have nothing in common, that is, they are mutually exclusive. Sometimes that will always be the case, as in the Product Preference example. It is not physically possible for a single worker to prefer more than one of the products, just as it is not possible for a single roll of a die to result in a 2 and a 3 at the same time. What happens when this is *not* the case, when it is possible that both of the events, A AND B, can occur?

Look at the example of rolling a single die. The sample space for the experiment is

$$S = \{1, 2, 3, 4, 5, 6\}$$

When we let A be the event that the number that came up was even and B be the event that the number that came up was a 3, these events had nothing in common. It is impossible for them to both occur on a *single trial* of the experiment. The law of probability worked just fine and $P(A \text{ OR } B) = \frac{3}{6} + \frac{1}{6} = \frac{4}{6}$.

Suppose that we define a new event, C, where

$$C = \text{the number that comes up is exactly divisible by 3}$$

This is the same as asking for the probability that A OR C will occur.

What is the probability that the number that comes up is even or exactly divisible by 3? Remember that our definition of the word OR is that one or the other or *both* of the events can occur. This means that we are looking at any outcomes in S that are even, or exactly divisible by 3, or both even AND exactly divisible by 3. Looking at the sample space, you can see that there are four outcomes that fit this description, 2, 3, 4, and 6. Thus, using the formula for probability, $P(A \text{ OR } C) = \frac{4}{6}$. What happens if we try to use the simple addition rule?

From the sample space you can see that $P(A) = \frac{3}{6}$ (there are three outcomes that are even) and $P(C) = \frac{2}{6}$ (there are two outcomes that are exactly divisible by 3). If we add these two probabilities we get: $\frac{3}{6} + \frac{2}{6} = \frac{5}{6}$. Wait! This is *not* the answer we got using the definition. What went wrong?

What went wrong is what usually goes wrong when mathematics leads to an incorrect answer—we violated the rules, in this case, the rules for using the simple addition rule. The simple addition rule is valid only if the events of interest are *mutually exclusive*. In this case, events A and C are *not* mutually exclusive; they have an outcome in common, the outcome of a 6.

When we calculated P(A), the outcome that the die turned up a 6 was included in that probability. Then, when we calculated P(C), the outcome of a 6 was included again. When we added the two probabilities together, the outcome got included twice, once for each event. Thus, the answer we obtained was too large by $\frac{1}{6}$, which is the probability that the die turns up a 6.

How can we adjust the simple addition rule of probability to work in situations when the events are NOT mutually exclusive? You can see from the example that the problem occurs when an outcome is included in *both* events, that is, in A AND C. The answer obtained by adding the individual probabilities was too large because the probability that both A AND C occur, P(A AND C), is included in both individual probabilities. If we consider this we can come up with a **general addition rule** for probability that considers the probability of A OR B when the events are not mutually exclusive:

$$P(A \text{ OR } B) = P(A) + P(B) - P(A \text{ AND } B)$$

Formula for general addition rule: P(A OR B)

EXAMPLE 7.8 Flipping Two Coins

Calculating P(A OR B)

In the example where Joe and Tom are matching coins, the sample space is

$$S = \{HH, HT, TH, TT\}$$

We let A be the event that both coins come up tails and we know that $P(A) = \frac{1}{4}$.

Suppose we define the event B = the coins match. What is the probability that the coins match or they both come up tails?

From the sample space we can see that two outcomes (HH and TT) correspond to the event A OR B. Thus, $P(A \text{ OR } B) = \frac{2}{4}$. To use the addition rule we need to look at events A and B and decide if they have any elements in common. We see that they do, since the outcome TT satisfies both the definition of A and the definition of B. We will have to use the general rule and remember to subtract P(A AND B). Thus, we find

$$P(A \text{ OR } B) = P(A) + P(B) - P(A \text{ AND } B)$$

or

$$P(A \text{ OR } B) = \frac{1}{4} + \frac{2}{4} - \frac{1}{4} = \frac{2}{4}$$

which we know is the correct answer. ∎

In Chapter 3 you learned that the first step in organizing and summarizing data for a single variable is to create a frequency table. The frequency table contains the number of occurrences of each value of the variable expressed as a count, fraction, decimal, or percentage. When data are collected on two *related* variables, they are organized by using a **cross-classification table** or **contingency table.**

A *contingency table* is a table whose rows represent the possible values of one variable and whose columns represent the possible values for a second variable. The entries in the table are the number of times that each *pair* of values occurs.

EXAMPLE 7.9   Credit Problems

The General Addition Rule

The size of a company cannot be determined by sales alone. It also depends on the price of the products to a large extent, so the Chamber of Commerce also asked a question about the number of employees that the company employed. The following table is a contingency table based on the questions about sales and number of employees:

Employees	No response	Under $1 m	$1–5 m	$6–10 m	$11–20 m	Over 20 m	Grand Total
0–5	0	44	7	7	0	0	58
6–10	0	12	9	3	3	0	27
11–50	1	4	4	14	3	3	29
51–150	0	0	1	8	5	7	21
151–250	0	0	0	6	5	6	17
Over 250	0	0	0	4	5	5	14
Grand total	1	60	21	42	21	21	166

Suppose that the Chamber of Commerce is interested in knowing the probability that the company chosen for further interviewing had sales over $20 million (A) or more than 250 employees (C). How can we find this probability, P(A OR C)?

If we look at the two events we see that $P(A) = \frac{21}{166}$ and $P(C) = \frac{14}{166}$. Now, because we are going to *add* the probabilities, we have to determine whether the two events have any outcomes in common. We need to know whether there are any businesses that have both sales over $20 million and more than 250 employees. Looking at the table, we see that 5 businesses are included in both events. We have to use the general addition rule:

$$P(A \text{ OR } C) = P(A) + P(C) - P(A \text{ AND } C)$$

or

$$P(A \text{ OR } C) = \tfrac{21}{166} + \tfrac{14}{166} - \tfrac{5}{166} = \tfrac{30}{166} \qquad ■$$

When you are looking for the probability of the event A OR B in a contingency table problem, it is not difficult to recognize when you need to use the general addition rule or when the simple addition rule will work. You can think about the outcomes that the events might have in common as the *intersection* of the two simple events in the table. If one event is represented by a row and the other by a column, then there will be a place where they intersect or overlap and you will need to use the general addition rule. If both of the simple events A and B are rows or both columns, then they cannot intersect, and the simple addition rule applies.

EXAMPLE 7.10   Student Status

Using the General Addition Rule

The School of Business is concerned about the number of upper level students who are enrolled in the Introductory Statistics course. The school collected data on 28 students and asked the students their year and whether they had transferred to the University. The data are shown in the following contingency table:

| | Status | | |
Year	Nontransfer	Transfer	Total
Fr	3	1	4
So	10	2	12
Jr	2	0	2
Sr	1	9	10
Total	16	12	28

Suppose that the committee decides to select a student at random and look at that student's records more closely. The sample space of the experiment is the twenty-eight students. Since every student is just as likely to be selected, the outcomes in S are equally likely. What is the probability that the student whose records are examined is a sophomore or a transfer student?

Let S represent the event that the student selected is a sophomore and T be the event that the student selected is a transfer student. We are looking for P(S OR T).

To start, we will find P(S) and P(T). From the table we can find that $P(S) = \frac{12}{28}$ and $P(T) = \frac{12}{28}$.

Next, we need to think about whether these events have any outcomes in common. That is, are there any students who are both sophomores AND transfer students? Again, looking at the table, we see that there are 2 students who are classified as sophomores and transfer students. These two students are included in both P(S) and P(T), so we will have to use the general addition rule. Thus,

In general, it is helpful to use letters to represent events that are coded to the actual data rather than just A or B.

$$P(S \text{ OR } T) = \frac{12}{28} + \frac{12}{28} - \frac{2}{28} = \frac{22}{28}$$

If we look at the table and use the cells that correspond to S OR T instead of the totals we get

$$P(S \text{ OR } T) = \frac{10 + 2 + 1 + 0 + 9}{28} = \frac{22}{28}$$

which is the same answer. ■

TRY IT NOW!

Quality Problems

Using the General Addition Rule

The company that manufactures cardboard boxes collected data on the defect type and production shift. The data are summarized in the contingency table below:

| | Shift | | | |
Defect	1	2	3	Total
Color	8	4	3	15
Printing	6	5	2	13
Skewness	0	2	0	2
Total	14	11	5	30

If a box has more than one defect, then it is classified by the more serious of the defects only.

Suppose that a box from the sample is selected at random and examined more closely. What is the probability that the box has a color defect?

What is the probability that the box was produced during the second shift?

Is it possible for the selected box to have a color defect and to have been produced on the second shift? If so, what is the probability?

What is the probability that the selected box will have a color defect or will have been produced on the second shift?

7.2.3 Probabilities as Relative Frequencies

At this point you may wonder why we need to study probability at all. It is a reasonable question if you think of probability only in terms of coins, dice, and spinners, or as random samples taken from known populations. These contexts are good for explaining how probabilities are calculated. The situations serve as models for examining probabilities for other situations.

In the previous chapters of this book you looked at methods for describing sample data taken from some population. We talked about the fact that the descriptions are exact for the sample, but are only estimates when applied to the whole population. How good those estimates are depends on many factors, such as the size of the sample and how well the sample represents the population of interest.

The same is true of probability models. When you collect sample data and calculate relative frequencies you can find, exactly, the probability that an item taken from the sample will have some characteristic(s) of interest. If the data are a good representation of the population, then we can also use the relative frequencies as estimates of the true probabilities for the population. Probabilities calculated in this way are often called **empirical** probabilities. The same rules of probability that you have already learned apply to these problems as well.

> An *empirical probability* is one that is calculated from sample data and is an estimate for the true probability.

ANS. $P(C) = \frac{15}{30}$, $P(2) = \frac{11}{30}$, YES, $\frac{4}{30}$, $\frac{22}{30}$.

EXAMPLE 7.11   Restaurant Survey

Probabilities as Relative Frequencies

A local marketing firm took a random sample of people in a large city to learn what kind of restaurants they think the city needs. The firm asked the customers to choose the type from the following list:

- Fast food
- Family restaurant
- Adult economical
- Adult moderate
- Adult upscale

As a result of their research they obtained the following data:

Type of restaurant	Number of people	Relative frequency (%)
Fast food	127	20
Family restaurant	234	36
Adult economical	158	24
Adult moderate	72	11
Adult upscale	56	9
Total	647	100.00

If we assume that the sample was a good representation of the adult residents of the city, what is the probability that a person selected at random in the city would favor some type of adult restaurant?

This question is really asking for the probability that a person would favor either an adult economical (AE), an adult moderate (AM), or an adult upscale (AU) restaurant. Since the events are mutually exclusive, we can use the simple addition rule to find the answer. What has changed is that our sample space is no longer the 647 original respondents to the survey. Instead, the sample space of the experiment becomes the possible responses that a person can give, or

$$S = \{\text{fast food, family, adult economical, adult moderate, adult upscale}\}$$

and the outcomes are no longer equally likely. We can, however, use the relative frequencies as estimates of the probabilities for each outcome:

$$P(\text{AE OR AM OR AU}) = 24\% + 11\% + 9\% = 44\% \qquad \blacksquare$$

7.2.4 Exercises—Learning It!

7.1 In 1997, the competition for local phone service increased significantly as a result of changes in federal law. A survey of over 10,000 people in a large metropolitan area asked whether they rated their current local phone service provider as Excellent, Very Good–Good, Satisfactory, or Poor–Very Poor. The results were tabulated and are shown in the table below:

Rating	Excellent	Very Good–Good	Satisfactory	Poor–Very Poor
% Responding	12	21	25	42

(a) What is the probability that a person selected at random from the area will rate his current phone service as Excellent or Very Good–Good?

(b) What is the probability that a person selected at random from the area will not rate his service as Poor–Very Poor?

7.2 A company was interested in looking at the way in which employees used the 2 floating holidays that were part of their benefits packages. They surveyed 300 employees and asked them what type of job they had and how they used the holidays. The answers were tabulated into a contingency table:

| Type of Job | *How Days Were Used* | | |
	Took Actual Holiday	**Added to Vacation**	**Took Random Days**
Professional	5	17	51
Clerical	13	46	32
Hourly	53	78	5
	71	141	88

(a) What is the probability that an employee surveyed took the floating holidays on the actual holiday?

(b) What is the probability that an employee surveyed was professional?

(c) What is the probability that an employee surveyed was clerical and added the floating holidays to his vacation?

(d) What is the probability that an employee surveyed took the floating holidays as random days off or was an hourly worker?

7.3 A computer magazine surveyed its readers to determine how likely it was that people who planned to purchase new computers in the near future would buy a portable/notebook or desktop model. The results are tabulated in the table below:

| Type of Computer | *When Purchase Will Be Made* | | |
	0–3 Months	**3–6 Months**	**6–12 Months**
Notebook/Portable	34	156	258
Desktop	56	346	128

(a) What is the probability that a person surveyed planned to buy a desktop model in the next 3–6 months?

(b) What is the probability that a person surveyed planned to buy a computer in the next 0–3 months or that the person was planning to buy a desktop model?

(c) What is the probability that a person was planning to buy a notebook computer and that the person planned to make the purchase in the next 0–3 or 3–6 months?

(d) What is the probability that a person did not plan to make a purchase in the next 6 months?

7.4 A marketing firm in the Northeast was looking at the type of medication that people with allergies took during the autumn allergy season. In particular, the firm wanted to know whether the person took medication daily and whether that medication was prescribed by a physician or purchased over the counter. The results of the survey are in the table:

| Frequency of Taking Medicine | *Type of Medication* | |
	Prescription	**Over the Counter**
Daily	86	43
As needed (sporadic)	23	156

(a) What is the probability that an allergy sufferer took prescription medication daily?

(b) What is the probability that an allergy sufferer took over-the-counter medication or that he took the medication daily?

(c) What is the probability that an allergy sufferer took medication daily and that he used over-the-counter medication?

7.5 A report by the Department of Justice on rape victims reports on interviews with 3721 victims. The attacks were classified by age of the victim and the relationship of the victim to the rapist. The results of the study are given in the next table.

Age of Victim	Relationship of Rapist		
	Family	Acquaintance or Friend	Stranger
Under 12	153	167	13
12 to 17	230	746	172
Over 17	269	1232	739

(a) What is the probability that a victim was under 12 years of age?

(b) What is the probability that a victim was between 12 and 17 and that the rapist was a member of the family?

(c) What is the probability that a victim was under 12 or that the rapist was an acquaintance or friend?

(d) What is the probability that the victim was not under 12 years of age?

(e) What is the probability that the rapist was not a family member, acquaintance, or friend?

7.3 RANDOM VARIABLES

When you learned the different techniques for summarizing and analyzing data, you learned that while qualitative data are important for understanding and interpreting the information obtained from a set of data, there is not much you can do with the data itself. Most statistical analyses use numerical or quantitative data as their basis. In much the same way, we also prefer to discuss probabilities for experiments whose outcomes are numerical.

You know from algebra that a variable is a quantity whose value can change or vary. In algebra the exact value that the variable takes on depends on the equation that includes it. We have also referred to the different kinds of data that can be collected from a population as variables. The exact value that a statistical variable will take on depends on the laws of chance or probability. When the variables in question are quantitative, they are known as **random variables.**

> A *random variable, X,* is a quantitative variable whose value varies according to the rules of probability.

When the outcome of an experiment is a random variable, then the elements of the sample space are all of the possible values that the variable can have. Random variables, like data, can be discrete (integers) or continuous (real numbers). In our initial discussion of random variables we will consider only the discrete case.

In the experiment of rolling a single die and recording the number of spots that are on the top face the sample space for the experiment can be written as

$$S = \{1, 2, 3, 4, 5, 6\}$$

The outcomes are numerical and the exact value that will turn up varies. We know from the rules of probability that each of the values is equally likely, or has an equal probability of happening. Thus, $X =$ the number of spots on the top face of a die is a random variable.

EXAMPLE 7.12 Tossing Two Coins

Defining a Random Variable

In the previous section we looked at an experiment in which two coins were tossed. The sample space of that experiment was

$$S = \{HH, HT, TH, TT\}$$

We always use capital letters, like X, to represent the random variable, and lowercase letters, such as x, to represent the values of the random variable.

This experiment does not involve a random variable, since the possible outcomes are not numerical. It is possible, however, to describe the outcomes of the experiment numerically by defining a random variable for the experiment. Let $X =$ the number of heads that appear on the two coins. The possible values of X are $x = 0, 1,$ and 2. Every time you perform the experiment you do not know what will happen, but the laws of probability give you some insight into what is likely to occur. ∎

7.3.1 Probability Distribution of a Discrete Random Variable

The rules of probability that describe the way a random variable behaves are known as the **probability distribution** of the random variable. The probability distribution of a discrete random variable assigns a probability to each of the possible values that can happen.

We read p(x) as "p of x."

> The *probability distribution* of a random variable, X, written as **p(x)**, gives the probability that the random variable will take on each of its possible values.

The notation that we use for the probability distribution of a random variable is similar to the notation for the probability of an event. For a random variable,

$$p(x) = P(X = x) \quad \text{for all possible values of } X$$

Most often the probability distribution is written in the form of a table.

EXAMPLE 7.13   Tossing Two Coins

Writing the Probability Distribution

In the experiment of tossing two coins we saw that $X =$ the number of heads is a random variable that can take on three values: 0, 1, and 2. Using the rules of probability that we learned, we can find the probability distribution of X, the number of heads in two tosses of a coin:

$$p(0) = P(X = 0) \text{ corresponds to only one outcome, TT, so } p(0) = \tfrac{1}{4}$$
$$p(1) = P(X = 1) \text{ corresponds to two outcomes, HT and TH, so } p(1) = \tfrac{2}{4}$$
$$p(2) = P(X = 2) \text{ corresponds to one outcome, HH, so } p(2) = \tfrac{1}{4}$$

In table form we would write this as

x	0	1	2
p(x)	$\frac{1}{4}$	$\frac{2}{4}$	$\frac{1}{4}$

∎

The rules for probability distributions are the same as the rules for probabilities. For each value of the random variable, X,

   1. $0 \le p(x) \le 1$ (probabilities are numbers between 0 and 1 inclusive)

   2. $\Sigma p(x) = 1$ for all values of x (the probabilities must sum to 1)

Notice that the outcomes of a random variable are *mutually exclusive.* This means that whenever we are interested in finding the probability that the random variable will take on one of its values OR another of its values we can use the simple addition rule.

EXAMPLE 7.14 Newspaper Sales

Finding Probabilities from a Probability Distribution

The number of copies of *The Wall Street Journal (WSJ)* that are sold daily by a convenience store is a random variable X, which is described by the probability distribution

x	0	1	2	3	4	5
$p(x)$	0.10	0.12	0.25	0.30	0.20	0.03

What is the probability that on any given day the convenience store sells exactly two copies of the *WSJ*?

We can obtain this answer directly from the table:

$$p(2) = 0.25$$

What is the probability that two or three copies are sold on any given day?

It is first important to recognize that we are trying to find $P(X = 2 \text{ OR } X = 3)$. Since the values of a random variable are mutually exclusive, we can simply add the probabilities to obtain the answer:

$$P(X = 2 \text{ OR } X = 3) = p(2) + p(3) = 0.25 + 0.30 = 0.55 \quad \blacksquare$$

Notation and Probabilities

Very often we are interested in the probability that a random variable takes on any one of a set of its possible values. In particular, we might be interested in finding the probability that the random variable takes on a value that is **"at least x,"** **"more than x,"** **"at most x,"** **"less than x,"** **"between x_1 and x_2,"** or **"between x_1 and x_2 inclusive."** There is nothing new that you need to know to find the probabilities. It just takes a little practice to be able to recognize each of the different problems and to write down what you are looking for. Since the words can get cumbersome, a standard notation is used.

As an example, we will use a random variable X that can take on values of $x = 0, 1, 2, 3, \ldots, n$. Table 7.1 provides a summary of each problem, what it means, and the correct notation.

TABLE 7.1 Probability Notation Summary

Find the probability that X takes on a value that is . . .	What it means	Notation
at least x	All of the values of the random variable that are the value x or larger (up to n)	$P(X \geq x)$
more than x	All of the values of the random variable that are larger than the value x (up to n)	$P(X > x)$
at most x	All of the values of the random variable that are the value x or less	$P(X \leq x)$
less than x	All of the values of the random variable that are smaller than the value x	$P(X < x)$
between x_1 and x_2 inclusive	All of the values of the random variable that start with the value x_1 and go up to and include the value x_2	$P(x_1 \leq X \leq x_2)$
between x_1 and x_2	All of the values of the random variable that are larger than the value x_1 and smaller than the value x_2	$P(x_1 < X < x_2)$

EXAMPLE 7.15   Newspaper Sales

Finding Interval Probabilities

The convenience store that sells the *WSJ* wants to know a bit more about the probability that it will sell the newspaper. The probability distribution of *X*, the number of copies of the *WSJ* sold per day, is given by

x	0	1	2	3	4	5
p(x)	0.10	0.12	0.25	0.30	0.20	0.03

In particular, the store needs to sell at least 3 copies per day to make a profit from the sales. What is the probability that the store will make a profit on any given day?

 We are looking for the probability that the store sells at least 3 copies, or $P(X \geq 3)$. This means $P(X = 3 \text{ OR } X = 4 \text{ OR } X = 5)$, which is calculated as

$$p(3) + p(4) + p(5) = 0.30 + 0.20 + 0.03 = 0.53$$

 What is the probability that the store will not make a profit?

 We can answer this question directly by determining that the store will not make a profit if it sells less than 3 copies of the newspaper and finding $P(X < 3)$. We can also do this problem by making use of the complement of an event. The events "make a profit" or "$X \geq 3$" and "do not make a profit" or "$X < 3$" are complements of each other. This means that their probabilities must sum to 1. Since we already know that $P(X \geq 3) = 0.53$, we can find $P(X < 3)$ by taking

$$1 - P(X \leq 3) = 1 - 0.53 = 0.47 \qquad \blacksquare$$

 **TRY IT NOW!**

Defective Diskettes

Finding Interval Probabilities

A company that sells computer diskettes in bulk packages for a warehouse club outlet knows that the number of defective diskettes in a package is a random variable with the probability distribution given below:

x	0	1	2	3	4	5	6
p(x)	0.30	0.21	0.12	0.10	0.10	0.09	0.08

Find the probability that a package of the diskettes will contain at least 3 defective disks.

(continued)

Find the probability that the package will contain between 2 and 5 defective diskettes.

Find the probability that the number of defective diskettes will be at most 2.

7.3.2 Probability Histograms

Random variables and their probability distributions are the models for the populations from which our sample data are taken. You learned in Chapter 3 that you can display quantitative data using a relative frequency table or a relative frequency histogram. In much the same way, a random variable can be displayed with a probability distribution table or a probability distribution histogram.

EXAMPLE 7.16 Newspaper Sales

Creating a Probability Histogram

The convenience store that sells the *WSJ* would like to see what the distribution of newspaper sales looks like. The store creates a probability histogram for the random variable:

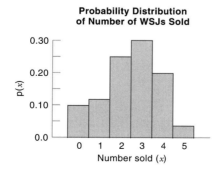

**Probability Distribution
of Number of WSJs Sold**

The histogram shows that the probability distribution of sales is not very variable, approximately symmetric, and centered at about 3 newspapers per day. ∎

In a probability histogram the area of each rectangle is equal to the probability that the random variable will take on the given value. This may not seem important right now, but we use this fact later in the chapter when we move from discrete to continuous random variables.

 *TRY IT NOW!*

Defective Diskettes
Creating a Probability Histogram

The company that sells computer diskettes in bulk packages for a warehouse club would like to have a picture of how the number of defective diskettes in a package behaves. The probability distribution is given below:

x	0	1	2	3	4	5	6
p(x)	0.30	0.21	0.12	0.10	0.10	0.09	0.08

Create a probability histogram for the number of defective diskettes.

Use the probability histogram to describe the distribution of the number of defective diskettes in a package.

7.3.3 Exercises—Learning It!

7.6   The number of employees who call in sick on any given day in a small business is a random variable with probability distribution:

x	0	1	2	3	4	5	6
p(x)	0.10	0.23	0.18	0.16	0.13	0.10	0.10

ANS.   **Probability Distribution of Number of Defective Diskettes**   SKEWED RIGHT, MUCH MORE LIKELY TO FIND A SMALL NUMBER OF DEFECTIVES

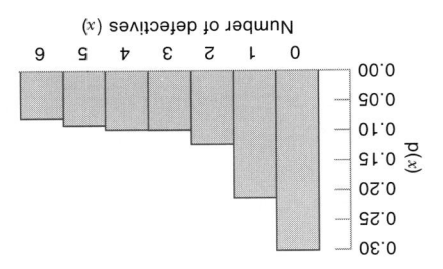

(a) What is the probability that on any given day at most 4 employees call in sick?

(b) What is the probability that between 2 and 4 employees call in sick?

(c) What is the probability that more than 4 employees call in sick?

7.7 The number of members who cannot get a tee time at a local country club is a random variable with probability distribution given by

x	0	1	2	3	4	5	6	7	8
$p(x)$	0.11	0.12	0.13	0.19	0.12	0.09	0.09	0.08	0.07

(a) What is the probability that between 2 and 5 members inclusive cannot get tee times?

(b) What is the probability that less than 3 cannot get tee times?

(c) What is the probability that at least 6 cannot get tee times?

(d) What is the probability that at most 4 cannot get tee times?

7.8 The number of times that a person gets a busy signal when calling the local cable television office is a random variable with the following probability distribution:

x	0	1	2	3	4
$p(x)$	0.23	0.34	0.17	0.15	?

(a) What is $p(4)$?

(b) What is the probability that a person will not get a busy signal?

(c) What is the probability that a person gets at least 1 busy signal?

(d) What is the probability that a person gets more than 2 busy signals?

7.9 A company that packages small items for resale is looking at the problems of incorrect packaging. The company finds that in a box that is supposed to contain two dozen items, the number of missing items is a random variable with probability distribution:

x	0	1	2	3	4	5	6
$p(x)$	0.13	0.17	0.26	0.30	0.07	0.05	0.02

(a) What is the probability that in a box of 2 dozen, exactly 3 are missing?

(b) What is the probability that the number of missing items is less than 4?

(c) What is the probability that there are exactly 20 items in the box?

(d) What is the probability that there are more than 20 items in the box?

7.10 A large airline keeps track of the number of no-shows for one of its most important commuter flights. Over time the airline has found that the number of ticketed passengers who do not show up is a random variable with the following probability distribution:

x	0	1	2	3	4	5	6	7	8
$p(x)$	0.05	0.08	0.13	0.23	0.18	0.13	0.08	0.06	0.06

(a) What is the probability that at least 3 ticketed passengers do not show up for the flight?

(b) What is the probability that between 2 and 5 passengers do not show up for the flight?

(c) What is the probability that not more than 6 passengers do not show up for the flight?

(d) The aircraft used for the flight has 35 seats. If the airline routinely overbooks the flight by 4 passengers, what is the probability that on any given day every ticketed passenger who shows up will get a seat?

7.4 THE BINOMIAL PROBABILITY DISTRIBUTION

In the previous section you learned the definition of a random variable and a probability distribution. Each of the random variables that you looked at was described by a probability distribution that was given in a table. The random variables represented many different types of data that might be collected in a statistical study. In fact, most of the random variables that we see in the real world fall into specific categories and can be described by a set of special models or probability distributions. We now look at one of these models, the binomial probability distribution, in detail.

7.4.1 The Binomial Model

One of the most common types of data that people collect is data on the number of times that some phenomenon occurs in a sample of given size. For example, you may be interested in the number of people in a sample who are in favor of certain legislation or in the number of people who like a new flavor of ice cream. The sample does not have to consist of people; it could be the number of defective diskettes in a box of 10, or the number of times a coin turns up heads in a certain number of tosses. The random variable in each case is the *number of times the phenomenon occurs* in the sample. All of these types of data are examples of a **binomial random variable.**

> A *binomial random variable* is the number of successes in n trials or in a sample of size n.

Certain characteristics define binomial random variables:

- There are a fixed number of identical trials of an experiment. *This is the same as taking a sample of size n. Just think of each trial as selecting the next item for the sample.*
- The outcome for each trial of the experiment can be classified in one of two ways: a *success,* S (when the phenomenon of interest happens) or a *failure,* F (when the phenomenon of interest does not happen).
- The probability that a success occurs in any sample element or on any trial of the experiment, π, is the same for each element or trial. This also means that the probability of a failure, which is $1 - \pi$, is also constant. *This means that nothing happens over the course of the experiment to change the probability of a success, such as a change in population.*
- The trials of the experiment are independent. *This means that the outcome from one trial does not affect the outcomes of subsequent trials.*
- The random variable is the number of successes that occur in the n trials of the experiment.

In finding real-world situations that exactly fit the characteristics of a binomial random variable when we are sampling from a population it is difficult to ensure that the trials are independent of each other and that π, the probability of a success on any trial or the proportion of successes in the population, remains constant. Keep in mind that the reason we want the trials to be independent is that if they are not, then the probability of a success will change from trial to trial. This is a violation of the third characteristic. In fact, the only way to ensure independence when sampling, is to sample from a finite population with replacement or from an infinite population.

The first problem, the idea of sampling with replacement, is not very appealing from a practical point of view. Since the purpose of statistics is to obtain information, sampling the same item or person repeatedly will not add to the information contained in the sample. The second problem, the idea of an infinite population, is more philosophical. Although there are no truly infinite populations, many populations are infinite for all practical purposes, such as the number of people in the world (or even in a large country for that matter) or all of the production, past and present, of a machine or factory.

What happens when the population from which we sample is finite and we do not sample with replacement?

EXAMPLE 7.17   CD Jewel Cases

Sampling Without Replacement

Suppose that you are inspecting a shipment of CD jewel cases for cracks in the cover. The box from which you are sampling contains 30 cases, of which 5 have cracked covers. The first time that you select an item from the population, the probability of getting a case with a cracked cover (the proportion of successes in the population) is $\frac{5}{30} = 0.167$. Now, you certainly are not going to sample with replacement in this situation. What would you gain by throwing a case you have already inspected back in the box, particularly if it is defective?

The next time you sample an item from the box, what is the probability of a success? If you are thinking. "It depends on what happened the first time," then you are absolutely correct! Look at the diagram below. You see that as you continue to sample from the box, the probability of obtaining a case with a cracked cover on *that trial* is different from the previous trial; that is, π changes from trial to trial. The trials are not independent because the outcome of one trial directly affects the outcome of the next!

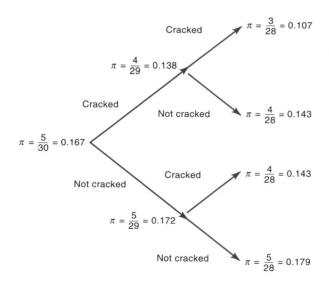

What could we do to fix the problem? The real issue is that when the denominator in the proportion is small and changing, the value of the fraction changes considerably. What happens if the denominator is much larger? Will the change still be noticeable?

Let's look at the same problem, but with everything increased proportionally. Suppose that the box contains 3000 jewel cases, of which 500 have cracked covers. In this case π, the probability or proportion of cracked cases, is still 0.167 when we begin sampling. Now what happens to π as items are sampled and not replaced in the population? The next figure illustrates the results:

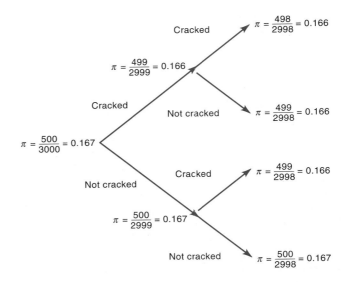

You see that although π changes, the changes are much less noticeable. In fact, if you were writing π to the nearest percent you would not see the change at all. ■

Clearly, as the population gets larger the problem gets smaller. The magnitude of the change in π depends on two things: the size of the population and the size of the sample that is drawn from the population. In general, if the population size is large and the sample size is small relative to the population size, then the random variable represented by the number of successes in the sample can be assumed to be binomial.

Usually the population must be at least 1000 to ensure that the changes do not occur before the third decimal place and the sample size should be no more than 10% of the population.

EXAMPLE 7.18 Credit Problems

Identifying a Binomial Random Variable

In addition to identifying variables, the Chamber of Commerce that is interested in the credit problems of small businesses asked the small businesses a series of questions about the problems they might be experiencing. One of these questions, to which the businesses responded Yes or No, was "Are you experiencing credit problems?" In previous applications of this questionnaire, the Chamber of Commerce found that approximately 10% of the businesses have reported having such problems.

Is the number of businesses who reported having credit problems in the sample of 166 a binomial random variable?

To answer this question we need to consider the five characteristics of the binomial random variable.

1. Yes. There are a fixed number of trials, in this case 166, and the selection was random.

2. Yes. Each trial can result in only one of 2 outcomes: a success (the company is having credit problems) or a failure (the company is not having credit problems).

3. Yes. If we assume that the time to return the survey was short enough, there would be no economic situations that would change the proportion from 10%.

4. Yes. Here we have to be careful. The sample of 166 is taken from a population of 1536 small businesses in the city. If 10% of the businesses do have credit problems, then that would correspond to 154 successes in the population. In

the worst case, the probability of a success could change to 0 (if all 154 successes in the population wound up in the sample of 166), but that is nearly impossible. (If it did happen, then either the sample was not random or else the estimate of 10% is not at all realistic.) Most likely it would change from 154/1536 (10.03%) at the beginning of the sampling to about 137/1370 (10%) when the entire sample of 166 was completed.

5. Yes. The data collected are the number of businesses in the sample that are experiencing credit-related problems. ■

In general, if the sample size is about 10% of the population then the assumption that the probability of a success does not change is a reasonable one.

EXAMPLE 7.19   Wearing Seatbelts

Identifying a Binomial Random Variable

The state of Connecticut has seatbelt laws for drivers and passengers. It is known that 70% of the drivers in Connecticut wear their seatbelts when they drive. A police trap selects a random sample of five drivers and counts the number that are wearing seatbelts. Is the number of drivers wearing seatbelts in the sample of five a binomial random variable?

To answer this question we need to consider the five characteristics of a binomial random variable.

1. Yes. There are a fixed number of identical trials if we assume that each driver is selected randomly.

2. Yes. Each trial can result in only one of two outcomes: a success (the driver is wearing a seatbelt) or a failure (the driver is not wearing a seatbelt).

3. Yes. If we assume that the sample is taken over a short enough period of time that driving habits will not have changed drastically, then the 70% will remain constant.

4. Yes. If we assume that there is no way that the driver of a car that is stopped can influence the outcome for the next driver then the trials are independent. The population of the drivers in Connecticut is very large and the sample size is definitely small relative to the population so the trials will be independent and the proportion of successes in the population will remain essentially constant.

5. Yes. The data that are collected are the number of people wearing seatbelts in the sample of five selected at random. ■

It is important that you learn to recognize problems to which the binomial distribution applies. It is easy enough to do in a textbook (after some practice) but when you are actually practicing statistics you will need to know what to look for. Keep in mind the five characteristics of a binomial random variable when you are looking at both problems in this book and at data that have been collected in real statistical studies.

 *TRY IT NOW!*

Loan Defaults
Recognizing a Binomial Random Variable

While the Chamber of Commerce is concerned about the problems of small businesses, it must also be sensitive to the problems that the lending institutions have when issuing credit. One of the problems that banks have with small businesses is default on loan payments. It is estimated that approximately 20% of all small businesses with less than 50 employees are at least six months behind in loan payments. The Chamber of Commerce that surveyed the small businesses of a city wants to look at this problem in more detail. It finds that of the 1536 small businesses in the city, 965 have less than 50 employees. It randomly selects 25 of these small businesses, check their credit histories, and count the number of companies in the sample of 25 that are at least six months behind in loan payments.

Does this qualify as a binomial probability distribution?

7.4.2 The Binomial Probability Distribution

Remember, we use parameters for populations, and a probability distribution is a model for a population.

Now that we understand binomial random variables, we need to know how to find their probability distribution. The probability distribution for the binomial random variable will depend on two quantities or parameters: n, the number of trials or the sample size, and p, the proportion of successes in the population. The random variable X is the number of successes in n trials of the experiment.

The probability distribution of X is determined by this formula:

Binomial probability distribution

$$p(x) = \frac{n!}{(x!)(n-x)!} \pi^x (1-\pi)^{n-x} \quad \text{for } x = 0, 1, 2, \ldots, n$$

The terms that contain the ! are known as factorials. $n!$ is equal to $n \times (n-1) \times \cdots \times 1$. The other factorial terms are calculated in a similar manner.

The formula considers the sample size, n, the probability of a success, π, and the value of X that you are interested in, x. The number of trials of the experiment defines the possible values that the random variable can have. If there are n trials of the experiment, then the smallest number of successes that can occur is 0 and the largest number of successes is n. Thus, a binomial random variable, X, can have values $x = 0, 1, 2, \ldots, n$. The probability of a success on any trial, π, will determine how the probabilities are distributed over the values of X. When π is large, then you will expect to see a lot of successes in the sample, so the higher values of X will have the larger probabilities. When π is small, the lower values of X will have the larger probabilities.

You do not really need to know how to use the formula to find the probability distribution of a binomial random variable. There are tables for this purpose for many different combinations of n and π. These tables are identical in use to the probability distribution tables that you used in the previous section. A good set of these tables is found in Appendix A at the end of this book.

Binomial Probability Tables

Most statistical software packages, such as Minitab and SPSS, and many spreadsheets, such as Excel, find binomial probabilities.

The probability distribution tables for the binomial distribution are classified according to the value of n, the number of trials of the experiment. There are tables for values of n from 5 to 30. Each table covers a range of values for π from 0.05 to 0.95. A sample of some of the table for $n = 5$ is shown in Figure 7.1.

$n = 5$						π						
x	**0.05**	**0.10**	**0.20**	**0.25**	**. . .**	**0.50**	**0.60**	**0.70**	**0.75**	**. . .**	**0.90**	**0.95**
0	0.774	0.590	0.328	0.237	. . .	0.031	0.010	0.002	0.001	. . .	0.000	0.000
1	0.204	0.328	0.410	0.396	. . .	0.156	0.077	0.028	0.015	. . .	0.000	0.000
2	0.021	0.073	0.205	0.264	. . .	0.313	0.230	0.132	0.088	. . .	0.008	0.001
3	0.001	0.008	0.051	0.088	. . .	0.313	0.346	0.309	0.264	. . .	0.073	0.021
4	0.000	0.000	0.006	0.015	. . .	0.156	0.259	0.360	0.396	. . .	0.328	0.204
5	0.000	0.000	0.000	0.001	. . .	0.031	0.078	0.168	0.237	. . .	0.590	0.774

FIGURE 7.1 Portion of binomial table for $n = 5$

The two columns that are shaded are the probability distribution table for $n = 5$ and $\pi = 0.70$ (or 70%). If you just take those columns and transpose as shown in Figure 7.2, then the result will look like every probability distribution table you have seen before!

Do not forget that decimals and percentages are equivalent!

Except for learning to recognize binomial random variables and using the tables, there is nothing new to learn about answering questions involving binomial random variables.

x	0	1	2	3	4	5
p(x)	0.002	0.028	0.132	0.309	0.360	0.168

FIGURE 7.2 Binomial probability distribution for $n = 5$ and $\pi = 0.70$

EXAMPLE 7.20 Wearing Seatbelts

Using the Binomial Probability Tables

The example about the Connecticut drivers identified the random variable for the problem as a binomial random variable with $n = 5$ and $\pi = 0.70$. What is the probability that, in the sample of 5 drivers, exactly 3 are wearing seatbelts?

Looking at the table in Figure 7.1 or 7.2, $p(3) = 0.309$.

What is the probability that in a sample of 5 drivers taken at most 2 are wearing seatbelts? Again, from the table in Figure 7.1 or 7.2,

$$P(X \le 2) = p(0) + p(1) + p(2) = 0.002 + 0.028 + 0.132 = 0.162 \qquad \blacksquare$$

You can see that there is really nothing new involved in actually solving problems that involve the binomial distribution. The parts that require some thinking and work are recognizing the problem as binomial, identifying the parameters n and π for the particular problem, and finding the correct table.

 *TRY IT NOW!*

Loan Defaults

Solving Binomial Probability Problems

The Chamber of Commerce that is checking credit problems of small businesses estimated that 20% of all small businesses were at least six months behind in loan payments. The Chamber of Commerce took a random sample of 25 small businesses and counted the number of the businesses that were at least six months behind in loan payments.

Define a success for this problem.

Describe the random variable, *X*, in words.

Find the parameters of the binomial distribution for this problem.

Find the probability that in the sample of 25 businesses less than 6 were at least six months behind in loan payments.

Find the probability that between 4 and 9 inclusive were at least 6 months behind in loan payments.

It is important to recognize that the definition of a success is critical to problem solving with the binomial distribution. A success is not necessarily always something good, nor does it have to stay the same in a given problem. To determine what the success is, you need to look at the question that needs to be answered, define a success, and use the appropriate value of π.

EXAMPLE 7.21 Parking Tickets

Calculating Binomial Probabilities

A local watchdog agency has been looking at parking problems outside the city court building. It estimates that 40% of all of the cars parked in the metered lot receive parking tickets for meter violations. The agency decides to take a random sample of 10 cars from the lot and check to see if they have a parking ticket on the windshield.

(a) Find the probability that, of the 10 cars sampled, exactly 6 have parking tickets.

For this problem a success is having a parking ticket and so we use the table with $n = 10$ and $\pi = 0.40$ and find that $P(X = 6) = 0.111$.

(b) Find the probability that between 4 and 7 inclusive do not have parking tickets.

For this problem a success is *not* having a parking ticket. Since 40% of the cars have parking tickets, we can use the definition of the complement to find that $100\% - 40\% = 60\%$ do not have parking tickets.

Use $n = 10$ and $\pi = 0.60$ to get $P(4 \leq X \leq 7) = 0.111 + 0.201 + 0.251 + 0.215 = 0.778$.

If the watchdog group found in its sample of 10 cars that none of the cars had parking tickets, would you think that the estimate of 40% was reasonable?

With $n = 10$ and $\pi = 0.40$ the probability that no cars have tickets, $P(X = 0)$, is 0.006, which is very small. Thus, it is highly unlikely that if 40% of the cars get tickets that the sample of 10 would find no cars with tickets. The estimate seems to be high.

∎

Caution! No matter what you may think, it is NOT easier to keep using $\pi = 0.40$ and try to change the question to be in terms of cars having parking tickets!

The previous example is a preview of what the next part of statistics that you will study is all about. In the first part of this book you learned how to describe data that are collected and how to calculate different sample statistics. Now you are learning about probability models and how they can be used to determine how likely it is that different events will occur when we assume some parameters for our population data. In the previous example we looked at how well what we observed (the data) fit the probability model (binomial with $\pi = 40\%$ and $n = 10$). We used the probability that such an event would happen to come to the conclusion that the model did not seem appropriate. In subsequent chapters you will learn the more formal methods of hypothesis testing to accomplish this same thing.

7.4.3 The Mean and Standard Deviation of the Binomial Distribution

Since probability distributions are models of populations, and random variables are numerical, it makes sense that just like quantitative sample data, they have means and standard deviations. There are some general formulas for calculating the mean and standard deviation of a random variable, but we will concentrate on the probability distributions that we study.

Since probability distributions are population models, their means and standard deviations are represented by the Greek letters μ and σ. In particular, for a binomial random variable, X, the **mean, μ,** and the **standard deviation, σ,** are found using the following formulas:

$$\mu = n\pi \quad \text{and} \quad \sigma = \sqrt{n\pi(1 - \pi)}$$

You see that the mean and standard deviation depend on the parameters of the probability distribution, n and π. The formula for the mean is actually quite intuitive if you think about it. If you knew, for example, that 40% of a certain population wore eyeglasses, and you took a sample of 10 people from that population, how many of the 10 would you *expect* to wear glasses? Instinctively you would take 40% of the 10, to get 4 people who wear glasses. From the formula for μ we would get

$$\mu = n\pi = (10)(0.40) = 4 \text{ people}$$

The formula for the standard deviation is really not at all intuitive and deriving it is beyond the scope of this text. To see how it is used, we can calculate the standard deviation of our example and find that

$$\sigma = \sqrt{n\pi(1 - \pi)} = \sqrt{(10)(0.40)(0.60)} = 1.55 \text{ people}$$

Thus, we know that if $n = 10$ and $\pi = 0.40$, then the number of people who wear glasses is a random variable with a mean of 4 and a standard deviation of 1.55.

EXAMPLE 7.22   Credit Problems

Calculating the Mean and Standard Deviation of a Binomial Random Variable

With a sample of $n = 166$ we could not use the tables to determine probabilities for the number of small businesses that reported having credit problems. Most statistical software packages will calculate probability distributions for binomial random variables for any values of n and π, but we can find the mean and standard deviation of this random variable without any problems.

The mean represents the expected number of businesses that are experiencing credit problems in the sample of 166. That is, with $n = 166$ and $\pi = 0.10$ the mean is:

$$\mu = n\pi = (166)(0.10) = 16.6 \text{ businesses}$$

The standard deviation of the number of small businesses in 166 that report credit problems is

$$\sigma = \sqrt{n\pi(1 - \pi)} = \sqrt{(166)(0.10)(0.90)} = 3.87 \text{ businesses} \qquad \blacksquare$$

EXAMPLE 7.23   Wearing Seatbelts

Finding the Mean and Standard Deviation of a Binomial Random Variable

The Department of Transportation in Connecticut would like to know how many drivers the state troopers should expect to find wearing their seatbelts at the checkpoint. It decides to calculate the mean of the binomial random variable.

In this case, $n = 5$ and $\pi = 0.70$, so it calculates that

$$\mu = n\pi = (5)(0.70) = 3.5 \text{ people}$$

The Department would also like to know the standard deviation of the number that would be wearing seatbelts. In this case

$$\sigma = \sqrt{n\pi(1 - \pi)} = \sqrt{(5)(0.70)(1 - 0.70)} = \sqrt{(5)(0.70)(0.30)} = \sqrt{1.05} = 1.02 \text{ people} \quad \blacksquare$$

Remember that the mean and standard deviation tell us something about how random variables behave. In particular, they tell us where the center of the probability distribution is located and how much the random variable will vary around that center.

 *TRY IT NOW!*

Loan Defaults

Calculating the Mean and Standard Deviation of a Binomial Random Variable

The Chamber of Commerce that was looking at the loan defaults for small businesses wants to know the mean and standard deviation for the binomial random variable with $n = 25$ and $\pi = 0.20$.

Find the mean and standard deviation of the number of small businesses in 25 that will default on their loans.

How can we use our knowledge of the mean and standard deviation? One thing we can do is to compare our knowledge of what *should* happen, to the reality of what *did* happen, to get some idea of how well the reality fits the theory. This is the basis for the work you will do in the remainder of this book, *inferential statistics*. Right now, we will look at this work in an informal way.

EXAMPLE 7.24 Credit Problems

Using the Mean and the Standard Deviation

The Chamber of Commerce wonders whether or not the current random sample of 166 businesses supports its belief that 10% of the small businesses in the city are experiencing credit problems. It already has the mean and standard deviation for its model, so it decides to compare the model to the actual data.

The mean of the binomial random variable with $n = 166$ and $\pi = 0.10$ is 16.6 and the standard deviation is 3.87. From the data, the Chamber of Commerce analysts find that the number of businesses that answered yes to the question of credit problems was 25. Clearly, 25 is not equal to 16.6, but how different is it?

ANS. $\mu = 5$, $\sigma = 2$

If they consider the standard deviation they can calculate the number of standard deviations that their data value is from the mean:

$$z = \frac{25 - 16.6}{3.87} = 2.17$$

They know that the empirical rule holds only for symmetric distributions and they have no idea whether that applies here, but they do know that 2.17 standard deviations away from the mean is at least possibly unusual.

As a second check the analysts use their data to estimate the probability that a small business will report having credit problems. They found 25 such businesses in their sample of 166, so they estimate the probability that a business will report having credit problems to be $25/166 = 0.1506$ or approximately 15%. This is different from the 10% they were expecting, but the difference may just be a result of sampling.

As a third check they use a statistical software package to calculate the probability of having 25 successes in a sample of 166 when $\pi = 0.10$. They find that the probability is 0.0108, which is only slightly more than 1%. It would seem that perhaps this sample does not fit their earlier beliefs. ■

EXAMPLE 7.25   Parking Tickets

Using the Mean and Standard Deviation

The watchdog agency that is looking at the problem of parking tickets would like to know what it should expect to see in the data it collects. The analysts in the agency decide to calculate the mean and standard deviation of the binomial random variable.

The mean is

$$\mu = (10)(0.40) = 4 \text{ cars}$$

and the standard deviation is

$$\sigma = \sqrt{n\pi(1 - \pi)} = \sqrt{(10)(0.40)(0.60)} = \sqrt{2.4} = 1.55 \text{ cars}$$

This tells the analysts that they should expect to find 4 cars with parking tickets. Wondering how much the number of cars might vary, they decide to find the probability that the number of tickets will be within 2 standard deviations of the mean. To do this they calculate $\mu \pm 2\sigma = 4 \pm (2)(1.55) = 4 \pm 3.1 = (0.9, 7.1)$. They want to find the probability that the random variable is between these two values, or $P(0.90 < X < 7.1)$.

How will they do this? The binomial tables certainly do not include numbers like 0.90 and 7.1! The analysts realize that they will have to convert the problem to the nearest integer values that satisfy the probability expression. That means they need to find $P(1 \leq X \leq 7)$. From the tables for $n = 10$ and $\pi = 0.40$ they find that the answer is 0.981. They interpret this to mean that 98.1% of the time the number of tickets they find will be between 1 and 7 inclusive. This gives them a good idea of what to expect in their data. ■

The idea of comparing what should happen to what does happen is a critical idea in statistics. In the next few chapters we develop a formal method for doing this called *hypothesis testing*.

Discovery Exercise 7.1
EXPLORING THE BINOMIAL DISTRIBUTION

Dear Mom and Dad: Send cash
According to USA Today, *70% of college students receive spending money from their parents when at school.*

For this exercise, you will need to simulate selecting 30 samples of 5 students from this population of college students and observe whether they receive spending money from their parents. Consider the successful outcome to be "receives money" with $\pi = 0.70$ and the failure outcome to be "does not receive money." If your instructor does not provide you with a method, you can take ten (small) pieces of paper and write an S on 7 of them and an F on 3 of them. Put the papers in a bag or other container and select one at random to simulate an observation. *Note:* Be sure to replace the paper each time or π will not always be 0.70.

1. Record an S when you select a student who receives money from his/her parents and an F when you select a student who does not receive spending money from his/her parents. For each sample, record the number of successes you sampled.

   In the last column compute a running estimate of π. Remember that π is the probability of a successful outcome. In this case, π is known to be 0.70. Let's see how close the estimate gets to 0.70 as the sample size increases. So, after the first sample is selected your estimate of π is simply the number of successes divided by 5. After the second sample is selected, your estimate of π is the number of successes in both samples divided by 10, and so forth.

Sample number	Observation number 1	2	3	4	5	X = Total number of successes in $n = 5$ trials	Running total number of successes	Running total of number sampled	Estimate of π
1								5	
2								10	
3								15	
4								20	
5								25	
6								30	
7								35	
8								40	
9								45	
10								50	
11								55	
12								60	
13								65	
14								70	
15								75	
16								80	
17								85	
18								90	
19								95	
20								100	
21								105	
22								110	
23								115	
24								120	

(continued)

Sample number	Observation number					X = Total number of successes in $n = 5$ trials	Running total number of successes	Running total of number sampled	Estimate of π
	1	2	3	4	5				
25								125	
26								130	
27								135	
28								140	
29								145	
30								150	

2. Using the results of the 7th column of the table, construct a relative frequency distribution for the number of successes in 5 trials.

X = number successes	Number of samples that had X successes	Observed relative frequency = $X/30$	Theoretical relative frequency
0			
1			
2			
3			
4			
5			
Totals:	**30**	**1.00**	**1.00**

3. Display the relative frequency as a bar chart.

4. Complete the theoretical relative frequency column in the table above by using the binomial table with $n = 5$ and $\pi = 0.70$.

5. Display the binomial distribution in a bar chart.

6. Compare the bar chart from step 3 to the binomial distribution displayed in step 5. How do they compare? Why are they different?

7. The graph below has a line at the theoretical value of π, 0.70. Graph your estimates of π for each sample on the same graph. What happens to your estimate as the sample size increases?

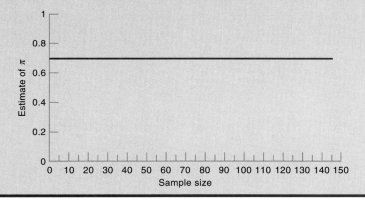

7.4.4 Exploring the Binomial Distribution

You have started to become familiar with using the binomial probability tables and solving binomial probability problems, but you still may not have an understanding what role the parameters n and π play in determining what the probability distribution of a binomial random variable looks like.

In Figure 7.3 you see the effects of changing the value of the parameter π for a fixed value of n. What you see agrees with our earlier thoughts that when π is small it

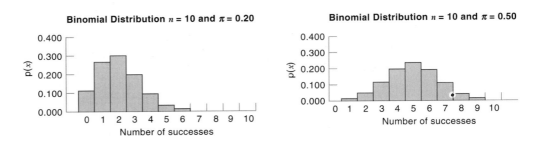

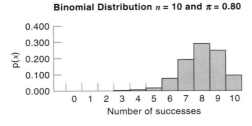

FIGURE 7.3   Effects of changing π when n is fixed

is more likely that the number of successes will be small and the distribution skews to the right. When π is large it is more likely that the number of successes will be large and the distribution is skewed to the left. You can also see that when π is equal to 0.50 the distribution is symmetric.

If you look at the binomial table for $n = 10$ you will notice, not surprisingly, that the probability distributions for $\pi = 0.20$ and $\pi = 0.80$ are mirror images of each other.

Where does the mean fit into the picture? If you remember, the mean is the value that you expect to occur, that is, it is the most likely outcome. If you look at Figure 7.3, you see that in the first probability histogram the highest bar is for $X = 2$ and the mean of a binomial random variable with $n = 10$ and $\pi = 0.20$ is $\mu = (10)(0.20) = 2$.

Figure 7.4 shows the effects of changing the value of the parameter n for a fixed value of π. The y scale for each histogram is approximately the same. From the pic-

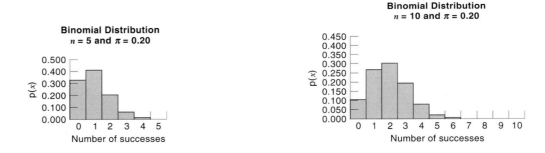

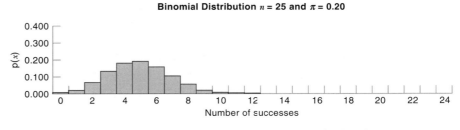

FIGURE 7.4   Effects of changing n for a fixed value of π

ture you can see that, for the most part, the shape of the distribution remains the same as n varies, but the number of possible values and the probability of each value of X (bars in the histogram) changes. In fact, as n gets larger, the individual probabilities get smaller. This makes sense, since we know that the probabilities must sum to 1 and if there are more values of X, then each value will get a smaller share of the total.

7.4.5 Exercises—Learning It!

7.11 The Board of Realtors of a small city reports that 80% of the homes that are sold have been on the market for more than 6 months. The Board takes a random sample of 15 homes that have recently been sold and counts the number that were on the market for more than six months. What is the probability that of the 15 houses in the sample

(a) less than 12 have been on the market for more than six months?

(b) between 8 and 13 have been on the market for more than six months?

(c) at least 10 have been on the market for more than six months?

(d) at most 4 have been on the market for less than six months?

7.12 The Department of Transportation for a city has found that 25% of all parking tickets that have been issued have not been paid within one month of issue. The Department takes a random sample of 20 parking tickets that were issued one month ago and counts the number that have not been paid. What is the probability that

(a) at most 5 have not been paid?

(b) between 4 and 8 inclusive have not been paid?

(c) more than 7 have not been paid?

(d) at least 6 have been paid?

7.13 A large construction firm estimates that 10% of the jobs that it manages are finished within the contracted time period. It looks at a random sample of 5 jobs that it has managed.

(a) What is the probability that 4 of the five jobs were not completed within the contracted time period?

(b) What is the mean number of jobs that are completed within the contracted time period?

(c) What is the standard deviation of the number of jobs in 5 that are completed within the contracted time period?

7.14 A large bank that issues many loans for General Motors estimates that 70% of the loans are approved within 24 hours of application. If a consumer group takes a random sample of 25 recent GM loan applications, what is the probability that

(a) at most 5 were not approved within 24 hours?

(b) between 10 and 17 are approved within 24 hours?

(c) What is the mean number of loans in 25 that will be approved within 24 hours?

(d) What is the standard deviation of the number of loans in 25 that will be approved within 24 hours?

(e) What is the probability that the number of loans in 25 that are approved within 24 hours is within 3 standard deviations of the mean?

7.15 Companies have been having problems with employees playing computer games at work for a long time. As the size and complexity of such games increases, computer system administrators find that network resources are being drained and that the games are using more and more hard disk space. A recent survey across various industries revealed that 30% of workers said that the last computer game they played had been played at work. If a random sample of 15 employees is taken, what is the probability that

(a) less than 6 said they played their last computer game at work?

(b) at least 4 said they played their last computer game at work?

(c) at most 10 said they did not play their last computer game at work?

(d) What is the expected number of employees in a sample of size 15 who played their last computer game at work? What is the standard deviation of the number of employees in 15 who played their last computer game at work?

(e) What is the probability that the number of employees in 15 who played their last computer game at work is within 2 standard deviations of the mean?

7.5   CONTINUOUS RANDOM VARIABLES

In Chapter 2 you learned that there are two types of quantitative data: discrete and continuous. Discrete data are integer and are often a count of the number of times that something happens. The binomial distribution that you just studied is one good probability model for discrete data.

Continuous data occur when the variable of interest can take on any one of an infinite number of values over some *interval* on the real number line. Much of the time, continuous data result from taking data on a *measurement*. Examples of this would be height of students or grade point averages (GPA). The actual number of values that you can obtain is limited by the measuring instrument (we choose to measure height to the nearest inch or report GPA to two decimal places) but any value in the interval is valid.

When you learned about discrete random variables you learned that to find probabilities associated with the random variable, you simply added the relevant individual probabilities. With continuous random variables you will have to shift your thinking a little bit to see how probabilities are calculated.

When we looked at the effects of changing the value of n in the binomial distribution, we saw that as n increased, the number of possible values of X increased and the individual probabilities got smaller. When $n = 10$ it is not that tedious to calculate $P(X \geq 4)$, but when $n = 20$ or $n = 25$ it is considerably more tedious to do it directly. You are adding many more terms and they are all very small numbers. If we let n get very large, say $n = 100$, it would be tiring indeed! Figure 7.5 looks at the binomial distribution for $\pi = 0.50$ and $n = 10, 25, 50,$ and 100.

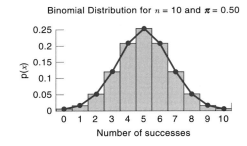

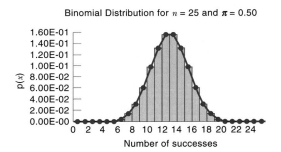

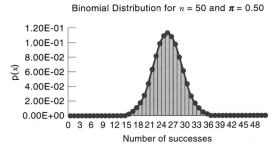

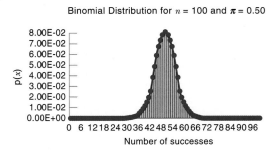

FIGURE 7.5   Binomial distribution for large values of n

You can see from the graphs that if we connect the tops of all of the histogram bars, as n gets larger, the curve becomes smooth. For continuous random variables, which can take on many more than 100 possible values, the probability distribution is called a **probability density function, $f(x)$,** and is represented by a smooth curve such as the one in Figure 7.6.

Probability Density Function for a Continous Random Variable

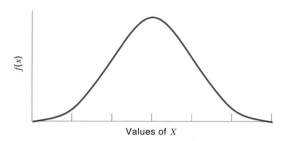

FIGURE 7.6 Probability distribution for a continuous random variable

A *probability density function, $f(x)$,* is a smooth curve that represents the probability distribution of a continuous random variable.

You might be wondering how this curve relates to probabilities, since there is no corresponding probability table with numbers involved. How do you find the probabilities that you are interested in? These are good questions.

When we increased the value of n for the binomial distribution, you saw that the individual probabilities got much smaller. In fact, if you look at the scales on the y axes for Figure 7.5 you will see that for $n = 100$ the *largest* number on the graph is 0.08! Again, this makes sense because the total probability is always equal to 1 and when you have to divide it up over more and more possible values, each individual probability gets closer and closer to 0.

In fact, for continuous random variables we can no longer talk about the probability that the random variable will assume one particular value, $P(X = x)$, because this is equal to 0. Instead, we can talk about the probability that the random variable will take on any one of the values over an *interval* of interest; that is, we can find $P(x_1 < X < x_2)$. This probability is represented by the area under the probability density curve as shown in Figure 7.7. Finding probabilities for a particular probability density involves the mathematics of calculus, but for certain distributions the probabilities have already been calculated and tabulated.

For continuous random variables $P(x_1 < X < x_2)$ is exactly the same as $P(x_1 \le X \le x_2)$, since $P(X = x) = 0$.

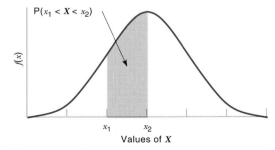

FIGURE 7.7 Probability represented by an area under the curve

In the next section we look at one particular continuous probability distribution that serves as a model for many natural phenomena. This is known as the *normal distribution.*

7.6   THE NORMAL DISTRIBUTION

7.6.1 The Normal Curve

The normal probability distribution is the symmetric, bell-shaped curve shown in Figure 7.8. It is usually referred to as the *normal curve*. This is the same curve we saw when we learned about the empirical rule.

In reality, most measurement data are modeled very well by this distribution. The actual formula for the probability density is

*Normal probability
density*

$$f(x) = \frac{1}{\sigma\sqrt{2\pi}} e^{-(x-\mu)^2/2\sigma^2} \qquad \text{for } -\infty < x < \infty$$

*Here, π is the familiar
constant, 3.141 . . . ,
not related to the bino-
mial distribution.*

where $e = 2.71828$ . . . and $\pi = 3.14159$. . . . If you look at the equation you can see that the probability density for a value, x, relies on two **parameters,** μ and σ.

> For a normal random variable, the parameter μ is the mean of the normal random variable, X, and σ is the standard deviation.

When we refer to a normally distributed random variable we often use a special notation

$$X \sim N(\mu, \sigma)$$

This shorthand is equivalent to saying that the random variable X is normally distributed with a mean of μ and a standard deviation of σ.

When we studied the binomial probability distribution we learned that the parameters, in that case n and π, defined how the probability was distributed over the set of possible values of X. That is, they defined the shape of the probability histogram. In a similar way, μ and σ define the way the normal distribution looks.

You have come across the mean in two other situations. When you learned about descriptive statistics, you learned that the mean was a measure of central tendency, or an estimate of the most typical data value. Then, when you studied the binomial distribution, we said that the mean is the most likely value, or the one that you expect to happen most often. For a normally distributed random variable, the mean is the center of the distribution. It is the value that determines the location of the probability density curve on the number line.

You also learned about the standard deviation. You know that the standard deviation measures how far the data are spread out around the mean. For a normally distributed random variable, σ determines how spread out the probability density is around its center. You may remember that in Chapter 6 we used the mean and the standard deviation to obtain a mental image of the distribution of the data. The empirical rule told us that for a symmetric distribution, virtually all of the data fall within three standard deviations of the mean. Since the normal distribution is the fundamental symmetric curve, we can use this same idea to get a picture of a normal probability distribution.

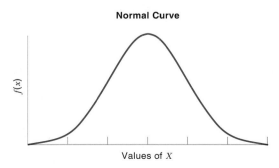

Normal Curve

FIGURE 7.8   Normal probability curve

Normal Curves with Equal Means and Different Standard Deviations

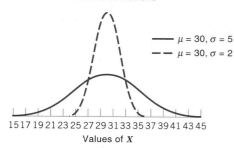

— $\mu = 30, \sigma = 5$
-- $\mu = 30, \sigma = 2$

15 17 19 21 23 25 27 29 31 33 35 37 39 41 43 45
Values of X

Normal Curves with Different Means and Equal Standard Deviations

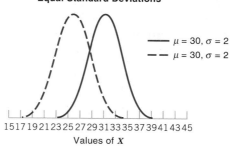

— $\mu = 30, \sigma = 2$
-- $\mu = 30, \sigma = 2$

15 17 19 21 23 25 27 29 31 33 35 37 39 41 43 45
Values of X

FIGURE 7.9   Effects of changing μ and σ on the normal curve

In Figure 7.9 you can see the effect that changes in μ and σ have on the way the normal curve looks.

As the standard deviation increases, the distribution gets more spread out around its center. For the curve with a mean of 30 and a standard deviation of 5 you see that the ends of the curve are around 15 and 45, three standard deviations away from the mean! As the mean increases, the distribution moves to the right (in the direction of increasing values of X) and when the mean decreases the distribution moves to the left. Notice that although the appearance of the normal curve does change when μ and σ are varied, the basic shape does not change. The curve is always bell-shaped and symmetric. This is an important feature of the normal probability distribution.

 *TRY IT NOW!*

Food Expenditures

Looking at the Normal Curve

The amount of money that a person working in a large city spends each week for lunch is a normally distributed random variable. For professional and management personnel the random variable has a mean of $35 and a standard deviation of $5. For hourly employees the mean is $30 with a standard deviation of $2.

Sketch the normal curves for each of the two random variables on the same graph.

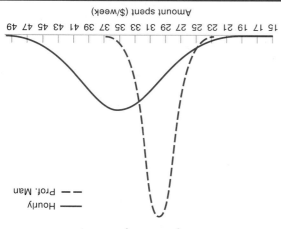

ANS:

7.6.2 The Standard Normal Curve

When we learned about continuous random variables we said that the probability that the random variable takes on any value in an interval is equal to the area under the probability density curve. We also said that finding these probabilities requires the use of integral calculus, which is beyond the scope of this book. How will we find the probabilities we are interested in?

Remember that when we studied the binomial probability distribution, the formula to find probabilities was also complicated, and we solved the problem by using tables for different sets of values for n and π. Since π is constrained to be between 0 and 1, it was not hard to get tables for many different values of n.

The idea of using probability tables for the normal probability distribution seems like a good one, but there is a practical limitation. Unlike n and π, the values of μ and σ are not at all constrained. Depending on the variable of interest, there are an infinite number of different pairs of μ and σ that we might be interested in. It is not feasible to have a table for every different problem we can think of.

The problem is solved by finding a way to transform every normally distributed random variable to a single, standard normal random variable. This standard normal random variable is known as **Z** and has a mean of 0 and a standard deviation of 1. We can use the shorthand notation to write that as $Z \sim N(0, 1)$. Once we transform the problem we can use a single normal probability table, a **standard normal table,** to solve it.

> A **Z random variable** is normally distributed with a mean of 0 and a standard deviation of 1, $Z \sim N(0, 1)$.

> A **standard normal table** is a table of probabilities for a Z random variable.

This is not mysterious, if you think about it for a minute. If the class average on an exam were a 70 and you wanted to make it a 75, you would ADD 5 points to everyone's grade. Similarly, to move the mean from μ to 0, you subtract μ from the values.

How does the transform work? Remember that the mean of a normally distributed random variable locates the normal curve on the number line. What we want to do is move any arbitrary normal random variable that is centered at μ down the number line to center at 0. It is not hard to see that this can be accomplished by subtracting the value of μ from the random variable, X.

The change from a standard deviation of σ to a standard deviation of 1 is not difficult either. Essentially you need to scale the probability distribution so that the total area under the curve stays equal to 1 (one of the rules of probability). A value of X that is two standard deviations away from μ for the original random variable must stay two standard deviations away from 0 when it is transformed to Z. To accomplish this we describe the distance that X is from μ, $X - \mu$, in terms of the number of standard deviations it is away from μ. The transform from X to Z is then given by

Formula for Z-score

$$Z = \frac{X - \mu}{\sigma}$$

This is the population version of the z-scores you calculated in Chapter 6.

where $X \sim N(\mu, \sigma)$ and $Z \sim N(0, 1)$. At this point, an example will help you see what is happening.

EXAMPLE 7.26   Aptitude Scores

Transforming from X to Z

Scores on an aptitude test given by the training department of a large company are normally distributed with a mean of 75 points and a standard deviation of 5 points.

The managers of the company are interested in knowing what proportion of the people that take the test score between 65 and 85 points.

The first thing we need to do is *draw a picture* representing the problem we are trying to solve.

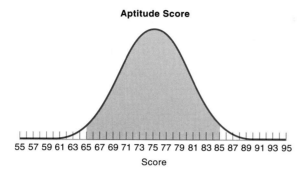

Note: It may seem to be a waste of time to you to draw the pictures. A little time spent here will avoid a LOT of time wasted later on.

The picture shows the distribution of the scores and the area under the curve that corresponds to the problem that management wants to solve. We know that to solve the problem we must transform it to a standard normal problem. To do this we must transform each of the values of X that we are interested in to corresponding values of Z. In this case,

$$\text{for } X = 65: \quad Z = \frac{65 - 75}{5} = -2$$

$$\text{for } X = 85: \quad Z = \frac{85 - 75}{5} = 2$$

Neither of these values of Z should be a surprise since you know that a Z-score measures the number of standard deviations that a value is from its mean. The score of 65 is 10 points, or exactly 2 standard deviations, *below* the mean of 75 and the score of 85 is exactly 2 standard deviations *above* the mean of 75. The sign on the value of Z indicates whether the value of X was above (+) or below (−) the mean.

If you look at a picture of the transformed problem you see that it does not look any different from the original picture except for the scale on the x axis. It is not necessary to draw both X and Z pictures.

Caution: It is easy to mix up X and μ when you use the formula for Z. Use the sign of the answer you get to check for errors. Make sure that the sign of your Z agrees with the location of X relative to μ.

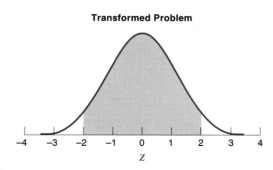

Transforming problems about normally distributed random variables to problems about standard normals is not difficult. Just as in all probability problems the hardest part is being able to understand what problems you need to solve.

TRY IT NOW!

Speed Reading

Translating from X to Z

The number of pages of a statistics textbook that a student can read in a given hour is a normally distributed random variable with a mean of 7 pages and a standard deviation of 1.5 pages. One of the professors who uses the book wants to know the probability that a randomly selected student can read more than 8.5 pages of the textbook in an hour.

Draw a picture that depicts the problem to be solved and find the Z values necessary to solve the problem.

The importance of drawing the picture that relates to the problem will be clear in the next section when we look at how to solve the probability problems.

7.6.3 The Standard Normal Tables

The tables for the binomial distribution that you used contained $P(X = x)$ for different combinations of n and π. To find quantities like $P(x_1 \leq X \leq x_2)$ you summed the probabilities for all of the values of X that were included in the interval. We know that $P(X = x) = 0$ for continuous random variables, so the probability tables for the standard normal probability distribution must contain some kind of **interval** probability.

> An **interval probability** gives the probability that a random variable will take on a value *between* two given values $P(x_1 \leq X \leq x_2)$.

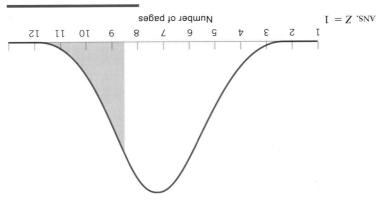

ANS. $Z = 1$

Number of pages

Pages of Statistics Textbook Read in an Hour

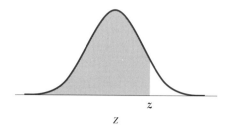

FIGURE 7.10   Probability given by the standard normal table

There are several different types of standard normal tables that can be used. They each give probabilities for different types of intervals. The tables that we use in this book give $P(Z < z)$; that is, the entries in the table are the probability that the random variable Z will take on a value that is less than some value z. A picture that corresponds to this is shown in Figure 7.10.

The table that is included in Appendix A has two halves, for both negative and positive values of Z. This makes solving problems involving the table less cumbersome.

All tables for the standard normal probabilities are set up the same way. The values of Z are defined to two decimal places, and the corresponding probability is located at the intersection of the appropriate row and column of the table. If you think of the Z value as always being in the form $X.XX$, the rows contain the first two digits $X.X$ while the columns contain the last digit $0.0X$. It sounds complicated but it is easy once you are familiar with it. A portion of the table is shown in Figure 7.11.

To look up a value of Z, say $Z = 1.14$ you first locate the row that corresponds to the first two digits, 1.1. This is the shaded row in Figure 7.11. The column you need is the one that corresponds to the last digit, 0.04, which is also shaded. The entry in the table at the intersection of the row and column, 0.8729, gives the probability of obtaining a value of the random variable Z that is less than 1.14, or the area under the curve to the left of $Z = 1.14$.

The half of the table for negative values of Z is used exactly the same way. It is important to be consistent about the way you calculate and look up Z values. Since the tables give the values of Z to two decimal places, you need to calculate Z values to two decimal places, rounding correctly, and look them up that way.

Caution: When looking up z values such as 0.08, the first two digits are 0.0 not 0.8.

Second decimal place

z	0.00	0.01	0.02	0.03	0.04	0.05	0.06	0.07	0.08	0.09
0.0	0.5000	0.5040	0.5080	0.5120	0.5160	0.5199	0.5239	0.5279	0.5319	0.5359
0.1	0.5398	0.5438	0.5478	0.5517	0.5557	0.5596	0.5636	0.5675	0.5714	0.5753
0.2	0.5793	0.5832	0.5871	0.5910	0.5948	0.5987	0.6026	0.6064	0.6103	0.6141
0.3	0.6179	0.6217	0.6255	0.6293	0.6331	0.6368	0.6406	0.6443	0.6480	0.6517
0.4	0.6554	0.6591	0.6628	0.6664	0.6700	0.6736	0.6772	0.6808	0.6844	0.6879
0.5	0.6915	0.6950	0.6985	0.7019	0.7054	0.7088	0.7123	0.7157	0.7190	0.7224
0.6	0.7257	0.7291	0.7324	0.7357	0.7389	0.7422	0.7454	0.7486	0.7517	0.7549
0.7	0.7580	0.7611	0.7642	0.7673	0.7704	0.7734	0.7764	0.7794	0.7823	0.7852
0.8	0.7881	0.7910	0.7939	0.7967	0.7995	0.8023	0.8051	0.8078	0.8106	0.8133
0.9	0.8159	0.8186	0.8212	0.8238	0.8264	0.8289	0.8315	0.8340	0.8365	0.8389
1.0	0.8413	0.8438	0.8461	0.8485	0.8508	0.8531	0.8554	0.8577	0.8599	0.8621
1.1	0.8643	0.8665	0.8686	0.8708	0.8729	0.8749	0.8770	0.8790	0.8810	0.8830
1.2	0.8849	0.8869	0.8888	0.8907	0.8925	0.8944	0.8962	0.8980	0.8997	0.9015

FIGURE 7.11   The standard normal table

For continuous random variables, probability is often referred to as the area under the curve.

You now know how to use the table to find $P(Z < z)$. How does the procedure change if you need $P(Z > z)$ or $P(z_1 < Z < z_2)$? The first problem is not difficult to figure out. Since the tables give $P(Z < z)$ and since the total area under the curve must always equal 1, we can obtain $P(Z > z)$ by using the definition of the complement of an event. That is,

$$P(Z > z) = 1 - P(Z < z)$$

So, to find the area under the curve above a given value of Z you look up the Z value, find the probability below Z, and subtract it from 1. Thus, using our earlier example,

$$P(Z > 1.14) = 1 - 0.8729 = 0.1271$$

Once you are comfortable using the normal probability table you can find upper area probabilities easily by relying on the symmetry of the normal distribution. Since the normal probability distribution is symmetric about the mean, it must be true that the area to the right of a Z value must be equal to the area to the left of the negative of that Z value, or

$$P(Z > z) = P(Z < -z)$$

This is illustrated in Figure 7.12.

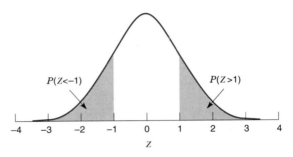

FIGURE 7.12   Comparison of upper and lower probabilities

Using the tables to find $P(z_1 < Z < z_2)$ is not difficult either. Figure 7.13 illustrates the process.

When you look up the larger Z value, in this case $Z = +2$, the table gives the entire area under the curve to the left of 2, which is 0.9772. Looking at the picture, you see that this includes the piece of the curve to the left of the lower value, in this case $Z = -2$, which is 0.0228. Since we do not want to include this lower portion we must subtract it from the first value. Thus, in general,

$$P(z_1 < Z < z_2) = P(Z < z_2) - P(Z < z_1)$$

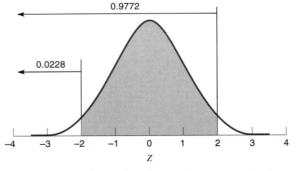

FIGURE 7.13   Finding the area between two Z values

and for this example,

$$P(-2 < Z < +2) = P(Z < +2) - P(Z < -2) = 0.9772 - 0.0228 - 0.9544$$

If you think about it, you already knew this! Remember that the Z random variable is $N(0, 1)$, so finding $P(-2 < Z < +2)$ is the same as finding the probability that the random variable will be within two standard deviations of the mean. The answer, not surprisingly, is 0.9544 or 95.44%. Remember the empirical rule? It told us that for a symmetric distribution, approximately 95% of the data will fall within two standard deviations of the mean. This is where the empirical rule comes from.

Remember that Z can also be thought of as the number of standard deviations that a value is from its mean.

Table 7.2 summarizes the rules for using the standard normal probability tables.

TABLE 7.2   Rules for solving normal probability problems

To Find...	Area Under the Curve...	Look Up...
$P(Z < z)$	below a value of Z	the Z value and use the table directly
$P(Z > z)$	above a value of Z	1. the Z value and subtract the value in the table from 1
		OR
		2. the negative of the Z value and use the table directly
$P(z_1 < Z < z_2)$	between two values of Z	both Z values and subtract the lower value from the higher value

 ***TRY IT NOW!***

The Standard Normal Table
Using the Table to Find Probabilities

For each of the following questions, *draw a picture* of what you are trying to find BEFORE you use the table to find it.

Find the probability that a Z random variable takes on a value that is less than 2.74.

Find the probability that a Z random variable is greater than 0.85.

Drawing a picture will help you use the table correctly.

(continued)

Find the probability that Z is between -1.36 and 1.87.

7.6.4 Solving Normal Probability Problems

Just as you found with the binomial distribution, once you understand how the random variable and the probability distribution work, solving problems is really not hard. The hardest part is understanding what the problem really means.

The following steps are used to solve problems that involve normally distributed random variables with mean μ and standard deviation σ:

Step 1: Write down the information about the random variable, i.e., $X \sim N(\mu, \sigma)$.

Step 2: Draw a picture that represents the problem and write down the probability statement, e.g., $P(X > 30)$.

Step 3: Transform the values of X that are involved in the problem into Z values.

Step 4: Look up the Z values on the standard normal table.

Step 5: Perform any additional calculations that need to be done, as described in Table 7.2.

EXAMPLE 7.27   Aptitude Scores

Solving Normal Probability Problems

The company that administered the aptitude test to its employees wants to know what proportion of the people who take the test score between 65 and 85 points.

We have essentially answered this question in the course of this section, but we can put it all together here.

We know that $X \sim N(75, 5)$ and that we are looking for $P(65 < X < 85)$ or

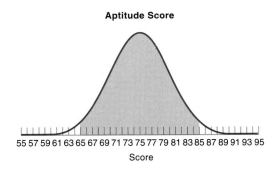

Aptitude Score

55 57 59 61 63 65 67 69 71 73 75 77 79 81 83 85 87 89 91 93 95

Score

We found the Z values of interest to be -2 and $+2$. Looking these values up in the normal tables, we find that the probability for $Z = -2$ is 0.0228 and for $+2$ it is 0.9772. From the picture we see that we are trying to find the area *between* these two values and so we subtract the probability values to find

$$0.9772 - 0.0228 = 0.9544$$

The company is also interested in knowing what proportion of the people who take the test score in what is considered to be the superior range, above 87 points.

We know that $X \sim N(75, 5)$. To answer their question we need to find $P(X > 87)$.

This is represented by the graph below:

Aptitude Scores

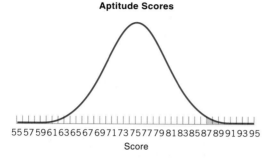

5557596163656769717375777981838587899193 95
Score

Transforming the X value of 87 to Z, we get

$$Z = \frac{87 - 75}{5} = \frac{12}{5} = 2.40$$

Notice that this answer makes sense, since it is positive and the X value of interest is to the right of the mean.

Looking the Z value of 2.40 up in the table, we get a probability of 0.9918.

Now we must decide whether there is anything left to do. The answer is most certainly YES! Common sense also tells you that 0.9918 cannot possibly be the answer. Look at how much of the area in the picture is shaded—it is very small, hardly over 99%! We are looking for the area *above* a value and the table gives the opposite area. We must subtract the value in the table from 1 to obtain

$$1 - 0.9918 = 0.0082 \text{ or } 0.82\%$$

Thus, the company sees that less than 1% of the people who take the test score in the superior range. ■

When doing problems, keep in mind that probabilities, proportions, and percentages are all ways of expressing the same quantity. Do not be misled by the way a question is asked. It is best to answer the question using the quantity type specified, but you are not wrong in using either decimals or percentages.

 *TRY IT NOW!*

Speed Reading
Solving Normal Probability Problems

The instructor who is interested in how many pages of the statistics text that students can read in an hour knows that the random variable is $N(7, 1.5)$.

Find the probability that a student could read more than 11.5 pages in an hour.

The instructor was worried about the percentage of students who could not finish reading a 5-page section in the given hour. What percentage of the students is this?

7.6.5 Finding Values That Correspond to Known Probabilities

In the previous section you learned how to use the standard normal probability tables to solve problems involving any normally distributed random variable. In these problems you were interested in finding the probability or the area under the curve that relates to a specific value of the random variable, X.

Another interesting type of problem which we can solve uses the normal distribution and the standard normal tables. Many people use the probabilities of the normal distribution to make decisions about what values of a variable define certain outcomes.

Suppose that a company wants to reward with an incentive bonus those salespeople whose sales are in the top 10%, or suppose that a city decides that only those people who score in the top 5% on a civil service exam will get job interviews. In many cases it does not make sense to simply take the top 5% of each group that takes the exam or the top 10% of each month's sales. If a particularly ill-suited group of people take the exam, the city could wind up interviewing people who are in no way qualified for the positions available. Similarly, the company could reward people in one month and deny people in the next who have the same sales figures.

ANS. 0.0013; 0.0918

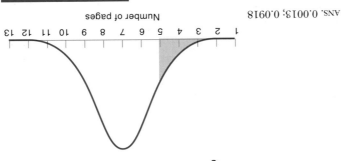

Pages of Statistics Textbook Read in an Hour

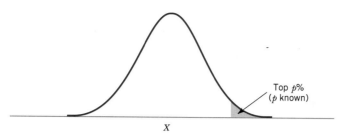

FIGURE 7.14 The "inverse" normal probability problem

Instead, these decision makers rely on the normal probability model to define the cutoff points for the variables of interest. How do they find the numbers that they are looking for?

You know how to use the normal probability tables to find the probability that corresponds to a given value of X or Z. The problem we just described is really the exact opposite of that problem. The decision makers know the percentage or area under the curve that they want to define. They do not know the value of Z (and subsequently X) that it corresponds to. This is illustrated in Figure 7.14.

Solving this type of problem involves using the normal probability tables "inside out." Previously you have located the Z value on the *outside* of the table and determined the answer to the question by looking on the *inside* of the table. For this type of problem you will start by looking on the *inside* of the table for the known probability and move to the *outside* to find the answer. It sounds more complicated than it really is. Again, you simply need to be systematic in your approach.

To get started, we will concentrate on the technique. Suppose you want to find the value of Z that has 10% of the area below it. This is illustrated in Figure 7.15.

Since we know that the normal tables give the area below the value of Z, and that is what we are looking for, we can look the probability up directly. At first glance it may seem like you are looking for a needle in a haystack, but with a little bit of common sense and insight you will see that it is really not difficult.

If you look at the picture we drew (and this is why we draw them), you can see immediately that the value of Z that we want must be negative. We have already halved our search! Also, if you look at the tables, you will see that the numbers in the table are in numerical order. Once you find a suitable starting place for your search, you simply have to search systematically for the number you are looking for. The section of the normal table that contains the value we are looking for is shown in Figure 7.16.

This is probably a good place to point out that since the tables give probabilities to four decimal places, we are not likely to find the value we are looking for exactly. However, if the value you want is not there, you will always be able to locate two adjacent values, one smaller and one larger than the value you are looking for. Examining the table you see that the numbers from top to bottom are increasing and that at the row for $z = -1.3$ they are too small and at the $z = -1.2$ row they are too large. The value you want must be in there somewhere! A more careful search will lead you to the two cells that are highlighted in the table, 0.0985 and 0.1003.

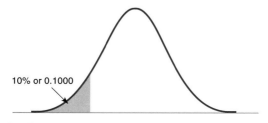

FIGURE 7.15 Bottom 10% of the normal distribution

z	0.00	0.01	0.02	0.03	0.04	0.05	0.06	0.07	0.08	0.09
−1.9	0.0287	0.0281	0.0274	0.0268	0.0262	0.0256	0.0250	0.0244	0.0239	0.0233
−1.8	0.0359	0.0351	0.0344	0.0336	0.0329	0.0322	0.0314	0.0307	0.0301	0.0294
−1.7	0.0446	0.0436	0.0427	0.0418	0.0409	0.0401	0.0392	0.0384	0.0375	0.0367
−1.6	0.0548	0.0537	0.0526	0.0516	0.0505	0.0495	0.0485	0.0475	0.0465	0.0455
−1.5	0.0668	0.0655	0.0643	0.0630	0.0618	0.0606	0.0594	0.0582	0.0571	0.0559
−1.4	0.0808	0.0793	0.0778	0.0764	0.0749	0.0735	0.0721	0.0708	0.0694	0.0681
−1.3	0.0968	0.0951	0.0934	0.0918	0.0901	0.0885	0.0869	0.0853	0.0838	0.0823
−1.2	0.1151	0.1131	0.1112	0.1093	0.1075	0.1056	0.1038	0.1020	0.1003	0.0985
−1.1	0.1357	0.1335	0.1314	0.1292	0.1271	0.1251	0.1230	0.1210	0.1190	0.1170
−1.0	0.1587	0.1562	0.1539	0.1515	0.1492	0.1469	0.1446	0.1423	0.1401	0.1379
−0.9	0.1841	0.1814	0.1788	0.1762	0.1736	0.1711	0.1685	0.1660	0.1635	0.1611
−0.8	0.2119	0.2090	0.2061	0.2033	0.2005	0.1977	0.1949	0.1922	0.1894	0.1867

FIGURE 7.16   Normal probability table containing the probability 0.1000

As we guessed, one of these is a bit too small and the other is too large. The value of 0.0985 corresponds to a value of Z of −1.29, while the 0.1003 corresponds to $z = -1.28$. Thus, the value of Z that we are looking for must be in between these two.

At this point we need to discuss how we will determine the Z values that we will use. Some people would suggest performing a linear interpolation between the two values to get an estimate of the correct value of Z. Even though the normal curve is certainly not linear, for small increments in Z this is not a bad approximation. The only problem is, it is not worth the trouble.

For our purposes we will use the value of the probability that is closest to the value we are looking for unless there is something in the problem that tells us to use the one that is smaller or larger. The something would be some directional description of the percentage we are looking for, like "at least 10%," in which case we would choose 0.1003, or "at most 10%," which would lead us to the 0.0985. When we look at an actual application of this technique you will see how little difference the exact answer makes.

Now on to solving real problems. We start with a simple example that uses what we have already figured out.

EXAMPLE 7.28   Aptitude Scores

Solving the Inverse Problem

As hard as it may be to take, downsizing and cost cutting are a reality of business life. The company that gives employees the aptitude test has decided that people who score in the bottom 10% of the test scores will not receive any additional job training. If there are to be layoffs, these people will be among the first to be cut. The question is, what cutoff score on the test should the company use?

We already know that the bottom 10% corresponds to a Z-score of −1.28. But clearly this cannot be the answer to the question. How do we translate the Z-score back into the realm of the variable of interest, the test scores?

You know that the Z value represents the number of standard deviations away from the mean that a value is. The negative sign on the Z value tells you to move down from the mean (subtract the number of standard deviations). Thus, we know that the cutoff score must be 1.28 standard deviations *below* the mean. That is,

$$X = 75 - (1.28)(5) = 68.6 \text{ points}$$

The company will have to decide whether to use 68 points (and affect less than 10% of the people) or 69 points (and affect more than 10%).   ∎

From this example you see why it is not necessary to interpolate to get the exact value of Z. If the answer is to be used to make a decision, the extra precision gained by getting the exact Z value does not usually translate into any practical information. The test is not scored to the nearest tenth of a point. Even if it were, the change in the Z value affects the hundredths place of the score.

We used our understanding of the Z value to determine the cutoff score, but the formula we used really comes from algebraically solving the definition of the Z value for X:

$$Z = \frac{X - \mu}{\sigma} \longrightarrow X = \mu + Z\sigma$$

When Z is positive you will add to μ and when Z is negative you will subtract from μ.

 **TRY IT NOW!**

Speed Reading
Solving the Inverse Problem

The instructor who is interested in how fast students can read the statistics textbook would like to identify the bottom 25% of the class, in terms of the number of pages that they can read in an hour.

Find the number of pages per hour that defines the bottom 25% of the students.

When solving problems that define the probability and looking for the value of the random variable, it is important to know when you are specifying an area above the value of interest or an area below the value. When the area is below the value you are looking for, you can look up the specified probability directly. When the area is above the value you are looking for you will either have to subtract the specified probability from 1 and look that up, or look up the given probability and rely on the symmetry of the table.

EXAMPLE 7.29   Aptitude Scores

Specifying Upper Area Probabilities

Although cost cutting and downsizing are negative aspects of the business world, the company with the aptitude test is also planning to give extra training to employees who score in the top 2% of those taking the test. The company would like to identify the score to use as the cutoff point.

If we draw a picture that represents the problem we see that the Z value we are looking for will most certainly be positive.

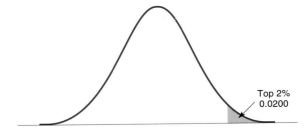

If you want to look up the value directly you will have to look up the bottom area or 0.9800. It is just as easy to use common sense and rely on the symmetry of the normal distribution. If you look up 0.0200 in the table you will find that it is not there, but that you can choose from 0.0197 and 0.0202. Since 0.0202 is closer to 2% (and slightly more generous to the employees) we will use the corresponding Z value of -2.05.

But we just said that the answer has to be positive. This is where common sense, symmetry, and the picture come together. We know the answer should be positive from the picture and common sense, but we used the symmetry feature to make the look-up easier. We will use $+2.05$ to find the correct answer.

The cutoff score for the top 2% is then

$$75 + (2.05)(5) = 85.25 \text{ points}$$

The company can use 85 (and train more than 2%) or 86 (and train less than 2%).

■

Inverse normal probability problems are an important application of the normal distribution. These types of problems are encountered many times in real life when standards or cutoff points need to be determined.

7.6.6 Exercises—Learning It!

7.16 The amount of money spent by students for textbooks in a semester is a normally distributed random variable with a mean of $235 and a standard deviation of $15.

(a) Sketch the normal distribution that describes the amount of money spent on textbooks in a semester.

(b) What is the probability that a student spends between $220 and $250 in any semester?

(c) What percentage of students spend more than $270 on textbooks in any semester?

(d) What percentage of students spend less than $225 in a semester?

7.17 On any given day, the number of leasable square feet of office space available in a small city is a normally distributed random variable with a mean of 850,000 square feet and a standard deviation of 25,000 square feet. The number of leasable square feet available in another small city is normally distributed with a mean of 900,000 square feet and a standard deviation of 25,000 square feet.

(a) Sketch the distribution of leasable office space for both cities on the same graph.

(b) What is the probability that the number of leasable square feet in the first city is less than 925,000 square feet?

(c) What is the probability that the amount available in the second city is less than 925,000?

7.18 The actual amount of a certain brand of orange juice in a container marked half gallon is a normally distributed random variable with a mean of 65 oz and a standard deviation of 0.35 oz.

(a) What percentage of the containers contain more than 64.5 oz?

(b) What percentage of the containers contain between 64 and 66 oz?

(c) If federal law says that 98% of all containers must be at or above the labeled weight, does this brand of orange juice meet the requirement?

7.19   The amount of money per month earned by an auditor with 10 years experience is a normally distributed random variable with mean $3500 and standard deviation $240.

(a) What percentage of auditors with 10 years experience earn more than $4000 per month?

(b) What percentage of auditors with 10 years experience earn less than $3200 per month?

(c) What is the probability that a randomly selected auditor earns between $3250 and $3800 per month?

(d) What monthly income defines the top 10% of all auditors with 10 years experience?

7.7 USING EXCEL TO GENERATE PROBABILITY DISTRIBUTIONS

Excel has built-in functions that allow you to calculate probabilities and generate random data from many different probability distributions. In this section, we will look at using Excel with the binomial and normal probability distributions.

7.7.1 Calculating Binomial Probabilities in Excel

So far, you have relied on tables to find probabilities for the binomial probability distribution. What would you do if you needed to solve a problem for values of n and π that were not in the tables? Excel has built-in functions that calculate binomial probabilities for any values of n and π.

Suppose that you want to calculate the probability that a random sample of $n = 50$ with $\pi = 0.45$ will have exactly 25 successes. You can use the Excel function BINOMDIST to calculate this probability. We will look at this function using the Function Wizard so that you understand how Excel functions work and how you can find out about Excel functions you have never used. We will also look at a macro included on the disk that accomplishes the same thing. Follow these steps:

1. To use a function, position the cursor in an empty cell in the worksheet.

2. From the main toolbar, click on the Function Wizard icon. The **Paste Function** dialog box opens. This box lists all of the different categories of functions available in Excel. Highlight **Statistical** for the function category and **BINOMDIST** for the function, as shown in Figure 7.17.

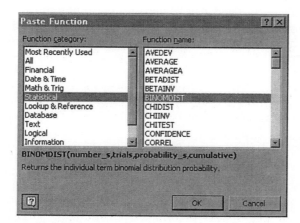

FIGURE 7.17   Choosing an Excel function

At the bottom you see the format for the function you are choosing and a description of what it does. Click **OK** and the dialog box for the BINOMDIST function opens.

As you place the cursor in each textbox, a description of the input for that box is described at the bottom of the dialog box. It is important that you read these descriptions so that you know what you are supposed to input.

3. Place the cursor in the textbox labeled **Number_s.** At the bottom you can see that this is where you enter the number of successes you are interested in. Type in the number "25."

4. Now, place the cursor in the textbox labeled **Trials** and enter "50," which is the number of trials that you have.

5. Place the cursor in the textbox labeled **Probability_s.** Notice that the bottom of the dialog box tells you that Excel is looking for the probability of a success on any trial, π. Type in "0.45."

6. The last textbox lets you indicate what kind of probability you want. Excel will calculate either $P(X = x)$ or $P(X \le x)$, the cumulative probability. In this case we want $P(X = 25)$ so we will set this value to False. The completed dialog box should look like the one in Figure 7.18.

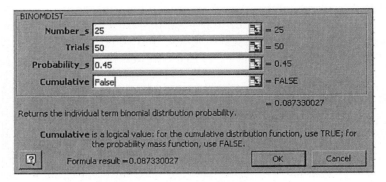

FIGURE 7.18   Completed BINOMDIST dialog box

7. Click on **OK** and the probability—in this case, 0.087330027—will appear in the cell in which you started.

You might have noticed that while you were filling in the dialog box, the status line in Excel was recording exactly what gets entered in the cell where you place the function, as shown in Figure 7.19. When you become familiar with particular functions, you can type the command in directly and bypass the Function Wizard. We will do that for the other functions in this chapter.

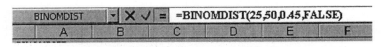

FIGURE 7.19   The BINOMDIST command

Remember, to use a macro you must open the MacDoIt.xls worksheet on the disk.

There is also a macro available for calculating binomial probabilities. To use this macro, choose **Macro** from the **Tools > Macro** menu and select **Binomial Probabilities** from the list of macros and click **Run.** The dialog box shown in Figure 7.20

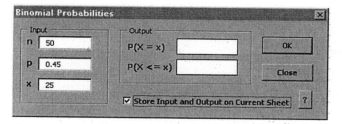

FIGURE 7.20 Binomial Probabilities macro

opens. You will need to input the same values that you did using the Excel function. The textbox labeled **p** is where the value of π goes. The macro gives both $P(X = x)$ and $P(X \le x)$ as output.

You can choose to have the output placed into the current worksheet if you want. If you do, then when you click on **OK,** the output table is placed in the location you specify, as shown in Figure 7.21. If you do not, then the output just appears in the dialog box.

Binomial Probabilities			
n	50	P(X=x)	0.08733
p	0.45	P(X<=x)	0.803369
x	25		

FIGURE 7.21 Binomial probability output

7.7.2 Calculating Normal Probabilities in Excel

Excel has functions either for calculating probabilities from a normal distribution with a given mean and standard deviation, or from a standard normal. There are also functions for solving inverse normal probability problems—that is, finding the value of Z that corresponds to a given probability. Table 7.3 lists all of the functions, describes what they do, and explains what values are expected as input.

TABLE 7.3

Function format:	What it does	Input
NORMDIST(x, mean, standard_deviation, cumulative)	Returns $P(X \le x)$ for a normally distributed random variable	x = value of interest; mean = mean of the random variable; standard_deviation = standard deviation of the random variable; cumulative = true for $P(X \le x)$
NORMINV(probability, mean, standard_deviation)	Returns the value of x that has the specified probability below it	probability = the area under the curve below the X value of interest; mean = mean of the random variable; standard_deviation = standard deviation of the random variable
NORMSDIST(x)	Returns $P(Z \le x)$ for a standard normal random variable	x = value of interest for Z
NORMSINV(probability)	Returns the value of Z that has the specified probability below it	probability = area under the standard normal curve below the Z value of interest

To calculate probabilities like $P(z_1 < Z < z_2)$ or $P(x_1 < X < x_2)$, you will need to calculate two probabilities and subtract them, just as you do with the tables. This is also true for $P(Z > z)$ and $P(X < x)$.

There is a macro available on the disk that will calculate any or all of these probabilities. To use it, select **Normal Probability** from the list of macros and click **Run.** The dialog box shown in Figure 7.22 opens. The input for the macro is very simple.

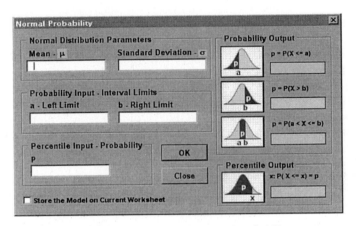

FIGURE 7.22   Dialog box for Normal Probability macro

To use the macro, follow these steps:

1. Fill in the mean, μ, and the standard deviation, σ, of the random variable in the appropriate textboxes.
2. Look at the pictures on the right of the box. If you want only a lower-tail probability, fill in the textbox for **a.** If you want only an upper-tail probability, fill in the textbox for **b.** If you want the probability between two values, fill in the textboxes for both **a** and **b.**
3. If you want to solve the inverse normal problem, type the area under the normal curve *below* the value of interest in the textbox labeled **Percentile Input.**
4. Click **OK** to obtain the output.

You can choose to have the output placed in the worksheet, by checking the box labeled **Store the Model on Current Worksheet.** If you do not do this, then the answers will just appear in the dialog box when you click **OK.**

7.7.3 Generating Random Data in Excel

In addition to calculating probabilities in Excel, you can also generate random data that come from populations with specific probability distributions. Such data are useful in simulations and other exercises. To generate data from different probability distributions, you use the **Data Analysis** tools. From the **Data Analysis** tools list, select **Random Number Generation.** The dialog box shown in Figure 7.23 opens.

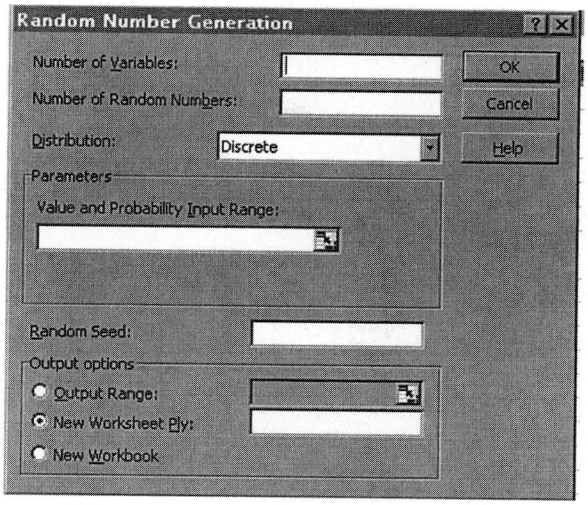

FIGURE 7.23   Random Number Generation dialog box

Excel will generate random data from several probability distributions, including the binomial and the normal. To see the list of available distributions, click on the drop down arrow in the textbox labeled **Distribution:** and the list appears as shown in Figure 7.24.

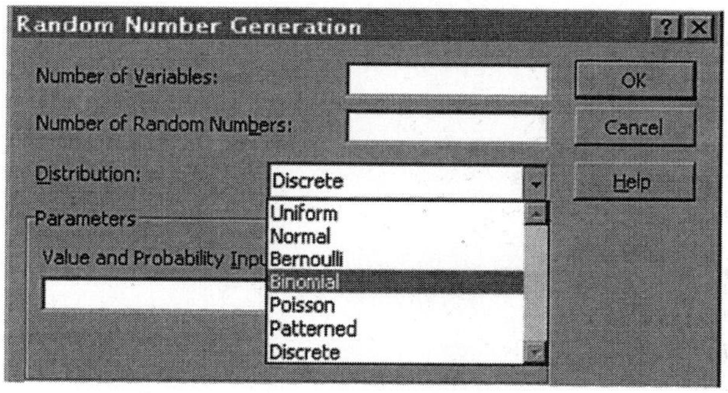

FIGURE 7.24   List of probability distributions

The top portion of the dialog box is the same for every probability distribution. In the textbox labeled **Number of Variables:** you enter the number of different random variables you want to generate. In most cases, this is 1. In the second textbox, labeled **Number of Random Numbers:** you input the number of observations from the probability distribution that you want to generate—usually either the sample size or the population size, depending on what you are generating.

We will generate 50 observations from a binomial distribution with $n = 20$ and $\pi = 0.70$. To do this, follow these steps.

1. Select **Binomial** from the list of distributions and the dialog box will change to allow input of the appropriate parameters.

2. In the textbox for **Number of Variables:** type "1" and in the textbox for **Number of Random Numbers:** type "50."

3. In the textbox labeled **p Value** enter the value for μ, 0.70. Finally, in the textbox for **Number of Trials:** enter 20.

4. Specify where you want the output to appear and click **OK.** The completed dialog box should look like the one in Figure 7.25 on page 310.

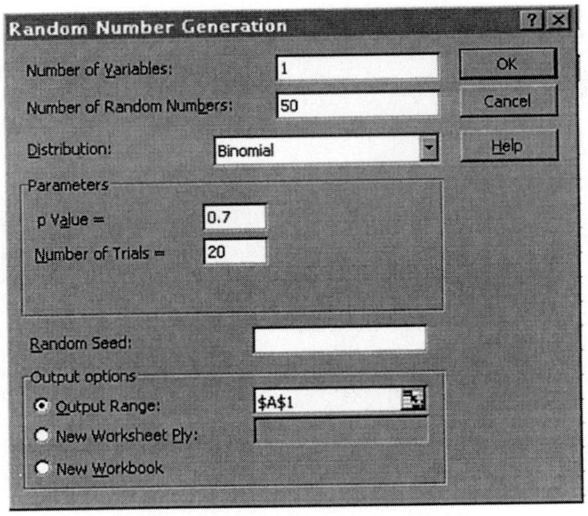

FIGURE 7.25   The Binomial Random Variable dialog box

The data will be output to the location you specify. Figure 7.26 shows a portion of the output. Notice that the values are all less than 20 (which you would expect) and that most of the values are higher than 10, since π is larger than 0.50.

	A
1	16
2	10
3	15
4	12
5	14
6	11
7	10
8	12
9	14
10	16
11	14
12	14
13	13
14	11
15	14

FIGURE 7.26   Random data from binomial distribution

The steps for generating data from a normal distribution are identical to those for the binomial distribution. The only difference is that the parameters you enter for the normal distribution are the mean and standard deviation, π and σ.

CHAPTER 7 SUMMARY

When many people think of probability they think of flipping coins, spinning spinners, or gambling. While these are certainly aspects of probability, and are useful for illustrating some of the basic principles of probability theory, they are by no means all there is to the subject. Probability is an important and interesting subject in its own right, but the study of random variables and probability distributions is important in the study and development of statistics.

Probability is the bridge between *descriptive statistics* and *inferential statistics.* In descriptive statistics we use different techniques to describe sample data. In inferential statistics we will test hypotheses about the populations from which these samples came. Probability is the tool that allows us to reconcile what happened (descriptive) with what we think is true by determining how likely the outcomes of the experiment we perform are.

Key Terms

Term	Definition	Page reference
A AND B	The event **A AND B** is the event that A and B both occur.	255
A OR B	The event **A OR B** describes the event when either A happens or B happens or they both happen.	255
Binomial random variable	A **binomial random variable** is the number of successes in n trials or in a sample of size n.	272
Complement	The **complement** of an event **A**, denoted **A**′, is the set of all outcomes in the sample space, **S**, that do not correspond to the event **A**.	253
Contingency table	A **contingency table** is a table whose rows represent the possible values of one variable and whose columns represent the possible values for a second variable. The entries in the table are the number of times that each *pair* of values occurs.	259
Empirical probability	An **empirical** probability is one that is calculated from sample data and is an estimate for the true probability.	262
Event	An **event, A,** is an outcome or a set of outcomes that are of interest to the experimenter.	251
Experiment	An **experiment** is any action whose outcomes are recordable data.	249
Interval probability	An **interval probability** gives the probability that a random variable will take on a value *between* two given values $P(x_1 \leq X \leq x_2)$.	294
Mutually exclusive	Two events, A and B, are said to be **mutually exclusive** if they have no outcomes in common.	256

Term	Definition	Page reference
Probability	**Probability** is a measure of how likely it is that something will occur.	249
Probability density function, $f(x)$	A **probability density function, $f(x)$,** is a smooth curve that represents the probability distribution of a continuous random variable.	289
Probability distribution, p(x)	The **probability distribution** of a random variable X, written as **p(x),** gives the probability that the random variable will take on each of its possible values.	266
Probability of an event, P(A)	The **probability of an event A, P(A),** is a measure of the likelihood that an event A will occur.	251
Random variable	A **random variable, X,** is a quantitative variable whose value varies according to the rules of probability.	265
Sample space, S	The **sample space, S,** is the set of all possible outcomes of an experiment.	249
Standard normal random variable, Z	A **standard normal random variable** is normally distributed with a mean of 0 and a standard deviation of 1, **$Z \sim N(0, 1)$.**	292
Standard normal table	A **standard normal table** is a table of probabilities for a Z random variable.	292

Key Formulas

Term	Formula	Page reference
Binomial mean and standard deviation	$\mu = n\pi$ and $\sigma = \sqrt{n\pi(1 - \pi)}$	280
Binomial probability formula	$p(x) = \dfrac{n!}{(x!)(n - x)!} \pi^x (1 - \pi)^{n-x}$ for $x = 0, 1, \ldots, n$	276
General addition rule	$P(A \text{ OR } B) = P(A) + P(B) - P(A \text{ AND } B)$	259
Normal probability density	$f(x) = \dfrac{1}{\sigma\sqrt{2\pi}} e^{-(x-\mu)^2/2\sigma^2}$ for $-\infty < x < \infty$	290
P(A)	$P(A) = \dfrac{\text{Number of ways that A can occur}}{\text{Total number of possible outcomes}}$	251
	$P(A) = \dfrac{n_A}{N}$	251
Probability of the complement, P(A$'$)	$P(A) + P(A') = 1$	253
Simple addition rule	$P(A \text{ OR } B) = P(A) + P(B)$	256
Z-score	$Z = \dfrac{X - \mu}{\sigma}$	292

CHAPTER 7 EXERCISES

Learning It!

7.20 In the survey about satisfaction with local phone service, those respondents who rated their current service as excellent and those who rated it Poor–Very Poor were asked what type of company their current local service provider was. The results are given in the next table.

Current Service Source	Excellent	Poor–Very Poor
Long distance company	264	1394
Local phone company	444	1318
Power company	131	485
Cable TV company	215	431
Cellular phone company	198	572

(a) What is the probability that a person selected from this group will rate his service as excellent and will have a long distance company as his current provider?

(b) What is the probability that a person selected from this group will use a power company or a cable TV company as his current provider?

(c) What is the probability that a person selected from this group will rate his current service as Poor–Very Poor or use a local phone company?

(d) What is the probability that a person selected from this group will rate his service as Poor–Very Poor and will use a cellular phone company?

7.21 In a study on college students and binge drinking, researchers were interested in looking at binge drinking and gender. Questions were asked about the number of times that a student had five or more drinks in the last two weeks and gender. The results are shown in the table below:

	Number of Times Had Five or More Drinks in a Row in Last Two Weeks					
Gender	**0**	**1**	**2**	**3**	**4**	**5**
Male	55	20	18	19	8	4
Female	75	14	12	24	2	0

(a) What is the probability that a student is a male and had five or more drinks in a row 2 times in the last two weeks?

(b) What is the probability that a student is female or had five or more drinks in a row 3 times in the last two weeks?

(c) What is the probability that a student had five or more drinks in a row in the last two weeks 3 or 4 times?

(d) What is the probability that a student did not have five or more drinks in a row in the last two weeks?

7.22 In the same study of binge drinking and college students, the researchers were also interested in the number of times that student experienced hangovers in the semester. The data they collected are found below:

	Hangover Since Beginning of Semester		
Gender	**Not at All**	**Once**	**Twice or More**
Male	61	23	40
Female	66	25	36

(a) What is the probability that a student is female and had a hangover twice or more during the semester?

(b) What is the probability that a student is male and has not had a hangover during the semester?

(c) What is the probability that a student had a hangover once or more during the semester?

7.23 Surveys indicate that 40% of all small businesses do not provide any kind of health insurance options for their employees. The Chamber of Commerce in a large city takes a random sample of 20 small businesses. What is the probability that in the sample of 20 small businesses

(a) at least 12 do not provide health insurance options for employees?

(b) at most 4 do not provide health insurance options for employees?

(c) between 8 and 13 do not provide health insurance options for employees?

(d) at least 5 provide health insurance options for employees?

7.24 Recent studies have shown that 30% of employees in the insurance industry telecommute four days a week. If a random sample of 15 insurance industry employees is taken, what is the probability that

(a) between 4 and 7 inclusive telecommute four days a week?

(b) less than 6 telecommute four days a week?

(c) at least 9 telecommute four days a week?

(d) between 5 and 10 do not telecommute four days a week?

7.25 A survey done at a state university in New England found that 40% of all seniors have encountered academic problems related to binge drinking. If a random sample of 25 seniors at the university is taken, what is the probability that of the 25

(a) at most 6 have encountered academic problems related to binge drinking?

(b) between 4 and 9 have encountered academic problems related to binge drinking?

(c) more than 15 have encountered academic problems related to binge drinking?

(d) Find the mean and standard deviation of the number of seniors in a sample of 25 who have encountered problems related to binge drinking.

7.26 The amount of office space allocated to production planners in consumer products companies is a normally distributed random variable with a mean of 120 square feet (sq ft) and a standard deviation of 6 sq ft.

(a) What percentage of production planners' offices have more than 135 sq ft?

(b) What percentage of production planners' offices have between 110 and 125 sq ft?

(c) What percentage of production planners' offices have less than 130 sq ft?

7.27 The size of a gift/specialty store in a regional super mall is a normally distributed random variable with a mean of 8500 sq ft and a standard deviation of 260 sq ft. What is the probability that a randomly selected gift/specialty store in a regional super mall is

(a) more than 8000 sq ft?

(b) between 8300 and 9000 sq ft?

(c) less than 9500 sq ft?

7.28 A recent study done at a university in New England found that 60% of all students have missed class in the semester because of drinking. A random sample of 20 students is taken. What is the probability that

(a) at least 15 have missed class because of drinking?

(b) between 12 and 17 inclusive have missed class because of drinking?

(c) less than 5 have not missed class because of drinking?

(d) at most 13 have not missed class because of drinking?

Thinking About It!

Requires Exercise 7.2 **7.29** The company that is looking at the way employees took floating holidays collected data from 300 employees. The data are shown again below:

| | *How Days Were Used* | | |
Type of Job	Took Actual Holiday	Added to Vacation	Took Random Days
Professional	5	17	51
Clerical	13	46	32
Hourly	53	78	5

(a) What is the probability that an employee selected at random is a clerical worker?

(b) Suppose you know that the person selected used the floating holidays as random days off. Now what is the probability that the person is a clerical worker?

(c) How do the probabilities in parts (a) and (b) compare? What is different about them? When you are given some information about the outcome of the experiment it reduces the size of the sample space. This type of probability is called a *conditional probability*. It is often described as the probability that an event occurs, given that you know another event has occurred.

(d) What is the probability that the person took the floating holidays on the actual holiday given that you know the person is a professional employee?

(e) If you know that the person selected is either a clerical worker or an hourly worker, what is the probability that the person took the days as added vacation days?

7.30 Consider the computer magazine survey about consumers' plans to buy computers. The data are given below:

Requires Exercise 7.3

	When Purchase Will Be Made		
Type of Computer	**0–3 Months**	**3–6 Months**	**6–12 Months**
Notebook/Portable	34	156	258
Desktop	56	346	128

(a) What is the probability that a person selected at random is planning to make a purchase in 0–3 months, given that the person plans to buy a notebook computer?

(b) What is the probability that a person selected is planning to buy a desktop computer given that the person plans to make the purchase in the next 6 months?

(c) If you know that the person selected is planning to buy a desktop computer, what is the probability that the person will make the purchase in the next 3–6 months?

(d) What is the probability that a person will buy a notebook computer, given that the person is not planning to buy it within the next 0–3 months?

7.31 Consider the study of binge drinking and college students. The data are found below:

	Hangover Since Beginning of Semester		
Gender	**Not at All**	**Once**	**Twice or More**
Male	61	23	40
Female	66	25	36

(a) Given that the student is female, what is the probability that she had a hangover twice or more during the semester?

(b) What is the probability that a student is male given that the student had a hangover once or less during the semester?

(c) What is the probability that a student has had two or more hangovers in a semester given that the student is male?

(d) Compare your answers to parts (a) and (c) and interpret the results for the researchers.

7.32 The Chamber of Commerce that is interested in whether small businesses provide health care options for employees wants a few more questions answered. The surveys indicated that 40% of employers did not provide health insurance options for employees.

Requires Exercise 7.23

(a) In the sample of 20, how many should the Chamber of Commerce expect to provide health insurance options for employees?

(b) What is the standard deviation of the number of small businesses in the 20 that provide health insurance options?

(c) Find the probability that the number of small businesses in a sample of 20 will be within 2 standard deviations of the mean.

(d) How does this probability compare to the percentage predicted by the empirical rule? If it does not agree, why not?

7.33 The Chamber of Commerce looking at health insurance options provided by small businesses finds 4 that provide health insurance options for employees and thinks this is unusual.

Requires Exercises 7.23, 7.32

(a) Do you agree that the findings are unusual? Why or why not?

(b) What would you tell the Chamber of Commerce this might mean?

7.34 The manager of a regional super mall wants to compare the gift/specialty stores in her mall to the size of the smallest 20% of the gift/specialty shops in similar malls. She knows that the size of such stores is normally distributed with a mean of 8500 sq ft and a standard deviation of 260 sq ft.

(a) What square footage defines the smallest 20% of such stores?

(b) Harriet's Gift Boutique has complained that it is much smaller than any similar stores in other malls. If the size of this store is 7800 sq ft, is the complaint reasonable?

Requires Exercise 7.28   **7.35** The Dean of Students at another university in New England read the report on binge drinking and decided to conduct a small survey on her own campus. She took a random sample of 20 students and found that 15 of them had missed class because of drinking. She decides that this indicates that her university is within the norm for this problem. Do you agree with her conclusion? Why or why not?

Doing It!

CHAMBER.XXX   **7.36** The Chamber of Commerce that is studying the credit problems of small businesses asked them 3 questions to classify their business and 7 questions related to the issue of credit problems. A portion of the datafile and an explanation of each variable are given below:

Size	Employees	Nature	Problem	Understands	Concerned	Call	Loan	Collateral	Access
2	2	1	1	2	1	2	2	0	2
1	2	3	1	2	2	0	2	2	1
4	3	1	2	1	2	2	2	2	0
1	1	1	2	1	2	2	2	2	2
1	2	1	2	0	0	0	0	0	0
3	1	5	2	0	2	2	2	2	1
3	3	2	2	1	1	2	2	2	1
2	2	3	2	1	2	2	2	2	2
3	5	2	2	1	2	2	2	1	2

- The variable *Size* refers to the annual sales of the company and is coded as follows:

  1   under \$1 million
  2   \$1–5 million
  3   \$6–10 million
  4   \$11–20 million
  5   over \$20 million

- The variable *Employees* refers to the number of employees that the company currently employs. This variable was coded as

  1   0–5 employees
  2   6–10 employees
  3   11–50 employees
  4   51–150 employees
  5   151–250 employees
  6   Over 250 employees

- The variable Nature refers to the type of business and is coded as

  1   Manufacturing
  2   Retail
  3   Service
  4   Real Estate
  5   Other

- The next seven variables contain the response to the questions or statements indicated and are coded as follows:

  1   Yes
  2   No

- *Problem*   "Are you experiencing credit related problems?"
- *Understd*   "The bank understands my problems."

- *Concern*   "I am concerned that my note might be recalled."
- *Call*   "The bank is planning to recall my loan."
- *Loan*   "The bank has called my loan."
- *Collateral* "The bank has demanded more collateral."
- *Access*   "Access to credit is affecting my business."

(a) Look at the variable *Problem* that was discussed in Example 7.24. Separate the respondents according to the type of company that they have. For each type of industry, generate a binomial probability distribution for $\pi = 0.10$ and $n =$ the number of that type of company. For each type of company find the probability that the number of people reporting credit problems is the actual number found in the sample.

(b) For each type of company estimate the value of π. Compare these to each other, to the overall estimated value of π and to the assumed value of 0.10. Report your conclusions.

(c) Repeat parts (a) and (b) for the different sizes of companies (both annual sales and number of employees).

(d) Now look at the variables related to loan recall, *Concern*, *Call*, and *Loan*. Use the data to estimate the percentage of small businesses that responded Yes to each of these questions. How does the number of nonresponses to the question affect your estimate of the percentage? Do you think that it is better to estimate the percentage of Yes respondents using the total of 166 companies or just the ones who responded to the question? Why?

(e) Investigate the variables *Concern*, *Call*, and *Loan* the same way you investigated *Problem*.

(f) Investigate the variable *Understd* the same way you investigated *Problem*.

(g) Investigate the variable *Access*.

(h) Prepare a report for the Chamber of Commerce about the credit problems and concerns of small businesses.

7.37 In Chapter 3 you learned about some of the customer complaints that a company that manufactures tissues can get. One of the categories of complaints that made up a large percentage of the total was Dispensing. In that category, Sheets Tear on Removal was a significant factor.   *TISSUES.XXX*

   The managers of the tissue company have decided to address this problem. They know that tensile strength is the factor that determines when a tissue will tear, and have decided that to solve the problem they will have to investigate the tensile strength of the tissues.

   As part of the Quality Control program at the company, facial tissue has certain product specifications, that is, criteria that must be met, for the product to be acceptable to consumers. One of the characteristics that is specified is tensile strength.

   The managers have decided to look at the current levels of tissue strength. They know the target values for the process and the parameters that should be met, and have decided to check to see if the process is meeting the current specifications. If it is not, then changes will need to be made to see that is does. If it is, then perhaps the process specifications will need to be changed. The managers will collect data on two variables:

- **Machine Direction (MD) Strength:** This is the strength in the direction that the machine pulls on the tissue during manufacture. It has to be high enough that the tissues do not break, causing machine down time.
- **Cross-Direction (CD) Strength:** This is the strength in the direction that the tissue is pulled out of the box. It is the variable that determines whether the sheets tear when you remove them from the box.

Samples were taken from tissue produced on a single tissue machine. The samples were taken over three different days and the results were recorded.

A portion of the datafile and an explanation of the variables are shown below:

Day	MDStrength	CDStrength
1	1006	422
1	994	448
1	1032	423
1	875	435
1	1043	445
1	962	464
1	973	472

- *Day* keeps track of the day on which the sample was taken and goes from 1 to 3.
- *MDStrength* measures machine-directional strength and is measured in lb/ream.
- *CDStrength* measures cross-directional strength and is measured in lb/ream.

According to the specifications, *MDStrength* is supposed to be normally distributed with a mean of 1000 and a standard deviation of 50 lb/ream. *CDStrength* should be normally distributed with a mean of 400 and a standard deviation of 25 lb/ream.

(a) Use a computer software package to create normal probability tables for the specified distributions of *MDStrength*. Use increments of 50 and go from 800 to 1200.

(b) Use the table to determine the probability that a tissue manufactured according to specifications will have an *MDStrength* of less than 850 lb/ream.

(c) Create a relative frequency histogram for the variable *MDStrength* that goes from 800 to 1200 in class intervals of 50 units.

(d) Do the data in the histogram appear to have the shape of a normal distribution? What is the center?

(e) Look at the relative frequencies on the histogram you just created, and calculate what percentage of the data are within one standard deviation of the mean value of 1000. According to the empirical rule what percentage should be within one standard deviation? How do the actual data compare with the empirical rule prediction?

(f) Use the procedure above to determine the percentage of *MDStrength* data that are within two and three standard deviations of the mean.

(g) The critical specifications for *MDStrength* are 850 on the low side and 1075 on the high side. Use the frequency distribution to determine the percentage of the tissues that actually do not meet these specifications.

(h) How does this compare to the percent defective expected by the product specifications?

(i) Do you think that the company should be concerned about the difference? Why or why not?

(j) Create a set of normal probabilities for the theoretical distribution of *CDStrength*.

(k) The critical values for *CDStrength* are 480 on the high side and 390 on the low side. *CDStrength* that is too high creates a stiff tissue. Since the cross direction is the one in which tissues are pulled from the box, a value of *CDStrength* that is too low can cause sheets to tear on removal from the carton. According to the specifications, what percentage of the tissues should have CD strengths that are too high? too low?

(l) Create a graph of the theoretical distribution of *CDStrength*.

(m) Generate a set of descriptive statistics for the variable *CDStrength*. Compare the mean and the median. Do you think that the distribution of *CDStrength* is symmetric?

(n) Create a relative frequency histogram for the variable *CDStrength*. Does it support the assumption of normality?

(o) Compare the percentage of CD strength measurements that are within 1, 2, and 3 standard deviations of the mean to the predictions of the empirical rule. Is the assumption of normality still reasonable?

(p) Prepare a report to management that indicates whether the process appears to be running to the product specifications. In this report include any changes that need to be made (in terms of mean and standard deviation) to bring the process back to target values.

WORKSHOP 4

CHAPTER 8

THE BUSINESS RESEARCH PROPOSAL

8.1   CHAPTER OBJECTIVES

In business, as in most areas, you cannot simply conduct a research study without first explaining what you want to do and how you will accomplish it. It might be that although you think that the research is a good idea, it does not satisfy an important need of the business. Also, the research you propose might involve costs that must be approved before the study begins. In order to start the research, you must first propose what it is you want to do and have the proposal approved by the business managers.

This chapter covers the following material:

- The steps of the research proposal
- How to conduct a literature review
- Ethics in business research

8.2   THE RESEARCH PROPOSAL

At this point, we have identified the needs for business research to solve business problems and we have outlined the process of business research. We have also learned some of the statistical tools that are used to carry out the research. However, how exactly do we move from *identifying* a problem that needs to be solved to the actual process? One thing that is used to accomplish this is the **research proposal.**

> The *research proposal* is a document that outlines the problem, motivates the need for the research, and outlines the methods that will be used to carry out the research.

It is important that the research proposal be complete and well thought out for several reasons. First, it is the selling point for the business research that you want to do. The proposal will, essentially, "state your case" to management. Second, the proposal serves as a "guide" when you actually do the research and afterward. With the proposal you will not forget important details of the study. In addition, you will be able to defend your study when people complain that you did not do one thing or another that you never proposed to do in the first place! Third, the research proposal is a foundation for the final report from the research.

A research proposal is made up of three or four parts, depending on how you divide it. We will describe a research proposal with four parts: the Introduction, the Literature Review, the Methodology, and the Timeline.

8.2.1 The Introduction

The introduction is, perhaps, the most important part of the proposal. It is here that you describe the business problem that you propose to study and justify the need for the study. This involves describing how solving the problem will contribute to the organization and how the study meets the needs of the organization. Essentially, the introduction is a formal statement of the first two steps of the Poet model: problem identification and statement of desired goal or outcome.

In the introduction, you motivate your study by explaining the problem that you want to solve. You state your hypothesis about what you believe the research will prove, the questions it will answer, and the objectives of the study. You also indicate *why* this study is important and what the measurable outcomes will be. Remember, this is where you *sell* your idea and so it is most important to do a good job. If the introduction is not compelling, it is likely that the remainder of the proposal will never even be read.

EXAMPLE 8.1 Turnover Problems

Writing the Introduction

You work for a software company in the Technical Support department. Lately, personal turnover is much higher than it has been. The length of time that people are staying at the job seems to be less than one year. The company spends a considerable amount of money in training and they should be concerned about the problem.

Some reasons for concern might be a less experienced workforce, which could lead to customer dissatisfaction and the cost and time of training new employees. Your statement might include information on the cost and time of training and an assessment of how long a technical support employee should stay on the job to make the training worthwhile. The difference between how long they stay and how long they should stay is the need of the organization.

The introduction might include a proposal to study the records of technical support personnel for the past three years and collect data on how long they have worked for the company. You would state your hypothesis that the average length of stay is less than one year or propose to answer the question of whether the length of stay of technical support employees has decreased in the past three years. You might also propose to answer questions about why the employees are leaving.

For employees who have left the company, information from the exit interview might also be used to collect data on why they are leaving. You could also propose to survey current employees to see how satisfied they are with their jobs and other issues.

The outcome of the study would be recommendations to increase the amount of time that technical support personnel stay at the company. ∎

8.2.2 The Literature Review

Reasons for Literature Survey

The purpose of the literature review is to ensure that no important variable is ignored that has in the past been found to have had an impact on the problem. It is possible that some of the critical variables are never brought out in the interviews, either because the employees cannot articulate them or are unaware of their impact, or because the variables seem so obvious to the interviewees that they are not specifically stated. If there are variables that are not identified during the interviews, but influence the problem critically, then research done without considering them would be an exercise in futility. In such a case, the true reason for the problem would remain unidentified even at the end of the research. To avoid such mishaps, the researcher needs to delve into all the important research work relating to the particular problem area.

The following example will help to highlight the importance of literature survey. In establishing employee selection procedures, a company might be doing the right things such as administering the appropriate tests to assess the applicants' analytical skills, judgment, leadership, motivation, oral and written communication skills, and the like. Yet, it might be consistently losing excellent MBAs hired as managers, within a year, despite being highly paid. The reasons for the turnover of MBAs may not be identified while conducting interviews with the candidates. However, a review of the literature might indicate that when employees have unmet job expectations (i.e., their original expectations of their role and responsibilities do not match actual experiences), they will be inclined to leave the organization. Talking further to the company officials, it might be found that *realistic job previews* are never offered to the candidates at the time of the interview. This might explain why the candidates experience frustration on the job and eventually leave. This important factor, significantly influencing the turnover of managerial employees, may not have come to light but for the literature survey. If this variable is not included in the research investigation, the problem may never be solved!

Sometimes an investigator might spend considerable time and effort to "discover" something that has already been thoroughly researched. A literature review would prevent such wastage of resources in reinventing the wheel. However, because every situation is unique, further research has to proceed taking into consideration the relevant variables applicable to each unique situation. Finally, a good literature survey could in itself be the basis for qualitative research, as for instance, tracing the origins and progress of technology and predicting where it is headed in the future.

A survey of the literature not only helps the researcher to include all the relevant variables in the research project, but also facilitates the creative integration of the information gathered from the structured and unstructured interviews with what is found in previous studies. In other words, it gives a good basic framework to proceed further with the investigation. A good literature survey thus provides the foundation for developing a comprehensive theoretical framework from which hypotheses can be developed for testing. The development of the theoretical framework and hypotheses is discussed in Chapter 10.

A good literature survey thus ensures that:

Advantages of a good literature survey

1. Important variables that are likely to influence the problem situation are not left out of the study.

2. A clearer idea emerges as to what variables would be most important to consider (parsimony), why they would be considered important, and how they should be investigated to solve the problem. Thus, literature survey helps the development of the theoretical framework and hypotheses for testing.

3. Testability and replicability of the findings of the current research are enhanced.

4. The problem statement can be made with precision and clarity.

5. One does not run the risk of "reinventing the wheel," that is, wasting efforts on trying to rediscover something that is already known.

6. The problem investigated is perceived by the scientific community as relevant and significant.

Conducting the Literature Survey

Based on the specific issues of concern to the manager and the factors identified during the interview process, a literature review needs to be done on these variables. The first step in this process involves identifying the various published and unpublished materials that are available on the topics of interest, and gaining access to these. The second step is gathering the relevant information either by going through the necessary materials in a library or by accessing the sources online; the third step is writing up the literature review. These are now discussed.

Identifying the Relevant Sources

Previously, one had to manually go through several bibliographical indexes that are compiled periodically, listing the journals, books, and other sources in which published work in the area of interest can be found. However, with modern technology, locating sources where the topics of interest have been published has become easy. Almost every library now has computer on-line systems to locate and print out the published information on various topics.

Global business information, published articles in newspapers and periodicals, and conference proceedings, among other sources, are all now available on databases. Computerized databases include bibliographies, abstracts, and full texts of articles on various business topics. Statistical and financial databases are also easily accessible. Computer hardware and software enable the storage, updating, and display of information on global activities. Economic indicators and other data for various countries can be tracked easily. Statistical abstracts, and other publications, now available on CD-ROM and on the Internet, bring to the researcher all the necessary information at the press of the relevant computer keys.

Basically, three forms of databases are useful in reviewing the literature, as indicated below.

1. The **bibliographic databases,** which display only the bibliographic citations, that is, the name of the author, the title of the article (or book), source of publication, year, volume, and page numbers. These have the same information as found in the Bibliographic Index books in libraries, which are periodically updated, and include articles published in periodicals, newspapers, books, and so on.

2. The **abstract databases,** which in addition provide an abstract or summary of the articles

3. The **full-text databases,** which provide the full text of the article.

Databases are also available for obtaining statistics—marketing, financial, and so on—and directories are organized by subject, title, geographic location, trade opportunities, foreign traders, industrial plants, and such.

On-line searches provide a number of advantages. Besides saving enormous amounts of time, they are comprehensive in their listing and review of references, and the researcher can focus on materials most central to the research effort. In addition, accessing them is relatively inexpensive.

Some of the important research databases available on-line and on the World Wide Web are provided in the appendix to this chapter. These can be accessed on-line or through the Internet. If a source of information is not known, the search strategies on the Internet help to find it. Databases include listings of journal articles, books in print, census data, dissertation abstracts, conference papers, and newspaper abstracts that are useful for business research. Details of some of these databases can be found in the appendix to this chapter.

Extracting the Relevant Information

Accessing the on-line system and getting a printout of all the published works in the area of interest from a bibliographical index will provide a comprehensive bibliography on the subject, which will form the basis for the next step. Whereas the printout could sometimes include as many as 100 or more listings, a glance at the titles of the articles or books will indicate which of these may be pertinent and which others are likely to be peripheral to the contemplated study. The abstract of such articles that seem to be relevant can then be obtained through the on-line system. This will give an idea of the articles that need to be examined, the full text of which can then be printed out. While reading these articles, detailed information on the problem that was researched, the design details of the study (such as the sample size and data-collection methods), and the ultimate findings could be systematically noted in some convenient format. This facilitates the writing up of the literature review with minimum disruption and maximum efficiency. While reading the articles, it is possible that certain other factors may be found that are closely related to the problem at hand. For instance, while reading articles on the effectiveness of Information Systems, the researcher might discover that the size of the company has also been found to be an important factor. The researcher might then want to know more about how the size of organizations is categorized and measured by others and, hence, might want to read materials on organization size. All the articles considered relevant to the current study can be then listed as references, using the appropriate referencing format.

Writing Up the Literature Review

The documentation of the relevant studies citing the author and the year of the study is called literature review or literature survey. The literature survey is a clear and logical presentation of the relevant research work done thus far in the area of investigation. As stated earlier, the purpose of literature survey is to identify and highlight the important variables, and to document the significant findings from earlier research that will serve as the foundation on which the theoretical framework for the current investigation can be based and the hypotheses developed. Such documentation is important to convince the reader that (1) the researcher is knowledgeable about the problem area and has done the preliminary homework that is necessary to conduct the research, and (2) the theoretical framework will be built on work already done and will add to the solid foundation of existing knowledge.

A point to note is that the literature survey should bring together all relevant information in a cogent and logical manner instead of presenting all the studies in chronological order with bits and pieces of uncoordinated information. A good literature survey also leads one logically to a good problem statement.

There are several accepted methods of citing references in the literature survey section and using quotations. The *Publications Manual of the American Psychological Association* (1994) offers detailed information regarding citations, quotations, references, and so on, and is one of the accepted styles of referencing in the management area. Other formats include *The Chicago Manual of Style* (1993), and Turabian's *Manual for Writers* (1996).

Examples of Two Literature Surveys

Let us take *a portion* of two literature reviews done and examine how the activity has helped to (1) introduce the subject of study, (2) identify the research question, and (3) build on previous research to offer the basis to get to the next steps of theoretical framework and hypotheses development.

EXAMPLE 8.2 Risk-Taking Behaviors and Organizational Outcomes

Literature Survey

Managers handle risks and face uncertainties in different ways. Some of these styles are functional and others adversely impact on corporate performance. Living in times of dramatic organizational changes (mergers, for instance), and with the company performance varying vastly in this turbulent environment, it is important to investigate the relationship between risk-taking behaviors of managers and organizational outcomes.

A vast body of knowledge exists regarding risk-taking behaviors in decision making. Some studies have shown that the context which surrounds the decision maker exerts an influence on the extent of risk the individual is prepared to take (Shapira, 1995; Starbuck and Milken, 1988). Other studies, such as those done by Sankar (1997) and Velcher (1998), indicate that the position of the risk taker, and whether the decision is taken by an individual or is the result of group effort, account significantly for the variance in risk-taking behaviors, and ultimately, to the performance of the organization. Schwartz (1994) has argued that the results of research done using subjects who participate in activities in a lab setting are different from those found in research done in organizational settings. Additionally, MacCrimmon and Wehrung (1984, 1986, 1990) suggest that the differences in the measurement tools used in research studies accounted for the differences in the findings of managerial risk attitudes. ■

You will note that the preceding example first introduces the subject of risk-taking behaviors and corporate performance, and why it is an important topic to be studied. Through the literature survey, it identifies the problem to be studied as one of investigating the factors that account for risk-taking behaviors. It also indicates the important factors to be considered in the research, which would enable the researcher to formulate a theory, based on which, hypotheses can be formulated and tested.

EXAMPLE 8.3 Organizational Effectiveness

Literature Survey

Organization theorists have defined organizational effectiveness (OE) in various ways. OE has been described in terms of objectives (Georgopolous & Tannenbaum, 1957), goals (Etzioni, 1960), efficiency (Katz & Kahn, 1966), resources acquisition (Yuchtman & Seashore, 1967), employee satisfaction (Cummings, 1977), interdependence (Pfeffer, 1977), and organizational vitality (Colt, 1995). As Coulter (1996) remarked, there is little consensus on how to conceptualize, measure, or explain OE. This should however, not come as a surprise since OE models are essentially value-based classification of the construct (the values being those of the researchers) and the potential number of models that can be generated by researchers is virtually limitless. Researchers are now moving away from a single model and are taking contingency approaches to conceptualizing OE (Cameron, 1996; Wernerfelt, 1998; Yetley, 1997). However, they are still limiting themselves to examining the impact of the dominant constituencies served, and the organization's life cycle on OE instead of taking a broader, more dynamic approach (Dahl, 1998, p. 25). ■

From the extract of the literature above, several insights can be gained. The literature review (1) introduces the subject of study (organizational effectiveness), (2) highlights the problem (that we do not have a good conceptual framework for understanding what OE is), and (3) summarizes the work done thus far on the topic in a manner that convinces the reader that the researcher has indeed surveyed the work done in the area of OE and wants to contribute to the understanding of the concept, taking off on the earlier contingency approaches in a more creative way. The scholar has carefully paved the way for the next step, which is to develop a more viable and robust model of organizational effectiveness. This model will be logically developed, integrating several streams of research done in other areas (e.g., cross-cultural management, sociology, etc.), which will be woven further into the literature review. Once the scholar has explicated the framework as to what constitutes OE and the factors that influence OE, the next step would be to develop testable hypotheses to see if the new model is indeed viable.

The literature survey thus provides the basis or foundation to develop a conceptual framework for looking at the problem in a more useful and/or creative way. This, in turn, helps to develop testable hypotheses that would substantiate or disprove our theory.

Examples of good literature survey can be found at the beginning of any article in the *Academy of Management Journal* and most other academic or practitioner-oriented journals.

8.2.3 Design and Methodology

So far we have motivated the study we want to do and have researched what other work has been done on problems like ours. Now the research proposal should explain *how* the business research will be carried out.

This section of the proposal should be very specific. It should explain how, when, and where data will be collected and exactly what variables will be included in the study. We talked about different ways to collect data in Workshop 2, Chapters 4 and 5. If you are using a survey, you should include a copy of the survey instrument. If interviews with employees are included, a copy of the interview script is appropriate.

The proposal should include, where possible, the statistical tools that will be used to analyze the data. Sometimes you do not know the exact test that might be used. For example, until you see the data, you will not know whether a nonparametric test or a z or t test is appropriate. You can state that a hypothesis test about the mean will be done and what the test might conclude. Sometimes the research will result in simple descriptive statistics. You should describe what variables you will look at and how you will summarize and display them.

The proposal should be as specific as possible with respect to sample sizes and sampling techniques. Sometimes you will find that you have to make adjustments during your data collection, but you should have a plan to work from.

EXAMPLE 8.4   Turnover Problems

Design and Methodology

For the study on technical support turnover, you might propose to examine the personnel records of all technical support staff who left the company in the past three years. If this is a very large number (depending on the size of the company) you might propose a sample of 30 or 60 records, stratified by year. You would state that you will collect data on length of stay, age, gender, and ethnicity. There might be other variables that the literature has suggested are factors in employee turnover.

Your proposal might state that after the data are collected you will look at descriptive statistics (mean, median, standard deviation) of length of stay for each of the demographic variables. You might propose to test whether the average length of stay in the past year was shorter than for the previous two years. You might want to know whether men stay longer, on the average, than women. You would propose a statistical technique like a z or t test (depending on your proposed sample size) and state the possible outcomes of the test.   ■

In addition to describing what you will do, it is useful to describe what the impact of the information is. For example, you can state that if the results indicate a certain outcome, then a further study might be appropriate. Or if a certain outcome is true, then it might be necessary to reexamine a policy or procedure. This is important because the end product of business research has to be a recommendation on what should be done about the problem you studied.

8.2.4 The Timeline

As is true with all aspects of business, it is important to state how long something will take and what it will cost. This enables management to evaluate the effectiveness of the study in a concrete way. Your timeline should include the milestones of the project—data collection, data analysis, and report writing. It should state when each phase of the research will begin and when it should end.

A careful time analysis will alert you to problems in your research proposal. Perhaps you want to study the effects of high humidity on a product that you make. This would limit the timeframe of the study to summer. If it is already July, the project just might not be feasible at this time.

A carefully written research proposal will, essentially, *sell* your idea to management. It is important that it be well written and well researched. In addition, it is *very* important to remember that you are committing yourself to doing this project and that you should not promise what you cannot produce.

8.3 ETHICS IN BUSINESS RESEARCH

Ethics is an important factor in all phases of business research. As you proceed through the business research process, you should consider the ethics of what you propose to study, how you propose to carry out the research, and what will be done with the results of your research.

8.3.1 Ethical Issues in the Preliminary Stages of Investigation

Once a problem is recognized and an investigation is decided on, it is important to inform all employees—particularly those who will be interviewed for preliminary data gathering through structured and unstructured interviews—of the proposed study. Although it is not necessary to acquaint them with the actual reasons for the study (since this might bias responses), knowing that the research is intended to help them in their work environment will enlist their cooperation. The element of unpleasant surprise will thus be eliminated for the employees. It is also necessary to assure them that their responses will be kept confidential by the interviewer(s) and that individual responses will not be divulged to anyone in the organization. These two steps make the employees comfortable with the research undertaken and ensure their cooperation. Attempts to obtain information through deceptive means should be avoided at all costs, because they engender distrust and anxiety within the system. In essence, employers have the right to gather information relating to work, and employees have the right to privacy and confidentiality, but respondent cooperation assures good information.

8.3.2 Ethics in Handling Information Technology

Although PC technology offers unbounded opportunities for organizations and facilitates decision making at various levels, it also imposes certain obligations on the part of its users. First, it is important that the *privacy* of all individuals is protected, whether they are consumers, suppliers, employees, or others. In other words, businesses have to *balance* their information needs with the individual rights of those they come in contact with and store data on. Second, companies also need to ensure that *confidential information* relating to individuals is protected and does not find its way to unscrupulous vendors. Third, care should be taken to ensure that *incorrect information* is not distributed across the many different files of the company. Fourth, those who collect data for the company should be honest, trustworthy, and careful in obtaining and recording the data. The responsibility of organizations rests in the fact that the use of PC technology should go hand in hand with the ethical practices followed by members of the organization as they pursue their daily ongoing business activities.

8.3.3 Ethical Issues in Experimental Design Research

It is appropriate at this juncture to briefly discuss a few of the many ethical issues involved in doing research, some of which are particularly relevant for conducting lab experiments. The following practices are considered *unethical:*

- Pressuring individuals to participate in experiments through coercion, or applying social pressure.
- Giving menial tasks and asking demeaning questions that diminish their self-respect.
- Deceiving subjects by deliberately misleading them as to the true purpose of the research.
- Exposing participants to physical or mental stress.
- Not allowing them to withdraw from the research when they want to.
- Using the research results to disadvantage the participants, or for purposes that they would not like.
- Not explaining the procedures to be followed in the experiment.
- Exposing respondents to hazardous and unsafe environments.
- Not debriefing participants fully and accurately after the experiment is over.
- Not preserving the privacy and confidentiality of the information given by the participants.
- Withholding benefits from control groups.

The last item is somewhat controversial as to whether or not it should be an ethical dilemma, especially in organizational research. If three different incentives are offered for three experimental groups and none is offered to the control group, then the control group has participated in the experiment with absolutely no benefit. Similarly, if four different experimental groups receive four different levels of training but the control group does not, the other four groups have gained expertise that the control group has been denied. But should this become an ethical dilemma *preventing* experimental designs with control groups in organizational research? Perhaps not, for at least three reasons. One is that several others in the system who did not participate in the experiment also did not benefit. Second, even in the experimental groups, some would have benefited more than the others (depending on the extent to which the causal factor is manipulated). Finally, if a cause-and-effect relationship is found, the system will in all probability implement the newfound knowledge sooner or later, and everyone will ultimately stand to gain. The assumption that the control group did not benefit from participating in the experiment may not be a sufficient reason not to use lab or field experiments.

Many universities have a "human subjects committee" to protect the rights of those participating in any type of research activity involving people. The basic function of these committees is to discharge the moral and ethical responsibilities of the university system by studying the procedures outlined in the research proposals and giving their stamp of approval to the studies. The human subjects committee might require the investigators to modify their procedures or inform the subjects fully.

8.3.4 Ethics in Data Collection

Several ethical issues should be addressed while collecting data. As previously noted, these pertain to those who sponsor the research, those who collect the data, and those who offer the same. The *sponsors* should ask for the study to be done to better the purpose of the organization, and not for any other self-serving reason. They should respect the confidentiality of the data obtained by the researcher, and not ask for the individual or group responses to be disclosed to them, or ask to see the questionnaires. Once the report is submitted, they should be open-minded in accepting the results and recommendations presented by the researchers.

Ethics and the Researcher

1. Treating the information given by the respondent as strictly confidential and guarding his or her privacy is one of the primary responsibilities of the researcher. If the confidentiality part of the survey has been communicated to the organizational executives before the survey starts, then, should the vice president or another top executive desire to take a look at the completed questionnaires, the prior understanding can be brought to their attention. Also, report on data for a subgroup of, say, fewer than 10 individuals, should be dealt with tactfully to preserve confidentiality. The data can be combined with others, or treated in another unidentifiable manner. It is difficult to sanitize reports to protect sources and still preserve the richness of detail of the study. An acceptable alternative has to be found, since preserving confidentiality is the fundamental goal.

2. Researchers should not misrepresent the nature of the study to subjects, especially in lab experiments. The purpose of the research must be explained to them.

3. Personal or seemingly intrusive information should not be solicited, and if it is absolutely necessary for the project, it should be tapped with high sensitivity to the respondent, offering specific reasons therefor.

4. Whatever be the nature of the data collection method, the self-esteem and self-respect of the subjects should never be violated.

5. No one should be forced to respond to the survey, and if someone does not want to avail of the opportunity to participate, the individual's desire should be respected. Informed consent of the subjects should be the goal of the researcher. This holds true even when data are collected through mechanical means, like recording interviews, videotaping, and the like.

6. Nonparticipant observers should be as nonintrusive as possible. In qualitative studies, personal values could easily bias the data. It is necessary for the researcher to make explicit his or her assumptions, expectations, and biases, so that informed decisions regarding the quality of the data can be made by the manager.

7. In lab studies, the subjects should be debriefed with full disclosure of the reason for the experiment after they have participated in the study.

8. Subjects should never be exposed to situations where they could be subject to physical or mental harm. The researcher should take personal responsibility for their safety.

9. There should be absolutely no misrepresentation or distortion in reporting the data collected during the study.

Ethical Behaviors of Respondents

1. The subject, once having exercised the choice to participate in a study, should cooperate fully in the tasks ahead, like responding to a survey or taking part in an experiment.

2. The respondent also has an obligation to be truthful and honest in the responses. Misrepresentation or giving information, knowing it to be untrue, should be avoided.

APPENDIX

SOME ON-LINE DATABASES USEFUL FOR BUSINESS RESEARCH

Databases contain raw data stored in disks or CD-ROM. Computerized databases which can be purchased deal with statistical data, financial data, texts, and such. Computer network links allow the sharing of these databases, which are updated on a regular basis. Most university libraries have computerized databases pertaining to business information which can be readily accessed.

On-line servers such as America Online, CompuServe, Prodigy, Delphi, and Microsoft Network provide, among other things, facilities of the electronic mail, discussion forums, real-time chat, business and advertising opportunities, stock quotes, on-line newspapers, and access to several databases. Some of the databases useful for business research are listed below.

1. **ABI/INFORM Global** and **ABI/INFORM** provide the capability to search most major business, management, trade and industry, and scholarly journals from 1971 onward. The information search can be made by keying in the name of the author, periodical title, article title, or company name. Full texts from the journals and business periodicals are also available on CD-ROMs and electronic services.

2. **INFOTRAC** has a CD-ROM with expanded academic, business, and investment periodicals index covering more than 1000 periodicals in social sciences and business, which are updated monthly.

3. **American Science and Technology Index** (ASTI) is available both on-line and on CD-ROM, and indexes periodicals and books.

4. The **Business Periodicals Index (BPI)** provides an index of more than 3000 business and management periodicals, and is available on-line and on CD-ROM.

5. **Human Resources Abstract** is a quarterly abstracting service that covers human, social, and manpower information.

6. The **Public Affairs Information Service (PAIS)** is available both on-line and on CD-ROM. This indexing service of books, periodicals, business articles, government documents in business, and so on, is a useful source of reference.

7. The **Wall Street Journal Index** is available in full text by using the Dow Jones News/Retrieval Service. This index covers corporate news as well as general economic and social news. The Dow Jones News/Retrieval Service offers full texts of articles.

Other Sources of Information

Dictionaries and encyclopedias are also accessible in the areas of accounting, business, finance, management, investing, international trade, business and management, marketing and advertising, and production and inventory management.

Information on books in print as well as book reviews are available on CD-ROM. Similarly, census data are also available on CD-ROM.

Some Reference Guides

American Statistical Index is available both on-line and CD-ROM as statistical master files.

Prompt-Predicasts provides an overview of markets and technology and offers access to abstracts and some full texts on industries, companies, products, markets, market size, financial trends, etc.

Other Databases

The following databases can also be accessed through the Internet:

Business and Industry Database (includes information on whether the company is private or public, description of business, company organization and management, product lines and brand names, financial information, stock and bond prices, and dividends, foreign operations, marketing and advertising, sales, R & D, and articles available on the company in newspapers and periodicals)

Guide to Dissertation Abstracts

Guide to Newspaper Abstracts

Conference Papers

Conference Proceedings

Operations Research/Management Science

Periodicals Abstracts

Personnel Management Abstract

Social Science Citation Index

STAT-USA

Conference Board Cumulative Index (covers publications in business, finance, personnel, marketing, and international operations)

Note: A cumulated annotated index to articles on accounting and in business periodicals arranged by subject and by author is also available. The Lexis-Nexis Universe provides specific company and industry information including company reports, stock information, industry trends, and the like.

On the Web

Some of the many Web sites useful for business research which can be accessed through a browser such as the Netscape Navigator or the Internet Explorer are provided below. Please note that each Web site can be accessed with the following references, but preceded in each case by http://. For example, the second reference for All Business Network will be: http://www.webcom.com/~garnet/labor/aa_eeo.html

1. Academy of Management   aom@academy.pace.edu
2. All Business Network   www. webcom.com/~garnet/labor/aa_eeo.html

    This site offers articles, publications, and government resources related to human resources management.
3. ASTD Home Page   www.astd.org

    ASTD (American Society for Training and Development) has information on shifting paradigms from training to performance.
4. AT&T Business Network   www.bnet.att.com

    This site gives access to good business resources and offers the latest business news and information.
5. Bureau of Census   www.census.gov
6. Business Information Resources   www.eotw.com/business_info.html

    Links small business researchers to magazines and journals, government and law, financial services, and other entrepreneurial organizations.
7. Business Management Home Page   www.lia.co.za/users/johannh/index.htm

    This page offers sources dealing with project management, total quality management (TQM), continuous improvement, productivity improvement, and related topics.
8. Business Researcher's Interests   www.brint.com/interest.html
9. Company Annual Reports   www.reportgallery.com/bigaz.htm
10. CNN Financial Network   http://cnnfn.com/index.html
11. Dow Jones Business Directory   www.Businessdirectory.dowjones.com
12. Entrepreneur Forum   http://upside.master.com/forum
13. Entrepreneur's Resources Center   www.herring.com/erc
14. Fidelity Investment   www.fid-inv.com
15. Harvard Business School Publishing   www.hbsp.harvard.edu
16. Human Resources Management on the Internet

    http://members.gnn.com/hrmbasics/hrinet.htm
17. Index of Business Topics   www1.usa1.com/~ibnet/iccindex.html

    Covers a vast range of subjects for companies engaged in international trade.

18. International Business Directory   www.et.byu.edu/-eliasone/main.html

    This site offered by BYU has valuable sources for international business.

19. I.O.M.A.   www.ioma.com/ioma/direct.html

    This site links to business resources which include financial management, legal resources, small business, human resources, and Internet marketing.

20. MBA Page   www.cob.ohio-state.edu/dept/fin/mba/htm

    Designed by Ohio State University to help MBA students.

21. Multinational Companies   http://web.idirect.com/~tiger/worldbea.htm

22. Operations Management   www.muohio.edu/~bjfinch/ominfo.html

23. Society for Human Resource Management   www.shrm.org

24. STAT-USA   www.stat-usa.gov

25. Systems Dynamics for Business Policy   http://web.mit.edu/15.87/www

26. Wall Street Journal   www.wsj.com

27. Wall Street Research Net   www.wsrn.com

SAMPLING DISTRIBUTIONS AND CONFIDENCE INTERVALS

THE DIAPER COMPANY

Most large manufacturing companies use some form of Statistical Quality Control (SQC) in the manufacture of their products. One form of SQC that is often used is a control chart. A control chart looks at variation in data from samples of products taken over time.

A large company that manufactures disposable diapers collects data from its machines at random times during the workday. One of the variables that is measured is diaper weight. Diaper weight is an important factor in the manufacturing process for two reasons. First, the material that contributes most to diaper weight is the most expensive component of the diaper. Thus, it is reasonable to want to provide enough of this material, but not an excessive amount. The second reason is that the weight of a diaper relates to the consumer's perception of how well that diaper will absorb liquid. The target values, previously identified by consumer research, for diaper weight are a mean of 55 g and a standard deviation of 0.55 g.

When diapers are collected they are grouped in samples of size 5. The sample number and the individual diaper weights and bulks are recorded. The sample averages are then calculated and plotted on charts. Machine operators use these charts to tell them whether the machine is behaving as expected or whether the machine needs adjustment. The first few lines of the data set are shown below.

Sample	Diaper	Weight	Bulk
1	1	55.87	0.419
1	2	55.35	0.380
1	3	54.50	0.365
1	4	53.97	0.406
1	5	54.29	0.360
2	1	54.85	0.397
2	2	54.79	0.405
2	3	54.65	0.393

9.1   CHAPTER OBJECTIVES

To properly make inferences about the population mean, you must understand the behavior of the sample mean, $\overline{X}$. Gaining an understanding of $\overline{X}$ is the focus of this chapter. In Chapter 7, we saw that the behavior of a quantitative variable can be described by its probability distribution, and this distribution in turn can be described by parameters such as the mean and the variance. In this chapter we make the link from the probability concepts you learned in Chapter 7 to the inferential statistics techniques covered in the remaining chapters in this book.

Figure 9.1 is a drawing you first saw in Chapter 2. The larger circle on the left represents the population you are studying. The smaller circle on the right represents the sample you have taken from this population. This figure is being used to help you see the relationship between probability and inferential statistics. In reality, the sample circle should sit inside the population circle since it is a piece of it.

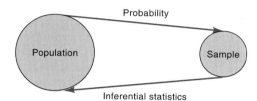

FIGURE 9.1   Relationship between probability and inferential statistics

Chapters 3 and 6 taught you the techniques of descriptive statistics which help you describe the sample. These techniques work on the smaller circle and are very helpful in gaining an understanding of the sample data, but they do not let you use the information in the sample to draw conclusions or inferences about the population. In other words, you do not yet have the ability to link that smaller circle to the larger one. In Chapter 7 you learned some tools of probability. In particular, you learned how to calculate the likelihood of a particular sample being selected from a population. You were thus making the trip across the top arc in Figure 9.1, from the population to the sample.

But this is not really our desired goal. We wish ultimately to make the trip from the sample to the population along the bottom arc in Figure 9.1. We wish to use the information in the sample to make probabilistic statements about the behavior of the population. This is the goal of inferential statistics. The key that unlocks this trip is found in this chapter.

More specifically, this chapter covers:

- Motivation for Point Estimators
- Common Point Estimators
- Desirable Properties of Point Estimators
- Distribution of the Sample Mean: Large Sample or Known σ
- The Central Limit Theorem—A More Detailed Look
- Drawing Inferences by Using the Central Limit Theorem
- Large-Sample Confidence Intervals for the Mean
- Distribution of the Sample Mean: Small Sample and Unknown σ
- Small-Sample Confidence Intervals for the Mean
- Confidence Intervals for Qualitative Data
- Sample Size Calculations

9.2 MOTIVATION FOR POINT ESTIMATORS

Remember: A parameter is a numerical descriptor of the population. Parameter values are typically unknown.

Since we most likely will not know the value of the parameters that describe the population, we must resort to using the information contained in the sample. It seems logical that if we can identify a numerical descriptor for the sample, then this statistic, called a point estimate, might be used to estimate the corresponding measure for the population. This is the right idea, which leads us to the definition of a **point estimate** and **point estimator.**

> A *point estimate* is a single number calculated from sample data. It is used to estimate a parameter of the population. A *point estimator* is the formula or rule that is used to calculate the point estimate for a particular set of data.

Sample statistics become point estimates.

Actually you have been working with point estimates without really knowing it. Up to this point, values calculated from the sample have been called sample statistics. Now we establish the link between sample statistics and the corresponding population parameters. In doing so we will have created point estimates out of these sample statistics. Let's take a look at the diaper company introduced at the beginning of the chapter.

EXAMPLE 9.1   The Diaper Company

Some Possible Point Estimates

Suppose a sample of 30 diapers was taken and the diaper weight was recorded in grams for each diaper. You decide to use the tools of descriptive statistics to describe this sample. The results of your analysis in Excel are shown below. Each of these values fits the definition of a point estimate. Each one is a single number calculated from the sample and it could be used to estimate an unknown population parameter. In the case of the sample mean, median, and mode, any of these values could be used as estimates of the true unknown mean weight of the population of all diapers manufactured.

Weight	
Mean	54.9833333
Median	54.86
Mode	#N/A
Standard deviation	0.63690162
Range	2.47
Minimum	53.97
Maximum	56.44
Count	30

Note: Excel uses #N/A to indicate that the Mode was unavailable. In most cases this means that there were no values that occurred more than once.   ∎

Before we go any further we must consider what parameters of the population we wish to estimate.

9.3 COMMON POINT ESTIMATORS

In Chapter 2 you learned that there are two major categories that can be used to classify variables: *qualitative* and *quantitative*. You also learned that the type of tools that can be used to analyze data depends on the type of data. Statisticians must know what kind of data they are analyzing before the analysis tools are selected and the inferences made.

Remember that a quantitative variable is numerical and a qualitative variable is descriptive.

9.3.1 Point Estimators for Quantitative Variables

If you are studying a *single quantitative variable,* then you typically wish to know the value of the *population mean* and the value of the *population standard deviation.* That is, you wish to know the center and the variability in the population. Remember that what you would like to know are the values of those parameters that describe the behavior of the population.

EXAMPLE 9.2   **The Diaper Company**

Parameters of Interest

For the diaper company, the population consists of all diapers made by this company. The characteristic that is being studied is the weight of the diapers. This is a quantitative variable and so we are interested in knowing the mean weight of all diapers made by this company and the standard deviation of the weights of all diapers made by this company. Knowing the mean and standard deviation of the diaper weights will allow us to set up a control chart as described in the chapter introduction. These are the population parameters of interest.

We know that we cannot weigh all the diapers in the company's current inventory but rather must base our estimates of the population mean and the population standard deviation on information contained in a sample of the product taken from the manufacturing line. ■

The population mean is always labeled with the Greek letter μ (pronounced mu) and the population standard deviation is labeled with the Greek letter σ (pronounced sigma). These values are typically unknown. We will use point estimates to estimate these values. In particular, we will use the sample mean, $\overline{X}$, to estimate μ and we will use the sample standard deviation, s, to estimate σ. It makes logical sense to use the sample mean to estimate the population mean and the sample standard deviation to estimate the population standard deviation. In the next section we discuss in more detail why these are the point estimates of choice.

Recall: This notation was introduced in Chapter 6.

Now suppose you wish to compare two quantitative variables. For example, you may wish to compare:

Comparing two populations

- the weight of diapers produced by two different machines
- the sales of a product at two different locations
- the time it takes to get your burger at McDonalds and Burger King
- the salaries for men and women in the same occupation

There are many such situations when you are studying two quantitative variables and wish to compare them. In these cases you often want to know how the true population means compare and how the amount of variation in one population compares with the amount of variation in the other population. For the purposes of keeping track of the different populations we will arbitrarily label one population as population 1 and the other as population 2.

To compare two numbers, what computation might you do? For example, if you wanted to compare your salary to your friend's salary, what would you do with the two

Use subtraction to compare two population means.

salary figures? If you said *subtract* one number from the other, then you are correct. By convention we always subtract the second mean from the first and thus we need to estimate the *true difference* between μ_1 and μ_2, or $\mu_1 - \mu_2$. It makes sense to estimate this true difference by using the actual difference in the two sample means or $\overline{X}_1 - \overline{X}_2$. If the two population means are the same, then the difference in the sample means should be close to zero. But remember that even if the two population means are the same and you select a sample from each population you will most likely get two slightly different sample means. This is due to the fact that you are looking at only a piece of each of the populations. This is what we have called sampling error. So just because the sample means are different, this does not at all imply that the population means are different.

Use a ratio to compare variability.

Another way to compare two numbers is to find the *ratio* of one to the other. If the numbers are the same, then the ratio will be 1. We will use ratios to compare the amount of variation in the first population to the amount of variation in the second population. So far in this book we have been using the standard deviation as a measure of variation. Unfortunately, for mathematical reasons, to compare the variation in two populations you must compare the variances. Recall that the variance is just the standard deviation squared. So if the standard deviation is 3 grams then the variance is 9 grams squared. The variance is not usually quoted because the units of measure are so odd—no one thinks in terms of grams squared. In this case what we wish to estimate is σ_1^2 / σ_2^2. It makes sense to estimate this by calculating the ratio of the sample variances or s_1^2 / s_2^2.

 **TRY IT NOW!**

Sales of Gizmos
Comparing Point Estimators Calculated From Samples Selected From Different Populations

Use a software package such as Excel (Excel instructions can be found at the end of this chapter) or Minitab to simulate picking a sample of size 10 from two different populations with mean 100 and standard deviation 3, or use the samples below.

Store 1		Store 2	
95	99	97	97
95	99	103	103
101	95	101	98
102	102	98	106
92	99	100	96

Suppose the first variable is daily sales of a new gizmo at store 1. The second variable is daily sales of the new gizmo at store 2. Select a sample of size 10 days from both stores. Assume that the daily sales at both stores are normally distributed with a mean of 100 and a standard deviation of 3.

Find $\overline{X}_1$, $\overline{X}_2$, and $\overline{X}_1 - \overline{X}_2$.

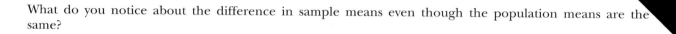

What do you notice about the difference in sample means even though the population means are the same?

Find s_1^2 (store 1 standard deviation squared), s_2^2, and s_1^2/s_2^2.

What do you notice about the ratio of the two sample variances?

9.3.2 Point Estimators for Qualitative Variables

If the variable you are studying is a *qualitative variable* then you typically wish to know what *proportion* or *percentage* of the population has a particular characteristic. For example, you may wish to know the percentage of defective items in the population, the percentage of people who will purchase a new product, or the percentage of people who are in favor of a particular political candidate or issue. The true unknown *population percentage* is labeled π. In this case we will use the sample proportion, p, to estimate π.

Since Greek letters (μ and σ) have been used to represent population parameters, the same convention will be followed with respect to population proportions. Some books use p for the population parameter and $\hat{p}$ for the sample proportion. This is done to avoid the use of π, which, of course, is traditionally reserved for the transcendental number 3.1415. The authors of this text feel that it makes sense to maintain the consistency of using Greek letters to denote population parameters.

The use of π as the notation for the population proportion

EXAMPLE 9.3   New Product Design

Sample Proportion Used to Estimate True Proportion

The company you work for is thinking about introducing a new product. You have been assigned the task of finding the percentage of consumers who would buy this new product if it were priced at $12.49. You cannot possibly survey all potential consumers to find the true percentage, but you know that taking a sample will provide you with the data to estimate this true percentage or proportion. Of the 100 people you survey, 62 indicate a willingness to buy this new product at this price. You find p = 62/100 = 0.62 or 62% as the estimate for the true proportion of consumers who would buy this new product. ∎

ANS. THE SAMPLE MEANS ARE 97.9 AND 99.9. THEY ARE NOT THE SAME AND THEREFORE THE DIFFERENCE IS NOT ZERO. THE SAMPLE VARIANCES ARE 11.88 AND 10.77. THEY ARE NOT THE SAME AND THEREFORE THE RATIO IS NOT EXACTLY ONE.

As we saw with quantitative variables, we often wish to compare two qualitative variables. For example, you may wish to compare

- the proportion of men who would buy a new product with the proportion of women who would buy it
- the proportion of defectives produced by the second shift with the proportion of defectives produced by the third shift
- the proportion of young people who like the new packaging with the proportion of older people who like it
- the proportion of bridges that are unsafe in Massachusetts compared with the corresponding proportion in Wyoming

When you are comparing two population proportions be sure that the proportions correspond to the same variable for the two different populations. Look at the first example in the list above. There are two populations here, men and women, and for both populations you are studying the proportion of the population that would buy the new product. You would not, however, compare the proportion of men who would buy the new product with the proportion of men who would not buy the product. This would be silly. These proportions would have to add up to one and so there is not anything to be learned by doing this.

Use subtraction to compare two population proportions.

As we did with the population means, we would like to compare the true population proportions π_1 to π_2 and we accomplish the comparison with a subtraction. So we wish to estimate the true difference in the population proportions, $\pi_1 - \pi_2$. It makes sense to estimate this true difference by using the difference in the two sample proportions or $p_1 - p_2$. If the two population proportions are the same, then the difference in the sample proportions should be close to zero. Remember that even if the two population proportions are both 0.20 and you select a sample from each population you will most likely get two slightly different sample proportions. This is again due to the fact that you are looking at only a piece of each of the populations. This is what we have called sampling error.

9.3.3 Summary of Commonly Used Point Estimators

The following table summarizes the most commonly used point estimators:

TABLE 9.1 Summary of Common Point Estimators

Type of Variable(s)	Population Parameter (unknown, what you wish to estimate)	Point Estimator or Sample Statistic (calculated from sample data)
A single quantitative variable	Population mean, μ	Sample mean, $\overline{X}$
	Population standard deviation, σ	Sample standard deviation, s
Two quantitative variables	Difference in population means, $\mu_1 - \mu_2$	Difference in sample means, $\overline{X}_1 - \overline{X}_2$
	Ratio of population variances, σ_1^2/σ_2^2	Ratio of sample variances, s_1^2/s_2^2
A single qualitative variable	Population proportion, π	Sample proportion, p
Two qualitative variables	Difference in population proportions, $\pi_1 - \pi_2$	Difference in sample proportions, $p_1 - p_2$

In the next section you will see that the estimators listed in column three of Table 9.1 are not the only estimators for the corresponding population parameters, but these are the most commonly used ones. The reason they are most commonly used is that, in a sense, they are the "best" estimators. In the next section we look at what makes them best.

9.3.4 Exercises—Learning It!

9.1 Rocky's Hardware has 10 stores located in western Massachusetts. To plan inventory and staffing, the owner has recorded the daily sales for 15 days for two of the stores. The data are shown below:

Store 1: $ 1981, 1802, 2053, 1877, 1647, 1931, 2065,
         $ 1988, 1694, 1848, 1736, 1951, 2111, 1957, 2120
Store 2: $ 2161, 2021, 1985, 2212, 2121, 2445, 1761, 2222,
         $ 2025, 1917, 2038, 1956, 1911, 1991, 2072

(a) What is your estimate of the average daily sales for store 1?

(b) What is your estimate of the average daily sales for store 2?

(c) What is the difference in the two estimates you found in parts (a) and (b)?

(d) These samples were both selected from populations with a population mean of $\mu = \$2000$. Explain why they have different sample averages.

9.2 Using the data for the hardware stores shown in Exercise 9.1, estimate the variability in the daily sales figures for both stores.

(a) What is your estimate of the variability of sales at store 1?

(b) What is your estimate of the variability of sales at store 2?

(c) What is the ratio of the two estimates you found in parts (a) and (b)?

(d) These samples were both selected from populations with a population variance of $\sigma^2 = (\$150)^2$. Explain why they have different sample variances.

9.3 Jiffy Burger does not want its customers to wait more than 3 minutes on the average for their food. A sample of the wait time in minutes for 20 customers is collected and shown below:

3.4	3.3	3.3	3.3	3.3	3.4	3.4	3.3	3.2	3.4
3.3	3.3	3.3	3.3	3.4	3.2	3.5	3.2	3.4	3.1

(a) What is your estimate of the average customer wait at Jiffy Burger?

(b) How much difference is there between your estimate and the target of 3 minutes?

9.4 A company is comparing the proportion of defective items produced by the second shift with the proportion of defective items produced by the third shift. In a sample of 100 items made by the second shift, 2 were found to be defective. In a sample of 100 items made by the third shift, 4 were found to be defective.

(a) What is your estimate of the proportion of defective items produced by the second shift?

(b) What is your estimate of the proportion of defective items produced by the third shift?

(c) What is the difference in the two estimates that you found in parts (a) and (b)?

9.5 A company that buys blank VHS tapes for video recording is concerned about the amount of time that the tapes are able to record. The tapes are rated at 120 minutes, but the company knows there is variation in the actual recording time. Data are collected on 25 randomly selected tapes and the actual recording times are listed below:

116	118	119	119	120
117	118	119	119	121
117	118	119	120	121
117	119	119	120	121
117	119	119	120	121

(a) What is your estimate of the average time available on the tape?

(b) Is your estimate different from 120 minutes? If so, by how much?

9.4 DESIRABLE PROPERTIES OF POINT ESTIMATORS

In this section, we develop the properties of point estimators. We do this by focusing on estimators for the population mean, μ. However, the resulting properties apply to any point estimator.

In Chapter 6 you learned how to calculate measures of the middle of the sample data. In particular, you learned about the sample mean, $\overline{X}$, the sample median, and the sample mode. The trimmed mean was also examined as a measure of the middle of the sample data. Based on what we have seen so far, it makes sense to consider using one or more of these statistics as our point estimator of the unknown value μ. Let's see how these might work for our diaper company.

EXAMPLE 9.4 The Diaper Company

Possible Point Estimates of μ

Suppose you took a sample of 5 diapers and recorded the diaper weight (in grams) of each of the 5 diapers. The data are shown below:

$$55.87 \quad 55.35 \quad 54.50 \quad 53.97 \quad 54.29$$

The sample mean is 54.80 and the sample median (the middle score) is 54.50. There is no mode for this data set since none of the values occur more than once. We could use either the value of 54.80 or 54.50 as our point estimate for the unknown true mean weight of the diapers being made. ■

It appears that the mean, the median, and the mode all fit the definition of a point estimator. In fact, we could dream up many other formulas that would fit the definition of a point estimator. For example, we could decide to average the minimum and maximum value in the sample and call this a point estimator. This is a legitimate point estimator, since the calculation is based only on sample data and yields a single number. Clearly, we need some criteria to judge which of these point estimators is the best one to use to estimate μ.

Let's think about the criteria that we could use to make this judgment. If the point estimate is to be used as our best guess of the value of the unknown population parameter, in this case μ, then we would like the point estimate to be close to μ. Ideally, we would like our point estimate to hit μ on the nose. But we know that the chance of that happening is pretty slim. In fact, we know from Chapter 2 that the sample is only a piece of the population and if we were to take a different sample, we would get a different sample mean, a different sample median, and a different sample mode.

Let's follow the diaper company an hour after our first sample of 5 diapers.

EXAMPLE 9.5 The Diaper Company

Second Sample Yields Slightly Different Values for Point Estimates

A second sample of 5 diapers is taken an hour after the first sample. The diaper weights, in grams, are as follows:

$$54.85 \quad 54.79 \quad 54.65 \quad 55.56 \quad 55.82$$

Some quick calculations yield

Sample mean = 55.13   Sample median = 54.85   Sample mode = N/A ■

The particular point estimates are different now. We used the same point estimators (formulas to calculate the sample mean and sample median) but we got different numbers because the data in this second sample were different from those of the first sample. This tells us that our point estimate depends on the particular sample we happen to choose and it changes from sample to sample. Combining this fact with the criterion of getting close to the unknown μ, we can see that we want to use the point estimator that comes closest to μ for most samples. That is, we would like a point estimator that does not yield radically different numbers from sample to sample.

Suppose for a minute that you dreamed up a formula for a point estimator for μ and you used it with two different samples from the same population. For the first sample your point estimate was calculated to be 24 and for the second sample it was calculated to be 245. How much faith would you have in your estimate? "Not much" should be your answer. Why? The feature of this point estimator that is troubling is that it yields wildly different estimates from sample to sample. The estimator has too much variability; that is, it jumps around too much from sample to sample.

In summary, what we really want is a point estimator with the following two properties:

1. The point estimator should yield a number close to the unknown population parameter.

2. The point estimator should not have a great deal of variability.

Properties of point estimators

Even though we have been considering point estimators for the unknown population mean, μ, these properties are desirable for any point estimator.

These two properties are more precisely stated as follows:

1. The point estimator should be unbiased.

2. The point estimator should have a small standard deviation.

The word *unbiased* in statistics has basically the same meaning that it has in ordinary language. When the word is used as a character trait of somebody, it generally implies that the person is fair and does not favor something or someone. We would like our point estimator to behave in this fashion as well. We do not want to use a point estimator that always overestimates μ or always underestimates μ. In this case the estimator would be biased or an unfair estimate. Sometimes our estimator $\overline{X}$ will be smaller than μ and sometimes it will be larger than μ, but we know that it will be close to μ most of the time.

This is basically the same definition you saw in Chapter 2 when we were learning about bias in a sample.

An **unbiased estimator** yields an estimate that is fair. It neither systematically overestimates the parameter nor systematically underestimates the parameter.

Unbiased estimators are fair. They do not systematically overestimate or underestimate the parameter.

Proving that an estimator is unbiased is the job of the mathematical statisticians. They have shown that $\overline{X}$ is an unbiased estimator and we will believe them! They have also shown that the median is an unbiased estimator only if the population has a normal distribution. If a proposed point estimator is wildly biased, it may be obvious simply from the formula. Let's look at one such point estimator for μ, one that is clearly ridiculous.

EXAMPLE 9.6 The Diaper Company

An Example of a Biased Estimator

Someone in the diaper company proposes using the smallest observation in the sample as the estimate of μ. For the first sample (shown in Example 9.4) this would yield an estimate of 53.97 g, and for the second sample (shown in Example 9.5) this formula would yield an estimate of 54.65 g. Clearly, this rule of selecting the minimum value as the estimate of μ is silly. It will always yield a number that is too small and is clearly an unfair estimate of the middle. ∎

The sample mean is the point estimator for μ with the smallest standard deviation.

With regard to showing that $\overline{X}$ has a small standard deviation, we will again leave that to the mathematical statisticians. They have shown that $\overline{X}$ has a smaller standard deviation than the median or the mode; that is, it jumps around less from sample to sample. If you think about the way the sample mean, the sample median, and the sample mode are found you can see that this makes sense. The sample mean incorporates all of the data in the sample and in doing so washes out the effect of any single number in the sample that happens to be a bit unusual. Both the median and the mode are based on only a portion of the sample and thus tend to change more from sample to sample. Thus, the sample mean, $\overline{X}$, is the best estimator of the unknown population mean, μ. The following example helps to demonstrate this fact.

EXAMPLE 9.7   The Diaper Company

Three More Samples

Let's follow the diaper company for 3 more hours. Each hour a sample of 5 observations is collected on the diaper weights (in grams). The observations are shown below along with the first two samples from Examples 9.4 and 9.5.

Recall that the sample mode is that value in the data set that occurs most frequently. Since none of the values repeat, there is no mode.

	Sample Mean	Sample Median	Sample Mode
Hour 1: 55.87 55.35 54.50 53.97 54.29	54.80	54.50	N/A
Hour 2: 54.85 54.79 54.65 55.56 55.82	55.13	54.85	N/A
Hour 3: 54.40 56.44 54.11 54.67 54.42	54.81	54.42	N/A
Hour 4: 54.69 54.55 54.56 54.22 54.87	54.58	54.56	N/A
Hour 5: 55.50 54.44 56.30 54.00 55.68	55.18	55.50	N/A

You can quickly see that the sample mode is not a very useful point estimator. In any given sample of 5 observations, no weight appears more than once.

As we examine the sample median and the sample mean let's think about the two properties of point estimators that we have been discussing. We must look at where the values of the point estimator are centered and how much the values jump around from sample to sample. In this example, which numbers have more variation: the sample means or the sample medians? How can we tell? We can calculate the average and the standard deviation of the sample means and we can calculate the average and the standard deviation of the column of sample medians.

The average of the sample means is 54.90 g, which is very close to the target mean of 55.00 g. This tells us that the sample means are clustered around the target population mean, μ. The standard deviation of the sample means is 0.251 g. For now, simply notice that this is quite a bit lower than the target standard deviation of 0.55 g. This is actually good news and we will return to this point in the next section.

The average of the sample medians is 54.77 g, which is also close to μ but not as close as the average of the sample means. The standard deviation of the sample medians is 0.441 g, telling us that the sample medians do jump around more than the sample means. That is, they have more dispersion. ∎

This small example should give you a sense of what we mean by saying that the point estimator yields values that are clustered around the true population parameter and do not vary much from sample to sample.

 *TRY IT NOW!*

The Diaper Company
Comparing the Variability of Two Point Estimators

Weights in grams for the next 5 hourly samples taken at the diaper company are shown below:

Hour 6:	54.89	55.06	55.45	55.23	55.75
Hour 7:	54.32	55.72	54.91	54.40	55.78
Hour 8:	54.14	55.18	55.78	55.37	55.69
Hour 9:	54.11	54.05	53.60	55.97	55.86
Hour 10:	55.21	55.40	53.87	55.09	55.70

For each sample, calculate the sample mean and the sample median.

Find the average of the sample means and the average of the sample medians.

Find the standard deviation of the sample means and the standard deviation of the sample medians.

Which point estimator has less variability?

ANS.

	MEAN	MEDIAN
HOUR 6:	55.28	55.23
HOUR 7:	55.03	54.91
HOUR 8:	55.23	55.37
HOUR 9:	54.72	54.11
HOUR 10:	55.05	55.21
AVERAGE	55.06	54.97
ST. DEV:	0.22	0.51

THE MEAN HAS LESS VARIABILITY.

9.5   DISTRIBUTION OF THE
SAMPLE MEAN, $\overline{X}$

As a result of your work in the previous section you have decided to use $\overline{X}$ as your point estimator for the true unknown population mean, μ. Your boss reminds you that the average diaper weight should be 55.00 g. You know that your sample mean $\overline{X}$ will not equal 55.00 g, but it should be close to 55.00 g if the manufacturing process is working properly. You take a sample at hour 11 and find a sample mean of 54.00 g. By looking at the last 10 values of $\overline{X}$ you realize that a sample mean of 54.00 g is a bit low, but you are not sure if this was caused simply by sampling error or if the machine is working improperly. How can you decide?

9.5.1 Putting Z-scores
and the Empirical Rule to Use

A Z-score measures the number of standard deviations that a data value is from the mean.

In Chapter 6 you learned about the *empirical rule* and *Z-scores*. These tools allowed you to decide if an individual observation was an outlier or unusual. To calculate the Z-score you used the following formula:

$$Z = \frac{\text{Distance between the data value and the average}}{\text{Standard deviation}}$$

Since we are trying to decide if an $\overline{X}$ value of 54.00 g is unusual, it seems like it would be a good idea to calculate the Z-score for the sample mean of 54.00 g. If 54.00 g is more than 3 standard deviations away from the target mean of 55.00 g, then we are pretty sure that there is a problem. Why? Because the empirical rule tells us that z-scores like this are very unusual and occur less than 1% of the time. This means that it would be *extremely unlikely* (less than 1% chance) that we would see a sample mean this small *if* the manufacturing process was producing diapers with a true mean weight of $\mu = 55.00$ g.

Now our data value is the observed sample mean, $\overline{X}$. This is a bit different from our previous use of Z-scores, but the idea is exactly the same. We are simply trying to decide if a particular $\overline{X}$ value is too unusual, whereas in Chapter 6 we were trying to decide if a particular individual observation was too unusual. We must rewrite the formula for Z to reflect this fact:

Z STATISTIC FOR $\overline{X}$

$$Z = \frac{\overline{X} - \mu_{\overline{X}}}{\sigma_{\overline{X}}}$$

Notice that the average is now the average of all of the possible $\overline{X}$'s and is therefore labeled $\mu_{\overline{X}}$. The standard deviation is the standard deviation of the $\overline{X}$'s and is labeled $\sigma_{\overline{X}}$. Although this calculation for the Z-score looks different, it is really the same beast we used in Chapter 6. It simply measures the number of standard deviations that a particular $\overline{X}$ is above the mean of all possible $\overline{X}$'s.

So, we should just calculate the Z-score for the sample mean of 54.00 g. But if we look at the formula we see that to calculate the Z-score for the sample mean of 54.00 g, we need to know the *average of the sample means* and *the standard deviation of the sample means*. In the Try It Now Exercise in Section 9.4, you calculated the average of 5 sample means and the standard deviation of 5 sample means. We need to expand these calculations beyond 5 sample means to all possible sample means. In other words, what we need now is the average of all possible sample means and the standard deviation of all possible sample means.

Clearly, we do not want to select all possible samples from the population. This would defeat the whole idea of taking one sample to learn about the population. Fortunately, we can once again rely on our friends the mathematical statisticians who have proved a theorem that will give us the results we need. This theorem is the Central Limit Theorem and it is the foundation for virtually all of inferential statistics. It is the key that unlocks the rest of the course.

9.5.2 The Central Limit Theorem

Let's look at the diaper company data again. In Section 9.4 we looked at 10 hourly samples taken from the manufacturing process of the diaper company. Each sample consisted of $n = 5$ observations of diaper weight in grams. We saw that the value of the sample mean, $\overline{X}$, varied from sample to sample. We called $\overline{X}$ a point estimator, but if we think about it we can see that it is also a *random variable*. In Chapter 7 you learned that a random variable has a mean, a standard deviation, and a probability distribution. Thus, the random variable $\overline{X}$ has a mean, a standard deviation, and a probability distribution.

Remember: A random variable is a quantitative variable whose value varies according to the rules of probability.

When the random variable is a point estimator, then the standard deviation and the distribution are labeled with slightly different words. These are detailed in the next two definitions.

> The **standard error** is the standard deviation of a point estimator. It measures how much the point estimator or sample statistic varies from sample to sample.

> The probability distribution of a point estimator or a sample statistic is called a **sampling distribution.**

We looked at the mean and standard deviation of $\overline{X}$ in the last section. This concept is recapped in the next example.

EXAMPLE 9.8   The Diaper Company

Illustration of $\overline{X}$ as a Random Variable with a Mean, a Standard Error, and a Sampling Distribution

In Section 9.4 we looked at 10 hourly samples. Each sample had $n = 5$ observations (remember sample size is always labeled as n). The 10 sample means (in grams) are shown below:

54.80   55.13   54.81   54.58   55.18   55.28   55.03   55.23   54.72   55.05

These 10 numbers are some of the possible values for $\overline{X}$ but not all of them. To list all of the possible sample means we would have to take all possible samples of size 5 from the population of all diapers manufactured by this company. This is not possible and, we will see, it will not be necessary.

We can begin to get a handle on how $\overline{X}$ behaves by looking at just these 10 values:

(a) The average of the 10 $\overline{X}$'s is 54.98 g.
(b) The standard deviation of the 10 $\overline{X}$'s is 0.24 g.
(c) The histogram, done in Excel, of these 10 $\overline{X}$'s looks like this:

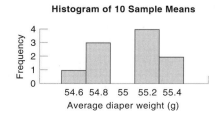

Histogram of 10 Sample Means

Remember: Excel displays the upper value of the class interval as though it were the middle of the class interval.

Before stating the theorem let's see if we can anticipate the results by looking at Example 9.8. It seems like the mean of the $\overline{X}$'s is pretty close to the population mean, $\mu = 55$ g. The standard deviation of the $\overline{X}$'s is about half of the population standard deviation, $\sigma = 0.55$ g. It makes sense that the sample means would be more similar to each other (have a smaller standard deviation) than the individual diaper weights since the averaging process washes out the individual values. Finally, the sampling distribution or the histogram of the $\overline{X}$'s doesn't seem to have much of a shape right at the moment, but remember that we have only 10 $\overline{X}$'s in this example.

Central Limit Theorem (CLT)

In random sampling from a population with mean μ and standard deviation σ, when n is large enough, the distribution of $\overline{X}$ is

- approximately normal with
- a mean, $\mu_{\overline{X}}$, equal to μ and
- a standard deviation, $\sigma_{\overline{X}}$, equal to $\sigma/\sqrt{n}$

The CLT holds for small samples when the population has a normal or symmetric distribution.

The Central Limit Theorem applies when you have a large enough sample size. A "large enough" sample size depends on how much the population distribution deviates from a normal distribution. Typically, if the sample size is larger than 30 then it is considered large enough. The larger the sample size, the better the normal approximation will be. If you are reasonably sure that the population you are sampling from has a normal distribution, then the results of the CLT hold even for small samples ($n \le 30$).

9.6 THE CENTRAL LIMIT THEOREM— A MORE DETAILED LOOK

Let's restate the Central Limit Theorem and examine each of the three points in the definition in terms of the diaper data.

9.6.1 The Shape of the Sampling Distribution of $\overline{X}$

The first point of the CLT: The distribution of all possible sample means is approximately normal.

The first point of the Central Limit Theorem is that if we took all possible random samples from an arbitrary population and calculated all the possible sample means, then the distribution of the sample means would be approximately normal. Remember that this applies if the sample size is large enough or if the underlying population distribution is normally distributed.

The diaper company is taking samples of size $n = 5$ every hour. Because each individual sample size is small ($n = 5$), to apply the CLT we will need to assume that the underlying distribution of diaper weights is normal or close to a normal shape. We can get a sense of the shape of the population distribution by examining a histogram of sample observations. The histogram of 260 diaper weights (representing 52 hours of sampling) is shown in Figure 9.2.

This histogram gives us a visual picture of the sample, not the population. But we remember that we can learn about the population by studying a representative sample like this one. If the sample looks reasonably bell shaped we can say that the population is probably also bell shaped. In Chapter 17 we will learn a more precise way to make an inference about the shape of the population.

We just examined the distribution of a sample of 260 diaper weights and, based on the histogram of the sample, we concluded that the population is reasonably close to a normal distribution. Thus, we expect the distribution of the sample means to be

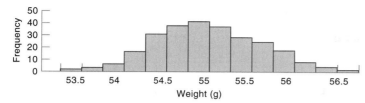

FIGURE 9.2 **The Diaper Company** Histogram of individual diaper weights

approximately normal even though the sample size is small ($n = 5$). This can be seen visually by examining the histogram of 52 sample means shown in Figure 9.3. The distribution of the sample means does indeed look approximately normal. The Central Limit Theorem tells us that the sampling distribution of $\overline{X}$ will be approximately normal if we are sampling from a normal distribution, as is the case for the diaper data.

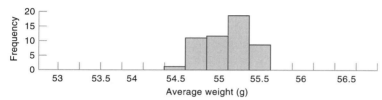

FIGURE 9.3 **The Diaper Company** Histogram of 52 sample means

If the sample size is sufficiently large, the Central Limit Theorem tells us that the sampling distribution of $\overline{X}$ will be approximately normal regardless of the shape of the underlying population distribution. This is great news because most of the time we will not know the shape of the population from which we are sampling.

9.6.2 The Mean of the Sampling Distribution of $\overline{X}$

The second point of the Central Limit Theorem is that the mean of the sampling distribution of $\overline{X}$ equals the mean of the population you are sampling from. This means the center of the histogram of the $\overline{X}$'s should be μ. This is consistent with what we have seen with the sample means from the diaper company. Most of the $\overline{X}$'s are close to μ. The theorem tells us that if we took all possible samples of size n, n sufficiently large, and calculated the sample mean for each of these samples and averaged the sample means, then we would get μ. In terms of our notation this means $\mu_{\overline{X}} = \mu$. For our diaper company this means that the average of all the $\overline{X}$'s should be close to the target mean of $\mu = 55.00$ g. In fact, the average of the 52 sample means is 54.994 g. This can also be seen visually by examining the histogram of the sample means shown in Figure 9.3. The sample means are all clustered around the target mean of $\mu = 55.00$ g.

Why is this important? Clearly, we are not going to take more than one sample. We will take one sample and have one sample mean, $\overline{X}$, which will be our estimate of μ. The Central Limit Theorem tells us that we can be reasonably sure that the sample mean that we get will be close to the true mean because the sample means cluster

The second point of the CLT: The average of all possible sample means is μ.

around the true mean, μ. Later in this chapter we see that we quantify the words "reasonably sure" and "close" by using what is known as a confidence interval.

9.6.3 The Standard Error of the Sampling Distribution of $\overline{X}$

The third point of the CLT: The standard error of $\overline{X}$ is $\sigma/\sqrt{n}$.

The third point of the theorem says that the standard deviation of the $\overline{X}$'s (also called the standard error) depends on two things: the amount of variability you start with in the population, σ, and the sample size, n.

Let's think about what this means. Suppose you had a population of students who were all the *same age* and you picked 10 different samples from this population and calculated the average for each of the 10 samples. How different would the sample means be from each other? They will all be the same since everyone in the population is the same age! There is no variability, $\sigma = 0$, in the population and so there will be no variability in the $\overline{X}$'s. Clearly, this is an extreme example.

Now consider one population that has a mean of 25 and a standard deviation of 1 and one that has a mean of 25 and a standard deviation of 5. Graphs of these distributions are shown in Figure 9.4.

Two Populations with Normal Distributions

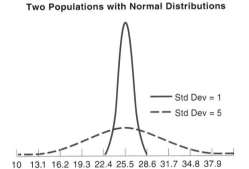

10 13.1 16.2 19.3 22.4 25.5 28.6 31.7 34.8 37.9
Sales

FIGURE 9.4 Graphs of two populations with the same mean but different standard deviations

Again consider taking 10 samples of size $n = 30$ from each population and calculate the 10 sample means. How would you expect the 10 sample means from the first population to behave compared to the 10 sample means from the second population? Let's examine the dotplots of the sample means shown in Figure 9.5.

In both cases, we see that the sample means cluster around $\mu = 25$, but the sample means from the first population appear to be more similar to each other (less dispersion) simply because the values in the first population are more similar to each other. Let's capture this in a result:

> As the standard deviation in the population increases, the standard error of $\overline{X}$ also increases.

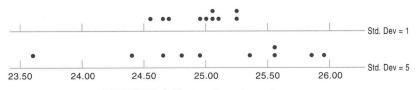

FIGURE 9.5 Dotplots of sample means

TRY IT NOW!

The Central Limit Theorem
Exploring the Third Point

Use a software package such as Excel (Excel instructions are found at the end of the chapter) or Minitab to simulate picking 10 samples each of size $n = 35$ from two different populations:

Population 1: normal distribution with a mean of $\mu = 25$ and a standard deviation of $\sigma = 1$
Population 2: normal distribution with a mean of $\mu = 25$ and a standard deviation of $\sigma = 5$

Alternatively you can use the samples shown below:

				Samples from population 1					
				Sample					
1	**2**	**3**	**4**	**5**	**6**	**7**	**8**	**9**	**10**
22.8	23.7	24.3	25.3	27.4	24.7	25.6	25.5	24.6	23.7
24.8	25.7	24.5	24.0	23.8	24.6	24.7	23.6	26.3	24.8
23.7	23.6	26.1	25.7	25.5	26.4	26.4	25.6	24.0	25.2
25.5	26.7	26.0	24.7	25.6	26.5	24.5	23.5	24.9	25.0
26.9	25.5	24.9	25.6	25.1	25.8	23.7	24.5	24.3	24.1
25.5	24.4	25.7	24.7	24.7	23.7	25.3	25.0	25.7	25.8
25.3	25.4	24.2	24.7	25.6	24.6	25.4	25.1	24.1	25.0
24.6	26.0	25.5	24.0	24.7	23.5	24.8	24.4	27.6	26.6
25.3	27.5	24.5	24.9	24.2	24.1	24.5	25.9	24.1	25.8
26.1	24.1	23.4	25.0	25.6	24.4	26.8	23.7	24.5	23.4
24.0	25.5	24.9	25.7	24.6	25.9	25.0	24.8	25.5	26.1
24.2	25.4	25.4	25.3	24.7	25.1	25.0	26.7	25.1	24.9
23.5	24.4	24.5	25.5	25.4	26.1	24.6	25.3	25.4	25.9
26.8	23.9	24.5	22.6	24.5	24.9	24.7	26.3	24.9	28.3
25.8	22.5	27.1	25.4	23.9	24.4	25.0	24.2	25.5	26.1
24.2	23.9	25.0	24.8	26.4	24.4	25.2	24.7	25.6	24.4
26.1	24.8	24.2	26.0	24.9	24.5	24.6	25.3	25.2	26.2
25.4	26.6	23.7	25.9	24.9	24.8	24.1	25.5	23.9	24.2
23.7	25.6	24.9	26.6	24.9	24.9	26.3	24.6	24.8	24.7
24.3	25.7	26.2	25.9	26.2	23.9	23.4	25.2	24.4	25.4
24.9	23.3	24.7	25.1	25.6	25.7	25.1	25.0	24.0	25.2
23.7	24.0	24.5	25.6	25.9	25.0	25.6	24.0	24.4	26.3
24.2	24.6	24.7	24.2	24.2	25.3	24.1	24.7	27.1	26.9
25.0	25.9	26.4	24.6	24.6	26.1	24.7	25.3	25.3	23.1
23.6	25.2	25.4	26.2	22.9	25.4	25.3	24.6	24.2	25.6
25.2	25.7	24.0	24.9	24.4	23.4	24.8	24.9	24.5	25.0
23.7	25.3	24.3	25.9	25.1	24.5	25.0	24.9	26.0	25.0
23.7	22.5	25.2	25.4	23.4	25.2	25.2	24.2	25.0	25.1
24.6	23.8	26.0	24.9	26.8	24.1	25.3	26.7	25.5	27.5
26.8	26.4	24.8	23.9	24.2	26.5	25.0	25.7	25.2	23.1
25.3	25.0	26.0	24.9	23.5	25.4	26.3	24.4	27.1	25.6
24.7	26.0	24.9	24.1	25.1	24.0	25.7	24.3	26.0	24.9
24.8	24.9	25.0	25.4	25.1	24.9	25.6	25.7	24.9	25.6
25.4	24.8	26.7	25.7	24.6	24.5	24.7	23.0	24.5	24.3
24.4	26.0	24.9	26.4	24.1	24.7	25.7	26.3	24.4	24.2

(continued)

Samples from population 2:

				Sample					
1	**2**	**3**	**4**	**5**	**6**	**7**	**8**	**9**	**10**
26.6	30.6	28.9	25.1	20.6	39.8	23.8	32.9	17.8	15.7
24.2	25.7	27.3	25.5	17.6	29.7	21.2	27.5	23.2	21.2
29.1	8.4	31.7	28.4	22.2	28.3	24.7	16.1	29.5	29.9
19.5	27.6	34.6	24.3	21.3	22.6	21.6	38.4	20.4	18.5
23.3	26.0	19.2	22.3	23.7	24.5	29.7	26.5	20.3	20.9
20.6	16.4	17.9	29.6	20.5	23.5	22.8	24.7	32.2	29.3
27.6	35.2	23.8	28.3	23.1	32.7	26.7	32.7	25.7	27.3
24.7	23.2	33.6	37.1	24.8	22.4	26.2	17.3	20.1	29.6
28.2	27.4	19.4	26.2	23.7	23.9	21.8	29.6	27.2	24.4
27.0	28.4	27.6	24.7	17.7	24.4	24.0	21.0	19.3	30.6
28.0	25.0	23.5	26.6	18.6	24.3	21.1	17.0	28.5	29.1
16.2	20.5	20.9	15.5	26.6	34.6	22.0	19.8	17.3	29.4
31.7	28.1	20.6	30.0	27.3	22.2	24.3	19.3	22.2	28.1
24.6	24.9	19.0	28.9	19.9	34.2	24.6	19.4	31.0	23.3
36.1	26.3	29.8	14.8	32.0	34.8	35.5	23.9	14.6	26.5
29.5	21.7	22.7	31.3	22.3	25.3	33.6	30.0	17.7	21.9
23.3	18.9	29.9	20.6	22.6	31.0	23.1	25.1	26.7	34.0
19.9	22.2	22.9	28.5	27.8	29.9	30.4	32.6	17.5	26.8
26.8	23.5	31.1	30.5	19.6	23.7	20.9	18.3	27.9	29.3
21.5	29.2	27.4	25.0	22.7	19.3	28.3	21.5	23.9	20.7
23.8	28.5	30.2	26.0	35.2	23.7	22.3	29.5	16.0	23.7
28.2	33.6	38.1	27.1	23.6	20.6	20.4	22.4	24.5	26.6
21.4	23.6	23.0	24.5	22.9	25.3	23.8	19.7	24.2	32.7
30.0	18.2	20.2	28.6	20.4	30.2	34.5	19.7	27.9	23.7
32.2	17.7	26.9	24.7	29.7	27.2	26.8	31.0	16.4	29.0
24.1	24.1	25.8	17.7	34.9	23.5	35.6	27.6	34.8	26.5
18.9	22.0	22.4	21.6	29.2	23.1	28.6	24.7	33.6	25.1
23.3	26.9	31.7	31.4	27.1	16.1	33.7	28.5	30.0	18.3
31.6	23.6	27.9	29.3	27.6	23.8	23.3	24.8	28.7	18.8
21.3	20.9	28.0	20.9	19.0	18.7	26.0	29.9	32.8	27.3
29.9	26.1	23.8	29.1	24.3	26.0	37.8	21.3	21.9	25.1
20.0	30.6	25.0	31.5	16.2	24.0	23.5	21.3	22.7	19.3
27.5	26.1	30.9	18.2	26.6	32.2	16.2	25.7	22.9	22.7
22.0	20.9	26.7	28.2	27.3	20.1	18.3	16.6	33.5	26.0
21.7	28.0	33.2	29.3	26.6	15.7	21.8	30.5	18.3	8.7

For each sample, calculate a sample mean, $\overline{X}$.

Find the average and standard deviation of the 10 $\overline{X}$'s from population 1 samples.

Find the average and standard deviation of the 10 $\overline{X}$'s from population 2 samples.

The second number that influences the size of the standard error of $\overline{X}$ is the sample size, n. This should also make sense. Let's start by looking at the smallest possible sample size: $n = 1$. Clearly, the sample mean of one number is just the single value you have selected and these sample means will have the same variation as the population since $\sigma/\sqrt{n} = \sigma/\sqrt{1} = \sigma$. Typically, your sample is more than a single observation and thus n is usually 2 or more. This means that no matter what the sample size is, the *amount of variation in the $\overline{X}$'s will be less than the amount of variation in the population you are sampling from.* This can be seen visually in the next example.

EXAMPLE 9. 9 **The Diaper Company**

Frequency Histogram of Diaper Weights and Average Weights, $\overline{X}$

The diaper company has taken 52 samples of size $n = 5$ over the last 52 hours. For each sample a sample mean, $\overline{X}$, has been calculated. The accompanying frequency histogram displays the distribution of the diaper weights ($5 \times 52 = 260$ observations) and the distribution of the 52 $\overline{X}$ values. Clearly, the distribution of the sample means has less variability.

ANS. THE AVERAGE OF THE $\overline{X}$'s SHOULD BE CLOSE TO $\mu = 25$ IN BOTH CASES. THE STANDARD DEVIATION OF THE $\overline{X}$'s FROM POPULATION 1 SAMPLES SHOULD BE SMALLER THAN THE STANDARD DEVIATION OF THE $\overline{X}$'s FROM POPULATION 2 SAMPLES.

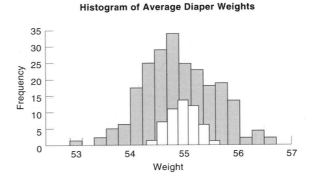

Histogram of Average Diaper Weights

Now let's see what the impact of increasing the sample size would be. Suppose we return to the population with a mean of 25 and a standard deviation of 1 and we take 10 samples of size 5 and calculate the 10 sample means. Now we take 10 samples of size 35 from this same population and calculate 10 more sample means. How would you expect the 10 sample means based on the samples with $n = 5$ to compare to the 10 sample means based on the samples with $n = 35$? We should expect that the sample means calculated from the bigger ($n = 35$) samples would be more similar (and thus have a smaller standard deviation) to each other since each sample mean contains more information and is thus a better estimate of μ.

The Central Limit Theorem says that the standard error of $\overline{X}$, $\sigma_{\overline{X}}$, equals $\sigma/\sqrt{n}$. So as n increases from 5 to 35 the denominator of the standard error gets bigger, which means we are dividing by a bigger number, yielding a smaller standard error. Let's follow this example and see what the numbers should be:

$\overline{X}$'s based on samples of size 5: The standard error should be $1/\sqrt{5} = 1/2.24 = 0.45$.

$\overline{X}$'s based on samples of size 30: The standard error should be $1/\sqrt{35} = 1/5.92 = 0.17$.

This is the second result:

> As the sample size increases, the standard error of $\overline{X}$ decreases.

 **TRY IT NOW!**

Central Limit Theorem

Impact of Sample Size on Standard Error

Using the random number table in Appendix A and the 350 values shown in the previous **Try It Now!**, select 10 samples of size 5 from population 1. Review Section 2.4 if you need a refresher on how to use the random number table.

For each sample, calculate a sample mean, $\overline{X}$.

Find the average and standard deviation of the 10 $\overline{X}$'s. Compare these values to the corresponding values that you found in the previous **Try It Now!** for population 1. In that case the sample size was $n = 35$.

9.6.4 Summary of Central Limit Theorem

Combining all three of the points of the Central Limit Theorem, we get Figure 9.6A, which displays the sampling distribution of $\overline{X}$, when n is sufficiently large.

We know from our work on the normal distribution that 68% of values will fall within one standard deviation of the mean, 95% will fall within two standard deviations of the mean, and 99.7% will fall within three standard deviations of the mean. With regard to the random variable $\overline{X}$ this means that 68% of time we will observe a sample mean that falls within one standard error of the unknown, population mean, μ. Similarly, 95% of the time we will observe a sample mean that falls within two standard errors of μ, and 99.7% of the time we will observe a sample mean that falls within three standard errors of μ. This idea leads to a concept known as a confidence interval or an interval estimate, which is developed in a later section of this chapter.

Remember that the standard deviation of $\overline{X}$ is called the standard error. It is equal to $\sigma/\sqrt{n}$.

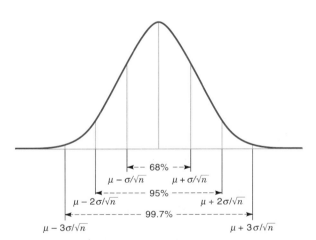

FIGURE 9.6A Sampling distribution of $\overline{X}$

$\overline{X}$'s FOR THE LARGER SAMPLES SHOULD BE CLOSE TO 0.17.
THE STANDARD DEVIATION FOR THE 10 $\overline{X}$'s FOR THE SMALL SAMPLE SHOULD BE CLOSE TO 0.45 AND THE STANDARD DEVIATION FOR THE 10
FOR THE LARGER SAMPLES.
ANS. THE AVERAGE OF THE 10 $\overline{X}$'s FOR THE SMALL SAMPLES SHOULD BE A LITTLE FARTHER AWAY FROM 25 THAN THE AVERAGE OF THE 10 $\overline{X}$'s

Discovery Exercise 9.1
THE CENTRAL LIMIT THEOREM IN ACTION

Part I: Draw a picture of a normal distribution with mean of 10 and standard deviation of 5. This is the population we will sample from.

Part II: Generate and examine 100 random samples.

(a) For this exercise you will need to generate 100 samples each consisting of 30 values selected from a normal distribution with a mean of 10 and a standard deviation of 5. If you do not know how to generate this data, your instructor will provide you with data.

Question: If you plotted the values from one of the samples, what shape would the histogram have? _____ Try it and see.

Question: What value would you expect the average of the 30 values from sample 1 to be close to? _____ Find the average of the first sample of 30 and see if it is close to the number you thought it would be.

Question: Do you expect the average of the 30 values in sample 2 to be close to the same number? _____ Why or why not? _____

Question: What value should the standard deviation of the first sample of 30 values be close to? _____ Find the standard deviation of the first sample of 30 values and check to see if you were right.

Question: Do you expect the standard deviation of the 30 values in sample 2 to be close to the same number? _____ How about sample 3? _____

Part III: Create a distribution of $\overline{X}$ for samples of size $n = 30$

(a) For each of the 100 samples find the sample mean, $\overline{X}_1$, $\overline{X}_2$, etc. You should have a column of 100 $\overline{X}$ values.

Question: Examine these $\overline{X}$ values. Are they all the same? _____ Why not?

Question: Are they all equal to 10? Why not? What is this called? _____

(b) Construct a histogram of the 100 $\overline{X}$ values.

Question: What shape does the distribution of $\overline{X}$ have? _____

Question: Will the shape of the $\overline{X}$'s always be normal or is it just normal because we sample from a normal population? _____

(c) Find the average of the 100 $\overline{X}$ values and the standard deviation of the 100 $\overline{X}$ values.

The CLT tells us that the average of the $\overline{X}$'s should be close to the mean of the distribution of the population.

Question: What is the mean of the population we sampled from? _____

Question: What is the average of the 100 $\overline{X}$'s? _____

Question: Are they close? _____

Question: How could you get them closer? _____

(d) The CLT tells us that the standard deviation of the $\overline{X}$'s should be about $\sigma/\sqrt{n}$.

Question: What is σ? _____

Question: What is n? _____

Question: So, what is the value that the standard deviation of the $\overline{X}$'s should be close to? _____

Question: What is the standard deviation of the $\overline{X}$'s? _____ Is it close to the value it was supposed to be? _____

Question: How could you get them closer?_____

9.7 DRAWING INFERENCES BY USING THE CENTRAL LIMIT THEOREM

Now we can return to the diaper company problem with which we started Section 9.5. We are ready to decide if an $\overline{X}$ value of 54.00 g is indeed unusual or if it is reasonable variation caused by sampling error when the process is running properly. Recall that the true mean diaper weight, μ, is 55.00 g and the true population standard deviation σ is 0.55 g when the manufacturing process is running properly.

9.7.1 Using the Central Limit Theorem

Let's create a picture similar to Figure 9.6A for the diaper manufacturing process. Recall that the samples each have 5 observations. Since the sample size is less than 30 we must assume that the underlying distribution of diaper weights is normally distributed. The $\overline{X}$'s should be centered at $\mu_{\overline{X}} = 55$ g and the standard error should be $\sigma_{\overline{X}} = \sigma/\sqrt{n} = 0.55/\sqrt{5} = 0.25$ g. Combining these results yields Figure 9.6B.

Create a histogram of the diaper weights to check the assumption of normality.

A sample mean of 54.00 g clearly falls in the tails of this distribution, telling us that if the process is running properly, the probability of seeing a sample mean this small is tiny. This conclusion has taken into account the fact the diaper weights themselves have some variability measured by σ and the sample size is $n = 5$. Thus, we

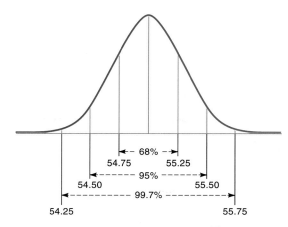

FIGURE 9.6B Sampling distribution for $\overline{X}$ when $\mu = 55.00$

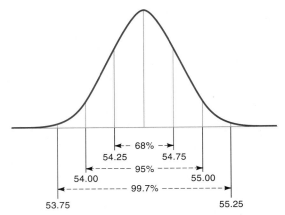

FIGURE 9.6C   Sampling distribution of $\overline{X}$ when $\mu = 54.50$

must conclude that something is not quite right with the process. We could also reach this conclusion by calculating the Z-score for our observed $\overline{X}$ value of 54 g:

$$Z = \frac{\overline{X} - \mu_{\overline{X}}}{\sigma_{\overline{X}}}$$

$$Z = \frac{54 - 55}{0.25} = -4.0$$

We know that a Z-score below -3 is extremely unusual. Thus, we conclude that we should check the machine.

Suppose for a moment that the process mean had shifted to 54.5 g. In this case the sampling distribution would look like Figure 9.6C.

Now a sample mean of $\overline{X} = 54$ g is not quite so unlikely. The Central Limit Theorem has given us the basis on which to judge whether the observed sample mean value of 54 g is too unusual and if so what could possibly have happened to the process that would result in such a low sample mean.

Let us examine some other problems that we can begin to tackle now that we have learned about the Central Limit Theorem.

EXAMPLE 9. 10   Car Batteries

Using the Central Limit Theorem

A company that produces car batteries warranties the batteries for 36 months. The battery lives have a standard deviation of 2 months. You and 8 of your friends purchased these batteries and you have collected data on the length of time each of them lasted. The data (in months) are shown below:

| 35.38 | 35.87 | 35.49 | 34.38 | 34.55 | 39.99 | 36.18 | 35.09 | 34.57 |

You calculate the average length of life of these batteries and find it to be 35.72 months. Do you have reason to be suspicious of the company claim on the basis of these data?

Let's calculate the Z-value for this sample mean value:

$$Z = \frac{\overline{X} - \mu_{\overline{X}}}{\sigma_{\overline{X}}}$$

$$Z = \frac{35.72 - 36}{2/\sqrt{9}} = -0.42$$

This Z-score tells us that the sample mean value of 35.72 months is within 1 standard deviation (Z-score between -1 and 1) of the warranty value of 36 months. We know that 68% of Z-scores are between -1 and 1. Thus, it is *not* an unusually low sample average and we have no basis to be suspicious of this company. ∎

In this example we used the value of the population standard deviation claimed by the company. What happens if we do not know the population standard deviation? This is not unlikely in many cases. If we need to estimate the mean, why would we *know* the standard deviation? If the sample is sufficiently large, $n > 30$, then we can use the point estimate s instead of σ in the calculation of the z-score. The next example demonstrates this situation.

EXAMPLE 9. 11   Manufacturer of Breakfast Cereals

Using the Central Limit Theorem

A manufacturer of breakfast cereals packs the cereal in 15-ounce boxes. Periodic quality checks must be made to ensure that the boxes are being filled adequately. A random sample of 35 boxes of the cereal are selected and weighed. The data are shown below:

14.91	15.23	15.34	15.40	15.48	15.59	15.62
14.96	15.24	15.34	15.42	15.49	15.60	15.62
15.11	15.28	15.36	15.45	15.50	15.60	15.63
15.21	15.30	15.38	15.46	15.52	15.61	15.64
15.22	15.30	15.39	15.48	15.58	15.61	15.67

You find the average and standard deviation to be $\overline{X} = 15.42$ oz and $s = 0.19$ oz. Is there any reason to believe that the boxes are not being filled properly?

Calculating the Z-score for this sample value, we find

$$Z = \frac{\overline{X} - \mu_{\overline{X}}}{\sigma_{\overline{X}}}$$

$$Z = \frac{15.42 - 15}{0.19/\sqrt{35}} = \frac{0.42}{0.03} = 14.00$$

This Z-score is extremely large, telling us that it would be extremely unusual to see an average of 15.42 oz (based on a sample of 35 boxes) if the boxes were indeed being filled to an average of 15 oz. Therefore, we have rather strong evidence to indicate that the boxes are not being filled to an average of 15 oz. ∎

In the preceding example you should have noticed that we used s, the sample standard deviation, as an estimate of σ in finding the Z-score, since σ was not given. This isn't precisely the correct procedure but it is close enough given the large sample size. In this case, the formula to compute the approximate Z-score is given as follows:

$$Z = \frac{\overline{X} - \mu_{\overline{X}}}{s_{\overline{X}}} = \frac{\overline{X} - \mu_{\overline{X}}}{s/\sqrt{n}}$$

SCORE WHEN σ IS UNKNOWN AND n IS LARGE

A more thorough discussion of this issue is addressed later in this chapter along with what to do if the standard deviation is unknown but the sample size is small ($n \leq 30$).

TRY IT NOW!

Cost of Books

Comparing the Sample Mean to the Claimed Population Mean

A university states that students spend an average of $225 per semester on books. Based on your own experience you feel that this is an underestimate of the true expenditure. You ask 30 of your friends how much they spent on textbooks this past semester and you obtain the following data:

214	236	241	247	248	253	258	262	269	277
233	239	244	248	249	254	260	263	274	277
234	241	245	248	250	254	262	265	276	281

Based on these data, do you have reason to tell the university that its statement is inaccurate?

9.7.2 Exercises—Learning It!

9.6 A manufacturer of pain relievers claims that it takes an average of 12.75 min for a person to be relieved of headache pain after taking its pain reliever. The time it takes to get relief is normally distributed with a standard deviation of 0.5 min. A sample of 12 people is taken and the data are shown below:

    12.9   13.2   12.7   13.1   13.0   13.1   13.0   12.6   13.1   13.0   13.1   12.8

(a) Find the sample mean.
(b) Find the standard error of $\overline{X}$.
(c) If the manufacturer claims that the mean is 12.75 min, find the Z-score of the sample mean.
(d) What do you think of the manufacturer's claim based on the Z-score?

9.7 A company that sells personal computers via mail order claims that its customers wait an average of 3 minutes on hold before reaching a technical support staff member. The on-hold time is normally distributed with a standard deviation of 0.5 min. A sample of the on-hold times for 20 different customers is shown below:

    3.21   3.13   3.72   2.88   2.84   3.49   3.04   2.98   2.56   2.91
    3.34   2.84   2.73   3.16   3.26   3.41   3.29   2.67   2.98   2.31

(a) Find the sample mean.
(b) Find the standard error of $\overline{X}$.
(c) If the claimed mean is 3 min, find the Z-score for the sample mean.
(d) What do you think of the company's claim based on the Z-score?

9.8 The U-Make-It Hardware chain has 10 different stores within a certain geographical region. The dollar value of customer sales is normally distributed with an average sale of $35.25 and a standard deviation of $2.50. You have recently been hired to manage one of these stores and under your leadership the average sales based on a sample of 100 customers is $36.50. You are very proud of the increased average sale and point this out to senior management.

(a) Find the Z-score for $36.50.
(b) Based on this Z-score, is your pride justified?

9.9 Scores on a national test are normally distributed with a mean of 75 and a standard deviation of 5 points. At a certain high school 50 students took this test and their average was 80. On the basis of these data, can this high school be considered unusually strong?

9.10 Starting salaries nationally among business school graduates follow a normal distribution with a mean of $25,000 and a standard deviation of $1500. The Career and Human Resource Director at College XYZ has taken a sample of 25 of the most recent graduates of the business school and found that their average salary is $24,500.

(a) Between what two values will 99.7% of the salary averages fall?
(b) Based on your answer to part (a), what can you conclude about the average starting salary of the most recent graduates of College XYZ?

WORKSHOP 5

9.8   LARGE-SAMPLE CONFIDENCE INTERVALS FOR THE MEAN

In this chapter, you have learned about point estimators, which give you a single number to be used as an estimate of the population parameter. A confidence interval takes the point estimate a step further and gives you a range of values and a probability. The probability value tells you the likelihood that you have an interval that actually includes the value of the unknown population parameter.

9.8.1 The Basics of Confidence Intervals

Let's return to the problem that was presented at the beginning of this chapter. The diaper company must decide if the manufacturing process is running to specification. In particular, the diaper company must decide if the population mean, μ, is 55.00 g. The company will decide this based on a sample of 30 observations that have a sample mean, $\overline{X} = 54.98$. We know that the sample mean, $\overline{X}$, is a reliable point estimate of μ. But we also know that it is not likely to hit μ on the nose. Initially, it looks as if $\overline{X}$ is not very helpful. It is an estimate of μ that is pretty likely to be wrong! Ideally, we would like to use $\overline{X}$ to create a range of values that would definitely contain μ. However, because of sampling error this is not possible. Instead, we can specify a high probability, say 0.90 or 0.95, that a particular range or interval covers the true mean, μ. For example, we would like to be able to say that based on the sample data, the interval from 54.50 to 55.50 g covers μ with a probability of 0.95. Rewriting this in a more mathematical form gives us

$$\text{Probability}(54.5 \leq \mu \leq 55.5) = 0.95$$

This is an example of a confidence interval. Let's examine the components of the confidence interval. First of all, it has a lower bound for μ. Call this L. In the example, L is 54.50. The confidence interval also has an upper bound for μ. Call this U. In the example, U is 55.50.

Finally, it has a probability value, which is called the confidence level and is labeled $1 - \alpha$. In the example, α is 0.05 and $1 - \alpha$ is 0.95. In this case we are constructing a $100(1 - 0.05)\% = 95\%$ confidence interval. This means that if we were to construct 20 intervals based on 20 different sample means, we would expect 95% of them or 19 intervals to actually contain μ. For any *individual* interval, μ is either in the interval or it is not. The problem is that, due to sampling error, we do not know for sure whether we have an interval which does include the value of μ. Thus the probability refers to the chance that we have one of the intervals that does indeed contain μ.

In general, a confidence interval for the population mean has the following form:

$$P(L \leq \mu \leq U) = 1 - \alpha$$

A *confidence interval* or an *interval estimate* is a range of values with an associated probability or *confidence level*, $1 - \alpha$. The probability quantifies the chance that the interval contains the true population parameter.

Think about the information such a confidence interval gives you. Instead of just being able to say that the true μ is somewhere close to $\overline{X} = 54.98$ now you can say the probability that the interval from 54.50 to 55.50 g covers μ is 0.95. This gives you a way to quantify the error you know is inherent in $\overline{X}$.

The next few sections provide the details for calculating the lower and upper bounds for confidence intervals for μ. These sections are followed by a section that focuses on the interpretation of confidence intervals.

9.8.2 Confidence Interval for μ: Normally Distributed Population and Known Standard Deviation

We start by developing a confidence interval for the mean of a population that is normally distributed and for which the population standard deviation, σ, is known. We will see what happens as we relax both the assumption of normality and the assumption of a known standard deviation in the next section.

Recall that an unbiased estimator of μ is a fair estimator. It does not systematically overestimate or underestimate μ.

As we start to look at how to calculate the lower and upper bounds for the confidence interval, think about where $\overline{X}$ should fall in the interval. Remember that $\overline{X}$ is an *unbiased* estimate of μ and so it makes sense to put $\overline{X}$ right in the middle of the interval as shown below:

The value of $\overline{X}$ should always go in the center of a confidence interval for μ.

Notice that the distance from the lower bound to the middle of the interval has been labeled e. The distance from the middle of the interval to the upper bound has also been labeled e. Thus, the width of the interval is $2e$. To find the lower bound we take $\overline{X}$ and subtract e, and to find the upper bound we take $\overline{X}$ and add the value of e. Now, all we need is the value of e, which stands for error.

The Central Limit Theorem tells us that the sampling distribution of $\overline{X}$ is a normal distribution.

Suppose we decide we want to construct a 95% confidence interval ($\alpha = 0.05$) for the population mean, μ. In a sense, we wish to "cover" the value of μ, 95% of the time. We know that $\overline{X}$ is the best point estimate we can use for μ and that $\overline{X}$ has a normal distribution when n is large enough. Remember that to cover 95% of the values that follow a normal distribution we need to go out about two standard deviations from the mean in both directions. Finally, we recall that the standard deviation of the sample mean, also known as the standard error, is equal to $\sigma/\sqrt{n}$. Bringing these three pieces of information together tells us that e should equal $2\sigma/\sqrt{n}$. This is just about right.

You may recall that in Chapter 7 you learned how to find Z-values given tail area probabilities. To get the correct value for Z, you must use that procedure with a tail area probability equal to $\alpha/2$. Label this value as $Z_{\alpha/2}$. For a 95% confidence interval we divide $\alpha = 0.05$ by 2 and find that the area in one of the tails is 0.025. This means we are given a probability and need the corresponding Z-value. Shown here is a portion of the Z table:

z	0.00	0.01	0.02	0.03	0.04	0.05	0.06	0.07	0.08	0.09
-2.1	0.0179	0.0174	0.0170	0.0166	0.0162	0.0158	0.0154	0.0150	0.0146	0.0143
-2.0	0.0228	0.0222	0.0217	0.0212	0.0207	0.0202	0.0197	0.0192	0.0188	0.0183
-1.9	0.0287	0.0281	0.0274	0.0268	0.0262	0.0256	0.0250	0.0244	0.0239	0.0233
-1.8	0.0359	0.0351	0.0344	0.0336	0.0329	0.0322	0.0314	0.0307	0.0301	0.0294
-1.7	0.0446	0.0436	0.0427	0.0418	0.0409	0.0401	0.0392	0.0384	0.0375	0.0367

We want to find the Z-value that has 0.025 in the lower tail. Look in the inside of the table for the number closest to 0.025 and see that it corresponds to a Z-value of -1.96. This gives us the value of $-Z_{\alpha/2}$. To get the positive cutoff value we can simply drop the negative sign because of the symmetry in the normal distribution. So $Z_{\alpha/2} = 1.96$. Figure 9.7 shows the normal distribution with 5% in the tail area.

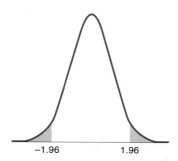

FIGURE 9.7 Normal distribution with 0.05 in the tails

So, the value of e is calculated as $1.96\ \sigma/\sqrt{n}$. This gives us the following formulas for the lower and upper bounds for a 95% confidence interval for μ:

$$L = \overline{X} - e = \overline{X} - \frac{1.96\ \sigma}{\sqrt{n}}$$

$$U = \overline{X} + e = \overline{X} + \frac{1.96\ \sigma}{\sqrt{n}}$$

Lower and upper bounds for a 95% confidence interval for μ

Let's use these to find a 95% confidence interval for the diaper manufacturer.

EXAMPLE 9.12 **The Diaper Company**

A 95% Confidence Interval for the Mean Weight

A sample of 30 diapers yielded a sample average weight of 54.98. Based on this sample, find a 95% confidence interval for μ.

Given information: $\overline{X} = 54.98$

$n = 30$

$\sigma = 0.55$ g   (from the specifications)

Calculations: Standard error $= \dfrac{\sigma}{\sqrt{n}} = \dfrac{0.55}{\sqrt{30}} = 0.10$

Error $= e = \dfrac{1.96\ \sigma}{\sqrt{n}} = (1.96)(0.10) = 0.20$

Lower bound $= \overline{X} - e = 54.98 - 0.20 = 54.78$

Upper bound $= \overline{X} + e = 54.98 + 0.20 = 55.18$

We can state that we are 95% confident that the true mean weight is between 54.78 and 55.18 g. This means that there is a 95% chance that we have constructed an interval that does indeed contain the true mean weight. Restating this mathematically gives us

$$P(54.78 \leq \mu \leq 55.18) = 0.95$$

Notice that the target value of 55.00 g is in between the lower and upper bound and thus a likely value for the population mean weight. ■

The formulas that we have used in this example are for a 95% confidence interval. Suppose instead that we wish to find a 90% confidence interval or a 99% confidence interval. The only adjustment that needs to be made to the formula is the value of Z that we multiply the standard error by to get e. For the 95% confidence level we used the value of Z that cut off a probability of 0.05 in the two tails of the normal distribution. The procedure for figuring out the correct Z-value for any confidence level is as follows:

STEPS FOR FINDING THE VALUE OF Z FOR A CONFIDENCE INTERVAL

Step 1.   Take the value of α and divide it by 2.

Step 2.   Look that value up in the body of the Z table (find the one closest to it).

Step 3.   Read off the corresponding Z-value. This gives you the value of $-Z_{\alpha/2}$.

Step 4.   Drop the negative sign to get the value of $Z_{\alpha/2}$.

Remember that Z tells you how many standard errors away from the mean that you need to go to get the coverage you desire.

In general, the formulas for a $100(1 - \alpha)\%$ confidence interval for μ when the population standard deviation, σ, is known are shown below:

CONFIDENCE INTERVALS FOR μ WHEN σ IS KNOWN

Error:

$$e = \frac{Z_{\alpha/2}\sigma}{\sqrt{n}}$$

Width of the interval: $w = 2e$

Lower bound: $L = \overline{X} - e$

Upper bound: $U = \overline{X} + e$

where $Z_{\alpha/2}$ is the value that cuts off $\alpha/2$ in the upper tail of the standard normal distribution and $\alpha/2$ in the lower tail of the standard normal distribution.

Some common values of α and the corresponding Z-values are shown in Table 9.2.

TABLE 9.2   Common Values of α and the Corresponding Z-values

	\multicolumn{5}{c}{Confidence Level $(1 - \alpha)$}
	0.90	**0.95**	**0.98**	**0.99**	**0.995**
α	0.10	0.05	0.02	0.01	0.005
$Z_{\alpha/2}$	1.645	1.96	2.33	2.58	2.81

As you increase the level of confidence, you lose precision.

Notice that as the level of confidence increases so does the value of Z used in the formula. Since the value of Z affects the size of e, the interval gets wider as the confidence level increases. Are wider intervals better? Suppose you had a choice between 2 intervals with the same level of confidence: one that stated the average age, μ, of the target population was between 2 and 95 years and one that stated the average age, μ, of the target population was between 25 and 35 years. Which interval gives you better information? Clearly, the interval that is narrower, the second one. This small example shows us that the narrower the interval, the more precise our estimate. Thus, as you *increase* the level of confidence you *decrease* the precision of the estimate.

 **TRY IT NOW!**

The Bottle-Filling Problem
Finding a Confidence Interval for μ

A sample of 36 bottles had a sample mean of $\overline{X} = 32.10$ oz. The population standard deviation, σ, was assumed to be 0.1 oz. Find a 95% confidence interval for μ. How wide is the interval?

Now find a 98% confidence interval for μ.

Which interval is wider?

The procedure just described depends on the knowledge that $\overline{X}$ has a normal distribution and the population standard deviation is known. The Central Limit Theorem tells us that $\overline{X}$ has a normal distribution when $n > 30$ or when the underlying population distribution is normally distributed (or close to normal). Therefore, if you have a *large sample* ($n > 30$), then you can use this procedure to find confidence intervals for μ regardless of the shape of the underlying distribution.

If the *sample size is small* ($n < 30$), we should visually check the shape of the distribution of the sample data to get a rough idea of the population shape. If the

Assumptions of confidence interval formulas

ANS.
P(32.07 ≤ μ ≤ 32.13) = 0.95
P(32.06 ≤ μ ≤ 32.14) = 0.98
THE 98% INTERVAL IS WIDER.

histogram or other plot of the data looks reasonably symmetric in shape, then you can use the formulas we just developed.

9.8.3 What Happens if σ Is Unknown or the Distribution Is Not Normal?

In many cases you do not know the population standard deviation, σ, and/or you do not have an underlying normally distributed population. First, consider what happens to the calculation of the confidence interval for μ if we do not know σ but we still have a normally distributed population. There are two cases: large sample size and small sample size.

If you have a large sample size then you can use s, the sample standard deviation, to estimate σ. Using s to approximate σ in the formulas for the upper and lower limits of the confidence interval does not appreciably affect the confidence level. This is the same approach we took in computing the Z-score when we did not know σ but the sample size was sufficiently large.

The formulas for a $100(1 - \alpha)\%$ confidence interval for the mean, μ, of a normally distributed population when the population standard deviation, σ, is unknown and n is sufficiently large are shown below:

CONFIDENCE INTERVAL FOR
μ WHEN σ IS UNKNOWN

Error:	$e = \dfrac{Zs}{\sqrt{n}}$
Width of the interval:	$w = 2e$
Lower bound:	$L = \overline{X} - e$
Upper bound:	$U = \overline{X} + e$

If you have a small sample size and the standard deviation is unknown then the sampling distribution of $\overline{X}$ is not normally distributed. The sampling distribution is called a t-distribution and is covered in the next section.

Finally, if you have a small sample size and the histogram of the data does not look roughly normal in shape and you do not know the population standard deviation, then you should not use this procedure for finding a confidence interval for μ. It may be possible to use a transformation to get normally distributed data or it may be necessary to use a different distribution or a nonparametric procedure. Nonparametric procedures are typically not as powerful but do not require the assumption of an underlying normal population. Transformations and nonparametric procedures are out of the scope of this text but can be found in the next level of statistics courses/books. Table 9.3 summarizes the different scenarios.

TABLE 9.3   Summary of Confidence Intervals for μ

Population	Standard deviation	Sample size	Confidence interval for μ
Normal	Known	$n \geq 1$	$L = \overline{X} - Z\sigma/\sqrt{n}$ $U = \overline{X} + Z\sigma/\sqrt{n}$
Normal	Unknown	$n > 30$	$L = \overline{X} - Zs/\sqrt{n}$ $U = \overline{X} + Zs/\sqrt{n}$
Normal	Unknown	$n \leq 30$	Use t-distribution covered in Section 9.9
Non-normal	Known	$n > 30$	$L = \overline{X} - Z\sigma/\sqrt{n}$ $U = \overline{X} + Z\sigma/\sqrt{n}$
Non-normal	Unknown	$n \leq 30$	Use transformation or nonparametric

9.8.4 Interpreting the Confidence Interval

The last major point of discussion for confidence intervals has to do with what the words "I am 95% confident" imply. Many people feel that this means that there is a 95% chance that the population mean is in the interval you have constructed. This is wrong! After all, μ is a number even if it is unknown to us. Therefore, there is no probability associated with it being between two other numbers; it is either in the interval or it is not.

So, then, what is the proper interpretation of those words? The correct interpretation of the 95% confidence level has to do with the chance that you have an interval that does in fact contain the population parameter. Remember that if you take 100 different samples from the same population, you will get 100 different sample means and therefore 100 different confidence intervals. If each of them is a 95% confidence interval then theoretically 95 out of the 100 intervals will contain μ and 5 of them will not.

Suppose that 10 samples of size $n = 30$ were selected from the population of diapers. For each sample, an average diaper weight was calculated and a 90% confidence interval was calculated based on this value of $\overline{X}$. These 10 confidence intervals are shown in Table 9.4. Let's assume that the manufacturing process was running properly so the population mean is 55 g. Remember that the population standard deviation is 0.55 g. According to the theory, 9 out of 10 of these intervals should include the value of $\mu = 55$ g.

The last row of Table 9.4 indicates whether the interval includes the value of 55 g. You can see in Figure 9.8 that there is only 1 interval that does not include the value of 55 g and 9 that do. This is just what we expected, since 90% of 10 intervals is 9 intervals that cover μ. The problem, of course, is that we don't know the actual value of μ and therefore we don't know if we have an interval that actually contains μ or not. The confidence level gives us the probability of our having an interval that does in fact contain the population parameter.

TABLE 9.4   Ten 90% Confidence Intervals

$\overline{X}$	55.21	54.94	54.99	55.05	55.06	54.97	55.11	54.92	55.03	55.10
Error	0.17	0.17	0.17	0.17	0.17	0.17	0.17	0.17	0.17	0.17
Lower	55.04	54.78	54.82	54.89	54.89	54.81	54.95	54.76	54.87	54.93
Upper	55.37	55.11	55.16	55.22	55.23	55.14	55.28	55.09	55.20	55.27
Covers μ?	No	Yes	Yes	Yes	Yes	Yes	Yes	Yes	Yes	Yes

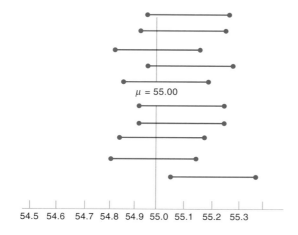

FIGURE 9.8   Comparison of confidence intervals and μ

To increase the probability that we have a "good" interval, we could widen the intervals. This is precisely what happens when you increase the confidence level. When you increase the confidence level from 90 to 95%, you are increasing the chance that you have a good interval, but you are losing precision by widening the interval. It is a tradeoff.

9.8.5 Exercises—Learning It!

9.11 A university wants to estimate the average amount of money that students spend on textbooks in a semester. It takes a random sample of 45 students and find that the average amount of money spent was $282 with a standard deviation of $21. Find a 98% confidence interval estimate for the true mean amount of money spent on textbooks in a semester.

9.12 The weights of jars of peanut butter are normally distributed with a standard deviation of 0.03 oz. A random sample of 12 jars has an average weight of 14.9 oz.

(a) Find a 95% confidence interval for the true mean weight of the jars.

(b) If the jars are labeled 15 oz, do you think that there is a problem?

9.13 Quality Foods' regional manager wants to determine the average fat content per pound of steak sold in Portland, Maine. She's considering quitting her job and starting a cattle ranch after reading the article "Queen of the Range" in the January 1989 issue of *Successful Farming*. Her ranch would produce low-fat beef for sale to upscale consumers. She purchased one steak from each of the 18 stores in the area and determined the fat content per pound. The data (in ounces) are shown below:

1.2	1.3
1.6	1.3
1.3	1.6
0.7	1.1
1.3	1.2
1.7	1.3
0.9	0.9
1.8	1.4
1.3	1.4

The standard deviation of fat content is known to be 0.30 oz.

(a) Assuming that the fat content is normally distributed, calculate a 95% confidence interval for the average fat content per pound in steaks sold in the Portland, Maine, area.

(b) Her research shows that the steaks from her ranch will average 1.0 oz of fat per pound. Are her steaks significantly lower in fat content than the ones available in the Portland area? Explain why or why not.

Data file: SALARY.XXX **9.14** Is it a lottery jackpot or a bonus? A salary or a life's savings? No, it's the annual compensation paid out to top executives of public stock corporations. The following table shows the 1996 salary and bonus for the first few of the 43 executives, the firm they work for, and the type of firm (manufacturing or nonmanufacturing).

Name	Firm	Type	1996 Salary and Bonus
Eugene Freedman	Enesco Giftware Group/Stanhome Inc.	M	$4,505,500
Thomas Wheeler	Massachusetts Mutual	NM	3,569,998
Terrence Murray	Fleet Financial Group	NM	3,909,200
Robert Shapiro	Monsanto	M	2,920,000
David E. Sams, Jr.	Massachusetts Mutual	NM	2,760,183
Charles Gifford	BankBoston	NM	2,600,000
George David	United Technologies	M	2,375,000

SOURCE: *Springfield Sunday Republican* (20 July 1997)

Assuming that these 43 executives represent a random sample of executives, find a 90% confidence interval for the mean salary of executives.

9.15 The United States Department of Labor publishes a great deal of data on a monthly basis. The following data on the average price of electricity were extracted from the World Wide Web:

[Bureau of Labor Statistics Data]
Data extracted on: January 4, 1997

Average	Price Data
Series	Catalog:
Series	ID APU0000702111
Area:	U.S. City Average
Item:	Average price per kwh of electricity

Jan-94	0.090	Jan-95	0.091	Jan-96	0.091
Feb	0.090	Feb	0.091	Feb	0.091
Mar	0.089	Mar	0.091	Mar	0.092
Apr	0.088	Apr	0.090	Apr	0.092
May	0.090	May	0.092	May	0.092
Jun	0.095	Jun	0.098	Jun	0.096
Jul	0.095	Jul	0.098	Jul	0.099
Aug	0.096	Aug	0.098	Aug	0.099
Sep	0.096	Sep	0.097	Sep	0.099
Oct	0.093	Oct	0.094	Oct	0.095
Nov	0.091	Nov	0.092	Nov	0.092
Dec	0.091	Dec	0.091	Dec	0.094

Treating these three years of data as a sample, find a 90% confidence interval for the average monthly price of electricity. Assume that the standard deviation is $.006 per kwh.

9.16 A national grocery chain is considering opening a store at a particular location. To be sure that enough traffic goes by that location, the grocery chain took a sample of vehicles crossing the intersection on 40 days. The results are shown in the table below:

Number of Cars Crossing Location/Day

1431	1540	1293	1340
1302	1700	1533	1402
1255	1840	1272	1467
1377	1642	1572	1220
1450	1139	1520	1477
1483	1227	1227	1515
1529	1684	1257	1242
1588	1782	1238	1350
1535	1491	1276	1367
1533	1513	1420	1375

(a) Find a 95% confidence interval for the average number of cars that pass this location on a daily basis. The standard deviation is assumed to be 165 cars.

(b) The company has decided to open a store at this location only if there is a daily average of at least 1400 cars passing this location. Based on your confidence interval, would you advise the company to open a store at this location? Explain why or why not.

9.17 The amount of time that it takes a student to complete an assignment in a statistics class has a standard deviation of 5 minutes. A group of 35 students are observed and it is found that the average time to complete the assignment was 55 minutes.

   (a) Find a 99% confidence interval for the true mean time to complete the assignment.

   (b) If the instructor believes that the students should be spending an average of 50 minutes on the assignment, what can you say about his belief?

9.18 Most companies have increased their dependence on computers and software. As a result, more employee time is spent on the telephone with technical support for the software. A sample of 22 times spent on hold for technical support is shown below:

Time on Hold (minutes)

8.6	12.9
12.7	11.4
8.7	10.3
12.2	7.9
11.7	9.2
7.3	10.5
9.8	11.8
14.5	10.9
13.0	11.5
12.9	10.6
11.4	11.7

Assume that the standard deviation is 2 minutes.

   (a) Since the sample size is less than 30 and you are not told that the population of time on hold is normally distributed, display the data in a histogram, and comment on the shape of the data. Is it reasonable to assume that the data come from a population that has a normal distribution? Why or why not?

   (b) Find a 99% confidence interval for the average amount of time spent on hold per call.

   (c) Find a 95% confidence interval for the average amount of time spent on hold per call.

   (d) If you are the manager of these employees and you are trying to argue for additional staff, which of the two confidence intervals would you use and why?

9.19 The symphony in a medium-size New England city is surveying the community to determine the average number of times during a year that a person would attend a concert at a particular price. The responses from 50 adults are shown below:

**Number of Times
Symphony Would Be
Attended in One Year**

1	3	2	2	2
4	4	4	4	4
3	2	4	4	4
2	1	4	4	4
2	3	3	3	3
2	3	3	3	3
3	5	5	5	5
2	4	4	4	4
4	3	2	2	2
3	3	2	2	2

   (a) Find a 90% confidence interval for the average number of times a year a person would attend a concert at the price studied.

   (b) How might this confidence interval be useful to the management of the symphony?

Discovery Exercise 9.2
EXPLORING CONFIDENCE INTERVALS FOR μ

From a population of college students across the United States, a sample was selected to find out how many hours per week a typical student spends playing sports.

Part I: A random sample of 2500 students was selected. The sample mean, $\overline{X}$, was found to be 12.5 hours. The population standard deviation, σ, is known to be 1.05 hours. Given this information, find

(a) a 90% confidence interval for μ

(b) a 92% confidence interval for μ

(c) a 94% confidence interval for μ

(d) a 96% confidence interval for μ

(*continued*)

(e) a 98% confidence interval for μ

(f) Discuss what happens to the size of the interval as the level of confidence increases.

Part II: A random sample of 2500 students was selected. The sample mean, $\overline{X}$, was found to be 10.5 hours. The population standard deviation, σ, is known to be 1.05 hours. Given this information, find

(a) a 90% confidence interval for μ

(b) a 92% confidence interval for μ

(c) a 94% confidence interval for μ

(d) a 96% confidence interval for μ

(e) a 98% confidence interval for μ

(f) Compare the intervals found in Part I with those found in Part II. Discuss what happened to the confidence interval due to the change in the value of the sample mean, $\overline{X}$.

(*continued*)

Part III: A random sample of 2500 students was selected. The sample mean, $\overline{X}$, was found to be 12.5 hours. Suppose you learn that the population standard deviation, σ, is actually 2.05 hours. Given this information, find

(a) a 90% confidence interval for μ

(b) a 92% confidence interval for μ

(c) a 94% confidence interval for μ

(d) a 96% confidence interval for μ

(e) a 98% confidence interval for μ

(f) Compare the intervals found in Part I with those found in Part III. Discuss what happened to the confidence intervals due to the change in the value of the population standard deviation, σ.

Part IV: A random sample of 2000 students was selected. The sample mean, $\overline{X}$, was found to be 12.5 hours. The population standard deviation, σ, is known to be 1.05 hours. Given this information, find

(a) a 90% confidence interval for μ

(b) a 92% confidence interval for μ

(*continued*)

(c) a 94% confidence interval for μ

(d) a 96% confidence interval for μ

(e) a 98% confidence interval for μ

(f) Compare the intervals found in Part I with those found in Part IV. Discuss what happened to the confidence intervals due to the change in the value of the sample size, n.

9.9 DISTRIBUTION OF THE SAMPLE MEAN: SMALL SAMPLE AND UNKNOWN σ

In the previous section we noted that if the standard deviation of a normally distributed population is unknown and the sample size is small ($n \leq 30$), then the sampling distribution of $\overline{X}$ does not follow a normal distribution. As we did with the large sample size situation, logic tells us to try using s in the formula for the Z-score instead of σ. When we do this a t-score is created instead of a Z-score. That is, the sampling distribution of $\overline{X}$ for small samples, when σ is unknown and the population is normally distributed, follows what is called *Student's t distribution*. The t-score is calculated as follows:

$$t = \frac{\overline{X} - \mu}{s/\sqrt{n}}$$

t-score

Notice that the calculation for t is just the same as for Z, with σ replaced with s.

The t distribution was first developed by a man named William Gosset who was working for Guinness Brewery in Ireland. The company did not allow research results to be published so he published his results under a pen name; Student was his pen name!

This distribution is not named for you, the student!

The graph of the t distribution looks very much like the standard normal distribution. It is symmetric, it has a bell shape, and it is centered at zero just like the Z distribution. The major difference between the Z and the t distributions has to do with the spread or variability of the distributions. The standard deviation of the t distribution is not 1 like it is for the Z distribution. Instead, the variability of the t distribution is related to a number that is called the degrees of freedom. Thus, there are many different t distributions, but they all have generally the same shape. A t distribution with 5 degrees of freedom is shown in Figure 9.9.

Let's examine the Z and t calculations. The two calculations are shown side by side below:

Each sample yields a different sample mean, $\overline{X}$, and a different sample standard deviation, s.

$$Z = \frac{\overline{X} - \mu}{\sigma/\sqrt{n}} \qquad t = \frac{\overline{X} - \mu}{s/\sqrt{n}}$$

Think about what causes the variability in the Z statistic. For a given sample size, say $n = 26$, the only thing that causes the Z values to change from sample to sample is $\overline{X}$. The other elements of the calculation stay the same from sample to sample. Now look at the t distribution. Both $\overline{X}$ and s are changing from sample to sample. This causes t to have more variability than Z. Based on this discussion we would correctly

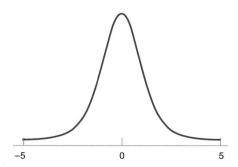

FIGURE 9.9 *t*-distribution with 5 degrees of freedom

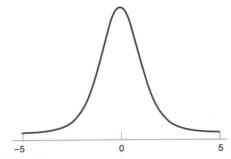

FIGURE 9.10 t distribution with 25 degrees of freedom

conclude that t will have "fatter" tails than Z and indeed it does. The t distribution for a sample size of $n = 26$ is shown in Figure 9.10.

You should ask, "Why do I need to know the value of n to draw the graph?" You might remember that as we take larger and larger samples (n gets larger) our estimate of the unknown population standard deviation gets better and better. Why is this so? Well, just think about the whole population as a big pie. When you take a small sample it is like taking a small piece of the pie and tasting it and trying to decide how the whole pie tastes based on that piece. As the piece you taste gets bigger you have a much better idea of how the whole pie tastes. Eventually, if you take a big enough sample you will eat the whole pie and you will know exactly how the whole pie tastes! This is the same as sampling the whole population. If the whole population were sampled you wouldn't need to estimate σ because you would know it.

So as you take a larger and larger sample you get a better and better estimate of σ, which, in turn, causes t to be less and less variable. All of this means that the graph of t depends on the size of your sample. In particular, the degrees of freedom associated with the t distribution is related to the sample size. The value for the degrees of freedom is $n - 1$, one less than the sample size. If you have taken a sample of size 26, then the t distribution will have 25 degrees of freedom. This is the graph shown in Figure 9.10. Examine the t distributions for various different degrees of freedom shown in Figure 9.11.

What do you notice about the graph of t as the number of degrees of freedom gets larger (i.e., as the sample size gets larger)? You should see that the t distribution gets tighter, meaning the variability of the distribution gets smaller. This is precisely what you should have expected to happen based on our discussion of what happens to our estimate s as we sample bigger pieces of the pie. In fact, eventually, if we use a large enough value for n, the graph of t will be indistinguishable from the standard normal distribution, Z. Typically, the value of n that is considered large enough is $n > 30$.

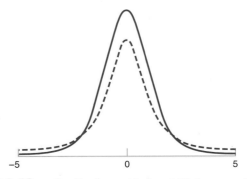

FIGURE 9.11 t distributions with 1 and 50 degrees of freedom

9.10 SMALL-SAMPLE CONFIDENCE INTERVALS FOR THE MEAN

Now that we have an understanding of the t distribution, the only thing that remains is to see how the confidence interval for μ changes for small samples. Replacing the Z table value with a t value in the formulas for the upper and lower limits of the confidence interval for μ gives us the following:

Error:	$e = t_{\alpha/2,n-1}s/\sqrt{n}$
Width of the interval:	$w = 2e$
Lower bound:	$L = \overline{X} - e$
Upper bound:	$U = \overline{X} + e$

CONFIDENCE INTERVAL FOR μ, SMALL SAMPLE, AND σ UNKNOWN

In addition, we must now assume that the underlying population of the variable we are estimating is normally distributed.

9.10.1 Using the t Table

The procedure for figuring out the correct t value for any confidence level is similar to that used to find the correct Z value. We need to find the t value so that there is a total probability of α in the tails of the distribution. Since the t distribution is symmetric, this means there will be $\alpha/2$ in each of the tail areas. Unlike with the Z table, there is no need to work the table backwards. Typically the t table is used to find the t value that corresponds to a certain probability in one of the tails of the distribution rather than to find probabilities under the t distribution. A portion of the t table is shown in Figure 9.12.

The complete t table is found in Appendix Table A.4. Notice that the table has several columns and several rows. The columns correspond to the tail area probability and the rows correspond to the number of degrees of freedom, which is one less than the sample size. The procedure for using the table is as follows:

Step 1: Take the value of α and divide it by 2.

Step 2: Use that column of the t table.

Step 3: Find the number of degrees of freedom by calculating $n - 1$. Use that row of the t table.

Step 4: Read off the corresponding t value at the intersection of the row and column you have identified. Label this value $t_{\alpha/2,n-1}$ to indicate that the tail area probability is $\alpha/2$ and there are $n - 1$ degrees of freedom.

STEPS FOR FINDING THE VALUE OF t FOR A CONFIDENCE INTERVAL

Let's look at an example to see how to use the t table.

Upper-Tail Areas

Degrees of Freedom	0.25	0.1	0.05	0.025	0.01	0.005
20	0.687	1.325	1.725	2.086	2.528	2.845
21	0.686	1.323	1.721	2.080	2.518	2.831
22	0.686	1.321	1.717	2.074	2.508	2.819
23	0.685	1.320	1.714	2.069	2.500	2.807
24	0.685	1.318	1.711	2.064	2.492	2.797
25	0.684	1.316	1.708	2.060	2.485	2.787
26	0.684	1.315	1.706	2.056	2.479	2.779

FIGURE 9.12 A portion of the t table

EXAMPLE 9.13

Using the t Table

Find the *t* value for a 95% confidence interval for a sample size of 25:

1. Since a 95% confidence interval is required, α is 0.05 and so $\alpha/2$ is 0.025.
2. Use the 0.025 column of the *t* table.
3. Since the sample size is $n = 25$, use the row corresponding to $n - 1 = 25 - 1 = 24$ degrees of freedom.
4. The *t* value is 2.064 in the table shown in Figure 9.12. Label it $t_{0.025,24}$. ■

Now let's use this *t* value to find a confidence interval for μ in the following example.

EXAMPLE 9.14   Assembly Time

Small-Sample Confidence Interval for μ

In pricing a new product, the manufacturer needs to determine the average number of minutes it takes to manufacture one item. A sample of 25 items is tracked and the number of minutes to make each unit is recorded below. The assembly time of a unit is known to have a normal distribution.

Time to Assemble Item

5.2	4.4	5.0	5.2	4.7	4.6	4.8	4.8	4.0	4.5
5.7	4.8	5.2	4.4	4.3	4.5	4.0	4.6	4.9	5.8
5.2	6.0	5.6	5.7	4.5					

To find the confidence interval for μ, we need to calculate $\overline{X}$ and s and find the appropriate *t* value from the table shown in Figure 9.12.

$$\overline{X} = 4.90 \text{ min}$$
$$s = 0.55 \text{ min}$$

The *t* value with 24 degrees of freedom and a tail area of 0.025 is $t_{0.025,24} = 2.064$.

The upper and lower 95% confidence limits for μ are calculated using the following formulas:

Error:	$e = t_{\alpha/2,n-1} \, s/\sqrt{n} = (2.064)(0.55)/\sqrt{25} = 0.23$
Width of the interval:	$w = 2e = (2)(0.23) = 0.46$
Lower bound:	$L = \overline{X} - e = 4.90 - 0.23 = 4.67 \text{ min}$
Upper bound:	$U = \overline{X} + e = 4.90 + 0.23 = 5.13 \text{ min}$

Notice that the *t* value for the 95% confidence interval with 24 degrees of freedom is 2.064. If we had known the value of σ, then we would have used a *Z* value of 1.96 instead of the *t* value. This would have made the value of *e* smaller, making the interval narrower. Confidence intervals constructed using *t* values are always wider than the corresponding interval would be if you could have used *Z*. This is a direct result of the increased variability due to having to estimate σ. ■

The *t* distribution is the correct sampling distribution to use whenever σ is unknown. However, when the sample size is large, $n > 30$, it is common practice to use the *Z* distribution instead of the *t* distribution because the graphs are virtually identical. The original reason for using Z instead of t for large samples has all but vanished but the practice continues. Prior to the use of statistical software, analysts had to rely on printed t tables. Since the t distribution depends on the sample size, very large t tables were needed to allow for many different sample sizes. Thus, it was generally agreed that Z could be used for sample sizes greater than 30, eliminating the need for t tables beyond 30 degrees of freedom. Software packages have eliminated the need for printed tables. Nevertheless, the convention of using Z instead of t for large samples has become entrenched, and in this book, when the sample size is larger than 30 we will use the Z distribution. Note that you would certainly be correct to use t whenever the standard deviation is unknown and that may be the approach taken by some software packages.

9.10.2 Exercises—Learning It!

9.20 The police department is concerned about the ability of officers to identify drunk drivers on the road. Before instituting a new training program they take a sample of 28 arrests and record the level of alcohol in the blood at the time of the arrest. Assume that the level of alcohol in the blood is normally distributed. The data are shown below.

Alcohol Level

92	204
93	182
108	173
173	105
194	153
133	150
207	180
127	209
256	141
184	151
253	133
159	147
101	209
133	252

(a) Find a 90% confidence interval for the average alcohol level in the blood at the time of arrest.

(b) Find a 95% confidence interval for the average alcohol level in the blood at the time of arrest.

9.21 A large amusement park has recently added 5 new rides, including a large roller coaster called the Mind Eraser. Management is concerned about the waiting times on the new roller coaster. A random sample of 10 people is selected and the time (in minutes) that each person waits to ride the Mind Eraser is recorded and shown below:

Mind Eraser

43	66
80	54
48	72
61	58
74	68

(a) Find a 95% confidence interval for the average waiting time for the Mind Eraser, assuming that the waiting time is normally distributed.

(b) The park management thinks that if customers have to wait more than 60 minutes for a ride, then the park should increase the staff to reduce the waiting time. Based on your confidence interval, does the park need to increase the staff? Explain why or why not.

9.22 Hospital administrators are paying increasing attention to length of stay of patients. A sample of 14 patient stays (in days) are shown below:

Patient Length of Stay

7	6
2	5
6	4
7	4
8	2
8	3
3	7

(a) Find a 95% confidence interval for the length of patient stay, assuming that the length of stay is normally distributed.

(b) Find a 90% confidence interval for the length of patient stay, assuming that the length of stay is normally distributed.

(c) Which of these intervals should the administrator use in negotiating with insurance companies and why?

9.23 The number of workers per vehicle at the top 10 car assembly plants is shown below:

Nissan Smyrna	2.22
Toyota Cambridge	2.35
Honda East Liberty	2.38
Toyota Georgetown #1	2.50
Chrysler Bramalea	2.54
Honda Marysville	2.57
Ford Atlanta	2.63
Ford Chicago	2.66
Chrysler Belvidere	2.68
GM Oshawa #1	2.68

SOURCE: *Manufacturing Engineering*, August 1997

Assuming that the number of workers per vehicle has a normal distribution, find a 95% confidence interval for the average number of workers per vehicle.

9.24 The percentage of foster children adopted into families varies widely from state to state according to an 8 August 1997 article in the *Springfield Union News*. Congress is preparing bills to encourage the process in some states. The number of children adopted in 1996, as a percentage of all children available for adoption that year, is shown for a sample of 14 states:

Alabama	18.46
Alaska	31.31
California	34.78
Kansas	29.67
Kentucky	30.60
Massachusetts	46.27
Minnesota	20.66
New Hampshire	51.61
North Dakota	96.73
Ohio	25.08
Rhode Island	44.83
Texas	28.93
Utah	53.45
Wyoming	50.00

Find a 95% confidence interval for the average adoption rate, assuming that the adoption rates are normally distributed.

9.11 CONFIDENCE INTERVAL FOR QUALITATIVE DATA

Often we are interested in estimating what proportion of the population has a particular characteristic or has a particular opinion. In fact, most of the data printed in newspapers and magazines are the result of surveys and often proportions or percentages are reported. This is particularly true at election times. For example, the candidate for mayor wishes to know what proportion or percentage of voters in the city favor him.

The increasing focus on quality and meeting the needs of the customer has led to increased data collection, much of which is qualitative data. A manufacturing company is clearly interested in the proportion of products manufactured that are defective. All businesses, both manufacturing and service industries, are interested in knowing if their products/services are meeting the needs of the customer. As a result you are often asked to fill out a questionnaire about how you liked the product and/or service you received. The data from questionnaires such as these are often qualitative data.

Remember: Qualitative data describe a particular characteristic of a sample item. They are most often nonnumerical in nature.

If the data we are analyzing are qualitative data, then we are most likely interested in estimating the proportion, π, of population members that have a certain characteristic (one of the categories of the nominal variable). Confidence intervals for the population proportion, π, have the same basic structure as those for μ. Remember that the sample proportion, p, is the best point estimate for π and so we should center the confidence interval at the value of p.

9.11.1 Finding the Confidence Interval for π

To develop the corresponding formulas for the lower and upper bounds of a $100(1 - \alpha)\%$ confidence interval for π, we need to know that the sampling distribution of p is a normal distribution. Again drawing on the properties of the normal distribution, we must extend the interval a certain number of standard errors to get the coverage we desire. The standard error of the point estimator, p, is $\sqrt{\pi(1 - \pi)/n}$. Notice that to calculate the standard error of p you must know the value of π. But you are trying to estimate π, so, clearly, you do not know π. It makes the most sense to use p as an estimate of π in the formula for the standard error.

The point estimate, p, should be at the center of any confidence interval for π.

Combining these pieces of information with the knowledge that we wish to place p at the center of the confidence interval, we arrive at the following formulas for the lower and upper bounds:

Error:
$$e = Z_{\alpha/2}\sqrt{\frac{\pi(1 - \pi)}{n}} \cong Z_{\alpha/2}\sqrt{\frac{p(1 - p)}{n}}$$

Width of the interval: $w = 2e$

Lower bound: $L = p - e$

Upper bound: $U = p + e$

FORMULA FOR CONFIDENCE INTERVAL FOR π

where $Z_{\alpha/2}$ is the Z value that cuts off $\alpha/2$ in the upper tail of the standard normal distribution and $\alpha/2$ in the lower tail of the standard normal distribution.

Let's look at an example.

EXAMPLE 9.15   Beverly Hills, 90210

Confidence Interval for π

Hard times have hit the *Beverly Hills, 90210* crowd. Nearly a third of teens polled by Teenage Research Limited said they have been personally affected by the recession.

Where do teens get their money? A survey of 2000 teens showed that 47% of them get some money from their parents. Find a 95% confidence interval for the proportion of teens who get some money from their parents.

Given information: $p = 0.47$

$n = 2000$

Calculations: Standard error $= \sqrt{\dfrac{p(1-p)}{n}} = \sqrt{\dfrac{0.47(1-0.47)}{2000}} = 0.011$

Error $= e = Z_{\alpha/2}\sqrt{\dfrac{p(1-p)}{n}} = (1.96)(0.011) = 0.022$

Lower bound $= p - e = 0.47 - 0.02 = 0.45$

Upper bound $= p + e = 0.47 + 0.02 = 0.49$

So we can state that we are 95% confident that the percentage of all teens who get money from their parents is between 45 and 49%. ∎

 **TRY IT NOW!**

Retirement Years
Confidence Interval for π

A survey shows that a growing number of Americans are willing to make sacrifices to become home owners despite increasing job and financial worries,. The Federal National Mortgage Association surveyed 1857 Americans and found that 67% would put off retirement for 10 years to own a home.

Find a 90% confidence interval for the proportion of all Americans who would put off retirement for 10 years to own a home.

9.11.2 Exercises—Learning It!

9.25 I asked 100 imaginary friends (only to avoid the time and cost of data collection) the following question: "Do you regularly watch MTV's *Beavis and Butthead*?" Of the 100 friends, 35 of them answered yes.

(a) Calculate a 95% confidence interval for the "viewership" of this show.

(b) MTV is considering canceling the show if less than one-third of the population regularly watches the show. Based on this information, what will MTV do?

9.26 Many companies have been experiencing downsizing. There is some feeling that when rumors of a downsize begin to float through an organization, employees begin to use their sick days at a faster rate than normal for fear of losing them. Fifty employees were surveyed and asked if they would use their sick days freely in the face of a potential downsizing. Of the 50 employees in the sample, 11 said they would.

(a) Find a 95% confidence interval for the proportion of employees who would begin to use their sick days freely when faced with a potential downsizing.

(b) Based on your confidence interval, what are some recommendations you might suggest to managers?

9.27 A poll of 450 registered Massachusetts voters found 46% of the Bay State voters opposed to allowing casino gambling in the state.

(a) Find a 95% confidence interval for the proportion of all Bay State voters opposed to allowing casino gambling in the state.

(b) Is 50% in the confidence interval? If so, what does this tell you? If not, what does this tell you?

(c) What would you recommend to the governor of Massachusetts, who is pushing for expanded gambling?

9.28 Thousands of years ago people hung the yellow blossoms of Saint-John's-wort over their doorways, hoping to ward off evil spirits. Today, German physicians write nearly 3 million prescriptions a year for pills made from extracts of the plant, meant to relieve depression. But doctors on the other side of the Atlantic haven't followed suit. Scientists in the United States say that there is no reliable evidence that the herb can help.

To find out more, researchers at the University of Munich and the University of Texas in San Antonio studied 1500 people with mild to moderate depression. One-third were treated with Saint-John's-wort, one-third were treated with antidepressant drugs, and one-third were treated with a placebo. The researchers found that after 6 weeks of treatment 64% of the people taking Saint-John's-wort pills felt markedly better, compared to 59% of those receiving synthetic antidepressant drugs.

(a) Construct a 95% confidence interval for the proportion of people who would feel markedly better using Saint-John's-wort pills.

(b) Construct a 95% confidence interval for the proportion of people who would feel markedly better using synthetic antidepressant drugs.

(c) Compare the two confidence intervals. Do they overlap at all? What is your recommendation to doctors?

9.29 In designing a new dormitory, a progressive university wishes to determine where students prefer to study in order to provide appropriate space. A survey of 100 randomly selected undergraduate students shows that 33% prefer to study in their room.

(a) Construct a 95% confidence interval for the proportion of students who prefer to study in their room.

(b) If the university has the option of making various sized rooms, what proportion of the rooms should be made larger to accommodate those who wish to study in their room?

9.30 A medium-size city hospital is concerned about the number of ventilator-acquired respiratory infections in each of the past three years. The random sample of infections were studied and the number of infections that were vent-related infections is shown in the table below:

Year	Number of Infections	Number of Vent-Related Infections
1995	13	10
1996	19	13
1997	14	11

(a) Find the sample proportion of vent-related infections for each of the three years.

(b) Construct a confidence interval for the proportion of vent-related infections for each of the three years.

(c) Should the hospital be concerned that the proportion of vent-related infections is increasing? Justify your answer using the confidence intervals found in part (b).

9.12 SAMPLE SIZE CALCULATIONS

Confidence intervals are easy to calculate and are commonly used to provide interval estimates for either the population mean or the population proportion. In fact, often a newspaper article will report the sample mean or sample proportion and a number called the sampling error or margin of error.

EXAMPLE 9.16   Political Polls

Illustration of Sampling Error

According to an article written by the Associated Press and published in the *Springfield Union News* on 23 October 1996, a CNN–USA Today–Gallup tracking poll put the national polling gap at 23 points, with Clinton favored by 55%, Dole by 32%, and Perot by 8%. The survey of 754 likely voters had a 4-point margin of error.   ■

This sampling error or margin of error is what we have labeled error. Thus, if you wanted to construct the confidence interval from such a report you simply need to add and subtract the sampling error to the point estimate.

EXAMPLE 9.17   Political Polls

Constructing the Confidence Interval

Continuing with the results of the CNN–USA Today–Gallup tracking poll, the proportion of voters who favor Clinton was 55% with a 4% margin of error. Thus, the percentage is between 51 and 59%. The article in the newspaper did not report the confidence level, but typically 95% confidence intervals are calculated.   ■

9.12.1 Understanding the Error

Sampling error is due to the fact that only a piece of the population has been studied.

We have not really examined why this distance between the center of the confidence interval and either endpoint is called an *error*. As we have seen many times before, the term sampling error does not imply that you made an error. It does indicate that you have imperfect information about the population, since you studied only a sample of that population. There is another reason why this value is called an error. Consider the confidence interval calculated in Example 9.12 for the diaper manufacturer. In that example we found a 95% confidence interval for the mean diaper weight in grams. It is shown below:

54.78   54.98   55.18

Remember that the value of 54.98 is the sample mean, $\overline{X}$. Suppose for the moment that we have found an interval that does contain μ but you don't know where in the interval μ is located. If you used $\overline{X}$ to estimate μ, what is the largest amount by which you could miss the value of μ? This *worst-case scenario* happens if μ is at either endpoint of the interval. If μ is at either of the endpoints, then our point estimate $\overline{X}$ is in error by the amount of $e = 0.20$ g. If μ is anywhere else in the interval, then our point estimate $\overline{X}$ is off by some amount less than 0.20 g. So we are 95% confident that the error is at most 0.20 g. Remember that a 95% confidence level tells you that

19 out of 20 intervals of this width contain the true population parameter. This is why we have labeled the distance from the middle of the interval to either endpoint as e standing for error. It is precisely this value that is often quoted in newspaper or research articles, although the newspaper rarely tells you what the corresponding level of confidence is. Typically, a 95% level of confidence has been used.

You might wonder if it is possible to specify the size of this error to achieve a certain level of accuracy. This is indeed often desirable. By specifying the amount of error that you can tolerate in a particular situation you are determining the sample size needed. Often you cannot achieve a particular level of accuracy because to do so would require a sample size that you cannot afford. This is the subject of the next subsection.

9.12.2 Determining the Sample Size

In Chapter 2, we identified the following factors that are important in determining the size of the sample needed:

- The amount of variation in the population
- The amount of error that can be tolerated
- The extent of resources available
- The size of the population

We now know enough statistics to develop a formula for the sample size that incorporates the first two of these factors. You should remember that these factors were identified as most important to the sample size determination.

Recall that the expression for e is one of the following, depending on whether you are estimating the mean, μ, or the proportion, π:

$$\text{Estimating } \mu: \qquad e = Z_{\alpha/2}\,\sigma/\sqrt{n}$$

$$\text{Estimating } \pi: \qquad e = Z_{\alpha/2}\sqrt{\frac{p(1-p)}{n}}$$

Each of these equations can be rewritten and solved for n, the sample size, by using some basic algebra. When this is done you get two equations for n that are algebraically equivalent to the two equations shown above:

$$\text{Estimating } \mu: \qquad n = \frac{Z_{\alpha/2}^2\,\sigma^2}{e^2}$$

$$\text{Estimating } \pi: \qquad n = \frac{Z_{\alpha/2}^2\,p(1-p)}{e^2}$$

Let's examine the equation for n if we are trying to estimate the population mean, μ. If we specify a certain level of confidence and a value for the maximum difference between $\overline{X}$ and μ, then we can calculate how large a sample we must select. Let's see how this formula could be used by the diaper manufacturer.

EXAMPLE 9.18   The Diaper Company

Sample Size Determination

The diaper manufacturer wants to be 95% confident that the estimate $\overline{X}$ is not in error by more than 0.10 g. That is, the diaper manufacturer wants to be sure that 19 out of 20 possible confidence intervals of width 0.20 g contain the true mean, μ. Recall that the manufacturing process has a standard deviation of 0.55 g. Thus, we have

$$e = 0.10 \text{ g}$$
$$Z_{\alpha/2} = 1.96$$
$$\sigma = 0.55 \text{ g}$$

Using this information to calculate n yields

$$n = \frac{Z_{\alpha/2}^2 \sigma^2}{e^2} = \frac{(1.96^2)(0.55^2)}{0.10^2} = 116.2$$

Since we can't sample a fractional diaper, we must round this value up to 117 diapers. ■

Always round up in sample size calculations.

For sample size calculations you should always round up to guarantee the level of confidence and error you have specified.

 **TRY IT NOW!**

Bottle Filling

Finding the sample size

How many bottles does the bottle manufacturer need to sample to be 98% confident that the error is at most 0.05 oz? Remember that the population standard deviation is 0.1 oz.

Ways to handle situations in which the standard deviation is unknown

Notice that we needed a value for the population standard deviation to calculate n. If this is unknown then we must estimate it. Unfortunately, often we need to take a sample to estimate σ but we need σ to figure out how large our sample needs to be. This is a classic "catch 22" situation. There appears to be no way out.

Fortunately, there are a couple of ways to handle this dilemma. First of all, we could take a small sample to get a rough estimate of the standard deviation and use this in the formula for n. Suppose that we took a sample of size 10 and we used this sample to calculate the sample standard deviation, s. This value of s can then be used as an estimate for σ and plugged into the formula for n. Suppose when we do this the formula tells us that we need a sample of size $n = 55$. Well, we already have data on 10 observations so we need an additional 45 to complete the sample. Another way is to use information about the variability of a similar product or process. This may not be perfect but it will give a rough idea of the value of σ which can, in turn, be used in the sample size calculation.

For you to be more confident, the sample size must be larger.

Very often the sample size determined from the formula cannot be collected because of limited resources (i.e., cost). If this happens then you must be willing to accept a lower level of confidence or tolerate a higher value of e or both. Let's examine the impact on n of varying the level of confidence. Before we do the calculations, let's see what we might anticipate. We know that as the level of confidence increases, the Z value also increases. Since Z is in the numerator of the formula for n, we would expect the sample size needed to increase as the level of confidence increases. This should make intuitive sense as well. For us to be more confident, the sample size must be larger.

EXAMPLE 9.19 The Diaper Company

Sample Sizes for Various Increasing Levels of Confidence
Require Increasing Sample Sizes

Suppose the diaper manufacturer considers several different levels of confidence. The sample sizes needed to achieve an error of at most 0.10 g for various levels of confidence are calculated below:

90% confidence level: $\quad n = \dfrac{Z_{\alpha/2}^2 \, \sigma^2}{e^2} = \dfrac{(1.645^2)(0.55^2)}{0.10^2} = 81.9$ rounded up to 82

95% confidence level: $\quad n = \dfrac{Z_{\alpha/2}^2 \, \sigma^2}{e^2} = \dfrac{(1.96^2)(0.55^2)}{0.10^2} = 116.2$ rounded up to 117

98% confidence level: $\quad n = \dfrac{Z_{\alpha/2}^2 \, \sigma^2}{e^2} = \dfrac{(2.33^2)(0.55^2)}{0.10^2} = 164.2$ rounded up to 165

99% confidence level: $\quad n = \dfrac{Z_{\alpha/2}^2 \, \sigma^2}{e^2} = \dfrac{(2.58^2)(0.55^2)}{0.10^2} = 201.4$ rounded up to 202 ■

If the sample size you calculate is too expensive, the other way you can cut costs is to increase the size of the error you can tolerate. Let's see what happens to the sample size as the maximum error is increased. Again, a quick look at the formula for n shows us that the value for e is in the denominator. So by increasing e we will be dividing by a larger number and hence the sample size needed will be smaller. Intuitively, this, too, makes sense. If you can tolerate a larger error, then you can take a smaller sample. Let's look at the diaper manufacturer from this perspective.

As the maximum error increases, the sample size decreases.

EXAMPLE 9.20 The Diaper Company

Impact of Larger Errors

Consider the following values for the maximum error that the diaper company is willing to tolerate in its estimate of μ. In all cases they wish to be 95% confident that the error is at most the specified value.

Error = 0.10 g $\quad n = \dfrac{Z_{\alpha/2}^2 \, \sigma^2}{e^2} = \dfrac{(1.96^2)(0.55^2)}{0.10^2} = 116.2$ rounded up to 117

Error = 0.15 g $\quad n = \dfrac{Z_{\alpha/2}^2 \, \sigma^2}{e^2} = \dfrac{(1.96^2)(0.55^2)}{0.15^2} = 51.6$ rounded up to 52

Error = 0.20 g $\quad n = \dfrac{Z_{\alpha/2}^2 \, \sigma^2}{e^2} = \dfrac{(1.96^2)(0.55^2)}{0.20^2} = 29.05$ rounded up to 30

Error = 0.25 g $\quad n = \dfrac{Z_{\alpha/2}^2 \, \sigma^2}{e^2} = \dfrac{(1.96^2)(0.55^2)}{0.25^2} = 18.6$ rounded up to 19 ■

Calculations of the sample size for situations when you wish to estimate the population proportion, π, are done in a similar manner. The only difference is the particular formula to be used. Consider the teens who are receiving money from their parents. We looked at this situation in Example 9.15.

EXAMPLE 9.21   *Beverly Hills, 90210*

Calculation of Sample Size for Proportions

How many teens need to be sampled to be 95% confident that our estimate of π is off by at most 5%? Remember that the point estimate is p, the sample proportion or percentage. If we want the sample proportion to be in error by at most 5%, then $e = 0.05$.

The formula for n is

$$n = \frac{Z_{\alpha/2}^2\, p(1 - p)}{e^2}$$

Plugging in the values we know gives us

$$n = \frac{(1.96^2)\, p(1 - p)}{0.05^2}$$

At this point we realize that we can't continue without a value for p but we need a sample to get a sample proportion. ∎

This is again a circular problem. We need p to find n but we need n to get p! There are two approaches you can take here. First of all, if you have any information about the value of p from previous samples or experience, then you should use that information as an estimate of p. If you have no information at all, then you should use a value of $p = 0.50$ in the formula for the sample size calculation. When you do this the sample size that is calculated is as large as it can be for the specified confidence level and error. It is the most conservative approach you can take and will often lead to a sample size larger than you really need. Let's finish Example 9.21 by using a value of $p = 0.50$.

EXAMPLE 9.22   *Beverly Hills, 90210 (cont'd)*

Sample Size Calculation for Proportion

Using a value of $p = 0.50$ in the formula for n gives us

$$n = \frac{(1.96^2)\, p(1 - p)}{0.05^2} = \frac{(1.96^2)(0.5)(1 - 0.5)}{0.05^2} = 384.16$$

which is rounded up to 385 teenagers. ∎

 **TRY IT NOW!**

Retirement Years

Sample Size Calculation for π

How many Americans must be sampled to determine the percentage who would put off retirement for 10 years to own a home? The estimate should not differ from the actual population proportion by more than 3% with a confidence of 90%.

The conclusions that we reached about what happens to the required sample size as the confidence level and the error vary are the same for proportions as for means. As the confidence level increases, the sample size needed to estimate the population proportion also increases. As the error that can be tolerated increases, the sample size needed decreases.

9.12.3 Exercises—Learning It!

9.31 How many stores must be sampled for the woman who wants to buy a ranch to be 95% confident that the error in estimating the average fat content per pound in steaks sold in the Portland, Maine, area is at most 0.05 oz?   *Requires Exercise 9.13*

9.32 How many months must be sampled for analysts to be 99% confident that the error in estimating the average monthly price per kwh of electricity is at most $.02?   *Requires Exercise 9.15*

9.33 How many days must be observed for the grocery store chain to be 90% confident that the error in estimating the average number of vehicles passing a certain location is at most 10 vehicles?   *Requires Exercise 9.16*

9.34 How many viewers must be surveyed to be 98% confident that the estimate of the viewership is in error by at most 3%?   *Requires Exercise 9.25*

9.35 How many employees must be sampled to be 95% confident that the error in estimating the proportion of employees who would begin to use sick days at a faster rate when rumors of downsizing are circulating is at most 0.02?   *Requires Exercise 9.26*

9.13   USING EXCEL TO FIND CONFIDENCE INTERVALS

Excel does not have a tool that automatically calculates confidence intervals. To do this, you will have to use some of the functions you already learned about in a formula. In this section, we will explain how to do so for a large-sample confidence interval for the mean. We will also explain how to use the Excel functions for the t distribution.

9.13.1 Large-Sample Confidence Intervals for the Mean

To find a confidence interval for a population parameter, you need a point estimate and a formula for the error of the estimate. For the population mean, when σ is known, or when n is large ($n > 30$), the formula for the confidence interval is

$$\overline{X} \pm Z_{\alpha/2} \frac{\sigma}{\sqrt{n}}$$

To find a confidence interval for the mean using Excel, you must have a point estimate, $\overline{X}$. You can obtain this from the Descriptive Statistics in the Data Analysis Tools or by using the AVERAGE function. You will also need a value for σ, or else a sample estimate, s, based on a large sample. The last thing that you need is the Z value, which you will obtain using the NORMSINV function discussed in Chapter 7.

Figure 9.13 shows a portion of an Excel worksheet that contains a sample of 50 diaper weights, and a set of summary statistics for the data.

To calculate a 95% confidence interval for the true weight of a diaper, follow these steps:

1. Position the cursor in the cell that you want to contain the upper limit of the confidence interval. In this case, we will use E1, which is just to the right of the summary data.

	A	B	C	D
1	*Weight*		*Weight*	
2	55.87			
3	55.35		Mean	54.993
4	54.50		Standard Error	0.096962964
5	53.97		Median	54.915
6	54.29		Mode	54.4
7	54.85		Standard Deviation	0.685631691
8	54.79		Sample Variance	0.470090816
9	54.65		Kurtosis	-0.848226678
10	55.56		Skewness	0.078374807
11	55.82		Range	2.84
12	54.40		Minimum	53.6
13	56.44		Maximum	56.44
14	54.11		Sum	2749.65
15	54.67		Count	50
16	54.56			
17	54.69			

FIGURE 9.13   Diaper data and summary statistics

2. Since you want to let Excel know that you are entering a formula, start the formula by typing an equal sign "=". All formulas in Excel must start with an equal sign.

What you want to calculate is $\overline{X} + Z_{.025}\dfrac{s}{\sqrt{n}}$. We will use the sample standard deviation because our sample size is large enough.

3. After you type the equal sign, position the cursor on the cell that contains the value of $\overline{X}$, in this case D3. If you look at the status line or the cell itself, you see that the cell D3 is entered in the formula. Since in our confidence interval we want to use that *specific* cell all the time for the sample mean, you need to change the cell reference to an absolute reference.

4. With the cursor still in the cell, press F4. The cell reference D3 will change to $D$3. The $ tells Excel to look *always* in column D, *always* in row 3 for this value. This is important to know when you copy formulas.

Remember: The NORM-SINV function returns the Z value for the lower tail probability.

5. Now, we want to add to $\overline{X}$ the quantity $Z_{.025}\dfrac{s}{\sqrt{n}}$. Type in + and then type NORMSINV(0.975).

Remember that the quantity $\dfrac{s}{\sqrt{n}}$ is referred to as the standard error of the mean. When you look at the summary statistics that Excel calculates, you see that the second value is indeed the standard error of the mean. This was not meaningful to us when we first looked at numerical descriptors, so we did not talk about it then. Now we will use it to finish our confidence interval.

6. Type in a multiplication sign, "*", and position the cursor in the cell that contains the standard error, in this case, D4. Again, press F4 to make the cell an absolute reference.

7. The finished formula is shown in Figure 9.14.

FIGURE 9.14   Formula for upper limit of confidence interval

8. Hit Enter and the result, 55.18304, will be displayed.

You can calculate the lower confidence limit by typing the same equation in the cell next to the upper limit. Be sure to change the Z value to a negative by using **NORMSINV(0.025).** To avoid retyping the whole equation, you can simply copy the one you just did and paste it into a new cell. Then, move to the display of the equation in the status bar and edit the **NORMSINV** parameter.

When you calculated the summary statistics, you might have noticed that there was an option available for Confidence Level for the Mean. Checking that box will add a new line to the end of the summary statistics output, which is supposed to be $Z_{.025} \dfrac{s}{\sqrt{n}}$. We have not used this option because the calculation is wrong (not drastically, but it is definitely not correct).

9.13.2 The *t* Distribution in Excel

The difference between large- and small-sample confidence intervals for the mean is in the sampling distribution of $\overline{X}$. You know that when σ is unknown and the sample size is not greater than 30, the Student-t distribution applies. The formula for the confidence interval is

$$\overline{X} \pm t_{\alpha/2, n-1} \frac{s}{\sqrt{n}}$$

To calculate small-sample confidence intervals for the population mean with Excel, you will use the **TINV** function to get the appropriate t values.

Just like other functions in Excel, the **TINV** function needs some input parameters. The format of the **TINV** function is **TINV(probability,deg_freedom).** The value that you input for **probability** is the value of α, the total tail area probability. The value for **deg_freedom** is $n - 1$, the degrees of freedom. As an example, for a 95% confidence interval for μ, based on a sample of size 25, the function would look like **TINV(0.05,24)** and would return the value 2.063898. There is no lower t value in Excel. To obtain the lower confidence limit, you simply subtract the quantity $t_{\alpha/2, n-1} \dfrac{s}{\sqrt{n}}$ from $\overline{X}$.

CHAPTER 9 SUMMARY

This chapter has covered the basics of estimating population parameters. In particular, you learned how to estimate the average of a numeric characteristic of a population, μ, and the proportion of a population that has a certain characteristic, π. The estimates are calculated from a sample selected from the population. Each sample would therefore yield a slightly different estimate of the population parameter. Thus, the estimators are themselves random variables. Just like the random variables you studied in Chapter 7, the estimators have a distribution. When the random variable is an estimator, the distribution is called a sampling distribution. The sampling distribution has a mean and a standard deviation, which is called the standard error.

You have learned how to use the sampling distribution of $\overline{X}$ to calculate probabilities and make inferences about μ. You have also learned how to create confidence intervals for μ and for π. Finally, you have learned how to calculate the required sample size to achieve a certain level of precision with a specified confidence.

Key Terms

Key Term	Definition	Page Reference
Point estimate	A **point estimate** is a single number calculated from sample data. It is used to estimate a parameter of the population.	336
Point estimator	A **point estimator** is the formula or rule that is used to calculate the point estimate for a particular set of data.	336
Unbiased estimator	An **unbiased estimator** yields an estimate that is fair. It neither systematically overestimates the parameter nor systematically underestimates the parameter.	343
Standard error	The **standard error** is the standard deviation of a point estimator. It measures how much the point estimator or sample statistic varies from sample to sample.	347
Sampling distribution	The distribution of a point estimator or a sample statistic is called a **sampling distribution.**	347
Central Limit Theorem	The **Central Limit Theorem** states that in random sampling from a population with mean μ and standard deviation σ, when n is large enough, the distribution of $\overline{X}$ is approximately normal with a mean, $\mu_{\overline{X}}$, equal to μ and a standard deviation, $\sigma_{\overline{X}}$, equal to $\sigma/\sqrt{n}$.	348
Confidence interval	A **confidence interval** or an **interval estimate** is a range of values with an associated probability or confidence level, $1 - \alpha$. The probability quantifies the chance that the interval contains the true population parameter.	363

Key Formulas

Description	Formula	Page Reference
Z-score for $\overline{X}$	$Z = \dfrac{\overline{X} - \mu_{\overline{X}}}{\sigma_{\overline{X}}}$	346
Mean of $\overline{X}$	$\mu_{\overline{X}} = \mu$	348
Standard error of $\overline{X}$	$\sigma_{\overline{X}} = \sigma/\sqrt{n}$	348
Confidence interval for μ when σ is known	Error: $e = \dfrac{Z_{\alpha/2}\sigma}{\sqrt{n}}$ Width of the interval: $w = 2e$ Lower bound: $L = \overline{X} - e$ Upper bound: $U = \overline{X} + e$	366
Large-sample confidence interval for μ when σ is unknown	Error: $e = \dfrac{Z_{\alpha/2}s}{\sqrt{n}}$ Width of the interval: $w = 2e$ Lower bound: $L = \overline{X} - e$ Upper bound: $U = \overline{X} + e$	368
Small-sample confidence interval for μ when σ is unknown, normal population	Error: $e = \dfrac{t_{\alpha/2,n-1}s}{\sqrt{n}}$ Width of the interval: $w = 2e$ Lower bound: $L = \overline{X} - e$ Upper bound: $U = \overline{X} + e$	381
Confidence interval for π	Error: $e = Z_{\alpha/2}\sqrt{\dfrac{\pi(1-\pi)}{n}} \cong Z_{\alpha/2}\sqrt{\dfrac{p(1-p)}{n}}$ Width of the interval: $w = 2e$ Lower bound: $L = p - e$ Upper bound: $U = p + e$	385
Sample size for estimating μ	$n = \dfrac{Z_{\alpha/2}^{2}\,\sigma^{2}}{e^{2}}$	389
Sample size for estimating π	$n = \dfrac{Z_{\alpha/2}^{2}\,p(1-p)}{e^{2}}$	389

CHAPTER 9 EXERCISES

Learning It!

9.36 The manufacturer of 15-oz breakfast cereal boxes samples and weighs 5 boxes every hour. The observations for seven hours are shown below: *Requires Example 9.11*

14.91	15.23	15.34	15.40	15.48	15.59	15.62
14.96	15.24	15.34	15.42	15.49	15.60	15.62
15.11	15.28	15.36	15.45	15.50	15.60	15.63
15.21	15.30	15.38	15.46	15.52	15.61	15.64
15.22	15.30	15.39	15.48	15.58	15.61	15.67

(a) Find the upper and lower confidence limits for the average weight of the breakfast cereals. Use $\alpha = 0.05$. Is the value of 15.00 oz contained in the interval?

(b) What does that tell you about the filling process?

9.37 A company that buys blank VHS tapes for video recording is concerned about the actual amount of time that the tapes are able to record. The tapes are rated at 120 minutes, but   *Requires Exercise 9.5* the company knows there is variation in the actual time that the tapes can be used. The company collects data on 25 randomly selected tapes and finds the recording times listed below:

116	118	119	119	120
117	118	119	119	121
117	118	119	120	121
117	119	119	120	121
117	119	119	120	121

(a) Find the average and the standard deviation of these data.

(b) Find a 95% confidence interval for the average recording time, assuming that the recording times are normally distributed.

(c) What do you conclude about the claim that the tapes have an average of 120 minutes of usable recording time? Use the confidence interval to support your conclusion.

9.38 A cross-country runner is evaluating his times against a team average of 18.5 minutes. The standard deviation of his times was 0.25 minute. This season the runner's average is 17.0 minutes for 10 races.

(a) Assuming that race times are normally distributed, calculate the Z-score for the runner's average time.

(b) Using the Z-score you found in part (a), decide if this runner is unusually fast compared to the team.

9.39 An insurance company is concerned about the amount of time it takes for operators who are responding to calls to the 800 number to access the database. It calls this time the response time. The Management Information Systems (MIS) department claims that the average response time is 10 seconds. A sample is taken and 40 response times are recorded. The results of generating descriptive statistics in Excel are shown below:

SUMMARY STATISTICS
FOR RESPONSE TIME

Mean	9.95656397
Standard Error	0.02942547
Median	9.9616274
Mode	#N/A
Standard Deviation	0.18610301
Sample Variance	0.03463433
Kurtosis	-0.29568133
Skewness	-0.31707278
Range	0.79489018
Minimum	9.49771882
Maximum	10.292609
Sum	398.262559
Count	40

(a) What are the average and standard deviation of the response times?

(b) Calculate the standard error of $\overline{X}$ using the standard deviation and n.

(c) Do you see the number calculated in part (b) anywhere in the output?

(d) Calculate the Z-score for the average response time.

(e) Based on the Z-score, do you think the MIS department's claim is correct?

9.40 A company concerned about the health of its employees has offered the employees free membership in a local health club. One year after this benefit was adopted, a survey of 50 employees was done to determine the average number of hours per week they exercise. The 50 employees surveyed exercised an average of 3.5 hours a week with a standard deviation of 0.5 hour. The national average number of hours of exercise per week is 3 hours a week.

(a) Find the Z-score for the average of 3.5 hours.

(b) Display the sampling distribution of $\overline{X}$.

(c) Based on this Z-score, do you think the employees of this company exercise more than the national average?

9.41 The Bureau of Labor Statistics publishes a great deal of data on a monthly basis. The following data on the average price of bread were extracted from the World Wide Web.

[Bureau of Labor Statistics Data]

Data extracted on: January 9, 1997

Average	Price Data
Series	Catalog:
Series	ID APU0000702111
Area:	U.S. City Average
Item:	Bread, White, pan (cost per pound/453 grams)

1994		**1995**		**1996**	
Jan	0.768	Jan	0.767	Jan	0.860
Feb	0.756	Feb	0.767	Feb	0.858
Mar	0.755	Mar	0.775	Mar	0.852
Apr	0.765	Apr	0.776	Apr	0.865
May	0.765	May	0.768	May	0.869
Jun	0.761	Jun	0.781	Jun	0.886
Jul	0.759	Jul	0.789	Jul	0.889
Aug	0.753	Aug	0.797	Aug	0.915
Sep	0.776	Sep	0.808	Sep	0.886
Oct	0.756	Oct	0.809	Oct	0.873
Nov	0.766	Nov	0.821	Nov	0.880
Dec	0.748	Dec	0.837	Dec	0.883

Treating these three years of data as a sample, find a 90% confidence interval for the average price of bread. Assume that the standard deviation is $0.110 per pound.

9.42 In an effort to improve the quality of the CD players that your company makes, you have started to sample the component parts that you purchase from an outside supplier. You will accept the shipment of parts only if there is less than 1% defectives in the shipment. Recognizing that you cannot test the entire shipment (or population), you select a sample of 25 components to test. You find 3 defectives in the sample of 25.

(a) Find a 90% confidence interval for the proportion of components in the population that are defective.

(b) Based on your confidence interval, should you accept the shipment? Why or why not?

9.43 A nationwide survey of practicing physicians will be taken to estimate μ, the true mean number of prescriptions written per day. The desired margin of sampling error is 0.75. A pilot study revealed that a reasonable planning value for the population standard deviation is 5.

(a) If the desired level of confidence was 99%, how many physicians should be contacted in the survey to estimate μ?

(b) If the desired confidence level were lowered to 95%, what would be the sample size?

(c) Discuss the results of parts (a) and (b).

9.44 A hotel is studying the proportions of rooms that are not ready when customers check in to the hotel.

(a) How many rooms must be in the sample for the hotel to be 95% confident that the margin of error is at most 1%?

(b) How many rooms must be in the sample for the hotel to be 95% confident that the margin of error is at most 3%?

Thinking About It!

9.45 Suppose you are studying the length of stay at a hospital and the population had a mean of 8 days with a standard deviation of 2.5 days. The length of hospital stay has a normal distribution.

(a) Display the population distribution of hospital stay.

(b) If you observed the hospital stay of 20 patients, find the standard error of $\overline{X}$. Display the sampling distribution of $\overline{X}$ on the same graph as part (a) using a different color pen or pencil.

(c) If you observed the hospital stay of 30 patients, find the standard error of $\overline{X}$. Display the sampling distribution of $\overline{X}$ on the same graph as (a) and (b) using a different color.

(d) If you observed the hospital stay of 80 patients, find the standard error of $\overline{X}$. Display the sampling distribution of $\overline{X}$ on the same graph as (a) and (b) using a different color.

(e) What happens to the sampling distribution of $\overline{X}$ as the sample size increases? What does this tell you about the accuracy of $\overline{X}$ as an estimator for μ?

9.46 Suppose you are studying the salaries of 3 different populations that all have the same average salary: $\mu = \$25,000$. The 3 populations have different standard deviations: accounting major, $\sigma = 1000$; computer information systems major, $\sigma = 2000$; and marketing major, $\sigma = 3000$. You select a sample of 30 from each population.

(a) Find the standard error of $\overline{X}$ for each of the three groups.

(b) What happens to the standard error as the population variability increases?

(c) What does this tell you about the accuracy of $\overline{X}$ as an estimator for μ?

Requires Exercise 9.14   **9.47** The 43 executives listed in Exercise 9.14 all work for firms that are headquartered or have significant operations in western Massachusetts.

(a) How does this affect the usefulness of the confidence interval for μ?

(b) These firms can be divided into 2 groups: manufacturing (M) and nonmanufacturing (NM). Compute a 90% confidence interval for the mean executive salary for each of the two groups assuming the salaries are normally distributed.

(c) Do these confidence intervals overlap? If so, what do you think that indicates? If not, what does that tell you?

9.48 According to a *New York Times* article of 31 December 1993, women are no more fully represented in French politics today than they were after universal suffrage was finally introduced 48 years ago. The article provides the percentage of seats held by women in several different countries and is shown below:

Country	% Seats Held by Women in Each Country's Main Legislative Assembly
France	6.1
Italy	8.1
Britain	9.1
U.S. House	11.0
U.S. Senate	7.0
Spain	15.7
Germany	21.6
Netherlands	27.3
Sweden	32.6
Denmark	33.5
Norway	38.0

(a) Leaving France out of the data set, construct a 95% confidence interval for the average percentage of seats held by women in international government.

(b) Does France appear to be atypical? Explain your answer using the confidence interval calculated in part (a).

9.49 The results of a survey conducted to see if a newly introduced product is meeting the customers' needs are available to the manager. The survey results indicate that 52% of those in the survey were very satisfied with the new product. The margin of error was 5%. The product manager argues that more than half of the customers are very satisfied. What is wrong with his argument?

9.50 The state of Hawaii had an adoption rate of 9.6% in 1996. Using the confidence interval you found in Exercise 9.24, what can you conclude about Hawaii's adoption rate in relation to the national average? *Requires Exercise 9.24*

Doing It!

9.51 The data in the file FCOJ.XXX have been collected over the past 11 years of futures trading of Frozen Concentrate Orange Juice (FCOJ). The FCOJ is traded on the New York Cotton Exchange. The unit of trading is 15,000 lb (one contract = 15,000 lb). The price quoted is in cents per pound. The data show average price for the month. The following is a brief description of data presented on each of the four sheets identified by name. *Datafile: FCOJ.XXX*

> *Popdata:* Population data for ALL 139 months (Jun 83–Dec 94).
> *Sample30:* A sample of 30 randomly selected months from the population. Fifty (50) samples are presented on this sheet.
> *Sample40:* A sample of 40 randomly selected months from the population. One sample is presented on this sheet.
> *Sample50:* A sample of 50 randomly selected months from the population. One sample is presented on this sheet.

(a) Using the 50 samples of size $n = 30$ (sheet *sample30*), calculate a 95% confidence interval for each of the 50 samples.

(b) Count how many of the 50 confidence intervals calculated in part (a) actually include the true population mean, μ. Is your count consistent with what you expected? Why or why not?

(c) Calculate a 90, 95, and 99% confidence interval for the single sample on the sheet *sample40*. How does the width of the confidence interval vary with the level of confidence?

(d) Calculate a 95% confidence interval for each of the following: (i) the first sample on the sheet *sample30*, (ii) the single sample on the sheet *sample40* (*note:* you did this one in part (c) above), and (iii) the single sample on the sheet *sample50*. How does the width of the confidence interval vary with the sample size?

9.52 The diaper company has collected data on 5 diapers per hour for a total of 52 hours. Thus, they have 260 rows of data. They have hired you to help them analyze these data. The data are located on the data disk in a file named DIAPER.XXX. A portion of the data is shown below: *Datafile: DIAPER.XXX*

Sample	Diaper	Weight	Bulk
1	1	55.87	0.419
1	2	55.35	0.380
1	3	54.50	0.365

Variable	Description
Sample	Keeps track of the sequence in which the samples were taken and goes from 1 to 52
Diaper	Keeps track of the individual diapers within a sample and goes from 1 to 5
Weight	The diaper weight in grams
Bulk	The diaper bulk in millimeters

In the chapter, the sampling distribution for the average diaper weight (Figure 9.3) was created when each sample had $n = 5$ observations. If you combine data from 2 samples and assume that a sample of size $n = 10$ was taken, you can examine the effects of a change in the sample size on the sampling distribution. Similarly, if you combine the data from 3 samples you can see the impact if the sample size had been $n = 15$. Finally if you combine the data from 4 samples you can see the impact if the sample size had been $n = 20$.

(a) Do this and display the histograms of the average weight. You will have 3 histograms in addition to Figure 9.3.

(b) For each set of average weights, find the mean and standard deviation and fill out the first three columns of the table below:

Sample Size	Average of the $\overline{X}$'s	Standard Deviation of the $\overline{X}$'s	Theoretical Standard Error $(\sigma/\sqrt{n})$
5			
10			
15			
20			

(c) What do you notice about the values for the average of the $\overline{X}$'s as n increases?

(d) What do you notice about the values for the standard deviation of the $\overline{X}$'s as n increases?

(e) How do the standard errors that you found (last column of the table) compare to what the Central Limit Theorem says they should be? Complete the last column in the table above.

Requires Exercise 9.52   **9.53**   Construct a histogram of the variable *Bulk* and calculate a set of summary statistics and a histogram for the variable *Bulk*.

(a) Do the data appear to be normally distributed?

(b) Find the sample average bulk for each of the 52 samples of size $n = 5$. Call these *AverageBulk*.

(c) Calculate summary statistics and make a histogram of these $\overline{X}$'s.

(d) Compare the histogram for the variable *Bulk* to the histogram of the variable *AverageBulk* found in part (b).

Requires Exercise 9.52   **9.54**   The product specifications for the diapers include a variable called *Density*, which is found by dividing the *Weight* of a diaper by its *Bulk*.

(a) Create the variable *Density*.

(b) Calculate a set of summary statistics for *Density* and make a histogram of *Density*.

(c) Does it appear to be normally distributed?

(d) For each sample of size $n = 5$, find the *Average Density*.

HYPOTHESIS TESTING: AN INTRODUCTION

THE TISSUE STRENGTH PROBLEM

A manufacturer of tissues has received numerous complaints about its products. One of the frequent complaints is about the strength of the tissue. As we have noted many times before in this book, all companies are paying increasingly more attention to the customer in an attempt to "totally delight the customer." Clearly, in this case, the customers are not totally delighted with the tissues made by this company. Management has asked the statisticians to investigate the source of this problem by taking a sample.

There are two possible explanations. One possibility is that manufacturing is producing tissues that are not as strong as the specifications. In this case, an adjustment may be needed to the manufacturing process. The second possibility is that despite the fact that manufacturing is making the tissues according to specifications, the customers prefer stronger tissues. In this case, an adjustment to the specifications may be in order.

Your boss has asked you to find out if the tissues are in fact as strong as they are designed to be. There are two measures of tissue strength: machine-directional strength *(MDStrength)* measured in lb/ream and cross-directional strength *(CD-Strength)* also measured in lb/ream. You decide to collect data on 3 different days. A portion of the data file is shown below.

Day	MDStrength	CDStrength
1	1006	422
1	994	448
1	1032	423
1	875	435
1	1043	445
1	962	464
1	973	472
1	1036	489
1	1084	440

The product specifications for *MDStrength* state that the mean should be 1000 lb/ream with standard deviation of 50 lb/ream. *CDStrength* is supposed to have a mean of 450 lb/ream with a standard deviation of 25 lb/ream.

10.1 CHAPTER OBJECTIVES

Remember: *A point estimate is a single number,
calculated from sample
data, which is used to
estimate a population
parameter.*

From Chapter 9 we learned that the sample mean is a good point estimate of the population mean. Suppose we took a sample of tissues and measured the strength of each tissue. We can use the data to calculate a sample mean strength. If that sample mean doesn't agree with the target mean strength, then we should tell the boss that the process needs to be adjusted. Simple, right? Well, not quite.

After a bit more thought, you remember that although the sample mean, $\overline{X}$, is a single number for any particular sample, if you pick a different sample you will probably get a different sample mean. In fact, there are many different possible values you could get for the sample mean, and virtually none of them will actually equal the true mean, μ.

EXAMPLE 10.1 Tissue Strength Problem

Impact of Sampling Error

Suppose that you take a sample of tissues and find the sample average *MDStrength* to be 1010 lb. Clearly, the sample average, $\overline{X} = 1010$, does not exactly equal the specification of 1000. Does this mean that the process is not running correctly? Probably not, but we will need a more precise way of deciding. An hour later you might take another sample and get a sample *MDStrength* of 995 lb. Again, you need to decide if the process is running correctly on the basis of that sample average, $\overline{X} = 995$ lb. Clearly, the sample average will be different each time you take a sample and will almost never be equal to 1000 even if the process is running correctly!

Remember that you are examining only a sample of the population of tissues, and each tissue manufactured has a slightly different MD strength. The variability in the tissue strengths combined with the fact that you have data on only a piece of the population leads to variability in the sample averages, the $\overline{X}$'s. Recall that the variability of $\overline{X}$ is called the standard error and it is based on the amount of variability in the population, σ, and the sample size, n. ∎

This example shows you that you should not simply compare the $\overline{X}$ value you get from the sample to the target mean strength, because even if the tissues were being made properly, the $\overline{X}$ that you observe will almost never be equal to that number. So, a simple point estimate will not do the job.

This chapter introduces the major concepts and philosophy of a technique called *hypothesis testing*, which will do the job. You might remember that when we talked about the need for inferential statistics we said that most of the techniques of inferential statistics could be classified as either estimation tools or hypothesis testing tools. This chapter lays the groundwork for hypothesis testing. The remaining chapters of the text rely heavily on the ideas developed in this chapter.

Specifically, in this chapter we will look at:

- What Is a Hypothesis Test?
- Overview of Hypotheses to Be Tested
- The Pieces of a Hypothesis Test
- Two-Tail Tests of the Mean: Large Sample
- Which Theory Should Go Into the Null Hypothesis?
- One-Tail Tests of the Mean: Large Sample
- What Error Could You Be Making?

10.2   WHAT IS A HYPOTHESIS TEST?

The word **hypothesis** has the same meaning in statistics as it does in everyday use. What does this word mean to you? Some possibilities are

- an idea
- an assumption
- a guess
- a theory

You actually work with many hypotheses on a daily basis without even realizing it. For example, you might think that exercising on a regular basis improves your overall ability to study. Your hypothesis, in this case, is that there is a relationship between exercising and effective studying. To decide if your hypothesis is correct you might use your recollection of past experiences.

> In statistics, a *hypothesis* is an idea, an assumption, or a theory about the behavior of one or more variables in one or more populations.

Once a hypothesis is formed, you must *test* it. You must decide whether to believe the hypothesis. The only information you have to help you decide is contained in your sample. Thus, to do a **hypothesis test,** you use the information in your sample data to help you decide if you should believe the hypothesis. Basically, you are trying to decide if the sample is consistent with the hypothesis (in which case you believe the hypothesis) or if the sample is inconsistent with the hypothesis (in which case you choose not to believe it or to reject it).

The sample data are your evidence and on the basis of these data you must make a decision.

> A *hypothesis test* is a statistical procedure that involves formulating a hypothesis and using sample data to decide on the validity of the hypothesis.

There are many different types of hypotheses that you could test and these are the subject of the next section. However, regardless of the specific hypothesis that you are testing, the basic procedure is the same. The purpose of this chapter is to explain the framework of the hypothesis testing procedure. The details of the test depend on the particular hypothesis that you are testing, but the purpose and general approach are the same for all tests. In fact, virtually all of the remaining chapters in this book are devoted to spelling out the details for the various tests you are likely to need for analyzing sample data and making informed business decisions. It is important that you not view these as separate and isolated chapters. They add flesh to the skeleton we will develop in this chapter. Think of each chapter as a variation on the same theme.

10.3   DESIGNING HYPOTHESES TO BE TESTED—AN OVERVIEW

Let's start by thinking about what type of variables we might be examining. In Chapter 2 you learned that two major categories can be used to classify variables, *qualitative* and *quantitative*. The distinction was made because different techniques are

Different types of data require different analysis tools.

used depending on the type of data you have. You have seen (Chapters 3, 6, and 9) that different *descriptive* tools and point estimators are used, depending on what kind of data you wish to evaluate. Now we will see that there are different *inferential* tools for different kinds of data.

10.3.1 Hypotheses About Quantitative Variables

Sample data that are inherently numerical in form are called quantitative data. Recall from Chapter 7 that if we are analyzing a quantitative variable then we know that it can be described by its distribution. The distribution in turn can be described by parameters. Thus, if we are constructing a theory or hypothesis about a quantitative variable it might be a statement about

Hypotheses about a quantitative variable

- The shape of the distribution of the variable in one population
- The mean value, μ, of the variable in one population
- How the mean value of the variable in one population, μ_1, compares to the mean value of the variable in a second population, μ_2
- The equality of the mean values of the variable in more than two populations
- The amount of variability, σ^2, of the variable in one population
- How the amount of variability of the variable in one population, σ_1^2, compares to the amount of variability of a variable in a second population, σ_2^2

All but the first bulleted item are generally referred to as tests on means and tests on variances. The first item requires a "goodness of fit" test. Goodness of fit tests are the subject of Chapter 17.

EXAMPLE 10.2   **Tissue Strength Problem**

Possible Hypotheses

For the problem that you are investigating for your boss, tissue strength is a quantitative variable. An example of each of the hypotheses just discussed is shown for this variable:

- The variable *MDStrength* has a normal distribution.
- The population mean *MDStrength* is equal to 1000 lb/ream.
- The mean *MDStrength* of tissues made by machine 1 is greater than the mean *MDStrength* of tissues made by machine 2.
- The mean *MDStrength* of tissues made by machines 1, 2, and 3 are equal.
- The variability of the tissue *MDStrength* is less than 625 (lb/ream)^2. (Recall that the variance is the standard deviation squared.)
- The tissues made by machine 1 have more variability in the *MDStrength* characteristic than those tissues made by machine 2. ■

Remember: Qualitative data describe a particular characteristic of a sample item.

Notice that a qualitative variable, in this case the number of the machine that made the tissue, was used to divide the data into two or more populations. This is often the major use of a qualitative variable, particularly nominal data. Recall that nominal data are one type of qualitative data. They are data that are created by assigning numbers to different categories when the numbers have no real meaning.

 *TRY IT NOW!*

7-11 Stores
Possible Hypotheses

7-11 stores are located all over the Northeast. Management is studying sales data. Develop a specific hypothesis for each of the different types of hypotheses that we have discussed. Here's one hypothesis to get you started:

> *Daily Sales are normally distributed.*

10.3.2 Hypotheses About Nominal Variables

In addition to using nominal variables as a way to divide the data into two groups, we are often interested in what percentage or proportion of a population has a particular characteristic. For example, we might be interested in what proportion of the product we are manufacturing is defective. In this case, the nominal variable is the quality status of the product, nondefective or defective, and we are interested in the percentage or proportion of the population that has the quality status "defective." If the data we are analyzing are *nominal data*, the hypothesis might be a statement about

- The value of the proportion, π, of population members that have a certain characteristic (one of the categories of the nominal variable)
- How the proportion who have a certain characteristic in one population, π_1, compares with the corresponding proportion in a second population, π_2

EXAMPLE 10.3   Cereal Manufacturer

Possible Hypotheses

A cereal company is considering a new package design. The company would like to know what percentage or proportion of the consumers like this new design. An example of each of the hypotheses just discussed is shown for this variable:

- At least 50% of the customers like this new design.
- A greater proportion of women than men like this new design.   ∎

ANS. (ANSWERS MAY VARY.) THE AVERAGE DAILY SALES IS $2500. THE AVERAGE SALES IN THE MA STORES IS GREATER THAN IN THE CT STORES. THE AVERAGE DAILY SALES FOR ALL COUNTIES IN MA ARE THE SAME. THE VARIABILITY OF STORES IN NY IS 100 DOLLARS^2. THE VARIABILITY OF DAILY SALES IN NY IS GREATER THAN IN THE CT STORES.

Notice that, in this example, we are interested in the proportion of the population that prefers the new design. In this case, the nominal variable is whether or not a consumer prefers the new design. We have used a second nominal variable (gender) to divide the consumers into two populations and have then constructed a hypothesis about the proportion of women who prefer the new design compared to the proportion of men who prefer the new design.

 *TRY IT NOW!*

The Sports Complex
Possible Hypotheses

A particular university is considering building a new sports complex. It wishes to know if the sports complex would be widely used by students. Develop a specific hypothesis for each of the different types of hypotheses that we have discussed.

10.3.3 Hypotheses About Ordinal Variables

Another way in which qualitative data can be expressed numerically is when the order in which the numbers are assigned has some meaning. When numbers are used to name ordered categories the data are called *ordinal*. If we are analyzing ordinal data, the hypothesis might be a statement about

- The mean value, μ, of the variable in one population
- How the mean value of the variable in one population, μ_1, compares to the mean value of the variable in a second population, μ_2
- The equality of the mean values of the variable in more than two populations

EXAMPLE 10.4   **The Soft Drink Manufacturer**

Hypotheses for Ordinal Data

Suppose you asked a group of people to rank five different versions of a new soft drink. The resulting data for one person might look like this:

Version	3	2	5	1	4
Ranking	1	2	3	4	5

In this case the numbers are not entirely meaningless since they indicate a relative position for each version of the product on a scale. But the variable ranking is not a quantitative variable because you cannot tell from this scale whether the person doing the ranking liked version 3 a lot and really hated the other 4 or whether versions 3 and 2 were similar and much superior to the remaining three versions. That is, the distances between the numbers assigned are not necessarily equal.

Possible hypotheses for this situation are:

- The average value for the ranking of version 3 is 1.5.
- Versions 2 and 5 have the same average ranking.
- There is no difference in the average rankings of the 5 different versions. ■

You can see that these hypotheses are very similar to the hypotheses we might use for a quantitative variable. However, the details of how to test these hypotheses when you have ordinal data are different. These tests are referred to as nonparametric tests. Although these tests are not presented in this text, the general hypothesis testing concepts that you will learn in this chapter and the next few chapters are also the basis for nonparametric tests.

10.3.4 Hypotheses About Two Variables

We started Section 10.2 with an example about exercise and effective studying. In this example we have two variables. How we investigate the relationship between these two variables depends on how the data are collected.

Suppose we have number of hours per day that a person exercises and a rating of the effectiveness of the person's studying ranging from 1 = highly effective to 5 = highly ineffective. In this case we could treat the number of hours per day that a person exercises as a quantitative variable and use the effectiveness rating to divide the data into 5 populations. We could then construct the hypothesis that the average number of hours exercised is the same for students with an effectiveness rating of 1, 2, 3, 4, and 5. This type of hypothesis was discussed in Section 10.3.1.

However, if the data are collected in such a way that both variables are qualitative variables, then we need to construct the hypothesis that says that the variable exercise rating and study effectiveness rating are independent. This is typically called a test for independence and is the subject of Chapter 17.

10.3.5 Summary of Kinds of Hypotheses

Table 10.1 summarizes the various kinds of hypotheses you are likely to need to analyze sample data and make informed business decisions.

TABLE 10.1   Summary of Various Hypotheses to Be Tested

	One Population	Two Populations (Often a Qualitative Variable is Used to Divide the Data into 2 Populations)	More Than Two Populations
Quantitative data	1. The shape of the distribution. 2. The mean value, μ, of the variable. 3. The amount of variability, σ^2, of the variable.	1. Compare the mean value of the variable in one population, μ_1, to the mean value of the variable in a second population, μ_2. 2. Compare the amount of variability of the variable in one population, σ_1^2, to the amount of variability of the variable in a second population, σ_2^2.	The equality of the mean values of the variable.
Nominal data	The value of the proportion, π, of population members that have a certain characteristic (one of the categories of the nominal variable).	Compare the proportion who have a certain characteristic in one population, π_1, with the corresponding proportion in a second population, π_2.	Test for independence.

10.4   THE PIECES OF A HYPOTHESIS TEST

To carry out any hypothesis test, you need to set it up according to some general guidelines.

10.4.1 The Null and Alternative Hypotheses

The first step is to take your idea or hypothesis and construct two opposing views. One of these is called the **null hypothesis** and the other is called the **alternative hypothesis.**

EXAMPLE 10.5   The Tissue Strength Problem

Setting Up the Null and Alternative Hypotheses

For the tissue strength problem that you are looking at you may choose to set it up this way:

Null hypothesis:	The true mean *MDStrength* is equal to 1000 lb/ream.
Alternative hypothesis:	The true mean *MDStrength* is not equal to 1000 lb/ream. ∎

Remember: You do not know μ! You are checking to see if μ is 1000. You are using the sample mean as your evidence.

Clearly, if you believe the null hypothesis then you cannot believe the alternative hypothesis. You must choose between them on the basis of the evidence in the sample.

Using the notation of hypothesis testing, you would rewrite this as

$$H_0: \quad \mu = 1000 \text{ lb/ream}$$
$$H_A: \quad \mu \neq 1000 \text{ lb/ream}$$

> The *null hypothesis* is a statement about the population(s). It is referred to as H_0.
>
> The *alternative hypothesis* is a statement about the population(s) that is opposite to the null hypothesis. It is referred to as H_A.

Sometimes the alternative hypothesis is labeled H_1. It doesn't matter whether you use H_1 or H_A as long as you are consistent. In this book we will always use H_A. You can see that the words have been rewritten in terms of the standard symbol for the population mean, μ.

You should notice that no value of μ is part of both the null and the alternative hypotheses. The null and alternative hypotheses cannot overlap. That is, the two hypotheses are mutually exclusive. Not only is that true but, in addition, when considered together, the null and the alternative hypotheses must cover all the possibilities. This means the two hypotheses are exhaustive. For this example, this means that all the possible values of μ are included in either the null or the alternative hypothesis but not both. Finally, note that the "=" sign is in the null hypothesis. This will always be the case.

EXAMPLE 10.6   The Bottle-Filling Problem

Reinforces the Two Opposing Views Concept

In any bottling process, a manufacturer will lose money if the bottles contain either more or less than the amount claimed on the label. Therefore, manufacturers pay close attention to the amount of their product that is dispensed by the bottle-filling machines. Suppose you are working for a soda company and each of the soda bottles is supposed to contain 32 oz of soda. You wish to set up a hypothesis test to decide if the bottle-filling machines are putting an average of 32 oz of soda into the bottles.

The two opposing views are, therefore,

$$H_0: \quad \mu = 32 \text{ oz}$$
$$H_A: \quad \mu \neq 32 \text{ oz} \qquad \blacksquare$$

Let's look at a situation in which the data to be analyzed are nominal data.

EXAMPLE 10.7   U.S. Unemployment

Null and Alternative Hypotheses for Test of Proportion

The U.S. government wishes to know what proportion of the workforce is unemployed. Last month the estimated proportion was 6% and the government wishes to set up a hypothesis test to decide if the proportion for this month has changed from 6%.

The two opposing views are, therefore,

$$H_0: \quad \pi = 0.06$$
$$H_A: \quad \pi \neq 0.06 \qquad \blacksquare$$

Now it is time for you to try your hand at setting up the null and alternative hypotheses.

 **TRY IT NOW!**

The Potato Chip Manufacturer
Setting Up the Null and Alternative Hypotheses

Many people eat chips with their soda. Suppose a potato chip manufacturer is concerned that the bagging equipment may not be functioning properly when filling 10-oz bags. You have been asked to set up a hypothesis test that will help determine if there is a problem with the bagging equipment. What null and alternative hypotheses would you use?

ANS. H_0: $\mu = 10$ oz
H_A: $\mu \neq 10$ oz

Discovery Exercise 10.1
FORMULATING HYPOTHESES

Consider the population of all M&M packages like the one you have in your hand.

Step 1: Identify as many different variables as you can. Be sure you have some quantitative and some qualitative variables. Record the values of these variables for your package. (*Hint:* You should carefully examine the package before you rip it open.)

Step 2: Select one of the quantitative variables and set up a null and alternative hypothesis for a parameter of this variable.

Step 3: Select one of the qualitative variables and set up a null and alternative hypothesis for a parameter of this variable.

Step 4: As a class, agree on several quantitative and qualitative variables that you feel are important. Record the data for all of the teams on each of these variables.

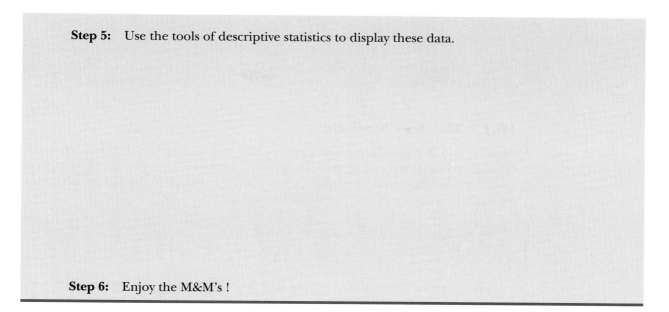

Step 5:   Use the tools of descriptive statistics to display these data.

Step 6:   Enjoy the M&M's !

10.4.2 The Rejection Region

The next step is to determine how you will decide between the null and the alternative hypotheses. To be more specific, you must decide to reject or not to reject the null hypothesis. Your decision is always phrased with regard to the null hypothesis. If you choose not to reject the null hypothesis this means that the sample data are consistent with the null hypothesis. If you choose to reject the null hypothesis this means that the sample data are sufficiently inconsistent with the null hypothesis.

To make this decision you must set up what is known as the **rejection region.**

> The *rejection region* is the range of values of the test statistic that will lead you to reject the null hypothesis.

The test statistic is discussed in more detail in the next section. For example, the rejection region might be all values of Z larger than 1.96. This is shown as the shaded region in Figure 10.1.

If this is the rejection region, we would reject the null hypothesis if the value of Z calculated from the data is greater than 1.96. If the value of Z calculated is less than or equal to 1.96, we would not reject the null hypothesis. The Z statistic is the same

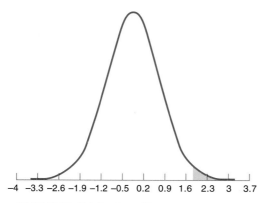

FIGURE 10.1   Possible rejection region

one you learned about in Chapter 9. Often, Z is the test statistic we need to use. This is discussed more in the next section.

The procedure for actually finding the rejection region depends on several things. The specifics are provided in the last section of this chapter and in upcoming chapters. For now, it is important that you understand the general concepts.

10.4.3 The Test Statistic

We have said that the information contained in your sample is the evidence that you will use to decide between H_0 and H_A. However, as you can see, the sample consists of n observations. Somehow you must capture the information in the sample into a single number. This number will be called the **test statistic.**

> A *test statistic* is a number that captures the information in the sample. It will be used to choose between the null and alternative hypotheses.

If you think about it, you will see that this term is entirely consistent with what we have already defined. In general, a statistic is any number calculated from the sample data. In this case, you will calculate a number from the sample data and use this number to decide between the null and the alternative hypotheses. In other words, you will use this statistic to test the hypothesis. Hence, it is called a test statistic.

10.5   TWO-TAIL TESTS OF THE MEAN: LARGE SAMPLE

Now we will put these pieces together to form a *5-step hypothesis testing procedure:*

Steps for any hypothesis test

Step 1:   *Set up the null and alternative hypotheses.*

Step 2:   *Pick the value of α and find the rejection region.*

Step 3:   *Calculate the test statistic.*

Step 4:   *Decide whether or not to reject the null hypothesis.*

Step 5:   *Interpret the statistical decision in terms of the stated problem.*

There are two different cases to consider when testing the population mean: tests of the mean when you know something about the *population* standard deviation and tests of the mean when all you have is the *sample* standard deviation. The first case is covered in this chapter. This case also includes the situation where you do not know the population standard deviation, σ, but you have a sufficiently large sample size, $n > 30$. When this happens you use the sample standard deviation, s, as an estimate of σ. For this reason, the tests in this chapter are referred to as "large-sample tests" but they are used whenever you know the population standard deviation regardless of the sample size.

10.5.1 Two-Tail Test of the Mean: A Detailed Example

The first step is to take your idea or hypothesis about the mean and construct the null and the alternative hypotheses.

In general, the procedure for finding the rejection region for a two-tail test is as follows:

STEPS FOR FINDING THE REJECTION RE-GION FOR A TWO-TAIL TEST

Step 1: *Take the value of α and divide it by 2.*

Step 2: *Look that value up in the body of the Z table (find the one closest to it).*

Step 3: *Read off the corresponding Z value. This gives you the value of* $-Z_{cutoff}$.

Step 4: *Drop the negative sign to get the value of* Z_{cutoff}.

Step 5: *The rejection region consists of all values of Z greater than* Z_{cutoff} *or less than* $-Z_{cutoff}$.

 **TRY IT NOW!**

The Tissue Company

Finding the Rejection Region

Suppose the tissue company decided to set α at 0.10. Find the rejection region.

The third step is to capture the information in the sample into a single number called the test statistic. It is the sample mean, $\overline{X}$, that is particularly relevant, since you are testing an idea about the true population mean, μ. Suppose the sample mean *MDStrength* of 36 tissue samples is found to be 980 lb/ream. On the surface it looks like maybe the process needs to be adjusted. After all, 980 is 20 units less than the target value of 1000. You must remember that this $\overline{X}$ value of 980 is based on a sample. If you took another 36 tissue samples you might get a sample mean of 1020. You cannot decide if a difference of 20 is really big or not until you compare it to the standard error! This is precisely what the *Z*-score calculation does.

Remember that $\overline{X}$ *is the best point estimator for* μ.

We can calculate *Z* using the following formula for standardizing $\overline{X}$:

$$Z = \frac{\overline{X} - \mu}{\sigma/\sqrt{n}}$$

It seems like we have a problem since we need to know μ to calculate the *Z*-statistic but we don't know μ. This is easily resolved at this point because μ is the *target* specification value. In general, the value of μ that you use is the number that you are testing your sample evidence against, the value in your null hypothesis.

Let's see how this test statistic would be calculated for the tissue strength problem that we have been following.

EXAMPLE 10.10   The Tissue Company

Step 3: Calculating the Test Statistic

Recall that the population standard deviation, σ, for the tissue manufacturer is 50. The Z-statistic is then

$$Z = \frac{\overline{X} - \mu}{\sigma/\sqrt{n}}$$

$$Z = \frac{980 - 1000}{50/\sqrt{36}} = \frac{-20}{8.333} = -2.40$$

Notice that you use the target value of 1000 as the value for μ in this calculation. ∎

The fourth step is to decide whether or not to reject the null hypothesis. If the Z value that was calculated at step 3 falls in the rejection region, then we reject H_0. If the Z value does not fall in the rejection region, then we fail to reject H_0. Let's do the fourth step for the tissue company.

EXAMPLE 10.11   The Tissue Company

Step 4: Using the Test Statistic to Decide Whether or Not to Reject the Null Hypothesis

The rejection region for this two-tail test is shown below:

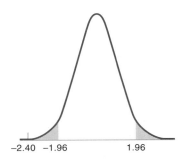

The calculated Z value of -2.40 is marked on this line and clearly falls in the rejection region. Thus, we reject H_0 and believe the alternative hypothesis. ∎

The fifth step is to interpret the statistical decision in terms of the stated problem. At this step you must "translate" the result of the statistical test into a recommended action, or decision, or reach a conclusion depending on the nature of the problem. Let's complete the hypothesis test for the tissue company

EXAMPLE 10.12   The Tissue Company

Step 5: Interpreting the Statistical Decision

At step 4 we rejected the null hypothesis. Therefore, we believe the alternative hypothesis. The alternative hypothesis was that the true mean was not equal to 1000 lb/ream. Therefore, on the basis of the evidence in the sample that is captured in the Z value of -2.4, you decide that the process needs adjustment.

     Your decision is based on the fact that it is unlikely that you would get such a large Z value ($Z = -2.4$) if the process is really OK. But is it possible that the process really does not need adjustment? You bet. Maybe you got a sample with a large $\overline{X}$, which gave you the large Z value. How likely is this? The chance of this happening is precisely α. In this case, that chance is at most 5%, since α was set to 0.05. ∎

Discovery Exercise 10.2

EXPLORING THE IMPACT OF
VARYING THE VALUE OF α

You may remember that in Chapter 6 we examined the data collected by a company that was concerned about the length of time its customers were on hold. The data shown below were first presented to you in Example 6.14. There are 50 observations on customer hold times:

0.6	4.6	5.6	6.3	6.8	7.5	7.8	8.3	8.9	9.6
2.9	4.7	6.0	6.3	6.9	7.5	7.9	8.4	9.2	10.1
3.4	5.2	6.0	6.6	6.9	7.6	8.0	8.4	9.2	10.2
3.8	5.5	6.1	6.6	7.0	7.6	8.1	8.6	9.4	10.7
4.5	5.5	6.1	6.7	7.2	7.8	8.2	8.6	9.4	11.1

The company wishes to test the hypothesis that the true mean hold time is 7 minutes. The standard deviation of hold times is known to be 2 minutes.

(a) Test this hypothesis using a value of $\alpha = 0.10$.

Step 1:

Step 2:

Step 3:

(continued)

Step 4:

Step 5:

(b) Now vary the value of α and complete the following table.

α	Rejection Region	Decision
0.10 (from part a above)		
0.05		
0.02		
0.01		

(c) What happened to your decision as the value of α changed?

10.5.2 Understanding, Calculating and Using p values

You may have noticed that the rejection region is tied to the value of α that we selected. You may therefore be wondering if selecting a different value of α would have led to a different conclusion. This is indeed a good point and worthy of consideration. Let's follow the details of the hypothesis-testing procedure through another example and then examine the impact of selecting a different value for α.

EXAMPLE 10.13   **The Bottle-Filling Problem**

Step 1: Setting Up the Null and Alternative Hypotheses

In Example 10.6 we looked at the hypothesis set up to test whether a soda machine was dispensing 32 oz of soda. Recall that we wished to know if the machine was underfilling or overfilling the cups. Since we are interested in detecting differences from 32 oz in either direction, we used a two-tail test. The null and alternative hypotheses are shown below:

$$H_0: \quad \mu = 32 \text{ oz}$$
$$H_A: \quad \mu \neq 32 \text{ oz}$$

Thus, step 1 has been done. ■

Continuing with this example, we will complete step 2 by selecting the value of α and finding the rejection region.

EXAMPLE 10.14　The Bottle-Filling Problem

Step 2: Selecting α and Finding the Rejection Region

Set $\alpha = 0.05$. The rejection region is the same as the one we found for the tissue company. Reject H_0 if $Z < -1.96$ or $Z > 1.96$. ■

Suppose you observe the machine filling 30 bottles and observe the following data:

32.12	32.04	32.04	32.00	32.01	32.12
32.13	32.05	32.05	31.72	32.08	32.03
32.02	32.22	32.22	32.22	31.89	32.20
31.73	31.95	31.95	32.07	32.13	32.13
31.98	32.21	32.21	31.97	32.09	31.97

You also know from the machine specifications that the population standard deviation is supposed to be 0.1 oz. What do you do with the sample information? Remember, step 3 is to capture the information in the sample into a single number called the test statistic. Let's complete step 3 for the bottle-filling problem.

EXAMPLE 10.15　The Bottle-Filling Problem

Step 3: Calculating the Test Statistic

To calculate the Z test statistic:

$$Z = \frac{\overline{X} - \mu}{\sigma/\sqrt{n}}$$

we clearly need $\overline{X}$. Calculating $\overline{X}$, we get 32.05 oz . The Z test statistic is then

$$Z = \frac{32.05 - 32}{0.1/\sqrt{30}} = \frac{0.05}{0.018} = 2.78$$ ■

Now we are ready to use the evidence in the sample, which has been captured in the Z test statistic, to decide if we should reject the null hypothesis or not. Step 4 for the bottle-filling problem is completed in the next example.

EXAMPLE 10.16　The Bottle-Filling Problem

Step 4: Using the Test Statistic

The value of $Z = 2.78$ clearly falls in the rejection region. So our statistical decision is to reject the null hypothesis and conclude that the true mean is not equal to 32 oz. ■

Step 5 requires that we interpret this statistical decision in terms of the problem statement.

EXAMPLE 10.17   The Bottle-Filling Problem

Step 5: Interpreting the Statistical Decision

We concluded that the true mean of the bottle-filling process was not 32 oz. This tells us that there is something wrong with the filling process. Possibly the true mean has shifted to a value other than 32 oz.   ∎

Let's consider a variety of different α values for the bottle-filling example and see what the impact is on the rejection region and ultimately on the decision.

Remember: To get the cutoff Z value you divide α by 2 and look up that value in the body of the Z table.

		Decision		
		$\leftarrow$───┼───┼───$\rightarrow$		
α	Z_{cutoff}	$-Z_{cutoff}$ 0 Z_{cutoff}		
0.10	1.64 or 1.65	$2.78 > 1.64$, so reject H_0		
0.05	1.96	$2.78 > 1.96$, so reject H_0		
0.01	2.58	$2.78 > 2.58$, so reject H_0		
0.005	2.81	$2.78 < 2.81$, so fail to reject H_0		
0.001	3.29	$2.78 < 3.29$, so fail to reject H_0		

Notice that when you change α from 0.01 to 0.005 your decision switches from rejecting H_0 to failing to reject H_0. Initially, this should make you a bit uncomfortable. After all, you need to tell your boss whether this machine is working correctly and you get a different answer depending on how you set α! There will always be some value for α where your decision switches. This is called the ***p* value.**

> The ***p* value** is defined to be the smallest value of α for which you can reject H_0.

For some value of α, the value of the test statistic (based on sample data) will be the same as the Z_{cutoff} that defines the rejection region. When that happens you have found the *p* value.

Calculating the p value is just like the calculations done in Chapter 9.

Remember that α is a probability and therefore so is the *p value*. In fact you have actually calculated *p* values in Chapter 9 without knowing it. In that chapter you calculated the probability of observing a particular *Z* value. This is exactly what needs to be done to get the *p* value. You must calculate the chance of observing the particular sample mean, $\overline{X}$, if the null hypothesis is true. Let's calculate the *p* value for the bottle-filling problem.

EXAMPLE 10.18   The Bottle-Filling Problem

Calculating the p Value

For the bottle-filling problem, we observed data and calculated a sample mean of $\overline{X} = 32.05$ oz and a *Z* statistic of 2.78. To calculate the *p* value we must determine the chance that we would have observed this $\overline{X}$ value if the null hypothesis was true. The Central Limit Theorem tells us that $\overline{X}$ has a normal distribution which we know is a continuous distribution. We also remember that there is no probability (zero area under the curve) of observing a specific value of a continuous random variable. Thus, we need to calculate the probability of observing the particular $\overline{X}$ value or one even more extreme. Since the test is a two-tail test we need to double the tail area probability. The *p* value is calculated by finding

$$p\text{ value} = 2P(\overline{X} > 32.05)$$

This calculation is done by converting the random variable $\overline{X}$ to a Z value and using the procedure developed in Chapter 9. But we already have the Z value from the test statistic so we know we can simply calculate

$$p \text{ value} = 2P(Z > 2.78)$$

Using the symmetry of the normal curve, we find that this is just two times the tail area probability for $Z = -2.78$. Using the standard normal table, we get

$$p \text{ value} = 2(0.0027) = 0.0054$$

This result is consistent with our previous analysis. We expected a p value less than 0.01. ∎

In this example the p value is very small. In particular, the p value is less than 0.01. This tells us that the smallest value of α for which we can reject H_0 is 0.0054. It also tells us that there is a chance of 54 in 10,000 that we would have observed a sample mean of $\overline{X} = 32.05$ oz, if the population mean was really 32 oz. Clearly, the data are very inconsistent with the null hypothesis. The data are telling you to reject the null hypothesis for virtually every possible value of α.

Consider another two-tail test where the calculated Z value is 1.80. The same type of analysis can be done to get an idea of what the p value is for this case.

α	Z_{cutoff}	Decision
0.10	1.64 or 1.65	1.8 > 1.64, so reject H_0
0.05	1.96	1.8 < 1.96, so do not reject H_0
0.01	2.58	1.8 < 2.58, so do not reject H_0
0.005	2.81	1.8 < 2.81, so do not reject H_0
0.001	3.29	1.8 < 3.29, so do not reject H_0

In this case, the decision switches from rejecting H_0 to not rejecting H_0 somewhere between $\alpha = 0.10$ and $\alpha = 0.05$. Thus, the p value is somewhere between 0.10 and 0.05. To calculate the p value for this situation we would find

$$p \text{ value} = 2P(Z > 1.8) = (2)(0.0359) = 0.0718$$

A p value between 0.05 and 0.10 tells you that the data are not strongly consistent with the null hypothesis, but neither are they strongly inconsistent with the null hypothesis. In such cases the data are "on the fence."

Finally, consider an example where the data are extremely consistent with H_0. Suppose you are doing a two-tail test of μ and the calculated Z statistic is 0.95. The same analysis is shown below.

α	Z_{cutoff}	Decision
0.10	1.64 or 1.65	0.95 < 1.64, so do not reject H_0
0.05	1.96	0.95 < 1.96, so do not reject H_0
0.01	2.58	0.95 < 2.58, so do not reject H_0
0.005	2.81	0.95 < 2.81, so do not reject H_0
0.001	3.29	0.95 < 3.29, so do not reject H_0

As you can see for all values of α less than or equal to 0.10, we would not reject H_0. See if you can find the value for α where the decision would switch to a rejection. It is clearly larger than 0.10. The calculation is shown below:

$$p \text{ value} = 2P(Z > 0.95) = (2)(0.1711) = 0.3422$$

You should also have realized that a Z value of less than 1 means that the observed sample mean is less than 1 standard error away from the hypothesized value and we know that such Z values are likely. This also points to the same conclusion. So in this case our p value is large (>0.10), which means we will almost always fail to reject the null hypothesis. The data are *highly consistent* with the null hypothesis.

In summary, we have the following table:

p Value	Data are . . .	Decision
<0.05	Highly inconsistent with H_0	Strong rejection of H_0
Between 0.05 and 0.10	"On the fence"	Collect more data (if possible)
>0.10	Highly consistent with H_0	Strong failure to reject H_0

 **TRY IT NOW!**

The Tissue Company

Finding the p Value

Find the p value for the tissue company's two-tail test of μ. Recall that the average *MDStrength* was found to be 980 lb/ream and the Z statistic was calculated to be -2.40.

10.5.3 Exercises—Learning It!

10.1   The School Committee members of a midsize New England city agreed that a strict discipline code has caused an increase in the number of student suspensions. The number of suspensions for a sample of schools in this city for the period September 1992–February 1993 are shown below:

Central	245	Putnam	1024
MCDI	1	Kiley	56
Chestnut	65	Central Academy	254
Duggan	133	Commerce	114
Kennedy	97	Bridge	7
Forest Park	149		

The average number of suspensions for the previous year was 130.5 with a population standard deviation of 158.2.

(a) Set up the null and the alternative hypotheses to test if the average number of suspensions has changed.

(b) Test your hypothesis using $\alpha = 0.05$.

(c) Find the p value.

(d) Display the data to see if it is reasonable to assume that the underlying population distribution is normal.

(e) Based on the p value, what can you conclude about the average number of suspensions?

10.2 The Educational Testing Service (ETS) designs and administers the SAT exams. Recently the format of the exam changed and the claim has been made that the new exam can be completed in an average time of 120 minutes. A sample of 50 new exam times yielded an average time of 115 min. The standard deviation is assumed to be 2 minutes.

(a) Set up the null and the alternative hypotheses to test if average time to complete the exam has changed from 120 minutes.

(b) Test your hypothesis using $\alpha = 0.05$.

(c) Find the p value.

(d) Based on the p value, what can you conclude about the average time to complete the new exam?

10.3 A vending machine that dispenses coffee into cups must fill the cups with 7.8 oz of liquid. Before selling the vending machine to a college or business, the company tests the machine to be sure it is dispensing an average amount of 7.8 oz of coffee. A sample of 20 amounts is shown below. The amount of coffee dispensed is assumed to be normally distributed with a standard deviation of 0.05 oz.

Ounces Dispensed

7.78	7.79
7.82	7.82
7.87	7.84
7.80	7.82
7.80	7.78
7.83	7.75
7.85	7.83
7.84	7.73
7.82	7.87
7.81	7.88

(a) Set up the null and the alternative hypotheses to test if average amount of coffee dispensed is different from 7.8 ounces.

(b) Test your hypothesis using $\alpha = 0.05$.

(c) Find the p value.

(d) Based on the p value, what can you conclude about the average amount of fluid dispensed by the machine?

10.4 According the data released by the United States Bureau of Labor, the average price of a half-gallon of ice cream in 1996 was $2.86. The monthly averages for the price of a half-gallon of ice cream at 12 stores in your town in 1996 are shown below.

Price for Half-Gallon of Ice Cream

$2.87	$2.93	$3.13	$2.92	$2.99	$3.01	$3.02	$3.01	$2.93	$3.08	$2.67	$2.89

Assume that the ice cream prices are normally distributed with a standard deviation of $0.10.

(a) Set up the null and the alternative hypotheses to test if the average price of a half-gallon of ice cream in your town is different from the 1996 national average.

(b) Test your hypothesis using $\alpha = 0.05$.

(c) Find the p value.

(d) What can you conclude about the average price of a half-gallon of ice cream in your town?

10.5 Your manager has just read an article stating that, on the average, employees spend one hour a day playing video games at work. She would like you to see if her employees are really spending this amount of work time playing games. A random sample of 35 employees played games an average of 55 minutes per day with a standard deviation of 5 minutes. You report this to your manager and she is relieved to learn that her employees are playing less than the national average.

  You quickly remind her that she has information on a sample of employees and suggest that she test the hypothesis that the population mean is 60 minutes. She is impressed with your knowledge of statistics and asks you to do the hypothesis test and report back to her.

(a) Set up the null and the alternative hypotheses to test if the average amount of time spent playing games at work is different from 60 minutes.

(b) Test your hypothesis using $\alpha = 0.05$.

(c) Find the p value.

(d) Based on the p value, are the employees at this company any different than the average employee nationwide?

10.6 A cereal manufacturer is concerned that the boxes of cereal not be underfilled or overfilled. Each box of cereal is supposed to contain 13 oz of cereal. A test is done to see if the machines are really putting an average of 13 oz of cereal into the boxes. A sample of 30 boxes is tested. The average weight is 12.58 oz and the standard deviation is 0.25 oz.

(a) Set up the null and the alternative hypotheses to test if average number of ounces in the cereal boxes is different from 12.58 oz.

(b) Test your hypothesis using $\alpha = 0.10$.

(c) Find the p value.

(d) Based on the p value, what can you conclude about the average number of ounces in the cereal boxes?

10.6   WHICH THEORY SHOULD GO INTO THE NULL HYPOTHESIS?

In Section 10.4.1 we said that the first step is to construct two opposing views. We called one of them the null hypothesis and one of them the alternative hypothesis. Because of the way the hypothesis testing procedure works, it is important to carefully consider which of the views you are going to call the null hypothesis and which one you will call the alternative hypothesis.

  Several different approaches may be taken to determine how the null and alternative hypotheses should be set up. These different approaches are the subject of the next two subsections. The first subsection introduces the concept of two-tail and one-tail tests, and the second subsection illustrates the conservative nature of the hypothesis testing procedure. Depending on the specifics of the problem you are trying to solve, one of these approaches will shed some light on resolving the question, Which theory should go into the null hypothesis?

10.6.1 Two-Tail Tests and One-Tail Tests

Much of our discussion so far has been about null and alternative hypotheses, which are called **two-tail tests.** The format is familiar to you by now and is shown in the following definition.

> A *two-tail test* of the population mean has the following null and alternative hypotheses:
>
> $$H_0: \quad \mu = \text{[a specific number]}$$
> $$H_A: \quad \mu \neq \text{[a specific number]}$$

The null hypothesis of a two-tail test claims that the mean (or whatever parameter you are testing) is actually *equal* to the particular number stated. The opposing view is clearly that the mean is *not equal* to that particular value, and this is the alternative hypothesis.

Remember: You could be testing a variance or a proportion instead.

When you use a two-tail test you are interested in seeing if the true mean is *different* from the number specified. You wish to know if the true mean is *higher than* the number or *lower than* the number. In other words, you want to test for deviations from the number in either direction—on the high side or the low side.

If you are doing a two-tail test, there is no decision to be made about how the null and the alternative hypotheses should be set up. The hypothesis testing procedure requires that the view with the equal sign be in the null hypothesis.

EXAMPLE 10.19   The Tissue Strength Problem

Setup as a Two-Tail Test of the Mean

For the tissue strength example, we could set up the following two-tail test:

$$H_0: \quad \mu = 1000 \ \text{lb/ream}$$
$$H_A: \quad \mu \neq 1000 \ \text{lb/ream}$$

Here the specific number is 1000.   ■

Both the soda bottle-filling problem and the potato chip-packaging problem discussed in the previous section are also two-tail tests. In each of these cases the view that the true mean, μ, was equal to 32 oz (for the soda) or 10 oz (for the chips) was the null hypothesis.

Consider another example.

EXAMPLE 10.20   Testing Supplies

Illustrating the Setup of a Two-Tail Test of a Proportion

Suppose you are purchasing 3.5-in. disks from Disks Are Us. The company claims that only 1% of the disks it manufactures are defective. As the purchasing agent you wish to check this claim. The null and alternative hypotheses are shown below:

$$H_0: \quad \pi = 0.01$$
$$H_A: \quad \pi \neq 0.01$$ ■

Notice that, in this example, we used a test on proportions, since we were interested in the percentage of the population of disks that were defective. The quality status of the defective can be considered a nominal variable.

 **TRY IT NOW!**

The Chapperel Steel Company
Setting Up the Null and Alternative Hypotheses for a Two-Tail Test of the Mean

Another recent management approach is to have employees become actual partners of the business. Chapperel Steel Company has done exactly this and the company feels that one of the benefits of this concept is that the average number of sick days will decrease. Prior to implementing this program, Chapperel had an average of 7.2 sick days per employee.

(continued)

Set up the null and alternative hypotheses to test if the average number of sick days per employee is different from 7.2.

Sometimes you really wish to see only if the population mean (or whatever parameter you are testing) is lower than the stated value. In this case, you are interested only in testing if the true mean, proportion, or variance is *less than* some number. Then you should use what is called a **lower-tail test.** A lower-tail test is one of two types of one-tail tests.

> A *lower-tail test* of a population mean has the following null and alternative hypotheses:
>
> $$H_0: \quad \mu \geq \text{[a specific number]}$$
> $$H_A: \quad \mu < \text{[a specific number]}$$

As with the two-tail test you could just as easily be testing a proportion, a variance, or the difference between two means or proportions.

EXAMPLE 10.21   The Tissue Strength Problem

Set Up as a Lower-Tail Test

Suppose the manufacturer of the tissues is worried only about the tissue strength being less than the 1000-lb/ream specified value. Then the test would be set up as follows:

$$H_0: \quad \mu \geq 1000 \text{ lb/ream}$$
$$H_A: \quad \mu < 1000 \text{ lb/ream} \qquad \blacksquare$$

Notice that the viewpoint that the manufacturer is interested in testing became the alternative hypothesis. You should also notice that the inequality sign with the equal sign attached to it (in this case $\geq$) became part of the null hypothesis.

EXAMPLE 10.22   The Chapperel Steel Company

Illustrating a Lower-Tail Hypothesis Test

Reconsider the Chapperel Steel Company example. The company is particularly interested in seeing if their partner idea has in fact reduced the average number of sick days from 7.2 per year. It would make more sense to set this test up as a lower-tail test:

$$H_0: \quad \mu \geq 7.2 \text{ days}$$
$$H_A: \quad \mu < 7.2 \text{ days} \qquad \blacksquare$$

You can see that we have placed the theory that management wishes to test into the alternative hypothesis and again the inequality with the equal sign (in this case $\geq$) is placed into the null hypothesis.

ANS. H_0: $\mu = 7.2$ DAYS; H_A: $\mu \neq 7.2$ DAYS

 *TRY IT NOW!*

The Bank Example
Lower-Tail Test

Suppose a bank knows that its customers are waiting in line an average of 10.2 minutes during the lunch hour. The branch manager has decided to add an additional teller during the 12–2 P.M. period and wishes to test the hypothesis that the average wait has decreased due to the additional teller. Set up the null and alternative hypotheses for the bank manager.

You may also wish to see if the true mean is greater than the stated value. In this case, you are interested only in testing if the true mean is *greater than* some number. This is also called an **upper-tail test.** An upper-tail test is the other kind of one-tail test. In general, such a test would look like this:

> An ***upper-tail test*** of a population mean has the following null and alternative hypotheses:
>
> $$H_0: \quad \mu \leq \text{[a specific number]}$$
> $$H_A: \quad \mu > \text{[a specific number]}$$

EXAMPLE 10.23   The Tissue Strength Problem

Set Up as an Upper-Tail Test

For example, suppose the manufacturer of the tissues is worried only about the tissue strength being greater than the 1000-lb/ream specified value. Then, the test would be set up as follows:

$$H_0: \quad \mu \leq 1000 \text{ lb/ream}$$
$$H_A: \quad \mu > 1000 \text{ lb/ream} \qquad \blacksquare$$

Here again, notice that the manufacturer's contention becomes the alternative hypothesis. The inequality with the equal sign (in this case $\leq$) is again placed into the null hypothesis.

EXAMPLE 10.24   Quality Tip Corp.

Set Up as an Upper-Tail Test of Proportions

Reconsider the cereal manufacturer from Example 10.3. One of the hypotheses that the company might be interested in studying was the hypothesis that a greater proportion of women (population 1) like the new packaging design than men (population 2). In this case we have two populations but it is still an upper-tail test. The appropriate null and alternative hypotheses are shown below:

$$H_0: \quad \pi_1 \leq \pi_2$$
$$H_A: \quad \pi_1 > \pi_2 \qquad \blacksquare$$

ANS. $H_0: \mu \geq 10.2$ MIN.; $H_A: \mu < 10.2$ MIN.

In Example 10.24 we see again that the theory management wishes to test is placed into the alternative hypothesis and the inequality with the equal sign (in this case $\leq$) is again placed into the null hypothesis.

 **TRY IT NOW!**

New Advertising Program
Setting Up an Upper-Tail Test

Suppose a company has implemented a new advertising program in the hopes of increasing sales from last year's annual average of $4.3 million. Set up the null and alternative hypotheses to test the theory that sales have increased.

Summary

A Two-Tail Test

- Is used to test if the parameter has shifted away from a certain number in either direction, increased or decreased.
- Must always be set up so the "=" theory is the null hypothesis.
- Is used when the problem statement has the key words *changed* or *different* in the problem statement.

One-Tail Tests
A Lower-Tail Test

- Is used to test if the parameter has shifted to a number less than a certain number.
- Must always be set up with the "=" as part of the null hypothesis.
- Is used when the problem statement has the key words *decreased, reduced,* or *less than.*
- The theory that you wish to "prove" is placed into the alternative hypothesis.

ANS. H_0: $\mu \leq \$4.3$ MILLION; H_A: $\mu > \$4.3$ MILLION.

An Upper-Tail Test

- Is used to test if the parameter has shifted to a number more than a certain number.
- Must always be set up with the "=" as part of the null hypothesis.
- Is used when the problem statement has the key words *increased, greater than*.
- The theory that you wish to "prove" is placed into the alternative hypothesis.

10.6.2 What View Requires No Action?

The next approach is particularly useful if you have decided to use a one-tail test. This is because, as we saw in the preceding section, there is really no choice about how to set up the null and alternative hypotheses once you decide to use a two-tail test.

This approach considers the question, What view requires that I take no action? Typically, this is the view that the population is behaving as it should be or as someone claims it should be. This view becomes the null hypothesis. Sometimes people call this the status quo.

EXAMPLE 10.25   The Tissue Strength Problem

Manufacturing Specification Becomes the Null Hypothesis

For the tissue strength problem that we have been examining, the view that requires no action (or adjustments) is to believe that the manufacturing specification is being met. In this case, the specification is that the mean *MDStrength* is greater than or equal to 1000 lb/ream. This becomes our null hypothesis. Thus, H_0: $\mu \geq 1000$ lb/ream. Once we have the null hypothesis, it is easy to construct the alternative hypothesis since it has to cover all the other cases. So we have H_A: $\mu < 1000$ lb/ream. This hypothesis test is shown below:

$$H_0: \quad \mu \geq 1000 \text{ lb/ream}$$
$$H_A: \quad \mu < 1000 \text{ lb/ream}$$

■

Now you should ask, why do it this way? The reason has to do with the fact that the hypothesis testing procedure is a conservative procedure. As such it behaves like a conservative person. A conservative person will take action only when he/she is very sure that action needs to be taken.

One way to help you understand what this means is to think about the belief in Santa Claus. Many young children believe in Santa Claus until they are about 7 or 8 years old. In this case, the null hypothesis would be H_0: Santa Claus exists and the alternative hypothesis would be H_A: Santa Claus does not exist. Children start off believing in Santa Claus (it is never proven to them) and they continue to believe that Santa Claus exists until there is overwhelming evidence to the contrary.

In a similar manner, the hypothesis testing procedure will tell us to believe the null hypothesis unless the evidence in the sample data overwhelmingly contradicts the null hypothesis. In other words, the status quo, or the view that implies that no action be taken, should be placed in the null hypothesis. It will be assumed to be correct until the data in the sample are *really* incompatible with it.

This approach is just like our judicial system. A person is assumed innocent until the evidence is so strong that it is impossible to continue to believe that the person is innocent. Remember that the instructions to the jurors are always to prove "beyond a reasonable doubt" that the person is guilty. This means that even if most of the evidence indicates that the person is guilty, if there is still some reasonable doubt, the jurors must find the person not guilty.

 **TRY IT NOW!**

Judicial System

Setting Up the Null and Alternative Hypotheses

If you think about the judicial system in terms of a hypothesis test, how would you set up the null and the alternative hypotheses?

Consider another example.

> **EXAMPLE 10.26** The Soda Machine Problem
>
> *Demonstrating the Approach, Which View Requires*
> *That No Action Be Taken? for a Two-Tail Test*
>
> A soda machine should dispense, on average, 8 oz of liquid into cups. Having purchased soda from this machine in the past, you recall that sometimes you get a small amount of soda and sometimes the cup overflows. Using the approach described in this section, you would ask, What view requires that no action be taken? Since the view that requires us to take no action is that the true mean fill is 8 oz, you should put this statement in the null hypothesis. Thus, you would get
>
> $$H_0: \quad \mu = 8 \text{ oz}$$
> $$H_A: \quad \mu \neq 8 \text{ oz}$$ ∎

TRY IT NOW!

VCR Manufacturer

Setting Up the Hypotheses so the Status Quo Is in the Null Hypothesis

Suppose a manufacturer of VCRs claims that the average life of his VCRs is at least three years. You have a VCR made by this company and have had problems with it, and so you question this claim. Set up the hypothesis test to investigate the manufacturer's claim.

ANS. $H_0: \mu \geq 3; H_A: \mu < 3$

ANS. $H_0:$ PERSON IS INNOCENT; $H_A:$ PERSON IS GUILTY

10.6.3 Exercises—Learning It!

10.7   Administrators at a small college are concerned that part-time evening students may not be familiar with all the services of the College. They wish to offer an orientation program to these students, but recognize that most of the part-time students work during the day and are generally very busy. They do not want to prepare an elaborate presentation if only a handful of part-time students will attend. They will conduct the orientation if more than 25% of the part-time students are interested in attending. Set up the null and alternative hypotheses to be used to decide if the administrators should offer the session.

10.8   A company is thinking about setting up an on-site day-care program for its employees. The CEO has stated that she will do so only if more than 80% of the employees favor such a decision. Set up the null and alternative hypotheses to be tested.

10.9   A human resource manager feels that men use e-mail less than women. Set up the null and alternative hypotheses that should be tested to determine if the human resource manager is correct.

10.10   In an attempt to improve quality many manufacturers are developing partnerships with their suppliers. A local fast-food burger outfit has partnered with its supplier of potatoes. The burger outfit buys potatoes in bags that weigh 20 lb. It wishes to set up the null and alternative hypotheses to test if the bags do weigh on the average 20 lb.

10.11   You are a connoisseur of chocolate chip cookies and you do not think that Nabisco's claim that every bag of Chips Ahoy cookies has 1000 chocolate morsels is correct. Set up the null and alternative hypotheses to test this claim.

10.12   Anti-lock brake systems (ABS) have been hailed as a revolutionary safety feature. A study by the National Highway Traffic Safety Administration looked at fatal accidents. The claim is that cars with ABS are in fewer fatal crashes than those without. Set up the null and alternative hypotheses to test this claim.

10.13   A college placement office wonders if the average entry level salary is different this year than last year's value of $25,000. Set up the null and alternative hypotheses to test this question.

10.14   The college placement office in Exercise 10.13 also wonders if there is a difference between the average salary of engineering graduates and business school graduates. Set up the null and alternative hypotheses to see if these averages are different.

Requires Exercise 10.13

10.15   Your new television has a 1-year warranty. You are given the option to buy a 3-year warranty and wonder if it is worth it. You wish to test the hypothesis that the average time before a problem occurs is more than 3 years. Set up the null and alternative hypotheses to test this belief.

10.16   It seems like you spend more money on groceries during the summer months when you eat more ice cream and drink more fluids. You know that you spend an average of $25/week on groceries during the winter months. Set up the null and alternative hypotheses to decide if, on the average, you spend more than this amount per week during the summer.

10.17   M&M/Mars claims that at least 20% of the M&M's in each package are the new blue color. Set up the null and alternative hypotheses to test this claim.

10.18   The computer center at a university claims that the average amount of time that students spend on-line has increased from last year's average of 1 hour per day. Set up the null and alternative hypotheses to test this claim.

10.7 ONE-TAIL TESTS OF THE MEAN: LARGE SAMPLE

In Section 10.5 you learned how to do a two-tail hypothesis test of the mean. In this section, we adapt that procedure to do a one-tail hypothesis test of the mean, μ. The procedure applies when you know the standard deviation of the population or if you have a sufficiently large sample. In the latter case, the only change in the procedure is to use the sample standard deviation, s, instead of σ. Therefore, we again use the label of "large sample" to describe these tests.

10.7.1 Lower-Tail Tests of the Mean

In Section 10.6.1 we saw that often we are interested in whether the mean has shifted in *one* direction. Reconsider the tissue strength problem. Most likely the company is interested in checking to see if the true average *MDStrength* has decreased from the specification of 1000 lb/ream. Thus, a lower-tail test would make more sense in this case. Let's use this example to illustrate the procedure for a one-tail hypothesis test of μ when the population standard deviation is known.

Fortunately, the steps for completing a one-tail test of the mean when the population standard deviation is known are quite similar to those we used in the previous section. In fact, the only thing that will change is the procedure for finding the rejection region in step 2. Recall that we have been using the following 5-step hypothesis testing procedure:

Steps for any hypothesis test

Step 1:	*Set up the null and alternative hypotheses.*
Step 2:	*Pick the value of α and find the rejection region.*
Step 3:	*Calculate the test statistic.*
Step 4:	*Decide whether to reject the null hypothesis.*
Step 5:	*Interpret the statistical decision in terms of the stated problem.*

Reconsider the problem that the tissue company faces as a one-tail test. The first step is to set up the null and alternative hypotheses.

> **EXAMPLE 10.27** The Tissue Company
>
> ### Step 1: Setting Up Null and Alternative Hypotheses
>
> Since the tissue manufacturer is checking to see if the true mean is *less than* 1000 lb/ream, this must go into the alternative hypothesis. So, we have the following setup.
>
> $$H_0: \quad \mu \geq 1000 \text{ lb/ream}$$
> $$H_A: \quad \mu < 1000 \text{ lb/ream}$$ ∎

We called this type of test a *lower-tail test*. The fact that the $<$ sign is in the alternative hypothesis makes it a lower-tail test. Next, you will see why it is called a lower-tail test as we find the rejection region.

Again, when we talk about "a lot lower" we must consider the difference in terms of the size of the standard error.

The second step is to select a value for α and find the rejection region. Again, let's think about what size Z values should lead us to reject the null hypothesis. We are interested in detecting if the true mean has decreased below 1000 lb/ream. So, values of $\overline{X}$ that are much lower than 1000 would lead us to be suspicious of the null hypothesis. Following the same logic that we used in the previous section, we can see that if $\overline{X}$ is a lot lower than 1000 then $\overline{X} - \mu$ will be negative and Z will be a negative number. Clearly, large negative Z values will then lead us to reject the null hypothesis.

How about large positive Z values? To get a positive Z value, $\overline{X}$ would have to be bigger than 1000. If this is the case, would we ever want to reject the null hypothesis? No! An $\overline{X}$ value of say, 1150, would certainly give us a large positive Z value but a sample mean of 1150 is consistent with the null hypothesis. Remember that we reject the null hypothesis only when the sample evidence is inconsistent with the null hypothesis. Thus, any $\overline{X}$ value larger than 1000 would lead us to continue to believe the null hypothesis. In fact, we wouldn't even have to do the hypothesis test in such cases!

So we have decided that only large negative Z values will lead us to reject the null hypothesis. We need only decide how large in the negative direction Z has to be before we reject H_0. Keep in mind that we arrived at this rejection region by simple logic! The particular value of the cutoff point for Z is again completely determined by the value of α.

Combining the idea that we will reject H_0 when the Z value is too large in the negative direction with the requirement that we incorrectly reject H_0 only a certain percentage (α) of the time, we arrive at the following picture:

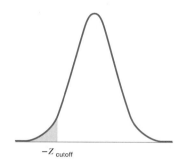

$-Z_{\text{cutoff}}$

Remember: The rejection region consists of those values of Z that lead you to reject H_0.

The shaded region is the rejection region. The rejection region is the *lower tail* of the normal distribution and hence the name *lower-tail test*. Remember that the rejection region consists of those values of Z, the test statistic, that lead you to reject the null hypothesis. Thus, a Z value smaller than $-Z_{\text{cutoff}}$ (or a large negative Z value) will lead you to reject the null hypothesis.

By using the standard normal table we can find this negative cutoff Z value so that the probability that we are in the rejection region is α. The rejection region for the tissue company can be found using this procedure.

EXAMPLE 10.28 The Tissue Company

Step 2: Selecting α and Finding the Rejection Region

Suppose α is set at 0.05. In this case, the shaded area must be 0.05. Since we are specifying the probability of 0.05, we look for 0.05 in the body of the standard Z table and see that the closest values are 0.0495 and 0.0505 corresponding to Z values of -1.65 and -1.64, respectively. It is OK to use either of these values or the average of these two numbers, which would be -1.645. Thus, we will reject H_0 if Z is smaller than -1.645 as shown in the shaded area below:

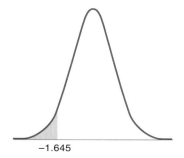

-1.645

■

You may remember that for the two-tail test when we set $\alpha = 0.05$, we found the rejection region to be

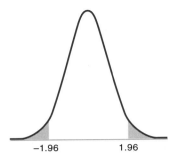

$$-1.96 \qquad 1.96$$

Can you explain why the negative Z_{cutoff} is different for the two-tail test? Remember that for the two-tail test, we split the value of $\alpha = 0.05$ in half before looking it up in the table. Thus, the cutoff value of -1.96 has an area of only 0.025 compared to the full value of 0.05 we used to get the cutoff value of -1.645.

 **TRY IT NOW!**

Finding the Rejection Region

Try to predict what will happen to the $-Z_{\text{cutoff}}$ value for the one-tail test if $\alpha = 0.025$. Now find it to confirm your guess.

In general, the procedure for finding the rejection region for a lower-tail test is as follows:

Steps for finding the rejection region for a lower-tail test

Step 1:   *Look up the value of α in the body of the Z table (find the one closest to it).*

Step 2:   *Read off the corresponding value: call it $-Z_{cutoff}$.*

Step 3:   *The rejection region consists of all values of Z that are less than $-Z_{cutoff}$ as shown below:*

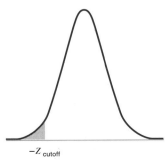

$$-Z_{\text{cutoff}}$$

The third step is to capture the information in the sample into a single number called the test statistic. It is the same calculation done in the previous section and is repeated here.

EXAMPLE 10.29   The Tissue Company

Step 3: Calculating the Test Statistic

Recall that the population standard deviation, σ, for the tissue manufacturer is 50 lb/ream and that the sample mean, $\overline{X}$, was observed to be 980 lb/ream. The Z statistic is then

$$Z = \frac{\overline{X} - \mu}{\sigma/\sqrt{n}}$$

$$Z = \frac{980 - 1000}{50/\sqrt{36}} = \frac{-20}{8.333} = -2.4$$ ■

Now we can finish the tissue strength problem. At step 4 the value of the test statistic is examined to see if it falls in the rejection region. Finally, step 5 requires that you interpret the statistical decision in terms of the problem statement.

EXAMPLE 10.30   The Tissue Company

Step 4: Deciding Whether the Null Hypothesis Should Be Rejected

Step 5: Interpreting the Statistical Decision

The rejection region for this one-tail, lower-tail alternative hypothesis test is shown below:

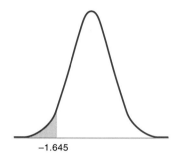

Clearly the calculated Z value of -2.40 falls in the rejection region. Thus, we reject H_0 and conclude that the true mean has decreased below 1000 lb/ream.

Remember that there is less than a 5% chance of observing a sample mean as low as 980 lb/ream if in fact the true mean really has not decreased below 1000! The precise calculation can be found based on the p value.

Let's look at another example.   ■

EXAMPLE 10.31   The Chapperel Steel Company

Lower-Tail Hypothesis Test

In Example 10.22 we set up the hypothesis test for the company that had adopted the partnering approach with its employees. Remember that the company was interested in seeing if this approach did in fact reduce the average number of sick days per year from 7.2. Suppose a sample of 20 employees of this company used an average of 6.9 sick days the first year after the partner approach was implemented. National personnel data indicate that the standard deviation is 0.5 day.

Is there enough evidence to conclude that the partnering approach did decrease the average number of sick days? Use $\alpha = 0.05$.

Following the hypothesis test procedure, step 1 requires you to set up the null and alternative hypotheses. Since the company is trying to show that the average number of sick days has decreased, that must be the alternative hypothesis. So, we have the following null and alternative hypotheses:

$$H_0: \quad \mu \geq 7.2$$
$$H_A: \quad \mu < 7.2$$

Step 2 requires you to set α and find the rejection region. Lacking any specific direction from the company, set $\alpha = 0.05$. Since it is a one-tail, lower-tail alternative test, you look up 0.05 in the body of the standard normal table and find $-Z_{cutoff}$ to be -1.645. The rejection region is shown below:

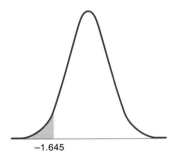

-1.645

Step 3 requires the calculation of the Z statistic:

$$Z = \frac{\overline{X} - \mu}{\sigma/\sqrt{n}} = \frac{6.9 - 7.2}{0.5/\sqrt{20}} = -2.68$$

Step 4 asks you to check to see if the value of the test statistic is in the rejection region. Clearly, $Z = -2.68$ is in the rejection region and we reject H_0 and conclude that $\mu < 7.2$.

Step 5 requires that you restate the statistical conclusion in terms of the problem. In this case, the data indicate that the average number of sick days used has been reduced. ■

 **TRY IT NOW!**

Frozen Foods
Lower-Tail Test of the Mean

Jake Bramhall can identify the make, model, and number of cylinders of any passing car but he can't tell the difference between stewed tomatoes and tomato paste. While more men are pushing shopping carts these days, many like Mr. Bramhall show little aptitude in the supermarket and display markedly different purchasing behavior than women. A study done by Consumer Network Inc. shows that the average amount of money spent by 100 single men on facial tissues was $7.38. On the basis of these data can you conclude that men spend less money on facial tissues than the average $8.19 spent by women on facial tissues? Use a population standard deviation of $3.50 and an α value of 0.05.

Step 1:

Step 2:

Step 3:

Step 4:

Step 5:

Are the results different if you use $\alpha = 0.01$?

10.7.2 Upper-Tail Test of μ

So far all of the one-tail tests that we have performed have been lower-tail tests. That is, the $<$ symbol was used in the alternative hypothesis. We must now consider what will be different in our procedure if we need to do an upper-tail test, that is, one with the $>$ symbol in the alternative hypothesis. Let's look at an example.

EXAMPLE 10.32 Frozen Foods

Step 1: Setting Up the Null and Alternative Hypotheses for an Upper-Tail Test

In the same study of supermarket behavior that you just looked at in the *Try It Now!* exercise, the average amount of money spent of frozen dinner/entrees by men was found to be $51.48. There were 100 men in the study, and the population standard deviation can be assumed to be $10. Is there any evidence to indicate that the men spend more than the national average for women of $40.71?

The major steps of the hypothesis testing procedure are the same. Here again, the only difference is in the rules for finding the rejection region used in step 2.

Since we are interested in showing that the men spend *more* on frozen dinner/entrees, we know it is a one-tail test with the $>$ symbol in the alternative hypothesis. That gives us the following null and alternative hypotheses:

$$H_0: \quad \mu \leq \$40.71$$
$$H_A: \quad \mu > \$40.71$$

That completes step 1. ■

Step 2 asks us to select a value for α and find the rejection region. If you look back at the logic that led us to the rejection region for the lower-tail test, you will see that in this case it is values of $\overline{X}$ that are much larger than 40.71 that will lead us to reject H_0. These values of $\overline{X}$ will yield large positive Z statistics. This observation coupled with what we have learned about the behavior of the normal distribution gives the following picture:

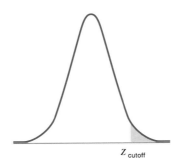

$$Z_{\text{cutoff}}$$

Here again the shaded region has an area of α and we will reject H_0 only when Z is bigger than Z_{cutoff}, thus falling into the rejection region.

EXAMPLE 10.33   Frozen Foods

Step 2: Selecting the Value for α and Finding the Rejection Region

Suppose we set $\alpha = 0.05$. In this case the shaded area must equal 0.05. Remember that the normal distribution table gives us only lower-tail areas as the probabilities. But, since we know that the distribution is symmetric we can look up 0.05 and simply drop the negative sign in order to get Z_{cutoff}. The value for Z_{cutoff} is arrived at by dropping the negative sign in front of either -1.64 or -1.65. Hence, use either 1.64 or 1.65 or the midpoint, 1.645.

The rejection region is shown below:

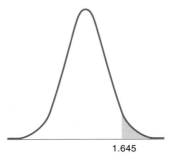

1.645 ■

This example has shown us the steps for the rejection region for an upper-tail test:

Steps for finding the rejection region for an upper-tail test

Step 1:   *Look up the value of α in the body of the Z table (find the one closest to it).*

Step 2:   *Read off the corresponding value and drop the negative sign. Call it Z_{cutoff}.*

Step 3:   *The rejection region consists of all values of Z that are greater than Z_{cutoff} as shown below:*

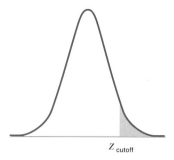

Z_{cutoff}

We are now ready to complete this example.

EXAMPLE 10.34   Frozen Foods

Steps 3,4, and 5

For step 3 we calculate the *Z* test statistic as follows:

$$Z = \frac{\overline{X} - \mu}{\sigma/\sqrt{n}} = \frac{51.48 - 40.71}{10/\sqrt{100}} = 10.77$$

Finishing this example, we do steps 4 and 5. Clearly, our *Z* value of 10.77 falls in the rejection region and we should reject the null hypothesis and conclude that men spend more than \$40.71 on average for frozen dinners/entrees!   ■

 **TRY IT NOW!**

Supermarket Survey
Upper-Tail Test of the Mean

From the same supermarket survey it is found that the 100 men spent, on the average, \$19.98 on low-calorie soft drinks. Is there enough evidence to conclude that men spend more than women, who, on the average, spend \$18.86? Assume that the population standard deviation is \$10 and use $\alpha = 0.05$.

10.7.3 Adjustments to *p* Value Calculation for One-Tail Tests

In Section 10.5.2 we calculated the *p* value for a two-tail hypothesis test of μ when σ was known. The procedure for one-tail tests is the same except that you do not double the tail area probability.

ANS. $Z = 1.12$; NO

> **EXAMPLE 10.35**   Frozen Foods
>
> *Calculating the p Value*
>
> Consider the frozen foods example (Example 10.34). This problem used an *upper-tail test of* μ with σ known. A sample of 100 men spent $51.48 on the average on frozen foods. The Z statistic was calculated to be 10.77. A quick look at this Z statistic tells us that the p value will be extremely small.
>
> The p value is calculated as follows:
>
> $$p \text{ value} = P(Z > 10.77) = \text{virtually } 0 \qquad \blacksquare$$

10.7.4 Exercises—Learning It!

10.19   A major manufacturer of glue products thinks it has found a way to make the glue adhere longer than the current average of 90 days. The manufacturer wishes to see if the glue products made this way have an average time to failure greater than 90 days. A sample of 30 tubes of the new glue yields an average of 93 days before failing. The failure time is normally distributed with a standard deviation of 3 days.

  (a) Set up the null and the alternative hypotheses to test if average time to failure is greater than 90 days.

  (b) Test your hypothesis using $\alpha = 0.05$.

  (c) Find the p value.

  (d) Based on the p value, what can you conclude about the average time to failure for the new product?

10.20   Recent medical research indicates that skin cancer patients who receive a new medication for skin cancer live longer than those who do not. The average length of life prior to the development of this medication was 18 months. The medical community wishes to test the claim made by the developers of this drug. A sample of 35 patients who received the medication lived an average of 21 months. The standard deviation is 5 months.

  (a) Set up the null and the alternative hypotheses to test if average length of life has increased from 18 months.

  (b) Test your hypothesis using $\alpha = 0.05$.

  (c) Find the p value.

  (d) Based on the p value, what can you conclude about the average length of life for patients who receive the vaccine?

10.21   An automobile company thinks that with new designs, its cars will last longer before having a problem. For this reason, the company wishes to extend the warranty that comes with the vehicle in hopes of attracting more customers. Before making this change, the idea is tested. Prior to the design changes, the cars lasted on the average 43 months before having a major problem. A sample consisting of 50 cars was tested. The cars lasted an average of 44 months before having a major problem. The standard deviation is 2 months.

  (a) Set up the null and the alternative hypotheses to test if average time before having a major problem is longer than 43 months.

  (b) Test your hypothesis using $\alpha = 0.05$.

  (c) Find the p value.

  (d) Based on the p value, what can you conclude about the average time before having a major problem?

10.22   A manufacturer of top-of-the line tennis rackets claims that its Smack Em tennis racket will change a player's game. A tennis pro currently serves the ball at an average speed of 115 mph with a standard deviation of 2.5 mph. The speeds are normally distributed. The tennis pro decides to test the company's claim and records the speed of his serve for 15 balls using the Smack Em racket. The data are shown in the following table:

Speed (mph)	
117.3	115.9
115.1	115.2
116.0	115.0
116.2	113.0
112.9	120.8
115.4	116.9
113.8	114.4
114.2	

(a) Set up the null and the alternative hypotheses to test if the average service speed has increased using the new racket.

(b) Test your hypothesis using $\alpha = 0.05$.

(c) Find the p value.

(d) Based on the p value, should the tennis pro invest in the new racket?

10.23 In an attempt to improve quality, many manufacturers are developing partnerships with their suppliers. A local fast-food burger outfit has partnered with its supplier of potatoes. The burger outfit buys potatoes in bags that weigh 20 lb. It does not wish to accept underweight bags of potatoes. A sample of 40 bags shows an average weight of 19.95 lb with a standard deviation of 0.1 lb.

(a) Set up the null and the alternative hypotheses to test if the average bag weighs at least 20 lb.

(b) Test your hypothesis using $\alpha = 0.05$.

(c) Find the p value.

(d) Based on the p value, should the burger outfit accept the shipment of potatoes?

10.8 WHAT ERROR COULD YOU BE MAKING?

Since you will be deciding between the null and the alternative hypotheses using only sample information, there is always a chance that you will be wrong. That is, you may choose to believe the null hypothesis when, in fact, it is not true. Alternatively, you may choose to believe the alternative hypothesis (rejecting the null hypothesis) when in fact it is not true. These are the two ways that you could be wrong when you perform a hypothesis test.

In Chapter 7 you learned the basics of probability. You saw that probabilities are a way to measure the chance that something occurs. In hypothesis testing, we use probabilities to measure the chance of being wrong. If we can state that there is a 5% chance that we have made an error, then we have a sense of how often we will be wrong.

Let's look in more detail at the two ways that you could be wrong.

10.8.1 Two Types of Errors

Four outcomes could result from the decisions you make in each hypothesis test situation. Look at Table 10.2 (page 444) and think carefully about how each of these outcomes could occur.

As you can see from Table 10.2, there are two ways that you could make the right choice and two ways that you could make a mistake. Unfortunately, you will not know whether you made the right decision or you made an error. This is because to know which category you are in you would have to know what the population actually looked like and thus if H_0 or H_A were true. If you knew that, you wouldn't be taking a sample!

TABLE 10.2 Possible Outcomes in a Hypothesis Test

Sample Indicates You Should	In Actuality:	
	H_0 is True	H_A is True
Believe H_0	You made the right choice—no error	*Error*
Believe H_A	*Error*	You made the right choice—no error

Let's look at these four possible outcomes from the perspective of the tissue manufacturer. They are presented in Table 10.3.

TABLE 10.3 Four Possible Outcomes for the Tissue Strength Problem

Sample Indicates You Should	In Actuality:	
	H_0 is True $\mu = 1000$	H_A is True $\mu \neq 1000$
Believe H_0: You believe that $\mu = 1000$ and so the product specs are being met.	You correctly decide that the product specs are being met.	You **incorrectly** decide that the product specs are being met when they are not.
Believe H_A: You believe that $\mu \neq 1000$ and so the product specs are not being met.	You **incorrectly** decide that the product specs are not being met when they are.	You correctly decide that the product specs are not being met.

To be clear which of the two possible errors we are talking about, we need to give them names. Look at Table 10.4, which shows the same four situations with each error named.

TABLE 10.4 Name Those Errors: Type I and Type II

Sample Indicates You Should	In Actuality:	
	H_0 is True	H_A is True
Believe H_0	You made the right choice—no error	*Type II error*
Believe H_A	*Type I error*	You made the right choice—no error

You can see that statisticians are uncreative at naming errors and they are simply called **Type I** and **Type II errors.** These errors are stated in the following definitions:

A *Type I error* is made when you reject the null hypothesis and the null hypothesis is actually true. In other words, you incorrectly reject a true null hypothesis.

A *Type II error* is made when you fail to reject the null hypothesis and the null hypothesis is false. In other words, you continue to believe a false null hypothesis.

EXAMPLE 10.36 The Bottle-Filling Problem

Type I and Type II Errors

Returning to the bottle-filling problem in Example 10.6, look at the Type I and Type II errors for this hypothesis test. We had set up the following null and alternative hypotheses:

$$H_0: \quad \mu = 32 \text{ oz}$$
$$H_A: \quad \mu \neq 32 \text{ oz}$$

Remember that a Type I error is made if you believe H_A but H_0 is true. For this problem that would mean that you believe the true mean is not equal to 32 oz (H_A) and therefore you believe the bottles are not being filled properly. You adjust the bottle filling machine when in fact the true mean is really 32 oz (H_0) and the machine did not need to be adjusted.

A Type II error is made if you believe H_0 is true when H_A is true. In this case, you would believe that the soda machine is properly filling the bottles when, in fact, it is either overfilling them or underfilling them. So, you do not adjust a machine that needs to be adjusted. ■

TRY IT NOW!

The Potato Chip Manufacturer
Examining the Type I and Type II Errors

Find the Type I and Type II errors for the hypothesis test that you set up for the potato chip manufacturer of this chapter.

10.8.2 Probability of Making an Error

Since we can't eliminate sampling error unless we sample the entire population, the next best thing we can do is try to control the chances that we are in error. The standard method of labeling the chance of making these errors is given in the following definition.

ANS. A TYPE I ERROR IS MADE IF YOU CONCLUDE THAT THE BAGS ARE NOT BEING FILLED PROPERLY WHEN THEY ARE BEING FILLED PROPERLY. A TYPE II ERROR IS MADE IF YOU CONCLUDE THAT THE BAGS ARE BEING FILLED TO 10 OUNCES BUT THEY ARE REALLY BEING OVER OR UNDER FILLED.

> The probability of making a Type I error is called **α (alpha).**
>
> The probability of making a Type II error is called **β (beta).**

Clearly, α and β must be numbers between 0 and 1 since they are probabilities. As the investigator, you will get to decide the value of α. This means that you can specify the chances of making a Type I error to be anything you wish. Once you set α, the value of β is completely determined. Your first thought might be to set α to be as small as possible so there is hardly any chance of making a Type I error. Well, there is a price to pay for making α really small. As α gets smaller, the size of β gets larger. So, just like most things in life, there is a trade-off. You can force the chance of making a Type I error to be really small but then you have to live with a greater chance of making a Type II error. This trade-off is pictured in Figure 10.3.

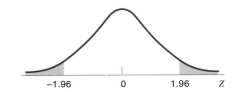

FIGURE 10.3   Two-sided rejection region with $\alpha = 0.05$

The area of the shaded region is equal to α. This is the chance that we reject H_0 when we should not have done so. If we look at Figure 10.4 we can see that the value of α has been reduced. This means we will be less likely to reject H_0, when we should not reject it. If there is less chance of rejecting the null hypothesis, then there must be a greater chance of continuing to believe the null hypothesis when we should not. This is the definition of β.

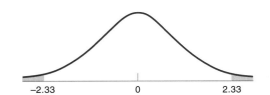

FIGURE 10.4   Two-sided rejection region with $\alpha = 0.02$

Thus, as $\alpha \uparrow \beta \downarrow$ and as $\alpha \downarrow \beta \uparrow$

Let's examine this trade-off in terms of the tissue strength problem.

EXAMPLE 10.37   **The Tissue Strength Problem**

Possible Value for α

In terms of the tissue strengths, suppose you decide to set the chance of making a Type I error, α, really small, say 0.01. This means that only 1% of the time will you incorrectly decide that the product specs are not being met, when really they are OK. This is the good news.

The trade-off is that the chance of making a Type II error increases. In this case, it means that the likelihood of deciding that the product specs are OK when they are not gets larger. How large the likelihood gets will depend on the real value of μ.   ∎

The standard value of α used to test hypotheses is 0.05. However, if making a Type I error is much more costly than making a Type II error, then you should consider testing at a value of α equal to 0.01. This will, of course, cause the value of β to increase. If, on the other hand, you feel that making a Type II error is particularly costly, then you should consider testing at a value of $\alpha = 0.10$. If both errors are equally costly then you should probably use the standard value of $\alpha = 0.05$.

Consider another example.

EXAMPLE 10.38   The Airplane Bolts Problem

Considering the Costs of the Errors

Suppose you are inspecting bolts that are used to fasten engines to airplanes. The mean strength of these bolts must be at least 100 pounds per square inch (psi). You need to decide if a shipment of these bolts is OK—that is, if they are strong enough. You realize that you cannot test them all because once you test a bolt it cannot be used in the airplane. This is what we call destructive sampling. If you test them all you will have none to use. You decide to take a sample of bolts, and on the basis of the mean strength of the sample you will decide if the shipment is OK.

One possible way to set up the hypothesis test is as follows:

$$H_0: \quad \mu \geq 100 \text{ psi} \qquad \text{(The shipment is OK.)}$$
$$H_A: \quad \mu < 100 \text{ psi} \qquad \text{(The shipment is not OK.)}$$

Remember: A Type I error is that you believe the alternative hypothesis when, in fact, the null hypothesis is true.

For this hypothesis test, what is a Type I error? In this case, you would believe that the shipment is not OK when in fact it is OK. So, you would send the bolts back for rework when they did not need to be reworked.

Now let's look at a Type II error for this problem. In this case, you would keep the bolts even though they are not strong enough.

Which of these errors is more costly?

Recall: A Type II error is that you believe the null hypothesis when, in fact, the alternative hypothesis is true.

- Send bolts back for rework when they did not need to be reworked

- Keep bolts that are not strong enough.

We would probably all agree that keeping bolts that are not strong enough is the more costly error (in terms of human life). This is a Type II error. Thus, we should try to minimize the value of β by testing at α equal to 0.10. ■

TRY IT NOW!

New Package Design:

Setting the Value of α

We have seen that a one-sided test is often used to investigate whether a new method of advertising or producing something is better than the existing method. Consider a company that is trying a new package design for its product. The average sales for this product are currently $1500/month. The null and alternative hypotheses would be

$$H_0: \quad \mu \leq \$1500$$
$$H_A: \quad \mu > \$1500$$

(continued)

In terms of the company's decision to adopt or not adopt this new design, what are the Type I and Type II errors?

What value of α would you suggest be used to conduct the test?

10.8.3 Exercises—Learning It!

10.24 Administrators at a small college are concerned that part-time evening students may not be familiar with all the services of the College. They wish to offer an orientation program to these students but recognize that most of the part-time students work during the day and are generally very busy. The administrators do not want to prepare an elaborate presentation if only a handful of part-time students will attend. Hence, they will conduct the orientation if more than 25% of the part-time students are interested in attending.

(a) State the consequences of a Type I error.

(b) State the consequence of a Type II error.

(c) Suggest a value for α, and justify your choice.

10.25 A company is thinking about setting up an on-site day-care program for its employees. The CEO has stated that she will do so only if more than 80% of the employees favor such a decision.

(a) State the consequences of a Type I error.

(b) State the consequence of a Type II error.

(c) Suggest a value for α, and justify your choice.

10.26 In an attempt to improve quality, many manufacturers are developing partnerships with their suppliers. A local fast-food burger outfit has partnered with its supplier of potatoes. The burger outfit buys potatoes in bags that weigh 20 lb. It does not wish to accept underweight bags of potatoes.

(a) State the consequences of a Type I error.

(b) State the consequence of a Type II error.

(c) Suggest a value for α, and justify your choice.

ANS. TYPE I ERROR IS TO CONCLUDE THE NEW DESIGN IS BETTER WHEN IT IS NOT. TYPE II ERROR IS TO CONCLUDE THE NEW DESIGN IS NOT BETTER WHEN IT IS. THESE ARE EQUALLY COSTLY ERRORS SO USE $\alpha = 0.05$.

10.27 You are a connoisseur of chocolate chip cookies and you do not think that Nabisco's claim that every bag of Chips Ahoy cookies has 1000 chocolate morsels is correct.

(a) State the consequences of a Type I error.

(b) State the consequence of a Type II error.

(c) Suggest a value for α, and justify your choice.

10.28 Anti-lock brake systems (ABS) have been hailed as a revolutionary safety feature. A study by the National Highway Traffic Safety Administration looked at fatal accidents. The claim is that cars with ABS are in fewer fatal crashes than those without.

Requires Exercise 10.12

(a) State the consequences of a Type I error.

(b) State the consequence of a Type II error.

(c) Suggest a value for α, and justify your choice.

10.29 A College Placement Office wonders if there is a difference between the average salary of engineering graduates and business school graduates.

Requires Exercise 10.13

(a) State the consequences of a Type I error.

(b) State the consequence of a Type II error.

(c) Suggest a value for α, and justify your choice.

10.30 Your new television has a 1-year warranty. You are given the option to buy a 3-year warranty and wonder if it is worth it. You wish to test the hypothesis that the average time before a problem occurs is more than 3 years.

(a) State the consequences of a Type I error.

(b) State the consequence of a Type II error.

(c) Suggest a value for α, and justify your choice.

10.31 M&M/Mars claims that at least 20% of the M&M's in each package are the new blue color.

(a) State the consequences of a Type I error.

(b) State the consequence of a Type II error.

(c) Suggest a value for α, and justify your choice.

10.32 A computer center is arguing for more computers in the lab for students at a mid-size college. The computer center at a university claims that the average amount of time that students spend on-line has increase from last year's average of 1 hour per day.

(a) State the consequences of a Type I error.

(b) State the consequence of a Type II error.

(c) Suggest a value for α, and justify your choice.

10.9 HYPOTHESIS TESTING IN EXCEL

Hypothesis tests from real data are sometimes inconvenient to do using a pencil and paper. While Excel has many statistical analysis tools already available, it does not have a tool for large-sample hypothesis tests for the mean. In this section we will look at a macro that allows you to do large-sample hypothesis tests for the population mean.

The macro provides output for both two-sided and one-sided hypothesis tests. You will use whatever output is pertinent to your test. We will test the tissue strength data to see if the population mean is really 1000 lb/ream, which is a two-sided test. The macro assumes that you have the data located in an Excel worksheet. A portion of the data and the summary statistics are shown in Figure 10.5 (page 450).

	A	B	C	D
1	Day	MDStrength	MDStrength	
2	1	1006		
3	1	994	Mean	990.8266667
4	1	1032	Standard Error	6.66764065
5	1	875	Median	991
6	1	1043	Mode	992
7	1	962	Standard Deviation	57.74346186
8	1	973	Sample Variance	3334.307387
9	1	1036	Kurtosis	-0.180965938
10	1	1084	Skewness	-0.264673668
11	1	1067	Range	261
12	1	939	Minimum	830
13	1	984	Maximum	1091
14	1	1001	Sum	74312
15	1	1076	Count	75

FIGURE 10.5   Tissue data and summary statistics

To perform a large sample hypothesis test for a mean, follow these steps:

1. Make sure that you have the MacDoIt.xls worksheet open and select **Tools** > **Macro** > **Macros** to obtain the list of macros. From the list, select **Z test** and click **Run.** A new worksheet is added to the current workbook and the table shown in Figure 10.6 appears. The table opens with values already in place. You will use Excel functions with your data to fill in the Input section of the table.

	A	B	C
1	Hypothesis Test for Mean - Z-test		
2	Input		
3	Sample Mean	m	990.8266667
4	Hypothesized Mean	μ_0	1000
5	Standard Deviation	σ	57.74346186
6	Sample Size	n	75
7	Significance Level	α	0.05
8	Output		
9	Standard Error	StdErr	6.6676
10	Z-Statistic	z	-1.3758
11	Lower-Tail Test for H_A: $\mu < \mu_0$		
12	Lower-Tail Critical Z	Z_L	-1.6449
13	Lower-Tail p-value	p-value$_L$	0.0844
14	Two-Tail Test for HA: $\mu \neq \mu_0$		
15	Two-Tail Absolute Critical Z	Z_T	1.9600
16	Two-Tail p-value	p-value$_T$	0.1689
17	Upper-Tail Test for HA: $\mu > \mu_0$		
18	Upper-Tail Critical Z	Z_U	1.6449
19	Upper-Tail p-value	p-value$_U$	0.9156

FIGURE 10.6   Table for Z test

If you get a #DIV/0! error, you switched to the z test sheet before typing the last parenthesis. It might be easier to copy the data to the sheet with the table to avoid this problem.

2. The first value that should be input is the sample mean. Position the cursor in the cell next to the one labeled m. You are going to type in the Excel function **AVERAGE,** which expects a data range as its input. Type **=average(** and switch to the worksheet that contains your data. Highlight the data and move the cursor to the status line of the worksheet that contains the data. Type in a right parenthesis ")" and hit Enter to finish the function. You will see the value of $\overline{X}$ appear in the table and the cursor will move to the next input cell.

3. Enter the hypothesized value for the population mean. In this case we will enter 1000 because that was the target value for *MDStrength*. Hit Enter to move the cursor to the next input cell.

4. Input the value of σ. Since you are doing a Z test, you are expected to know the population standard deviation. If you do not, but you have a large sample, then you can use the sample standard deviation, s. To do this, use the Excel function **STDEV** in this cell. This function works exactly like the one for average.

5. Hit Enter and input the sample size—in this case, 75.

6. Hit Enter again and enter the level of significance for the test.

When you hit Enter the last time, the table will update and give the test output as shown in Figure 10.7.

	A	B	C
1	**Hypothesis Test for Mean - Z-test**		
2	**Input**		
3	Sample Mean	m	98
4	Hypothesized Mean	μ_0	100
5	Standard Deviation	σ	10
6	Sample Size	n	30
7	Significance Level	α	0.05
8	**Output**		
9	Standard Error	StdErr	1.8257
10	Z-Statistic	z	-1.0954
11	Lower-Tail Test for H_A: $\mu < \mu_0$		
12	Lower-Tail Critical Z	Z_L	-1.6449
13	Lower-Tail p-value	p-value$_L$	0.1367
14	Two-Tail Test for HA: $\mu \neq \mu_0$		
15	Two-Tail Absolute Critical Z	Z_T	1.9600
16	Two-Tail p-value	p-value$_T$	0.2733
17	Upper-Tail Test for HA: $\mu > \mu_0$		
18	Upper-Tail Critical Z	Z_U	1.6449
19	Upper-Tail p-value	p-value$_U$	0.8633

FIGURE 10.7 Z test table with input and output

The output section of the Z test includes the standard error of the mean and the value of the Z test statistic. In addition, it includes information for all three different ways that the null and alternative hypotheses can be set up. In this case you want to look at the section labeled Two-Tail Test. You see that the critical value for the test is ± 1.96. Comparing the test statistic to this value, you would fail to reject H_0. The p value for the test is also provided in the output.

CHAPTER 10 SUMMARY

In this chapter you have learned the key steps involved in doing any hypothesis test. You first formulate two opposing viewpoints called the null and alternative hypotheses. These hypotheses are typically theories or ideas about the value of one or more population parameters. The technique of hypothesis testing helps you decide between these opposing hypotheses using the sample data as the evidence upon which to base your decision. In doing any hypothesis test there are two possible errors you can make. These are called Type I and Type II errors. The probabilities of making these errors are labeled α and β, respectively. It is desirable to make both of these probabilities small, but there is a tradeoff.

In this chapter you have also learned the procedure for doing large-sample tests of the mean. This procedure applies whenever you know the population standard deviation or you have a sufficiently large sample size, $n > 30$. Thus, the tests are called large-sample tests.

Key Terms

Term	Definition	Page Reference
Hypothesis	A **hypothesis** is an idea, an assumption, or a theory about the behavior of one or more variables in one or more populations.	405
Hypothesis test	A **hypothesis test** is a statistical procedure that involves formulating a hypothesis and using sample data to decide on the validity of the hypothesis.	405
Null hypothesis	The **null hypothesis** is a statement about the population(s). It is referred to as H_0.	410
Alternative hypothesis	The **alternative hypothesis** is a statement about the population(s) that is opposite to the null hypothesis. It is referred to as H_A.	410
Rejection region	The **rejection region** is the range of values of the test statistic that will lead you to reject the null hypothesis.	413
Test statistic	A **test statistic** is a number that captures the information in the sample. It will be used to choose between the null and alternative hypotheses.	414
p value	The **p value** is defined to be the smallest value of α for which you can reject H_0. This is also called the level of significance of the test.	422
Two-tail hypothesis test	A **two-tail test** of the population mean has the following null and alternative hypotheses: H_0: $\mu =$ [a specific number] H_A: $\mu \neq$ [a specific number]	426

Term	Definition	Page Reference
Lower-tail hypothesis test	A **lower-tail hypothesis test** of the population mean has the following null and alternative hypotheses: $H_0: \mu \geq$ [a specific number] $H_A: \mu <$ [a specific number]	428
Upper-tail hypothesis test	An **upper-tail hypothesis test** of the population mean has the following null and alternative hypotheses: $H_0: \mu \leq$ [a specific number] $H_A: \mu >$ [a specific number]	429
Type I error	A **Type I error** is made when you reject the null hypothesis and the null hypothesis is actually true. In other words, you incorrectly reject a true null hypothesis.	444
Type II error	A **Type II error** is made when you fail to reject a false null hypothesis. In other words, you fail to reject H_0 when you should have rejected it.	445
Alpha (α)	The probability of making a Type I error is called $\boldsymbol{\alpha}$.	446
Beta (β)	The probability of making a Type II error is called $\boldsymbol{\beta}$.	446

Key Formulas

Term	Formula	Page Reference
Z test statistic	$Z = \dfrac{\overline{X} - \mu}{\sigma/\sqrt{n}}$	415

CHAPTER 10 EXERCISES

Learning It!

10.33 The manufacturer of an over-the-counter pain reliever claims that its product brings pain relief to headache sufferers in an average of 3.5 minutes. To be able to make this claim in its television advertisements, the manufacturer was required by a particular television network to present statistical evidence in support of the claim.

(a) Is this a one-tail test or a two-tail test?

(b) Set up the null and the alternative hypotheses.

(c) What is a Type I error?

(d) What is a Type II error?

(e) A sample of 40 headache sufferers is used. They report that it took an average of 3.3 minutes to get some relief. If the standard deviation is 0.5 minute, perform the hypothesis test using $\alpha = 0.02$.

10.34 Pharmaceutical companies spend millions of dollars annually on research and development of new drugs. After a new drug is formulated, the pharmaceutical company must subject it to lengthy and involved testing before receiving the necessary permission from the Food and Drug Administration (FDA) to market the drug. The pharmaceutical company must provide substantial evidence that a new drug is safe prior to receiving FDA approval, so that the FDA can confidently certify the safety of the drug.

(a) Set up the null and alternative hypotheses.

(b) What is a Type I error?

(c) What is a Type II error?

10.35 Suppose a quality manager for a catsup company is interested in testing whether the mean number of ounces of catsup per family-size bottle differs from the labeled amount of 20 oz.

(a) Is this a one-tail or a two-tail test?

(b) Set up the null and the alternative hypotheses.

(c) What is a Type I error?

(d) What is a Type II error?

(e) A sample of 30 catsup bottles is checked. The average number of ounces in the sampled bottles is 19.97 oz. If the standard deviation is 0.1 oz, perform the hypothesis test using $\alpha = 0.05$.

(f) Find the p value. What conclusion can you draw about the mean number of ounces in catsup bottles?

10.36 The LEGO Group is an international company which makes the LEGO blocks that many of us have played with at some time. Many of its products require that the production process perform according to specifications. One of the products is Little People and the diameter of the neck of each of the Little People must be 0.5 inch so that it can be attached to the head properly. LEGO is interested in testing to see if this process is performing according to specifications.

(a) Is this a one-tail test or a two-tail test?

(b) Set up the null and the alternative hypotheses.

(c) What is a Type I error?

(d) What is a Type II error?

(e) A sample of 30 blocks is tested. The average diameter of the sampled necks was 0.48 inch. If the standard deviation is 0.05 inch, perform the hypothesis test using $\alpha = 0.02$.

(f) Find the p value and make a recommendation to LEGO.

10.37 Another concept from the TQM (total quality management) movement is the idea of building better relationships with your vendor. In doing so the quality of the incoming raw material is improved. If you buy raw materials from me you may wish to sample some of the material I sell you to test the quality claims that I have made. Suppose I claim that the weight of the material I sold you was on the average 10 lb. You wish to test this claim.

(a) Is this a one-tail test or a two-tail test?

(b) Set up the null and alternative hypotheses.

(c) What is a Type I error?

(d) What is a Type II error?

(e) A sample of 30 yields an average weight of 10.03 lb. If the standard deviation is 0.03 lb, perform the hypothesis test using $\alpha = 0.05$.

(f) Find the p value and decide if you would like to continue to purchase material from one supplier.

10.38 An up-and-coming restaurant chain is trying to decide whether to locate in the town of Longmortgage. It will locate there only if the average number of days of the week that people eat out is three or more. The company wishes to investigate the town of Longmortgage as a potential location.

(a) Is this a one-tail test or a two-tail test?

(b) Set up the null and the alternative hypotheses.

(c) What is a Type I error?

(d) What is a Type II error?

Thinking About It!

Requires Exercise 10.2

10.39 Would it be possible to switch the null and the alternative hypotheses for the ETS problem? Explain why or why not.

Requires Exercise 10.19

10.40 For the glue manufacturer, which of the possible errors is the more costly error? Consider the consequences of making a Type I error and the consequences of making a Type II error.

Requires Exercise 10.20

10.41 Consider the case of the medical research described in Exercise 10.20.

(a) What position would you be taking if you had made the null hypothesis: $H_0: \mu \geq 18$?

(b) What factors other than the medication might influence the length of time a cancer patient lives?

10.42 What are the implications of the car company setting the warranty too long?   *Requires Exercise 10.21*

10.43 Consider the cereal manufacturer.   *Requires Exercise 10.6*
   (a) From the cereal manufacturer's perspective, what is the more costly error?
   (b) From the consumer's perspective, what is the more costly error?

10.44 For the television network, what are the implications and consequences of a Type I error? A Type II error?   *Requires Exercise 10.33*

10.45 Rewrite the hypothesis test for the FDA problem by switching the null and alternative hypotheses.   *Requires Exercise 10.34*
   (a) What is a Type I error for this new setup?
   (b) What is a Type II error for this new setup?
   (c) Compare your answers with those you found in Exercise 10.34. What has happened to the errors?
   (d) Which setup do you feel is better and why?

10.46 If you took a sample of catsup bottles and observed a sample mean, $\overline{X} = 19.7$ oz, does this automatically mean that the bottle-filling machine is not working properly? Explain why or why not.   *Requires Exercise 10.35*

10.47 What are the implications for LEGO if the process that manufactures the necks for the Little People is not producing necks with the specified diameter?   *Requires Exercise 10.36*

10.48 As the receiver of raw material, explain why you would not want to test every component of raw material (i.e., why not test the population instead of just examining a sample).   *Requires Exercise 10.37*

10.49 From the perspective of the restaurant chain, what are the consequences of a Type I error? a Type II error? Which error is more costly?   *Requires Exercise 10.38*

Exercises—Doing It!

10.50 Let us return to the tissue manufacturer presented at the beginning of this chapter. Recall that the manufacturer was concerned about customer complaints involving sheets tearing on removal. It decided to look at the manufacturing process to see how it compared to the product specification and to see if any changes needed to be made. Making changes to the manufacturing process is a big job and before proceeding the manufacturer would like to be a little more certain that the changes need to be made. A sample of 225 was taken from a single tissue machine. The samples were taken over three different days and the results are stored in the data file TISSUES.XXX. A portion of the data file is shown below: *Datafile: TISSUES.XXX*

Day	MDStrength	CDStrength
1	1006	422
1	994	440
1	1032	423
1	875	435
1	1043	445

The variable *Day* keeps track of the day on which the tissue was produced and ranges from 1 to 3.
The variable *MDStrength* measures the machine-directional strength and is measured in lb/ream. The product specifications for *MDStrength* state that the measurement is normally distributed with $\mu = 1000$ lb/ream and $\sigma = 50$ lb/ream.
The variable *CDStrength* measures the cross-directional strength and is measured in lb/ream. The product specifications for *CDStrength* state that the measurement is normally distributed with a mean of 450 lb/ream and a standard deviation of 25 lb/ream.
After giving the matter some thought the engineers looking at the tissue manufacturing process wondered if the fact that the measurements were taken on three different days might bias the analysis. Since operating conditions are not always the

same on any given day, they wondered if the process was off target on all three days, or if it was off target on one or two days. They decided to check each day individually.

(a) Perform a hypothesis test at the 0.05 level of significance to determine whether *MD-Strength* was 1000 on each of the three days separately. Use the target standard deviation of 50 and write your results in the table below:

Day	Z Statistic	Decision (reject H_0/fail to reject H_0)
1		
2		
3		
All 3 days together		

(b) Using these results, do you think the company can assume that *MDStrength* is running on target? Why or why not?

(c) Using a 0.05 level of significance, do the sample data indicate that the variable *CD-Strength* was running according to target specifications on each of the three days? Do you conclude the same for the three-day period as a whole?

Day	Z Statistic	Decision (reject H_0/fail to reject H_0)
1		
2		
3		
All 3 days together		

(d) Using these results, do you think the company can assume that *CDStrength* is running on target? Why or why not?

(e) Since the consumer problem that prompted the study was sheets tearing on removal, the company decides that it is really interested in knowing only whether the average *CDStrength* is less than the target of 450. If it is not, the company will not make any adjustments. Redo the tests in part (c) as a one-tail test to reflect this change. What is the result?

(f) One of the other variables that needs to be considered is some measure of total strength, which is related to both the dispensing defects and another important tissue variable, softness. Often strength and softness are tradeoffs. Using both *MD-Strength* and *CDStrength*, the company can calculate the Geometric Mean Tensile (GMT) strength, which is equal to the square root of the product of the two variables. This variable is also subject to process specifications.

Unfortunately, the group is divided on the subject of GMT. The operation's specialists think that since the company has critical specifications for the measurement, they need to check if those specifications are being met. The machine operators say that since GMT is calculated from two other variables that have been tested, they do not need to perform a hypothesis test concerning these data.

(g) What do you think? Is it possible for both *MDStrength* and *CDStrength* to be running according to specs but yet have GMT out of specs? Experiment with the data and see.

(h) Using the results of your analysis prepare a report to management about the current tissue manufacturing process. Make a recommendation on whether the process needs to be adjusted or whether it is running on target. Remember, if it is not running on target, the management will be considering making changes to the specifications to reduce customer complaints about dispensing. Address this point in your report.

10.51 A manufacturer of electronic telecommunications equipment was receiving complaints from the field about low volume on long-distance connections. Aunt Sue in California couldn't hear Cousin Fred in Florida.

A string of amplifiers manufactured by the company was being used to boost the signal at various points along the way. The boosting ability of the amplifiers (the "gain") was naturally the prime suspect in the case.

The design of the amplifiers calls for a gain of 10 decibels (dB). This means that the output from the amplifier should be about 10 times stronger than the input signal. This amplification makes up for the natural fading of the signal over long-distance connections. Because it is difficult to make every amplifier with a gain of exactly 10 dB, the design allows amplifiers to be considered acceptable if the gain falls between 7.75 and 12.2 dB. These permissible minimum and maximum values are sometimes called the specification (or spec) limits. The average value is to be 10 dB. Because there are literally hundreds of amplifiers boosting the signal on a long connection, low-gain amplifiers should be balanced by high-gain amplifiers to give an acceptable volume level.

The quality improvement team investigating the "couldn't hear" condition arranged to test the gain of 120 amplifiers. The results of the tests are listed below and are found on the disk file named AMPLIF.xxx.

Gain of 120 Tested Amplifiers   *Filename: AMPLIF.XXX*

8.1	10.4	8.8	9.7	7.8	9.9	11.7	8.0	9.3	9.0
8.2	8.9	10.1	9.4	9.2	7.9	9.5	10.9	7.8	8.3
9.1	8.4	9.6	11.1	7.9	8.5	8.7	7.8	10.5	8.5
11.5	8.0	7.9	8.3	10.0	9.4	9.2	10.7	9.0	8.7
9.3	9.7	8.7	8.9	8.6	9.5	9.4	8.8	8.3	8.2
8.4	9.1	10.1	7.8	8.1	8.8	9.2	8.4	7.8	8.0
7.9	8.5	9.2	8.7	10.2	7.9	9.8	8.3	9.0	9.6
9.9	10.6	8.6	9.4	8.8	8.2	10.5	9.7	9.1	8.0
8.7	9.8	8.5	8.9	9.1	8.4	8.1	9.5	8.7	9.3
8.1	10.1	9.6	8.3	8.0	9.8	9.0	8.9	8.1	9.7
8.6	9.2	8.5	9.6	9.0	10.7	8.6	10.0	8.8	8.6
8.5	8.2	9.0	10.2	9.5	8.3	8.9	9.1	10.3	8.4

SOURCE: *Quality Progress,* September 1990

(a) How many of the amplifiers fell within the specification limits?

(b) Does this mean that the data were of little value?

(c) Generate a histogram of the data to get a better "picture" of the data. What shape does it have and what does that tell you about the problem?

(d) Calculate the standard summary statistics for these data: mean, median, trimmed mean, standard deviation, variance, and range.

(e) Although the amplifiers were designed to have 10-dB gain, very few of them actually had a measured gain of 10 dB. Furthermore, very few amplifiers had exactly the same gain. Speculate about what could be causing that variation.

(f) Test the hypothesis that the true average gain is 10 dB.

(g) Find a 95% confidence interval for the true average gain, μ.

(h) Based on your analysis in parts (a)–(f), what are your recommendations to the manufacturer of the amplifiers?

RES 342: RESEARCH & EVALUATION II

WORKSHOP 1

NEEDS ASSESSMENT/ANALYSIS AND EVALUATION

INTRODUCTION

Although you have been learning many statistical techniques in the last few chapters, remember that we began this book with the idea that statistics is a tool that can be used to assist managers in doing business research. Before continuing with some additional statistical techniques, it is perhaps a good time to revisit the steps involved in doing research. Recall that the six steps for doing applied research were identified in Chapter 1 as follows:

Step 1. Problem Identification

Step 2. Statement of Desired Goal or Outcome

Step 3. Research Evidence and Hard Data

STEPS FOR BUSINESS RESEARCH

Step 4. Outcomes

Step 5. Identification of Possible Cause(s)

Step 6. Proposed Solutions

Oftentimes we rush to collect data and propose solutions without carefully considering what exactly the problem is that we are trying to solve and what is our desired goal; these are our steps 1 and 2. This part of the process is known as a **needs assessment.**

A *need* is a gap between the current state and a specified future state.

The size, scope, significance, and relative priority of the gap are the subject matter of needs assessment.

11.1 CHAPTER OBJECTIVES

This chapter summarizes and synthesizes the methodologies used to conduct needs assessments. You will learn:

- The three-level approach to needs assessment
- Data collection methods used for needs assessment
- Four types of needs assessments, classified according to the type of problem being addressed and the conditions under which the problem occurs
- Some guidelines that may be applied to almost any needs assessment

11.2 LEVELS OF ANALYSIS

Needs assessments have traditionally been classified as focusing on three levels of analysis: organization, job or task, and individual or person. The three-level approach to needs assessment suggests that assessors should start by analyzing the organization to determine what results are not occurring and should be, and what organizational factors are contributing to that condition. The assessors should then analyze work, jobs, or tasks to determine what performance should be occurring. Finally, assessors should study individuals to determine who needs learning to accomplish those jobs or tasks. Figure 11.1 shows how these corresponding levels can be viewed as a U-shaped process of needs assessment leading to evaluation.

By starting at the organizational level, needs assessment is most likely to lead to well-designed interventions with a very good chance of solving real performance problems. Evaluation after the program is easy because the criteria for each level were determined before the interventions were designed.

Assessors are usually cautioned to be sure that a needs assessment encompasses these three traditional levels (organization, work or task, and individual). Failure to collect assessment data at all levels is likely to lead to misdirected resources and low-impact interventions that are difficult to evaluate.

It should be pointed out, however, that not all needs assessors are given the opportunity to perform an all-encompassing three-level analysis. Kaufman and English (Kaufman, R. & English, F. W., *Needs assessment: Concept and application.* Englewood Cliffs, NJ: Educational Technology Publications, 1979) classified needs assessments as ranging from *alpha* to *zeta*, depending on the breadth of questions that the assessor is allowed to ask, with an *alpha* assessment having the broadest scope, encompassing all levels of analysis. The levels of analysis included in an assessment depend on such fac-

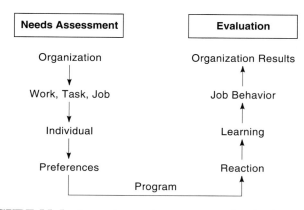

FIGURE 11.1   Integrating needs assessment and evaluation processes

tors as the status of the assessor in the organization, available time and resources, management policy, and the assessor's biases.

11.3 METHODS EMPLOYED

Needs assessors have long been known for employing a wide variety of methods. Table 11.1 summarizes the methods used.

TABLE 11.1 Data Collection Methods Used

Category	Methods
Quantitative	Survey
	Task analysis
	Benchmarking
Qualitative	Interview
	Focus group
	Subject matter expert
	Committee
Blended	Nominal group technique
	Action research
	Observation
	Work sampling
	DACUM (developing a curriculum)
	Subject matter analysis
Extant data	Job descriptions
	Internal reports
	Performance appraisal
	Personnel records
	Industry data
	Annual report
	Literature

Some notable and encouraging trends evident in needs assessment include the following:

- *Multimethod strategies*—Although surveys and interviews are the most popular methods, many others are used as well. In fact, many needs assessors use multiple methods. Most combine qualitative and quantitative methods. Weaving together multiple data sources significantly enhances the assessments, particularly in multilevel analyses.

- *Integration of qualitative methods*—Needs assessors have sometimes been criticized as not being rigorous enough because they use so many qualitative methods. In many cases, however, their use is a strength, not a weakness. Interviews and focus groups provide open-ended information that enrich an assessment. True revelations tend to occur more frequently with qualitative methods, largely because quantitative data collection instruments constrain the range of responses. Interviews are often the primary way to explore problems at the organizational level.

- *Appropriate use of surveys*—Needs assessment has also been criticized for an over-reliance on surveys. The message to assessors is this: Surveys are powerful tools, but they are best used in conjunction with other methods, and they are not a necessary part of needs assessment, even with a large group.

11.4 FOCUS OF NEEDS ASSESSMENT

The previous two sections have classified needs assessments according to two traditional approaches: level of analysis and methods used. These two approaches fail to capture another dimension of needs assessment that points out a critical issue facing needs assessors. Rothwell and Sredl (Rothwell, W. J. & Sredl, H. H., *The ASTD reference guide to professional human resource development roles and competencies,* 2d ed. Amherst, MA: HRD Press, 1992) described needs assessments as being either deficiency or opportunity oriented. Traditionally, needs assessment had been deficiency oriented, designed to identify and address existing deficiencies or gaps in performance. A deficiency approach is, by definition, focused in the present. Opportunity-oriented needs assessment is future oriented, identifying performance gaps likely to occur in the future and proactively implementing solutions to prevent them. And if organizations are to handle today's fast-changing business environments, needs assessments must be more focused on the future.

A useful classification scheme comes from the father of action learning, R. Revans (keynote presentation at the 1994 Academy of Human Resource Development annual meeting) although it has not been applied to needs assessment before now. He has suggested that organizations face four types of change, varying along two dimensions. One dimension is the type of problem—known or unknown. The other dimension is the conditions under which the problem occurs—also divided into known and unknown. The four-cell matrix that results (see Figure 11.2) can be used to describe the following four types of needs assessments:

- *Corrective needs assessments* are those that analyze existing problems in existing circumstances to identify performance problems. These are the traditional deficiency-oriented assessments, usually using discrepancy performance models, such as Mager and Pipe's model (Mager, R. F. & Pipe, P., *Analyzing performance problems.* Belmont, CA: Lake Publishing, 1984).

- *Adaptive needs assessments* occur in organizations that find themselves performing under new conditions, but facing the same job demands as in the past. For example, an organization that restructures to work teams but faces essentially unchanged customer demands should conduct an adaptive needs assessment.

- *Developmental needs assessments* are those designed to improve an organization's ability to deal with additional problems in the existing environment and conditions. For example, an organization may seek to improve its ability to deal with unknown problems in the current environment by strengthening its ethical practices.

- *Strategic needs assessments* require the anticipation of unknown problems or deficiencies that are likely to occur in the future under changing, but unknown, conditions. Such assessments usually occur in organizations facing fundamental and rapid change.

	Problem	
	Known	**Unknown**
Known	Corrective	Developmental
Unknown	Adaptive	Strategic

FIGURE 11.2 Four types of needs assessments

Traditional needs assessment and performance analysis methodologies are strong tools for corrective needs assessments, limited tools for adaptive and developmental needs assessments, and weak tools for strategic needs assessments. Needs assessors have traditionally avoided strategic needs assessments because they require a great deal of "crystal ball" analysis. However, needs assessors must develop the capability to assist organizations with noncorrective assessments. Corrective assessments, by definition, place the human resource development (HRD) person conducting the assessment in a reactive mode. But no organization can afford to wait for performance problems to occur in order to figure out how to prevent them. To link HRD practice to the strategic goals of an organization requires needs assessors to have the ability to conduct strategic needs assessments. Experience indicates that the basic assessment methods appear to work in strategic assessments with minimal adaptation. Perhaps the barriers lie more in a lack of strategic thinking in HRD, not in the lack of methodologies.

11.5   PLANNING AHEAD

Plan to use data.   Someone once said, "Planning is a substitute for good luck." The data you collect in a needs assessment can be useful long after you've used it to determine training needs. You should plan to make your needs-assessment data an integral part of your whole training project. The data can keep you from having to "reinvent the wheel" at every stage of the training, as new needs arise.

Establish criteria or goals.   What do you expect your needs assessment to find? List your expectations and make a note of conditions prior to the assessment.

Research the topic, task, or focus.   Become familiar with the subject that is targeted by your needs assessment. Interview people in the field. Review books, journals, and reports on the topic.

Create a guidance group.   Get together a group of people whose function is to make decisions and keep the needs assessment on track.

The group should include some of the people who requested the needs assessment, some subject matter experts, and some stakeholders—people who may be affected by the outcome.

Before starting the assessment, get everyone's commitment to working as a group and achieving the assessment's goals.

Identify your data sources.   With the group's help, determine the best sources for the data you'll need. Even while collecting data, continue looking for other sources of information. Ask the people who provide you with information to recommend additional resources.

There are two kinds of data: hard and soft. Hard data are factual. They can take the form of production reports, defective-parts reports, recall reports, and absentee reports. Such data provide real numbers that you can count, analyze, and translate into statistics.

Soft data are pieces of information that are obtained through such means as group discussions, interviews, questionnaires, and literature reviews. Soft data are subjective—similar to opinions and beliefs—but they can often be supported by hard data.

Design an interview format.   A trial run can help you refine the content of your interviews with the people on your list of sources and stakeholders. The more refined an interview is, the more information you're likely to get. Carefully planned interviews also take less time to conduct.

Construct a questionnaire—either on paper or on a computer disk—to use as the basis for your interviews. For the trial run, ask several people in your guidance group to answer the questionnaire.

In both the trial and actual interviews, describe for participants the goals of your needs assessment. Tell them who authorized it. You also should tell them why they were selected, how you plan to use the information they provide, how much you value their participation, what kinds of feedback they'll receive, and how they might benefit from the outcome of the assessment.

11.6   COLLECTING DATA

Interviews.   Interviewing is an excellent way to get information. The best interview subjects tend to be those who are professionals in the area you're assessing and who may be directly affected by the outcome of the assessment. But remember that the information gleaned from interviews is subjective; it must be weighed with other data.

Let an interviewee know in advance how much time you expect the interview to take. You should have some idea from your trial interviews. When you begin an interview, ask for permission to record the session. Warn people that you may ask probing questions. Invite them to ask questions as well.

Feel free to stray from your present format. For example, interviewees may elaborate immediately on some points that you'd planned to ask later. Just glance through the questions at the end to ensure that all of them have been addressed.

When you can't interview people in person, consider sending them questionnaires to fill out instead. If you're gathering data from a lot of people, sending a questionnaire on computer disks might be cost-effective. One study shows that disk-based surveys reap better response rates than written surveys.

Afterwards, be sure to send interviewees thank-you notes that emphasize how valuable their input will be to the outcome of the needs assessment.

Group discussion.   A group discussion can bring out all kinds of data; an interactive environment tends to elicit information that individuals might not bring up on their own.

It's important to select a diverse group of people, to follow an agenda, and to keep the number of members manageable. Group size can affect the amount and kinds of information you get. If a group is too large, some people may not join the discussion. If a group is too small, it may not provide the richness or variety of information you seek.

The agenda should lead the discussion, but as the facilitator, you should assert some control. Don't let one or two people in the group dominate. And don't let anyone stray from the topic.

Hard data.   If your needs assessment requires hard data, make sure the data are relevant to and accurately reflect work conditions that are observed during the assessment. Also, it's important that the data address the concerns of people who requested the needs assessment.

11.7 ANALYZING THE DATA

Compile the data.   This stage of the needs assessment can be tricky. By now, you probably have a lot of information. What are you going to do with it?

It's up to you and the group to decide which data are relevant. In other words, does the information pertain to what you wanted the assessment to measure?

To pinpoint areas in which you may need more information, try dividing the data into major topics or focus points.

Provide statistics.   If any of the hard data include statistics, a statistical analysis can help clarify the results. And sometimes, it may be necessary to provide statistical documentation in order to obtain resources for projects. The data used for statistics also may be used to monitor any changes that are implemented.

Prepare a report.   Your report should flow. It should lead readers step-by-step through the needs assessment to the results.

The report should contain a statement of purpose and goals, dates, methodologies, data sources, analysis techniques, and results—as well as a list of the needs you and the group identified. Include charts and graphs for clarification. Use the data you have collected as documentation for the needs you list.

The report provides a solid basis for making changes and for measuring the effectiveness of those changes. You may even want to go ahead and make some recommendations at this point.

By now, the wide array of needs assessments may not appear to be so confusing, even to a first-time needs assessor. Using the guidelines, you should be able to apply a cohesive framework to any kind of needs assessment. At the very least, following the recommendations may help prevent some unpleasant surprises.

CHAPTER 11 SUMMARY

Clearly, needs assessment must be the first step in any business research plan. It is the gap identified between your current situation and where you would like to be. As you move to step 3, research evidence and hard data, you will need some additional statistical tools. The next few chapters will present these tools.

INFERENCES: ONE POPULATION

THE HOSPITAL

The health-care industry is as concerned with quality as manufacturing companies. With the rising costs of health care, many hospitals are looking for ways to contain costs without sacrificing quality of care. Studies have shown that a simple way to cut down on the spread of bacteria is to be sure that all employees wash their hands adequately. The literature indicates that a minimum of 5 seconds is needed.

To be sure that the medical staff was washing their hands long enough, a large hospital observed a sample of employees washing their hands. The employees did not know they were being observed. In total, 28 employees were observed. The average hand-washing time was found to be 2.556 seconds with a sample standard deviation of 3.755 seconds. On the basis of these data, the hospital must decide if the employees should participate in a training session to learn about the consequences of inadequate hand washing.

A portion of the data set is shown below:

Observation	Unit	Time 1
1	CCU	3
2	CCU	2
3	CCU	0
4	CCU	5
5	CCU	2
6	CCU	0

12.1   CHAPTER OBJECTIVES

The hospital described above does not know the standard deviation of hand-washing times and the sample size is small ($n \leq 30$). Therefore, the hypothesis testing procedure introduced in Chapter 10 cannot be used. Now it is time to examine what to do if the population standard deviation is unknown and the sample size is small. In addition to handling small-sample tests of the mean, we will also see how to perform a hypothesis test for other population parameters.

If we are analyzing a *quantitative variable,* such as time an employee washes his/her hands, then we know that it can be described by its distribution. The distribution, in turn, can be described by parameters. Thus, if we are constructing a theory or hypothesis about a *quantitative variable* it might be a statement about

- The mean value, μ, of the variable in one population
- The amount of variability, σ^2, of the variable in one population

If the data we are analyzing are *nominal data,* such as whether or not an employee is satisfied with the job, the hypothesis might be a statement about

- The value of the proportion, π, of population members that have a certain characteristic (one of the categories of the nominal variable)

In this chapter we look at the numerical details of the test statistics and rejection regions for hypothesis tests about a single population parameter. We will be studying only one population. For example, we will look at the average sales of a particular product or the proportion of people who favor legalizing marijuana. In Chapter 13 we extend these results to compare the behavior of two populations. This might entail comparing the average sales of two products or comparing the views of men and women on the issue of legalizing marijuana.

This chapter covers the following material:

Remember: Both μ and π describe the behavior of the population not the sample. They are typically unknown.

- Hypothesis Test of the Population Mean, μ: Small Sample
- Hypothesis Test of the Population Variance, σ^2
- Hypothesis Test of the Population Proportion, π
- The Relationship Between Hypothesis Testing and Confidence Intervals

12.2   HYPOTHESIS TEST OF THE MEAN: SMALL SAMPLE

In Chapter 10 you learned to perform a large-sample hypothesis test of the mean. In our examples, we have seen that sometimes there is information about the population standard deviation from either the product specs or from some previous study. However, more often than not, the population standard deviation is not known. In this case the sample standard deviation, s, must be used to calculate an estimate of the unknown population standard deviation, σ. If the sample is sufficiently large, $n > 30$, then you can use the Z test statistic from Chapter 10 to do a hypothesis test on the mean. However, very often you have a small sample.

Remember: To find the sample variance you use:
$$s^2 = \frac{n \sum x^2 - (\sum x)^2}{n(n-1)}$$

Now let us think about the implications of having a small sample on our hypothesis test. The steps of any hypothesis test are listed below:

Step 1:   *Set up the null and alternative hypotheses.*

Step 2:   *Pick the value of α and find the rejection region.*

Steps for any hypothesis test

Step 3: *Calculate the test statistic.*

Step 4: *Decide whether or not to reject the null hypothesis.*

Step 5: *Interpret the statistical decision in terms of the stated problem.*

Step 1 will not be affected since we are just setting up the null and the alternative hypotheses. If you guessed that steps 2 and 3 are affected then you were right. In Chapter 10, we calculated a Z statistic as our test statistic in step 3. But the formula for Z uses σ, which is now unknown, and our estimate is based on a small sample. Logic tells us simply to use s in the formula instead of σ. However, when we do this the test statistic no longer follows a normal distribution. Instead, it has a t *distribution*.

You used the t distribution in Chapter 9 when you learned how to find a confidence interval for the mean of a normally distributed population when σ was unknown and the sample size was small. We will use our knowledge of the t distribution to do a hypothesis test of μ for precisely this situation.

The test statistic is calculated as follows:

$$t = \frac{\overline{X} - \mu}{s/\sqrt{n}}$$

Notice that the calculation for the t statistic is just the same as Z with σ replaced with s.

Remember that a test statistic is a single number based on the sample which allows you to decide between the null and the alternative hypotheses.

t test statistic

Remember: The population must have a normal distribution in order to use the t statistic.

12.2.1 Two-Tail Test of the Mean: Small Sample

Let's first look at two-tail tests of the mean when σ is unknown.

EXAMPLE 12.1   The Hospital

Two-Tail Test of μ: Small Sample

The table shown below gives the hand-washing times (in seconds) for the 28 employees who were observed:

3	3	0	6	2	0	1	0	5	2
2	0	0	0	0	6	15	0	5	0
0	1	0	10	2	0	8	0		

The hospital wishes to test if the population average washing time is different from 5 seconds. Use $\alpha = 0.05$.

Step 1: *Set up the null and alternative hypotheses.* Since the company is interested in detecting shifts from the value of 5 seconds in either direction, we know that this is a two-tail test. So we have

$$H_0: \quad \mu = 5 \text{ s}$$
$$H_A: \quad \mu \neq 5 \text{ s}$$

Step 2: *Select a value for α and find the rejection region.* The value of α has been specified as 0.05. From our work in the previous section, we know that we will reject H_0 if the calculated t statistic is too large or too small.

We need to use the t table to find the value of t_{cutoff}. Since this is a two-tail test we know that we must divide α in half. Thus, we want $0.05/2 = 0.025$ in the upper-tail area and 0.025 in the lower-tail area. Using the column labeled 0.025 and the row for $n - 1 = 28 - 1 = 27$ degrees of freedom, we find the t_{cutoff} value to be 2.052.

Since the t distribution is symmetric, the value for $-t_{cutoff}$ is -2.052 and the rejection region is shown as the shaded area below:

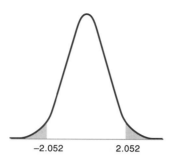

-2.052   2.052

Do you remember what the value for the z_{cutoff} was when $\alpha = 0.05$? It was 1.96. So you can see that the t_{cutoff} value is just a bit larger than the corresponding z_{cutoff} value. This will always be the case, and as the degrees of freedom get larger, the t_{cutoff} value will get closer and closer to 1.96. Look down the column for $\alpha = 0.025$ and see that, as the degrees of freedom increase (due to larger sample sizes) the values for t_{cutoff} get closer and closer to 1.96.

The t_{cutoff} value will always be a bit larger than the corresponding z_{cutoff} value.

Step 3: *Calculate the test statistic and the* p *value.* From the data we can calculate $\overline{X}$ and s and find them to be $\overline{X} = 2.536$ s and $s = 3.687$ s and we know $n = 28$. With this information we can calculate the t test statistic:

$$t = \frac{\overline{X} - \mu}{s/\sqrt{n}} = \frac{2.536 - 5}{3.687/\sqrt{28}} = -3.54$$

Find the p value. To get this probability you will most likely need to use some software support such as Minitab or Excel. This is because the t table is not designed to be used to find probabilities but rather to find the cutoff values for the rejection region.

$$p \text{ value} = 2P(t < -3.54) = (2)(0.00075) = 0.0015$$

The output shown below is from MINITAB. You can see the t statistic labeled T. In addition, the p value is calculated by Minitab and shown in the last column labeled P.

```
T-TEST OF THE MEAN
Test of mu = 5.000 vs mu not = 5.000

Variable     N     Mean    StDev    SE Mean      T        P
Hand was    28    2.536    3.687     0.697     -3.54    0.0015
```

Step 4: Here we check to see if the t statistic calculated at step 3 falls in the rejection region. Since our t statistic from step 3 falls in the rejection region, we reject H_0. We can also reach this conclusion by examining the p value from Minitab. Since the p value is considerably smaller than 0.05, we would reject the null hypothesis.

Step 5: We translate this decision in terms of the problem statement and conclude that the average employee hand-washing time is different from 5 seconds. ∎

Let's look at another example.

EXAMPLE 12.2   **The Cereal Company**

Two-Tail Test of μ When σ Is Unknown

We have seen that packaging the correct amount is critical to most companies. Underfilling the package will clearly yield customer complaints and overfilling the packages will cost the company money. Suppose you are working for a cereal manufacturer and each box of cereal is supposed to contain 13 oz of cereal. The weight of the cereal boxes is assumed to be normally distributed. The company has taken a sample of 25 cereal boxes and found the following weights:

12.985	12.976	13.107	13.006	12.910
12.755	12.938	13.139	13.015	13.033
13.029	12.887	13.049	12.823	12.910
13.073	13.024	13.088	13.061	13.111
13.050	13.141	13.008	13.149	12.907

Is there evidence that the true mean is not equal to 13 oz? Use $\alpha = 0.05$.

> **Step 1:** *Set up the null and alternative hypotheses.* Since the company is interested in detecting differences from the value of 13 ounces in either direction, we know that this is a two-tail test. So we have
>
> $$H_0: \quad \mu = 13 \text{ oz}$$
> $$H_A: \quad \mu \neq 13 \text{ oz}$$

> **Step 2:** *Select a value for α and find the rejection region.* The value of α has been specified as 0.05. From our work in the previous section, we know that we will reject H_0 if the calculated t statistic is too large or too small.

We need to use the t table to find the value of t_{cutoff}. Since this is a two-tail test we know that we must divide α in half. Thus, we want $0.05/2 = 0.025$ in the upper-tail area and 0.025 in the lower-tail area. Using the column labeled 0.025 and the row for $n - 1 = 25 - 1 = 24$ degrees of freedom, we find the t_{cutoff} value to be 2.064. Since the t distribution is symmetric, the value for $-t_{\text{cutoff}}$ is -2.064 and the rejection region is shown as the shaded area below:

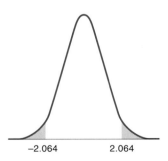

−2.064   2.064

> **Step 3:** *Calculate the test statistic and the* p *value.* From the data we can calculate $\overline{X}$ and s and find them to be $\overline{X} = 13.007$ oz and $s = 0.1004$, and we know $n = 25$. With this information we can calculate the t test statistic:

$$t = \frac{\overline{X} - \mu}{s/\sqrt{n}} = \frac{13.007 - 13}{0.1004/\sqrt{25}} = 0.35$$

Alternatively, we can use a software package such as Minitab to find the test statistic and the *p* value. The output from Minitab is shown at the top of page 473.

```
T-TEST OF THE MEAN
Test of mu = 13.0000 vs. mu ≠ 13.0000

Variable     n      Mean     StDev    SEMean      T      P
Cereal      25     13.0070   0.1004   0.0201     0.35   0.73
```

Step 4: Here we check to see if the t statistic calculated at step 3 falls in the rejection region. Since our t statistic from step 3 does not fall in the rejection region, we do not reject H_0. Using the p value to make this same decision, we see that the p value is larger than 0.05, leading us to fail to reject H_0.

Step 5: We translate this decision in terms of the problem statement and conclude that the cereal boxes are being filled properly. ■

Let's summarize the procedure for finding the rejection region for a two-tail test of μ when σ is unknown:

Step 1: *Divide the value of α in half and use that column of the* t *table.*

Step 2: *Find the number of degrees of freedom by calculating* n $-$ 1 *and use that row of the* t *table.*

Step 3: *Find the value of* t_{cutoff} *in that row and column.*

Step 4: *The value of* $-t_{cutoff}$ *is the same as* t_{cutoff} *with a negative sign.*

Steps for finding the rejection region for a two-tail test of μ when σ is unknown

 **TRY IT NOW!**

The Soda Machine

Test of Population Mean When σ Is Unknown

In Chapter 10 we looked at the hypothesis test to see if a soda machine was correctly dispensing 32 oz of soda. The amount dispensed is assumed to be normally distributed. The machine was not working properly if the bottles were overfilled or underfilled. You observed the machine filling 30 bottles and collected the following data:

32.12	32.04	32.04	32.00	32.01	32.12
32.13	32.05	32.05	31.72	32.08	32.03
32.02	32.22	32.22	32.22	31.89	32.20
31.73	31.95	31.95	32.07	32.13	32.13
31.98	32.21	32.21	31.97	32.09	31.97

This time assume that you do not have any information about the population standard deviation and so you must calculate the sample standard deviation.

Is there any evidence to indicate that the machine is not filling the bottles properly? Use $\alpha = 0.05$.

Step 1:

Step 2:

Step 3:

Step 4:

Step 5:

12.2.2 One-Tail Test of the Mean: Small Sample

In the previous section we learned that the two-tail hypothesis testing procedure for μ is affected in two major ways by the lack of knowledge about σ. The test statistic becomes a t test instead of a Z test statistic, and the rejection region cutoff values must be found from the t table rather than the Z table. The same can be said about one-tail tests of the mean when σ is unknown.

The only step in the procedure that we need to update is finding the rejection region using the t table for one-tail tests. The form of the rejection region is the same as when σ is known. The only difference is in finding the cutoff values. Remember that we constructed the rejection regions by following a series of logical arguments as to what values of $\overline{X}$ would lead us to reject the null hypothesis. These arguments still apply.

For a one-tail test we want to reject H_0 if the calculated t statistic is too small. Thus, we have the rejection region shaded in Figure 12.1.

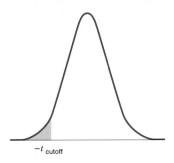

$-t_{\text{cutoff}}$

FIGURE 12.1   Rejection region for a lower-tail test of μ

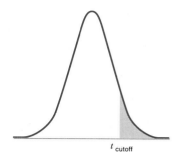

FIGURE 12.2   Rejection region for an upper-tail test of μ

For an upper-tail test we want to reject H_0 if the calculated t statistic is too large. The rejection region for this type of test is shaded in Figure 12.2.

The cutoff values are found in a manner similar to that used for the two-tail test except that there is no reason to split the value of α in half. These steps are summarized here:

Step 1: *Use the column of the* t *table that corresponds to the value of α you have selected.*

Step 2: *Find the number of degrees of freedom by calculating* n $-$ 1 *and use that row of the* t *table.*

Step 3: *For upper-tail tests the desired* t_{cutoff} *value is found at the intersection of that row and column. For lower-tail tests, place a negative sign in front of the value to get the value of* $-t_{cutoff}$.

Steps for finding the rejection region for one-tail tests of μ: small sample

Let's look at some examples.

EXAMPLE 12.3   The Hospital

One-Tail Test of μ: Small Sample

The hospital is most likely interested in testing whether the mean hand-washing time is less than 5 seconds. This is a one-tail test of μ.

Step 1: *Set up the null and alternative hypotheses.* Since the hospital is interested in seeing if average hand-washing times are *less than 5 seconds,* this is a lower-tail test. The null and alternative hypotheses are shown below:

$$H_0: \quad \mu \geq 5 \text{ s}$$
$$H_A: \quad \mu < 5 \text{ s}$$

Step 2: *Select a value for α and find the rejection region.* The value of α has been set at 0.05. The specific value of t_{cutoff} is found by looking in the t table. Use the column labeled 0.05 and the row corresponding to 27 degrees of freedom ($28 - 1 = 27$). The table value at the intersection of this column and row is -1.703. This is the value of $-t_{cutoff}$.

Step 3: *Calculate the test statistic and the* p *value.* Since the value of σ is not known we must calculate both $\overline{X}$ and s from the sample data. The output from Minitab is shown below.

```
T-TEST OF THE MEAN
Test of mu = 5.000 vs mu < 5.000

Variable    N    Mean    StDev   SE Mean     T         P
Hand was   28   2.536    3.687    0.697    -3.54    0.0007
```

Step 4: *Decide whether to reject the null hypothesis.* Since the calculated test statistic of $t = -3.54$ is in the rejection region and the p value is less than 0.05, we reject H_0.

Step 5: *State the statistical decision in terms of the problem.* Based on these data, we can conclude that the average hand-washing time is less than 5 seconds. ∎

Let's look at one more example.

EXAMPLE 12.4   New Marketing Plan

Upper-Tail Test of μ When σ Is Unknown

A company is trying out a new marketing plan and wishes to evaluate its success. Prior to the new advertising scheme the average store sales per week was $4000. The new method is tried on a random sample of 15 stores and the following sales data were collected:

$4128	$4132	$4163
$4148	$4157	$4039
$4028	$4146	$4174
$4190	$4054	$4181
$4088	$4069	$4099

On the basis of these data, can the company conclude that the new marketing plan worked? The weekly sales are assumed to be normally distributed. Use $\alpha = 0.05$.

Step 1: *Set up the null and alternative hypotheses.* Since the company is interested in seeing if average sales have *increased,* this is an upper-tail test. The null and alternative hypotheses are shown below:

$$H_0: \quad \mu \leq \$4000$$
$$H_A: \quad \mu > \$4000$$

Step 2: *Select a value for α and find the rejection region.* The value of α has been set at 0.05. The specific value of t_{cutoff} is found by looking in the t table. Use the column labeled 0.05 and the row corresponding to 14 degrees of freedom ($15 - 1 = 14$). The table value at the intersection of this column and row is 1.761. This is the value of t_{cutoff}.

Step 3: *Calculate the test statistic and the p value.* Since the value of σ is not known we must calculate both $\overline{X}$ and s from the sample data:

$$\overline{X} = \$4119.7 \qquad s = \$53.4 \qquad t = \frac{4119.7 - 4000}{53.4/\sqrt{15}} = 8.69$$

```
T-TEST OF THE MEAN
Test of mu = 4000.0 vs. mu > 4000.0

Variable    n      Mean     StDev    SE Mean     T        P
sales      15     4119.7    53.4      13.8      8.69    0.0000
```

Step 4: *Decide whether to reject the null hypothesis.* Since the calculated test statistic of $t = 8.69$ is in the rejection region, we reject H_0. The p value of 0.0000 also tells us to reject H_0.

Step 5: *State the statistical decision in terms of the problem.* Based on these data, we can conclude that the advertising scheme has indeed increased sales. ∎

 *TRY IT NOW!*

Diameter of Washers

Lower-Tail Test of μ: Small Sample

Your company purchases washers. It is important that the diameter of the hole not be more than 0.5 inch. If the hole in the center of the washer is too large then your company will not be able to use the washer. The diameters are assumed to be normally distributed. You have just received a shipment of 10,000 of these washers. You decide to sample 20 to check to be sure that the diameters are not more than 0.5 inch, on the average. Your measurements (in inches) are

0.5053	0.5098	0.4606	0.4606
0.4711	0.4627	0.4800	0.4800
0.4672	0.5642	0.5495	0.5495
0.4672	0.5346	0.5745	0.5745
0.5340	0.3767	0.3933	0.3933

Should you accept the shipment? Use $\alpha = 0.05$.

Step 1:   *Set up the null and the alternative hypotheses.*

Step 2:   *Select a value for α and find the rejection region.*

Step 3:   *Calculate the test statistic and the p value.*

(continued)

Step 4:   *Decide whether to reject the null hypothesis.*

Step 5:   *State the statistical decision in terms of the problem.*

12.2.3 Summary of Tests of the Mean: Small Sample

In the last two sections we have seen several more examples of the five-step hypothesis testing procedure. In each case, regardless of whether it was a two-tail or one-tail test, the same five steps were utilized. There were two major differences from the tests of μ when σ is known: (1) a t test statistic was used instead of Z and (2) the cutoff values for the rejection region were found by using the t table instead of the Z table.

The rejection regions are summarized below:

Type of Test	Rejection Region
Two-tail test of μ	Reject H_0 if $t < -t_{\text{cutoff } \alpha/2}$ or if $t > t_{\text{cutoff } \alpha/2}$
$\quad H_0: \mu =$ [a specific number]	
$\quad H_A: \mu \neq$ [a specific number]	
Lower-tail test of μ	Reject H_0 if $t < -t_{\text{cutoff } \alpha}$
$\quad H_0: \mu \geq$ [a specific number]	
$\quad H_A: \mu <$ [a specific number]	
Upper-tail test of μ	Reject H_0 if $t > t_{\text{cutoff } \alpha}$
$\quad H_0: \mu \leq$ [a specific number]	
$\quad H_A: \mu >$ [a specific number]	

ANS. $t = -0.72$, FAIL TO REJECT THE NULL HYPOTHESIS, ACCEPT THE SHIPMENT.

12.2.4 Exercises—Learning It!

12.1 The cost of common goods and service in 5 cities is shown in the table below *(USA Today)*:

City	Aspirin (100)	Fast food (hamburger, fries, soft drink)	Woman's haircut/blow dry	Toothpaste (6.4 oz)
Los Angeles	$7.69	$4.15	$20.11	$2.42
Tokyo	$35.93	$7.62	$76.24	$4.24
London	$9.69	$5.80	$44.35	$3.63
Sydney	$7.43	$4.53	$29.93	$2.08
Mexico City	$1.16	$3.63	$17.94	$1.08

(a) You have just returned from a business trip and you lost your receipt for the aspirin you purchased but would like to be reimbursed by your company (since you had take the aspirin after a stressful business meeting!). You guesstimate a cost of $10.00. Your boss claims that the average cost of aspirin is less than $10.00. Using these data, can you "prove" your boss wrong? Conduct the necessary hypothesis test. Assume that all costs are normally distributed.

(b) Based on these data, is there enough evidence to support your submitting a cost of $10.00 for the fast-food meal on your trip?

(c) If you remove Tokyo from the data set do your answers to parts (a) and (b) change? What does this tell you about the effect of outliers on the hypothesis test of μ when you have a small sample?

12.2 The marketing material for a New England ski resort advertises that they can make snow whenever the temperature is 32°F or below. To demonstrate how often this happens their material includes the following line graph of the weekly average temperatures.

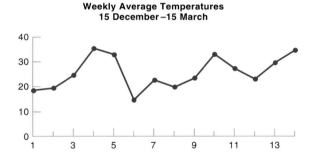

**Weekly Average Temperatures
15 December–15 March**

The data that generated this graph are shown below:

Week	1	2	3	4	5	6	7	8	9	10	11	12	13	14
Temperature	18	19	24	35	33	14	22	20	23	33	27	23	30	35

Is there enough evidence for the ski resort to claim that the average weekly temperature is less than 32°F? Assume that the average weekly temperature is normally distributed.

12.3 While job prospects for nurses were once good, many now face an uncertain future as hospitals cut staff and train unlicensed workers to do some of their jobs. The change is reflected in enrollment at several area nursing schools. The following table shows the enrollment at nursing schools in Western Massachusetts:

School	1995	1996
Holyoke Community College	60	45
Berkshire Community College	96	96
Greenfield Community College	90	79
Springfield Technical Community College	146	135
American International College	189	180
Elms College	298	300
Umass–Amherst	411	429
Baystate Medical Center	150	130

Using the 1996 data, test the hypothesis that the average enrollment in nursing programs has decreased from the average of 180 in 1995. Assume that enrollment in nursing programs is normally distributed.

12.4 If you like shopping for the best deal on long-distance phone service, then you'll enjoy sorting through offers from 10 different marketers vying to be your energy supplier. Residents of 16 communities will be the first in Massachusetts to wade into the coming nationwide experiment in deregulation of the natural gas industry. The average customer uses 1232 therms of natural gas, for which the average cost has been $520.24. The table below shows the proposed costs to deliver 1232 therms of natural gas from 10 competitors:

Company	Cost ($)
All Energy Marketing Co.	478.66
Broad Street/Energy One	450.24
Global Energy Services	468.16
Green Mountain Energy Partners	471.24
KBC Energy Services	435.53
Louis Dreyfus Energy Services	472.24
National Fuel Resources	468.22
NorAm Energy	442.20
WEPCO Gas	443.52
Western Gas Resources	457.81

Is there enough evidence to conclude that the average cost to the customer from the competitors is less than $520.24? Assume that the costs are normally distributed.

12.5 Computer centers at universities and colleges are certainly aware of the increased number of web surfers. To begin to understand the demands that will be made on the computer center resources, one school studied 25 children in grades 7 to 12. The number of hours that these children spent on the Internet in one week is shown below:

5.0	4.4	5.7	5.6	5.5
5.2	5.0	4.8	3.6	4.1
4.6	4.9	4.0	6.7	5.5
5.4	6.7	5.8	5.4	4.8
5.9	5.1	3.8	4.1	6.7

Is there enough evidence to indicate that children spend more than an average of 5 hours per week web surfing? Assume that the time spent web surfing is normally distributed.

12.3 χ^2 TEST OF A SINGLE VARIANCE

Remember: The variance is the standard deviation squared.

Hypothesis tests of the population variance, σ^2, follow the same basic steps that were used to do a hypothesis test of the population mean and the population proportion.

There are two types of situations for which you might be interested in doing a hypothesis test of the population variance. First, you may wish to see if a manufacturing process is running to the specified standard deviation. You cannot test the standard deviation but must instead test the population variance. This is the case for the tissue manufacturer. Second, to perform some other statistical analysis of the data, you may need to know the population variance. In this case you would use sample data to test to see if the population variance is, in fact, a particular value.

Just like with tests of the mean, you can do either two-sided or one-sided tests of the variance. To decide which type of test you need, you must determine what you are trying to learn from your study. The key words and guidelines are found in Chapter 10.

12.3.1 Two-Tail Hypothesis Test of the Variance

To learn how to do a two-sided test of the variance we will complete the five steps of the hypothesis testing procedure for the tissue manufacturer looked at in Chapter 10.

EXAMPLE 12.5   The Tissue Manufacturer

Step 1: Setting Up the Null and Alternative Hypotheses for a Test of the Population Variance

The variable *MDStrength* should have a standard deviation of 50 lb/ream. Also, the population variance should be $50^2 = 2500$ (lb/ream)^2. This gives us the following null and alternative hypotheses:

$$H_0: \quad \sigma^2 = 2500 \text{ (lb/ream)}^2$$
$$H_A: \quad \sigma^2 \neq 2500 \text{ (lb/ream)}^2 \qquad \blacksquare$$

The next step in the hypothesis testing procedure is to select a value for α and find the rejection region. To do this we need to know what test statistic will be used in step 3 of the procedure. This is the main difference between a hypothesis test of the population mean, μ, and the population variance, σ^2. In testing means, we used the sample mean as the basis for our decision to reject or fail to reject the null hypothesis. Because the Central Limit Theorem told us that $\overline{X}$ has an approximately normal distribution for sufficiently large sample sizes, the appropriate test statistic for a large-sample test of the mean is a Z statistic and thus the rejection region is determined by Z values. As we have seen, in the small-sample case, $\overline{X}$ follows a t distribution for normally distributed populations.

If we are testing the population variance then the sample variance, s^2, will be used as the basis for deciding between H_0 and H_A. Relying once again on the mathematical statisticians to do the theoretical work, we learn that the sampling distribution associated with the sample variance, s^2, is called the chi-square (χ^2) distribution. The rejection region will be determined by critical values from the chi-square distribution. An example of a chi-square distribution is shown in Figure 12.3.

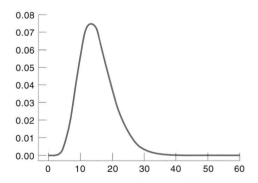

FIGURE 12.3   A chi-square distribution

The chi-square distribution, just like any distribution, describes how the random variable behaves. Recall from Chapter 7 that a distribution tells you the most likely values of the random variable (where there is the most area under the curve) and the least likely values of the random variable (where there is little to no area under the curve). You will see the chi-square distribution again in Chapter 17.

If the variable being studied is assumed to be normally distributed, then the statistic to test whether or not the population variance is equal to a particular value is calculated as follows:

Chi-square test statistic

$$\chi^2 = \frac{(n-1)s^2}{\sigma^2}$$

where

n = sample size

s^2 = the sample variance

σ^2 = the hypothesized value of the population variance under the null hypothesis

The chi-square test assumes that the underlying population distribution is normal.

Notice that this test statistic is basically comparing the variability contained in the sample and reflected in the sample variance to the value of the population variance that is being tested. In this case, the comparison takes the form of a ratio. In testing means, the comparison of the sample evidence, reflected in $\overline{X}$, to the value of the population mean being tested is accomplished by a subtraction. In both cases the idea is the same: Is the sample evidence consistent with the null hypothesis?

This test statistic has been labeled with the Greek letter χ (chi) and a squared symbol, hence the name chi-square test statistic. Like the t distribution, the shape of the chi-square distribution is determined by the number of degrees of freedom. This test statistic has $n - 1$ degrees of freedom. Unlike the Z and t distributions, the χ^2 distribution is not symmetric.

Recall that the rejection region is the set of those values of the test statistic that would lead you to reject the null hypothesis. Let's think a minute about what values of the chi-square test statistic would lead us to reject the null hypothesis. Clearly, if the sample variance exactly equals the population variance then the ratio of s^2 to σ^2 will be one and the test statistic will be equal to $n - 1$. In the case of the tissue manufacturer, this would be $36 - 1 = 35$. In this situation the sample evidence is clearly consistent with the null hypothesis and we would fail to reject H_0.

Remember: The total area of the rejection region is α.

If the sample variance is quite different from the population variance being tested then it will be either larger or smaller than σ^2. If the sample variance is a great deal lower than the population variance being tested then the ratio of s^2 to σ^2 will be a fraction and the test statistic will be some value less than $n - 1$. A similar analysis tells us that if the sample variance is a great deal larger than the population variance being tested then the ratio of s^2 to σ^2 will be greater than one and the test statistic will be larger than $n - 1$. These are the situations that would lead us to reject H_0. This gives us the basic structure for the rejection region. We will reject H_0 if the value of the test statistic is much less than $n - 1$ or much greater than $n - 1$. Since this is a two-sided rejection region, the area in each tail must be $\alpha/2$. A typical rejection region is shown as the shaded region in Figure 12.4.

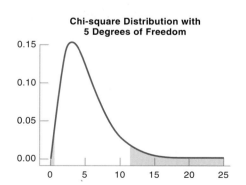

Chi-square Distribution with 5 Degrees of Freedom

FIGURE 12.4   Rejection region for a two-sided test of the variance

TABLE 12.1   A Portion of the Chi-square Table

Degrees of Freedom	Upper-Tail Areas											
	0.995	0.99	0.975	0.95	0.9	0.75	0.25	0.1	0.05	0.025	0.01	0.005
32	15.134	16.362	18.291	20.072	22.271	26.304	36.973	42.585	46.194	49.480	53.486	56.328
33	15.815	17.073	19.047	20.867	23.110	27.219	38.058	43.745	47.400	50.725	54.775	57.648
34	16.501	17.789	19.806	21.664	23.952	28.136	39.141	44.903	48.602	51.966	56.061	58.964
35	17.192	18.509	20.569	22.465	24.797	29.054	40.223	46.059	49.802	53.203	57.342	60.275

The specific values for χ^2_{lower} and χ^2_{upper} must be found from the chi-square table. A portion of this table is shown in Table 12.1. The complete table is found in Appendix A.5.

Let's use this table to complete step 2 of the hypothesis testing procedure for the tissue company.

EXAMPLE 12.6   The Tissue Company

Rejection Region for Variance Test

Since the sample size was $n = 36$, we must use the row with $n - 1 = 35$ degrees of freedom. To get the value for χ^2_{upper}, use the column labeled 0.025 in the upper tail.

The table value at the intersection of the correct row and column is 53.203. You may have noticed that the column values correspond to upper-tail areas. There is no need to have another table for lower-tail areas. To have the area of the lower tail be 0.025, the upper tail must be 0.975.

To get the value for χ^2_{lower} use the column labeled 0.975 in the upper tail. The table value at the intersection of the row with 35 degrees of freedom and 0.975 in the upper tail is 20.569.   ■

To get a sense of how different the sample variance must be from the population variance in order to reject H_0, let's examine this rejection region for a moment. The upper cutoff value is 53.203. If we divide this by 35, the number of degrees of freedom, we get 1.52. Thus, even if the sample variance is 1.5 times larger than the population variance you are testing it against, you would not yet be in the rejection region. Likewise if we take the lower cutoff value of 20.569 and divide it by 35 we get 0.588. So, if the sample variance is, for example, 60% (0.60) of the population variance you are testing it against, you would not reject the null hypothesis. This tells us that the sample evidence must be "pretty different" from the population variance you are testing it against in order to reject the null hypothesis. The chi-square table quantifies "pretty different" for us.

Now we can complete step 3 of the hypothesis testing procedure. Remember that at this step the test statistic is calculated.

EXAMPLE 12.7   The Tissue Company

Calculating the Chi-square Test Statistic

A sample of 36 tissues had a sample standard deviation of 60.1 lb/ream. The test statistic is

$$\chi^2 = \frac{(n - 1)s^2}{\sigma^2}$$

$$= \frac{(36 - 1)(60.1)^2}{2500} = 50.57$$ ■

The fourth step of the hypothesis testing procedure is to decide whether to reject the null hypothesis on the basis of the test statistic. The fifth step is to interpret this decision in terms of the original problem statement. Let's complete the test of the variance for the tissue manufacturer.

EXAMPLE 12.8 The Tissue Company

Steps 4 and 5 of the Hypothesis Testing Procedure

Since the value of 50.57 does not fall in the rejection region, we fail to reject the null hypothesis. This tells us that the manufacturing process is running according to specifications for the variability of the variable *MDStrength*. ■

 TRY IT NOW!

The Cereal Company
Testing the Variance

A cereal manufacturer wishes to test if the population variance of the weight of the boxes is equal to 0.0500 oz². A random sample of 20 boxes has a standard deviation of $s = 0.25$ oz. Use $\alpha = 0.05$.

Step 1:

Step 2:

Step 3:

Step 4:

Step 5:

12.3.2 One-Sided Tests of the Variance

If you are testing the population variance, most of the time you are interested in do-ing a two-sided test. However, it is possible to do a one-sided test of the variance. The only change in the procedure needed to complete a one-sided test of the variance is in step 2. A one-sided rejection region is used in this case. The rejection regions for the two possibilities are shown in Figure 12.5.

H_0: $\sigma^2 \geq$ [a specific number] H_0: $\sigma^2 \leq$ [a specific number]
H_A: $\sigma^2 <$ [a specific number] H_A: $\sigma^2 >$ [a specific number]

Lower-tail test of the variance

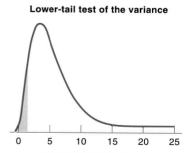

Upper-tail test of the variance

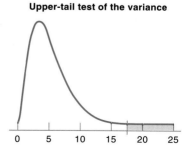

FIGURE 12.5 Rejection regions for one-sided tests of the variance

EXAMPLE 12.9   The Tissue Company

One-Sided Test of the Variance

Let's reconsider the variability for *MDStrength* for the tissue manufacturer. Suppose that the company wishes to be sure that the variance is at most 2500 lb/ream. The key words "at most" tell us this is an upper-tail test of the variance.

Step 1:

$$H_0: \quad \sigma^2 \leq 2500 \text{ lb/ream}$$
$$H_A: \quad \sigma^2 > 2500 \text{ lb/ream}$$

Step 2: Set $\alpha = 0.05$. Use the chi-square table with 35 degrees of freedom to find the rejection region. The table value at the intersection of the row for 35 degrees of freedom and the column with 0.05 in the upper-tail area is 49.802. Note that we did not divide α in half because this is a one-sided test with a one-sided rejection region.

Step 3: The test statistic is calculated below:

$$\chi^2 = \frac{(n-1)s^2}{\sigma^2}$$
$$= \frac{(36-1)(60.1)^2}{2500} = 50.57$$

Step 4: This test statistic falls in the rejection region, so we reject the null hypothesis.

Step 5: We conclude that the population variance is greater than 2500 lb/ream and the process should be adjusted. ■

A note on ethics in data analysis

You should notice that we got two different results based on the same sample of data. When we used a two-sided test of the variance we failed to reject the null hypothesis and concluded that the process was running properly. But when we used an upper-tail test of the variance we concluded that the population variance was greater than 2500 lb/ream and that the process was not running properly. You must decide what you are interested in detecting before you do the test. This is an example of how to "lie with statistics." You might say that the data do not lie. You are right about that in one sense. However, by manipulating the type of test that is done and the value of α that is used, you can change the conclusions of the analysis. If you are using the results of someone else's data analysis you must ask the right questions to be sure that the data have not been unfairly manipulated to support a particular position.

12.3.3 Summary of Tests on the Variance

In this section all of the tests used the same test statistic in step 3. The purpose of the test statistic is always the same: It gives you a single number that summarizes the sample information and helps you to decide between the null and alternative hypotheses. The test statistic for tests on variances is

$$\chi^2 = \frac{(n-1)s^2}{\sigma^2}$$

At step 2 we did see that the rejection region is different depending on what type of test we are doing. These differences are summarized in the next table.

Type of Test	Rejection Region
Two-tail test of σ^2 H_0: $\sigma^2 =$ [a specific number] H_A: $\sigma^2 \neq$ [a specific number]	Reject H_0 if $\chi^2 < \chi^2_{lower}$ or if $\chi^2 > \chi^2_{upper}$
Lower-tail test of σ^2 H_0: $\sigma^2 \geq$ [a specific number] H_A: $\sigma^2 <$ [a specific number]	Reject H_0 if $\chi^2 < \chi^2_{lower}$
Upper-tail test of σ^2 H_0: $\sigma^2 \leq$ [a specific number] H_A: $\sigma^2 >$ [a specific number]	Reject H_0 if $\chi^2 > \chi^2_{upper}$

12.3.4 Exercises—Learning It!

12.6 A company that sells mail-order computer systems has been planning inventory and staffing based on an assumption that the variance of their weekly sales is 180 (1000^2). The weekly sales are normally distributed. The company selects 15 weeks at random from the past year and obtains the data (in thousands of dollars) shown below:

Weekly sales 191 222 222 223 223 225 227 228 229 232 234 234 236 244 253

(a) What is the sample variance for these data?

(b) Set up the hypotheses to test whether the population variance is different from 180.

(c) At the 0.05 level of significance, what can you conclude about the company's assumption?

12.7 In manufacturing, the amount of material that is wasted or lost during a process is very important. In preparing financial estimates, a company assumes that the percent material lost for its new process has a variance $10\%^2$. After the new process has been running for a month and appears to be stable, the cost analyst looks at the percent material lost and finds the following data:

Daily Loss   10   12   12   13   14   14   18   19   19   20

(a) What is the sample variance for these data?

(b) Set up the hypotheses to test whether the actual variance is greater than the value the company has been assuming. Assume that the daily loss is normally distributed.

(c) At the 0.05 level of significance, what can you conclude?

12.8 Lead time for an order is a critical factor in inventory planning. A company has noticed that inventory for assembling its new product line has been consistently short in the past few months. It has been assumed that lead time is normally distributed with a standard deviation 2 days. The company decides to look at the last 20 orders placed, and finds that the lead times (in days) are

1	4	6	7
4	5	6	8
4	5	7	8
4	6	7	10
4	6	7	11

(a) What is the sample variance for these data?

(b) Set up the hypotheses to test whether or not the actual variance is greater than the value the company has been assuming. Be sure you use the variance here.

(c) At the 0.05 level of significance, what can you conclude?

12.9 In an effort to understand the cost overruns by a particular department in a company, data were collected on the amount of the overrun ($) for 24 different days:

87.3	93.7	96.8	98.4	100.9	107.8
89.9	94.9	97.0	99.6	101.3	109.7
91.5	96.5	97.1	100.3	105.7	111.5
93.6	96.7	97.3	100.4	107.7	114.2

(a) Calculate the sample variance for the daily cost overrun.

(b) At the 0.05 level of significance, do the data agree with the assumption that the variance of the daily overrun is 50 (dollars)^2. Assume that the cost overruns are normally distributed.

12.4   TEST OF A SINGLE PROPORTION

All of the tests that we have done thus far in the chapter have been tests on μ. However, very often it is not the average value that we are interested in but rather some proportion of the population that behaves in a certain manner. If the data we are analyzing are *nominal data*, the hypothesis might be a statement about

- The value of the proportion, π, of population members that have a certain characteristic (one of the categories of the nominal variable)

For example, the hypothesis might be a statement about the proportion of

- Residents of Swingfield who are in favor of a casino
- Students who are interested in graduate school
- Vaccinated patients who remain cancer free
- CEOs who use computers as a major tool
- People who are unemployed

In these cases, the parameter of interest is a proportion. In Chapter 9 you learned that an estimator for the true unknown proportion, π, is the corresponding sample proportion, p. For example, if 100 people are surveyed and 58 are in favor of a casino, then the sample proportion, p, is 58/100 or 0.58.

Like $\overline{X}$, the sample proportion, p, rarely if ever actually equals the true population proportion, π. Remember that this is not because you did something wrong but rather because you are looking at only a piece of the population and not the entire population. This is what we have called sampling error. It is a fact of life in statistics. You cannot eliminate sampling error unless you examine the entire population.

Since the sample proportion is not likely to equal the population proportion, we again see that the estimate alone is inadequate for making decisions. Suppose, for example, that more than 60% of the voters must be in favor of a casino and the sample proportion comes out to be 0.58. Does this automatically mean that the casino should be killed? Is it possible that more than 60% of the population favors the casino but only 58% of those sampled favor the casino? Possibly. How possible again depends on the standard error. So we need to develop the hypothesis testing procedure for tests on proportions. We start by looking at two-tail tests of proportions.

12.4.1 Two-Tail Tests of Proportions

Fortunately much of what we have already learned about hypothesis testing can be easily adapted to tests of proportions. In fact, we will use precisely the same five steps that we have been following for any hypothesis test:

Steps for any hypothesis test

Step 1:   *Set up the null and alternative hypotheses.*

Step 2:   *Pick the value of α and find the rejection region.*

Step 3:   *Calculate the test statistic.*

Step 4:   *Decide whether or not to reject the null hypothesis.*

Step 5:   *Interpret the statistical decision in terms of the stated problem.*

Let's use an example to see how these steps work for a two-tail test of proportions.

EXAMPLE 12.10 Beverly Hills, 90210

Two-Tail Test of Proportions

Hard times have hit the *Beverly Hills, 90210* crowd. Nearly a third of teens polled by Teenage Research Limited said they have been personally affected by the recession. Where do teens get their money? A survey of 2000 teens showed that 47% of them get some money from their parents. Suppose you wish to do a hypothesis test to see if half or 50% of all teens get some money from their parents.

Step 1: *Set up the null and alternative hypotheses.* The same general guidelines from Chapter 10 apply to setting up the null and the alternative hypotheses for a test of proportions. Since we are interested in seeing if the true proportion *differs* from 50% in either direction, we use a two-tail test. For this example, the null and alternative hypotheses are

$$H_0: \quad \pi = 0.50$$
$$H_A: \quad \pi \neq 0.50$$

Step 2: *Pick the value of α and find the rejection region.* In Chapter 9 you learned that the sampling distribution for p is the normal distribution. Therefore, the test statistic for proportions is a Z statistic. Thus, the form of the rejection region is exactly the same as the rejection region for a two-tail test on the mean when the standard deviation is known. If we set $\alpha = 0.05$, then the rejection region is as follows: reject H_0 if $Z > 1.96$ or $Z < -1.96$.

Step 3: *Calculate the test statistic and the* p *value.* Now we must consider what information in the sample will help us make a decision between the null and the alternative hypotheses. Clearly, we must use the sample proportion, p. But we know that we cannot simply use the value of p alone. We must take into account the fact that p will vary from sample to sample. As we have repeatedly seen, this is taken into account by dividing by the standard error of the estimator. The appropriate test statistic is

$$Z = \frac{p - \pi}{\sqrt{\pi(1 - \pi)/n}}$$

Z test statistic for proportions

where π is the hypothesized value of the population proportion.

This test statistic has the same basic form as the Z statistic we used in testing μ. In that case we used

$$Z = \frac{\overline{X} - \mu}{\sigma/\sqrt{n}}$$

In the numerator we are calculating the difference between the value of the population proportion, π, and the sample proportion, p. This is equivalent to the calculation that we did in the numerator of the Z calculation for μ. In the denominator of both Z statistics we use the standard error of the estimator. In the case of proportions, this is $\sqrt{\pi(1 - \pi)/n}$.

For this example the Z statistic is

$$Z = \frac{0.47 - 0.50}{\sqrt{0.50(1 - 0.50)/2000}} = -2.68$$

Remember that the p value is the smallest value of α for which you can reject H_0. To complete the hypothesis test you should calculate the p value. The p value for a test of proportions is calculated the same way as it was for the test of means. Since this is a two-tail test we must double the tail area probability. The Z statistic for this example was found to be -2.68. So, the p value is found as follows:

$$p\,\text{value} = 2P(Z > 2.68) = (2)(0.0037) = 0.0074$$

If you select an α of 0.05 as we have done in this example, then you will reject H_0.

Step 4: *Decide whether or not to reject H_0.* Since the value of $Z = -2.68$ is in the rejection region, we reject H_0.

Step 5: *Interpret the statistical decision in terms of the problem.* We conclude that the true proportion of teenagers who receive spending money from their parents is different from 0.50. ∎

 **TRY IT NOW!**

Poll of Americans

Test of Proportion

"People believe the country is still very, very screwed up," said Charles Cook of 10 political experts who analyzed the polling data upon its release at a Capitol Hill luncheon. The poll was conducted by sampling 2001 adults between 16 March and 21 March 1994. In the poll only 24% of Americans are satisfied with the country's economic course. Is there evidence that 25% of Americans are satisfied with the country's economic course? Use $\alpha = 0.05$.

Step 1:

Step 2:

Step 3:

Step 4:

Step 5:

Find and interpret the p value.

12.4.2 One-Tail Test of Proportions

Finally, we move to one-tail tests of proportions. Here again the five-step hypothesis testing procedure is identical to the one we have been using. The test statistic is the same as that used for a two-tail test of proportions and the rejection regions are the same as those used for one-tail tests of the mean.

Step 1: *Set up the null and alternative hypotheses.* There are two possible ways to set up a one-tail test of proportions. We have already looked at the issues that should be addressed in deciding how to set up the hypothesis test. Here we will simply present the two forms again using the terms upper-tail and lower-tail test.

ANS. $H_0: \pi = 0.25$, $H_A: \pi \neq 0.25$; CRITICAL VALUE $= \mp 1.96$; $Z = -1.03$; FAIL TO REJECT H_0; p VALUE $= 0.303$

Remember that the specific number must be a number from 0 to 1.

Upper-Tail Test	Lower-Tail Test
H_0: $\pi \leq$ [a specific number]	H_0: $\pi \geq$ [a specific number]
H_A: $\pi >$ [a specific number]	H_A: $\pi <$ [a specific number]

Step 2: *Select the value of α and find the rejection region.* Since the test statistic is a Z test, the rejection regions for the one-tail tests are precisely the same as the ones we used in testing means.

Step 3: *Calculate the test statistic.* We have seen in the previous section that the appropriate test statistic is

$$Z = \frac{p - \pi}{\sqrt{\pi(1 - \pi)/n}}$$

Steps 4 and **5** remain the same.

Consider an example.

EXAMPLE 12.11   Testing Raw Materials

An Upper-Tail Test of Proportions

Your company has been building a long-term relationship with one of its suppliers in the spirit of TQM. To validate the supplier's claim that at most 5% of its products are defective, you take a sample of 100 items and find six defectives. Should you believe the supplier's claim? Test using $\alpha = 0.05$.

Step 1: *Set up the null and alternative hypotheses.* Since you are interested in seeing if there is at most 5% defectives, this is a one-tail test. You will believe the supplier's claim unless the data indicate otherwise. This is an upper-tail test with the following setup:

H_0: $\pi \leq 0.05$ (Supplier's claim is correct.)

H_A: $\pi > 0.05$ (Supplier's claim is incorrect.)

Step 2: *Select a value for α and find the rejection region.* Since $\alpha = 0.05$, the rejection region is as shown below:

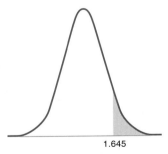

1.645

Step 3: *Calculate the test statistic.* The sample proportion was found to be $p = 6/100 = 0.06$ and the sample size was 100. Using this information we can calculate the test statistic:

$$Z = \frac{p - \pi}{\sqrt{\pi(1 - \pi)/n}} = \frac{0.06 - 0.05}{\sqrt{0.05(1 - 0.05)/100}} = 0.46$$

Notice that 0.05 is used as the value for π since that is the hypothesized value of the population proportion under H_0.

Step 4: *Decide whether or not to reject H_0.* Since $Z = 0.46$ does not fall in the rejection region, we fail to reject H_0.

Step 5: *Interpret the statistical decision in terms of the problem.* You cannot conclude with confidence that the vendor has more than 5% defectives in the entire shipment.

To finish the example we should calculate the p value. This is an upper-tail test so we do not double the tail area probability. The p value is found as follows:

$$p \text{ value} = P(Z > 0.46) = 0.3228$$

The large p value tells us that the data are highly consistent with the null hypothesis. This gives us greater confidence in accepting the shipment. ∎

 **TRY IT NOW!**

The Soft Drink Company
One-Tail Test of Proportion

The Coca-Cola Company is interested in entering the fruit drink market. Before bringing its new product, Fruitopia, to the market the company wishes to be sure that it will capture more than 20% of the fruit drink market. A survey of 1000 people shows that 210 respondents prefer Fruitopia to other fruit drinks. Is there enough evidence to allow Coca-Cola to proceed with the new product? Use $\alpha = 0.05$.

Step 1:

Step 2:

Step 3:

(continued)

Step 4:

Step 5:

12.4.3 Summary of Tests on Proportions

In this section all of the tests use the same test statistic in step 3. The purpose of the test statistic is always the same: It gives you a single number that summarizes the sample information and helps you to decide between the null and alternative hypotheses. The test statistic for tests on proportions is

$$z = \frac{p - \pi}{\sqrt{\pi(1 - \pi)/n}}$$

At step 2 we did see that the rejection region is different depending on what type of test we are performing. This is summarized below:

Type of Test	Rejection Region
Two-tail test of π	Reject H_0 if $Z < -Z_{\text{cutoff } \alpha/2}$ or if $Z > Z_{\text{cutoff } \alpha/2}$
$H_0: \pi = $ [a specific number between 0 and 1]	
$H_A: \pi \neq $ [a specific number between 0 and 1]	
Lower-tail test of π	Reject H_0 if $Z < -Z_{\text{cutoff } \alpha}$
$H_0: \pi \geq $ [a specific number between 0 and 1]	
$H_A: \pi < $ [a specific number between 0 and 1]	
Upper-tail test of π	Reject H_0 if $Z > Z_{\text{cutoff } \alpha}$
$H_0: \pi \leq $ [a specific number between 0 and 1]	
$H_A: \pi > $ [a specific number between 0 and 1]	

12.4.4 Exercises - Learning It!

12.10 Companies are increasingly concerned about employees playing video games at work. In addition to reducing productivity, this habit slows down networks and uses valuable storage space. A recent article stated that 80% of all employees play video games at work at least once a week. A large company that employs many engineers wonders if its employees are as bad as the article claims. If they are, the company will install software that detects and removes video games from the network. The company surveys (anonymously) 100 employees and finds that 85 of the employees surveyed had played video games at work in the past week.

(a) Set up the null and alternative hypotheses to test whether the proportion of the company's employees that play video games is greater than the proportion stated in the article.

(b) At the 0.05 level of significance, test the hypotheses.

(c) What do you recommend that the company do?

12.11 An alumni office is interested in serving their alumni better in order to encourage more donations to the college. A survey of 200 alumni was conducted to determine whether half-day training sessions offered on the campus were of interest. If more than 75% of the alumni were interested, the college would start a program. The survey showed that 160 of the alumni surveyed were interested in such a program.

(a) Set up the null and alternative hypotheses to test whether the college should implement the program.

(b) At the 0.05 level of significance, test the hypotheses.

(c) What do you recommend that the college do?

12.12 A company that makes computer keyboards has specifications that allow it to produce a product that has a maximum of 3% defective. The company has been receiving more customer complaints than usual. A sample of 50 keyboards has 2 defectives.

(a) Set up the null and alternative hypotheses to test whether the proportion defective keyboards has exceeded the amount allowed by the specifications.

(b) At the 0.05 level of significance, test the hypotheses.

(c) What do you recommend that the company do?

12.13 A university in the Northeast claims in its brochures that it has an acceptance rate of 60%. A sample of 300 high school seniors who applied to this university shows that 148 of them were accepted.

(a) Set up the null and alternative hypotheses to test whether the acceptance rate is what the university claims.

(b) At the 0.05 level of significance, test the hypotheses.

(c) Is there a need for the university to change its literature?

12.14 "Computer Jobs Increase by 6%" is the headline on an article in a national newspaper. A study of the classified job ads in that same newspaper indicates that out of 2202 advertisements in a typical Sunday edition, 502 were for computer related jobs. Statistics released by the Bureau of Labor Statistics for the last 12 months show that 15% of available jobs were computer related.

(a) Set up the null and alternative hypotheses to test whether the headline is correct.

(b) At the 0.05 level of significance, test the hypotheses.

(c) What conclusion can you reach?

Discovery Exercise 12.1
EXPLORING THE CONNECTION BETWEEN
CONFIDENCE INTERVALS AND HYPOTHESIS TESTING

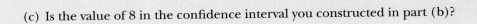

Part I

A recent survey is offering the first evidence that PCs are replacing TVs as the primary source of home recreation, information, and entertainment. The survey was conducted among 1200 homes nationwide. The average computer user spends 9.5 hours per week in front of the PC but only 8 hours per week watching prime-time TV. Assume that the standard deviation of hours spent in front of a PC is 3 hours/week.

(a) Test the hypothesis that the average number of hours per week a computer user spends in front of a PC is different from 8 hours. Use $\alpha = 0.05$.

(b) Using the same data, construct a 95% confidence interval for μ, the population average time spent in front of a PC.

(c) Is the value of 8 in the confidence interval you constructed in part (b)?

(d) Did you reject the null hypothesis in part (a)?

Part II

The Casual Businesswear Employee Survey was conducted to assess the attitudes and behavior of white-collar employees whose companies allow casual dress on some basis. The study was national in scope and the sample size was 752 people. Of those in the sample, 609 agree that allowing casual dress improves morale.

(a) Test the hypothesis that the proportion of white-collar employees who agree that allowing casual dress improves morale is different from 80%. Use $\alpha = 0.05$.

(b) Construct a 95% confidence interval for π.

(c) Is the value of 0.80 in the confidence interval?

(d) Did you fail to reject the null hypothesis?

Part III

Summarize.

Based on these two situations, speculate how you would complete the following statements:

If the value of the parameter being tested (the one in the null hypothesis) is not in the confidence interval then I will _____ the null hypothesis.

If the value of the parameter being tested (the one in the null hypothesis) is in the confidence interval then I will _____ the null hypothesis.

12.5 CONNECTION BETWEEN HYPOTHESIS TESTING AND CONFIDENCE INTERVALS

Now you have learned about both confidence intervals and hypothesis testing. These two techniques are actually closely related even though they are used for different purposes. You can see the relationship between these techniques easily if we reconsider one of the hypothesis testing examples from this chapter and calculate the corresponding confidence interval. Let's look at the tissue company.

EXAMPLE 12.12 The Tissue Company

Relationship of Confidence Interval to Hypothesis Test

A sample of 36 tissues was taken and the average *MDStrength* was 980 lb/ream. The standard deviation of the process is 50 lb/ream. In Example 10.11, we tested the hypothesis that the population mean *MDStrength* was 1000 lb/ream. The following hypothesis test was done:

$$H_0: \quad \mu = 1000 \text{ lb/ream}$$
$$H_A: \quad \mu \neq 1000 \text{ lb/ream}$$

The value of the test statistic was $Z = -2.4$. Based on this Z value and an α value of 0.05, the null hypothesis was rejected, leading us to conclude that the population mean, μ, was not 1000 lb/ream.

Now construct a 95% confidence interval for μ using these data:

Upper $= \overline{X} + Z(\sigma/\sqrt{n}) = 980 + (1.96)(50/\sqrt{36}) = 980 + 16.33 = 996.33$ lb/ream

Lower $= \overline{X} - Z(\sigma/\sqrt{n}) = 980 - (1.96)(50/\sqrt{36}) = 980 - 16.33 = 963.67$ lb/ream ∎

Notice that the value of μ of 1000 lb/ream is not in the confidence interval. This means that 1000 is not a likely value of μ. Using the same data we rejected the null hypothesis, concluding that μ was not 1000 lb/ream. You can see that the results from the confidence interval are consistent with the conclusions drawn from the hypothesis test.

This result can be stated in general for any hypothesis test, whether it be a test of μ, σ^2, or π. If the value of the parameter being tested (the one in the null hypothesis) is in the confidence interval, then you will fail to reject the null hypothesis of a two-tail test. If the value of the parameter being tested is not in the confidence interval, then you will reject the null hypothesis. Of course, the same value of α must be used for both the confidence interval construction and the hypothesis test, and the hypothesis test must be a two-tail test.

12.6 HYPOTHESIS TESTING IN EXCEL

One of the things you learned in this chapter is that all hypothesis tests have the same basic form. What changes from test to test is how to calculate the test statistic and which sampling distribution applies. It is most important to understand when each test is appropriate.

This is also true about using Excel to perform hypothesis tests. All that changes for the different tests is the input.

12.6.1 Small-Sample Tests for the Population Mean

Since Excel does not have any analysis tools for single population hypothesis tests, we have written a macro that will do this analysis. Suppose that we were looking at the hand washing data, and that we want to perform the test to see if the mean time spent washing hands by the 28 employees is different from 5 seconds. Since we do not have the population standard deviation and the sample size is small ($n < 30$), the t test is appropriate.

You should have the data you want to analyze in an Excel worksheet and have the MacDoIt.xls worksheet open so that the macros are available. From the list of macros, choose **t Test.** The table shown in Figure 12.6 opens in a new worksheet. You can see that it is very similar to the one for the Z test.

	A	B	C
1	Hypothesis Test for Mean - t-Test		
2	Input		
3	Sample Mean	m	95
4	Hypothesized Mean	μ_0	100
5	Standard Deviation	s	10
6	Sample Size	n	16
7	Significance Level	α	0.05
8	Output		
9	Standard Error	StdErr	2.5
10	Degrees of Freedom	df	15
11	t-Statistic	t	-2
12	Lower-Tail Test for H_A: $\mu < \mu_0$		
13	Lower-Tail Critical t	t_L	-1.7531
14	Lower-Tail p-value	p-value$_L$	0.0320
15	Two-Tail Test for HA: $\mu \neq \mu_0$		
16	Two-Tail Absolute Critical t	t_T	-2.1315
17	Two-Tail p-value	p-value$_T$	0.0639
18	Upper-Tail Test for HA: $\mu > \mu_0$		
19	Upper-Tail Critical t	t_U	1.7531
20	Upper-Tail p-value	p-value$_U$	0.9680

FIGURE 12.6   Table for t test macro

Follow these steps to carry out the test:

1. Position the cursor in the Input section in the row for the Sample Mean. Type **=AVERAGE(** and highlight the range that contains the data. Remember to type in the right parenthesis *before* you switch back to the sheet with the table.

2. After you hit Enter, you must type in the value for the hypothesized mean. In this case, we are testing whether the mean washing time is different from 5 seconds, so type in "5" and hit Enter.

3. Now you must enter the value for the sample standard deviation using the **STDEV** function.

4. Enter the sample size, in this case 28, and hit Enter again.

5. Now enter the level of significance for the test. Let's test the hypothesis at the 0.05 level of significance.

After you hit Enter the last time, the table will update. The results are shown in Figure 12.7.

	A	B	C
1	Hypothesis Test for Mean - *t*-Test		
2	Input		
3	Sample Mean	m	2.428571429
4	Hypothesized Mean	μ_0	5
5	Standard Deviation	s	3.655828137
6	Sample Size	n	28
7	Significance Level	α	0.05
8	Output		
9	Standard Error	StdErr	0.690886578
10	Degrees of Freedom	df	27
11	*t*-Statistic	t	-3.721925791
12	Lower-Tail Test for H_A: $\mu < \mu_0$		
13	Lower-Tail Critical *t*	t_L	-1.7033
14	Lower-Tail *p*-value	p-value$_L$	0.0005
15	Two-Tail Test for HA: $\mu \neq \mu_0$		
16	Two-Tail Absolute Critical *t*	t_T	-2.0518
17	Two-Tail *p*-value	p-value$_T$	0.0009
18	Upper-Tail Test for HA: $\mu > \mu_0$		
19	Upper-Tail Critical *t*	t_U	1.7033
20	Upper-Tail *p*-value	p-value$_U$	0.9995

FIGURE 12.7   Output from *t* test macro

From the output section, we see that the value of the *t* statistic is -3.722. Since this is a two-tailed test, look at that section of the output and compare the test statistic to the critical value of -2.0518. The decision is to reject H_0 and conclude that the mean hand washing time is not 5 seconds. We also see that the *p* value of the test is 0.0009, which is much smaller than the specified level of significance.

Recall that one of the assumptions of the *t* test for the mean is that the data are normally distributed. Although you cannot test this assumption formally yet, you can (and most certainly should) create a graphical display of the data to see whether the assumption of normality is valid. Using the boxplot macro in Excel produces the plot shown in Figure 12.8 on page 500.

From the plot you can see that the data are really rather skewed to the right and that the assumption might not be a good one for these data.

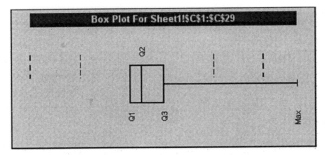

FIGURE 12.8   Boxplot of hand washing data

12.6.2 χ^2 Tests for the Variance in Excel

There is no built-in tool or macro for the χ^2 test for the variance, but the test can be performed by using some built-in Excel functions. Most of these functions should be familiar to you.

We will look at the problem of the tissue manufacturer. When we did the test on the mean tissue strength in Chapter 10, we used the target value for the standard deviation, 50 lb/ream. Now we will test to see whether it was reasonable to assume that the population variance is 2500 (lb/ream)^2, using a χ^2 test. The data are the same data that we used in Chapter 10.

Since we want to know whether the true population variance is different from the target value of 2500 (lb/ream)^2, we will do a two-tailed test. To do the test, we will need to calculate the value of the test statistic and the critical value for the test. We will also calculate the p value for the test.

The test statistic for a χ^2 test for the variance is

$$\chi^2 = \frac{(n-1)s^2}{\sigma^2}$$

To calculate this using Excel, the following steps are used:

1. To organize the output, put labels in worksheet cells to identify all of the output values. Then when you are done, you will know which value is which. In this case, label cells for s^2, σ^2, the sample size, the test statistic, the critical value of the test, and the p value of the test. An example of this is shown in Figure 12.9.

E	F
s^2	
σ^2	
n	
Test Statistic	
Critical Value	
p value	

FIGURE 12.9   Setup for χ^2 test

2. In the cell adjacent to the one labeled s^2, use the **STDEV** function to calculate the sample standard deviation, and then square it. The formula will look like the one shown in Figure 12.10.

$$= \quad \text{=STDEV(B1:B76)^2}$$

FIGURE 12.10 Formula for variance

As an alternative to this, you can use the **VAR** function, which calculates the sample variance. The input for this function is the data range, just as in the **STDEV** function.

3. In the cell adjacent to the one labeled σ^2, type in 2500, and type 75 in the one adjacent to n.

4. To calculate the test statistic, we will need the three values just input. In the cell adjacent to Test Statistic, type in the following formula:

$$= \quad \text{=(F3-1)*F1/F2}$$

Note: You need parentheses for $n - 1$.

These are the cell locations from the worksheet we are using as an example. Be sure to use the cell locations from *your* worksheet if they are different. You could enter the numbers directly in the formula, but it really is easier to use the cell locations.

5. We will use the **CHIINV** function to find the two critical values for the test. The **CHIINV** function works the same way as the **TINV** function. The input values are the upper tail probability and the degrees of freedom. Since α for this test is 0.05 and it is a two-tailed test, the probability for the lower critical value will be 0.975, and 0.025 for the upper value. The degrees of freedom are $75 - 1 = 74$. The upper tail formula is shown below:

$$= \quad \text{=CHIINV(0.025,74)}$$

The output value is 99.67838. Place the formulas for the upper and lower critical values in the two cells adjacent to the one labeled Critical Value.

6. Remember that the p value of the test is associated with the test statistic. To find the p value of the test, we will use the **CHIDIST** function. This function accepts as input the value of the χ^2 variable and the degrees of freedom, and outputs the tail area probability. In the cell adjacent to the one labeled p value, type

$$= \quad \text{=CHIDIST(F4,74)*2}$$

The cell F4 is the reference of the cell that contains the test statistic value. We multiply the p value by 2 because this is a two-tailed test.

The complete output is shown in Figure 12.11.

E	F	G
s^2	3334.307	3334.307
σ^2	2500	
n	75	
Test Statistic	98.6955	
Critical Value	99.67838	52.10282
p value	0.058305	

FIGURE 12.11 Output for χ^2 test

Comparing the test statistic to the critical value, we find that we cannot reject H_0. That is, it is reasonable to assume that the population variance is 2500 $(\text{lb}/\text{ream})^2$. The decision is a rather close one, as you can see from the p value.

Remember that the χ^2 test, just like the t test, assumes that the population is normally distributed. Before you report any results, you should check this assumption with some type of graphical display.

Any hypothesis test can be done in Excel using the same type of procedure outlined here, although it is easier if there are built-in tools or macros available.

12.6.3 Hypothesis Tests for a Single Proportion in Excel

Although Excel does not have a built-in analysis tool for the single population proportion test, we have a macro that will accomplish this task. The macro is very similar to the ones for the Z and t tests.

Suppose we want to look at the data on the proportion of teenagers who get money from their parents. We want to determine whether the true proportion of teenagers who get money from their parents is different from 0.50 or 50%. The data for this problem have been partially analyzed, and we know that in the sample of 2000 teenagers, the proportion who receive money from their parents is $p = 0.47$ or 47%.

To carry out the test, follow this procedure:

1. Select **Tools > Macro > Macros** from the menu and run the **SinglePropTest** macro from the list. The table shown in Figure 12.12 opens in a new worksheet.

	A	B	C	D	E	F
1						
2		Test Statistic for Proportion				
3						
4		Sample Proportion	p$_s$	0.5		
5		Hypothesized Proportion	π	0.5		
6		Sample Size	n	100		
7						
8		Test Statistic	Z	0.0000		
9		Significance Level	α	0.05		
10		*Critical Values*				
11		Lower Tail Test	Z1	-1.6449	p-value1	0.5000
12		Upper Tail Test	Z2	1.6449	p-value2	0.5000
13		Two Tailed Test	Z3	-1.9600		

FIGURE 12.12   Table for a single proportion test

2. Fill in each of the cells in the top portion of the table with the relevant data: $p = 0.47$, $\pi = 0.50$, and $n = 2000$.

3. Fill in the significance level of the test; in this case, we will use $\alpha = 0.05$.

The table will update and the result is shown in Figure 12.13.

From the output, we see that the two-tailed critical value is -1.96 and that the test statistic is -2.6833. The decision is to reject H_0 and conclude that the true proportion is different from 50%. Also, we see that the p value of the test is much smaller than the level of significance.

	A	B	C	D	E	F
1						
2		Test Statistic for Proportion				
3						
4		Sample Proportion	p,	0.47		
5		Hypothesized Proportion	π	0.5		
6		Sample Size	n	2000		
7						
8		Test Statistic	Z	-2.6833		
9		Significance Level	α	0.05		
10		Critical Values				
11		Lower Tail Test	Z1	-1.6449	p-value1	0.0036
12		Upper Tail Test	Z2	1.6449	p-value2	0.0036
13		Two Tailed Test	Z3	-1.9600	p-value 3	0.00729

FIGURE 12.13   Output for single proportion test

You might wonder how you would proceed with the test if you had raw data and not the value of p available. In this case you would use the Pivot Table tool to obtain a frequency table. The steps to do this were explained in Chapter 3. As part of the pivot table, you can either display the output as a count or as a percentage of the total.

CHAPTER 12 SUMMARY

In this chapter you have learned how to do a hypothesis test of the population mean, population variance, and population proportion. The five-step hypothesis test procedure is the same for any hypothesis test. The differences in the tests are in the rejection regions and the calculation of the test statistic. For a test of the population mean, you use a Z test if the population standard deviation is known; otherwise, you use a t test. For a hypothesis test of the population variance use a chi-square test, and for a test of proportions use a Z test.

You have seen that in any hypothesis test the rejection region and thus the final decision depend on the value of α. You can alter the outcome of the test by adjusting the value of α. To handle the potential unethical use of hypothesis testing and to provide management with more information than simply whether or not the null hypothesis was rejected, you learned about p values. You have seen that the p value frees you from specifying a value of α for the test. By reporting the p value for the test, you put the decision to reject or fail to reject the null hypothesis in the hands of management. The decision maker must weigh the costs of Type I and Type II errors and make a decision using these costs and the p value. The final section of this chapter ties the results of a hypothesis test to the corresponding confidence interval calculation.

Key Formulas

Term	Formula	Page Reference
t test statistic	$t = \dfrac{\overline{X} - \mu}{s/\sqrt{n}}$	470
Chi-square statistic	$\chi^2 = \dfrac{(n-1)s^2}{\sigma^2}$	482
Z test statistic for proportions	$Z = \dfrac{p - \pi}{\sqrt{\pi(1-\pi)/n}}$	489

CHAPTER 12 EXERCISES

Learning It!

12.15 Most traffic lights are set so that there is enough time for pedestrians to cross the road safely. A recent study indicates that a large number of elderly cannot get across the road in the usual 15 seconds allowed. To determine the average amount of time it takes senior citizens to cross the street, a study was taken of 25 seniors. On the average, it took them 19.5 seconds to cross the street, with a sample standard deviation of five seconds. Assume the time to cross the road has a normal distribution.

(a) Set up the null and alternative hypotheses to see if the data show that it does indeed take seniors longer than 15 seconds to cross the street.

(b) In terms of traffic flow, what are the implications of a Type I error?

(c) What are the implications of a Type II error?

(d) Find the value of the test statistic.

(e) If $\alpha = 0.05$, what is the rejection region?

(f) What is your recommendation to the city?

12.16 Kids think Moms are doing a great job! According to a recent poll conducted by Massachusetts Mutual Insurance Company, 90% of children feel their mothers spend enough time with them. A similar survey was conducted in New York to see if children there felt the same way. Of 1000 children surveyed in New York 880 felt that their mothers spend enough time. Do these data indicate that less than 90% of the New York children feel their mothers spend enough time with them?

(a) Set up the null and alternative hypotheses.

(b) What is the value of the test statistic?

(c) If $\alpha = 0.10$, what is the rejection region?

(d) What do you conclude about the children of New York?

(e) What might be causing this difference?

12.17 In a new advertising campaign, Coca-Cola pokes fun at the decision to revive its famous contoured bottle. Atlanta-based Coke faced significant technical hurdles in bringing out the well-known bottle in a new material: plastic. Coke also conducted extensive marketing tests to make sure that one of the world's best-known packages, seldom seen since the 1970s, would be a hit in the 1990s.

Coca-Cola wishes to be sure that at least 60% of the consumers prefer the contoured bottle. Of 3,000 consumers nationwide, 1900 prefer the new bottle. Is this sufficient evidence to give the new bottle the go-ahead?

(a) Set up the null and alternative hypotheses for Coca-Cola.

(b) In terms of Coke's decision to introduce the new bottle or not, what is a Type I error?

(c) In terms of Coke's decision to introduce the new bottle or not, what is a Type II error?

(d) What is the value of the test statistic?

(e) If $\alpha = 0.05$, what is the rejection region? What is your decision?

12.18 You purchased your home in 1987 for $165,000, which was the average price at that time. Now you are thinking of selling your home. You take a sample of homes sold in your neighborhood during the past 2 months and find the following sale prices in $:

| 130,000 | 135,500 | 136,000 | 140,000 | 160,000 |
| 167,000 | 168,000 | 174,500 | 177,400 | 180,000 |

(a) Based on these data, conduct the hypothesis test to see if the average selling price has increased from $165,000. Assume that the selling price of homes in this town is normally distributed.

(b) What advice would you give this homeowner regarding selling his home at this time? Explain your answer in terms of the results of the hypothesis test.

12.19 According to the *Springfield Union News*, the New England Blizzard may not yet be ready to take a place among the American Basketball League's elite, but the Hartford–Springfield franchise is in a whole other dimension when it comes to getting people through the turnstiles. The next table shows the 1997 ABL attendance figures for the other ABL teams:

Team	Dates	Attendance
Portland	4	19,856
San Jose	5	23,775
Seattle	3	9,233
Philadelphia	5	15,308
Colorado	3	9,176
Columbus	4	11,239
Atlanta	4	10,776
Long Beach	3	5,372

(a) Calculate the average attendance for the ABL league without the Blizzard.

(b) The Blizzard had an attendance figure of 49,879 in 6 games. Find the average attendance for the sample.

(c) Test the hypothesis that the Blizzard had a higher average attendance than the ABL league. Assume that attendance figures are normally distributed.

(d) What conclusions can you share with the owners of the Blizzard franchise?

(e) Upon further investigation you discover that the Blizzard plays at both the Hartford Civic Center and the Springfield Civic Center. What additional information would you like to know about the attendance figures to help the Blizzard increase attendance?

12.20 In the wake of a large number of repeat crime offenders, many states are considering the death penalty. A sample of 100 citizens of a New England state was taken and 56 of the 100 people favored the death penalty.

(a) Legislatures in favor of the death penalty are using the results of this survey to say that the majority of the people in the state are in favor of the death penalty. Conduct the appropriate hypothesis test to determine if these legislatures are justified in their remark.

(b) What conclusion can be reached from this survey about the preference for citizens of this state toward the death penalty?

Thinking About It!

Requires Exercise 12.17

12.21 Coca-Cola has asked you your recommendation about their new bottle. Basically, the company wants to know if you are secure in your decision before it makes any drastic changes. You decide to test how much the procedure you chose affected your decision.

(a) Suppose you had used $\alpha = 0.10$. What is your decision?

(b) Now use $\alpha = 0.01$. What is your decision?

(c) Find the p value for this problem.

(d) Based on this new information, what would you tell the Coca-Cola company to do?

Requires Exercise 12.15

12.22 Look again at your recommendation to the city about the length of its walk light signal. In writing your report, you start to think about whether or not the study you did was adequate.

(a) What other factors might be useful in determining the length of time to allow people to cross the street safely?

(b) What are the drawbacks of making decisions such as this based on the mean?

(c) What might be a better statistic to use in this case?

12.23 Many states are considering allowing casino gambling in an attempt to revitalize urban areas. Opponents say gambling would be a magnet for crime and would shift spending away from entertainment and other industries. A recent poll conducted by the Boston *Sunday Herald* found that 46% of the Bay State voters are opposed to gambling, while 36% support the proposal. Eighteen percent of those surveyed had no opinion or were neutral. The poll of 450 registered voters across the state was conducted between a Tuesday and a Friday. Legislatures will endorse the casinos if less than 50% of the voters oppose it. Do the data indicate that this is the case?

(a) Set up the null and the alternative hypotheses.

(b) What should you do with the 18% who had no opinion or were neutral?

(c) Find the value of the test statistic.

(d) If the members of the legislature wish to be conservative, how should they set α?

(e) If the members of the legislature wish to be more aggressive on this issue, how should they set α?

(f) Using the value of α set in part (d), find the rejection region.

(g) What is your recommendation to your local representative?

(h) Using the value of α set in part (e), find the rejection region.

(i) What is your recommendation now to your local representative?

12.24 Remember the problem where you were trying to convince your boss that your expense report for the aspirin purchase was justified? Your boss just asked you for the data you used to justify the expense. He points out that the price of aspirin in Tokyo seems a bit high and asks you to reduce your estimate. The data are repeated below: *Requires Exercise 12.1*

City	Aspirin (100)	Fast food (hamburger, fries, soft drink)	Woman's haircut/blow dry	Toothpaste (6.4 oz)
Los Angeles	$7.69	$4.15	$20.11	$2.42
Tokyo	$35.93	$7.62	$76.24	$4.24
London	$9.69	$5.80	$44.35	$3.63
Sydney	$7.43	$4.53	$29.93	$2.08
Mexico City	$1.16	$3.63	$17.94	$1.08

(a) If you remove Tokyo from the data set should you change your expense report?

(b) What does this tell you about the effect of outliers on the hypothesis test of μ when you have a small sample?

Doing It!

12.25 A portion of the data file for the hospital that collected the data on hand washing is shown below: *Data file: HOSPITAL.XXX*

Observation	Unit	Time 1	Time 2
1	CCU	3	16
2	CCU	2	7
3	CCU	0	5
4	CCU	5	8
5	CCU	2	15
6	CCU	0	15
7	CCU	2	20
8	CCU	3	16
9	CCU	0	18
10	IMCU	1	16
11	IMCU	2	8

You can see that information was also collected on the unit of the employee and a second time, labeled Time 2, was also recorded. Assume that the hand-washing times are normally distributed and are recorded in seconds.

(a) Find the sample means and sample standard deviations for the first hand-washing time, Time 1, for each unit.

(b) Test to see if any of the units had an average hand-washing time, Time 1, less than 5 seconds.

(c) The second hand-washing time, Time 2, was recorded after the employees received some training on the effects of not washing their hands long enough. Find the sample means and sample standard deviations for the second hand-washing time, Time 2, for each unit.

(d) Test to see if any of the units had an average hand-washing time after training, Time 2, of less than 5 seconds.

(e) Display all the initial hand-washing times, Time 1, in a histogram. Is the assumption of normality reasonable?

(f) Display all the second hand-washing times, Time 2, in a histogram. Is the assumption of normality reasonable?

(g) Test the hypothesis that the variance of the initial hand-washing time is 14 seconds.

(h) Test the hypothesis that the variance of the hand-washing times after training is greater than 14 seconds.

(i) Find the sample proportion of employees who increased their hand-washing time after training. Test the hypothesis that the proportion of employees who increased their time after training is greater than 0.80.

(j) Based on your analysis, write a report to the manager of the hospital.

WORKSHOP 2

COMPARING TWO POPULATIONS

TRAINING

As the number of management consultant firms rapidly grows, more and more companies are relying on company outsiders to provide different types of training. While using company personnel might be more cost effective, management feels that employees attach a stigma to in-house trainers (often their peers or even subordinates) and feels that outside consultants provide a more credible service.

In an attempt to determine whether there are any real differences in the results achieved by in-house and external trainers, a large corporation decided to try an experiment with its quality inspectors. The corporation took two groups of employees at a large manufacturing location who were training for the American Society for Quality (ASQ) certification exam as Certified Mechanical Inspector (CMI). One group used the on-site statistician to provide training, while the other group used an outside consultant.

The company measured several variables, including a "pretraining" score for the exam, an evaluation survey for the training course, and whether the engineer passed the actual certification examination. Gender and age of the employees were also noted. A portion of the data is shown below:

Employee	Training	Gender	Age	Pre Test	Post Test	Certification Exam
1	I	M	43	27	64	F
2	I	M	40	33	71	P
3	I	M	36	41	60	F
4	I	M	34	45	51	F
5	I	M	43	38	65	F
6	I	M	29	48	63	F
7	I	M	38	55	58	P

13.1 CHAPTER OBJECTIVES

In the previous chapter you learned how to do hypothesis tests for many different population parameters. In each case you tested hypotheses about whether the sample data could have come from a population whose parameter was equal to a specific value. While these tests provide valuable information to decision makers, they are not the only questions that you might want answered. For example, suppose you wanted to know the answer to the following questions:

- Who uses electronic mail more frequently, senior managers or midlevel managers?
- Is it better to promote a person from within the organization or hire from outside the organization?
- Do men and women behave the same way in the supermarket?
- Does one design of a golf ball travel farther than another?
- Do college graduates actually earn more money than high school graduates?

Although the questions listed above cover widely different topics, they do have a common theme. In each case a *comparison* is being made. In this chapter we will extend the hypothesis testing procedure that we learned in Chapter 12 to handle the *comparison* of two population means, two population proportions, or two population variances. The basic five-step hypothesis testing procedure is the same. The only difference is the test statistic to be used.

In Chapter 12 we looked at hypothesis tests of a mean, proportion, or variance from a single population.

This chapter covers the following material:

- Collecting Data From Two Populations
- Large-Sample Test of the Difference in Two Population Means
- Small-Sample Test of the Difference in Two Population Means
- Test of the Difference in Two Population Means—Paired Data
- Hypothesis Test of the Difference in Two Population Proportions
- Hypothesis Test of the Difference in Two Population Variances

13.2 COLLECTING DATA FROM TWO POPULATIONS

When you are comparing characteristics of two different populations, you must have a sample from each of the populations. These samples are usually selected independently of each other. In other words, the selection of one sample should not have any effect on the selection of the second sample. We will label all of the parameters of one population with a subscript 1 and all the parameters of the second population with a subscript 2. It does not matter which population you label 1 or 2. The populations and samples are shown in Figure 13.1.

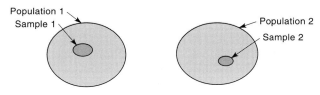

FIGURE 13.1 Two populations and two samples

Consider the question posed about whether men or women spend more money on frozen foods. We could label the population of males as population 1 and the population of females as population 2. If we do this then the parameters and statistics corresponding to the male population will be identified with a subscript 1 and those describing the female population will carry a subscript 2. These are shown in the following lists.

	Population 1: Males	Population 2: Females
Size of population	N_1	N_2
Population mean	μ_1	μ_2
Population standard deviation	σ_1	σ_2
Sample size	n_1	n_2
Sample mean	$\overline{X}_1$	$\overline{X}_2$
Sample standard deviation	s_1	s_2

For this example, we would select a sample of men shoppers and a separate sample of women shoppers. We would ask all members of the sample how much money they spent on frozen foods in the past week. It is not necessary that the sample sizes be equal, but if possible it is desirable to have both sample sizes (n_1, n_2) greater than or equal to 30. The reason for this stems from the fact that the Central Limit Theorem generally applies when the sample size is 30 or greater. Remember that we developed the Z test statistic based on the knowledge that the sample mean, $\overline{X}$, has a normal distribution.

Qualitative variables are often used to identify two populations for comparison.

However, often a single sample is selected and a qualitative variable is used to identify two populations for comparison. For the food shoppers example, we might select one sample of shoppers and then record the gender of the respondent as part of the data. This means that the data can then be divided into the two comparison populations after the data have been collected. If at the same time you collect data on the age of the person as "under 40" or "40 and over" then you can also compare the average frozen food expenditure for younger buyers to that of older buyers. Clearly, spending differences that are identified by gender or age could be of great assistance in developing a marketing strategy.

13.3 HYPOTHESIS TEST OF THE DIFFERENCE IN TWO POPULATION MEANS—OVERVIEW

Population standard deviations are known or the sample sizes are large.

There are several different cases to consider if you are testing a hypothesis of the difference in two population means. To determine into which case your data fit, you must first see if you have information on the population standard deviation of both populations. As you saw in Chapter 12, sometimes you know the value of the population standard deviation but most of the time you do not. If you know the value of the standard deviation for each of the two populations that you are comparing, then you will use a Z test statistic. In this case, it does not matter how large the samples are. The underlying populations must be normally distributed. This is covered in Section 13.4.

Remember that when $n \geq 30$ the Central Limit Theorem generally applies.

If you do not know the value of the standard deviation for each of the two populations, then you must check the size of the sample. If each of the samples has 30 or more observations then you will be in what is commonly referred to as the "large-sample" case. This case is very similar to the situation when you know the standard deviations.

If you do not know the population standard deviations and one or both of your samples has less than 30 observations, then you will be in the "small-sample" case. You must then decide if the variances in each of the two populations are the same, even though they are unknown, or different. Section 13.5 covers the case when you do not know the value of the standard deviations but you do know that the populations are normally distributed.

Population standard deviations are unknown and sample sizes are small.

Finally, sometimes the data are collected in such a way that the samples are not independent. This happens quite often when you are comparing two different medical procedures or treatments. You wish to be sure that the differences that you observe are not simply due to the fact that you have different patients with different medical histories. In this case you might use the same patients in both samples. If your samples are dependent then you must use what is known as a paired *t* test to compare the population means. This is the subject of Section 13.6.

Paired data must be analyzed differently.

In summary, three sections will be devoted to hypothesis testing of the difference in means. Section 13.4 will cover tests when the standard deviation is known, Section 13.5 will cover the cases when the standard deviation is not known, and Section 13.6 will cover the test for paired data. Regardless of the particular case, each hypothesis test will follow the five-step procedure outlined in Chapters 10 and 12. It is repeated here for you:

Step 1: Set up the null and alternative hypotheses.

Step 2: Pick the value of α and find the rejection region.

Step 3: Calculate the test statistic and the p value.

Step 4: Decide whether or not to reject the null hypothesis.

Step 5: Interpret the statistical decision in terms of the stated problem.

Steps for any hypothesis test

Since you are already familiar with hypothesis testing, in this chapter we often combine the first two steps, which define the test procedure, into a single step.

13.4 LARGE-SAMPLE TESTS OF THE DIFFERENCE IN TWO POPULATION MEANS

13.4.1 Large-Sample Tests of Two Means With Known Standard Deviations

The basic test concerning two population means occurs when we want to know whether the two samples come from populations with equal means and we *assume that the population standard deviations are known*. Although this may not seem reasonable, there are cases where a historical or specified standard deviation is appropriate.

EXAMPLE 13.1 The Tissue Manufacturer

Identifying the Two Populations

The company that is looking at tissue strength wants to look at the problem again. Since tissue strength seems to vary quite a bit, your boss has asked you to find out if there is any difference in the strength of the tissues made on one day compared to another day. The company uses two measures of tissue strength: machine-directional strength *(MDStrength)* measured in lb/ream and cross-directional strength *(CD-Strength)* also measured in lb/ream. Right now he wants you to look at the *MD-Strength.*

The product specifications for *MDStrength* state that the mean should be 1000 lb/ream with standard deviation of 50 lb/ream. You have collected data on 75 sheets for each of three different days. For each sheet you recorded the *MDStrength* and the *CDStrength.*

The first thing to do is to identify the two populations to be compared. The tissue manufacturer wants to compare the strength of tissues made on day 1 to the strength of tissues made on day 2. Population 1 is all tissue sheets made on day 1 and we have a sample of size $n_1 = 75$ from this population. All tissue sheets made on day 2 comprise population 2 and we have a sample of $n_2 = 75$ from this population. ■

Use the techniques of Chapter 12 to test the variance for each of the populations.

In this case, we will use the specified values of the population variance to perform the test. Of course, this is not something you want to do blindly. Just because you have specifications does not mean that you have compliance. Still, for our example we already established that the specified values for the population standard deviation are valid.

Once you can establish that the values of the population standard deviations for both populations are known, you know what case we are dealing with and we are ready to proceed with the hypothesis test.

- *The first step of the procedure is to construct the null and alternative hypotheses.*

As with tests of a single mean, there are three different ways to set up the hypothesis test. The guidelines for determining which of these tests you should use are the same as presented in Chapter 12. The three possible setups are shown below:

Remember the key words different, less than, *and* greater than *tell you which test to use:*

Two-Sided Test

$H_0:\ \mu_1 = \mu_2$ or	$H_0:\ \mu_1 - \mu_2 = 0$	Use this test if you wish to test if the mean of
$H_A:\ \mu_1 \neq \mu_2$	$H_A:\ \mu_1 - \mu_2 \neq 0$	population 1 is *different* from the mean of population 2.

Lower-Tail Test

$H_0:\ \mu_1 \geq \mu_2$ or	$H_0:\ \mu_1 - \mu_2 \geq 0$	Use this test if you wish to test if the mean of
$H_A:\ \mu_1 < \mu_2$	$H_A:\ \mu_1 - \mu_2 < 0$	population 1 is *less than* the mean of population 2.

Note: *You should never have $\overline{X}$'s in the statement of the hypotheses. There is no need to formulate a hypothesis about the sample means because you know the values. You do not know the population means.*

Upper-Tail Test

$H_0:\ \mu_1 \leq \mu_2$ or	$H_0:\ \mu_1 - \mu_2 \leq 0$	Use this test if you wish to test if the mean of
$H_A:\ \mu_1 > \mu_2$	$H_A:\ \mu_1 - \mu_2 > 0$	population 1 is *greater than* the mean of population 2.

Notice that the equals sign is always part of the null hypothesis. Also notice that the hypotheses are statements about the *relationship* between the size of the mean of population 1 and the mean of population 2. The tests do not give you information about the value of the means, only about how the value of μ_1 compares to the value of μ_2.

You can see that there are two ways to state each of the hypotheses. For each setup shown above the first way of writing the test makes a statement about the relative value of μ_1 to μ_2. The second way of writing the same test makes a statement about the value of the difference, $\mu_1 - \mu_2$. They are equivalent to each other. Look at the two-sided test. Clearly, if $\mu_1 = \mu_2$ then the difference between them must be zero. It is also possible to test for differences of values other than zero. For example, you could test that the difference between the two means is 10 or any other value you like. The procedure is basically the same with only a slight difference in the test statistic formula.

EXAMPLE 13.2   The Tissue Manufacturer

Setting Up the Test

The tissue manufacturer wishes to see if the mean tissue strengths are the same for day 1 and day 2 tissues. The variable to be examined is *MDStrength*. The null and alternative hypotheses are

$$H_0: \quad \mu_1 = \mu_2$$
$$H_A: \quad \mu_1 \neq \mu_2$$

In addition to setting up the hypotheses, the test is defined by choosing a level of significance, α.

Suppose for this test we choose $\alpha = 0.05$. The test statistic to be calculated in step 3 is a Z statistic so we know how to find the rejection region for the two-sided test. Since we know that the total area of the rejection region must be equal to α and $\alpha = 0.05$, we split it in half and require the area of each tail of the normal distribution to be 0.025. From the standard normal tables we find the critical values for the test to be ± 1.96. ■

The rejection regions are found using the procedures from Chapter 10.

Once the test has been defined, the data are collected and processed. This allows us to proceed with the third step of the procedure, calculating the value of the test statistic.

Remember that the test statistic is calculated from sample data and is used to decide between the null and the alternative hypotheses. If we are trying to decide if $\mu_1 = \mu_2$, then it makes sense to look at the size of the difference between $\overline{X}_1$ and $\overline{X}_2$. We know that even if the two population means are exactly the same, we will almost never get two equal sample means. So we don't expect the difference $\overline{X}_1 - \overline{X}_2$ to be zero even if the null hypothesis is true. However, if the null hypothesis is true, then the difference $\overline{X}_1 - \overline{X}_2$ should be small relative to the size of the standard error. Clearly, large differences in the sample means would lead us to be suspicious of the null hypothesis. The appropriate test statistic is a Z statistic and is calculated as follows:

$$Z = \frac{(\overline{X}_1 - \overline{X}_2) - 0}{\sqrt{\sigma_1^2/n_1 + \sigma_2^2/n_2}}$$

Formula for test statistic for difference between two population means

Remember: A point estimator is calculated from sample information only and is used to estimate an unknown population parameter.

The numerator of the test statistic is simply comparing the evidence, $\overline{X}_1 - \overline{X}_2$, to the difference in the population means if H_0 is true. For the hypotheses we are using, if H_0 is true then the true difference is zero, so you see that the comparison is between $\overline{X}_1 - \overline{X}_2$ and zero. As we saw in Chapter 12, the size of the difference is always measured in terms of the standard error. Thus, the denominator is the standard error of the point estimator. In this case we are trying to estimate $\mu_1 - \mu_2$ and the natural point estimator is $\overline{X}_1 - \overline{X}_2$. The standard error of $\overline{X}_1 - \overline{X}_2$ is a natural extension of the standard error of $\overline{X}$, $\sqrt{\sigma^2/n}$, which we often wrote as $\sigma/\sqrt{n}$.

When we are testing for a difference of zero, the last part of the numerator is often left out entirely. This makes sense, but it is important that you know what you are really testing. When you need to test for a difference equal to some value other than zero, then you need to put that value in the formula for the test statistic in place of the zero. For example, if you wished to test to see if the difference in the populations means was equal to ten, then you would change the null and alternative hypotheses to the following:

$$H_0: \quad \mu_1 - \mu_2 = 10$$
$$H_A: \quad \mu_1 - \mu_2 \neq 10$$

You would then use the value of 10 in the formula for the test statistic instead of zero. This is called the hypothesized difference and is labeled d.

EXAMPLE 13.3   The Tissue Manufacturer

Calculation of Test Statistic

Now we are ready to calculate the test statistic for the tissue manufacturer. The relevant data are shown below:

Population 1: Day 1 Tissues	Population 2: Day 2 Tissues
$n_1 = 75$	$n_2 = 75$
$\overline{X}_1 = 990.8$	$\overline{X}_2 = 977.0$
$\sigma_1 = 50 \text{ lb/ream}$	$\sigma_2 = 50 \text{ lb/ream}$

We can calculate the value of the test statistic Z as follows:

$$Z = \frac{(990.8 - 977.0) - 0}{\sqrt{50^2/75 + 50^2/75}} = 1.69$$

Using the standard normal table, we can find the p value of the test by looking up the test statistic of 1.69. We find that the area to the right of $Z = 1.69$ is 0.0455. Since this is a two-sided test, we double that and find that the p value is 0.09.   ■

Although much of the work is done at this point, the test procedure is not. The fourth step of the hypothesis testing procedure requires you to reject or fail to reject the null hypothesis on the basis of the test statistic. At the fifth step of the procedure you interpret this decision in terms of the problem. Let's finish the tissue manufacturer problem now.

EXAMPLE 13.4   Tissue Manufacturer

Finishing the Test

The test statistic for the comparison of tissue strengths was 1.69. The rejection region for the test had critical values of ± 1.96. Since 1.69 is not outside the critical values, that is, it is not in the rejection region of the test, we fail to reject the null hypothesis. Our conclusion is that there is no evidence that the mean *MDStrength* of tissues made on day 1 is different than that of those made on day 2.

From previous experience in hypothesis testing, you know that you should also examine the p value for these data. The p value is less than 0.10. This means that if you set α larger than the p value, say at 0.10, then you would reject the null hypothesis of equal means. The decision, in this case, depends very much on the choice of α. We might say that the decision is "on the fence" with regard to these two hypotheses. Perhaps we should compare day 2 to day 3 to get a better understanding of what is happening.   ■

▨ *TRY IT NOW!*

The tissue manufacturer also recorded values of *MDStrength* for 75 tissues made on day 3. The sample mean *MDStrength* for that day was found to be 1000.32 lb/ream. Is there any evidence that the average *MDStrength* is different on day 2 than day 3? Use $\alpha = 0.05$.

Identify

population 1 :
population 2 :

Step 1:

Step 2:

Step 3:

Step 4:

(*continued*)

Step 5:

Once the conclusion of the test is reached, the real work in terms of decision making begins. The statistical analysis simply confirms or fails to confirm a hypothesis. It does not help you make decisions about the problem you are trying to solve.

EXAMPLE 13.5 **Tissue Strength**

Interpreting and Using the Results of a Test

As a result of our tests, we have concluded that the average *MDStrength* of the tissues made on day 1 and day 2 are equal but the average *MDStrength* of the tissues made on day 2 and day 3 are different. Clearly, something is not quite right with this process. How can we interpret this? What should we do next?

What we know as a result of this test is that the tissue strength is not behaving entirely as it is supposed to. It is not clear *how* the mean strength differs, or whether it is *always* different or just *sometimes different*. It is also not clear on which, if any, of the three days the *MDStrength* is correct.

This would be a good time to graph the data and see what is happening. This would help us look for any patterns in the data.   ■

To answer the question we started with, we would like to compare all 3 days. That is, we would like to do a hypothesis test that looks like this:

$$H_0: \quad \mu_1 = \mu_2 = \mu_3$$
$$H_A: \quad \text{At least one of them is different.}$$

This is a test of more than 2 means and you cannot do such a test using the techniques of this chapter. You need to use a technique called analysis of variance, which is covered in Chapter 14.

13.4.2 Large-Sample Tests of Two Means With Unknown Standard Deviations

The large-sample test for the difference between two population means requires that *both* of the sample sizes be greater than 30. Remember that the only differences in the various hypothesis tests are in the calculation of the test statistic.

Large-sample test of the difference between two means

The test statistic for this case is very similar to the case when you know the standard deviations. Since the sample sizes are large, each individual sample standard deviation is a good estimate of the corresponding unknown population standard deviation. So, we simply use each of the sample standard deviations in the formula instead of the corresponding values of σ.

ANS. $Z = -2.86$ AND PVALUE $= 0.0042$. REJECT H_0 AND CONCLUDE THAT THE AVERAGE *MDStrength* IS DIFFERENT ON DAY 2 AND DAY 3.

The test statistic becomes

$$Z = \frac{(\overline{X}_1 - \overline{X}_2) - 0}{\sqrt{s_1^2/n_1 + s_2^2/n_2}}$$

Finding the rejection region is the same as it is for any hypothesis test involving the Z distribution—it depends on α and whether the test is one-sided or two-sided.

EXAMPLE 13.6 Training Issues

Setting Up the Hypotheses

Recall the corporation that was interested in whether there was a difference in the results of in-house vs. outside training programs. The employees involved in the training completed a sample copy of the CMI certification examination at the end of each of the respective training programs. the company wants to know if the outside trainers achieve better results. In this case the two populations are:

Population 1: Employees trained by an outside consultant
Population 2: Employees trained in-house

Since the company is interested in whether one group is *better* than the other, it will use a one-sided test and the hypotheses are

$$H_0: \quad \mu_1 \leq \mu_2$$
$$H_A: \quad \mu_1 > \mu_2$$ ∎

Once the hypotheses are set up and you know the population parameters you are testing, you need to make a decision about which specific type test you will use. To do this you need to assemble the data you have and look at standard deviation and sample size. With tests about the mean, the first question is whether the population standard deviation is known. If it is, then you would perform the Z test defined in the previous section. If it is not, then you need to look at the sample sizes to make the next decision. When both sample sizes n_1 and n_2 are large (≥ 30), then the test is also a Z test with the test statistic just described.

EXAMPLE 13.7 Training Issues

Setting Up the Test

The company looking at training issues assembled and processed its data as shown below:

	Population 1 (Outside Consultant)	Population 2 (In-house Training)
Sample size	$n_1 = 50$	$n_2 = 47$
Sample mean	$\overline{X}_1 = 75.46$	$\overline{X}_2 = 63.55$
Sample standard deviation	$s_1 = 4.12$	$s_2 = 7.69$

You see that the company had 50 employees in the group receiving outside training and 47 people in the group being trained in-house. Since the study was a new one, the company had only sample measures of variation. It will be able to use a Z test, because of the large sample sizes.

Once the company knows it will use a Z test, it can find the rejection region. In this case there is no real reason to choose a level of significance other than $\alpha = 0.05$, so that is what is used.

The one-sided rejection region is found using the techniques of Chapter 12. The value of Z_{cutoff} is found to be 1.645. That is, the rejection region is to the right of the value 1.645. ∎

Steps 1 and 2 of the hypothesis test define the test procedure to be used. The only other computational portion of the test is to calculate the test statistic. This is completely determined by the test procedure and so it is a matter of using the correct formula.

EXAMPLE 13.8 Training Issues

Performing the Test

The corporation looking at training issues knows that it is using a Z test for its data. The test statistic is calculated as follows:

$$Z = \frac{(75.46 - 63.55) - 0}{\sqrt{4.12^2/50 + 7.69^2/47}}$$

$$Z = \frac{11.91}{1.2640} = 9.42$$

To find the p value of the test, we look up the Z score of 9.42 on the standard normal table and find that the p value is 0.0000. ■

The remainder of the hypothesis testing procedure is the same for this test as it has been for any of the tests we have done. So let's complete the test by doing steps 4 and 5.

EXAMPLE 13.9 Training Issues

Conclusions and Interpretation

Step 4: *Decide whether to reject the null hypothesis or fail to reject it based on the test statistic.* Since $Z = 9.42$ is most definitely in the rejection region, we reject the null hypothesis.

Step 5: *Interpret the decision in terms of the original problem statement.* Since we reject the null hypothesis, we conclude that the mean post-test score for the group trained by an outside consultant is higher than the group trained in-house. ■

Remember that the standard error is the standard deviation of the point estimator. It is the yardstick by which we judge all differences.

You are probably not surprised at this conclusion since the average for the outside group was higher than the one for the inside group. However, this is not sufficient reason to conclude that they are different. You cannot simply look at the two $\overline{X}$ values and conclude that employees trained by an outside consultant did better on the post-test than those trained by an in-house person. Yes, $\overline{X}_1$ is higher than $\overline{X}_2$ if you look only at the observed difference. It is possible that such a difference might occur due to the random nature of the sampling and testing procedure. The only way you can tell if there is a "significant" difference is to compare the observed difference in the sample means to the size of the standard error. This is precisely what the test statistic does for you.

📖 TRY IT NOW!

The management of the corporation looking at training issues are not convinced that they should switch to outside consultants for all of their training programs. They want to look at how the two groups compared on the pretest.

The relevant data are given below:

	Population 1 (Outside Consultant)	Population 2 (In-house Trainer)
Sample size	$n_1 = 50$	$n_2 = 47$
Mean score on pretest	$\overline{X}_1 = 43.52$	$\overline{X}_2 = 47.66$
Standard deviation of pretest score	$s_1 = 14.80$	$s_2 = 12.80$

Test to see if there is a difference in the mean pretest scores. Use $\alpha = 0.05$.

Step 1:

Step 2:

Step 3:

Step 4:

(*continued*)

Step 5:

13.4.3 Exercises—Learning It!

13.1 Many studies have been done comparing consumer behavior of men and women. One such on-going study concerns take-out food. In particular, the study focuses on whether there is a difference in the mean number of times per month that men and women buy take-out food for dinner. The most recent results of the study are shown below:

	Population	
	Men	**Women**
Sample size	$n_1 = 34$	$n_2 = 28$
Sample mean	$\overline{X}_1 = 25.6$	$\overline{X}_2 = 21.2$
Population standard deviation	$\sigma_1 = 4.2$	$\sigma_2 = 3.8$

Because the study has so much historical data, information is known about the population standard deviations.

(a) Set up the hypotheses to test whether there is a difference in the mean number of times per month that a person buys take-out food for dinner for men and women.

(b) Use the Z test with known population variances to set up and perform the test. Use a level of significance of 0.05.

(c) Find the p value for the test.

(d) Do the data provide evidence that the mean number of times per month for men differs from that for women?

(e) Does the choice of α in this case affect the decision?

13.2 Professional employees who work for large corporations often contend that the mean salary paid by a company differs by location in the United States. To test that claim, data were collected on financial analysts working for a large corporation at locations in New England and in the upper Midwest. Because there is an extensive history of salary data, the population standard deviations are available. The study found the following results:

	Population	
	New England	**Upper Midwest**
Sample size	$n_1 = 25$	$n_2 = 18$
Sample mean	$\overline{X}_1 = \$38,348$	$\overline{X}_2 = \$36,782$
Sample standard deviation	$s_1 = \$2336$	$s_2 = \$2258$

(a) Set up the appropriate hypotheses to test whether the company's analysts in New England were paid more, on the average, than those working in the upper Midwest.

(b) Use the Z test with known population variances to set up and perform the test. Use a level of significance of 0.05.

(c) Find the p value for the test.

(d) Do the data support the contention that the mean pay for analysts in New England is higher than that of analysts in the upper Midwest?

13.3 A study at a university in New England focuses on binge drinking by students. The people who administered the study wanted to determine whether students who lived on campus had more episodes of binge drinking per semester on the average than those who lived at home. They surveyed all students who were enrolled in a required health course and obtained the following data:

	Population	
	On Campus	**At Home**
Sample size	$n_1 = 220$	$n_2 = 196$
Sample mean	$\overline{X}_1 = 37.3$	$\overline{X}_2 = 35.6$
Population standard deviation	$\sigma_1 = 8.6$	$\sigma_2 = 10.1$

The study used the same survey instrument used in many other similar studies and so they used the population standard deviation.

(a) Set up the appropriate hypotheses to test whether students who live at home have fewer episodes of binge drinking per semester, on the average, than those who live on campus.

(b) Use the Z test with known population variances to set up and perform the test. Use a level of significance of 0.05.

(c) Find the p value for the test.

(d) Do the data indicate that students who live at home binge less on the average?

13.4 A company uses two suppliers to provide paper for the copy machines. The company has experienced an excessive number of paper jams and wonders if there is a difference in the paper provided by each supplier. The company collects data on the number of jams per ream of paper (500 sheets) over a period of 2 months. The summary data are

	Population	
	Supplier 1	**Supplier 2**
Sample size	$n_1 = 45$	$n_2 = 52$
Sample mean	$\overline{X}_1 = 8.1$	$\overline{X}_2 = 9.3$
Population standard deviation	$\sigma_1 = 1.1$	$\sigma_2 = 1.2$

(a) Set up the appropriate hypotheses to test whether paper from the two suppliers results in the same average number of jams per ream.

(b) Use the Z test with known population variances to set up and perform the test. Use a level of significance of 0.05.

(c) Find the p value for the test.

(d) Do the data indicate that there is a difference between suppliers on average?

13.5 You are considering selling your house and need to choose between 2 real estate agencies. You collect data on the number of weeks that a house is on the market for both agencies. The summary statistics are shown below:

	Population	
	Agency A	**Agency B**
Sample size	$n_1 = 25$	$n_2 = 20$
Sample mean	$\overline{X}_1 = 22.3$	$\overline{X}_2 = 18.5$
Sample standard deviation	$s_1 = 1.5$	$s_2 = 2.2$

(a) At the 0.05 level of significance, test to see if houses listed with Agency A are on the market on the average longer than those listed with Agency B.

(b) Find the p value for the test.

(c) Which agency would you use and why?

13.5 SMALL-SAMPLE TESTS OF THE DIFFERENCE IN TWO POPULATION MEANS

When the population standard deviation is unknown and the sample size is large, the Z test can be used. If the sample size is not large, then it is necessary to use a t test, just as we did in the one-population tests.

13.5.1 Small-Sample Tests of Two Means With Unknown But Equal Standard Deviations

The reason that 30 is used as the cutoff point has to do with the fact that the Central Limit Theorem applies for sample sizes of 30 or more.

Unfortunately, we do not always have the luxury of large samples. Remember that data collection is expensive, so often the sample size for one or both of the samples is less than 30. This is called the "small-sample" case.

Just as in the one-population case, the small-sample tests for two population means will involve the t distribution. The t test carries with it the same assumption that the populations involved in the test are normally distributed. In addition, though, the two-population case requires that you know whether or not the two population variances can be considered equal. In this section, we look at both cases.

EXAMPLE 13.10   Training and Gender

Setting Up the Hypotheses

In addition to looking at the results of training in-house vs. using an outside consultant, the company decided to look at gender issues in training. It took the group of people that had been trained in-house and differentiated them according to gender. The company wanted to know if there was any difference in the post-test scores for males vs. females.

We can call the females population 1 and the males population 2. Since the company is interested only in whether the groups are different, it can use a two-sided test. The hypotheses are

$$H_0: \quad \mu_1 = \mu_2$$
$$H_A: \quad \mu_1 \neq \mu_2$$

∎

The first approach is to assume that even though you don't know the variability in the two populations, you know enough to believe that there is the same amount of variability in each population. At first this may seem a bit odd to you but actually it is not such a ridiculous assumption. Remember that you are comparing the means of two populations. Since you are measuring the same variable in both populations, it is quite possible that the two populations have a normal distribution with the same shape but with different means. Figure 13.2 illustrates such a situation.

There is, of course, a way to tell by doing a hypothesis test to compare two variances. This is the subject of Section 13.8.

FIGURE 13.2   Two normally distributed populations with the same variance but different means

For now, we will assume that equal variance has been confirmed through a statistical procedure. We have two estimates of this variance, one from each sample. We have s_1^2 and s_2^2, which are two different estimates of the same number. Instead of using two different estimates of the unknown but common population variance, it makes more sense to combine the data and find one better estimate of the variance. That is, instead of using s_1^2 and s_2^2 in the formula for the test statistic, we will "pool" the data from the samples and get one estimate of the common variance. This is called the **pooled variance.**

One way to estimate the pooled variance is simply to put all the data together from both samples and calculate the sample variance of this "big" sample. There are two reasons for not doing that. One reason is that if, indeed, the two population means are different, then the consolidated data will have a mean that is not from *either* population. This will result in a distorted measure of variance. In addition, at this point most people have already calculated the sample variances for each sample individually. Thus, we will find the pooled variance by building from the individual sample variances. Your first thought might be to average the two sample variances. This is the right idea, but instead we will use a weighted average. If, for example, one of the samples sizes is 10 and the other one is 25, then the sample variance from the sample based on 25 observations is more accurate and should carry a greater weight in calculating the pooled variance. The formula for the pooled variance, s_p^2, is shown below:

$$s_p^2 = \frac{(n_1 - 1)s_1^2 + (n_2 - 1)s_2^2}{n_1 + n_2 - 2}$$

Formula for the pooled variance

This formula weights the sample variances by one less than the sample size. If the sample sizes are the same, this formula is the same as a simple average of the two sample variances.

EXAMPLE 13.11   Training and Gender

Calculating the Pooled Variance

The data for the post-tests on the in-house group are given below:

	Population 1 Females Trained In-House	Population 2 Males Trained In-House
Sample size	$n_1 = 20$	$n_2 = 27$
Sample average	$\overline{X}_1 = 63.95$	$\overline{X}_2 = 63.26$
Sample standard deviation	$s_1 = 8.86$	$s_2 = 6.87$

The calculation of the pooled variance of the in-house post test scores is then

$$s_p^2 = \frac{(20 - 1)(8.86^2) + (27 - 1)(6.87^2)}{20 + 27 - 2}$$

$$s_p^2 = \frac{1491.4924 + 1227.1194}{45} = \frac{2718.6118}{45} = 60.4136$$

$$s_p = \sqrt{60.4136} = 7.77$$ ∎

As a double check, be sure the number you end up with for the pooled standard deviation is in between the two sample standard deviations. If it is not, recheck your work, as you have made a mistake.

Now we are ready to consider the test statistic for the small-sample case when the population variances are equal. The test statistic in this case has a t distribution with $(n_1 + n_2 - 2)$ degrees of freedom. Remember that for simpler notation we can write the t value as $t_{\alpha, n-1}$. The test statistic is calculated as follows:

$$t = \frac{\overline{X}_1 - \overline{X}_2}{\sqrt{s_p^2/n_1 + s_p^2/n_2}}$$

Test statistic using pooled variance

or equivalently

$$t = \frac{\overline{X}_1 - \overline{X}_2}{s_p\sqrt{1/n_1 + 1/n_2}}$$

Knowing the degrees of freedom and the formula for the test statistic, we can complete the hypothesis test.

EXAMPLE 13.12 Gender and Training

Conducting the Hypothesis Test

Now the company looking at training issues can find the rejection region for the test and calculate the test statistic. Since it is doing a two-tail test, the cutoff values will be the t statistic with 0.025 in the tail area and 45 degrees of freedom, $t_{0.025, \, 45}$. From the t table the critical values are found to be ± 2.014. The test statistic is calculated as

$$t = \frac{63.95 - 63.26}{7.77\sqrt{\frac{1}{20} + \frac{1}{27}}}$$

$$t = \frac{0.69}{7.77\sqrt{0.08704}} = \frac{0.69}{(7.77)(0.2950)} = 0.3010$$

The company uses a statistical software package to look up the t statistic and finds that the area outside the test statistic is 0.3824. Since this is a two-sided test, it doubles that to find that p value $= 0.7648$.

The results of the test show that since 0.3010 is not in the rejection region of the test, the company cannot reject H_0. There is not enough evidence to say that the mean test score for the male in-house group is different than the mean post-test score for the female in-house group. ■

TRY IT NOW!

The company also wants to look at gender differences in the test scores for the group of employees trained by the outside consultant. It assembles the relevant data and finds the following:

	Population 1 Females Trained by Outside Consultant	Population 2 Males Trained by Outside Consultant
Sample size	$n_1 = 21$	$n_2 = 29$
Sample mean	$\overline{X}_1 = 75.9$	$\overline{X}_2 = 75.2$
Sample standard deviation	$s_1 = 3.9$	$s_2 = 4.4$

Find the pooled variance for the data.

Test to see if there is evidence that the mean post-test score for the females trained by the outside consultant is different than the mean post-test score for the males trained by the outside consultant. Use a level of significance of 0.05.

Step 1:

Step 2:

Step 3:

(*continued*)

Step 4:

Step 5:

13.5.2 Small-Sample Tests of Two Means With Unknown and Unequal Standard Deviations

Second way to handle small samples

If you test for equality of variances and find that the population variances are not equal, then you cannot pool the data. This is known as the Behrens–Fisher problem. Figure 13.3 depicts this situation. In this case both of the populations have a normal distribution with different variances.

FIGURE 13.3   Two normally distributed populations with unequal variances

The test statistic is called a separate-variance t test. As the name indicates, you must use the individual sample variances in the test statistic. In fact, the formula for calculating the test statistic looks exactly like the Z statistic for the large-sample case when the variances are unknown. It is calculated as follows:

Small-sample test with unknown standard deviations

$$t = \frac{(\overline{X}_1 - \overline{X}_2) - d}{\sqrt{s_1^2/n_1 + s_2^2/n_2}}$$

It does not follow a Z distribution because the estimates of the unknown variances are based on small samples. We know from our work in Chapter 12 that the appropriate distribution for small-sample situations is the t distribution. This is the case here as well, but the number of degrees of freedom is found according to a rather more complicated formula than you have seen thus far. This test statistic can be approximated by a t distribution with ν (Greek letter pronounced "new") degrees of freedom. The formula for ν is given as follows:

$$\nu = \frac{\left(\dfrac{s_1^2}{n_1} + \dfrac{s_2^2}{n_2}\right)^2}{\dfrac{(s_1^2/n_1)^2}{n_1 + 1} + \dfrac{(s_2^2/n_2)^2}{n_2 + 1}} - 2$$

Formula for degrees of freedom for small samples and unequal variances

When you calculate ν you will most likely get a fractional result. But we know that degrees of freedom are integer values. Simply drop the fractional portion of the result to get just an integer value for the degrees of freedom. You may wonder why we do this. The reason is that the number that results for the degrees of freedom in this case will be smaller than the one used in the pooled variance case. The number of degrees of freedom in the test defines the test's ability to detect significant differences in the population means. That is, when you assume that the population variances are *not* equal, your hypothesis test becomes less *powerful* and the chances that you will make a Type II error for a given value of α increase. By not pooling variances you are more likely to miss true differences in the population means.

Note: There are several equivalent formulas for calculating the degrees of freedom in this case.

EXAMPLE 13.13   Training and Gender

Small-Sample Test with Unequal Variances

Suppose we repeat the test on gender in the in-house group, this time assuming that the population variances are not equal. How does this change the analysis? The relevant data are again:

	Population 1 Females Trained In-House	Population 2 Males Trained In-House
Sample size	$n_1 = 20$	$n_2 = 27$
Sample average	$\overline{X}_1 = 63.95$	$\overline{X}_2 = 63.26$
Sample standard deviation	$s_1 = 8.86$	$s_2 = 6.87$

Step 1: Construct the null and the alternative hypotheses.

Section 13.8 will cover the procedure for testing for equal variances.

$$H_0: \quad \mu_1 = \mu_2$$
$$H_A: \quad \mu_1 \neq \mu_2$$

Step 2: Select α and find the rejection region.

To do this we need to find the degrees of freedom to use. Substituting in the formula we get

$$\nu = \frac{\left(\dfrac{8.86^2}{20} + \dfrac{6.87^2}{27}\right)^2}{\dfrac{(8.86^2/20)^2}{20 + 1} + \dfrac{(6.87^2/27)^2}{27 + 1}} - 2$$

$$= \frac{32.18308 \ldots}{0.84272 \ldots} - 2$$

$$= 38.18939 - 2 = 36.189$$

Taking the integer portion gives us 36 degrees of freedom. The value for t_{cutoff} is $t_{0.025, 36}$, found using a computer software package to be ± 2.0281. ∎

Many statistical packages offer you the option of doing the t test assuming equal variances or assuming unequal variances. To make an informed choice about whether you should pool the data, you must do a test for the equality of the variances. This is the subject of Section 13.8.

13.5.3 Exercises—Learning It

13.6   The Board of Realtors for Greater Bridgeport, CT, wants to know whether the average price of a single family home in the area has increased in the past two years. They take a random sample of homes sold in 1995 and 1996 and calculate the following statistics:

	Population	
	1995	**1996**
Sample size	$n_1 = 25$	$n_2 = 25$
Sample mean	$\overline{X}_1 = \$151{,}166$	$\overline{X}_2 = \$160{,}669$
Sample standard deviation	$s_1 = \$5332$	$s_2 = \$6468$

(a) Set up the appropriate hypotheses to test whether there has been an increase in the mean selling price of a home in Greater Bridgeport.

(b) Calculate the pooled variance for the data.

(c) Assuming that the data are normally distributed, use the small-population test with equal variances to test the hypotheses. Use a level of significance of 0.10.

(d) Do the data provide evidence that the average price of a home in 1996 is higher than it was in 1995?

13.7   The cost of shipping is of great concern to mail-order businesses. Traditionally, mail-order companies have used United Parcel Services (UPS) for their shipping. The U.S. Postal Service has been promoting their parcel post service as a competitor for this service. To determine whether parcel post is better than UPS ground, a mail-order company sent 20 packages via UPS ground and 20 packages via parcel post and tracked the number of working days for the packages to reach their intended destinations. A summary of the data is given below:

	Population	
	UPS Ground	**U.S. Postal Service**
Sample size	$n_1 = 20$	$n_2 = 20$
Sample mean	$\overline{X}_1 = 7.3$	$\overline{X}_2 = 7.1$
Sample standard deviation	$s_1 = 0.9$	$s_2 = 2.3$

(a) Set up the appropriate hypotheses to test whether the U.S. Postal Service is better than UPS Ground.

(b) Assuming that the data are normally distributed, use the small-population test with unequal variances to test the hypotheses. Use a level of significance of 0.05.

(c) Do the data provide evidence that the average delivery time for Parcel Post is less than UPS Ground?

13.8   After months of working overtime, you have saved for some money for a set of new golf clubs and you want to make sure that you are buying the best. You can get a really good deal on Brand X clubs, but would make the sacrifice to buy Brand Z if they really improve your game. The salesperson allows you to take the number 3 wood from each brand and use them to hit balls on a driving range. The data for your efforts are summarized below:

	Population	
	Brand X	**Brand Z**
Sample size	$n_X = 15$	$n_Z = 15$
Sample mean	$\overline{X}_X = 255$	$\overline{X}_Z = 271$
Sample standard deviation	$s_X = 8.7$	$s_Z = 9.1$

(a) Set up the appropriate hypotheses to test whether Brand Z clubs are better than Brand X.

(b) Calculate the pooled variance for the data.

(c) Assuming that the data are normally distributed, use the small-population test with equal variances to test the hypotheses. Use a level of significance of 0.05.

(d) Should you spend the extra cash or save it for golf balls?

13.9 Few things are more frustrating than needing information and having to wait for it. The time spent on hold waiting for technical support to resolve software problems is one of the largest drains on productivity experienced by small businesses. Before a company will consider switching from its current word processing software (Microsoft Word for Windows 95) to a competitor (WordPerfect for Windows) the company will have to be convinced that it will spend less time on hold. The company decides to perform a test by placing 10 calls to each technical support line and recording the amount of time (in minutes) that the caller is on hold. The summary data are

	Population	
	MS Word	**WordPerfect**
Sample size	$n_1 = 10$	$n_2 = 10$
Sample mean	$\overline{X}_1 = 11.2$	$\overline{X}_2 = 9.7$
Sample standard deviation	$s_1 = 3.4$	$s_2 = 1.6$

(a) Assuming that the data are normally distributed, use the small-sample test with unequal variances to test the hypotheses that the company will spend less time per call, on the average, with WordPerfect technical support.

(b) Should the company switch software?

13.10 Are all discount mail-order companies the same? In an attempt to answer this question, data were collected on the price of software for the top ten business software packages reported by *PC Magazine,* August 1997. Two well-known mail-order companies were asked for prices on each piece of software. The data are shown below:

Top Ten Business Software Packages	Computability Price ($)	PC Connection Price ($)
Windows 95 Upgrade	88	95
Norton Anti-Virus	59	70
McAfee ViruScan	49	60
First Aid 97 Deluxe	54	58
Clean Sweep III	37	37
Norton Utilities	68	75
Netscape Navigator	45	40
MS Office Pro 97 Upgrade	300	310
First Aid 97	32	35
Win Fax Pro	95	95

(a) For each vendor, calculate the average price and the standard deviation of the prices.

(b) Based on these statistics, do you think that the companies are the same or different? Why?

(c) Set up the hypothesis test to determine whether the average price for software is different at the two mail-order companies.

(d) Assume that the variances are equal and calculate the pooled variance.

(e) Assuming that the data are normally distributed, use the t test with equal variances to perform the test at the 0.05 level of significance.

(f) What can you conclude about software prices at the two companies?

(g) What assumption was violated in this example?

13.6 SUMMARY OF TESTS OF THE DIFFERENCE IN TWO POPULATION MEANS—INDEPENDENT SAMPLES

The previous two sections have presented several different situations in which you might find yourself when testing the difference in two population means. This is a good place to summarize what you have learned so far. Table 13.1 lists the criteria for each of the tests.

TABLE 13.1 Summary of Tests

Populations	Variances	Sample Sizes	Test
Independent and normal	Known	No restrictions	Z test
Independent and non-normal	Unknown	n_1 and $n_2 \geq 30$	Z test
Independent and normal	Unknown but assumed equal	n_1 and $n_2 < 30$	t test with a pooled variance
Independent and normal	Unknown and assumed not equal	n_1 and $n_2 < 30$	Modified t test (Behrens–Fisher problem)

For all but one of these situations the populations must be normally distributed. It is unlikely that you will know information about the distribution of the populations, so you must use the sample data to decide if this assumption is reasonable. To determine whether the data are normally distributed you should use a statistical test such as a chi-square goodness of fit test or a normal probability plot. The chi-square test is presented in Chapter 17 and normal probability plots are discussed in Chapter 15. Right now, you can consider the assumption of normality by using the descriptive techniques you learned in Chapters 3 and 6. You can look at boxplots, histograms, or dotplots of the data. Since the assumption of normality is an issue in small-sample tests, histograms often do not work well as visual tools. While these graphs are certainly not rigorous tests, unless there are large departures from normality (very skewed or non-mound-shaped distributions) the test will be valid. You could also apply the empirical rule for normality, which was covered in Chapter 6.

EXAMPLE 13.14   Training and Gender

Checking for Normality of Data

The company looking at gender issues in training knows that it needs to check the data to see if it is reasonable to assume that the population scores (males and females) are normally distributed. The company decides to make boxplots of the data and finds the following:

Post-Test Scores for In-house Training

From the boxplots the data appear to be reasonably symmetric—the medians are located near the centers of the boxes and the whiskers are approximately the same length. It would appear that normality is a reasonable assumption. ∎

A quick look at the summary table tells you that all of these tests are for *independent samples*. Recall that this means that you select a sample from population 1 and it has no impact on the selection of the sample from population 2. This is typically what is done. However, sometimes it is necessary or desirable to use dependent samples instead. This is the subject of the next section.

13.7 TEST OF TWO POPULATION MEANS—DEPENDENT SAMPLES

13.7.1 Tests With Dependent Populations

In many cases, such as the example on training programs, data are collected *before* and *after* some experiment is performed. This type of data is known as **pretest** and **post-test** data. When data of this type are collected, the samples come not from *independent* populations, but from *dependent* populations. Often, in this type of situation, we are interested in knowing whether the experimental procedure has an effect on the variable being studied.

> **Pretest** and **post-test** conditions exist when data are collected on the same sample elements before and after some experiment is performed.

This is not the only way samples can be **dependent.** Sometimes each member of the sample from population 1 is matched or paired with a member of the sample from population 2. For example, in studying the effect on sales of a new product display, you would want to compare stores that are either similar in size or in similar locations. Otherwise, the apparent increase in sales that is observed might simply be due to the fact that a larger, busier store was randomly chosen to try out the new display. In this case, it is possible that the sales did not really increase due to the new display.

> **Dependent samples** are related to each other. The members of one sample are identical to or matched or paired with the members in the other sample according to some characteristic.

Discovery Exercise 13.1
INTRODUCTION TO EXPERIMENTAL DESIGN

Did you ever wonder why the restrooms in restaurants have a sign instructing all employees to wash their hands before returning to work? The owner of a large restaurant chain wanted to increase the amount of time that employees washed their hands after reading about the health implications of washing for only a few seconds. A random sample of 10 employees was selected. The amount of time each employee washed was recorded. The data were collected in such a way that the employees did not know they were being observed.

These employees were then educated on the benefits of hand washing. They watched a health video that detailed the benefits of increasing the amount of time they washed. One week after the training, employee hand washing was timed again. The times (in seconds) are shown in the next table:

(continued)

Employee	Before Training	After Training
1	3	3
2	3	3
3	2	4
4	4	4
5	3	5
6	3	4
7	4	4
8	3	4
9	5	6
10	2	3

(a) Use a two-population paired t test to test the hypothesis that no learning has occurred. Use a pooled variance.

$$H_0: \quad \mu_1 = \mu_2$$
$$H_A: \quad \mu_1 \neq \mu_2$$

(b) What is your conclusion?

(c) Why is this surprising? (*Hint:* Of the 10 employees, how many of them washed at least as long after the training?)

(d) Explain how these data are different from most of the two-sample data sets you have looked at up to this point in the chapter. Why might this be important?

(e) Suppose 10 different employees were used for the second timing. Would this make a difference? Explain why or why not.

In the situation described above, one of the characteristics that could be used to pair or match the sample members is the size of the store. A second is the location of the store. This is the first departure we have seen from random sampling. In fact, you are *trying* to match the items that are sampled in an attempt to control other variables that influence the data. Clearly, the size of the store will influence sales. In a sense, you match sample elements to keep the comparison "fair." Similarly, it would not be fair to compare the average miles per gallon of two cars driven by two people with very different driving styles. You would want either the same person testing both cars or people matched on the basis of their driving styles. There may be more than one characteristic that you wish to use to match the members of the two samples. This discussion is really the beginning of a topic known as experimental design. In designing the statistical experiment, you are trying to get the most information for your money. Paired data are the simplest case of a designed experiment. It fits naturally here because often paired data are collected to compare two population means and yet the data are incorrectly analyzed using the methods discussed in the last section.

You should *not* analyze these paired or matched data using the small-sample t test that was described in the previous section. If you do, you might miss differences that exist but are masked by improper analysis. In this section we see how to analyze dependent or paired data.

13.7.2 The Paired Sample Test

The problem with using the tests for independent populations is that if there is a large amount of variation among the sample elements, then the standard errors of the estimates will be high. Since we divide by the standard error to obtain our test statistic, this value is often smaller than it should be. Real differences in the populations are *hidden* by the amount of variation among the sample elements.

TABLE 13.2 Data set up for paired difference test

	Variable		
Observation	Pretest X_1	Post-test X_2	Difference $d = (X_2 - X_1)$
1	x_{11}	x_{21}	d_1
2	x_{12}	x_{22}	d_2
3	x_{13}	x_{23}	d_3
$\vdots$	$\vdots$	$\vdots$	$\vdots$
n	x_{1n}	x_{2n}	d_n
Average			$\bar{d}$

How can we fix this problem? Well, if you think about it, the real question we are asking is whether the difference between the pretest and post-test measurements is significantly different from zero. In that case, we can look at the differences as our measurements and perform a test about them. The hypotheses for this test are then

$$H_0: \quad \mu_d = 0 \qquad H_0: \quad \mu_d \leq 0 \qquad H_0: \quad \mu_d \geq 0$$
$$H_A: \quad \mu_d \neq 0 \qquad H_A: \quad \mu_d > 0 \qquad H_A: \quad \mu_d < 0$$

Table 13.2 shows how the data for such a test are set up. Each observation in the sample is a row in the table. The second and third columns are the values of the pretest (population 1) and post-test (population 2). For example, in the second column, the entry x_{13} is the pretest value for the third sample element. The fourth column, labeled Difference, is calculated by subtracting the pretest score from the post-test score. These differences are labeled $d_1, d_2, \ldots, d_n$. Finally, the average of these difference is found and labeled $\bar{d}$. That is,

Average difference

$$\bar{d} = \frac{\sum\limits_{i=1}^{n} d_i}{n}$$

EXAMPLE 13.15 Training Issues

Setting Up the Hypotheses for a Paired Difference Test

The company that was looking at the results of two different types of training programs wonders if there is really an improvement realized from any type of training program. It decides to analyze the in-house training program it has and look at the difference in the pretest and post-test scores for the people who have gone through the training. The company wants to know if the training increases the scores. The company knows it must use a paired difference test because the two populations are dependent. Since it wants to know if the program *increases* scores, it will use a one-sided test. The question is, in what direction do each of the hypotheses go?

If the training program does increase scores, then the differences (post-test − pretest) will be positive. In this case, the test should be an upper-tail test and the appropriate hypotheses are

$$H_0: \quad \mu_d \leq 0$$
$$H_A: \quad \mu_d > 0$$

∎

You see that the direction of the test depends on what the experimenter thinks the effect of the experiment will be and the order in which you subtract the pretest and post-test values. Most paired difference tests are one-sided.

EXAMPLE 13.16   Sleep Apnea

Setting up Hypotheses

Because of skyrocketing health-care costs, many hospital administrators are working to contain costs. Studies are being done to see if diagnosis and treatment of some conditions can be done at home at a significantly reduced cost. One such area of study is in the diagnosis and treatment of obstructive sleep apnea syndrome (OSAS), which affects 2 to 5% of the adult male population and is characterized by extremely heavy snoring. As a result of OSAS, breathing is suspended either partially or entirely, which can lead to suffocation.

Once OSAS is diagnosed, the traditional treatment calls for hospitalization to begin nasal continuous positive airway pressure (NCPAP). In an effort to reduce costs, a study was done where NCPAP was initiated at home.

One of the variables observed for each patient before and after treatment was the number of obstructions. If the treatment worked, then the average number of obstructions post-treatment should be less than the pretreatment average. The hospital running the study wants to know if home treatment of OSAS was effective.

Since the number of obstructions should decrease after NCPAP if treatment is effective, the analyst knows that the test should be one-tailed, and the differences (post-test − pretest) should be negative. This means that the test should be a lower-tail test and the appropriate hypotheses are

$$H_0: \quad \mu_d \geq 0$$
$$H_A: \quad \mu_d < 0$$ ∎

The hypothesis test is now essentially a test of a single mean. We know how to do this test from Chapter 12. The paired difference test is considered to be a small-sample test, so the test statistic will be a t statistic. The only difference is that the mean we are testing is the mean difference between two populations rather than the mean of a single population. Otherwise, the procedure is the same.

We need to define the test statistic for the test. For a paired difference test, the test statistic has a t distribution with $n - 1$ degrees of freedom. Remember that in this test n is the *number of pairs* and not the number of data points:

$$t = \frac{\bar{d}}{s_d/\sqrt{n}}$$

Paired difference test

where s_d is the standard deviation of the differences and is found in the usual way. Since the test is a t test, we must, as always, have normally distributed populations.

Figure 13.4 shows typical data for a paired sample test. Since the data are plotted by observation, you can clearly see that the values after treatment are higher than the values before treatment for every observation in the sample. The statistical test will see if the differences are statistically significant.

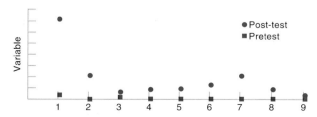

FIGURE 13.4   Pre- and post-test data for similar observations

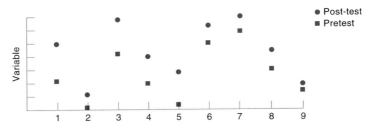

FIGURE 13.5   Pre- and post-test data for highly variable observations

If we used the small, independent sample t test we would average all of the values in the first sample and all of the values in the second sample. In doing so we would lose some of the information contained in the sample. To see this effect, let's look at some different data to make a point. When you look at Figure 13.5, it is not obvious that the post-test data have a higher average if you do not view the data in pairs. The variation from sample element to sample element hides the differences between the pretest and post-test data. This is what happens when you simply average all of the pretreatment values and all of the post-treatment values and compare the sample means using a t test.

EXAMPLE 13.17   **Training Issues**

Performing the Paired Sample Test

The company that wants to know the value of training programs has 47 people who participated in the in-house training program. It analyzes the data and finds the following summary statistics:

$$n = 47$$
$$\bar{d} = 15.89$$
$$s_d = 8.45$$

The company decides to test with $\alpha = 0.05$. Since there are 47 pairs of data, the t test will have $n - 1 = 46$ degrees of freedom and the critical values will be $t_{0.05,\ 46} = 1.679$, obtained from statistical software.

The test statistic is calculated to be

$$t = \frac{15.89}{8.45/\sqrt{47}} = 12.89$$

Clearly, this has a p value of 0.0000.   ■

EXAMPLE 13.18   **OSAS Data**

Doing the Paired Sample Test

Let's complete the OSAS example using a level of significance of 0.05. If the treatment was effective, then the average difference should be significantly less than zero. If the treatment was not effective, then the pretest and post-test values would be close and the differences small, resulting in an average difference close to zero. The actual data for the $n = 9$ patients are shown in the next table.

Patient	Number of Obstructions		
	Pretreatment	**Post-Treatment**	**Difference**
1	365	22	− 343
2	107	1	− 106
3	28	12	− 16
4	40	8	− 32
5	48	3	− 45
6	64	8	− 56
7	109	0	− 109
8	55	4	− 51
9	20	0	− 20
$\bar{d}$			− 86.44
s_d			101.83

The test statistic follows a t distribution with $n - 1$ degrees of freedom. The value of n is equal to the number of pairs or the number of differences. In this case, $n = 9$ so there are 8 degrees of freedom. We need to find $t_{0.05,\, 8}$. From the t table, the critical value is -1.860.

The test statistic looks just like the t statistic for a single population mean. It is calculated as

$$t = \frac{-86.44}{101.83/\sqrt{9}}$$
$$= -2.546$$

The p value of the test is found by looking up the test statistic in the t distribution with 8 degrees of freedom. We find that the p value is 0.0172. ■

Once the testing procedure is done we require a decision and then an interpretation.

EXAMPLE 13.19   Does Training Work?

Deciding What the Test Results Mean

The company looking at whether training programs are effective looks at the results of the test. It sees that the test statistic is in the rejection region, so they will reject H_0.

The company concludes that the average difference in score before and after the training program is greater than zero. This means that the training program is effective. ■

EXAMPLE 13.20   OSAS Data

Conclusions

In the study looking at the effectiveness of home treatment of sleep apnea, we see that the test statistic of -2.546 is in the rejection region, so we reject H_0.

The hospital administration can conclude that the average difference between the number of obstructions after and before home treatment is less than zero—that is, it has gone down. Therefore, the home treatment is effective. ■

 *Try It Now!*

In addition to measuring the number of obstructions before and after home treatment in the OSAS study, the number of hypopnea both pre- and post-treatment for each patient is shown below.

| Patient | Hypopnea | | Difference |
	Pretreatment	Post-Treatment	
1	208	22	
2	297	0	
3	126	17	
4	150	13	
5	133	0	
6	201	10	
7	272	7	
8	137	1	
9	310	2	
$\overline{d}$			
s_d			

Complete the table above.

Is there evidence that the treatment has reduced the number of hypopnea at the 0.05 level of significance?

Step 1:

Step 2:

Step 3:

Step 4:

Step 5:

13.7.3 Exercises—Learning It!

13.11 Having learned about the paired t test you realize that you really should have used the test for the data on software price comparison. The data are repeated below:

Top Ten Business Software Packages	Computability Price ($)	PC Connection Price ($)
Windows 95 Upgrade	88	95
Norton Anti-Virus	59	70
McAfee ViruScan	49	60
First Aid 97 Deluxe	54	58
Clean Sweep III	37	37
Norton Utilities	68	75
Netscape Navigator	45	40
MS Office Pro 97 Upgrade	300	310
First Aid 97	32	35
Win Fax Pro	95	95

(a) Calculate the differences between the prices for each type of software package. Just looking at the differences, do you think that one company charges more than the other? Why or why not?

(b) Calculate the average difference and the standard deviation of the differences.

(c) Set up the hypotheses to test whether the mean difference in price between the two companies is zero.

(d) Assuming that the data are normally distributed, at the 0.05 level of significance, is there a difference in the mean price of software for the two companies?

(e) Did these results differ from the last time you analyzed the data? Why do you think this happened?

13.12 A hospital administrator is concerned about the length of time that the nursing staff washes their hands. A recent study in health care showed that longer washing greatly reduces the spread of germs. The hospital observed the amount time that a sample of nine nurses in the Cardiac Care Unit (CCU) washed their hands. The data were collected in such a way that the employees did not know that they were being observed. The hospital then showed the nurses an educational video on the negative effects of shortened time spent hand washing. After the video, the hospital again watched and timed the group of nurses washing their hands. The data are shown below:

Observation	Unit	Time 1 (s)	Time 2 (s)
1	CCU	3	16
2	CCU	2	7
3	CCU	0	5
4	CCU	5	8
5	CCU	2	15
6	CCU	0	15
7	CCU	2	20
8	CCU	3	16
9	CCU	0	18

(a) Calculate the differences between the times for each nurse. Just looking at the differences, do you think that, on the average, they washed their hands longer the second time? Why or why not?

(b) Calculate the average difference and the standard deviation of the differences.

(c) Set up the hypotheses to test whether there was an increase in the average amount of time spent washing hands.

(d) Assuming that the data are normally distributed, at the 0.05 level of significance, what can you conclude?

(e) Can you conclude that the video caused the nurses to wash their hands longer? Why or why not?

13.13 While it cannot be denied that advertising impacts sales, it is also true that advertising is expensive and companies want to advertise in ways that have the greatest benefit for the amount of money spent. A company that sells snack food designs two different advertising strategies, one focusing on print media and the other on radio. It ran the campaigns in a total of 16 different cities, paired on population size and measured the sales ($1000) in the week directly following the beginning of the campaign. The data are given below:

Print Sales	Radio Sales
28.3	22.1
24.6	19.1
23.1	20.3
21.0	24.4
25.7	22.4
22.5	19.2
32.0	22.8
23.5	20.3
24.3	25.5
25.2	22.6
23.3	24.9
25.3	29.7
22.2	22.2
23.4	28.5
23.9	28.2
25.7	21.6

(a) Calculate the differences in sales for each pair of cities.

(b) Based on these differences, do you think there is a difference in mean sales for the two types of advertising campaigns? Why or why not?

(c) Calculate the average difference and the standard deviation of the differences.

(d) Set up the hypotheses to test whether there is a difference in mean sales due to type of advertising.

(e) Assuming that the data are normally distributed, at the 0.01 level of significance, what can you conclude?

13.14 How does time spent using the computer impact the speed with which you work? A software company ran a study that looked at the effectiveness with which a person uses a mouse. It selected 10 people, matched on computer skills, and measured the speed with which they moved a mouse at the beginning of a long session of computer use and after two hours of use. The data (in hundredths of a second) are shown below:

67	57
64	53
69	71
88	61
72	73
80	50
85	53
116	80
77	63
78	41

(a) Calculate the differences between the times for each person. Just looking at the differences, do you think that there was a change in the average speed after two hours? Why or why not?

(b) Calculate the average difference and the standard deviation of the differences.

(c) Set up the hypotheses to test whether there was a change in the mean speed with which the people moved the mouse.

(d) Assuming that the data are normally distributed, at the 0.05 level of significance, what can you conclude?

13.15 Does gender impact the use of electronic mail (e-mail)? An insurance company studied the use of e-mail in its organization by counting the number of business-related e-mails generated by 10 men and 10 women matched on job position in a day. The data are given below:

Men	Women
82	48
77	61
78	56
83	59
82	58
78	56
81	60
74	64
86	59
76	63

(a) Calculate the differences in the number of business-related e-mails for each male/female pair. Just looking at the differences, do you think that men use e-mail more? Why or why not?

(b) Calculate the average difference and the standard deviation of the differences.

(c) Set up the hypotheses to test whether the average number of business-related e-mails generated by men is greater than that generated by women.

(d) Assuming that the data are normally distributed, at the 0.05 level of significance, what can you conclude?

13.8 HYPOTHESIS TEST FOR THE DIFFERENCE IN TWO POPULATION PROPORTIONS

13.8.1 Two-Population Tests With Qualitative Data

One of the things you should realize by now is that different types of data require different statistical techniques. So far, all of the two-population tests that we have looked at involve quantitative data. In this section, we look at the two-population tests for a kind of qualitative data—population proportions or percentages.

A lot of data are available in the form of proportions or percentages. You see this type of data all the time in the newspaper. Here are some examples taken from the newspaper.

- A study of Americans in 1987 found that 71% said the government should take care of people who can't take care of themselves. In 1994 the percentage was 57%. Is there evidence that Americans have become more cynical and less compassionate?

- A recent nationwide poll of 1225 adults showed that blacks and whites differ in their opinions of the causes and solutions of society's problems. But do they really? Seventy percent of blacks feel progress has been made in easing racial tension in the past decade compared to 65% of whites. Based on these data, do blacks and whites really feel differently about this issue?

- Researchers in San Francisco using the diseased cells of melanoma patients have developed a vaccine that they say dramatically reduces the recurrence of the deadliest form of skin cancer. After three years, 70% of those vaccinated remained cancer-free, compared to 20% of patients treated with surgery alone.

- A study of 1049 men and women aged 18 to 65 shows that a greater percentage of women (86%) find it difficult to have sex without emotional involvement compared to men (71%).

- A study conducted by an on-line service found that 30% of respondents under age 45 drove sports cars compared to 17% of the 45 or over population.

Each of the studies cited above has a structure similar to those we have already looked at in this chapter. In each case two populations are being compared and we have taken a sample from each population. What has changed is that the parameter being analyzed is no longer the mean, but the population proportion. For each sample, the percentage of the sample that has a certain characteristic is found. These percentages then need to be compared.

13.8.2 The Test for Two-Population Proportions

We know that even if the percentage of two populations that have a certain characteristic were exactly the same, we would almost never get exactly the same percentage in two samples from the populations. This is due to sampling error. The question then becomes, how large a difference in the percentages is large enough to declare the difference statistically significant? That is, how large does the difference have to be for us to be reasonably sure that there really is a difference in the behavior of the two populations, that something more than just sampling error is causing the difference. This is the same question we addressed in comparing means. This time the population parameters being compared are proportions and the evidence from the sample is summarized by the sample proportions. Our notation will follow the pattern established for the tests of means.

	Population 1	Population 2
Size of population	N_1	N_2
Sample size	n_1	n_2
Population proportion	π_1	π_2
Number in sample that have the characteristic being studied	x_1	x_2
Sample proportion	$p_1 = x_1/n_1$	$p_2 = x_2/n_2$

As in the tests for comparing two population means, there are two different but equivalent ways of stating the null and alternative hypotheses. You can also test for a difference in proportions other than zero. The null and alternative hypotheses for the upper-tail and lower-tail tests of differences in proportions are shown below.

Lower-Tail Test

H_0: $\pi_1 \geq \pi_2$ or H_0: $\pi_1 - \pi_2 \geq 0$ Use this test if you wish to test if the pro-
H_A: $\pi_1 < \pi_2$ H_A: $\pi_1 - \pi_2 < 0$ portion of population 1 is *less than* the proportion of population 2.

Upper-Tail Test

H_0: $\pi_1 \leq \pi_2$ H_0: $\pi_1 - \pi_2 \leq 0$ Use this test if you wish to test if the propor-
H_A: $\pi_1 > \pi_2$ H_A: $\pi_1 - \pi_2 > 0$ tion of population 1 is *greater than* the proportion of population 2.

EXAMPLE 13.21   Is Training Really Better?

Null and Alternative Hypotheses

The company that is looking at training programs has learned that the mean score on the sample CMI certification exam is higher for the group of employees that was trained by the outside consultant than for the group that was trained in-house. Some upper managers are ready to abandon all in-house training and move to outside training immediately. Other managers, however, are not so sure. They reason that the end result is really whether or not the employees pass the actual certification exam. If the percentage of employees that pass is the same in both groups, then why bother with the more expensive route? They decide to look at the proportion of employees who pass the exam in each group. If the percentage that pass for the outside consultant is higher than the percentage for the in-house group, then they will consider switching.

      The ASQ does not publish scores for people who take the exam; scores are simply classified as "Pass" or "Fail." The data for the two groups in question are found to be

	Population 1 In-House	Population 2 Outside Consultant
Sample size	47	50
Population proportion	π_1	π_2
Number that passed CMI	13	20
Sample proportion	$p_1 = \dfrac{13}{47} = 0.277$ or 27.7%	$p_2 = \dfrac{20}{50} = 0.400$ or 40.0%

Since the managers want to know if the proportion for the outside trained group is higher than that of the in-house group, they want to do a one-tail test. They set up their hypotheses as

$$H_0: \pi_1 \geq \pi_2 \quad \text{or} \quad H_0: \pi_1 - \pi_2 \geq 0$$
$$H_A: \pi_1 < \pi_2 \quad\quad\quad H_A: \pi_1 - \pi_2 < 0$$

This may seem strange to you, since the managers want to know if the outside group has a higher percentage of passes than the in-house group, but if you think about it, that is the same as saying the in-house group has a lower percentage of passes. The managers used the latter because of the way they labeled their populations. ∎

After you become proficient at hypothesis testing you might want to change from the standard subscripts of 1 and 2 to denote populations. It is often easier to use subscripts that reflect the two populations, such as M and F for comparisons of males and females or O and I for outside and in-house. In fact, we will do that in the next example.

EXAMPLE 13.22   E-Mail Usage

Null and Alternative Hypotheses

Although most companies have implemented some form of electronic mail (e-mail) to increase the frequency and timeliness of information, very little information is available on the impact that e-mail has had on organizational behavior. A study of 5 different companies was conducted to learn how e-mail is being used at different levels of the organization. A total of 70 people responded to the survey. The respondents provided information on their title (VP/Director, Manager, Staff/Administrator), their gender, and the degree to which they used e-mail for various purposes. All companies in this study had utilized e-mail for at least one year. In this case, the study focused on the differences in the proportion of males and females that used e-mail to communicate decisions.

	Population 1 Men	Population 2 Women
Sample size	35	35
Population proportion	π_M	π_W
Number that use e-mail	13	27
Sample proportion	$p_M = \dfrac{13}{35} = 0.371$ or	$p_W = \dfrac{27}{35} = 0.771$ or
	37.1%	77.1%

Remember that sample proportions must always be between 0 and 1 or between 0 and 100 if expressed as a percentage.

We wish to decide if there is evidence of different use of e-mail by men and women. Since we wish to detect differences in the two population proportions we know to use a two-sided test. The null and alternative hypotheses are shown below:

$$H_0: \quad \pi_M = \pi_W \quad \text{or} \quad H_0: \quad \pi_M - \pi_W = 0$$
$$H_A: \quad \pi_M \neq \pi_W \qquad H_A: \quad \pi_M - \pi_W \neq 0$$

Recall that the test statistic of a single-population proportion was a Z statistic. The extension to comparing two-population proportions is also a Z test statistic. Therefore, we already know how to find the rejection region.

The test statistic for this test is constructed using logic similar to what we have used before. The two sample proportions are compared by subtracting one from the other. To determine if this difference is "large," it is compared to the standard error of the estimate. In this case, the estimate of the true difference in the population proportions, $\pi_1 - \pi_2$, is the difference in the sample proportions, $p_1 - p_2$. The standard error of this estimate is similar to the standard error for a single-sample proportion. The test statistic is then given by the formula

Test statistic for comparing two population proportions

Remember that the test statistic is calculated under the assumption that the null hypothesis is true.

$$Z = \frac{p_1 - p_2}{\sqrt{\bar{p}(1 - \bar{p})\left(\dfrac{1}{n_1} + \dfrac{1}{n_2}\right)}}$$

Notice that $\bar{p}$ is used in the calculation of the standard error. The value of $\bar{p}$ is similar to the pooled variance in the sense that it combines all of the sample data. If the null hypothesis is true and the population proportions are equal, then the best estimate of the common but unknown population proportion is obtained by pooling the data. The formula for $\bar{p}$ is then

$$\bar{p} = \frac{x_1 + x_2}{n_1 + n_2}$$

Common population proportion

EXAMPLE 13.23 Is Training Really Better?

Performing the Test

After thinking about the problem, the company testing to see if the proportion of passes from the in-house training group is lower than that of the outside training group decides that a Type I error will be more costly. The managers reason that if they reject H_0 in error and conclude that the in-house group has a significantly lower percentage of passes, they will wind up making changes and implementing expensive programs in error. They decide to test at the 0.01 level of significance.

The critical value for the test is $Z = -2.33$.

They calculate the value of $\bar{p}$ to be

$$\frac{13 + 20}{47 + 50} = 0.340 \text{ or } 34\%$$

The value of the test statistic is then

$$Z = \frac{0.277 - 0.400}{\sqrt{(0.340)(0.660)\left(\dfrac{1}{47} + \dfrac{1}{50}\right)}}$$
$$= -1.28$$ ■

EXAMPLE 13.24 E-Mail Usage

Doing the Test

For the study of e-mail usage, there is no reason to test at any level other than 0.05. The rejection region for a two-tail Z test is defined by the critical values ± 1.96.

The value of $\bar{p}$ is calculated as follows:

$$\bar{p} = \frac{13 + 27}{35 + 35} = 0.571$$

and the test statistic is found to be

$$Z = \frac{0.371 - 0.771}{\sqrt{(0.571)(0.429)\left(\dfrac{1}{35} + \dfrac{1}{35}\right)}}$$
$$= -3.39$$ ■

As in any hypothesis test, we are not finished until we make a decision about H_0 and interpret that decision in terms of the problem we are trying to solve.

EXAMPLE 13.25 Is Training Really Better?

Conclusions

The group of managers who are looking at training programs found that the test statistic for their data was -1.28. Since this is not in the rejection region, they fail to reject H_0. Thus, they conclude that there is not enough evidence to say that the group trained in-house has a lower percentage of passes on the certification exam than the group trained by the outside consultant.

Does this end the issue? Not by a long shot. The managers who are promoting the idea of using outside consultants are still not convinced that it is not a better route. They argue that even if the percentage of passes *is* the same in both groups there are other factors to consider. They want to look at employee perceptions about the training programs. They also argue that the higher test scores on the post-test must mean something. Perhaps the true difference will be evident in subsequent job performance.

What you see here is that statistics *cannot* solve the problem. Both groups have legitimate arguments. What the group doing the study has to decide on is *what* measurable outcome they really want to see. If it is the percentage of their employees who pass the certification exam, then in-house training is probably sufficient. If they are interested in employee attitudes and job performance issues, then they have a lot more work to do before they can make a good decision.   ■

EXAMPLE 13.26   E-Mail Usage

Conclusions

For the question about using e-mail we see that the test statistic is in the rejection region, so we reject the null hypothesis. We conclude that a different proportion of men and women use e-mail to communicate decisions. Although this tells the company something, the results of the test do not indicate whether men or women use e-mail more. If this is what the company wanted to know, then it should have done a one-sided test.   ■

As you can see, testing the difference in two population proportions is not very different from testing the difference in two population means. This test has been developed parallel to the procedure for testing the difference in two population means. It has also been developed as a natural extension of the test of a single-population proportion.

 *Try It Now!*

E-mail Usage

Tests of Two Proportions

The company looking at use of e-mail is also interested in the use of e-mail for personal messages. After seeing the results of the last test, it decides that what it really wants to know is if a higher proportion of women than men use e-mail to send personal messages. Use $\alpha = 0.05$. The data are shown below:

	Population 1 Men	Population 2 Women
Sample size	35	35
Population proportion	π_1	π_2
Number that use e-mail	15	18
Sample proportion		

Finish the table and then complete the hypothesis test.

Step 1:

Step 2:

Step 3:

Step 4:

Step 5:

13.8.3 Exercises—Learning It!

13.16 People who watch major league baseball may have noticed a decrease in the percentage of successful stolen bases. The increase in slide-step pitching motion and players who focus on homers may be behind the decline, according to an article in *The New York Times*, 24 May 1994. In 1993 there were 957 stolen bases in 1446 attempts, while in 1994 there were 807 stolen bases in 1193 attempts.

(a) Calculate the sample proportion of stolen bases for each year.

(b) Set up the hypotheses to test whether there has been a decrease in the percentage of successful stolen bases.

(c) At the 0.05 level of significance, has the percentage really decreased?

13.17 Does television match reality? A recent study looked at the percentage of television characters who survived cardiopulmonary resuscitation (CPR) compared to those who survive in real life. The study looked at 34 instances where CPR was applied to a television character and found that 25 survived, while data provided by Emergency Medical Technicians found that in 40 similar cases, only 16 survived.

(a) Calculate the sample proportion of people who survive CPR for television and for reality.

(b) Set up the hypotheses to test whether there is a difference in the proportion of people who survive CPR in the two groups.

(c) At the 0.05 level of significance, does television reflect reality?

13.18 Women who smoke suffer an increased risk of dying of breast cancer, according to a recently published study. In the study, out of 319,000 women who never smoked there were 468 deaths from breast cancer, while out of 120,000 smokers, there were 187 deaths.

(a) Calculate the sample proportion of women who died of breast cancer for smokers and nonsmokers.

(b) Set up the hypotheses to test whether the proportion of women who die of breast cancer is higher for smokers than for nonsmokers.

(c) At the 0.05 level of significance, can you conclude that smoking causes breast cancer? If not, what can you conclude?

13.19 Selling personal computers is big business and consumers are becoming increasingly aware of vendor reputation. A recent study of two vendors of desktop personal computers reports on the units that need repair for Dell Computer and Gateway 2000. Of 1584 computers manufactured by Dell Computer, 427 needed repair, while for Gateway 2000, 825 of 2662 computers needed repair.

(a) Calculate the sample proportion of computers needing repair for each company.

(b) Set up the hypotheses to test whether the proportion of computers needing repairs is different for the two companies.

(c) At the 0.05 level of significance, what can you conclude?

13.20 How punctual are Amtrak trains? One of the reasons that passenger trains are late is that almost all of the tracks are owned and maintained by freight companies who control the schedules for use. Amtrak's definition of on time is based on trip length. After customer complaints about late trains, Amtrak has tried to improve and claims that it has. Of 100 runs of the Metroliner between New York City and Washington, DC, in March 1993, 87 were on time, while in the same period in 1994, 90 were on time.

(a) Calculate the sample proportion of on-time trips for each time period.

(b) Can Amtrak really claim it has improved ($\alpha = 0.05$)? Why or why not?

13.9 HYPOTHESIS TEST OF THE DIFFERENCE IN TWO POPULATION VARIANCES

13.9.1 The *F* Test for Comparing Population Variances

Remember the test to compare population means when the samples are small and the standard deviations are unknown? In working with small samples, we had to decide whether the populations had a common but unknown variance. If they did then we pooled the data and calculated a pooled estimate of the variance, s_p^2. We did not, however, discuss how to decide if the populations had a common variance. Now it is time to learn how to determine this. A hypothesis test of variances follows the same five steps that we have repeatedly used. Once again, the only difference is the test statistic.

To decide if we should pool the data we need to test to see if two population variances are *equal*. Thus, we should use a two-sided test. The null and alternative hypotheses are shown below:

Two-Sided Test

$H_0: \sigma_1^2 = \sigma_2^2$ or $H_0: \sigma_1^2/\sigma_2^2 = 1$ Use this test if you wish to test if the variance
$H_A: \sigma_1^2 \neq \sigma_2^2$ $H_A: \sigma_1^2/\sigma_2^2 \neq 1$ of population 1 is *different* from the variance of
population 2.

Remember: the key words different, less than, and greater than tell you which test to use.

As with the tests to compare two population means or proportions, it is also possible to do one-sided tests. These are shown below.

Lower-Tail Test

$H_0: \sigma_1^2 \geq \sigma_2^2$ or $H_0: \sigma_1^2/\sigma_2^2 \geq 1$ Use this test if you wish to test if the variance of
$H_A: \sigma_1^2 < \sigma_2^2$ $H_A: \sigma_1^2/\sigma_2^2 < 1$ population 1 is *less than* the variance of population 2.

Upper-Tail Test

$H_0: \sigma_1^2 \leq \sigma_2^2$ $H_0: \sigma_1^2/\sigma_2^2 \leq 1$ Use this test if you wish to test if the variance of
$H_A: \sigma_1^2 > \sigma_2^2$ $H_A: \sigma_1^2/\sigma_2^2 > 1$ population 1 is *greater than* the variance of population 2.

Because of the difficulty in finding the rejection region for a lower-tail test, most one-sided tests are conveniently done as upper-tail tests.

Since we are trying to decide how two population variances compare, it makes sense to compare the sample variances. We have seen two ways of comparing numbers. One way is to subtract one number from the other and see how close the difference is to zero. If the two numbers are estimates of the same unknown population parameter, then their difference will be close to zero. The other way to compare two numbers is to divide one number by the other and see how close the ratio is to one. If the two numbers are estimates of the same unknown population parameter, then their ratio will be close to one. Notice that the second way of writing each of the variance tests shown above uses a ratio and a comparison to one. When comparing variances, you should always use a ratio to compare them and not a difference.

We know from our work in Chapter 9 that the best estimator of the population variance is the sample variance. Extending this idea to two populations, the point estimate for the ratio of the population variances is the ratio of the sample variances. This is also the test statistic. It is shown below:

The test statistic for comparing variances is an F statistic.

$$F = \frac{s_1^2}{s_2^2}$$

Formula for test statistic for comparing two variances

Notice that this ratio is labeled F. This means that the test statistic follows an F distribution, if the two original populations are normally distributed. The F distribution was briefly introduced to you in Chapter 9. It is named after the famous statistician R. A. Fisher. Like the χ^2 distribution, which we used to test a single-population variance, the specific shape of the F distribution is determined by its degrees of freedom. But the F distribution has not one, but two values that determine its shape. One of these is called the degrees of freedom in the numerator and it is equal to one less than the sample size on which s_1^2 is based, $n_1 - 1$. The other one is called the degrees of freedom in the denominator and is equal to one less than the sample size on which s_2^2 is based, $n_2 - 1$.

The F distribution is determined by two sets of degrees of freedom.

The procedure for finding the rejection regions is similar to that which we have used for other tests. If you are a doing a two-sided test, then the rejection region is two-sided; if you are doing a one-sided test, then the rejection region is one-sided. These are shown in Figure 13.6.

The critical values that define the rejection region are labeled $F_{\text{upper,df1,df2}}$ and $F_{\text{lower,df1,df2}}$ rather than F_{cutoff} and $-F_{\text{cutoff}}$, the notation you might have expected. To find the values for $F_{\text{upper,df1,df2}}$ and $F_{\text{lower,df1,df2}}$, we need to notice that the shape of the F distribution is not symmetric and the distribution is not centered at zero. Therefore, the absolute values of F_{upper} and F_{lower} are not the same and they will always be greater than zero. This is why they are not labeled F_{cutoff} and $-F_{\text{cutoff}}$. It happens that there is a relationship between the upper and lower F values. In particular, $F_{\text{lower,df1,df2}}$ can be found from an upper-tail value as follows:

The symbols df_1 and df_2 stand for the number of degrees of freedom in the numerator and the denominator, respectively.

Formula for finding the lower critical value for an F test

$$F_{\text{lower,df1,df2}} = \frac{1}{F_{\text{upper,df2,df1}}}$$

That is, the lower critical value is found by taking the reciprocal of the upper critical value *with the degrees of freedom reversed.* Therefore, we need table values only for F_{upper}.

Since the F distribution is determined by two sets of degrees of freedom, labeled df_1 and df_2, we will need an entire table to specify the values for F_{upper} that cut off a certain amount of probability in the upper tail of the distribution. Unlike the other tables that we have encountered so far, each upper-tail area probability requires a separate table. For instance, Table A.6a shows the values for F_{upper} that cut off an area of 0.01 in the upper tail. The row and column we need to use depends on the number of degrees of freedom in the numerator and the denominator. However, if we need 0.05 in the upper-tail area, then we must use Table A.6b. This is the first time we have encountered a different table for each value of α. When we used the t or the χ^2 distribution, each column corresponded to a different value of α.

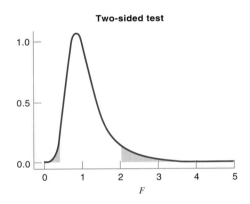

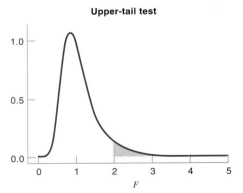

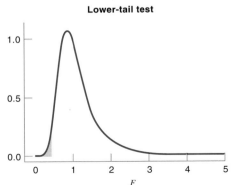

FIGURE 13.6 Rejection regions for test of variances

EXAMPLE 13.27   Training and Gender

Setting Up the Test

When the company looking at training programs wanted to know if there was a difference in the post-test scores for males versus females, we made an assumption about whether or not the population variances were equal. The relevant data from the problem were

	Population 1 Females Trained In-House	Population 2 Males Trained In-House
Sample size	$n_1 = 20$	$n_2 = 27$
Sample average	$\overline{X}_1 = 63.95$	$\overline{X}_2 = 63.26$
Sample standard deviation	$s_1 = 8.86$	$s_2 = 6.87$

To really determine whether this assumption is reasonable we need to test the hypotheses:

$$H_0 : \sigma_1^2 = \sigma_2^2$$
$$H_A : \sigma_1^2 \neq \sigma_2^2$$

With these hypotheses, if we reject H_0, then we will conclude that the variances *are* different and we will know that we cannot pool the variances. For the moment we will assume that the populations are normally distributed and we will do the test at the 0.10 level of significance, which means that we need to use the F table with the upper-tail area of 0.05, Table A.6b. A portion of that table is shown below:

		\multicolumn{10}{c}{Numerator Degrees of Freedom (df_1)}
		18	19	20	21	22	23	24	25	26	27
	18	2.217	2.203	2.191	2.179	2.168	2.159	2.150	2.141	2.134	2.126
	19	2.182	2.168	2.155	2.144	2.133	2.123	2.114	2.106	2.098	2.090
	20	2.151	2.137	2.124	2.112	2.102	2.092	2.082	2.074	2.066	2.059
Denominator	21	2.123	2.109	2.096	2.084	2.073	2.063	2.054	2.045	2.037	2.030
degrees of	22	2.098	2.084	2.071	2.059	2.048	2.038	2.028	2.020	2.012	2.004
freedom	23	2.075	2.061	2.048	2.036	2.025	2.014	2.005	1.996	1.988	1.981
(df_2)	24	2.054	2.040	2.027	2.015	2.003	1.993	1.984	1.975	1.967	1.959
	25	2.035	2.021	2.007	1.995	1.984	1.974	1.964	1.955	1.947	1.939
	26	2.018	2.003	1.990	1.978	1.966	1.956	1.946	1.938	1.929	1.921
	27	2.002	1.987	1.974	1.961	1.950	1.940	1.930	1.921	1.913	1.905

To find the upper critical value we will use the numerator degrees of freedom as $20 - 1 = 19$ and the denominator degrees of freedom as $27 - 1 = 26$. The value of $F_{upper,19,26}$ is 2.003. Now, the lower critical value that we want is $F_{lower,19,26}$, so using the relationship that

To find $F_{lower,df1,df2}$ you must reverse the degrees of freedom and find the corresponding F_{upper} value and then take the reciprocal.

$$F_{lower,19,26} = \frac{1}{F_{upper,26,19}}$$

we find that

$$F_{lower,19,26} = \frac{1}{2.098} = 0.3356 \qquad \blacksquare$$

This example has illustrated the procedure for finding the rejection region for a two-sided test. Let's summarize the steps for finding a two-sided rejection region and note the differences for a one-sided rejection region.

Two-sided rejection region:

   Step 1:   Divide the value of α in half.

   Step 2:   Find the F table that corresponds to those values that cut off $\alpha/2$ in the upper tail of the distribution.

Step 3:   Find $F_{\text{upper,df1,df2}}$ at the intersection of the column corresponding to $df_1 = n_1 - 1$ and the row corresponding to $df_2 = n_2 - 1$.

Step 4:   Find F_{lower} by first finding $F_{\text{upper,df2,df1}}$ with $n_2 - 1$ degrees of freedom in the numerator and $n_1 - 1$ degrees of freedom in the denominator. Then take the reciprocal of this number.

Step 5:   Reject H_0 if the F statistic is larger than F_{upper} or less than F_{lower}.

If we are using a one-sided test of the variances, the main difference in finding the rejection region is that you do not split α in half. In addition, you need to find only one of the critical values: either F_{upper} or F_{lower}.

Upper-tail rejection region:

Steps for finding the rejection regions for the F distribution

Step 1:   Find the F table that corresponds to those values that cut off α in the upper tail of the distribution.

Step 2:   Find $F_{\text{upper,df1,df2}}$ at the intersection of the column corresponding to $df_1 = n_1 - 1$ and the row corresponding to $df_2 = n_2 - 1$.

Step 3:   Reject H_0 if the F statistic is larger than F_{upper}.

Lower-tail rejection region:

To avoid the tedious problem of finding the lower-tail rejection region for a one-tail test, it is easier to just set the test up as an upper-tail test by defining population 1 as the one with the larger of the two sample variances.

Step 1:   Find the F table that corresponds to those values that cut off α in the upper tail of the distribution.

Step 2:   Find F_{lower} by first finding $F_{\text{upper,df2,df1}}$ with $n_2 - 1$ degrees of freedom in the numerator and $n_1 - 1$ degrees of freedom in the denominator. Then take the reciprocal of this number.

Step 3:   Reject H_0 if the F statistic is less than F_{lower}.

The remainder of the hypothesis testing procedure involves calculating the test statistic and making a decision both in terms of the null and alternative hypotheses and the original problem. We can now finish the example.

EXAMPLE 13.28   **Training and Gender**

Concluding the Test

For the training data we can calculate the test statistic as

$$F = \frac{8.86^2}{6.87^2} = 1.663$$

Since the critical values for the test were 0.3356 and 2.003, we see that we cannot reject H_0. This means that there is not enough evidence to say that the variances are different and so the assumption that they are equal is reasonable.   ■

 **TRY IT NOW!**

Training and Gender

Test for Equality of Variances

When you looked at gender differences for the group of employees trained by the outside consultant, you did the test using the pooled variance. That is, you assumed the population variances were equal. The relevant data were as shown in the table.

	Population 1 Females Trained by Outside Consultant	Population 2 Males Trained by Outside Consultant
Sample size	$n_1 = 21$	$n_2 = 29$
Sample mean	$\overline{X}_1 = 75.9$	$\overline{X}_2 = 75.2$
Sample standard deviation	$s_1 = 3.9$	$s_2 = 4.4$

Assume that the data are normally distributed and perform a test to see whether the assumption of equal variances was reasonable. Use a level of significance of 0.10.

Step 1:

Step 2:

Step 3:

Step 4:

(*continued*)

Step 5:

13.9.2 Exercises—Learning It!

13.21 Consider the problem in which The Board of Realtors for Greater Bridgeport, CT, was looking at the average selling prices of homes. The data are given again below:

	Population	
	1995	**1996**
Sample size	$n_1 = 25$	$n_2 = 25$
Sample mean	$\overline{X}_1 = \$151,166$	$\overline{X}_2 = \$160,669$
Sample standard deviation	$s_M = \$5332$	$s_W = \$6468$

(a) Assuming that the populations are normally distributed, set up the hypotheses to test whether the population variances are equal at the 0.10 level of significance.

(b) Was the decision to test using the pooled variance justified?

13.22 In your quest for the perfect golf clubs you made an assumption about the population variances when you tested your hypotheses. The data you collected are given below:

	Population	
	Brand X	**Brand Z**
Sample size	$n_X = 15$	$n_Z = 15$
Sample mean	$\overline{X}_X = 255$	$\overline{X}_Z = 271$
Sample standard deviation	$s_X = 8.7$	$s_Z = 9.1$

(a) Set up the appropriate hypotheses to test whether the variance of Brand Z clubs is the same as the variance for Brand X.

(b) Assuming that the populations are normally distributed, at the 0.10 level of significance was your decision to pool the variances a good one?

(c) In general, would a difference in variation between the clubs be a factor in your purchase decision?

13.23 The members of the Chamber of Commerce of a small city in Fairfield County, CT, are looking at the amount of vacant office space in their city compared to a similar city. The data are given below:

	Population	
	Our City	**Their City**
Sample size	$n_1 = 12$	$n_2 = 12$
Sample mean	$\overline{X}_1 = 210,700$	$\overline{X}_2 = 167,607$
Sample standard deviation	$s_1 = 2200$	$s_2 = 2100$

Assuming that the populations are normally distributed, at the 0.02 level of significance, should they have pooled the variances?

13.24 A company has two different production lines that make the plastic cards used for credit cards and ATM cards. Both lines use $\overline{X}$ control charts to make sure that they run to the target specification, and both have been in control for the past six weeks. Recently, however, the quality manager has noticed that one of the machines (machine A) has many more items being rejected for the measurement on the width of the card. Since both machines are running to target, he decides the problem must be with the variability and decides to run a test. He samples 40 items from each production line and calculates the following summary statistics:

	Population	
	Machine A	**Machine B**
Sample size	$n_A = 40$	$n_B = 40$
Sample standard deviation	$s_A = 1.1$ mm	$s_B = 0.62$ mm

(a) Set up the hypotheses to test whether the variance of machine A is greater than the variance of machine B.

(b) Assuming that the populations are normally distributed, perform the test at the 0.10 level of significance.

(c) Is the quality manager correct in his perception that machine A is more variable?

13.25 A large utility company is considering two sites for locating a large-scale wind energy conversion system (windmill). In selecting a site both the average speed and the variation in speed are important: the more consistent the wind speeds, the more efficient the energy conversion. The variability for the two sites is measured using a sample of 30 wind speed observations at each site. The summary data are given below:

	Population	
	Site 1	**Site 2**
Sample size	$n_1 = 30$	$n_2 = 30$
Sample standard deviation	$s_1 = 1.8$ mph	$s_2 = 2.62$ mph

(a) Set up the hypotheses to test whether the variability in wind speed at the two sites is the same.

(b) Assuming that the populations are normally distributed, perform the test at the 0.10 level of significance.

(c) What can you tell the utility company about the variability at the two sites?

13.10 TWO POPULATION HYPOTHESIS TESTS IN EXCEL

Excel has built-in data analysis tools for two population hypothesis tests. Once you know which test to use, the procedure is not very different from test to test.

13.10.1 Large-Sample Tests of Two Means

For a large-sample test of two population means, either the population standard deviations must be known or else the sample size from each population must be large—that is, greater than 30. Suppose that we want to consider the data from the company that was looking at in-house vs. outside consultant training. They want to know whether the results obtained by the outside consultant are better than those of the in-house trainers. This is a one-tailed test, and the hypotheses are

$$H_0 : \mu_1 \leq \mu_2$$
$$H_A : \mu_1 > \mu_2$$

To use Excel to perform the hypothesis test, the data must be in a spreadsheet. If you do not know the population standard deviations and are using the sample standard deviations, you must first calculate these using either the Descriptive Statistics tool or the **STDEV** or **VAR** functions. The value you will actually need is the variance, so if you calculate the standard deviation, you will have to square it to find the variance.

In our example, the population standard deviations are not known, but the sample sizes are $n_1 = 50$ and $n_2 = 47$, respectively. To perform the test, open the **Tools > Data Analysis** menu and from the list choose **z Test: Two Sample for Means.** This will open the dialog box for the test as shown in Figure 13.7.

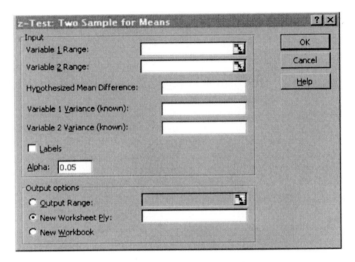

FIGURE 13.7 Dialog box for *Z* test for two means

Use the following procedure to perform the test:

1. Position the cursor in the textbox next to **Variable 1 Range:** and highlight the range that contains the post-test data for the first population—in this case, the group trained by the outside trainer.

2. With the cursor in the textbox for **Variable 2 Range:** highlight the location of the data for the second population, the in-house group.

3. Position the cursor in the textbox for **Hypothesized Mean Difference** and enter 0. Since we want to know whether $\mu_1 > \mu_2$, this is the same as $\mu_1 - \mu_2 > 0$.

Note: You must enter the value of the variance itself. You cannot enter a formula or highlight a data range here.

4. Now, put the cursor in the textbox for **Variable 1 Variance (known)** and enter the *value* of the variance for population 1. For the training example, the variance in post-test scores for the outside group is $(4.12)^2 = 16.97$.

5. Repeat this for **Variable 2 Variance (known).**

6. If there were labels in any of the data ranges you highlighted, check **Labels** checkbox.

7. Enter the level of significance for the test. In this case, we will use the default value of $\alpha = 0.05$.

8. Finally, indicate where you want the output from the test to appear.

When you have entered all the data, the dialog box should look like the one in Figure 13.8.

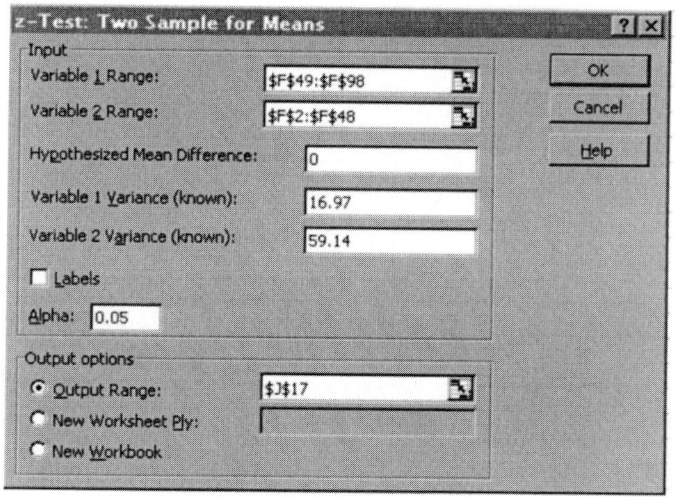

FIGURE 13.8 *Z* test dialog box

9. Hit Enter to perform the test. The resulting output is shown in Figure 13.9.

z-Test: Two Sample for Means		
	Variable 1	Variable 2
Mean	75.46	63.55319
Known Variance	16.97	59.14
Observations	50	47
Hypothesized Mean Difference	0	
z	9.419937922	
P(Z<=z) one-tail	0	
z Critical one-tail	1.644853	
P(Z<=z) two-tail	0	
z Critical two-tail	1.959961082	

FIGURE 13.9 Output from z Test for two sample means

From the output, we see that the value of the test statistic, labeled z in the output, is 9.419937922 (which is the same as the 9.42 we obtained by hand). Since the critical value for a one-tailed test is 1.644853, we can reject H_0 and conclude that the mean for the outside group is higher than the mean for the in-house group.

The procedure gives the p value for the test, but does not label it as such. The value labeled P(Z <= z) one-tail is the p value for the one-tailed test, and P(Z <= z) two-tail is the p value for the two-tailed test.

13.10.2 Small-Sample Tests for Two Population Means

When you do not know the population standard deviations and the sample sizes are not large, the appropriate test is the *t* test. To choose the correct *t* test, you must decide whether the two population variances are the same or whether they differ from each other. Excel provides *t* test procedures for both cases.

Let's look at the training data for the outside consultant. Suppose that now we want to know whether there is a difference in post-test scores for males and females. Since there were only 50 people in the original sample, it is not possible that both sample sizes are larger than 30. Therefore, we must use a t test. We will use the t test that assumes the two population variances are equal.

From the **Data Analysis Tools** menu, select **t-Test: Two Sample Assuming Equal Variances.** The dialog box that opens should look very similar to the one for the Z test. The only difference is that you are not required to put in the population variances. The t test assumes that they are unknown and the procedure will calculate them from the data. Fill in the dialog box, just as you did for the Z test. An example of the completed box is shown in Figure 13.10.

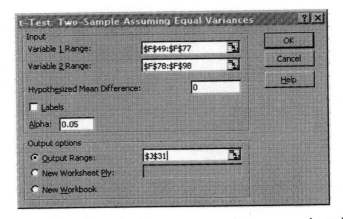

FIGURE 13.10 t Test for two means, variances assumed equal

Click **OK;** the output is placed in the location you specified. The output for this test is shown in Figure 13.11.

From the output, we see that the value of the t statistic is -0.575819805. Since we wanted to determine whether there was a difference between males and females, it is a two-sided test and the critical values are ± 2.01063358. Comparing the test statistic and the critical value, we fail to reject H_0 and conclude that there is no difference in the scores for males and females in the outside group. The output from the t test does include the p value of the test, although it is not labeled as such. The value labeled P(T $<=$ t) two-tail is the p-value for the two-tailed test. Notice that it is twice the value for the one-tailed test.

t-Test: Two-Sample Assuming Equal Variances		
	Variable 1	*Variable 2*
Mean	75.17241379	75.85714
Variance	18.93349754	14.82857
Observations	29	21
Pooled Variance	17.22311166	
Hypothesized Mean Difference	0	
df	48	
t Stat	-0.575819805	
P(T<=t) one-tail	0.283713515	
t Critical one-tail	1.677224191	
P(T<=t) two-tail	0.567427031	
t Critical two-tail	2.01063358	

FIGURE 13.11   Output for t test assuming equal variances

What if we cannot assume that the variances are equal? In this case, the two variances are 18.93 and 14.83, which do not appear to be that different. However, we can do the same test, assuming that the variances are not equal, and see what changes. From the list of **Data Analysis** tools, select **t Test: Two Sample Assuming Unequal Variances.** There are no differences in the dialog box. This makes sense, because there are no differences in the test inputs, just the outputs. Entering the same data as in the previous test produces the output shown in Figure 13.12.

t-Test: Two-Sample Assuming Unequal Variances		
	Variable 1	Variable 2
Mean	75.17241379	75.85714
Variance	18.93349754	14.82857
Observations	29	21
Hypothesized Mean Difference	0	
df	46	
t Stat	-0.58736595	
P(T<=t) one-tail	0.2799151	
t Critical one-tail	1.678658919	
P(T<=t) two-tail	0.559830201	
t Critical two-tail	2.012893674	

FIGURE 13.12   Output from *t* test assuming unequal variances

From the output, you see that the only real difference is in the value for degrees of freedom. The test assuming unequal variances is a more conservative test. Since the degrees of freedom have changed, the critical values will also change. The conclusion of the test in this example is the same.

Do not forget that all of the *t* tests *assume the populations are normally distributed.* You need to check this assumption before reporting any results.

13.10.3 Paired Difference Test in Excel

When the two populations in a hypothesis test are related, we must use a paired difference test. There is an Excel data analysis tool for the paired difference test.

Suppose that we want to look at the training scores to determine whether training really changes test scores. In this case, the two populations are related; they are the same people, so we must use a paired difference test. We will look at the in-house group, using the same data as for the previous examples. The hypothesis test is a one-sided test, because we want to know whether training increases test scores. We will define the difference in scores to be post-test − pretest.

From the list of data analysis tools, select **t Test: Paired Two Sample** for Means. The dialog box is the same as the dialog boxes for the other *t* tests that we have looked at. In this case, the first variable is the post-test scores for the in-house group, and the second variable is the pretest scores for the same group. The output from the test is shown in Figure 13.13 on page 562. By now, the output should look familiar.

From the output, you can see that the value of the test statistic is 7.17031. Comparing this to the one-tail critical value of 1.678658919, we reject H_0 and conclude that the test scores have increased. The *p* value of the test is 0.0000.

t-Test: Paired Two Sample for Means		
	Variable 1	Variable 2
Mean	63.55319149	47.65957
Variance	59.20906568	163.7946
Observations	47	47
Pearson Correlation	-0.040209148	
Hypothesized Mean Difference	0	
df	46	
t Stat	7.170308992	
P(T<=t) one-tail	2.53155E-09	
t Critical one-tail	1.678658919	
P(T<=t) two-tail	5.0631E-09	
t Critical two-tail	2.012893674	

FIGURE 13.13   Output from paired difference test

13.10.4 Two Population Tests of Proportions

Excel does not have a built-in test for comparing two population proportions, but we have supplied a macro that will perform the test. The macro assumes that you have the proportion of frequency of successes calculated for each population. You can use pivot tables to do this.

The company looking at training scores wants to determine whether the proportion of people who pass the test from the outside training group is greater than the proportion for the in-house group. An example of the pivot table is shown in Figure 13.14. The counts for each cell are displayed as percentage of row.

Count of Certificaton Exam	Certificaton Exam		
Training	F	P	Grand Total
I	0.723	0.277	1.000
O	0.600	0.400	1.000
Grand Total	0.660	0.340	1.000

FIGURE 13.14   Pivot table for training data

From the list of MacDoIt macros, choose **TwoPropTest** and click **Run.** A table will open in a new worksheet, just as in the macros for the one-population tests. We will let the outside group be population 1 and the in-house group be population 2. In the top portion of the table, type in the sample proportions and sample sizes for each group. The table will update automatically. The results of the test are shown in Figure 13.15.

	A	B	C	D	E
2					
3			Proportion	Sample Size	
4	Sample 1	p1	0.4000	n1	50
5	Sample 2	p2	0.2770	n2	47
6	Pooled	pp	0.3404		
7					
8	Test Statistic	Z	1.2777		
9	Significance Level	α	0.05		
10	*Critical Values*				
11	Lower Tail	Z1	-1.6449	p-value1	0.100684
12	Upper Tail	Z2	1.6449	p-value2	0.100684
13	Two Tail	Z3	-1.9600	p-value3	0.201367

FIGURE 13.15   Table for two proportion test

The test statistic is 1.2777. The test is an upper-tail test so the critical value is 1.6449 with a p value of 0.1007. From the results, we conclude that we cannot reject H_0. There is not enough evidence to say that the proportion that passes from the outside group is higher.

13.10.5 *F* Test for Comparing Two Variances

When we looked at the small-sample tests for comparing population means, we had to make a decision about whether the population variances were equal. Excel has an analysis tool to test hypotheses about population variances.

We can use this test to determine which t test was correct. From the list of Data Analysis tools, select **F Test Two Sample for Variances.** The dialog box is the same one we have seen for the other two population tests.

The only difference is not obvious from the dialog box—the test is done as an upper-tail test. So, to test whether the variances are not equal, you must enter *half* the value of α that you want to use and you must enter the larger variance as population 1. Finish filling out the dialog box and click **OK.** The output is shown in Figure 13.16.

F-Test Two-Sample for Variances		
	Variable 1	*Variable 2*
Mean	75.17241379	75.85714286
Variance	18.93349754	14.82857143
Observations	29	21
df	28	20
F	1.27682546	
P(F<=f) one-tail	0.289003794	
F Critical one-tail	2.365666774	

FIGURE 13.16   Output from two sample variance test

From the output, the test statistic is labeled F and is 1.27682546. Comparing this to the critical value of 2.365666774, we conclude that we cannot reject H_0 and that it is reasonable to assume that the variances are equal. The p value of 0.289003794 given is the one-tailed p value.

CHAPTER 13 SUMMARY

Key Terms

Term	Definition	Page Reference
Dependent samples	**Dependent samples** are related to each other. The members of one sample are identical to or matched or paired with the members in the other sample according to some characteristic.	533
Pretest and post-test	**Pretest** and **post-test** conditions exist when data are collected on the same sample elements before and after some experiment is performed.	533

Key Formulas

Term	Formula	Page Reference
Formula for average difference, $\overline{d}$	$\overline{d} = \dfrac{\sum\limits_{i=1}^{n} d_i}{n}$	536
Formula for common population proportion, $\overline{p}$	$\overline{p} = \dfrac{x_1 + x_2}{n_1 + n_2}$	547
Formula for degrees of freedom for small-sample, unequal variances test	$\nu = \dfrac{\left(\dfrac{s_1^2}{n_1} + \dfrac{s_2^2}{n_2}\right)^2}{\dfrac{(s_1^2/n_1)^2}{n_1 + 1} + \dfrac{(s_2^2/n_2)^2}{n_2 + 1}} - 2$	529
Variances test formula for F_{lower}	$F_{\text{lower,df1,df2}} = \dfrac{1}{F_{\text{upper,df2,df1}}}$	552
Formula for pooled variance, s_p^2	$s_p^2 = \dfrac{(n_1 - 1)s_1^2 + (n_2 - 1)s_2^2}{n_1 + n_2 - 2}$	525
Test for paired difference	$t = \dfrac{\overline{d}}{s_d/\sqrt{n}}$	537
Test for two population means:		
• Variances known	$Z = \dfrac{(\overline{X}_1 - \overline{X}_2) - 0}{\sqrt{\sigma_1^2/n_1 + \sigma_2^2/n_2}}$	515
• Variances unknown, $n_1, n_2 \geq 30$	$Z = \dfrac{(\overline{X}_1 - \overline{X}_2) - 0}{\sqrt{s_1^2/n_1 + s_2^2/n_2}}$	519
	$t = \dfrac{(\overline{X}_1 - \overline{X}_2) - d}{\sqrt{s_1^2/n_1 + s_2^2/n_2}}$	528
• Variances unknown but equal, $n_1, n_2 < 30$	$t = \dfrac{(\overline{X}_1 - \overline{X}_2)}{s_p\sqrt{1/n_1 + 1/n_2}}$	526

Term	Formula	Page Reference
Test for two population proportions	$Z = \dfrac{p_1 - p_2}{\sqrt{\bar{p}(1 - \bar{p})\left(\dfrac{1}{n_1} + \dfrac{1}{n_2}\right)}}$	546
Test for two population variances	$F = \dfrac{s_1^2}{s_2^2}$	551

CHAPTER 13 EXERCISES

Learning It

13.26 The members of the Chamber of Commerce of a small city in Fairfield County, CT, are wondering whether they need to worry about the amount of vacant office space in the city. They would consider lobbying for tax incentives for businesses if they find that there is more vacant office space in their city than in a comparable area. They take weekly data on the number of square feet of vacant office space, for the second quarter of 1996 for each city and find the following:

	Population	
	Our City	**Their City**
Sample size	$n_1 = 12$	$n_2 = 12$
Sample mean	$\overline{X}_1 = 210{,}700$	$\overline{X}_2 = 167{,}607$
Sample standard deviation	$s_1 = 2200$	$s_2 = 2100$

(a) Calculate the pooled variance for the data.

(b) Assuming that the data are normally distributed, at the 0.01 level of significance, should they lobby for tax incentives?

13.27 A recent study of consumer behavior focused on the amount of money spent monthly on frozen foods. The study wanted to determine whether there was a difference in the average amount of money spent for men and women. Data were collected on samples of 50 men and 50 women and the following information was found:

	Population	
	Men	**Women**
Sample size	$n_M = 50$	$n_W = 50$
Sample mean	$\overline{X}_M = \$72.24$	$\overline{X}_W = \$67.44$
Sample standard deviation	$s_M = \$8.23$	$s_W = \$8.12$

(a) Set up the appropriate hypotheses to test whether men spend more per month, on the average, for frozen foods than women.

(b) Use the large-population test with unknown variances to test the hypotheses. Use a level of significance of 0.10.

(c) Do the data provide evidence that the average amount spent per month on frozen food by men is greater than by women?

(d) What is the p value of the test?

13.28 The nurses who were part of the hand-washing experiment are still not convinced that the length of time spent washing hands makes that much difference. They design their own study and decide to have each nurse in the CCU wash his or her hands twice, once for 2.5 seconds and once for 15 seconds. After each washing they do a bacteria culture and measure the number of bacteria that remain on the person's hands. The data are shown in the next table.

Observation	Culture 1 2.5 s	Culture 2 15 s
1	66	78
2	132	115
3	120	93
4	187	48
5	190	77
6	17	3
7	33	12
8	92	12
9	1000	146

(a) Calculate the differences between the number of bacteria for each nurse. Just looking at the differences, do you think that washing longer decreased the amount of bacteria? Why or why not?

(b) Calculate the average difference and the standard deviation of the differences.

(c) Set up the hypotheses to test whether there was a decrease in the average amount of bacteria after washing longer.

(d) Assuming that the data are normally distributed, at the 0.05 level of significance, what can you conclude?

(e) Suppose that you were told that the second episode of hand washing was done right after the first. Would that change the way you interpret the results of the study?

Thinking About It

13.29 It has been a widely held belief that the switch to participative management would increase employees' buy-in to the company. One of the benefits that should be realized is a reduction in the number of sick days that employees use. A company that has made the switch in some departments wonders if this has been true. It decides to sample 25 employees from each of two manufacturing departments. The first has been using a participative management style for almost two years and the second is still using a traditional management style. The data on the number of sick days used by each employee in the past 12 months are found below:

Participative					Traditional				
1	3	5	5	6	0	5	6	7	9
1	4	5	6	7	3	5	7	7	9
2	4	5	6	8	4	6	7	7	10
2	4	5	6	8	4	6	7	8	11
3	4	5	6	8	5	6	7	8	11

At the 0.05 level of significance, do the data provide enough evidence to say that employees who use participative management styles use, on the average, fewer sick days than those who use a traditional management style? Be sure to justify any assumptions you make in selecting the test procedure you use.

13.30 The software company that is looking at the time to failure of the diskettes (hours) it uses decides to look at an alternative supplier of the product. The data for the current supplier and for the new supplier are shown below:

Current Supplier				Alternative Supplier			
486	494	502	508	489	492	495	498
490	496	504	510	489	492	496	499
491	498	505	514	491	493	497	502
491	498	506	515	492	493	497	503
494	498	507	527	492	494	497	505

The company has decided that if the mean time to failure for the new supplier is longer than it is for the current supplier, it will switch suppliers.

(a) What level of significance would you suggest the company use? Justify your choice.

(b) Assuming that the data are normally distributed, should the company switch suppliers? Use the level of significance you chose in part (a).

(c) What impact did your choice of α have on the decision?

13.31 You are still wondering about the results of the test on the difference in software prices. You wonder why the two tests came to different conclusions and figure that it must be the amount of variability in the software prices for the packages chosen.

Requires Exercise 13.11

Top Ten Business Software Packages	Computability Price ($)	PC Connection Price ($)
Windows 95 Upgrade	88	95
Norton Anti-Virus	59	70
McAfee ViruScan	49	60
First Aid 97 Deluxe	54	58
Clean Sweep III	37	37
Norton Utilities	68	75
Netscape Navigator	45	40
MS Office Pro 97 Upgrade	300	310
First Aid 97	32	35
Win Fax Pro	95	95

(a) Look at the data again. Does any software have a price that seems to be very different from the others? If so, which one?

(b) Drop the data for the unusual observation and perform the hypothesis test again using the test for independent samples.

(c) Does anything change from the last time you did the test? If so, what?

(d) Does dropping the observation change the decision?

(e) Do you think this was the right test to use? Why or why not?

13.32 A study was recently completed by an insurance company concerning a particular surgical procedure. The study looked at the hospital records of 40 patients at two different hospitals and compared the length of patient stay. The data were analyzed using Minitab. Output showing the descriptive statistics for the data is given below:

```
DESCRIPTIVE STATISTICS

Length of Stay    N      Mean     Median   Trim Mean   St.Dev.    SE Mean
Hospital 1        40     7.725    7.500    7.667       2.562      0.405
Hospital 2        40     10.350   10.000   10.222      3.340      0.528

Length of Stay    Min    Max      Q1       Q3
Hospital 1        2.000  14.000   6.000    9.000
Hospital 2        5.000  18.000   8.000    13.500
```

(a) Set up the hypotheses necessary to test whether the patients at hospital 1 had, on the average, a shorter stay than those at hospital 2.

(b) What type of test would you use to make this decision? Why?

(c) Perform the appropriate hypothesis test. Use a level of significance of 0.05.

(d) Do the data provide evidence that the mean stay at hospital 1 is shorter than at hospital 2?

13.33 After looking at the results of the data analysis, the Director of Human Resources at the company looking at sick days and type of management writes a memo to the Vice President of Human Resources suggesting that the company change all units over to Participative Management. He cites the results of the test and states that the data "provide evidence that Participative Management causes people to take fewer sick days." Since you did the analysis, he gives you the memo to read before he sends it.

Requires Exercise 13.29

(a) Do you agree with the Director of Human Resources? Why or why not?

(b) Write a memo to the Director of Human Resources explaining your reaction. Include plans for further study if you think it is warranted.

Requires Exercise 13.27

13.34 Reconsider the study of the amount of money spent monthly on frozen foods. Data were collected on samples of 50 men and 50 women:

	Population	
	Men	**Women**
Sample size	$n_M = 50$	$n_W = 50$
Sample mean	$\overline{X}_M = \$72.24$	$\overline{X}_W = \$67.44$
Sample standard deviation	$s_M = \$8.23$	$s_W = \$8.12$

(a) Do you think your decision was sensitive to the value chosen for α? Why or why not?

(b) Suppose that you were interested only in whether the average spent by men was different than the average spent by women. How would this have changed the setup of the test? Would it have changed the conclusion?

Requires Exercises 13.11, 13.31

13.35 After looking at the software price data again, and based on the results of the paired test, you are considering buying your software from Computability. You decide to check the ads for each company one more time to see if there are any hidden catches and you notice that shipping charges for PC Connection are $5 while for Computability they are $16.95.

At the 0.05 level of significance, who will you buy your software from?

13.36 Since the data were available, the Nursing Supervisor was interested in knowing whether there was a difference in the average amount of time that nurses from two different departments spent washing their hands. She was not sure whether to pool the variances, and Minitab does not do an F test on variances, so she decided to run the t test both ways. The Minitab output is shown below:

```
Two Sample T-Test and Confidence Interval

Two sample T for C9
C8       N     Mean    StDev    SE Mean
IMCU    10     1.00    1.89      0.60
N4       8     4.87    5.79      2.0
95% CI for mu (IMCU) — mu (N4 ): (−8.79, 1.0)
T-Test mu (IMCU) = mu (N4 ) (vs not =): T = −1.82 P = 0.11 DF = 8
```

```
Two Sample T-Test and Confidence Interval

Two sample T for C9
C8       N     Mean    StDev    SE
IMCU    10     1.00    1.89     0.60
N4       8     4.87    5.79     2.0
95% CI for mu (IMCU) — mu (N4 ): (−7.98, 0.2)
T-Test mu (IMCU) = mu (N4 ) (vs not =): T = −2.00 P = 0.063 DF =
16
```

(a) Interpret the results of the output for the first test, without pooling the variances. How many degrees of freedom are there for this test? If you use a level of significance of 0.10 what would you conclude about the two departments?

(b) Interpret the results of the output for the second test, pooling the variances. How many degrees of freedom are there for this test? If you use a level of significance of 0.10 what would you conclude about the two departments?

(c) How does this example confirm what you learned about the effects of pooling the variances?

Doing It!

Datafile: BOSSSAL.XXX

13.37 Every year the *Wall Street Journal* has a feature on compensation of CEOs of different companies. The data include type of company, amount of salary, amount of bonus, % change from the previous year, and several other compensation forms. A sample of the data is shown in the table.

Company	1996 Salary (000)	1996 Bonus (000)	% Change from 1995	Options Gains	Other	Total Direct Compensation (000)	Present Value of Options Grants (000)
Basic materials							
Air Products	$738.10	$473.00	− 8.70%	$560.60	$390.50	$2162.10	$887.00
Alcoa	750	810	− 22	6113.60	0	7673.60	13,905.60
Alumax	800	797.1	27.3	0	670	2267.10	8,491.10
Armco	559.2	0	− 42.2	0	0	559.2	422.7
Asarco	811.7	485	− 20.5	0	539.8	1836.40	571.6

(a) Look at the 1996 salary data for cyclical and noncyclical companies. Create a plot of salaries for each group. Based on the graphs do you think the data are normally distributed? Why or why not?

(b) Perform the appropriate hypothesis test to determine whether the variances in salary for the two types of companies are equal. Based on this test, can you assume equal variances?

(c) Based on the results of your answers to parts (a) and (b) select the appropriate test procedure to test whether or not the mean salary for cyclical and noncyclical companies is the same.

(d) Perform the test at the 0.05 level of significance. What is your conclusion?

(e) Repeat the procedure you used to answer parts (a) to (d) to determine whether the mean bonus for the two types of companies is different.

(f) Look at the entire set of data. One variable that is reported is the change in compensation level from the previous year (%). Calculate the proportion of companies whose CEOs received increases in compensation. Do the same thing for the proportion who received decreases in compensation.

(g) At the 0.05 level of significance, is the proportion of those receiving positive increases greater than the proportion of those receiving decreases?

(h) Do the necessary analysis to compare the mean salaries of the technology companies to those of the industrial companies.

EXPERIMENTAL DESIGN AND ANOVA

THE AIRSPACE PROBLEM

Have you ever pulled the first tissue out of the box and had it tear? Have you ever opened a box of tissues and in trying to get one tissue out, ended up with several tissues? The problem causing both of these annoying things to happen is that there is not enough airspace in the box. This problem became an important issue for a large manufacturer of tissues when an unusually high number of complaints were registered.

Airspace is defined to be the amount of space between the top of the tissues and the top of the box. It is measured in millimeters and it should be at least 9 mm. Even if there is 9 mm of airspace when the box is manufactured, there may not be enough airspace by the time the customer opens the box. This is due to a phenomenon called "growback." As the box sits in the warehouse or on the supermarket shelf, the tissues, which were heavily compressed when they were put into the box, begin to expand or "grow back." Thus, the airspace is reduced.

The tissue company needs to estimate how much growback will occur in order to determine how much airspace should be left in the box at the time of manufacturing. A portion of the data is shown below:

Position	Time	Airspace
1	1	23
1	1	25
1	1	23
1	1	23
1	1	23

14.1   CHAPTER OBJECTIVES

Throughout most of this book we have assumed that the samples have been selected at random from the population we are studying. These types of samples are called simple random samples. They were first defined in Chapter 2. The reasons we have studied primarily simple random samples are twofold. First, very often data are collected this way. Second, data analysis is simplest when the samples are simple random samples, so it makes sense to present these samples first when learning the data analysis techniques.

However, we are now at a point where it makes sense to think about *how* the sample is selected in order to get the most information for our money. Remembering that a statistical experiment is any action whose outcomes are recordable data, it is time to learn how to set up or design that statistical experiment. In doing so we will specify how the data are to be collected. You may recall that in Chapter 2 we recognized the fact that the amount of resources we have available does constrain the size of our sample. However, the formulas for determining the sample size that we learned about in Section 9.7 do not explicitly take this into account. So, it seems that we have, in effect, ignored the fact that we have limited resources. This chapter attempts to address that issue.

By now, we know that the techniques used to analyze data must take into account the type of data being analyzed and how the data were collected. Thus, this chapter also introduces a technique to analyze data that result from a designed experiment. This technique is called Analysis of Variance, which is abbreviated ANOVA.

In particular, this chapter covers the following topics:

- Motivation for Using a Designed Experiment
- Analysis of Data From One-Way Designs
- Assumptions of ANOVA
- Analysis of Data From Blocked Designs
- Analysis of Data From Two-Way Designs
- Other Types of Experimental Designs

A simple random sample is a sample that has been selected in such a way that all members of the population have an equal chance of being selected.

14.2   MOTIVATION FOR USING A DESIGNED EXPERIMENT

To help the tissue company, you might suggest that the company take a random sample of tissue boxes and measure the amount of airspace in the box. Let's follow that idea for a minute. When should you measure the airspace? If you suggested measuring the airspace at the time that the tissue box is manufactured, then you are partway to the right answer. These are called in-process data, since they are taken at the time the box is manufactured.

Suppose there is an average of 12 mm of airspace in the tissue boxes right after they are manufactured. Is this enough to ensure that, even after growback, there is an average of 9 mm of airspace by the time the customer opens the tissue box? It looks like we need to check the airspace in the box at some future point in time, say 2 weeks after the box is made. Can we use the same boxes and measure the airspace 2 weeks later? Clearly, we cannot since the boxes need to be opened to measure the airspace right after they are manufactured. This is an example of destructive inspection. The product is, in a sense, destroyed once we open it and measure the airspace. We certainly cannot sell opened tissue boxes! We need to use another set of boxes that have been sitting on the shelf for 2 weeks.

Initially, it looks as if we could use the two population tests of Chapter 13 to compare the mean airspace of two populations: the population of boxes right after manufacturing is completed and the population of boxes that have been sitting on the shelf for 2 weeks. If these samples were selected randomly from the two populations then this would work. What might happen if we use this approach? Well, the boxes that we check 2 weeks after manufacturing might have been manufactured on a different day, under different conditions than those we check right after manufacturing. How can we tell if the differences in airspace are due to different manufacturing conditions or the "growback" during the 2-week time delay? If we decided that the boxes checked after 2 weeks should be selected at the same time as the ones checked immediately then we will have addressed this issue. We would have samples that are, in fact, comparable. They are matched in terms of the manufacturing conditions. This is an example of the simplest type of experimental design.

Remember: In dependent samples the members of one sample are identical to or matched with members in the other sample.

Actually, you have already been introduced to the idea of a designed experiment in Section 13.7. In that section, we recognized that in some situations, other characteristics will influence the variable being observed. One example we looked at was the effect on sales of a new product display. Clearly, the size of the store influences sales and to do a fair comparison we compared stores of similar size. The stores were matched on the size characteristic. These types of data were called paired data or dependent samples. The technique to analyze these types of data was a paired t test. It was presented in Chapter 13 because in that chapter we were learning how to draw conclusions about two population parameters.

If we were to just compare the average airspace in process to the average airspace 2 weeks later then we could use a t test. But how do we know that 2 weeks is the right time to check the airspace? Perhaps we should examine the airspace at several different times. This is precisely what was done at the tissue company. A total of 4 cases were sampled, one case every hour for 4 hours. One case was used to obtain in-process measurements. The other 3 cases were saved for observations taken after 24 hours, 2 weeks, and 4 weeks. Each case contained 24 cartons of tissue boxes or 120 tissue boxes.

A portion of the data is shown in Figure 14.1. The variable *Time* indicates when the measurement was taken and is coded as follows: 1 = in process, 2 = after 24 hours, 3 = after 2 weeks, and 4 = after 4 weeks. The variable *Airspace* is the measurement of the airspace measured in millimeters. Figure 14.2 shows the sample mean and sample standard deviation for each of the 4 time periods.

Time	Airspace
1	23
1	25
1	23
1	23
1	23

FIGURE 14.1   A portion of the tissue company dataset

Time	Sample Mean	Sample Standard Deviation
1	21.01	1.53
2	17.46	1.48
3	15.78	2.08
4	15.02	2.32

FIGURE 14.2   Summary statistics for the tissue company

As you can see, we are now comparing the mean airspace of four different samples to draw a conclusion about four different populations: the population of boxes at the time of manufacturing (in process), the population of boxes that have been sitting on the shelf for 24 hours, the population of boxes that have been sitting on the shelf for 2 weeks, and the population of boxes that have been sitting on the shelf for 4 weeks. The average amount of airspace seems to be decreasing the longer the tissue box sits on the shelf. But we know from our work in earlier chapters that just because the sample means are different that does not automatically mean that the population means are different. We would like to be able to tell the tissue company if the mean airspace is the same for all four of these populations. Thus, our null and alternative hypotheses are as follows:

H_0: $\mu_1 = \mu_2 = \mu_3 = \mu_4$

H_A: At least one of the population means is different from the others.

We need a technique that extends the work we did in Chapter 13 beyond two populations. If the population means test different, that is, if we reject the null hypothesis, then we would like a model to explain how the airspace decreases as the box sits on the shelf for a longer period of time. Both of these issues are addressed in the next section.

14.3 ANALYSIS OF DATA FROM ONE-WAY DESIGNS

14.3.1 One-Way Designs: The Basics

The tissue company is interested in comparing the characteristics of four populations that differ on the basis of one **factor:** the amount of time that has elapsed since the box was manufactured. In this experiment the factor of interest has four different **levels** or possible values.

> A *factor* is a variable that can be used to differentiate one group or population from another. It is a variable that may be related to he variable of interest. A *level* is one of several possible values or settings that the factor can assume.

The variable that is being studied is referred to as the **response variable.** This is just a further description of the term "variable" which you have been working with throughout this text. The descriptor word "response" in front of the word variable indicates that what you are measuring or observing may respond to the experimental conditions, that is, the setting of the factor(s).

> The *response variable* is a quantitative variable that you are measuring or observing.

Many situations involve populations that differ on the basis of one factor. In fact, all of our work in Chapter 13 involved comparing populations that were different on the basis of one factor and that factor had only two levels. In Chapter 2 we said that a qualitative variable is often used to divide a large population into two smaller groups for comparisons. A commonly used qualitative variable is gender, which clearly has two levels: male and female. Thus, using the techniques of Chapter 13, we would compare the average salary of males and females. The factor would be gender and we would be comparing average male salaries to average female salaries.

The natural extension of this situation is to handle a factor that has more than two levels. For example, a manufacturer of diapers might wish to study the differences in the amount of fluid the diaper can absorb when different filler materials are used. Here, the factor is the filler material. If the diaper company is considering six different materials for the filler, then there are six levels of this factor. The response variable is the amount of fluid the diaper can absorb. A university career office might wish to study differences in starting salaries for different majors. Here, the factor is the academic major and the response variable is starting salary. If the career office is studying engineering, business, and humanities majors, then there are three levels of this factor. A soft drink manufacturer might be interested in studying the taste test of 4 different versions of a new drink. In this case, the factor is the version of the drink and there are 4 levels. The response variable is the taste test score. A medical researcher may wish to study the amount of time it takes for five different experimental drugs to work. In this case, the factor is the drug and there are five different levels. The response variable is the time it takes for the drug to work. A manufacturer may wish to study the life of products made at 3 different plants. In this case the plant location is the factor and there are three levels. The response variable is the life of the product.

Notice that, in each of these situations, the response variable is a quantitative variable (airspace, absorbency, salary, tastiness, time to work, product life) and the factor takes on one of several possible values. These are all examples of **one-way** or **completely randomized designs.**

> An experiment has a *one-way* or *completely randomized design* if there are several different levels of one factor being studied and the objects or people being observed/measured are randomly assigned to one of the levels of the factor.

The term *one-way* refers to the fact that the groups differ with regard to the one factor being studied. The term *completely randomized* refers to the fact that individual observations are assigned to the groups in a random manner. For example, in the case of the soft drink manufacturer, people would be assigned to sample one of the 4 versions of the drink in a random manner.

 **TRY IT NOW!**

One-Way Designs
Designing a Simple Study

Select a population that you might be interested in studying and identify a quantitative variable that you might wish to analyze.

Now, specify a factor that might be of interest. This should be some characteristic that you think might influence the variable you are analyzing.

Indicate the various levels of the factor.

14.3.2 Understanding the Total Variation

When the data are the result of a one-way design and certain assumptions are met, then the proper tool to analyze the data is called **analysis of variance (ANOVA).**

> *Analysis of variance (ANOVA)* is the technique used to analyze the variation in the data to determine if more than two population means are equal.

R. A. Fisher originated the technique of ANOVA in England in connection with agricultural experiments. These experiments studied the yield of crops when the farmland was treated differently. In this case the factor was the treatment of the farmland and the response variable was the yield of the crop. The word **treatment** has remained a part of the language of designed experiments. It need not refer to treatment of the farmland, but more generally it refers to a particular setting or combination of settings of the factor(s) being studied.

> A *treatment* is a particular setting or combination of settings of the factor(s) being studied.

Your initial reaction to the name of this technique is most likely that it seems to be misnamed. After all, we are really interested in determining if the groups, which correspond to differing levels of the factor of interest, have different means. What does the variance have to do with this technique and why are we analyzing the variance and not examining the means? This is completely normal thinking at this point. Interestingly enough, it is through an analysis and breakdown of what is causing the variation that we see in the data that we reach conclusions about the group means! Let's see how this works.

Suppose we think about *all* of the data that were collected for the tissue company—that is, 120 observations per time period times the 4 time periods, or 480 observations of airspace. We are interested in deciding if the airspace changes, so we might ask the question, How different are these airspace measurements from each other? We know from our work in descriptive statistics that we measure how spread out the data values are by using a measure of spread or variation. Thus, we could calculate the variance of the entire dataset. The variance is calculated using the formula for the sample variance that we learned in Chapter 6:

Remember that the units of measure on a variance calculation are the original units squared.

$$s^2 = \frac{\sum_{i=1}^{n}(x_i - \overline{x})^2}{n-1}$$

Sample variance formula

In this case $n = 480$ and the value of $\overline{x}$, the overall average for the dataset, is found to be 17.32 mm. Since the data were collected in 4 different groups corresponding to the 4 levels of the factor time, it is convenient to label each observation with a double subscript rather than just a single subscript ranging from 1 to 480. We use the letter c for the number of different levels of the factor. In this case $c = 4$. Each observation can be written as x_{ij}, where the first subscript, i, tells you how the observation is numbered within the group and the second subscript, j, tells you what group the observation is in.

EXAMPLE 14.1   Airspace Data

Setting Up the Notation

In the case of the tissue company, x_{11} would refer to the first observation in the first group, which was the in-process data. Similarly, x_{14} would refer to the first observation in the fourth group, the data collected 4 weeks after the time of manufacturing. ■

 **TRY IT NOW!**

Airspace Data
Getting Used to the ANOVA Notation

Using the information in Figure 14.1, what is the value of x_{31}?

What notation would be used to refer to the 120th observation of the data taken 2 weeks after the time of manufacturing?

What is the range of values for the first subscript for group 1?

What is the range of values for the second subscript?

ANS: $x_{31} = 23$ MM; $x_{120,3}$, 1–120, 1–4

Carrying the benefits of this notation further, we notice that the number of observations in each group does not have to be the same. Thus, n, which is the total number of observations in the experiment, can be written as follows:

$$n = n_1 + n_2 + \cdots + n_j + \cdots + n_c$$

where each n_j is the sample size of group j and c is the number of different groups or levels of the factor.

EXAMPLE 14.2   **Airspace Data**

Sample Size

For the tissue company, there are 4 time periods or 4 levels of the factor, so $c = 4$. There are 120 observations at each time period so $n_1 = n_2 = n_3 = n_4 = 120$ and

$$n = 120 + 120 + 120 + 120 = 480 \qquad \blacksquare$$

Finally the **overall mean** is relabeled $\bar{\bar{x}}$. This is read as "x double bar." For the tissue company $\bar{\bar{x}}$ is 17.32 mm. It is based on all 480 of the observations and therefore is sometimes called the **grand mean.**

> The *grand mean* or the *overall mean* is the sample average of all the observations in the experiment. It is labeled $\bar{\bar{x}}$.

We can now rewrite the variance calculation as follows:

$$s^2 = \frac{\sum_{j=1}^{c} \sum_{i=1}^{n_i} (x_{ij} - \bar{\bar{x}})^2}{n - 1}$$

Formula for sample variance with double subscript notation

This formula yields the same result as the formula for the variance we learned in Chapter 6. The numerator is simply rewritten to reflect the fact that each observation belongs to some level of the factor.

The numerator of this formula is called the **total variation** or **Sum of Squares Total (SST).**

> The *total variation* or *sum of squares total (SST)* is a measure of the variability in the entire data set considered as a whole.

SST is calculated as follows:

$$\sum_{j=1}^{c} \sum_{i=1}^{n_i} (x_{ij} - \bar{\bar{x}})^2$$

Formula for sum of squares total (SST)

The technique of ANOVA breaks down this total variation into several components and examines the contribution each component makes to the total variation. So instead of analyzing the sample variance, we will actually be analyzing the total variation, which is the numerator of the sample variance.

You could calculate SST by hand, but it can be a long and tedious calculation, even for small datasets. The next example shows you the first few terms of the calculation of SST for the tissue company.

EXAMPLE 14.3   **Airspace Data**

Calculation of SST for the Tissue Company

Using the information in Figure 14.1 and the fact that $\bar{\bar{x}} = 17.32$ mm, we can set up the first five terms of the calculation of SST for the tissue company:

Time	Airspace, x_{ij}	$(x_{ij} - \bar{\bar{x}})^2$
1	23	$(23 - 17.32)^2 = 32.3$
1	25	$(25 - 17.32)^2 = 59.0$
1	23	$(23 - 17.32)^2 = 32.3$
1	23	$(23 - 17.32)^2 = 32.3$
1	23	$(23 - 17.32)^2 = 32.3$

The calculation shown in column 3 of the table would be done for all 480 observations and then totaled. This would give you a value of SST = 4249.87. ∎

Any statistical package that you use will calculate SST for you and produce what is known as an ANOVA table. The ANOVA tables produced by Excel and MINITAB are shown in the next example.

EXAMPLE 14.4   Airspace Data

Output from Excel and Minitab

ANOVA table from Excel and Minitab for the tissue company.

ANOVA

Source of Variation	SS	df	MS	F	P-value	F crit
Between Groups	2554.75	3	851.58333	239.13025	1.342E-94	2.6236364
Within Groups	1695.116667	476	3.5611695			
Total	4249.866667	479				

Excel output: SST is shown in the line labeled Total under the column Source of Variation.

```
          Analysis of Variance for Airspace

Source    DF      SS        MS        F        P
Time       3    2554.75   851.58    239.13   0.000
Error    476    1695.12     3.56
Total    479    4249.87
```

Minitab output: SST is shown in the line labeled Total under the column heading Source. ∎

Other values are printed in the ANOVA table besides SST. The next section explains what the other numbers mean and how to use them.

14.3.3 Components of Total Variation

If you look at the ANOVA tables shown in Example 14.4 you will see that there is a column labeled SS in both the Excel and the Minitab outputs. The SS label stands for sum of squares. So far we have seen what the SST or total sum of squares term means and how it is calculated. A small amount of detective work leads you to believe that the two numbers above SST add up to SST. This is exactly correct. The technique of ANOVA focuses on the relative sizes of these components of the total variation.

To start, we will use the Excel output terminology and then relate the discussion to the Minitab output. Looking at the words in the first column of the ANOVA table,

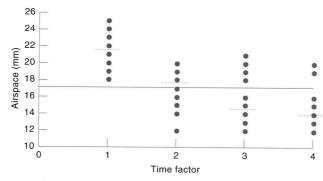

FIGURE 14.3 Dotplot of airspace by time for the tissue company

we can see that one of the components of SST is called the **Between Groups Variation** and the other component is called the **Within Groups Variation.** Let's try to understand what each of these components measures.

The between groups variation is also called the **Sum or Squares Between** or the **Sum of Squares Among** and it measures how much of the total variation comes from actual differences in the treatments. The dotplot shown in Figure 14.3 displays the sample average for each of the four time treatments. These are called **treatment means.**

> A ***treatment mean*** is the average of the response variable for a particular treatment.

By examining the graph we can see that the treatment means are "pretty different" from the grand mean of 17.32 mm. Thus, we might expect that a substantial portion of the total variation is due to the fact that the airspace measurements were taken at four different times. This is what the between groups variation component measures.

Note: The term "pretty different" will be quantified once we get a handle on the concept.

In computing this component, you look at differences between the treatment means ($\overline{x}_j$, shown as dotted lines) and the overall mean ($\overline{\overline{x}}$, shown as a solid line). Those differences are weighted by the sample size of each group (n_j).

> ***Between Groups Variation*** measures how different the individual treatment means are from the overall grand mean. It is often called the ***sum of squares between*** or the ***sum of squares among (SSA).***

The formula for sum of squares among (SSA) is

$$SSA = \Sigma n_j(\overline{x}_j - \overline{\overline{x}})^2$$

Formula for sum of squares among (SSA)

Let's verify the calculation of SSA for the tissue company.

EXAMPLE 14.5 **Airspace Data**

Calculation of SSA

The calculation of SSA for the tissue company is shown in the following table:

Column A Time	Column B Sample Mean	Column C (Sample Mean − Grand Mean)^2	Column D Column C × Group Sample Size
1	21.01	13.63	1635.41
2	17.46	0.02	2.41
3	15.78	2.35	282.13
4	15.02	5.29	634.80
		SSA	2554.75

The calculation of SSA is not as long and tedious as that of SST because there are only c terms, but SSA is always part of an ANOVA table regardless of what statistical package you are using. There is really little value in calculating it by hand. Since you are learning the technique of ANOVA for the first time, it may help you to understand what the number SSA measures. For this reason it is time to try your hand at a small example.

 **TRY IT NOW!**

Career Office
Calculation of SSB

The career office is interested in studying starting salaries for 3 different majors: Engineering, Business, and Humanities. The overall average starting salary was $\bar{\bar{x}} = \$28,200$. There were 30 students in each group and the averages are shown below. Find SSA.

Column A Major	Column B Sample Mean	Column C (Sample Mean − Grand Mean)^2	Column D Column C × Group Sample Size
Engineering	$38,100		
Business	$25,600		
Humanities	$20,900		
Total			

Notice that the between group variation is labeled with the word "time" in the Minitab output under the column Source (of variation). This is because the label "time" was used when the data were entered into Minitab to indicate that time was the factor being studied.

The second component of the total variation is labeled "within groups variation" in the Excel output and it is labeled "error" in the Minitab output. These are the most common labels used to identify this component of variation. This component is also called experimental error, which explains why the word "error" is used in the Minitab output. Once again, this is not an error indicating that you have made a mistake but rather an indication of the fact that you are studying a sample.

If you take another look at the dotplot shown in Figure 14.3 you can see the within groups variation there as well. Concentrate for the moment on time period 1 or one group. Clearly, the depth of airspace in the boxes checked at the time of manufacturing differ from each other and from the treatment mean for that time period ($\bar{x}_1 = 21.01$ mm). It is this variation that is captured in the within groups component of the total variation. The difference between each observation and the mean of the group it is in is squared and these squared differences are accumulated into the **Sum of Squares Within,** which is also called the **Sum of Squares Error (SSE).**

> Within groups variation measures the variability in the measurements within the groups. It is often called sum of squares within or *sum of squares error (SSE).*

The sum of squares error is calculated as follows:

Formula for sum of squares error (SSE)

$$\textbf{SSE} = \sum_{j=1}^{c} \sum_{i=1}^{n_i} (x_{ij} - \bar{x}_j)^2$$

Let's set up the first few terms for the calculation of SSE for the tissue company. To show the detailed calculation of SSE for the tissue company we would have to show all 120 squared differences. It is not particularly helpful to do this.

EXAMPLE 14.6   Airspace Data

First 5 Terms of SSE

Using the information in Figure 14.1 and the fact that $\bar{x}_1 = 21.01$ mm, we can set up the first five terms of the calculation of SSE for the tissue company:

$$SSE = (23 - 21.01)^2 + (25 - 21.01)^2 + (23 - 21.01)^2 + (23 - 21.01)^2 + (23 - 21.01)^2 + \cdots$$

We know from the computer output that SSE = 1695.12. ∎

Since SSA and SSE are the only two components of SST and you already know two of these numbers you can find SSE by simple subtraction. Thus,

$$SSE = SST - SSA$$

This formula shows us that SSE measures whatever variation is not a result of the fact that the data were collected in were several different groups.

EXAMPLE 14.7   Airspace Data

Finding SSE by Subtraction

We can find SSE for the airspace data by subtraction:

$$SSE = SST - SSA = 4249.87 - 2554.75 = 1695.12$$ ∎

Let's summarize what we know so far about the airspace dataset. From the dotplot and the relative size of SSA compared to SSE, it seems that much of the total variation in the data is due to the time factor being studied. To draw the inference that the means of the four populations are in fact different, we must perform a hypothesis test using the information in the ANOVA table. This is discussed in the next section.

Before moving on, one more example is presented. This example was first introduced in Chapter 12. The same scenario was reconsidered in Section 13.4 when you learned how to compare the mean of 2 populations. However, there are really 3 populations in this situation. Now you have the right tool to do the complete analysis.

EXAMPLE 14.8   Tissue Strengths

Finding SST, SSA, and SSE

You might remember from Chapter 12 that customer complaints about the tissues tearing led the company to look at the strength of the tissues. Two tissue strength measurements were studied: machine-directional strength *(MDStrength)* and the cross-directional strength *(CDStrength)*. They are both measured in lb/ream. We will look at only the *MDStrength* in this example. The data were collected over 3 days and the company would like to know if the average *MDStrength* is the same for all 3 days.

In the complete dataset there are 75 observations per day. In this example we will look at only five observations per day so you can see the computational details. At the end of the chapter you will have the opportunity to do a complete analysis of this dataset.

The five MD strengths for each of the 3 days are shown below.

	Day 1	**Day 2**	**Day 3**
	1006	951	993
	994	994	1093
	1032	1017	939
	875	965	966
	1043	966	992
Averages	990	978.6	996.6
Grand mean	988.4		

A scatter plot of the data was created in Excel and is shown below:

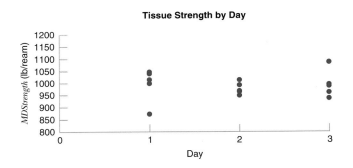

There are $c = 3$ levels of the factor "day" and $n_1 = n_2 = n_3 = 5$.

The treatment means have been calculated as well as the grand mean and are shown in the table: $\bar{x}_1 = 990$ lb/ream, $\bar{x}_2 = 978.6$ lb/ream, $\bar{x}_3 = 996.6$ lb/ream, and $\bar{\bar{x}} = 988.4$ lb/ream.

Recall that the formula for the sum of squares total is

$$\text{SST} = \sum_{j=1}^{c} \sum_{i=1}^{n_j} (x_{ij} - \bar{\bar{x}})^2$$

The Excel spreadsheet on page 583 displays the 15 squared differences, which added together give you SST.

Day	MDStrength	(MDStrength − Grand Mean)	(MDStrength − Grand Mean)²
1	1006	17.6	309.76
1	994	5.6	31.36
1	1032	43.6	1,900.96
1	875	− 113.4	12,859.56
1	1043	54.6	2,981.16
2	951	− 37.4	1,398.76
2	994	5.6	31.36
2	1017	28.6	817.96
2	965	− 23.4	547.56
2	966	− 22.4	501.76
3	993	4.6	21.16
3	1093	104.6	10,941.16
3	939	− 49.4	2,440.36
3	966	− 22.4	501.76
3	992	3.6	12.96
		SST	**35,297.6**

Recalling that the formula for SSA or sum of squares among is

$$SSA = \Sigma n_j (\bar{x}_j - \bar{\bar{x}})^2$$

and $\bar{x}_1 = 990$ lb/ream, $\bar{x}_2 = 978.6$ lb/ream, $\bar{x}_3 = 996.6$ lb/ream, and $\bar{\bar{x}} = 988.4$ lb/ream, we can easily compute this component as follows:

$$SSA = (5)(990 - 988.4)^2 + (5)(978.6 - 988.4)^2 + (5)(996.6 - 988.4)^2 = 829.2$$

Finally, the sum of squares error or SSE can be found by subtracting SSA from SST:

$$SSE = 35,297.6 - 829.2 = 34,468.4$$

Alternatively, SSE can be calculated using the formula

$$SSE = \sum_{j=1}^{c} \sum_{i=1}^{n_i} (x_{ij} - \bar{x}_j)^2$$

The details of the calculation using the formula for SSE are shown below in an Excel spreadsheet.

If you are using Excel to generate the ANOVA table you must have the data organized in columns as displayed at the beginning of this example.

	Day 1	SSE calculation	Day 2	SSE calculation	Day 3	SSE calculation	
	1006	256	951	761.76	993	12.96	
	994	16	994	237.16	1093	9292.96	
	1032	1764	1017	1474.56	939	3317.76	
	875	13225	965	184.96	966	936.36	
	1043	2809	966	158.76	992	21.16	
Averages	990		978.6		996.6		
Sum of SSE contribution		18070		2817.2		13581.2	34468.4 SSE

All of our computations are confirmed in the following ANOVA table.

ANOVA

Source of Variation	SS	df	MS	F	P value	F crit
Between groups	829.2	2	414.6	0.1443409	0.867073	3.8852903
Within groups	34,468.4	12	2872.3667			
Total	35,297.6	14				

■

Clearly, you want to use a statistics package or a spreadsheet package that has some statistical tools in order to find SST, SSA, and SSE. Even this small example ($n = 15$) took two pages to perform by hand.

The next sections explain what the other numbers in the ANOVA table mean and how to do the hypothesis test to determine if any of the population means are different from the others.

14.3.4 The Mean Square Terms in the ANOVA Table

Remember that the technique of ANOVA analyzes the variances to make an inference about the equality of the population means. What we have found in the previous sections is the numerator of three variances. To turn these sums of squares into variances we must divide by a number that is typically one less than the number of observations in the sample. This number is called the degrees of freedom. It is typically labeled "df" or "DF." You have already encountered this term in conjunction with the t distribution, the F distribution, and the chi-square distribution.

The most obvious formula is the degrees of freedom associated with SST. Since this term measures the total variability in the dataset and uses all n observations, it has $n - 1$ degrees of freedom associated with it. Following this line of thinking, we know that there are c levels of the factor being compared and therefore c terms are added together to find SSA. Thus, SSA has $c - 1$ degrees of freedom. This leaves us with $n - c$ degrees of freedom for SSE. There are two ways to think about why this is correct. First, just like the SS column, the degrees of freedom column must add up correctly. If the column labeled degrees of freedom must total $n - 1$ and we have used $c - 1$ degrees of freedom for SSA, then that leaves us with $(n - 1) - (c - 1)$ or $n - c$ degrees of freedom for SSE. Another way to think about this is to realize that each of the c levels contributes $n_j - 1$ degrees of freedom to SSE. Summing these, we get the right number for the degrees of freedom for SSE:

$$\sum_{j=1}^{c} (n_j - 1) = n - c$$

If we divide each of the sum of square (SS) terms by the appropriate degrees of freedom we will have three variances or **mean square terms.**

> The *mean square among* is labeled MSA. The *mean square error* is labeled MSE and the *mean square total* is labeled *MST*.

The formulas for the mean squares are

Mean square formulas

$$\text{MSA} = \frac{\text{SSA}}{c - 1} \qquad \text{MSE} = \frac{\text{SSE}}{n - c} \qquad \text{MST} = \frac{\text{SST}}{n - 1}$$

By looking at the output from Excel and Minitab you can see that these values are found in the column labeled MS. Notice that typically only MSA and MSE are printed.

EXAMPLE 14.9   Airspace Data

Calculating MSA and MSE

For the airspace data the mean square calculations are

$$MSA = \frac{2554.75}{3} = 851.58$$

$$MSE = \frac{1695.12}{476} = 3.56$$

∎

TRY IT NOW!

Tissue Strengths

Finding MSA and MSE

Find MSA and MSE for the tissue strength data shown in Example 14.8. You should do the calculations using the formula and then find those values in the computer output.

14.3.5 Testing the Hypothesis of Equal Means

So far we have examined the first 4 columns of the ANOVA table. Typically, there are 2 more columns in the ANOVA table. These are labeled *F* and *P* or *p* value. Excel provides one additional column labeled *F* crit, which stands for the critical value of the *F* distribution. These columns are based on the mean square column and allow us to do the hypothesis test that we set out to do, that is, test to see if at least one of the population means is different.

In general, the null and alternative hypotheses for a one-way designed experiment are shown below:

H_0: $\mu_1 = \mu_2 = \mu_3 = \cdots = \mu_c$

H_A: At least one of the population means is different from the others.

EXAMPLE 14.10   Airspace Data

Null and Alternative Hypotheses

For the airspace data there are $c = 4$ levels of the factor, so the null and alternative hypotheses are as follows:

H_0: $\mu_1 = \mu_2 = \mu_3 = \mu_4$

H_A: At least one of the population means is different from the others.   ∎

Think for a moment about what it means if the null hypothesis is true. In the case of the airspace data, if the null hypothesis is true then the amount of time the box sits on the shelf does not affect the average airspace in the tissue box. In this case we say that there is no *treatment effect*.

ANS. MSA = 414.6
MSE = 2872.3667

If there is no treatment effect, then we conclude that the factor does not affect the variable being studied. Even in this case, we would not expect every observation in the dataset to be identical because there is some natural variation inherent in the process. Remember that SSE measures the variation within the groups and so MSE is certainly an estimate of this unknown population variability, σ^2.

If the null hypothesis is true and there is no treatment effect, then not only is MSE an estimate of σ^2, but all three of the mean square values, MSA, MSE, and MST, are estimates of the natural variability in the data. In this case we have 3 estimates of the same parameter, σ^2. We know from our work in Section 13.9 how to test for equality of variances of two normally distributed populations. To do this we used an F test. We use this same F test here to see if the MSA and MSE are actually two different estimates of the same parameter. If MSA and MSE test equal, then we will conclude that there is no treatment effect and the population means are equal. We will fail to reject the null hypothesis.

The formula for the F test statistic from Section 13.9 is calculated by taking the ratio of the two sample variances: $F = s_1^2 / s_2^2$. In ANOVA, MSA and MSE are our two sample variances. So the F statistic is calculated as

Formula for F test statistic for ANOVA

$$F = \frac{\text{MSA}}{\text{MSE}}$$

The value for F critical is obtained from the F table and the procedure for reading this table was explained in Section 13.9.

The hypothesis test is easily done by determining if the F value is in the rejection region. Excel provides the critical F value as part of its ANOVA output. It is labeled F crit. If the F test statistic is "too large," that is, larger than the cutoff value shown in F critical, then you conclude that MSA and MSE are not estimates of the same number and there is a treatment effect. If the software you are using does not provide you with the critical F value, you can look it up in the F table in the Appendix. The F statistic has $c - 1$ degrees of freedom in the numerator (from the MSA term) and $n - c$ degrees of freedom in the denominator (from the MSE term).

An easier way to perform the hypothesis test is to use the p value. Remember that we used p values to do all of the hypotheses tests in this book. Both Minitab and Excel provide you with a p value to use. Most statistical packages will typically output both the F test statistic and the corresponding p value as part of the ANOVA table.

Now we can complete the airspace data analysis.

EXAMPLE 14.11 Airspace Data

The Hypothesis Test

The F test statistic for the airspace data is

$$F = \frac{\text{MSA}}{\text{MSE}} = \frac{851.58}{3.56} = 239.13$$

Looking at the ANOVA table from Excel or Minitab shown in Example 14.4, we find this F value printed under the label F. This means that the ratio of the two mean squares is about 240:1, or MSA is 240 times larger than MSE. Clearly, the ratio is not even close to 1 and these two values are not likely to be estimates of the same number.

The p value of 0.000 shown in the Minitab output and the p value of 1.342×10^{-94} (equivalent to a decimal point followed by 93 zeros and then 1342) confirms this line of thinking and tells us to reject the null hypothesis.

Thus, we conclude that at least one of the population means is different from the others and there is a treatment effect—the time the box sits on the shelf does affect the amount of airspace in the tissue box.   ■

That's it! We have looked at all of the numbers in the ANOVA table and learned what they measure, how to calculate them, and how to use them to draw an inference about the population means. You should have noticed that the technique of ANOVA ties together a number of concepts and techniques that have been developed in earlier chapters. For this reason, the technique of ANOVA is in a sense a "capstone" technique for this book. It is a very commonly used tool and it is an interesting technique since it analyzes the variability in the data to see if there are any differences in the population means. There are some assumptions that we have glossed over up to this point and it is now time to take a look at these assumptions. People tend to use ANOVA without checking to see if it is actually appropriate, which could lead to erroneous conclusions. Section 14.4 discusses the assumptions necessary to use ANOVA.

Before moving on you should complete the analysis for the tissue strength data that was started in Example 14.8.

 ***TRY IT NOW!***

Tissue Strength Data
Completing the Analysis

Set up the null and the alternative hypotheses for the tissue strength example.

Using the ANOVA table shown in Example 14.8, find the *F* statistic and decide if you should reject the null hypothesis or fail to reject it.

What do you conclude about the variable *MDStrength* and what is your recommendation to the company?

14.3.6 A Summary of One-Way Designs

In this section you have learned how to examine a set of data that has resulted from a one-way designed experiment. There is a variable of interest and you wish to know if a particular factor affects the average level of this variable. So, you have one factor, which can be "set" to various values or levels. The variable of interest is repeatedly observed or measured with the factor set at each of the levels.

Once you have the data you quickly see that not all of the values of the variable are the same! No surprise here. The technique of ANOVA allows you to decide what is causing the variation that you see. Is the variation largely due to the fact that you

ANS: $H_0: \mu_1 = \mu_2 = \mu_3$ H_A: AT LEAST ONE MEAN IS DIFFERENT.
$F = 0.87$; FAIL TO REJECT H_0. TELL MANAGEMENT THAT THE *MDStrength* IS THE SAME FOR ALL 3 DAYS.

have treated the observations differently by setting the factor at different levels? If so, you will conclude there is a treatment effect. Or is the variation largely due to natural or inherent variation in the variable being studied? If so, you will conclude that there is no treatment effect. You break the total variability in the data into two components and then examine the relative sizes of these two estimates of variability.

All of the computations necessary to carry out a one-way ANOVA are summarized in the ANOVA table shown below:

Source of Variation	SS	df	MS	F	p Value
Between groups	SSA	$c - 1$	$MSA = \dfrac{SSA}{c - 1}$	$F = \dfrac{MSA}{MSE}$	
Within groups	SSE	$n - c$	$MSE = \dfrac{SSE}{n - c}$		
Total	SST	$n - 1$			

You can calculate each of the values in the ANOVA table by hand, but it is much more likely that you will be using a statistical software package or a spreadsheet package that has some data analysis tools built-in. The calculations are done from left to right across the columns of the table with the p value being the final value to be calculated. Based on the p value you make a decision to reject or fail to reject the null hypothesis. If you fail to reject the null hypothesis, then you conclude that the factor you studied does *not* affect the average value of the variable. If you reject the null hypothesis, then you conclude that there is a treatment effect and at least one of the population means is different from the others. In this case, although you cannot tell at this point which population mean(s) are different, you do know that the factor you have investigated is important to the response variable. This is one of the major focuses of the technique of ANOVA. It tells you whether anything interesting is going on with respect to this factor. It tells you whether you need to further investigate this factor or if it can simply be ignored because it does not affect the average level of the variable being studied.

14.3.7 Building the Model for the Response Variable

Remember when we started looking at the airspace data at the beginning of the chapter we wanted to be able to tell the tissue company if the mean airspace differed when the box was left on the shelf different amounts of time. We said that if the population means were different, then we wanted to be able to tell the tissue company how the airspace changes as the box sits on the shelf. Now that we know there is a treatment effect let's consider how to estimate the airspace.

Just as the overall variability in the data can be split into component parts so can a single observation be divided into its component parts. Each observation can be thought of as follows:

$$\text{Response} = \text{Grand mean} + \text{Treatment effect} + \text{Error}$$

Notice that if you concluded that there was no treatment effect (by failing to reject H_0), then each response would be equal to the overall mean plus some random variation called error.

We must use the data to estimate the grand mean and the treatment effect. As you will see in Section 14.4 on the assumptions of ANOVA, the error term is assumed to have an average of zero. The grand mean can be estimated by $\bar{\bar{x}}$ and the treatment effect is the adjustment that must be made to the grand mean to predict a response. The treatment effect can be estimated by taking the difference between the treatment mean and the overall mean, $\bar{x}_j - \bar{\bar{x}}$. Thus, the estimate of the response is

$$\text{Response} = \bar{\bar{x}} + (\bar{x}_j - \bar{\bar{x}}) + \text{Error} = \bar{x}_j + \text{Error}$$

or just the treatment mean + error.

EXAMPLE 14.12   Airspace Data

A Prediction Model

For the airspace data consider $x_{11} = 23$ mm. Writing this in terms of its components gives us

$$\text{Response} = \text{Grand mean} + \text{Treatment effect} + \text{Error}$$
$$23 = 17.32 + (21.01 - 17.32) + \text{error}$$
$$23 = 17.32 + 3.69 + 1.99$$

Our model would predict an average of 21.01 mm of airspace in process. In this case we would be in error by 1.99 mm. ∎

 TRY IT NOW!

Tissue Strength Data
Prediction Model

For the tissue strength data, write x_{11} in terms of its components.

Using the first treatment mean as your prediction, what is your error for this particular observation?

14.3.8 The Next Step—Multiple Comparisons

If you have rejected the null hypothesis of equal population means, as we have in the case of the airspace data, you can say that there is sufficient evidence in the data to state that not all the population means are the same. Clearly, this is only the first step. To decide what type of experiment to run next to further investigate the effect this factor has on the variable of interest, we should try to learn a little bit more from this dataset.

Let's see what tools we have already learned that might help us with our detective work. We started off by simply looking at the sample mean for each of the levels

Remember that your goal is to understand how the airspace changes as the tissue box sits on the shelf.

ANS. 1006 = 988.4 + (990 − 988.4) + 16

ERROR = 16 LB/REAM

```
Analysis of Variance

Source     DF          SS        MS        F         P
Time        3     2554.75    851.58   239.13     0.000
Error     476     1695.12      3.56
Total     479     4249.87

                                         Individual 95% CIs For Mean
                                         Based on Pooled StDev
Level       N        Mean     StDev    -------+---------+---------+---------
  1       120      21.008     1.526                                  (-*-)
  2       120      17.458     1.483                          (*-)
  3       120      15.783     2.083              (-*-)
  4       120      15.017     2.319          (-*-)
                                         -------+---------+---------+---------
Pooled StDev =     1.887                    16.0      18.0      20.0
```

FIGURE 14.4   Confidence intervals for airspace data

of the factor. But we know from our work in Chapter 9 that the sample mean is a point estimate and is never likely to actually "hit" the population mean right on the money. To get an idea of how far off our point estimate was likely to be, we constructed confidence intervals for the population mean. It seems that this might be a useful tool at this point. In fact, most statistical software packages calculate the individual sample means and the corresponding confidence intervals when you run ANOVA. This portion of the Minitab output for the airspace data is shown in Figure 14.4.

By this point in the text, you have seen that you can learn a great deal by examining graphs and noticing patterns. As you examine these confidence intervals you can see that some of them overlap and some do not. For example, the confidence interval for the mean airspace in process (population or level 1) seems particularly different from the confidence interval for the mean airspace after 4 weeks (level 4). However, the confidence intervals for the mean airspace for levels 3 (2 weeks) and 4 (4 weeks) overlap a bit. Although this is not a formal statistical test, you can certainly intelligently speculate about whether there is one population mean that is causing the F value in ANOVA to be large (leading to a rejection of the null hypothesis) or whether all of the means are different from each other. In the case of the airspace data it seems that there is a great difference between the in-process data and the 24-hour data. Perhaps it would be a good idea to look at the differences in the sample means. Table 14.1 shows the amount of growback (in mm) in the tissues from the time of manufacturing until 4 weeks after manufacturing. These values are easily found by subtracting pairs of sample means. For example, the difference between $\bar{x}_1$ and $\bar{x}_2$ is $(21.01 - 17.46 \text{ mm}) = 3.55 \text{ mm}$.

This table tells us that the airspace decreases 3.55 mm between the time of manufacturing (level 1) and 24 hours later (level 2), it decreases an additional 1.68 mm between the 24-hour mark and 2 weeks later, and it decreases only an additional 0.76 mm between 2 weeks and 4 weeks. In total, the tissues "grow back" about 6 mm from the time of manufacturing until the customer opens the box 4 weeks later. Clearly, our idea that the most of the growback occurs during the first 24 hours after manufacturing seems to be correct. In fact, the tissues are greatly compressed at the time

TABLE 14.1   Airspace Data

Level	1	2	3
1			
2	3.55		
3	5.23	1.68	
4	5.99	2.44	0.76

of manufacturing to get them into the box, so perhaps focusing on the first 24 hours is a good approach for your next experiment.

This analysis also points out how much easier it is to do data analysis when you are actually involved in the manufacturing process or the situation that generated the data. If you understand the situation you are much more likely to be able to make sense of the data and make relevant recommendations.

The intuitive, visual approach we have taken here works well at pointing you in the right direction and will not differ much from the conclusions you could draw from formal techniques if the sample sizes are equal, as they were in the airspace data. In fact, the most common formal statistical test done at this point uses the difference between two of the sample means as the center of a confidence interval, which estimates the difference in the corresponding population means. Several techniques could be used to do this. Alternatively, a hypothesis test of the difference in the two population means can be done. It might seem that a t test or a paired t test from Chapter 13 would work here. It is natural to think so since we now wish to compare two population means. Although the setup of the null and alternative hypotheses is the same here, unfortunately we cannot use either a t test or a paired t test to do the test. The reason for this has to do with the fact that you are making multiple pairwise comparisons from a single dataset and in addition to that you are doing the comparison after the data have been analyzed.

There are several techniques that could be used and most statistical software packages will have several options for doing the pairwise comparisons. For example, if you are using Minitab then you can use any one of four techniques to do the pairwise comparisons. Sometimes these techniques lead to conflicting conclusions. Many issues need to be addressed to properly explain these techniques and they will not be addressed here. However, by examining the confidence intervals for the mean of each level of the factor, looking at how they overlap and how much the sample means change as you change levels of the factor, you will detect a good deal about how the factor influences the variable you are studying. This will help guide the design of your next experiment.

14.3.9 Exercises—Learning It!

14.1 A diaper company is considering 3 different filler materials for their disposable diapers. Eight diapers were tested with each of the 3 filler materials and 24 toddlers were randomly given a diaper to wear. As the child played, fluid was injected into the diaper every 10 minutes until the product failed (leaked). The amount of fluid (in grams) at the time of failure was recorded for each diaper. The data are shown below:

Material 1	Material 2	Material 3
791	809	828
789	818	814
796	803	855
802	781	844
810	813	847
790	808	848
800	805	836
790	811	873

(a) What is the response variable and what is the factor?

(b) How many levels of the factor are being studied?

(c) Is there any difference in the average amount of fluid the diaper can hold using the 3 different filler materials? If so, which ones are different?

(d) What is your recommendation to the company and why?

14.2 In Chapter 6 you looked at a large company that sells software. The company had received complaints that the disks provided by the company failed to work properly after *Datafile:* *DISKFAIL.XXX* extended use. The company has decided to investigate the complaint. Data are collected on the time to failure for disks from their current supplier and two alternative suppliers. The data are shown below:

Current Supplier	Alternative 1 Supplier	Alternative 2 Supplier
486	489	508
490	489	510
491	491	517
491	492	506
494	492	515
494	492	520
496	492	503
498	493	524
498	493	515
502	494	503
504	495	507
505	496	515
506	497	509
507	497	518
508	497	495
510	498	499
514	499	510
515	502	503
527	503	533
498	505	517

(a) What is the response variable and what is the factor?

(b) How many levels of the factor are being studied?

(c) Is there any difference in the average time to failure of the disks from the 3 different suppliers? If so, which ones are different?

(d) What is your recommendation to the company and why?

Datafile:
HOMEWORK.XXX

14.3 Grading homework is a real problem. It takes an enormous amount of time and many students do not do a very good job or copy answers from other students or the back of the book. A teacher of elementary statistics decided to conduct a study to determine what effect grading homework had on her students' exam scores. She taught 3 sections of Elementary Statistics and randomly assigned each class one of three conditions: (1) no homework given, (2) homework given but not collected, and (3) homework given, collected, and graded. After the first exam, she collected the data (exam scores). They are shown below:

No Homework	Homework, Not Collected	Homework, Collected
69	73	83
69	63	97
92	68	72
84	79	79
79	57	84
84	68	76
76	72	91
63	74	76
76	49	83
82	84	88
89	79	91
72	71	96
72	80	68
65	74	99
73	71	89
47	63	80

(*continued*)

No Homework	Homework, Not Collected	Homework, Collected
92	88	79
71	83	91
83	89	83
81	82	83
92	69	76
80	92	90
64	79	79
72	81	67
84	76	86
79	81	86
74	81	82
81	75	84

(a) What is the response variable and what is the factor?

(b) How many levels of the factor are being studied?

(c) Is there any difference in the average exam scores using the 3 different approaches to homework? If so, which ones are different?

(d) What is your recommendation to the teacher and why?

14.4 Two tissue strength measurements were studied: machine-directional strength *(MD-Strength)* and the cross-directional strength *(CDStrength)*. They are both measured in lb/ream. We will look at only the *CDStrength* in this exercise. The data were collected over 3 days and the company would like to know if the average *CDStrength* is the same for all 3 days.

In the complete dataset there are 75 observations per day. In this example we will look at only five observations per day. At the end of the chapter you will have the opportunity to do a complete analysis of this dataset.

The five CD strengths for each of the 3 days are shown below:

Day	CDStrength	Day	CDStrength	Day	CDStrength
1	422	2	436	3	473
1	448	2	468	3	440
1	423	2	419	3	441
1	435	2	458	3	443
1	445	2	459	3	400

(a) Display the data as a dotplot.

(b) Find the grand mean and the treatment means.

(c) Mark the grand mean and the treatment means on the dotplot.

(d) Examine the first observation for day 1: 422 lb/ream. Find the difference between the first treatment mean and the grand mean. This is the treatment effect for the first treatment. On the dotplot mark off this difference.

(e) Continue to examine the first observation for day 1: 422 lb/ream. Find the difference between this observation and the first treatment mean and the grand mean. This is the error for this observation. On the dotplot mark off this difference.

(f) Analyze these data using one-way ANOVA. Is there a significant treatment effect?

14.5 The sports industry is a large and competitive market. A manufacturer of golf balls is considering 4 new ball designs. A sample of 36 balls from each of the 4 models is tested and the distance the balls carry (in yards) is recorded. The balls are hit by a machine. The data are shown below.

Datafile: BALLDESN.XXX

M1 Model	M2 Model	M3 Model	M4 Model
257	256	244	250
255	255	243	255
256	258	241	251
255	257	243	249

(continued)

M1 Model	M2 Model	M3 Model	M4 Model
255	257	240	250
256	257	249	251
255	258	244	251
258	258	249	248
252	256	248	247
256	258	243	253
253	258	242	250
254	256	245	255
260	259	253	250
258	255	250	256
258	259	251	260
257	257	249	259
259	256	249	252
257	255	249	253
255	255	255	257
256	257	249	255
259	260	251	251
261	255	250	255
260	261	248	253
257	262	250	256
260	258	258	262
262	260	258	258
258	260	257	263
260	256	258	259
258	255	254	256
258	256	256	261
258	257	255	258
256	251	254	258
260	253	255	256
260	254	260	263
261	254	254	264
257	255	253	257

(a) What is the response variable and what is the factor?

(b) How many levels of the factor are being studied?

(c) Is there any difference in the average distance the ball carries?

(d) What is your recommendation to the golf manufacturer and why?

14.4 ASSUMPTIONS OF ANOVA

We have already mentioned that ANOVA is a very commonly used technique. Unfortunately, many times ANOVA is used when it is not the appropriate technique. This may be partly due to the fact that it is relatively easy to run ANOVA using a statistical package and most software tools simply do the calculations but do not check that it is the appropriate technique. This is your job as the data analyst. As with many of the tools presented in this book, the tool of ANOVA gives you valid conclusions if your data meet certain assumptions. If your data do not meet these assumptions and you use ANOVA, you could easily draw the wrong conclusions. Thus, you should always check that the three major assumptions of ANOVA are met *before* you use this technique.

The three major assumptions of ANOVA are as follows:

1. The errors are random and independent of each other.

2. Each population has a normal distribution.

3. All of the populations have the same variance.

14.4.1 Assumption About the Errors

The first assumption talks about the behavior of the errors. What are these errors? The term error refers to the difference between any observed value and the sample mean of its group. We looked at these differences when we developed the formula for SSE in Section 14.3.3. Remember that SSE is found by adding up the squared differences of the form $x_{ij} - \bar{x}_j$. Each term in the sum for SSE is considered an "error." Thus, the first assumption says that these differences should be random and the error for one observation should not influence the error for another observation. If you have randomly selected a sample from each of the populations then typically this assumption is met. The most common situation in which this assumption is violated is when you have time series data. In this case, the observation at time period t may influence the observation at time period $t + 1$ (in fact, it is precisely this dependency that you exploit when you build a time series model). If the observations are dependent, then the errors will be dependent. A violation of this assumption could seriously affect the conclusions reached using ANOVA. If you are suspicious of a bias in the sampling process or if you have time-dependent data, then you should consult a more advanced textbook.

14.4.2 Assumptions About the Underlying Populations

The second and third assumptions of ANOVA refer to the behavior of the variable being collected. The second assumption says that the variable should have a normal distribution in each of the populations (each level of the factor). You can check this assumption formally using a chi-square goodness of fit test, which is the subject of the next chapter. At this point you should visually check the shape of the data in each sample by displaying a histogram of the data.

EXAMPLE 14.13   Airspace Data

Looking at Histograms for Normality

Histograms for all 4 levels of the factor "time" for the airspace data are shown below:

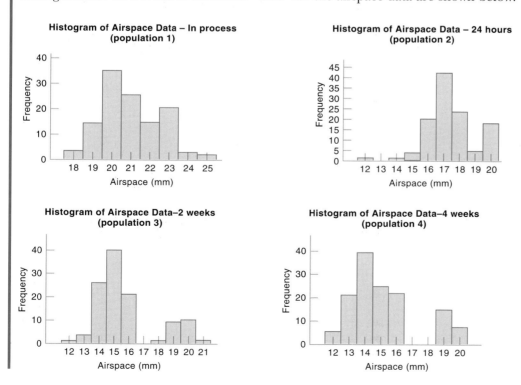

A look at these graphs tells us that we may be in violation of the assumption of normality for at least one of the populations.   ■

Remember: When the data are normally distributed, the plot will be a straight line.

Another way to test for the assumption of normality is to use a normal probability plot. You were introduced to this concept in the discussion of the assumptions of simple linear regression in Section 15.6.1. A normal probability plot for the in-process data is shown in Figure 14.5. It was generated by Minitab.

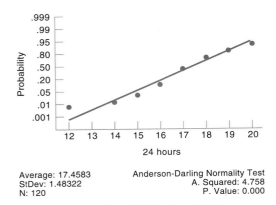

Average: 17.4583
StDev: 1.48322
N: 120

Anderson-Darling Normality Test
A. Squared: 4.758
P. Value: 0.000

FIGURE 14.5   Normal probability plot of 24-hour airspace data

Fortunately, the technique of ANOVA is not particularly sensitive to violations of the normality assumption. This means that if your data follow a distribution that is not extremely different from a normal distribution, then your conclusions from ANOVA are probably fine, especially for a large sample size.

Remember: You are using ANOVA to test equality of means.

The third assumption of ANOVA also has to do with the behavior of the distribution of the variable in each of the populations. Once you have checked to see that the second assumption has been met, or at least is not severely violated, then you must check to see that the amount of variability in each of the populations is the same. That is, the variance within each population should be equal. This is necessary in order to combine the data from the various samples to get an estimate of the inherent variability measured by SSE.

At a minimum, you should compare the size of the sample variances and do a visual check of the data by looking at a boxplot of the data to see if the spread in each sample looks about the same. Neither of these approaches constitutes a test for equal variances and, of course, there are such tests, but they will certainly identify glaring violations of this assumption.

EXAMPLE 14.14   **Airspace Data**

Checking for Equal Variances Using a Boxplot

The sample variances for the airspace data are shown in the table below:

Time	Sample Standard Deviation	Sample Variance
1	1.53	2.34
2	1.48	2.19
3	2.08	4.33
4	2.32	5.38

The boxplots of airspace by time are shown on the next page.

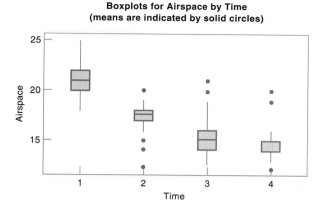

Boxplots for Airspace by Time
(means are indicated by solid circles)

The largest sample variance (5.38) is almost 2.5 times larger than the smallest sample variance (2.19). This represents a large difference in the variances given the size of this sample. These data may be in violation of the third assumption of ANOVA.   ■

There are formal statistical procedures to test the assumption of equal variances. The statistical software you are using will most likely have more than one way to test this assumption. If the data are in violation of the third assumption, you will need to use an appropriate data transformation to normalize the data and stabilize the variance. Alternatively, you could use a nonparametric technique.

14.4.3 Exercises—Learning It!

14.6 Check the ANOVA assumptions for the disposable diaper company interested in testing   *Requires Exercise 14.1*
different materials.

  (a) Is there any reason to believe that the errors are not independent?

  (b) Construct histograms of the amount of fluid in the diapers at the time of failure for each of the 3 different materials. Do they look normally distributed?

  (c) Calculate and compare the size of the sample variances and do a visual check of the data by looking at boxplots of the data to see if the spread in each sample looks about the same.

  (d) Comment on the validity of the assumptions of ANOVA for these data.

14.7 Check the ANOVA assumptions for the software company.   *Requires Exercise 14.2*

  (a) Is there any reason to believe that the errors are not independent?

  (b) Construct histograms of the time to failure for each of the 3 different suppliers. Do they look normally distributed?

  (c) Calculate and compare the size of the sample variances and do a visual check of the data by looking at boxplots of the data to see if the spread in each sample looks about the same.

  (d) Comment on the validity of the assumptions of ANOVA for these data.

14.8 Check the ANOVA assumptions for the statistics teacher investigating the effect of   *Requires Exercise 14.3*
homework on exam scores.

  (a) Is there any reason to believe that the errors are not independent?

  (b) Construct histograms of the exam scores for each of the 3 different methods. Do they look normally distributed?

  (c) Calculate and compare the size of the sample variances and do a visual check of the data by looking at boxplots of the data to see if the spread in each sample looks about the same.

  (d) Comment on the validity of the assumptions of ANOVA for these data.

14.9 Check the ANOVA assumptions for the golf ball manufacturer.   *Requires Exercise 14.5*

(a) Is there any reason to believe that the errors are not independent?

(b) Construct histograms of the distance the balls carried for the 4 different ball designs. Do they look normally distributed?

(c) Calculate and compare the size of the sample variances and do a visual check of the data by looking at boxplots of the data to see if the spread in each sample looks about the same.

(d) Comment on the validity of the assumptions of ANOVA for these data.

14.5   ANALYSIS OF DATA FROM BLOCKED DESIGNS

14.5.1 Motivation for Block Designs

In this chapter we have looked at two issues for the tissue company: the effect of time on the airspace in the box and the strength of the tissues being manufactured as measured by the variable *MDStrength*. In both of these situations there was one factor of interest and the company was able to collect a sample of *comparable* observations from each population.

Often there is one factor of interest but the observations in the sample may not be comparable due to some other factor. Consider the effect on sales of product display. One factor is being studied here: the product display. Suppose we are considering 3 different product displays; then there are 3 levels for this factor. To use a one-way ANOVA we would need to collect sales data from a sample of stores using display method 1, a sample of stores using display method 2, and a sample of stores using display method 3. This is not a problem, but the question that arises is, Are the sales from these stores different due to some reason other than the display method being used? It is likely that the answer to this question is yes. The stores are probably different in size and in different geographic locations. These factors of size and location would not matter if they did not affect the variable we were studying: sales. If they did not affect sales, then we could happily use a one-way ANOVA, randomly assigning stores to use one of the 3 product display methods. But they most likely do affect sales. We are not going to find a sample of like size stores in the same location! So we cannot get a sample of *comparable* stores to sample.

In this example, it seems as though we actually have 3 factors: product display method, size, and location of the store. But really we are interested only in the effect of the product display method on sales. The other 2 factors are in a sense "nuisance factors." They need to be taken into account, but we are not directly interested in their effect on sales at this point. The factors size and location need to be blocked out so that we may see the effect of the product display on sales. Otherwise, there is no way to tell if the differences we see in sales are a result of the differing product displays or the differing size and/or location of the store.

There are two possible ways to handle the data collection for this example. We could find 3 stores that are similar in size and location (large city, small city, rural, etc.) and randomly assign a product display method to each store. Then we would need to find another 3 stores that are similar in size and location and do the same thing. Each set of 3 stores becomes what is called a **block.** We would use as many blocks as feasible given the amount of time, the amount of money, and the availability of stores.

Alternatively, we could observe a set of 20 stores for 3 weeks. During the first week, each store would randomly be assigned to use one of the 3 product display methods, during the second week, the store would use a second product display method, and during the third week, the store would use the remaining product display method. In this case each store is a **block.**

> A **block** is a group of objects or people that have been matched. An object or person can be matched with itself, meaning that repeated observations are taken on that object or person and these observations form a block.

If the realities of data collection lead you to use blocks, then you must take this into account in your analysis. Your experimental design is called a **randomized block design.** Instead of using a one-way ANOVA you must use a block ANOVA.

> An experiment has a **randomized block design** if several different levels of one factor are being studied and the objects or people being observed/measured have been matched. Each object or person is randomly assigned to one of the c levels of the factor.

14.5.2 Partitioning the Total Variation

The analysis of data from a block design is conceptually the same as that of a one-way design. You essentially still have just one factor of interest. You are still interested in deciding if there is a treatment effect. That is, you are interested in seeing if the population means are all equal or if at least one is different. The null and alternative hypotheses remain as follows:

$$H_0: \quad \mu_1 = \mu_2 = \mu_3 = \cdots = \mu_c$$
$$H_A: \quad \text{At least one of the means is different.}$$

Like the approach we took with data from a one-way design, the idea is to take the total variability as measured by SST and break it down into its components. With a block design there is one additional component: the variability between the blocks. It is called the **sum of squares blocks** and is labeled **SSBL.**

> The **sum of squares blocks** measures the variability between the blocks. It is labeled **SSBL.**

The difference between each observation and the mean of its block is squared and these squared differences are accumulated into the sum of squares blocks. This component measures the portion of the variation that is attributable to the fact that the data were collected in blocks. Since we used blocks because we thought that there were some "interfering factors," we would expect this component to be of significant size. If it is not, then we probably worried about factors that do not, in fact, affect the variable of interest.

For a block design, the variation we see in the data is due to one of three things: the level of the factor, the block, or the error. Thus, the total variation is divided into three components:

$$SST = SSA + SSBL + SSE$$

Formula for SST in terms of other components in a block design

The terms SSA and SSE have the same formulas and definition as for a one-way ANOVA. The effect of partitioning the total variation into another component is to reduce the error term, SSE. Recall that SSE is used to find MSE, which becomes the denominator of the F test statistic. If you can reduce SSE by the use of blocks, then the MSE term will be smaller, making the F test statistic larger. Since "big" F values lead you to reject the null hypothesis of equal means, the use of blocks allows you to see differences in the treatment means that you might not have seen otherwise.

14.5.3 Using the ANOVA Table in a Block Design

Consider a randomized block design with r blocks and c groups. There are a total of $n = rc$ observations. The ANOVA table for such a block design looks just like the ANOVA table for a one-way design with an additional row. It is shown below:

Source of Variation	SS	df	MS	F	p Value
Between groups	SSA	$c - 1$	$MSA = \dfrac{SSA}{c - 1}$	$F = \dfrac{MSA}{MSE}$	
Between blocks	SSBL	$r - 1$	$MSBL = \dfrac{SSBL}{r - 1}$		
Within groups	SSE	$(r - 1)(c - 1)$	$MSE = \dfrac{SSE}{(r - 1)(c - 1)}$		
Total	**SST**	$rc - 1$			

The last 2 columns of the ANOVA table are used to do the hypothesis test of the equality of the population means. The hypothesis test is easily done by determining if the F value is in the rejection region. If the F test statistic is "too large," that is, larger than the critical F value from the table, then you conclude that MSA and MSE are not, in fact, estimates of the same number and there is a treatment effect. If the software you are using does not provide you with the critical F value, you can look it up in the F table in Appendix A. The F statistic has $(c - 1)$ degrees of freedom in the numerator (from the MSA term) and $(r - 1)(c - 1)$ degrees of freedom in the denominator (from the MSE term).

Let's see how this works by following the sales and product display example.

EXAMPLE 14.15   Product Display Location

A Block Design

A national chain of stores is investigating the effect of the product display on sales of the product. The analysts are interested in 3 product display locations: on the shelf (group 1), at the end of the aisle (group 2), and at the entrance to the store (group 3). They have matched stores with respect to size and location and have created 10 blocks of stores. The stores in the blocks were randomly assigned a product display location and the monthly sales were recorded. The data are shown below:

Block	Shelf	End of Aisle	Front of Store
1	4457	4500	5800
2	4400	4370	5290
3	4310	4300	5000
4	4600	4400	5600
5	5000	5000	6000
6	4500	4500	5300
7	4700	5100	5900
8	4590	4280	5460
9	8510	8670	10600
10	6470	6500	8410
Average	**5153.7**	**5162**	**6336**

The ANOVA table and the group sample means and confidence intervals generated by Minitab are shown below:

```
Analysis of Variance for Sales

Source     DF        SS        MS
Display     2    9253927   4626964
Block       9   60738440   6748716
Error      18    1371972     76221
Total      29   71364339
```

```
                        Individual 95% CI
Display          Mean    ------+---------+---------+---------+-----
Shelf            5154        (----*---)
End of Aisle  5162      (----*----)
Front of Store   6336                                    (---*----)
                        ------+---------+---------+---------+-----
                        5200      5600      6000      6400
```

Looking at the sample means and the confidence intervals for the individual population means, we can see that locating the product at the front of the store generates considerably more sales.

To complete the hypothesis test we need to find the F value and then use the F table to find the rejection region. Minitab does not calculate the F or p values for a block design.

The null and alternative hypotheses are shown below:

$$H_0: \quad \mu_1 = \mu_2 = \mu_3$$
$$H_A: \quad \text{At least one population mean is different.}$$

The F statistic is $MSA/MSE = 4{,}626{,}964/76{,}221 = 60.7$. The critical value for the rejection region is found from the F table with 2 $(c-1)$ and 18 $(r-1)(c-1)$ degrees of freedom. If $\alpha = 0.05$, then the critical F value is 3.55. Clearly, $F = 60.7$ is larger than the critical F value and we reject H_0.

The null hypothesis is rejected and we conclude that the location of the product display affects the average sales. Our scatter plot tells us that it is probably group 3 (front of store) that has the different mean, since the other two sample means are quite similar and the confidence intervals overlap substantially. ■

All of the assumptions of one-way ANOVA pertain to the block design. Thus, you should be sure to check these assumptions for your dataset.

 *TRY IT NOW!*

Participative Management
Block Design

In Chapter 13 you looked at the number of sick days used by employees in the past 12 months by employees who were traditionally managed and those who were involved in a participative management style. Management is now considering a third approach which is a mixture of the old structure with the team-based approach. The study involves 75 employees matched by age, type of work, and gender.

Use the Minitab output shown below to determine if there are any differences in the mean number of sick days used by the different populations.

```
Analysis of Variance for Sick Day

Source   DF       SS       MS
Style     2   57.680   28.840
Block    24  336.213   14.009
Error    48   24.987    0.521
Total    74  418.880

                      Individual 95% CI
Style   Mean    -------+---------+---------+---------+----
Part.   4.72    (----*----)
Hybrid  6.56                                (----*----)
Trad    6.60                                (----*----)
                -------+---------+---------+---------+----
                    4.80      5.40      6.00      6.60
```

14.5.4 Benefits of Blocking

The purpose of blocking is to allow you to see a treatment effect with a smaller sample size. However, it is often difficult to set up the blocks and so it is interesting to examine whether the effort to set up the blocks is worth it. The following formula gives you a measure of the relative efficiency of the block design compared to a one-way design:

Formula for relative efficiency of block design

$$\text{Relative efficiency} = \frac{(r-1)\text{MSBL} + r(c-1)\text{MSA}}{(n-1)\text{MSE}}$$

The terms in this formula are all found in the ANOVA table from a block design. The value of n is the total sample size. The relative efficiency value tells you by what factor the sample size would have to be increased to see the treatment effect without using a block design. Clearly, if the relative efficiency is a low number such as 1.5, then perhaps it is not worth the effort to set up the blocks. Let's see how this formula works for the product display location.

THE CONFIDENCE INTERVALS, IT APPEARS THAT PARTICIPATIVE GROUP IS DIFFERENT FROM THE OTHER TWO.

ANS. $F = 55.35$; $F_{crit} = 3.23$;USING 2 AND 40 DEGREES OF FREEDOM AND $\alpha = 0.05$;REJECT H_0. AT LEAST ONE GROUP IS DIFFERENT. FROM

EXAMPLE 14.16 Product Display Location

Efficiency of Block Design

The ANOVA table for the product display location exercise is shown below:

Source	DF	SS	MS
Display	2	9,253,927	4,626,964
Block	9	60,738,440	6,748,716
Error	18	1,371,972	76,221
Total	**29**	**71,364,339**	

The relative efficiency of this design is calculated as follows:

$$\text{Relative efficiency} = \frac{(10-1)6,748,716 + 10(3-1)4,626,964}{(30-1)76,221} = 69.3$$

This number tells us that we would have needed 69 times as many observations to see the treatment effect if we did not use a block design. Clearly, the blocking was worth it in this case. ■

 **TRY IT NOW!**

Participative Management

Efficiency of Block Design

Calculate the relative efficiency of the block design for the sick days data. Comment on the usefulness of the blocking.

14.5.5 Exercises—Learning It!

14.10 A manufacturer of electronic components was concerned about the reliability of one of its products. The company wished to see if the input current that the component would draw from a power source such as a battery changed over time. Twenty components were pretested at the time of manufacture, 9 days later, and 58 days later. The data (in milliamps) are shown below:

Datafile: AMPS1.XXX

Number	Pretest	9 Days	58 Days
1	26.1	27.0	27.5
2	26.2	27.4	27.8
3	26.3	27.8	28.5
4	27.4	28.1	28.3
5	27.8	29.5	29.8
6	26.8	27.9	27.8
7	28.2	28.9	29.3
8	28.2	29.3	29.3

(*continued*)

Number	Pretest	9 Days	58 Days
9	27.1	28.1	28.9
10	26.9	27.9	28.0
11	27.8	28.9	29.0
12	27.4	28.8	28.8
13	27.1	28.0	27.7
14	27.4	29.0	29.5
15	26.5	27.7	27.8
16	27.0	27.6	27.7
17	28.3	29.1	29.6
18	28.2	29.1	29.1
19	27.0	27.5	28.0
20	28.1	29.7	30.4

(a) What is the factor and how many levels are there in this design?

(b) What is the blocking factor?

(c) Is there a significant difference in the power drawn over time?

(d) Calculate the relative efficiency of the blocking. Is the blocking effective?

Datafile:
ALGORTH.XXX

14.11 Many techniques for finding the "best" or optimal solution to a problem require that you provide the computer software with a starting solution. There are often different ways to get this starting solution. Four algorithms for generating starting solutions are run on 30 different problems. The number of iterations required to move from the starting solution to the optimal solution was recorded. A portion of the data is shown below:

Problem	Method	Iteration
1	1	5
2	1	4
3	1	4
4	1	4
5	1	2
6	1	2

(a) Is there any difference in the number of iterations required to find the optimal solution that is due to the difference in the method used to generate the starting solution?

(b) Which method would you recommend and why?

14.12 No. 1 Foods contracts with colleges and universities to provide the food services for the school. It has 3 different cafeteria layouts, which it is evaluating in terms of the time (in minutes) it takes for a "typical" student to get his/her food. To do a fair comparison No. 1 Foods has blocked on the size of the school. Is there a difference in the average time spent getting food for the 3 different layouts? If so, which layout is best? The data are shown below:

	Layout 1	Layout 2	Layout 3
1	6.5	2.9	6.9
2	5.8	4.5	6.1
3	5.8	4.1	7.0
4	6.7	4.2	7.4
5	6.4	5.1	6.9
6	5.8	4.7	7.7
7	6.2	5.2	7.8
8	5.7	4.9	8.1
9	6.9	4.8	7.8
10	5.0	4.3	8.2
11	5.8	4.2	7.1
12	4.8	4.1	7.4
13	6.7	4.7	7.2
14	6.6	4.3	7.6
15	4.7	4.4	7.5

14.13 Should you listen to classical music, watch TV, or just have silence when you study? To answer this question a group of 10 undergraduate students were randomly selected from a large university. For one semester they listened to classical music while they studied, for the next semester they watched TV while they studied, and for the third semester they studied in the library. Their GPA for each semester was recorded. The data are shown below:

Student	Music	TV	Library
1	3.62	3.35	3.23
2	3.38	3.23	3.07
3	3.09	2.24	2.70
4	2.69	1.88	2.41
5	3.29	2.31	3.01
6	2.97	2.75	2.82
7	2.60	1.98	2.14
8	3.56	3.26	3.07
9	3.50	2.74	3.23
10	3.74	3.19	3.55

(a) What is the response variable?

(b) What is the factor? How many levels are there?

(c) What is the blocking factor?

(d) Use block ANOVA to determine if there are any differences in the average GPA scores for the three treatment groups.

(e) What other factors would you suggest be included in a subsequent study?

14.14 What can I expect as a starting salary if I major in Business? Engineering? Education? This is a commonly asked question. The career office at a university decided to see if there were any differences in the starting salaries of these 3 majors. Analysts selected 50 students from last year's graduating class and recorded the starting salary for each student. To do a fair comparison, the students were matched on the basis of the overall grade point average. The data are shown below:

Datafile: GRADSAL.XXX

Business	Engineering	Education
22521	25698	20577
22482	25423	20609
22822	25662	20483
22503	25512	20531
22300	25453	20530
22418	25322	20710
22574	25715	20290
22545	25792	20730
22406	25499	20215
22521	25532	20491
22249	25543	20512
22398	25509	20540
22144	25451	20570
22615	25504	20662
22399	25367	20564
22565	25335	20500
22512	25461	20385
22469	25650	20672
22532	25399	20510
22516	25618	20412
22558	25481	20680
22414	25592	20587
22608	25479	20719
22327	25442	20390
22699	25192	20324
22545	25501	20418

(*continued*)

Business	Engineering	Education
22431	25761	20593
22622	25514	20482
22467	25208	20430
22064	25740	20309
22388	25743	20452
22805	25867	20689
22600	25622	20600
22446	25386	20523
22667	25568	20647
22655	25604	20298
22500	25340	20441
22285	25601	20374
22218	25466	20601
22471	25394	20328
22176	25479	20713
22403	25489	20004
22293	25072	20548
22670	25627	20610
22531	25250	20577
22208	25412	20228
22550	25434	20527
22483	25204	20413
22501	25409	20754
22810	25396	20439

(a) What is the response variable?

(b) What is the factor? How many levels are there?

(c) What is the blocking factor?

(d) Use block ANOVA to determine if there are any differences in the average starting salary for the three groups.

(e) What other factors would you suggest be included in a subsequent study?

Datafile:
WEIGHT.XXX

14.15 Surely you have heard of the "freshman ten," referring to the average weight gain of 10 lb by students during the freshman year. Does the same phenomenon occur to graduate students? A sample of 15 students was taken and their weight (in lb) was recorded in September, February of the following year, July of the following year, and then July two years later. The data are shown below:

Person	Sept	Feb	July	July, 1 Year Later
1	110	115	115	111
2	140	144	146	141
3	165	169	171	165
4	123	128	131	124
5	158	167	169	159
6	180	187	187	181
7	173	179	179	173
8	178	185	190	180
9	195	203	207	196
10	145	154	158	145
11	136	137	137	137
12	170	179	180	170
13	166	168	171	168
14	180	184	185	181
15	130	136	138	130

A new student of statistics decided to use a one-way ANOVA for these data and the results from Minitab are shown at the top of the next page.

Analysis of Variance for Weight

Source	DF	SS	MS	F	P
Month	3	634	211	0.36	0.784
Error	56	33108	591		
Total	59	33742			

(a) Using this ANOVA table, what would you conclude about the average weights at the four times the data were collected?

(b) Explain why this analysis is incorrect and why it must be analyzed as a block design.

(c) What are the blocks?

(d) The Minitab output for the block ANOVA is shown below. What can you conclude about the average weights for the 4 time periods?

Analysis of Variance for Weight

Source	DF	SS	MS
Person	14	32990.73	2356.48
Month	3	633.87	211.29
Error	42	117.13	2.79
Total	59	33741.73	

Discovery Exercise 14.1
THE BENEFITS OF BLOCKING

In manufacturing electronic products such as loudspeakers, it is important that connections are strong and hold so that they do not disconnect. The same is true for commercial heaters. One of the customers of a manufacturer of heaters complained that the pull poundage was too low. Thus, the customer did not have confidence in the terminal connection. New wires with the terminals attached were made and a pull test was done on a sample of 25. The terminals were connected and pulled apart. The pull poundage at which the terminals would disconnect from each other was recorded. Then the same terminals were reconnected and the test was repeated a total of 6 times. The object was to determine if the pull poundage changed over time. The minimum pull poundage required by the customer was 5 lb. The data (datafile HEATERS.XXX) are shown below:

Trial 1	Trial 2	Trial 3	Trial 4	Trial 5	Trial 6
11.0	12.3	9.8	12.3	10.8	8.5
8.5	10.8	7.5	11.0	8.0	8.0
9.5	10.0	9.0	9.0	9.0	7.8
10.0	7.5	8.0	9.8	10.5	8.5
10.5	7.5	10.3	9.0	9.0	9.8
11.8	11.8	10.0	9.0	9.5	10.3
11.5	10.0	14.0	11.3	14.0	13.0
10.0	9.5	10.0	8.0	8.3	8.0
11.3	12.0	11.0	11.3	10.5	10.3
10.3	10.5	9.8	9.8	11.8	10.5
11.0	15.0	13.5	11.5	9.8	10.0
11.3	10.8	9.8	10.3	10.0	9.5
10.5	10.0	10.5	10.5	10.0	9.5
11.8	10.8	10.0	11.8	9.8	9.5
10.0	10.0	9.5	11.8	9.0	9.3
11.5	11.0	11.0	9.8	9.0	9.8
10.5	9.0	12.5	11.3	11.0	11.8

(*continued*)

Trial 1	Trial 2	Trial 3	Trial 4	Trial 5	Trial 6
11.3	11.0	9.5	9.0	10.0	8.0
11.0	11.3	11.3	10.0	11.3	9.5
11.0	8.0	10.5	7.0	7.0	12.0
11.8	10.3	12.3	11.8	12.5	11.0
12.0	12.3	11.5	12.0	12.3	11.0
11.5	12.0	11.0	10.0	11.5	11.5
12.0	9.5	12.5	9.5	9.0	9.3
10.0	7.5	10.5	10.3	9.5	9.0

Part I: One-Way ANOVA

Use a one-way ANOVA to analyze these data.

(a) What is the response variable and what is the factor?

(b) How many levels of the factor are being studied?

(c) Is there any difference in the average pull poundage among the trials? If so, which ones are different?

(d) What is your recommendation to the company and why?

Part II: Block Design

Since the same terminals were connected and pulled apart 6 times, each row is actually a block. Reanalyze these data using a block design.

(a) Is there any difference in the average pull poundage among the trials? If so, which ones are different?

(b) Explain why you got a different answer when you analyzed the data as a one-way design.

(c) Calculate the sample size needed to see the treatment effect using a randomized one-way design.

(d) Now, what is your recommendation to the company?

14.6   ANALYSIS OF DATA FROM TWO-WAY DESIGNS

After learning about block designs and thinking a bit more about the results of the airspace analysis, you wonder if there might be some other factor that influences the airspace in the box (something other than the length of time that the box sits on the shelf). You return to talk to the people who work in manufacturing to ask them about how the tissue boxes are made.

14.6.1 Motivation for a Factorial Design Model

In Figure 14.6 you see a picture of a hardroll, which is a large roll of raw paper stock. Each box of 250 tissues is made from 25 hardrolls, each of which has 10 different slit positions. Four such slit positions are shown in the diagram. One tissue is taken from each slit and they are pressed into a tissue box and sealed. A hardroll lasts for 4 hours in production.

The manufacturing personnel feel that the mean airspace might differ by position in the hardroll from which the tissues were made. This hypothesis was proposed because of two facts:

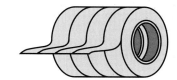

FIGURE 14.6   Schematic of a hardroll

1. As the hardroll sits in the warehouse, the outside collects moisture. Thus boxes made from tissues taken from the outside of the roll might have less growback.

2. The core (very center) of the roll is very compressed (a hardroll weighs 1000 pounds). Thus, boxes of tissues made from the core have the "life stretched out of them."

As a result of this discussion you have identified another potential factor, the position in the hardroll from which the box was produced. This is not a blocking factor because you are interested in knowing whether this factor affects the airspace in the box. Hence, it is not a nuisance factor. You return to look at your data and realize that a total of 4 cases of tissue boxes were sampled, one case every hour for four hours. One case was used to obtain in-process measurements. The other 3 cases were saved for observations taken after 24 hours, 2 weeks, and 4 weeks. Each case contained 24 cartons of tissue boxes or 120 tissue boxes. This means that the tissues that were allowed to sit on the shelf for 24 hours were made from a different part of the hardroll than the ones that sat on the shelf for 2 weeks or 4 weeks. You cannot tell from this dataset if the position in the hardroll is indeed a significant factor because you did not account for this factor in your experimental design.

Based on this new information you suggest the following sampling plan:

1. Sample 5 times through the hardroll. Since a hardroll lasts for 4 hours the 5 observations were taken at the beginning and once an hour thereafter. This results in observations at the top, the core, and 3 in the middle of the hardroll.

2. Each sample consists of a case of tissues (24 cartons or 120 tissue boxes).

3. A total of 4 cases is sampled every hour. One was used to obtain in-process measurements. The other 3 cases are saved for observations taken after 24 hours, 2 weeks, and 4 weeks.

Each airspace measurement is taken on a tissue box made from a certain position in the hardroll and allowed to sit on the shelf a certain amount of time before measurement. Thus, we have two factors: position and time. The position indicates from what part of the hardroll the sample was taken. The values for position are 1, 2, 3, 4, and 5. Position 1 indicates that it was taken from the outside of the roll and position 5 indicates it was taken from the core. The factor time indicates when the measurement was taken: 1 = in process, 2 = after 24 hours, 3 = after 2 weeks, and 4 = after 4 weeks.

This experimental design is an example of a **factorial design with two factors.** The two factors do not need to have the same number of levels but we will assume that the sample size is the same for each combination of the levels. For the airspace data this means that the number of tissue boxes sampled from position 1 and tested in process is the same as the number of tissue boxes sampled from position 3 and allowed to sit 2 weeks on the shelf and this is the same as the number of tissue boxes sampled from each of the 5 positions and allowed to sit for each of 4 different time possibilities. In this design, a sample of 1 case (24 cartons or 120 boxes) is selected from each of the $5 \times 4 = 20$ populations. This is referred to as **equal replication.**

Each observation within a population is referred to as a **replicate.** If you have unequal sample sizes, you must refer to a more advanced text.

> An experimental design is called a ***factorial design with two factors*** if there are several different levels of two factors being studied. The first factor is called ***factor A*** and there are r levels of factor A. The second factor is called ***factor B*** and there are c levels of factor B.

EXAMPLE 14.17   Airspace Data

Two-Factor Design

For the airspace data, we can label the position factor as factor A and the time factor as factor B. There are $r = 5$ levels of factor A: positions 1, 2, 3, 4, and 5. Position 1 refers to tissues made from the very outside of the hardroll and position 5 refers to tissues made from the core of the hardroll. There are $c = 4$ levels of the factor B, as we had before. ∎

> The design is said to have ***equal replication*** if the same number of objects or people being observed/measured are randomly selected from each population. The population is described by a specific level for each of the two factors. Each observation is called a ***replicate.*** There are n' observations or replicates observed from each population. There are $n = n'rc$ observations in total.

EXAMPLE 14.18   Airspace Data

Number of Replicates

For the airspace data, one case or 120 tissue boxes is selected for each combination of position and time. There are 20 combinations of position and time ($rc = (5)(4) = 20$). Thus, n' is 120 and there are $(20)(120) = 2400$ observations in total. ∎

The tissue company wishes to know if there a difference in airspace due to position in the hardroll and if there a difference in airspace by time tested. To answer these questions we must extend our analysis technique to handle this second factor.

14.6.2 Partitioning the Variation

The underlying concept in analyzing data from a two-factor design is the same as the approach we took in analyzing data from a one-way design. Remember that we looked at the relative sizes of the components of variation. For a two-factor design the variation we see in the data is due to one of four things: factor A, factor B, the interaction between factor A and B, and error. The idea is to take the total variability as measured by SST and break it down into 4 component sum of squares.

The names and labels for these are **sum of squares due to factor A (SSA), sum of squares due to factor B (SSB), sum of squares due to the interacting effect of A and B (SSAB)**, and **sum of squares error (SSE)**. Although the names are a bit different from those we used in a one-way ANOVA, they are an extension of the same concept.

> The *sum of squares due to factor A* is labeled **SSA**. It measures the squared differences between the mean of each level of factor A and the grand mean.

> The *sum of squares due to factor B* is labeled **SSB**. It measures the squared differences between the mean of each level of factor B and the grand mean.

> The *sum of squares due to the interacting effect of A and B* is labeled **SSAB**. It measures the effect of combining factor A and factor B.

> The *sum of squares error* is labeled **SSE**. It measures the variability in the measurements within the groups.

Thus, the total variation is divided into four components:

$$SST = SSA + SSB + SSAB + SSE$$

Formula for SST in terms of other components in a two-way design

The terms SSA and SSE have the same formulas and definition as for a one-way ANOVA. The SSB and SSAB terms are calculated in a similar fashion. If you are using a two-way experimental design, you will need to use software to do the calculations. The formulas for the sum of squares terms are long and tedious and any statistical software will generate the ANOVA table for a two-way design.

14.6.3 Using the ANOVA Table in a Two-Way Design

Consider a factorial design with r levels of factor A and c levels of factor B with n' replicates in each combination of factor A and B. There are a total of $n = n'rc$ observations. The ANOVA table for such a design looks just like the ANOVA table for a one-way design with two additional rows. It is shown below:

Source of Variation	SS	df	MS	F	p Value
Factor A	SSA	$r - 1$	$MSA = \dfrac{SSA}{r - 1}$	$F = \dfrac{MSA}{MSE}$	
Factor B	SSB	$c - 1$	$MSBL = \dfrac{SSB}{c - 1}$	$F = \dfrac{MSB}{MSE}$	
Interaction of A and B	SSAB	$(r - 1)(c - 1)$	$MSAB = \dfrac{SSAB}{(r - 1)(c - 1)}$	$F = \dfrac{MSAB}{MSE}$	
Error	SSE	$rc(n' - 1)$	$MSE = \dfrac{SSE}{rc(n' - 1)}$		
Total	**SST**	$rcn' - 1$			

The last 2 columns of the ANOVA table are used to do the hypothesis tests. In a two-way ANOVA, three hypothesis tests should be done. (1) To test the hypothesis of no difference due to factor A we would have the following null and alternative hypotheses:

H_0: There is no difference in the population means due to factor A.

H_A: There is a difference in the population means due to factor A.

This hypothesis test is easily done by determining if the F value in the row of the ANOVA table corresponding to factor A is in the rejection region. If the F test statistic is "too large," that is, larger than the critical F value from the table, then you conclude that factor A is significant. If the software you are using does not provide you with the critical F value, you can look it up in the table of F distribution values. The F statistic has $(r-1)$ degrees of freedom in the numerator (from the MSA term) and $rc(n'-1)$ degrees of freedom in the denominator (from the MSE term).

(2) To test the hypothesis of no difference due to factor B we would have the following null and alternative hypotheses:

H_0: There is no difference in the population means due to factor B.

H_A: There is a difference in the population means due to factor B.

This hypothesis test is easily done by determining if the F value in the row of the ANOVA table corresponding to factor B is in the rejection region. If the F test statistic is "too large," that is, larger than the critical F value from the table, then you conclude that factor B is significant. If the software you are using does not provide you with the critical F value, you can look it up in the F distribution table. The F statistic has $c-1$ degrees of freedom in the numerator (from the MSB term) and $rc(n'-1)$ degrees of freedom in the denominator (from the MSE term).

(3) To test the hypothesis of no difference due to the interaction of factors A and B we would have the following null and alternative hypotheses:

H_0: There is no difference in the population means due to the interaction of factors A and B.

H_A: There is a difference in the population means due to the interaction of factors A and B.

This hypothesis test is easily done by determining if the F value in the row of the ANOVA table corresponding to the interaction of factors A and B is in the rejection region. If the F test statistic is "too large," that is, larger than the critical F value from the table, then you conclude that the interaction is significant. If the software you are using does not provide you with the critical F value, you can look it up in the F distribution table. The F statistic has $(r-1)(c-1)$ degrees of freedom in the numerator (from the MSAB term) and $rc(n'-1)$ degrees of freedom in the denominator (from the MSE term). The interaction effect is discussed further in the next section. But first let's look at the airspace data.

EXAMPLE 14.19 Airspace Data

Two-Way ANOVA

The output from Minitab for the airspace data is shown below. Factor A is position and factor B is time.

```
Two-Way Analysis of Variance

Analysis of Variance for Airspace

Source        DF        SS         MS
Position       4     5942.98    1485.74
Time           3    12773.75    4257.92
Interaction   12      865.52      72.13
Error       2380     1667.08       0.70
Total       2399    21249.33
```

To test the hypothesis of no difference due to factor A we would have the following null and alternative hypotheses:

H_0:   There is no difference in the population means due to position.

H_A: There is a difference in the population means due to position.

The F value is MSA/MSE = 1485.74/0.70 = 2122.5. With such a large F value we really don't need an F critical to tell us that our F value is bigger than F critical. However, for completeness the table F value is found for $\alpha = 0.05$ with 4 degrees of freedom in the numerator and 2380 degrees of freedom in the denominator. F critical = 2.37, so, clearly, position is a significant factor.

To test the hypothesis of no difference due to factor B we would have the following null and alternative hypotheses:

H_0:   There is no difference in the population means due to time.

H_A: There is a difference in the population means due to time.

The F value is MSB/MSE = 4257.92/0.70 = 6082.7. With such a large F value we really don't need an F critical to tell us that our F value is bigger than F critical. However, for completeness the table F value is found for $\alpha = 0.05$ with 3 degrees of freedom in the numerator and 2380 degrees of freedom in the denominator. F critical = 2.60 so, clearly, time is a significant factor.

To test the hypothesis of no difference due to the interaction of factors A and B we would have the following null and alternative hypotheses:

H_0:   There is no difference in the population means due to the interaction of position and time.

H_A: There is a difference in the population means due to the interaction of position and time.

The F value is MSAB/MSE = 72.13/0.70 = 103.0. With such a large F value we really don't need an F critical to tell us that our F value is bigger than F critical. However, for completeness the table F value is found for $\alpha = 0.05$ with 12 degrees of freedom in the numerator and 2380 degrees of freedom in the denominator. F critical = 1.75, so, clearly, the interaction is significant.

At this point we know from the analysis that both position in the hardroll from which the tissue box is made and the length of time the box sits on the shelf affect the average airspace in the box. We also know that these two factors interact.   ■

 **TRY IT NOW!**

Participative Management
Two-Way Design

In a previous "Try It Now" you looked at the number of sick days used in the past 12 months by employees who were traditionally managed, employees who were involved in a participative management style, and employees who were involved in a management style that was a mixture of the old structure with the team-based approach. Management is wondering whether the employee's department is a factor as well.

Use the accompanying Minitab output to determine if there are any differences in the mean number of sick days by management style and by department and if there is any interaction effect.

(*continued*)

Analysis of Variance for Sick Day

Source	DF	SS	MS
Department	4	298.187	74.547
Mgt Style	2	51.440	25.720
Interaction	8	4.293	0.537
Error	60	42.000	0.700
Total	74	395.920	

14.6.4 Understanding the Interaction Effect

In the previous section we performed a hypothesis test to determine if the two factors had a significant interaction effect. The easiest way to understand this effect is to look at a graph of the sample averages for each of the possible combinations of the two factors. The line graph shown in Figure 14.7 displays the 20 sample means for airspace.

From this graph you can see that the mean airspace decreases the longer the box sits on the shelf, regardless of from what position in the hardroll the box was made. However, the line connecting the sample means for those made from position 2 crosses the line connecting the sample means for those made from position 5. This indicates that the amount of growback in the tissue box between the 24-hour mark and the 2-week mark depends on where in the hardroll the tissue came from. Thus, the airspace behavior is affected by the interaction of the time on the shelf and the position in the hardroll from which it was made. This is what we expected based on our hypothesis test.

The information needed to look at the interaction effect graphically is the average of the observed variable for each group. Such information can be generated by using a statistical software package. Sometimes it is part of the output of ANOVA and sometimes it is a separate command.

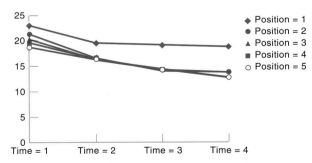

FIGURE 14.7 Average airspace based on time on shelf for different positions

EXAMPLE 14.20 Airspace Data

Sample Means for All Combinations for Position and Time Tested

In Minitab the analysis of means command generates output as shown below for the airspace data.

```
Rows: Position  Columns: Time

            1         2         3         4         All
1          120       120       120       120        480
          23.00     19.75     19.58     19.33      20.42
          0.648     0.523     0.705     0.473      1.613

2          120       120       120       120        480
          20.83     16.88     15.08     14.13      16.73
          0.803     1.171     0.762     0.668      2.713

3          120       120       120       120        480
          20.25     17.08     14.54     14.08      16.49
          0.664     1.120     0.916     0.705      2.605

4          120       120       120       120        480
          19.29     16.67     15.04     13.29      16.07
          0.679     0.853     0.938     0.938      2.371

5          120       120       120       120        480
          21.67     16.92     14.67     14.25      16.88
          1.286     0.816     0.690     0.882      3.096

All        600       600       600       600       2400
          21.01     17.46     15.78     15.02      17.32
          1.521     1.478     2.076     2.311      2.976

Cell Contents—

    Airspace:N

        Mean

        StDev
```

■

If there were no interaction effect, the lines connecting the sample means would be parallel. This would tell us that the growback behavior of the tissues in the box was the same no matter what position in the hardroll the tissues were made from. The graph might look something like the one shown in Figure 14.8.

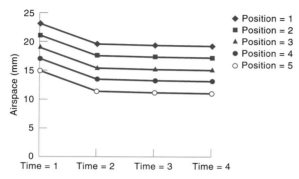

FIGURE 14.8 Hypothetical average airspace if there was no interaction effect

14.6.5 Exercises—Learning It!

14.16 The manufacturer of batteries is designing a battery to be used in a device that will be subjected to extremes in temperature. The company has a choice of 3 materials to use in the manufacturing process. An experiment is designed to study the life of the battery when it is made from materials A, B, and C and is exposed to temperatures of 15, 70, and 125°F. For each combination of material and temperature, 4 batteries are tested. The lifetimes in hours of the batteries are shown below:

	Temperature		
	15°F	**70°F**	**125°F**
Material A	130,155,74,180	34,40,80,75	20,70,82,58
Material B	150,188,159,126	126,122,106,115	25,70,58,45
Material C	138,110,168,160	174,120,150,139	96,104,82,60

(a) Calculate the average life for each of the 3 materials.

(b) Calculate the average life for each of the 3 temperatures.

(c) Calculate the average life for each of the 9 treatment groups.

(d) Plot the 9 treatment means on a graph with temperature factor on the x axis, and the life of the battery in hours on the y axis. Use a different color for each of the 3 materials and connect the averages for those of the same material. What do you speculate about the interaction effect based on the graph?

(e) Confirm your suspicions by doing a two-way ANOVA and testing to see if there is a significant interaction effect.

(f) What material do you recommend to this manufacturer and why?

Requires Exercise 14.5

14.17 The golf ball data that you analyzed as a one-way ANOVA in Exercise 14.5 were actually collected at three different times of day. To see if the time of day is a factor, reanalyze the data using a two-way ANOVA. The first 12 observations of each model were taken early in the morning, the second 12 observations were taken mid-day, and the last 12 observations were taken late in the afternoon.

(a) Is there any difference in the average distance the ball carries by design?

(b) Is there any difference in the average distance the ball carries by time of day?

(c) Is there any interaction effect? If so, interpret this in terms of the application.

(d) Check the assumptions of ANOVA for these data.

Requires Exercise 14.10
Datafile: AMPS2.XXX

14.18 The company that tested the power supplies in Exercise 14.10 wished to see if the input current that the component would draw from a power source, such as a battery, changed over time. Twenty components were pretested at the time of manufacture, 9 days later, and 58 days later. The company was also concerned about the possible differences in the product performance from lot to lot. The entire experiment was repeated for a sample taken from another lot. The data (in milliamps) are shown below:

Number	Pretest	9 Days	58 Days
1	25.6	27.3	28.5
2	27.9	30.0	30.5
3	25.7	27.1	27.8
4	25.9	27.6	29.1
5	25.2	26.6	27.5
6	26.1	27.9	29.4
7	26.5	28.8	29.4
8	29.9	32.1	33.2
9	27.9	29.1	30.3
10	27.3	29.8	32.0
11	25.1	27.2	28.5
12	25.7	27.2	27.9
13	27.0	28.3	29.2
14	25.6	26.2	26.8
15	26.2	27.1	27.6

(continued)

Number	Pretest	9 Days	58 Days
16	26.5	28.3	28.9
17	24.7	25.8	26.8
18	29.5	33.1	34.7
19	25.3	27.2	28.2
20	24.9	26.2	27.4

(a) With the additional lot this can now be considered a two-factor design. What is factor A? Factor B? How many replicates are there?

(b) Analyze both lots as a two-factor ANOVA.

(c) Is there any difference in the amount of power drawn from the battery over time?

(d) Is there any difference in the amount of power drawn from the battery by lot?

(e) Is there any interaction effect? If so, interpret this in terms of the application.

14.19 With the increased use of computers at all levels of the organization, many managers are finding themselves faced with new and different problems. One manager wished to study the effect of the employees' efficiency measured by how long it took to complete a particular task (in seconds). Twenty employees were timed at the beginning of the day and at 2, 4, and 6 hours after the employee had been doing intensive work. In each case the employee used his/her right hand to complete the task 20 times and the left hand 20 times. The data are shown below. *Datafile: MOUSE1.XXX*

Beginning of Day		After 2 h Intensive Work		After 4 h Intensive Work		After 6 h Intensive Work	
Right	**Left**	**Right**	**Left**	**Right**	**Left**	**Right**	**Left**
67	162	84	63	52	89	57	114
64	86	78	71	53	53	53	61
69	88	74	86	56	122	71	61
88	99	91	111	66	88	61	61
72	83	70	99	59	93	73	53
80	88	73	88	77	116	50	43
85	113	86	121	64	104	53	56
116	69	71	66	62	53	80	111
77	88	76	56	54	58	63	38
78	101	76	56	65	119	41	40
68	73	61	48	71	96	63	58
51	61	62	51	92	104	41	41
54	76	94	123	71	146	53	84
75	53	63	56	50	81	63	58
71	71	70	83	71	53	61	136
64	61	63	63	58	58	46	93
86	63	66	86	77	73	68	71
98	101	71	63	53	56	64	68
103	86	53	61	81	136	49	73
91	71	81	171	70	83	70	74

(a) Is there any difference in the average time to complete the task after differing amounts of computer intensive work?

(b) Is there any difference in the average amount of time to complete the task for the right hand versus the left hand?

(c) Is there a significant interaction effect?

(d) Display the treatment means to get some additional insight into the interaction effect. What does the graph tell you that the hypothesis test does not tell you?

(e) What is your recommendation to this manager?

14.20 A manufacturer of adhesive products designed an experiment to compare a new adhesive product to a competitor's product. The adhesive product, or glue, is used by automobile manufacturers. The response variable was the strength of the glue measured by tensile strength in pounds per square inch (psi). The ability to adhere to oil-contaminated surfaces under different humidity conditions was studied. There were 2 levels for factor A: no oil or oil. Oil contamination was applied by hand dipping the samples in an oil solution and allowing them to air dry at room temperature for 2 hours. There were 2 levels for factor B: 50% humidity and 90% humidity. Three samples were tested for each of the combinations of factor A and factor B. The tensile values (psi) for the new product are shown below:

Humidity	No Oil	Oil
50%	175,100,175	43,42,44
90%	95,115,85	95,105,116

(a) Does the product behave significantly differently if the surface is oil contaminated?

(b) Does the product behave significantly differently at different humidity levels?

(c) Is there any significant interaction effect present?

14.7 OTHER TYPES OF EXPERIMENTAL DESIGNS

Often you wish to investigate more than 2 factors. The analysis technique of ANOVA extends directly to what is known as a factorial design with more than 2 factors. The interaction terms become a bit more complicated and the required sample size increases as you consider additional factors. However, the basic approach that we have taken works and the concepts you have learned extend directly to handle more than 2 factors.

We can see from our work in the previous section that the number of observations that you need increases very quickly as you increase the number of factors. For the airspace data we ended up with 20 groups from 2 factors. This means your total sample size increases. Suppose, for example, that the tissue company wished to consider a third factor which had 3 levels. We would have (20)(3) or 60 groups to study. If we selected a sample of 5 from each of these populations we would need a total sample of size of 300. You often do not have the resources to collect this many data.

With the addition of a third factor you could perform more hypothesis tests. You would have 3 hypothesis tests for the effect of the individual factors; 3 hypothesis tests to test for interactions between 2 of the variables (factor A and factor B; factor B and factor C; factor A and factor C), and 1 hypothesis test to check for interaction between all 3 variables. Although there are more tests to be performed, each one is done using an F test just as we have seen in this chapter.

The two-way interaction terms are often relatively easy to interpret, as we have seen in the previous section. However, the higher order interactions are typically difficult to interpret and often do not tell you much. Thus, instead of using a full factorial design as we have done with two factors, we use a technique that helps keep the sample size down while still giving information about the main effect of the factors on the variable we are studying. These are called fractional factorial designs. In the case of the tissue company faced with 60 groups to study, the fractional design may call for a sample to be selected from only 30 groups. Thus, half the number of groups is studied, hence the term fractional design. You want to select the 30 groups to be studied judiciously so that you get information on the 3 factors and some of the important interaction effects. You lose information on some of the higher order interactions, but at least you have information on the major factors individually and some of the two-way interactions. This is better than not doing the experiment at all because of cost reasons. If you are in a situation where you wish to study more than 2 factors you should consult an advanced text on experimental design.

14.8   ANOVA IN EXCEL

Although Excel is not really a statistical software package, it does have some built-in tools for performing the basics of Analysis of Variance. We will look at how to use these tools for one- and two-way designs.

14.8.1 One-Way Analysis of Variance in Excel

To perform a one-factor ANOVA using Excel, the data must be in a worksheet with the data for each factor level in separate, adjacent columns. A portion of the airspace data is shown in Figure 14.9.

	A	B	C	D
1	Time 1	Time 2	Time 3	Time 4
2	23	20	19	19
3	25	19	19	19
4	23	20	20	19
5	23	20	20	19
6	23	20	19	19
7	23	19	20	19
8	23	20	20	19
9	23	20	19	19
10	22	20	19	19
11	22	20	20	20
12	23	20	19	20
13	23	19	19	19
14	22	20	19	19
15	22	20	21	19
16	24	20	20	19
17	23	20	18	20
18	23	20	19	20

FIGURE 14.9   Airspace data in Excel

To perform the analysis to see whether there is a difference in mean airspace for the different times, we will use the **ANOVA: Single Factor** tool from the **Tools > Data Analysis** menu. The dialog box shown in Figure 14.10 opens.

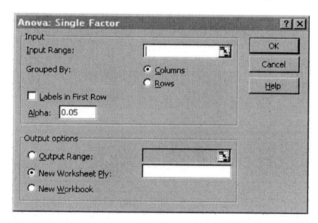

FIGURE 14.10   ANOVA: Single Factor dialog box

The input is similar to most Excel data analysis tools.

1. Enter the range that contains the data in the **Input Range:** textbox. If your data range contains labels, click on the textbox for **Labels in First Row.**

2. Since each factor is in a different column, click the radio button for **Grouped By: Columns.** This is the default value.

3. In the textbox for **Alpha,** enter the level of significance that you want to use for the test.

4. Finally, indicate where you want Excel to put the output from the analysis.

5. Click **OK;** the output will appear in the location specified. The output for the airspace data is shown in Figure 14.11. As with the output from the regression analysis, you might want to select **Format > Column > AutoFit Selection** while the output is still highlighted.

F	G	H	I	J	K	L
Anova: Single Factor						
SUMMARY						
Groups	Count	Sum	Average	Variance		
Time 1	120	2521	21.00833333	2.327661064		
Time 2	120	2095	17.45833333	2.199929972		
Time 3	120	1894	15.78333333	4.339215686		
Time 4	120	1802	15.01666667	5.377871148		
ANOVA						
Source of Variation	SS	df	MS	F	P-value	F crit
Between Groups	2554.75	3	851.5833333	239.1302467	1.34182E-94	2.623636419
Within Groups	1695.116667	476	3.561169468			
Total	4249.866667	479				

FIGURE 14.11 Output from ANOVA on airspace data

Excel provides some summary statistics for each factor and the ANOVA table as output. From the ANOVA table, you see that the F value for **Between Groups** is 239.13, compared to the critical value of 2.6236. The p value of the test is zero for all practical purposes. Thus, the conclusion is that there is a difference in the mean airspace values for the different time periods.

Excel does not provide any tools for doing multiple comparisons to determine which levels are different, nor does it provide any diagnostic plots for checking assumptions. Since the fitted values are not provided, there is no way to find the residuals and perform the analysis yourself.

14.8.2 Two-Way ANOVA Designs in Excel

The data analysis tools in Excel allow you to analyze two-way designs for situations where there are no replications, and for situations where there are replications. The

	A	B	C	D
1	Store	Shelf	End of Aisle	Front of Store
2	1	4457	4500	5800
3	2	4400	4370	5290
4	3	4310	4300	5000
5	4	4600	4400	5600
6	5	5000	5000	6000
7	6	4500	4500	5300
8	7	4700	5100	5900
9	8	4590	4280	5460
10	9	8510	8670	10600
11	10	6470	6500	8410

FIGURE 14.12 Display/Sales data

no-replication tool is similar to the method you would use to analyze a blocked design, whereas the replications tool allows you to analyze a full two-way design.

To analyze a blocked design, we will look at the data on the effects of display type on sales. The data should be in a spreadsheet with the factors (display type) in columns and the blocks (stores) in rows, as shown in Figure 14.12.

1. From the analysis tools list, select **ANOVA: Two Factor Without Replication** and the dialog box shown in Figure 14.13 opens.

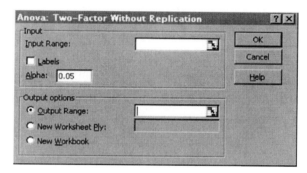

FIGURE 14.13   Two-way without replication dialog box

2. Enter the worksheet range for the data and fill in the other textboxes with the appropriate information.

3. Click **OK;** the output will be placed in the location specified. The output for the store display data is shown in Figure 14.14.

F	G	H	I	J	K	L
Anova: Two-Factor Without Replication						
SUMMARY	Count	Sum	Average	Variance		
1	3	14757	4919	582583		
2	3	14060	4686.667	273233.3		
3	3	13610	4536.667	161033.3		
4	3	14600	4866.667	413333.3		
5	3	16000	5333.333	333333.3		
6	3	14300	4766.667	213333.3		
7	3	15700	5233.333	373333.3		
8	3	14330	4776.667	374233.3		
9	3	27780	9260	1353100		
10	3	21380	7126.667	1235433		
Shelf	10	51537	5153.7	1782646		
End of Aisle	10	51620	5162	1970196		
Front of Store	10	63360	6336	3148316		
ANOVA						
Source of Variation	SS	df	MS	F	P-value	F crit
Rows	60738440.03	9	6748716	88.5418	4.73E-13	2.456282
Columns	9253927.267	2	4626964	60.70484	9.97E-09	3.554561
Error	1371972.067	18	76220.67			
Total	71364339.37	29				

FIGURE 14.14   Output for blocked design

The output includes summary information for each factor level and block in addition to the ANOVA table. We see from the p values that both the blocks (located in the Rows) and the display types (located in the Columns) were significant.

The method for analyzing a full two-way design with replications is very similar to the method for the one-way and two-way without replications. The only difference is in the way Excel expects the data to be stored in the worksheet.

Excel expects to find the data in table format, with the levels of one factor in columns and the levels of the other factors in rows. Each replicate (observation) for a pair of levels is placed in a cell under the previous one. An explanation of this data setup from the Excel documentation is shown in Figure 14.15.

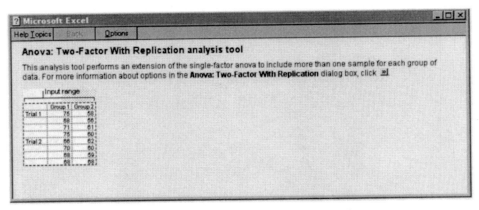

FIGURE 14.15   Data setup for two-way with replications

A subset of the airspace data is shown in Figure 14.16. In this example there are 24 boxes taken from each of five positions. The airspace is measured at four different times.

A	B	C	D	E	F
1		Time1	Time2	Time 3	Time 4
2	Position 1	23	20	19	19
3		25	19	19	19
4		23	20	20	19
5		23	20	20	19
6		23	20	19	19
7		23	19	20	19
8		23	20	20	19
9		23	20	19	19
10		22	20	19	19
11		22	20	20	20
12		23	20	19	20
13		23	19	19	19
14		22	20	19	19
15		22	20	21	19
16		24	20	20	19
17		23	20	18	20
18		23	20	19	20
19		23	20	20	19
20		23	18	20	20
21		23	20	20	19
22		23	20	20	19
23		23	20	21	20
24		24	20	20	20
25		23	19	19	20
26	Position 2	20	18	14	14
27		20	17	15	14
28		21	17	15	14
29		20	16	16	14
30		21	17	14	15
31		21	18	15	14
32		20	17	15	14

FIGURE 14.16   Airspace data for two-way analysis

The completed dialog box for the two-way with replications analysis tool is shown in Figure 14.17.

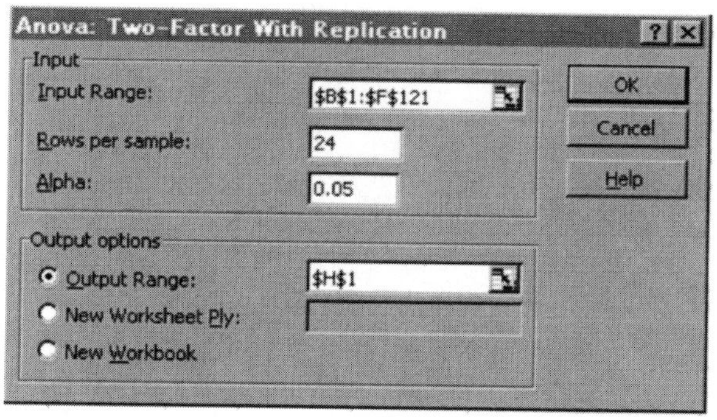

FIGURE 14.17 Two-way with replications dialog box

The only difference in the input is that you need to tell Excel how many replications there are by filling in the textbox labeled **Rows per sample:.** In this case there were 24. You *must* highlight the row with the first factor labels and the column with the second factor labels. Excel does not give you a choice here. The output is shown in Figure 14.18. It is very similar to the output from the other ANOVA tools.

Anova: Two-Factor With Replication						
	H	I	J	K	L	M	N
SUMMARY		Time1	Time2	Time 3	Time 4	Total
Position 1						
Count		24	24	24	24	96
Sum		552	474	470	464	1960
Average		23	19.75	19.5833333	19.3333333	20.4166667
Variance		0.43478261	0.2826087	0.51449275	0.23188406	2.6245614
Position 2						
Count		24	24	24	24	96
Sum		500	405	362	339	1606
Average		20.8333333	16.875	15.0833333	14.125	16.7291667
Variance		0.66666667	1.41847826	0.60144928	0.46195652	7.42061404
Position 3						
Count		24	24	24	24	96
Sum		486	410	349	338	1583
Average		20.25	17.0833333	14.5416667	14.0833333	16.4895833
Variance		0.45652174	1.29710145	0.86775362	0.51449275	6.84199561
Position 4						
Count		24	24	24	24	96
Sum		463	400	361	319	1543
Average		19.2916667	16.6666667	15.0416667	13.2916667	16.0729167
Variance		0.47644928	0.75362319	0.91123188	0.91123188	5.6683114
Position 5						
Count		24	24	24	24	96
Sum		520	406	352	342	1620
Average		21.6666667	16.9166667	14.6666667	14.25	16.875
Variance		1.71014493	0.6884058	0.49275362	0.80434783	9.66842105
Total						
Count		120	120	120	120	
Sum		2521	2095	1894	1802	
Average		21.0083333	17.4583333	15.7833333	15.0166667	
Variance		2.32766106	2.19992997	4.33921569	5.37787115	

ANOVA						
Source of Variation	SS	df	MS	F	P-value	F crit
Sample	1188.59583	4	297.148958	409.963072	3.869E-150	2.39132447
Columns	2554.75	3	851.583333	1174.89128	3.56E-215	2.62429012
Interaction	173.104167	12	14.4253472	19.9020037	4.7221E-35	1.77324111
Within	333.416667	460	0.72481884			
Total	4249.86667	479				

FIGURE 14.18 Output from airspace two-way design

CHAPTER 14 SUMMARY

In this chapter you have learned the tool of analysis of variance (ANOVA), which is used to draw inferences about more than 2 population means. It directly extends the two-population tests to compare means which you learned in Chapter 13. You typically find yourself comparing more than 2 populations when you are trying to determine what factors influence the variable you are studying.

Regardless of the specific number of factors you are investigating, the technique of ANOVA partitions the variation in the data into components and compares the relative sizes of these components. So it is through an analysis of the variation in the data that we learn something about the means. It is a clever approach and one that draws on many of the basic concepts and tools that you have learned in earlier chapters.

The results of the hypothesis tests that you can perform will guide you in your design of subsequent experiments as you try to understand the factors that cause your variable of interest to vary. Ultimately you are interested in building a model to predict or control this variation. Remember that unexpected variation is the enemy according to Deming, and customers are demanding higher quality in products and services. It is only by understanding variation that we can produce quality products.

Key Terms

Term	Definition	Page Reference
Response variable	The **response variable** is a quantitative variable that you are measuring or observing.	573
Factor	A **factor** is a variable that can be used to differentiate one group or population from another.	573
Level	A **level** is one of several possible values or settings that the factor can assume.	573
One-way or completely randomized design	An experiment has a **one-way** or **completely randomized design** if several different levels of one factor are being studied and the objects or people being observed/measured are randomly assigned to one of the c levels of the factor.	574
ANOVA	**Analysis of variance (ANOVA)** is the technique used to analyze the variation in the data to determine if more than two population means are equal.	575
Treatment	A **treatment** is a particular setting or combination of settings of the factor(s) being studied.	575

Term	Definition	Page Reference
Grand mean	The **grand mean** or the **overall mean** is the sample average of all the observations in the experiment. It is labeled $\bar{\bar{x}}$.	577
Sum of squares total (SST)	The **total variation** or **sum of squares total (SST)** is a measure of the variability in the entire data set considered as a whole.	577
Treatment mean	A **treatment mean** is the average of the response variable for a particular treatment.	579
Sum of squares among (SSA)	**Between groups variation** measures how different the individual treatment means are from the overall grand mean. It is often called the **sum of squares between** or the **sum of squares among (SSA).**	579
Sum of squares error (SSE)	Within groups variation measures the variability in the measurements within the groups. It is often called **sum of squares within** or **sum of squares error (SSE).**	581
Mean squares	The **mean square among** is labeled **MSA**. The **mean square error** is labeled **MSE** and the **mean square total** is labeled **MST.**	584
Block	A **block** is a group of objects or people that have been matched. An object or person can be matched with itself, meaning that repeated observations are taken on that object or person and these observations form a block.	599
Randomized block design	An experiment has a **randomized block design** if there are several different levels of one factor being studied and the objects or people being observed/measured have been matched. Each object or person is randomly assigned to one of the c levels of the factor.	599
Sum of squares blocks (SSBL)	The **sum of squares blocks** measures the variability between the blocks. It is labeled **SSBL.**	599
Factorial design with two factors	An experimental design is called a **factorial design with two factors** if there are several different levels of two factors being studied. The first factor is called **factor A** and there are r levels of factor A. The second factor is called **factor B** and there are c levels of factor B.	610
Replicate	The design is said to have **equal replication** if the same number of objects or people being observed/measured are randomly selected from each population. The population is described by a specific level for each of the two factors. Each observation is called a **replicate.** There are n' observations or replicates observed from each population. There are $n = n'rc$ observations in total.	610
Sum of squares due to factor A (SSA)	The **sum of squares due to factor A** is labeled **SSA.** It measures the squared differences between the mean of each level of factor A and the grand mean.	611
Sum of squares due to factor B (SSB)	The **sum of squares due to factor B** is labeled **SSB.** It measures the squared differences between the mean of each level of factor B and the grand mean.	611
Sum of squares due to the interacting effect of A and B (SSAB)	The **sum of squares due to the interacting effect of A and B** is labeled **SSAB.** It measures the effect of combining factor A and factor B.	611

Key Formulas

Term	Formula	Page Reference
SST, one-way design	$\sum_{j=1}^{c} \sum_{i=1}^{n_j} (x_{ij} - \bar{\bar{x}})^2$	577
SSA, one-way design	$SSA = \sum n_j (\bar{x}_j - \bar{\bar{x}})^2$	579
SSE, one-way design	$\sum_{j=1}^{c} \sum_{i=1}^{n_j} (x_{ij} - \bar{x}_j)^2$	581
MSA, one-way design	$\dfrac{SSA}{c-1}$	584
MSE, one-way design	$\dfrac{SSE}{n-c}$	584
MST, one-way design	$\dfrac{SST}{n-1}$	584
F statistic, one-way design	$\dfrac{MSA}{MSE}$	586
SST, block design	$SST = SSA + SSBL + SSE$	599
Relative efficiency of block design	$\dfrac{(r-1)MSBL + r(c-1)MSA}{(n-1)MSE}$	602
SST, two-way design	$SST = SSA + SSB + SSAB + SSE$	611

CHAPTER 14 EXERCISES

Learning It!

14.21 A manufacturer of watches is concerned about the defective rate. There have been numerous customer complaints. The watches are made on 3 shifts and the company wishes to see if there is any difference in the average percent defective on the 3 shifts. The defective rate for each shift for 10 days is recorded and shown below:

	Shift 1	Shift 2	Shift 3
Day 1	0.020	0.037	0.036
Day 2	0.016	0.032	0.032
Day 3	0.018	0.028	0.037
Day 4	0.021	0.029	0.033
Day 5	0.020	0.031	0.038
Day 6	0.022	0.034	0.035
Day 7	0.019	0.028	0.034
Day 8	0.019	0.032	0.027
Day 9	0.019	0.031	0.036
Day 10	0.021	0.030	0.034

(a) What type of design is this?

(b) Is there any difference in the percent defective by shift? If so, which one(s)?

(c) Check the assumptions of ANOVA for these data.

Datafile:
DISHWASH.XXX

14.22 A manufacturer of dishwashers was concerned about the consistency of measuring the resistance of the heater in the dishwasher product from one meter to another. A sample of 30 heaters was selected. The resistance of each heater was measured by 4 meters. The data (in ohms) are shown in the table.

Sample	Meter 1	Meter 2	Meter 3	Meter 4
1	14.5	14.4	14.3	14.63
2	14.8	14.8	14.7	14.98
3	14.8	14.8	14.7	14.93
4	14.6	14.6	14.5	14.80
5	15.8	15.7	15.6	15.95
6	14.6	14.6	14.5	14.80
7	14.8	14.8	14.7	14.99
8	14.4	14.4	14.3	14.62
9	14.6	14.5	14.4	14.71
10	14.3	14.2	14.2	14.43
11	14.3	14.3	14.2	14.49
12	14.6	14.6	14.5	14.75
13	14.3	14.2	14.2	14.44
14	15.0	14.9	14.8	15.11
15	14.5	14.5	14.4	14.68
16	14.7	14.7	14.5	14.86
17	14.6	14.6	14.5	14.76
18	14.7	14.6	14.5	14.81
19	14.7	14.7	14.6	14.85
20	14.6	14.5	14.4	14.71
21	14.5	14.4	14.3	14.58
22	14.4	14.3	14.3	14.50
23	15.3	15.3	15.2	15.44
24	14.4	14.4	14.3	14.54
25	14.5	14.4	14.3	14.58
26	14.8	14.7	14.6	14.86
27	14.6	14.5	14.4	14.72
28	14.5	14.4	14.3	14.57
29	14.4	14.3	14.2	14.50
30	14.6	14.6	14.5	14.75

(a) What type of design is this?

(b) Is there any difference in the measurements taken by the different operators? If so, which one(s)?

(c) Check the assumptions of ANOVA for these data.

14.23 The dishwasher company wondered whether it mattered if the meters had time to warm up before being used. In addition to the 4 meters used in Exercise 14.22, another meter of the same type as meter 4 was allowed a 2-hour warm-up period before being used. The readings on this meter for the same 30 sample products are shown below: *Requires Exercise 14.22*

Samples 1–5	14.62	15.00	14.95	14.78	15.93
Samples 6–10	14.79	15.01	14.61	14.70	14.45
Samples 11–15	14.48	14.76	14.45	15.09	14.67
Samples 16–20	14.86	14.77	14.79	14.84	14.71
Samples 21–25	14.58	14.50	15.45	14.55	14.59
Samples 26–30	14.85	14.72	14.58	14.51	14.76

(a) Rerun the ANOVA including these data as a 5th treatment.

(b) Do any of your conclusions change?

14.24 Many on-line service companies have a technical support telephone service to assist users who are having difficulty getting connected or using the service. Recently there has been a concern about the amount of time that a customer has to wait "on hold" before hearing a human voice. In an effort to decide which on-line service to purchase, a study was conducted. Thirty calls were made to 4 on-line services during 3 times of day: early morning (6–8 AM), midday (noon to 2 PM), and late afternoon (4–6 PM). The amount of time spent on hold in minutes was recorded. A portion of the data is shown in the next table.   *Datafile: ONHOLD.XXX*

	Company 1	Company 2	Company 3	Company 4
Early AM	11	11	7	3
	8	11	9	8
	7	11	7	6
	8	9	9	9
	8	9	8	11
	7	6	8	9
	8	9	5	9

(a) Is there any difference in the average time on hold among the four companies?

(b) Is there any difference in the average amount of time on hold by time of day?

(c) Is there any interaction between the time of day and the four companies?

(d) If you could use more than one of the companies, which company would you use during each of the three time periods? Explain your reasoning.

(e) If you cannot use more than one company, which company would you select? Explain your reasoning.

(f) What other factors might influence the time on hold?

14.25 One of the benefits of implementing a Total Quality Management program is the decreased time it takes to go from product design to market sales. This is often referred to as the cycle time. An automobile company recently implemented a TQM program companywide. The cycle times (in months) for 3 new cars before TQM and one year after TQM are shown below:

Before TQM	After TQM
26, 29, 22	17, 9, 13

(a) Based on these data, what can you conclude about the average cycle time before and after the implementation of TQM?

(b) What else would you like to know about these products to be certain that the reduction in cycle time is a result of the TQM program and not something else?

Datafile: SCORES.XXX

14.26 There has been increasing concern about the declining reading and math levels of American students. A study was conducted to see the effect of income on four response variables: behavior problems index, reading score, math score, and vocabulary score. All three of the scores were from 0 to 100. The behavior index ranges from 0 to 100 with low scores indicating fewer behavior problems. One hundred children aged 5 to 7 were tested from families with varying income levels. Three income levels were used: an annual income of $15,000 or less, an annual income of $25,000, and an annual income of $40,000. The results are found in the datafile named SCORES.xxx.

(a) For each of the response variables, test to see if there is a difference in the average score for the 3 levels of the income factor.

(b) Have any of the assumptions of ANOVA been violated?

(c) What recommendations can you make based on your analysis?

Requires Exercise 14.19
Datafile:
MOUSE2.XXX

14.27 The manager from Exercise 14.19 is also concerned about the accuracy of her employees after many hours of intensive computer work. At the same time as the data on time to complete the task was taken, a measure of accuracy (% correct) was also recorded. These data are shown below:

Beginning of Day		After 2 h Intensive Work		After 4 h Intensive Work		After 6 h Intensive Work	
Right	**Left**	**Right**	**Left**	**Right**	**Left**	**Right**	**Left**
95.88	89.70	93.60	90.78	87.96	82.54	79.88	92.38
91.94	90.00	98.00	94.17	96.39	87.96	84.70	86.58
92.79	85.68	90.10	94.17	98.59	95.76	93.68	94.17
95.76	95.53	97.00	89.70	96.00	91.00	93.92	95.00
91.46	81.97	87.96	96.84	84.35	83.24	97.76	96.84
92.72	90.15	95.53	90.78	96.39	95.00	74.68	84.48
93.60	92.72	96.84	92.72	94.90	94.61	86.40	89.23

(continued)

Beginning of Day		After 2 h Intensive Work		After 4 h Intensive Work		After 6 h Intensive Work	
Right	**Left**	**Right**	**Left**	**Right**	**Left**	**Right**	**Left**
94.61	83.45	80.76	91.00	94.61	87.96	99.00	90.00
99.00	98.00	95.53	92.19	79.38	79.60	94.61	86.11
97.76	88.30	92.00	88.82	89.70	85.00	76.65	82.91
100.00	93.60	96.39	97.17	93.68	96.84	98.59	94.17
89.18	89.23	80.90	85.58	98.59	90.00	81.13	90.00
87.47	97.00	85.68	94.61	92.93	91.75	86.96	95.88
94.17	96.39	93.29	91.94	85.68	94.90	92.72	83.00
91.40	91.94	96.00	84.97	93.60	87.27	91.94	94.90
95.53	84.19	94.90	90.15	90.57	89.70	80.69	94.61
93.68	96.39	94.34	89.00	96.39	84.70	91.75	87.79
94.90	84.44	92.19	87.79	87.96	95.53	95.00	95.00
97.76	94.17	95.00	91.40	94.90	93.92	85.58	82.97
97.00	93.29	98.59	91.40	95.53	90.57	96.84	88.69

(a) Is there any difference in the average accuracy level after differing amounts of computer intensive work?

(b) Is there any difference in the average accuracy level to complete the task for the right hand versus the left hand?

(c) Is there a significant interaction effect?

(d) What is your recommendation to this manager?

14.28 In an effort to reduce crime, several cities have implemented a plan that encourages members of the police force to live in the city they patrol. These police officers have been given the opportunity to purchase homes at half-price, the other half of the cost being funded by the city budget. To see if this program has had the desired effect, the yearly number of murders in 5 cities has been recorded the year before the program was initiated, one year after the program started, and two years after the program was initiated. The data are shown below:

City	Year Before Program Started	1 Year After Program Started	2 Years After Program Started
Los Angeles	44	43	43
Chicago	35	34	33
Detroit	41	39	38
Springfield	21	20	19
Philadelphia	29	28	27

(a) What type of experimental design was used in this situation?

(b) Has the program been effective? Why or why not?

(c) Have any of the assumptions of ANOVA been violated?

(d) A government publication states that the average number of murders has dropped from 34 to 31.8 for the cities that have adopted this program. The publication uses this statement to propose expanding this program to other cities. Do you agree with this analysis? Why or why not? Should this program be adopted by other cities or should it be scrapped?

14.29 The adhesive company in Exercise 14.20 was interested in comparing its new product with that of its competitor. The same experimental conditions were used to test the competitor's product. The tensile values (psi) for the competitor's product are shown below: *Requires Exercise 14.20*

Humidity	No Oil	Oil
50%	437,437,450	9,10,6
90%	115,115,50	105,87,105

(a) Does the competitor's product behave significantly differently if the surface is oil contaminated?

(b) Does the competitor's product behave significantly differently at different humidity levels?

(c) Is there a significant interaction effect for the competitor's product?

(d) Summarize your findings about the new product and the competitor's product in terms of the response to oil-contaminated surfaces and varying levels of humidity. Should this company continue to develop the new product?

Thinking About It!

Requires Exercises 14.22, 14.23

14.30 Upon further thought about the dishwasher company, you have decided to compare just 2 of the columns of data: meter 4 and meter 4 after a warm-up period.

(a) Use a paired t test to do this comparison. Is there any difference in the readings if you let the meter warm up?

(b) Run the appropriate ANOVA using just meter 4 and meter 4 after warm-up as the dataset. Does the ANOVA indicate any difference in the treatment means?

(c) Should you have expected the same answer using the paired t test and the ANOVA? Explain why or why not.

(d) Take the value of the t statistic from the paired test and square it. Compare it to the F statistic from the ANOVA. What do you notice? Do you think this will always happen or is this just a coincidence for this dataset?

Requires Exercise 14.3

14.31 Reconsider the teacher of statistics who was wondering whether to assign and grade homework. The teacher used 3 different classes to investigate the 3 different possible ways of dealing with homework. The teacher used a one-way design thinking that students randomly select sections of a course to take and so they should be comparable. Is there any reason to think that a block design should have been used in this situation instead of the one-way design that was used? Explain why or why not.

Requires Exercise 14.1

14.32 When the diaper company tested absorbency the product was actually placed on a child and the child was allowed to play during the testing. As the child played, fluid was injected into the diaper every 10 minutes until the product failed (leaked). The amount of fluid (in grams) at the time of failure was recorded for each diaper. This experiment was set up as a one-way design. Can you think of any reason why a block design might be better?

14.33 In any ANOVA you test the null hypothesis of equal means using an F statistic. If you fail to reject the null hypothesis and you construct a confidence interval for the difference in two of the treatment means, what number should you expect to find in the confidence interval? Explain why.

Requires Exercise 14.18

14.34 In Exercise 14.18 you analyzed a dataset with two factors: time and lot. Since then, two more months have gone by and data have been collected on two more lots. Unfortunately, for the third lot the experiment was run only for the pretest and 9 days later. For the fourth lot, data were collected during the pretest and 9 days and 16 days later (instead of 58 days). The data are shown below:

Number	Pretest	9 Days	Number	Pretest	9 Days	16 Days
1	26.1	31.5	1	30.8	34.6	35.0
2	26.8	29.8	2	27.5	30.9	31.4
3	25.3	28.7	3	26.8	31.3	31.8
4	28.1	32.2	4	29.4	32.8	33.5
5	25.3	30.3	5	28.7	32.6	32.0
6	27.3	31.7	6	24.0	26.5	26.9
7	26.5	30.4	7	26.4	30.9	31.2
8	24.8	28.0	8	28.3	31.0	31.2
9	26.1	28.6	9	30.0	32.4	32.8
10	26.2	29.1	10	25.0	27.3	27.4
11	25.3	27.4	11	26.8	30.3	30.8
12	28.0	31.6	12	26.7	30.9	31.2
13	25.5	29.1	13	26.1	27.9	28.2
14	24.8	28.2	14	27.6	31.5	31.7
15	27.2	31.0	15	27.4	29.7	30.0
16	25.1	27.1	16	26.3	29.0	29.5
17	26.2	28.2	17	28.0	31.5	32.0
18	26.8	29.5	18	25.3	27.1	27.5
19	27.3	29.4	19	26.4	28.3	28.4
20	27.1	28.8	20	27.2	30.0	30.7

(a) Explain why you cannot simply add two more levels to the lot factor and reanalyze the entire dataset using ANOVA.

(b) What analysis do you suggest be done to use these additional data?

14.35 You have decided to use your newly acquired data analysis skills in your coaching of a 6th-grade girls' basketball team. You want to see if there is any difference in the average number of points your team scores when you use different defense strategies. You suggest the following design: use "man-to-man" defense strategy for the first third of the season and record the number of points scored for each game. Switch to full-court press strategy for the second third of the season and record the number of points scored for each game. During the last third of the season use a mixture of defense strategies.

(a) Will you be able to tell which strategy is best from the data collected from this experiment?

(b) Suggest an alternative experimental design.

14.36 In Exercise 14.15 you concluded that the average weights for the students were different at the four different times the data were collected. Suppose you now learned that these students were enrolled in a one-year weekend MBA program starting in September of one year and ending a year later and that they were all working at full-time jobs during this time.   *Requires Exercise 14.15*

(a) What might be causing the weight differences that you observed?

(b) Support your theory by taking another look at this dataset and analyzing it differently.

14.37 In Exercises 14.19 and 14.27 you analyzed the amount of time to complete a task and the accuracy level after varying amounts of time of intensive computer work. What other factors would you suggest be studied in a subsequent experiment on this subject?   *Requires Exercises 14.19, 14.27*

Doing It!

14.38 A manufacturing company that makes paper products is having trouble with its paper towel line. When a roll of paper towels is manufactured, the last step before the roll is put in a case is to wrap it with the poly wrapper in which the consumer sees it on the store shelf. A large amount of product is scrapped because it is not being wrapped properly.

The problem appears to be that the diameters of the towel rolls are too large and the wrapper does not go all the way around and seal properly. There are many factors in the manufacturing process that would appear to affect roll diameter, but the engineers involved were not sure exactly how the different machine settings really impacted roll diameter. In fact, they were not convinced that all of the factors made a difference! The towel machine team designed a study to determine the effect that different machine settings actually had on roll diameter. The specifications for the diameter of the roll of towels is 5.35 inches. Several different people on the team expressed a concern that some settings that would result in a good roll diameter might have an adverse impact on another towel roll characteristic, roll firmness.

If a towel is not firm its wrapping will also be affected, and if it is too firm consumers will still react negatively. Fixing the roll diameter problem at the expense of firmness was not an option.

The team decided to look at two machine factors:

1. Embosser roll gap: the mechanism that puts the pattern in the towel

2. Speed: the speed at which the machine winds the rolls of towels

The machine was run at two different settings for speed and two different sizes of the embosser roll gap. The response variables were diameter of the roll of paper towels in inches and firmness measured on a specialized scale. Lower numbers indicate softer rolls, while larger numbers indicate firmer rolls. A portion of the dataset is shown below:

Speed	Embosser	Diameter	Firmness
0	0	5.43307	0.271667
0	0	5.39370	0.336667
0	0	5.47244	0.260333
0	0	5.39370	0.261333
0	0	5.43307	0.297000
1	0	5.39370	0.312000
1	0	5.47244	0.294667

(*continued*)

Speed	Embosser	Diameter	Firmness
1	0	5.43307	0.342333
1	0	5.47244	0.306000
1	0	5.43307	0.365667

Datafile: TOWEL.XXX

The variable *Speed* is a 0–1 variable that indicates the machine speed. A 0 indicates that the slower machine speed was used and a 1 indicates that the faster machine speed was used. The variable *Embosser* is a 0–1 variable that indicates the size of the embosser roll gap. A 0 indicates that the smaller gap measurement was used, while a 1 indicates that the larger gap measurement was used.

(a) Is there a difference in the diameter of the paper towel roll due to the speed at which the machine is run?

(b) Is there a difference in the diameter of the paper towel roll due to the size of the embosser roll gap?

(c) Is there any interaction of the two factors with respect to the diameter of the paper towel roll?

(d) Is there a difference in the firmness of the paper towel roll due to the speed at which the machine is run?

(e) Is there a difference in the firmness of the paper towel roll due to the size of the embosser roll gap?

(f) Is there any interaction of the two factors with respect to the firmness of the paper towel roll?

(g) Check the ANOVA assumptions for both response variables.

(h) Prepare a report for management explaining the effects that each of the machine settings have on *Diameter* and *Firmness*. Indicate which settings result in acceptable values for roll diameter. Make a recommendation on machine settings, if you can.

14.39 In several of the worked examples in this chapter we looked at a portion of a dataset measuring tissue strength. The management of the tissue company was concerned about complaints involving sheets tearing on removal. One possibility was the lack of airspace in the box. We have studied that dataset extensively in this chapter. Another possible explanation is that the tissues are not strong enough. Two tissue strength measurements were studied: machine-directional strength *(MDStrength)* and the cross-directional strength *(CDStrength)*. They are both measured in lb/ream. The data were collected over 3 days and the company would like to know if the average *MDStrength* and the average *CDStrength* are the same for all 3 days. There are 75 observations per day. A portion of the dataset is shown below:

Datafile: TISSUE.XXX

Day	MDStrength	CDStrength	Day	MDStrength	CDStrength
1	1006	422	1	962	464
1	994	448	1	973	472
1	1032	423	1	1036	489
1	875	435	1	1084	440
1	1043	445			

The specifications indicate that *MDStrength* should have a mean of 1000 lb/ream and a standard deviation of 50 lb/ream. The specifications indicate that *CDStrength* should be normally distributed with a mean of 450 and a standard deviation of 25 lb/ream.

(a) Is there any difference in the average *MDStrength* for the 3 days? If so, which day(s) are different?

(b) Is the *MDStrength* running to specification?

(c) Is there any difference in the average *CDStrength* for the 3 days? If so, which day(s) are different?

(d) Is the *CDStrength* running to specification?

(e) Check the assumptions of ANOVA for both response variables.

(f) Using the results of your analysis, prepare a report to management telling them about the current tissue manufacturing process. Make recommendations to them on whether the process needs to be adjusted or whether it is running to target. Remember, if it is running to target they will be considering making changes to the specifications to reduce customer complaints about dispensing.

WORKSHOP 3

15

REGRESSION ANALYSIS

IS TQM WORKING?

A company that manufactures computer storage media has had a Total Quality Management (TQM) program in place for the past two years. The program incorporates Quality Circles, Statistical Quality Control, and Team-Based Decision Making in the production of removable storage media, i.e., floppy diskettes. Before the company decides to expand the program to other departments it would like to assess whether the program has been effective. According to the literature, when TQM programs are used they should result in increased productivity and quality, and decreased waste and delay.

The company asked the production team to assemble some data so that it can decide whether the program is successful. The team decided to collect data on five different variables, **machine speed, waste, delay, rate of operation,** and **average outgoing quality** for the two-year period that the program has been in place. A sample of the data is shown below:

Month	Speed	Waste (%)	RateOper	Delay (h)	Quality
1	375	8.9	14.5	6.2642	95
2	334	9.9	12.8	6.4854	93
3	356	8.7	12.7	6.8372	94
4	378	9.5	13.9	5.7134	93
5	373	9.8	13.7	6.3136	94
6	381	8.8	14.7	6.2034	92

In particular, the company managers would like to know how production rate, waste, and delay are related to the speed at which a machine runs. They know that they have been able to run the production process at a faster speed in the past two years, but they are not sure that the increase in speed translates to an increase in productivity or a decrease in waste and delay.

15.1 CHAPTER OBJECTIVES

In this chapter you will learn about relationships between quantitative variables. In particular, you will learn that when the relationship is linear you can use a method called *least squares* to find the equation to describe the relationship.

For example, a company might want to predict sales of a particular product. The company knows that there are many different variables that might affect sales, but it does not know which variable(s) are most important or exactly how they relate to sales. The company needs to find a model that will enable it to predict sales as a function of the other variables. In *regression analysis*, this model is an *equation*.

In the first part of the chapter we look at *simple linear regression*, which predicts the value of Y as a linear function of a single independent variable, X. Once we have established the basics of regression analysis in the simple linear case we will look at *multiple regression models*, which predict the value of Y as a function of a set of independent variables.

In this chapter you will learn how to

- Find the linear regression equation for a dependent variable Y as a function of a single independent variable X
- Determine whether a relationship between X and Y exists
- Analyze the results of a regression analysis to determine whether the simple linear model is appropriate

15.2 THE SIMPLE LINEAR REGRESSION MODEL

15.2.1 Deterministic and Statistical Relationships

In some cases where two variables, x and y, are related, the relationship is *deterministic*, or *functional*. This means that when a value of x is selected, the value of y is uniquely determined. For example, if we were interested in the relationship between the total cost of an order, y, and the number of items ordered, x, we can describe this relationship by an equation such as

$$y = \$50 + \$1.20x$$

where the value of $50 might represent an ordering cost and the $1.20 is the cost per item ordered. Figure 15.1 illustrates this type of relationship.

If a person were to order $x = 100$ items, then the corresponding cost would be $y = \$50 + (\$1.20)(100) = \$170$. Every person who orders 100 items will incur the cost of $170. That is, the value of y is *unique*, for a given value of x.

FIGURE 15.1 Deterministic relationship between total order cost and number of items ordered

While many real problems are described by this type of relationship, we are interested in a different situation when we study *linear regression*. When we look at the relationship between two variables, X and Y, we are interested in situations where the value of Y varies for a given value of X. That is, we are interested in the *statistical relationship* between two variables.

Suppose that we are looking at the relationship between dollars spent in advertising and the revenues from sales. Clearly, we expect the two variables to be related, but we do not expect that every time a company spends $\$x$ in advertising it will always have $\$y$ in revenues. We know that there are other *factors,* or *variables,* such as the type of product, location, and various economic factors that will affect the value of Y for a given value of X.

When we collect our data we are collecting *pairs* of observations on the two variables, X and Y. Thus, we will have a set of n data pairs:

$$(x_1, y_1), (x_2, y_2), \ldots , (x_n, y_n)$$

A plot of the data might look like the one in Figure 15.2. You can see that the two variables are related, but that a particular value of advertising expenditures can result in more than one value for revenue. This type of plot, a scatter plot, is of primary importance in exploring relationships between variables, and should be done *before* any type of statistical analysis is performed.

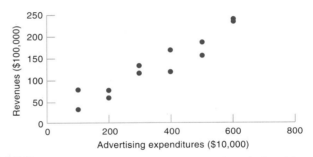

FIGURE 15.2   Statistical relationship between revenue and advertising expenditures

EXAMPLE 15.1   Is TQM Working?

Plotting the Data to Look for a Linear Relationship

The company that is looking at its TQM program has decided that if the program is effective then the observed increase in machine speed should be accompanied by an increase in productivity and a decrease in waste and delay.

The company decides to look at productivity and machine speed to see if there is any relationship between them that can be investigated further. The machine speed is measured in items per minute that the machine is set to produce. The measure chosen for productivity is Rate of Operation, which is a measure of the usable throughput of the machine per hour (in thousands of units).

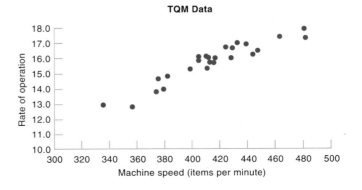

After looking at the plot it seems that a linear relationship is a reasonable model, and so the company decides to proceed with a simple linear regression analysis. ∎

In the case of simple linear regression we would like to find an equation, or model, that will allow us to predict a value for a variable, *Y*. Ideally, since the value of *Y* depends on many different factors, we would like to find some variable *X* that does a good job of predicting *Y* with a linear equation. By a "good" job of predicting, we mean that the prediction we obtain is *useful* for purposes of planning or problem solving. In regression analysis, the *X* variable is assumed to be controlled or at least controllable. This means that its values can be fixed by the person collecting the data.

EXAMPLE 15.2   HMO Health

Looking at the Relationship Between Variables

As approaches to health-care coverage change, Health Maintenance Organizations (HMOs) are growing in popularity. Some business analysts wondered how the increase in popularity affected the financial health of the HMOs. They collected data on revenue and number of members for 10 different HMOs:

HMO	Members (million)	Revenue ($ billion)
United HealthCare	4.24	5.49
Humana	3.19	4.63
FHP International	1.83	3.86
PacifiCare Health systems	1.62	3.60
U.S. Healthcare	2.07	3.43
WellPoint Health Network	2.30	2.91
Health Systems International	1.83	2.74
Foundation Health	2.15	2.40
Oxford Health Plans	0.97	1.71
Physician Corp. of America	0.89	1.20

Source: The Economist, 6 April 1996.

In this case the dependent variable is revenue, measured in billions of dollars and the independent variable is number of members, in millions. The analysts looked at a scatter plot of the data to see if a relationship existed:

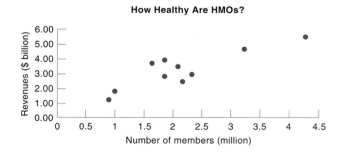

From the plot it appeared that a linear relationship existed between revenue and number of members. This meant that a straight line should provide a good model for predicting revenue for an HMO as a function of the number of members of the HMO. ∎

As you learn additional and more sophisticated statistical tools, you should not lose sight of the descriptive tools you learned first. Plotting data is essential to any statistical analysis and should be one of the first things you do so that you can determine what analysis, if any, is appropriate.

 **TRY IT NOW!**

Increasing Capacity
Plotting Data to Look at the Relationship

An oil company is trying to determine how the number of refining sites available for refining crude oil relates to the overall refining capacity. It would use this information to determine whether or not expansion will provide the increase in capacity that it wants or whether other steps to increase capacity will be necessary. The company collects data on other competitive companies and finds the following:

Oil Company	Number of Sites	Refining Capacity (million tons per year)
Royal Dutch/Shell	13	81.82
Exxon	10	81.82
Agip	13	58.18
BP	8	43.64
Repsol	5	40.00
Total	7	36.36
Turkish Petroleum	4	34.55
Elf	8	32.73
Mobil	7	29.09
Petrofina	3	25.45

Source: *The Economist,* 15 July 1995.

Use the grid below to create a scatter plot of the data.

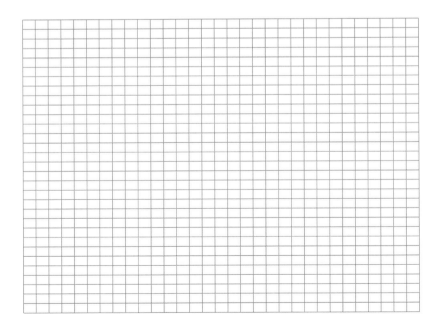

Do you think that a linear model is a good one?

15.2.2 The Simple Linear Regression Model

The objective of simple linear regression is to find a linear equation that describes the existing relationship between two variables, X and Y. The equation we find will be based on data taken from some population. Just as in any problems involving sampling from a population, we are trying to find an *estimate* for the true **regression model** between the two variables.

> The true relationship between the variables X and Y, the **simple linear regression model,** can be described by the equation
>
> $$y = \beta_0 + \beta_1 x + \varepsilon$$

This equation says that for a given value of the variable $X = x$, the actual value of Y will be determined by the expression $\beta_0 + \beta_1 x$, plus some random variation, ε, due to other, unmeasured, factors. Thus, if we knew the values of β_0, the true population intercept, and β_1, the true population slope, we could predict the value of Y to within some random error, ε. Figure 15.3 shows the population model for a linear regression.

*At this point we are assuming that a plot of the data has determined that a **linear model** is appropriate.*

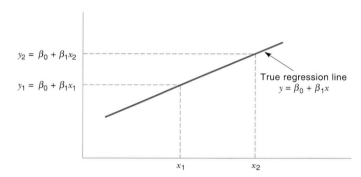

FIGURE 15.3 The true regression model showing how Y varies for a given value of X

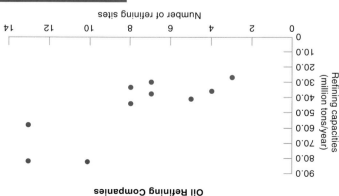

Oil Refining Companies

You can see that for a given value of $X = x_1$, the values of Y vary around the regression line. This variation is measured by the error term, ε. One of the assumptions of regression analysis is that the ε terms are normally distributed with a mean of 0 and a standard deviation of σ. We discuss this assumption along with some others later in the chapter.

How can we find estimates for the population values β_0 and β_1? We would like to find values that do the best job of describing the relationship between the variables. If you look at the data on advertising and revenue in Figure 15.2 you can probably imagine a straight line that you might draw that captures the relationship. Figure 15.4 shows such a line along with the data.

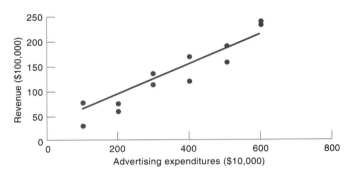

FIGURE 15.4   Straight line approximating the relationship between advertising and revenue

The equation of the line that we draw will be:

$$\hat{y} = b_0 + b_1 x$$

where $\hat{y}$ is the predicted value of Y for a particular value of $X = x$. The quantities b_0 and b_1 are the *estimates* of the population values β_0 and β_1. This line is called the *regression line of y on x* or the estimate of the *simple regression model*.

The problem is that there are many different lines that we might draw depending on what our criterion for "best" is. A list of possible criteria might be

- Hit as many points as possible.
- Have an equal number of points above and below the line.
- Pick two representative points and connect them.
- Connect the first and the last points.

The trouble with these criteria is that they do not produce a unique line. That is, there are many lines that hit 3 or 4 or 5 points and many lines that have an equal number of points above and below the line. (The last criterion actually produces a unique line, but it is not a very good criterion.) Figure 15.5 illustrates the problem. In Figure 15.5a both of the lines drawn have six points above and six points below. Figure 15.5b has several different lines, all of which go through exactly three of the data points.

You may wonder why it is important that we find a unique line to fit the data. Remember that we want to use the model (the equation) to predict values for Y, the dependent variable, for different possible values of X, the independent variable. One of the important features of a good model is that it be consistent. If the technique we use to find the line can produce many different models, then how will the user know which model or prediction to use? If everyone is allowed to choose the model he or

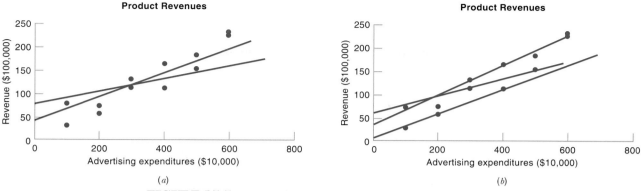

FIGURE 15.5   A single criterion can produce many different lines

she likes best, then the technique is, for all practical purposes, useless. You might as well just pick a number out of a hat.

15.2.3 The Least-Squares Line

At this point, we are sure about two things. We want the method we use for finding the equation of the line to produce one that is unique, and we want it to be a *good* representation of reality. While there are certainly many ideas about what is good, we can agree that the line should be close to as many of the data points as is possible. Figure 15.6 shows a set of data and a line drawn to represent the relationship between the variables. Although the line does not actually go through any of the data points, it is very close to most of them. The distance from each data point to the line is shown. These distances are called the **deviations** or **errors** of the line. We would like to find a line that somehow minimizes the overall deviation of the data points from the line.

> The distance between the predicted value of Y, $\hat{y}$, and the actual value of Y, y, is called the **deviation** or **error.**

When you learned about the variance in Chapter 6, you learned that when deviations are both positive and negative (in this case because the data points fall both above and below the line) you cannot simply look at the *sum* or *total* deviation, because it will be 0. For the variance, you solved the problem by squaring the deviations and then adding them together. We will use a similar approach to solving this problem.

The technique that we will use to find the line that best fits the data will be the **least-squares method.**

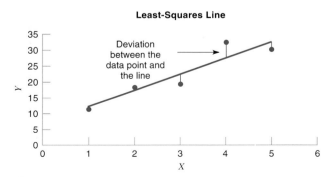

FIGURE 15.6   Deviations between the data points and the line

> The technique that finds the equation of the line that minimizes the total or sum of the squared deviations between the actual data points and the line is called the **least-squares method.**

You may wonder why we always square negative quantities to make them positive instead of just taking the absolute value. One reason is that mathematically the square is a nicer quantity to work with than the absolute value. The other reason, in the case of fitting a line to a set of data, is that using the square of the error makes the line try to fit *all* of the points. This is because the penalty for avoiding certain points is much larger. Suppose that the actual distance from a data point to the line is -10 units. If we simply wanted to minimize the *absolute* deviation, then the penalty attached to missing that point is $|-10| = 10$. When we minimize the square of the distance, the penalty for missing that point becomes $(-10)^2 = 100$ and so the line tries to get closer to that point.

The least-squares method finds the equation of the line

$$\hat{y} = b_0 + b_1 x$$

that minimizes

$$\sum_{i=1}^{n} (y_i - \hat{y}_i)^2$$

the total of squared deviations from the data points to the line. The values for b_0 (the intercept of the line) and b_1 (the slope of the line) are found by using the following equations:

Formulas for the slope and intercept of a least-squares line

$$b_1 = \frac{n \sum_{i=1}^{n} x_i y_i - \sum_{i=1}^{n} x_i \sum_{i=1}^{n} y_i}{n \sum_{i=1}^{n} x_i^2 - \left(\sum_{i=1}^{n} x_i\right)^2} \quad \text{or} \quad \frac{n \sum xy - \sum x \sum y}{n \sum x^2 - \left(\sum x\right)^2}$$

and

$$b_0 = \frac{\sum_{i=1}^{n} y_i}{n} - b_1 \frac{\sum_{i=1}^{n} x_i}{n} \quad \text{or} \quad b_0 = \bar{y} - b_1 \bar{x}$$

Although these equations may seem a bit complex, if you look at them carefully you will see that only five quantities have to be calculated. Four of the quantities involve the data values for X and Y, and the fifth is simply the number of observations in the sample, n. For example, look at the quantity $\sum xy$. This is simply the sum of the products of the x and y values for each data point. The easiest way to look at what is involved in the calculations is to make a table with a column for each sum needed. The table will look like the one in Table 15.1.

TABLE 15.1   Table for Calculating the Least-Squares Line

Observation Number	x	y	xy	x^2
1	x_1	y_1	$x_1 y_1$	$x_1 x_1$
2	x_2	y_2	$x_2 y_2$	$x_2 x_2$
.	.	.	.	.
.	.	.	.	.
.	.	.	.	.
n	x_n	y_n	$x_n y_n$	$x_n x_n$
Totals	$\sum x$	$\sum y$	$\sum xy$	$\sum x^2$

EXAMPLE 15.3   Is TQM Working?

Finding the Least-Squares Line

The manufacturing company that is trying to assess the effectiveness of its TQM program is looking at the relationship between productivity as measured by rate of operation (Y) and machine speed (X). It has plotted the data and thinks that the plots indicates that the linear model is a good one. To find the equation of the least-squares line, the managers of the company have assembled the following information:

Do not round until you have done the calculations.

$$\Sigma x = 10{,}340 \qquad \Sigma y = 391.4 \qquad \Sigma xy = 162{,}951.9 \qquad \Sigma x^2 = 4{,}306{,}918 \qquad n = 25$$

They first calculate b_1, the estimate of the slope:

$$b_1 = \frac{(25)(162{,}951.9) - (10{,}340)(391.4)}{(25)(4{,}306{,}918) - (10{,}340)^2} = \frac{26{,}721.5}{757{,}350} = 0.035282894$$

and then use that estimate to find b_0, the estimate of the y intercept of the line:

Although we would never use the estimate of b_1 to this many decimal places in the final equation, again, you should not round until after you use the value in the equation for b_0. In the final equation, the values for b_0 and b_1 should be rounded to reflect the precision of the original data.

$$b_0 = \frac{391.4}{25} - (0.035282894)\left(\frac{10{,}340}{25}\right) = 1.062995$$

The equation of the regression line relating rate of operation to machine speed is

$$\hat{y} = 1.06 + 0.035x$$

This means that there is a positive relationship between machine speed and productivity, that is, as machine speed increases, so does productivity. In fact, the equation tells the managers that for every unit increase in machine speed, the productivity goes up by 0.035. This is the definition of the slope of a line.

Although the positive relationship is encouraging, the people looking at the equation wonder if it really means anything. That is, they wonder if the increase in productivity realized by increasing speed is *significant*. The equation of the line alone cannot tell them that. They need to do further analysis. ∎

The equation of the least-squares regression line does give some information about the relationship between the independent variable Y and the dependent variable, X. The *sign* of the slope estimate tells whether the relationship is positive or negative. The *value* of the slope gives the change that will occur in Y when X is changed by one unit.

It is more difficult to explain the interpretation of the intercept. By definition, the intercept of a line is the value of Y when $X = 0$. In some situations, this number can be thought of as the value of the dependent variable that is related to *other factors* that are not considered in this model. For example, when we look at the relationship between sales and advertising, it is possible that for $0 spent on advertising, there will still be sales of a product.

In many cases, however, it does not make sense for the value of Y to be nonzero when $X = 0$. In the TQM example, if the machine speed is 0 there *is no productivity*, that is, Y must be 0. We discuss this problem in more detail as we proceed.

EXAMPLE 15.4   HMO Health

Calculating the Least-Squares Line

After they plotted the data and looked at the graph, the analysts who were interested in the relationship between HMO revenues (Y), and number of members (X), decided to use least squares to find the equation of the regression line relating the two variables. They assembled the data shown in the next table.

Although finding the equation of the regression line by hand is not difficult, it is tedious, even for small data sets. Many calculators have the capability to find the equation of the least-squares line. If you have such a calculator you might want to learn how to use it now.

Observation Number	Members, X (million)	Revenue, Y ($ billion)	XY	X²
1	4.24	5.49	23.2776	17.9776
2	3.19	4.63	14.7697	10.1761
3	1.83	3.86	7.0638	3.3489
4	1.62	3.60	5.8320	2.6244
5	2.07	3.43	7.1001	4.2849
6	2.3	2.91	6.6930	5.29
7	1.83	2.74	5.0142	3.3489
8	2.15	2.40	5.1600	4.6225
9	0.97	1.71	1.6587	0.9409
10	0.89	1.20	1.0680	0.7921
Totals	**21.09**	**31.97**	**77.6371**	**53.4063**

Substituting into the equations they found

$$b_1 = \frac{(10)(77.6371) - (21.09)(31.97)}{(10)(53.4063) - (21.09)^2} = \frac{102.1237}{89.2749} = 1.143924$$

$$b_0 = \frac{31.97}{10} - (1.143924)\left(\frac{21.09}{10}\right) = 0.7844643$$

Thus, the regression equation of HMO revenues on number of members is

$$\hat{y} = 0.78 + 1.14x$$

They realize that the equation means that for an increase in members of 1 million, the revenues will increase by $1.14 billion. The intercept of the line is 0.78, which would mean that when an HMO has no members, it will still generate $0.78 billion in revenues. This is a case where interpretation of the intercept does not make sense. ∎

Because least-squares analysis is mechanical, the equation alone cannot tell us whether the relationship is real or whether, in fact, the two variables are unrelated. You can find a least-squares line relating any two variables, but that certainly does not mean that the two variables are really related. To make that decision, further statistical tools are needed.

Most spreadsheet programs and all statistical software also perform regression analysis. It is not necessary to use the computer just to find the equation of the regression line, but further analysis can really only be done using a computer package. Although it is not necessary now, the remainder of this chapter assumes that you have access to some statistical software package. At the end of this section we look at regression output from different software packages. As the chapter progresses we will use the output extensively to make decisions about the problems we are trying to solve.

 **TRY IT NOW!**

Increasing Capacity
Finding the Equation of the Least-Squares Regression Line

The oil company that is looking at increasing refining capacity has decided that a linear relationship is appropriate.

Fill in the table below or use some other means to find the equation of the least-squares line:

Observation Number	Number of Sites, X	Capacity, Y	XY	X^2
1	13	81.82		
2	10	81.82		
3	13	58.18		
4	8	43.64		
5	5	40.00		
6	7	36.36		
7	4	34.55		
8	8	32.73		
9	7	29.09		
10	3	25.45		
Totals				

Interpret the meaning of the estimate of the slope of the line. Does the y intercept make sense for these data?

15.2.4 Using the Computer to Do Regression Analysis

Any statistical software that you might be using is capable of performing regression analysis. In addition, most spreadsheet packages also do regression. The output from the analyses will look different, but they have many common elements. We will start by identifying the estimates for the parameters of the regression equation in the output for several software packages. As we progress in the chapter we will continue to use the output in conjunction with other analyses.

In Figure 15.7 you see the regression output from two different computer packages for the TQM problem. The output from any software package contains three sections: (1) general information about the regression line, (2) information about the regression coefficients, and (3) ANOVA information. For each output, the coefficients of the regression equation are highlighted.

As you can see from the output, the equation is identified in a slightly different way for each package. You can also see that there is a lot more to the output of a regression analysis than just the equation. It is important that you understand the language of regression so that you can identify the output from whatever software package that you are using.

15.2.5 Using the Regression Equation to Make Predictions

Once we have the equation of the regression line, we can use the equation to predict values of the dependent variable, Y, for different values of the independent variable. We do this by substituting different values of X into the regression equation and calculating the corresponding value of $\hat{y}$. Remember in Figure 15.3 we saw that the values of Y vary around the true regression line.

ANS. $\hat{y} = 9.03 + 4.79X$ CAPACITY WILL INCREASE BY 4.79 M TONS FOR EVERY NEW SITE ADDED. NO.

```
Regression Analysis

The regression equation is
RateOper = 1.06 + 0.0353 Speed

Predictor        Coef      Stdev   t-ratio        p
Constant        1.063      1.084      0.98    0.337
Speed        0.035283   0.002611     13.51    0.000

s = 0.4544   R-sq = 88.8%   R-sq(adj) = 88.3%

Analysis of Variance

SOURCE        DF        SS        MS        F        p
Regression     1    37.712    37.712   182.64    0.000
Error         23     4.749     0.206
Total         24    42.462

Unusual Observations
Obs.   Speed   RateOper        Fit   Stdev.Fit   Residual   St.Resid
  2      334    12.8000    12.8475      0.2268    -0.0475    -0.12 X
  3      356    12.7000    13.6237      0.1757    -0.9237    -2.20R

R denotes an obs. with a large st. resid.
X denotes an obs. whose X value gives it large influence.
```

Minitab

SUMMARY OUTPUT

Regression Statistics	
Multiple R	0.942419659
R Square	0.888154813
Adjusted R Square	0.883291979
Standard Error	0.454404799
Observations	25

ANOVA

	df	SS	MS	F	Significance F
Regression	1	37.7124744	37.7124744	182.6413924	1.9909E-12
Residual	23	4.749125596	0.206483722		
Total	24	42.4616			

	Coefficients	Standard Error	t Stat	P-value	Lower 95.0%	Upper
Intercept	1.062994916	1.083622206	0.980964501	0.336817039	-1.178645356	3.304
Speed	0.035282894	0.002610746	13.51448824	1.9909E-12	0.029882162	0.040

Excel

FIGURE 15.7   Regression output for Minitab and Excel

> The value of $\hat{y}$ that we find is really a prediction of the mean value of Y for a given value of X.

EXAMPLE 15.5   Is TQM Working?

Prediction Using the Regression Equation

The company that is looking at effectiveness of its TQM program found that the regression equation for its sample data was

$$\hat{y} = 1.06 + 0.035x$$

The managers would like to know what the productivity will be if the machine runs at 350 items per minute, since this is a speed that they have been able to run for quite some time. They substitute into the equation and find that

$$\hat{y} = 1.06 + 0.035(350) = 13.31 \text{ thousand units per hour}$$

They would also like to know what the productivity will be when they run at 450 items per minute, since this is a speed that they have achieved only since TQM was instituted. They find

$$\hat{y} = 1.06 + 0.035(450) = 16.81 \text{ thousand units per hour}$$

From this it would seem that productivity does increase with speed. ∎

When we first looked at the regression equation and talked about what the parameters mean, we noticed that the slope of the line does not always make sense. What happens if we use the regression equation to predict Y when $X = 0$?

EXAMPLE 15.6 **Is TQM Working?**

Using the Regression Equation When X = 0

One of the workers who is involved in the TQM study thinks that there is something wrong with the model. He uses the equation to predict productivity for a machine speed of 0 items per minute and finds

$$\hat{y} = 1.06 + 0.035(0) = 1.06 \text{ thousand parts per minute}$$

If TQM can accomplish this it must be pretty good, since productivity is impossible when the machine is not running! He is not so sure about using the model to predict productivity, if it makes errors like this. ∎

When using a regression model we must remember that the model is constructed only from sample data. It is really relevant only *over the range of observed values*. The equation tells you about the relationship between the dependent and independent variables, only in this range. There are two kinds of predictions that you can do, **interpolation** and **extrapolation.**

Using the equation to predict values of Y within the range of the X data is called **interpolation.** Predicting values of Y for values of X outside the observed range is called **extrapolation.**

Extrapolation is risky and really should almost never be done. Without data you have no idea what the relationship beyond your "boundaries" is like. Over a larger range of the dependent variable, the relationship might change shape and be nonlinear. It might even change direction, making your predictions totally inappropriate.

 **TRY IT NOW!**

Increasing Capacity
Using the Regression Equation to Predict the Value of Y

Use the equation of the regression line you found earlier to predict the refining capacity for each of the observed values of X, the number of sites.

(continued)

Observation Number	Number of Sites, X	Capacity, Y	$\hat{y} = 9.03 + 4.79x$
1	13	81.82	
2	10	81.82	
3	13	58.18	
4	8	43.64	
5	5	40.00	
6	7	36.36	
7	4	34.55	
8	8	32.73	
9	7	29.09	
10	3	25.45	

15.2.6 Calculating Residuals

We used least squares to find the line for regression analysis, because we wanted the line we obtained to be close to the actual data points. How can we assess how well the regression line accomplishes this?

If we plot the regression line on the same plot with the data we can get a *visual* or *graphical* idea of the connection between the model and the data. The proximity of the data points to the line will give an overall picture of how well the regression line describes the relationship between the variables.

EXAMPLE 15.7 Is TQM Working?

Plotting the Regression Line and Calculating Residuals

The company assessing its TQM decided to plot its data and the regression equation together to try to get a visual idea of what the least-squares line did.

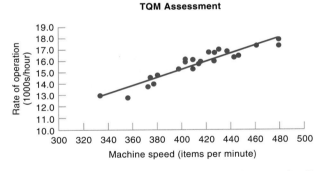

TQM Assessment

The graph shows that most of the data points are very close to the line. The company felt that the line it obtained did a pretty good job of describing the relationship between machine speed and productivity. ■

EXAMPLE 15.8 HMO Health

Plotting the Regression Line

The analysts who are looking at the relationship between revenues and number of members of HMOs created a plot of the data and the regression equation that they calculated.

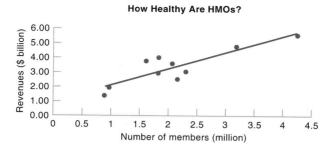

How Healthy Are HMOs?

They see that some of the points are very close to the line, but some others are farther away. Still, it looks like the line does a pretty good job of describing the relationship between revenues and number of members. The analysts wonder if they can get a numerical measure of the differences between the predicted values and the actual data. ■

A *numerical* measure of the agreement between the line and the data is obtained by looking at the differences between the observed and expected values of Y for each value of X. In regression analysis, these differences are known as the **residuals.**

> The difference between the observed value of Y (y_i), and the predicted value of Y from the regression equation ($\hat{y}_i$), for a value of $X = x_i$ is called the *i*th **residual,** e_i.

The *sign* of the residual, positive or negative, tells whether the prediction was lower or higher than the actual data value. The *size* of the residual gives an idea of how much the actual data vary around the line.

EXAMPLE 15.9 Is TQM Working?

Calculating Residuals

To obtain a better idea of how well the regression line agreed with their data, the workers decided to calculate the residuals for two predicted values close to the ones that they were interested in. They looked at their data and found that they had data for machine speeds of 373 and 447 items per minute, which were close to the 375 and 450 speeds. They used the regression equation to find the predicted values for $X = 373$:

$$\hat{y} = 1.06 + 0.035(373) = 14.1$$

and for $X = 447$:

$$\hat{y} = 1.06 + 0.035(447) = 16.7$$

and calculated the residuals, $y - \hat{y}$, for $X = 373$:

$$y - \hat{y} = 13.7 - 14.1 = -0.4$$

and for $X = 447$:

$$y - \hat{y} = 16.4 - 16.7 = -0.3$$

In each case, the difference between the observed data and the line was negative. The magnitude of the deviation was about 0.3 or 0.4, which is about 300 or 400 items per hour. ■

The residuals play a very important role in regression analysis. As you just saw, they give a numerical value to the differences between the model and the actual data for each of the data points in the sample. We have already discussed that, for a statistical relationship, the actual values of Y will vary for a fixed value of X. The residuals also provide a means to measure the overall variation in Y for any value of X. In addition, as we will see later on in this chapter, the residuals are used to determine the appropriateness of a linear model.

EXAMPLE 15.10   HMO Health

Calculating Residuals

For the HMO problem, the analysts wanted to use the model to predict revenues for different size HMOs. They were interested in predicting Y for $X = 1.0$ (1 million members) and 2.0 (2 million members). To have a way to compare the predicted values to the actual data, they chose the data points with $X = 0.97$ and 2.07, since these points were closest to the values of interest. The analysts calculated the predicted values using the equation $\hat{y} = 0.78 + 1.14x$ for each value and then calculated the residuals. The results were

Observation Number	x_i	y_i	$\hat{y}_i = 0.78 + 1.14x_i$	$e_i = y_i - \hat{y}_i$
9	0.97	1.71	$\hat{y}_9 = 0.78 + 1.14(0.97) = 1.89$	$e_9 = 1.71 - 1.89 = -0.18$
5	2.07	3.43	$\hat{y}_5 = 0.78 + 1.14(2.07) = 3.14$	$e_5 = 3.43 - 3.14 = 0.29$

The differences between the predicted revenues and the actual revenues were $180 million and $290 million or about 10%. ■

Although residuals are important, they cannot tell you everything about how good the model will be for predicting values of Y. Just because the residual for a certain data point is large, it does not mean that the regression equation does a poor job of predicting Y for values of X near that point. It may be that the data point used for the residual is unusually far from the regression line and that the prediction is actually more representative of the population. For this reason it helps to look at the residuals in context. Graphing the regression line with the data points provides additional insight about individual residuals.

EXAMPLE 15.11   HMO Health

Looking at the Residuals Graphically

When the HMO analysts looked at the plot of the data and the regression line, they saw that the data points that they used to calculate the residuals were actually very close to the regression line.

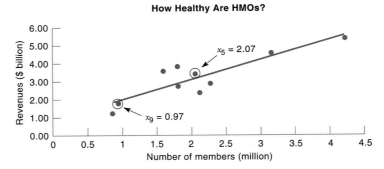

They felt that the errors for the number of members that they were interested in were not unusually large. ■

TRY IT NOW!

Increasing Capacity

Calculating the Residuals

The oil company that is looking at the relationship between refining capacity and the number of refining sites wants to get a better idea of how the regression line relates to the actual data. It decides to calculate the residuals for each observed value of X, the number of sites. Find the residuals and fill in the table below:

Observation Number	Number of Sites, X	Refining Capacity Y	$\hat{y}_i = 9.03 + 4.79x$	$e_i = y_i - \hat{y}_i$
1	13	81.82	71.3	
2	10	81.82	56.9	
3	13	58.18	71.3	
4	8	43.64	47.4	
5	5	40.00	33.0	
6	7	36.36	42.6	
7	4	34.55	28.2	
8	8	32.73	47.4	
9	7	29.09	42.6	
10	3	25.45	23.4	

Source: *The Economist,* 15 July 1995.

To get a picture of how the residuals and the regression line fit together, the company also decides to graph the regression line on a plot of the data. Graph the regression line on the data plot. How well do you think the line represents the data?

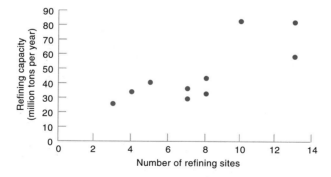

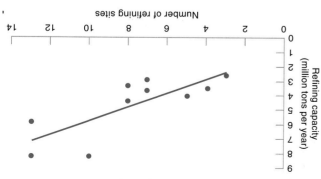

Oil Refining Companies

ANS. THE LINE DOES A PRETTY GOOD JOB EXCEPT FOR A FEW POINTS

15.2.7 The Standard Error of the Estimate

Since each residual corresponds to a single data point we know that we cannot use any single residual to determine how much the data differ from the model. Still, we would like some way to measure how much the data points vary around the regression line. This measure is known as the **standard error of the estimate,** $s_{y|x}$.

> The **standard error of the estimate,** $s_{y|x}$, is a measure of how much the data vary around the regression line.

The quantity that we are trying to measure is the *overall* or *average* variation of the data around the line. Remember that the residuals measure the *individual* variation of each data point from the line. We can obtain an overall measure of this variability by using something similar to the formula we used to measure the variability of the data from its mean in Chapter 6. In fact,

$$s_{y|x} = \sqrt{\frac{\Sigma(y - \hat{y})^2}{n - 2}} = \sqrt{\frac{\Sigma e^2}{n - 2}}$$

There are several different computational formulas for finding the standard error of the estimates. The one we have shown makes it easiest to understand what the quantity really measures. It is most convenient if you have already calculated the residuals. We will not look at any other formulas, because, as we already said, beyond finding the least-squares equation, regression analysis should be done using a computer.

EXAMPLE 15.12 HMO Health

Calculating the Standard Error of the Estimate

Since they knew that any single residual might be unusually large or small, the analysts decided to calculate the standard error of the estimate for the regression line. They calculated the residuals and squared residuals for each data point as

Observation Number	Members, X	Revenue, Y	$\hat{y}$	$e = y - \hat{y}$	e^2
1	4.24	5.49	5.61	−0.12	0.01
2	3.19	4.63	4.42	0.21	0.04
3	1.83	3.86	2.87	0.99	0.98
4	1.62	3.60	2.63	0.97	0.94
5	2.07	3.43	3.14	0.29	0.08
6	2.30	2.91	3.40	−0.49	0.24
7	1.83	2.74	2.87	−0.13	0.02
8	2.15	2.40	3.23	−0.83	0.69
9	0.97	1.71	1.89	−0.18	0.03
10	0.89	1.20	1.79	−0.59	0.35
Total					**3.38**

Then, the standard error of the estimate is

$$s_{y|x} = \sqrt{\frac{3.38}{8}} = \sqrt{0.423} = 0.65$$

This means that the typical deviation or distance of each data point from the line is 0.65 or $650 million. ■

The standard error of the estimate is analogous to the standard deviation of data from the mean and to the standard error of the mean, which we have studied before. From our knowledge of what these quantities measure, and the Empirical Rule, we can get an intuitive sense of what the quantity $s_{y|x}$ means. We can get a *mental image* of how the data are spread out around the line. Later on in the chapter we develop some more formal ways to use the standard error of the estimate to look at predictions.

In Section 15.2.4 we looked at the computer output for the HMO problem for several different software packages. Figure 15.8 shows the portion of the output from each of the packages that contains the standard error of the estimate. You will notice that in the Excel output, it is called the "standard error" while in the Minitab output it is simply labeled "S".

SUMMARY OUTPUT

Regression Statistics	
Multiple R	0.879773847
R Square	0.774002022
Adjusted R Square	0.745752275
Standard Error	0.65297731
Observations	10

Excel Output

Regression Analysis

The regression equation is
Rev \$bn = 0.784 + 1.14 Members (m)

Predictor	Coef	StDev	T	P
Constant	0.7845	0.5050	1.55	0.159
Members	1.1439	0.2185	5.23	0.000

S = 0.6530   R-Sq = 77.4%   R-Sq(adj) = 74.6%

Minitab Output

FIGURE 15.8   Computer output showing the standard error of the estimate

15.2.8 Exercises—Learning It!

15.1 In a study about postage tariffs and revenue, the European Economic Community wanted to look at the relationship between the number of postal employees in a country and the amount of domestic mail that was processed. They collected data for six different countries and found

Country	Number of Staff	Domestic Traffic (billions of pieces)
Germany	342,413	18.32
France	289,156	23.87
Italy	221,534	6.62
Britain	189,000	16.75
Spain	65,355	4.06
Sweden	52,251	4.21

Source: *The Economist,* 15 June 1996.

(a) Create a scatter plot of the data and determine whether a linear relationship is appropriate.

(b) Find the regression equation for the data.

(c) The Netherlands employs 53,560 postal workers. Use the regression equation to predict the domestic mail traffic for the Netherlands.

(d) Calculate the values of the residuals for France and Germany.

15.2 In trying to look at the effects of shopping center expansion, the Commerce Department decided to look at the relationship between the number of shopping centers and the retail sales for different states in the same region. It collected the data for the North Central states in the United States and found the following:

**North Central States
Retail Sales—1994**

State	Number of Shopping Centers	Retail Sales ($ billion)
Ohio	1559	34.8
Indiana	848	18.0
Illinois	1961	34.1
Michigan	967	21.0
Wisconsin	588	12.0
Minnesota	436	11.5
Iowa	277	6.3
Missouri	834	18.9
North Dakota	84	1.8
South Dakota	51	1.1
Nebraska	245	4.8
Kansas	458	9.7

Source: *Statistical Abstract of the United States 1995.*

(a) Create a scatter plot of the data.

(b) Find the regression equation relating retail sales and number of shopping centers.

(c) Plot the regression line on the same plot as the data. Do you think the line fits the data well? Why or why not?

(d) Use the regression line to predict retail sales for each state.

(e) Calculate the residuals for each state. Which state has the largest residual? Which state has the smallest? Do the residuals support your answer to part (d)?

(f) Find the standard error of the estimate.

15.3 As part of an international study on energy consumption, data were collected on the number of cars in a country and the total travel in kilometers. The data for 12 of the countries are shown below:

Country	Total Cars (millions)	Travel (billion km)
United States	142.35	3140.29
Finland	1.82	34.66
Denmark	1.66	30.76
Britain	21.32	352.76
Australia	8.53	138.22
Sweden	3.32	53.21
Netherlands	5.53	83.69
France	23.27	348.20
Norway	1.59	23.54
Italy	26.12	367.85
Germany	43.75	608.52
Japan	40.25	439.30

Source: *The Economist*, 22 June 1996.

(a) Create a scatter plot of the data. Do you think that there is a linear relationship between number of kilometers traveled and the number of cars?

(b) Find the least-squares regression line for the data. Interpret the value of the slope.

(c) Does the intercept make sense for these data? Why or why not?

(d) Plot the regression line on the same plot with the data. Does the line make you feel confident about predicting travel as a function of the number of cars?

(e) Use the regression line to predict the number of kilometers traveled for Sweden and Japan. How well do the predictions agree with the original data?

15.4 Large companies are always looking for expanding markets. Media companies in particular are looking at foreign markets for expansion. To obtain some information about expansion possibilities in Central Europe, a major communications company collected data on revenues from television advertising and the percentage of households with cable or satellite television for seven countries.

Country	% Cable	TV Advertising Revenues ($ million)
Hungary	55	170
Slovakia	47	28
Poland	33	369
Czech Republic	28	105
Romania	27	40
Russia	15	250
Bulgaria	9	23

Source: *The Economist*, 6 July 1996.

(a) Create a scatter plot of the data. Does it appear that advertising revenues are related to the percentage of households that have cable or satellite TV?

(b) Find the equation of the regression line for the data.

(c) Plot the regression line on the same plot as the data. Do you think that the line does a good job of predicting TV advertising revenues? Why or why not?

(d) From the plot, which country do you think will have the largest residual? Which will have the smallest?

(e) In the Ukraine, 8% of the households have cable or satellite television. Use the line to predict TV advertising revenues for the Ukraine.

(f) Find the residuals for each country. Do the values support your answer to part (d)?

(g) Calculate the standard error of the estimate for the regression line.

15.5 As part of the anti-crime bill passed in 1995, the U.S. government granted money to different cities to hire new beat-patrol officers. Nine cities in New Jersey were given grants and hired officers. In trying to assess the program, government analysts wanted to look at the relationship between the number of officers hired and the amount of the grant. The data are

Grant ($)	Number of Officers
150,000	2
375,000	5
471,125	6
70,967	1
450,000	6
525,000	7
375,370	7
750,000	10
1,000,000	12

Source: *Justice Department*.

(a) Create a scatter plot of the data.

(b) Find the equation of the linear regression line.

(c) Plot the regression line on the same plot as the data.

(d) How well do you think the line fits the data?

(e) Use the regression equation to predict the number of officers hired for each city.

(f) Calculate the residuals and the standard error of the estimate.

15.6   For many countries, tourism is an important part of revenues. In trying to predict tourism revenues, one of the independent variables that is considered important is the number of foreign visitors to the country. Data for six different countries were collected:

Country	Number of Visitors (million)	Tourism Receipts ($ billion)
France	60	27.3
Spain	48	25.1
US	45	58.4
Italy	30	27.1
Britain	23	17.5
Germany	15	11.9

Source: *The Economist,* July 1996.

(a) Create a scatter plot of the data and find the equation of the linear regression line.

(b) Use the regression line to predict tourism revenues for Italy and the United States.

(c) For which country does the regression line do a better job of predicting tourism receipts?

(d) Calculate the residuals and the standard error of the estimate.

15.7   As communications and the media change and become more important in the economy, the radio industry has become an area of concern. One of the major communications companies wanted to look at the relationship between the number of radio stations a company owned and the revenues generated by radio. It collected the following data:

American Radio Companies

Company	Number of Stations	Revenues ($ billion)
Westinghouse/Infinity	83	1.05
Jacor	57	0.31
Clear Channel	104	0.31
Evergreen	35	0.30
Disney/ABC	21	0.29
American Radio Systems	63	0.23
SFX	67	0.22
Chancellor	41	0.21
Cox	38	0.21
Bonneville	20	0.12

Source: *The Economist,* 29 June 1996, *Duncan's American Radio.*

(a) Create a scatter plot of the data. Do you think that there is a relationship between the number of stations that a company owns and the revenues generated by radio?

(b) Find the linear regression line for the data.

(c) Use the regression line to predict the radio revenues for Chancellor and Westinghouse/Infinity. Which prediction is better?

(d) Leave out the data point for Westinghouse/Infinity and recalculate the regression line. Do you think that this line will do a better job of predicting radio revenues? Why or why not?

15.3   INFERENCES ABOUT THE LINEAR REGRESSION MODEL

The tools you have learned for finding the simple linear regression equation are descriptive. The method of least squares allows you to find the equation of the line that best describes the sample data. Looking at the residuals allows you to get some idea about how well the line fits the data. Residuals are useful to some degree but, as in most cases, we would really like to be able to make *inferences* about the population based on our sample data.

In particular, we would like to know

- if the relationship described by the regression equation is *meaningful*.
- how well the equation enables us to predict values of the dependent variable.
- how useful the predictions are for decision making.

To accomplish these tasks we need to learn some inferential methods related to regression analysis.

15.3.1 Hypothesis Testing About the Slope, β_1

Saying that there is a linear relationship between X and Y means that as X changes, Y changes in some predictable, corresponding way. The parameter that describes the magnitude of the relationship is the slope of the line. Remember that the definition of slope is the change in Y for a unit change in X. Thus, if the variables X and Y are related, the slope of the line will be some number. If there is no relationship between X and Y, then the slope of the line is zero. That is, we say that as X changes, Y *does not change* in a related way (Figure 15.9). Thus, one of the ways that we can determine if the relationship between the two variables is real is to decide whether the slope of the regression line is equal to zero.

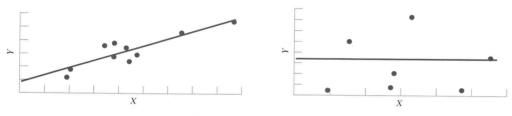

FIGURE 15.9   (a) Line with nonzero slope; (b) line with zero slope

While looking at data gives you an idea about what might be true in the population, you know quite well by now that you cannot simply look at the numbers to make these decisions. If the data consist of small numbers, the slope and intercept values, b_1 and b_0, will naturally be small. In the same way, if the data are large numbers, the corresponding slope and intercepts might or might not be large. You need to consider the *standard error* of the statistic in question to make decisions involving the values. We will use hypothesis testing to decide whether the slope of the regression line is *significantly different from zero*.

The first step in testing a hypothesis is to set up the appropriate hypotheses. In this case we want to test

$$H_0: \quad \beta_1 = 0$$
$$H_A: \quad \beta_1 \neq 0$$

If the test results in rejecting the null hypothesis, then we will conclude that the slope of the regression line is not equal to zero, and that the relationship between the X and Y variables is real.

Our estimate of β_1 is b_1, and to proceed with the steps of the hypothesis test we need to know about the sampling distribution of b_1. It turns out that the sampling distribution associated with the least-squares estimate of the slope is the Student t distribution. The test statistic for our hypothesis test is therefore

$$t = \frac{b_1 - \beta_1}{s_{b_1}}$$

You know that if you took a different sample your calculated regression line would not have exactly the same slope and intercept values. Thus, these estimates are sample statistics and have sampling distributions.

Formula for t *statistic*

which has a t distribution with $n - 2$ degrees of freedom. In the formula, s_{b_1} is the **standard error of the slope** b_1 and is calculated by

Formula for standard error of b$_1$

$$s_{b_1} = \frac{s_{y|x}}{\sqrt{\Sigma x^2 - (\Sigma x)^2 / n}}$$

This does not mean that the values cannot be calculated by hand. Once you find the regression coefficients and the standard error you have all of the pieces you need to find the test statistic.

Although the equation may look complicated we have said that for any purposes other than finding the regression line, we will be using statistical software to find the values we need. Since the test is a two-sided test, once the significance level of the test, α, is chosen, the critical values of the test are $\pm t_{\alpha/2, n-2}$. We now have the test set up. All that remains is to perform the test and make a decision.

EXAMPLE 15.13   Is TQM Working?

Hypothesis Test About the Slope Coefficient

The company looking at its TQM program has determined that the equation of the regression line relating productivity to machine speed is

$$\hat{y} = 1.06 + 0.035x$$

When the regression line and the data are graphed, it appears that the regression line does a good job of representing the data. Still, the company looks at the value of the slope, 0.035, and wonders if it really means anything. The company decides to perform a hypothesis test at the 0.05 level of significance, to see if the relationship is real.

The hypotheses are

$$H_0: \quad \beta_1 = 0$$
$$H_A: \quad \beta_1 \neq 0$$

Since the value of α is 0.05 and there are 25 data points, the critical values of the test are $\pm t_{0.025, 23}$ or ± 2.069.

From previous analyses the company knows that $\Sigma x = 10,340$, $\Sigma x^2 = 4,306,918$, and $s_{y|x} = 0.4544$. The value of the test statistic is

Remember that the hypothesized value for β_1 is zero.

$$t = \frac{0.035 - 0}{\dfrac{(0.4544)}{\sqrt{4,306,918 - (10,340)^2 / 25}}} = \frac{0.035}{0.00261} = 13.51$$

Comparing the value of the test statistic to the critical values, the company sees that 13.51 is definitely beyond the critical values and so it will reject H_0 and conclude that the slope of the regression line is not equal to zero. That is, there is a significant linear relationship between productivity and machine speed. ■

When we first talked about computer output, we said that there are usually three sections in the output. We are interested in the section with information about the regression coefficients. Figure 15.10 shows the relevant sections of the computer output for two different statistical packages.

In the output you are given all of the information needed to perform the hypothesis test about the slope coefficient. You can either look up the critical value for the level of significance that you have chosen and compare it to the t statistic, or use the p value of the test, which is also included in the output. In all cases, the p value of the test is 0.000. Since this is less than our chosen α, we reject H_0.

Predictor	Coef	StDev	T	P
Constant	1.063	1.084	0.98	0.337
Speed	0.035283	0.002611	13.51	0.000

Minitab

	Coefficients	Standard Error	t Stat	P-value
Intercept	1.062994916	1.083622206	0.980964501	0.336817039
Speed	0.035282894	0.002610746	13.51448824	1.9909E-12

Excel

FIGURE 15.10 *t*-test portion of computer output

EXAMPLE 15.14 HMO Health

Hypothesis Test About the Slope of the Regression Line

The analysts who were looking at the relationship between HMO revenues and number of members used statistical software to perform the regression analysis and to determine whether the relationship is significant. They want to test, at the 0.05 level of significance, the hypotheses

$$H_0: \quad \beta_1 = 0$$
$$H_A: \quad \beta_1 \neq 0$$

At the 0.05 level of significance with $10 - 2 = 8$ degrees of freedom, the critical values for the test are ± 2.306.

They obtained the following output:

```
Regression Analysis

The regression equation is
Rev $bn = 0.784 + 1.14 Members (m)

Predictor    Coef    StDev    T      P
Constant    0.7845   0.5050   1.55   0.159
Members     1.1439   0.2185   5.23   0.000
```

From the output line that corresponds to the slope estimate, they see that the coefficient is 1.1439 and the standard error of the slope is 0.2185. The corresponding t statistic is 5.23. Since the critical value for the test is 2.306, they reject H_0 and conclude that the slope of the regression line is not zero. That is, there is a significant linear relationship between revenues and number of members.

They can also use the p value of the test to make the decision. From the output, the p value of the test is 0.000. Since that is less than the specified α of 0.05, they know that they should reject H_0. ■

You might have noticed that in the computer output there is corresponding information about b_0, the intercept of the regression line. It is tempting to want to test hypotheses about β_0, the intercept, in particular whether it is equal to zero. *Do not do this*. Remember that the regression line is valid only in the range of X that has been observed in the data. If you do not have X data for $X = 0$, then you cannot make a determination about whether the intercept should remain in the equation. Removing the intercept from the equation when the slope is not equal to zero means that the regression line will move downward, parallel to itself, and will no longer be the line of best fit for the data.

 **TRY IT NOW!**

Increasing Capacity
Testing for Significance of the Regression Model

The oil company that is looking at increasing capacity wants to determine whether the relationship between refining capacity and number of refining sites that it calculated is significant.

    Write down the hypotheses that the company needs to test.

The company decides to use a 0.01 level of significance for the test. Find the critical values for the test.

Remember that you need to find the degrees of freedom, in this case $n - 2$.

It used a computer software package to run the analysis and obtained the following output:

	Coefficients	Standard Error	tStat	P-value	Lower 95%	Upper 95%
Intercept	9.028295455	11.04587353	0.817345539	0.437393793	−16.44355105	34.50014196
Number of sites	4.786628788	1.307226853	3.661666509	0.006385827	1.77215631	7.801101266

From the computer output, find the slope of the regression line, the standard error of the slope, and the value of the t statistic.

Perform the hypothesis test and make a decision about the regression line.

Find the p value of the test from the output and explain how you could use the p value on the output to make the same decision.

    Once we have determined that the relationship between X and Y is significant, we can perform some additional analyses to see if the predictions we obtain are useful for the purposes of decision making and to determine the strength of the relationship.

15.3.2 Partitioning the Variance in Linear Regression

In statistics, we are always interested in looking at the *variation* in data. In Chapter 6 you learned about the standard deviation and how that can be used to measure the variation of data around its mean value. When we are looking at the relationship between two variables, Y (the dependent variable) and X (the independent variable), we are interested in how Y *varies*, considering the values of the variable, X.

When two variables, X and Y, are linearly related, the variation of the y values from the overall mean of Y can be attributed to two different factors: its relationship with X and pure chance. In fact, the variation in the y values from the overall mean of Y can be partitioned, or divided, into parts called **sums of squares:**

> **SST**—the total variation in the y values around the mean $\bar{y}$
> **SSR**—the variation in Y that is caused by Y's relationship with X
> **SSE**—the variation in Y that remains unexplained

The quantities **SST, SSR,** and **SSE** are known as the **sums of squares:** SST is the total sum of squares, SSR is the regression sum of squares, and SSE is the error sum of squares. Figure 15.11 illustrates how these quantities relate to the data and to the regression line. From the picture you see that SST represents the distance of the data value y_i from the overall mean $\bar{y}$. SSR represents the distance of the predicted value $\hat{y}$ from the overall mean $\bar{y}$ and SSE represents the distance of a data value y_i from the predicted value $\hat{y}$. The formulas for calculating these quantities are

Notice that SSE is related to the residuals that you learned about previously.

$$\text{SST} = \sum (y - \bar{y})^2 \qquad \text{SSR} = \sum (\hat{y} - \bar{y})^2 \qquad \text{SSE} = \sum (y - \hat{y})^2$$

In addition you see that

$$\text{SST} = \text{SSR} + \text{SSE}$$

When X and Y are related, SSR is a large part of the total variation. This implies that a major reason that Y varies so much is *because* it is related to X. When this is true, the SSE component of the variation is small and is the variation in Y that happens "naturally" or entirely due to chance.

When X and Y are not related, the regression line is horizontal ($\beta_1 = 0$) and the SSR component of the variation disappears. The SSE part of the variation becomes dominant and we say that we cannot really explain the variation in Y using the linear model with X.

We can use these ideas to create a formal test for the significance of the linear

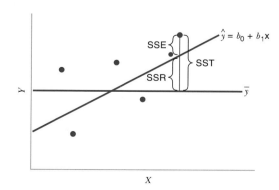

FIGURE 15.11 Components of the variation in y value

regression model. The test is referred to as an *analysis of variance (ANOVA)* test, because it is based on looking at the *variation* in the Y variable.

The hypotheses that we test are

H_0:   The linear regression model is not significant.

H_A: The linear regression model is significant.

The test statistic for this test uses the **mean squares,** which are obtained by dividing the sums of squares, SSR and SSE, by their respective degrees of freedom:

$$MSR = \frac{SSR}{1} \quad \text{and} \quad MSE = \frac{SSE}{n-2}$$

You saw the F *distribution when you looked at the test for comparing the variances of two normal populations in Chapter 13.*

The test looks at the ratio of the regression mean square (MSR) to the error mean square (MSE). As we just saw, when the regression model is really significant (that is, when X and Y are really linearly related), SSR (and therefore MSR) will be much larger than SSE (and therefore MSE). The ratio of the two mean squares has an F distribution with 1 and $n - 2$ degrees of freedom. If the ratio is larger than the critical F value, then the model is significant. Thus, the ANOVA test is a one-sided or one-tail test.

Calculation of SST, SSR, and SSE is not difficult, but for even small data sets it can be computationally exhausting. If you have any two of the components, you can obtain the third by subtraction. You can see from the formulas that SSE is really the sum of the squared residuals that you used to find $s_{y|x}$, and that SST is really the numerator portion of the variance of Y.

If you have been doing the analyses by hand or using a calculator, you probably have already calculated most of the quantities you need. However, if you are doing this much analysis you really should be using a statistical software package to do the job. You remember that we learned that output from a statistical analysis program includes a section known as the ANOVA section. Figure 15.12 shows the ANOVA output for Excel and Minitab.

You might wonder why you need another method for testing significance of a linear regression model. In reality you need only one method, since both tests (the t test and the F test) yield the same decision. The difference is, to some degree, a matter of preference. Other statistical analyses use the ANOVA approach and some people prefer that type of test.

ANOVA

	df	SS	MS	F	Significance F
Regression	1	37.7124744	37.7124744	182.6413924	1.9909E-12
Residual	23	4.749125596	0.206483722		
Total	24	42.4616			

Excel Output

Analysis of Variance

Source	DF	SS	MS	F	P
Regression	1	37.712	37.712	182.64	0.000
Error	23	4.749	0.206		
Total	24	42.462			

Minitab Output

FIGURE 15.12   Computer ANOVA output for regression analysis

The quantities SST, SSR, and SSE generated during the ANOVA approach also yield additional information about how good the linear regression model is. One of the reasons that we would like to know whether two variables are related is so that we can use the information to predict one given the other. Another reason is so that we can control or reduce the variation in one quantity by controlling the variation in the other. The regression sum of squares (SSR) measures the amount of the variation in the Y variable that can be accounted for or *explained by* Y's relationship with X. If you look at SSR as a portion of SST, then you can determine the amount of the variability in Y that can be explained, or accounted for. This value is called the **coefficient of determination, R^2**:

$$R^2 = \frac{\text{SSR}}{\text{SST}} \, 100\%$$

Formula for coefficient of determination

R^2 is usually part of the general information in the output from statistical packages.

The coefficient of determination gives you a measure of how much the variation in Y could be reduced if X were controlled to a single value. This is a way of measuring how useful a model is for planning purposes. Even if a linear regression model is significant, it may not explain enough of the variation in the Y variable to be useful for planning. When this is the case, you may want to use a regression model with more than one independent or *explanatory* variables. These models are called multiple regression models and are discussed in many standard statistical references.

15.3.3 Exercises—Learning It!

15.8 The data on the number of postal employees and the amount of mail processed were analyzed using Minitab. A portion of the output is given below:

```
Predictor         Coef        StDev       T       P
Constant         0.955        4.608     0.21    0.846
_ Staff       0.00005872   0.00002087   2.81    0.048

S = 5.461    R-Sq = 66.4%    R-Sq(adj) = 58.1%
```

(a) Set up the hypotheses to test whether the slope coefficient is equal to 0.

(b) There were 6 observations in the sample. At the 0.05 level of significance what are the critical values of the test?

(c) From the output, find the value of the coefficient, the standard error of the slope, and the value of the t statistic.

(d) Compare the t statistic to the critical value and determine whether the relationship between the number of postal employees and the amount of mail processed is significant.

15.9 Consider the data from Exercise 15.4 which looked at the relationship between the percentage of households with cable TV in Central Europe and TV Advertising revenues. The regression model is

$$\hat{y} = 127.2 + 0.442x \qquad s_{y|x} = 143.2 \qquad \Sigma x = 214 \qquad \Sigma x^2 = 8142$$

(a) Find the standard error of the slope.

(b) Set up the appropriate hypotheses and find the value of the test statistic.

(c) At the 0.10 level of significance, test to see if the slope coefficient is significantly different from 0.

15.10 In Exercise 15.6 you looked at the relationship between the number of visitors to a country, X, and the receipts from tourism, Y. The regression equation is

$$\hat{y} = 9.95 + 0.487x \qquad s_{y|x} = 15.50 \qquad \Sigma x = 221 \qquad \Sigma x^2 = 9583$$

(a) Calculate the standard error of the slope.

(b) Set up the hypotheses to determine whether the regression equation is significant at the 0.05 level.

(c) Find the critical values of the test and calculate the test statistic.

(d) Is the relationship between number of visitors and tourism receipts significant?

15.11 Consider the data from Exercise 15.3 on the number of kilometers of travel, Y, and the number of cars, X, for different countries. The results of a regression analysis using Microsoft Excel are given below:

	Coefficients	Standard Error	t Stat	P-value
Intercept	−106.212984	55.14747208	−1.92598102	0.082984278
Total Cars	2.15816E-05	1.194E-06	18.07496921	5.75839E-09

(a) From the output, find the slope coefficient, the standard error of the slope, the t statistic, and the p value of the test.

(b) At the 0.01 level of significance use the p value to determine if the relationship between number of kilometers traveled and the number of cars is significant. (The p value in Excel is a one-sided p value.)

(c) There were 12 observations in the data set. Find the critical values for the test and verify your answer to part (b).

15.12 In Exercise 15.5 you looked at the relationship between the amount of money, X, that a city was granted and the number of new police officers hired, Y, for nine different cities in New Jersey. The Minitab output from the regression analysis is shown below:

Predictor	Coef	StDev	T
Constant	0.7077	0.5360	1.32
Grant ($)	0.00001191	0.00000100	11.88

(a) Use the output to find the regression equation relating X and Y.

(b) Use the output to test whether the relationship between the amount of the grant and the number of new officers is significant.

15.13 Data on the number of radio stations owned by 10 different companies, X, and the revenues generated by radio, Y, were examined in Exercise 15.7. The data were analyzed using Microsoft Excel and the output is given below:

	Coefficients	Standard Error	t Stat	P-value
Intercept	0.085444208	0.176860232	0.483117131	0.641956743
# of Stations	0.004546468	0.003006584	1.512170824	0.16894211

Use the output to determine whether there is a significant relationship between the number of radio stations owned and the revenues generated by radio for a company.

15.14 Consider the data from Exercise 15.2 on the number of shopping centers and retail sales for the 12 North Central states. The data were analyzed with Minitab and the output is shown here

Predictor	Coef	StDev	T	P
Constant	1.331	1.006	1.32	0.215
No. Shop	0.019021	0.001129	16.84	0.000

At the 0.01 level of significance, is the relationship between the number of shopping centers and retail sales significant?

15.4 PREDICTION AND CONFIDENCE INTERVALS

We have discussed the fact that in a *statistical* relationship the values of the dependent variable will vary for any single value of the independent variable. No matter how strong the relationship between X and Y is, there is some amount of inherent variation that will exist in Y for any given value of X. This natural variation is identified in the population regression model $y = \beta_0 + \beta_1 x + \varepsilon$ as the quantity ε.

When the relationship between X and Y is strong, the variation in Y for a given value of X is small and can be thought of simply as *natural* variation. If you look at a plot of the data and the regression line, the points will be very close to the line. That is, the variation that exists is entirely due to chance and we should not waste time trying to figure out why it happens. In this case, the predictions we obtain from our estimate of the regression line will be fairly precise and can be used for the purposes of making decisions.

When the relationship between X and Y is not so strong, the variation in Y for a given value of X will be variation present because our linear model is *not adequate*. In this case, the data points will not be close to the regression line, the relationship might not be linear, or there may be other factors that cause Y to vary. In this case, even if X and Y have a statistically significant relationship, the predictions obtained from the regression model might not be precise enough for decision making purposes.

In Section 15.2 you learned that the standard error of the estimate, $s_{y|x}$, is a measure of the overall variation in Y for any value of X. In this section we put that value in context and see how it can be used to evaluate the usefulness of a regression model.

15.4.1 Confidence Intervals for Regression Analysis

The value of $\hat{y}$ obtained from the regression model is an estimate of the *mean value of Y for a given value* of X. That is, the value $\hat{y}$ is an *estimate* of the true mean, $\mu_{y|x}$. If the variable X represented advertising expenditures, then the regression model relating X to sales of a product, Y, would predict the *average* sales for a given level of advertising. Because our estimate is based on sample data, we know that there is some error in that estimate.

In Chapter 12 you learned how to find confidence intervals for the mean of a population, μ, based on a sample estimate, $\overline{X}$. We can use the same ideas to create a confidence interval for $\mu_{y|x}$ based on our sample estimate, $\hat{y}$.

To find a confidence interval for a population parameter we need to know the sampling distribution of the estimate and the standard error of the sample statistic. The estimate $\hat{y}$ has a Student t distribution with $n - 2$ degrees of freedom. The standard error is related to the standard error of the estimate, $s_{y|x}$. The formula for a $(1 - \alpha)100\%$ **confidence interval** for the mean value of Y for a given value of $X = x_i$, $\mu_{y|x_i}$, is

$$\hat{y}_i - t_{\alpha/2,n-2}s_{y|x}\sqrt{\frac{1}{n} + \frac{(x_i - \overline{x})^2}{\Sigma x^2 - (\Sigma x)^2/n}} \leq \mu_{y|x_i} \leq \hat{y}_i + t_{\alpha/2,n-2}s_{y|x}\sqrt{\frac{1}{n} + \frac{(x_i - \overline{x})^2}{\Sigma x^2 - (\Sigma x)^2/n}}$$

A **confidence interval** provides an estimate for the mean value of Y ($\mu_{y|x}$) at a particular value of X.

Again, the expression looks formidable, but it does not contain anything we have not already dealt with!

Previously, when you found confidence intervals, you found that the width of the interval was related to the level of confidence that you wanted. This is also true for the confidence interval for $\mu_{y|x}$, but the width of the interval also depends on how far the selected value of X is from the average value of X. The farther the value of interest is from the average X value, the wider (and less precise) the confidence interval becomes! Figure 15.13 (page 666) illustrates how the confidence interval is related to the values of the X variable.

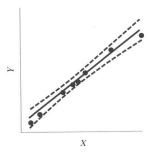

Y

X

FIGURE 15.13   Confidence interval for the mean estimate

If you look at the formula for the confidence interval again, you will see that when $X = \overline{X}$, the part of the formula in the square root reduces to $1/\sqrt{n}$ and the formula looks exactly like the confidence interval for the mean that you learned previously.

EXAMPLE 15.15   Is TQM Working?

Calculating a Confidence Interval for the Mean Response

The manufacturing group that is looking at the effectiveness of TQM wants to know something about the accuracy of the predictions from the regression model. To accomplish this, it decides to find 95% confidence intervals for two machine speeds that it runs frequently, 400 and 450 items per minute. The regression model is

$$\hat{y} = 1.06 + 0.035x$$

The first step is to find the predicted value for the 400 speed:

$$\hat{y} = 1.06 + 0.035(400) = 15.06$$

To actually calculate the interval it is probably easiest to calculate the ± part and then add and subtract that value from the prediction, $\hat{y}$.

From previous work the manufacturers know the predicted value for 450 is 16.81 and $s_{y|x} = 0.4544$, $\Sigma x = 10{,}340$, and $\Sigma x^2 = 4{,}306{,}918$. They calculate that $\overline{X} = 413.6$.

For a 95% confidence interval the t values are $\pm t_{0.025,23} = \pm 2.069$, so they calculate

$$(2.069)(0.4544)\sqrt{\frac{1}{25} + \frac{(400 - 413.6)^2}{4{,}306{,}918 - (10{,}340)^2/25}}$$

$$= (2.069)(0.4544)\sqrt{0.0461054 \ldots} = 0.202$$

So, the confidence limits are

$$15.06 \pm 0.20 = (14.86,\ 15.26)$$

or

$$(14.86 \leq \mu_{y|400} \leq 15.26)$$

The manufacturers see that the estimate of the mean productivity is somewhere between 14.86 (or 14,860 units) per hour and 15.26 (or 15,260 units) per hour. The width of the interval is

$$15.26 - 14.86 = 0.40 \text{ or } 400 \text{ units/hr.}$$

which is fairly precise for their purposes.

They do a similar calculation for a speed of 450 items per minute and obtain:

Notice that the only change in the calculation is substituting 450 for 400 in the square root part of the formula.

$$(2.069)(0.4544)\sqrt{\frac{1}{25} + \frac{(450 - 413.6)^2}{4,306,918 - (10,340)^2/25}}$$

$$= (2.069)(0.4544)\sqrt{0.0837367 \ldots} = 0.272$$

So the 95% confidence interval for a speed of 450 items per minutes is

$$16.81 \pm 0.27 = (16.54, 17.08)$$

or

$$(16.54 \leq \mu_{y|450} \leq 17.08)$$

That means that at a speed of 450 items per minute, average hourly production will be between 16,540 and 17,080 items. ■

It is not really difficult to calculate confidence intervals for one or two estimates, but if you want intervals for more estimates, using a computer is much easier.

EXAMPLE 15.16 HMO Health

Finding Confidence Intervals for the Mean Estimate

The analysts who are looking at the relationship between HMO revenues and number of members would like to know how accurate the estimates from the regression model are. They use Minitab to find 95% confidence intervals for the mean for each value of X in the data set. The output (edited) from the analysis is shown below:

X Value	Fit	95.0% CI
0.89	1.803	(1.025, 2.580)
0.97	1.894	(1.148, 2.640)
1.62	2.638	(2.101, 3.174)
1.83	2.878	(2.381, 3.374)
1.83	2.878	(2.381, 3.374)
2.07	3.152	(2.676, 3.629)
2.15	3.244	(2.767, 3.721)
2.30	3.415	(2.930, 3.901)
3.19	4.434	(3.710, 5.157)
4.24	5.635	(4.460, 6.809)

They see that the uncertainty in the estimated revenue (the width of the interval) goes from a low of $0.953 billion to a high of $2.349 billion. For the middle X values (close to the mean) the width of the interval is about $1 billion. They feel that the confidence intervals are very wide, especially considering the size of the actual estimates. It might not be practical to use this model to predict mean revenues. ■

Most statistical software packages will calculate the confidence intervals for either a single X value or for an entire set of X values. If you are using a spreadsheet program, entering the formula is not a difficult task.

TRY IT NOW!

Increasing Capacity
Finding Confidence Intervals for the Mean Predicted Value

After calculating the regression model and deciding that the model is significant, the analysts at the oil company would like to know about the accuracy of the estimates from the model. They decide to calculate 95% confidence intervals for $X = 8$ and 13 sites. They know from previous work that for the set of 10 observations in the model, $s_{y|x} = 13.43$, $\Sigma x = 78$, and $\Sigma x^2 = 714$.

Find 95% confidence intervals for the mean estimates.

Do you think that these estimates would be useful for planning purposes? Why or why not?

15.4.2 Prediction Intervals for Regression Analysis

A confidence interval for the mean estimate from the regression model gives the decision maker an idea of just how accurate the estimate is *over a long period of time*. That is, it is an interval estimate for the mean. This is certainly of importance for decisions that involve long-time horizons, but what about models that are important for predictions in a single instance?

If a company is looking at the relationship between sales and advertising, a confidence interval will tell it that, for a specific level of advertising, the *average* sales will vary between these two amounts. This is certainly useful information, but some additional information might also be helpful. The company might be interested in the accuracy of a *single observation*. For example, it might like to have an interval estimate for next month's sales for some specific advertising outlay. In a problem related to cash flow, this information would be more important than an overall mean. Intervals for individual estimates are called **prediction intervals** in regression analysis.

> A **prediction interval** gives an estimate for an individual value of Y at a particular value of X.

ANS. FOR $X = 8$ (37.51, 57.14) AND FOR $X = 13$ (52.77, 89.74).

The formula used to calculate a $(1 - \alpha)100\%$ prediction interval for a regression model is almost identical to the one for the confidence interval:

$$\hat{y}_i - t_{\alpha/2, n-2} s_{y|x} \sqrt{1 + \frac{1}{n} + \frac{(x_i - \bar{x})^2}{\Sigma x^2 - (\Sigma x)^2/n}} \leq y \leq \hat{y}_i + t_{\alpha/2, n-2} s_{y|x} \sqrt{1 + \frac{1}{n} + \frac{(x_i - \bar{x})^2}{\Sigma x^2 - (\Sigma x)^2/n}}$$

You see that the only difference is the "1" in the square root part of the formula. The impact of this change is that the prediction intervals for a specific value of X will be *wider* than the corresponding confidence interval. If you think about it, a wider interval for prediction makes sense. You know from the ideas of sampling and the Central Limit Theorem that the distribution of individual observations is always more variable than the distribution of the sample means.

EXAMPLE 15.17   Is TQM Working?

Finding a Prediction Interval for a Regression Model

The company that is looking at the TQM program thinks that the estimates for the mean production are useful, but the analysts would like to know how much variability there is in the estimate for the hourly productivity. They decide to find 95% prediction intervals for the estimates for speeds of 400 and 450 units per min. The relevant values for the 25 observations are $\hat{y}_{400} = 15.06$, $\hat{y}_{450} = 16.81$, $s_{y|x} = 0.4544$, $\Sigma x = 10,340$, and $\Sigma x^2 = 4,306,918$. The t value has not changed from the confidence interval calculations and so the analysts calculate the interval as follows:

$$(2.069)(0.4544)\sqrt{1 + \frac{1}{25} + \frac{(400 - 413.6)^2}{4,306,918 - (10,340)^2/25}}$$
$$= (2.069)(0.4544)\sqrt{1.0461055 \ldots} = 0.962$$

The 95% prediction interval is then

$$15.06 \pm 0.96$$

or

$$(14.10 \leq y \leq 16.02)$$

So, in a given hour, at 400 units per minute, production might vary from 14,100 to 16,020 units. Similarly for the 450 speed they find

$$(2.069)(0.4544)\sqrt{1 + \frac{1}{25} + \frac{(450 - 413.6)^2}{4,306,918 - (10,340)^2/25}}$$
$$= (2.069)(0.4544)\sqrt{1.0837367 \ldots} = 0.979$$

The prediction interval is

$$16.81 \pm 0.98$$

or

$$(15.83 \leq y \leq 17.79)$$

At 450 units per minute, hourly production will vary from 15,830 to 17,790 units.

They see that the variability for an individual month's productivity is greater than the variability for average production. The width of the intervals is about 0.97 or 970 items, which is still useful for planning purposes. ∎

Prediction intervals are important when the individual values for the Y variable are critical to the decision process. If, for example, the company looking at TQM is interested in planning warehouse capacity based on the amount of usable product it has, then a long-term estimate might not make sense.

**TRY IT NOW!**

Increasing Capacity

Calculating Prediction Intervals for Regression Estimates

The oil company analysts decide to calculate 95% prediction intervals for the two X values that they are interested in. The relevant values from the set of 10 observations are $s_{y|x} = 13.43$, $\Sigma x = 78$, and $\Sigma x^2 = 714$.

Find 95% prediction intervals for $X = 8$ and $X = 13$ refining sites.

Do you think that confidence intervals or prediction intervals would be more appropriate for the oil company's purposes?

15.4.3 Exercises—Learning It!

Requires Exercise 15.5   **15.15** For the data on federal grant money and police officers hired the relevant data are

$$\Sigma x = 4,167,462 \qquad \Sigma x^2 = 2,571,647,717,614 \qquad s_{y|x} = 0.8033 \qquad n = 9$$

(a) Find a 95% confidence interval for the mean number of officers hired when the amount of the grant is $350,000.
(b) Find a 95% prediction interval for the number of officers hired when the amount of the grant is $350,000.

Requires Exercise 15.1   **15.16** For the data on the number of postal workers in a country and the amount of mail processed, the relevant data are:

$$\Sigma x = 1,159,709 \qquad \Sigma x^2 = 292,657,611,087 \qquad s_{y|x} = 5.4614 \qquad n = 6$$

(a) Find 95% confidence intervals for the mean amount of mail processed when the number of postal workers is 50,000 and 250,000.
(b) Find 95% prediction intervals for the same two values of X.

15.17 For the data on the number of shopping centers and retail sales for the North Central States, the relevant data are: *Requires Exercise 15.2*

$$\Sigma x = 8308 \qquad \Sigma x^2 = 9,517,766 \qquad s_{y|x} = 2.1918 \qquad n = 12$$

(a) Find 98% confidence intervals for the mean retail sales when the number of shopping centers is 1000 and 1500.

(b) Which interval is wider? Why?

(c) Find 98% prediction intervals for the same values of X.

(d) Do you think that prediction or confidence intervals are more appropriate for these data?

15.18 For the data on the number of automobiles in a country and the total number of kilometers traveled, the relevant data are   *Requires Exercise 15.3*

$$\Sigma x = 319.51 \qquad \Sigma x^2 = 25,598.8975 \qquad s_{y|x} = 156.0981 \qquad n = 12$$

(a) Find 99% confidence intervals for the number of kilometers traveled for $X = 1.5$ and 10 (million) cars.

(b) Find 99% prediction intervals for the two values of X from part (a).

15.5 CORRELATION ANALYSIS

In many instances, when people talk about regression analysis they also talk about *correlation analysis*. The two topics are related, but not interchangeable. Both regression and correlation analyses deal with bivariate quantitative data and the relationship between the two variables. The main purpose of regression analysis is to find an equation or model that allows the decision maker to predict the value of the dependent variable. On the other hand, *correlation analysis* simply *measures the strength* of the relationship between two quantitative variables. The output of the analysis is a single number. In correlation analysis, there is no need to identify which variable is dependent and which is independent, since prediction is not the end result.

15.5.1 The Correlation Coefficient

We have talked about the types of linear relationships that can exist between two variables. In Figure 15.14 you see three types of relationships: perfect negative, none, and perfect positive.

If we simply want to measure the strength of a relationship between two variables, we use the **correlation coefficient.**

> The **correlation coefficient** is used as a measure of the strength of a linear relationship. A correlation of -1 corresponds to a perfect negative relationship, a correlation of 0 corresponds to no relationship, and a correlation of $+1$ corresponds to a perfect positive relationship.

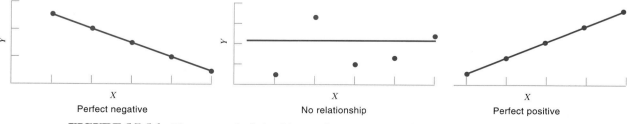

FIGURE 15.14   Three types of relationships: perfect negative, no relationship, and perfect positive

The **correlation coefficient, r,** is calculated using the formula

$$r = \frac{\Sigma xy - (\Sigma x)(\Sigma y)/n}{\sqrt{\Sigma x^2 - (\Sigma x)^2/n}\ \sqrt{\Sigma y^2 - (\Sigma y)^2/n}}$$

Again, we have a complicated formula, but it uses all of the familiar elements. Any scientific calculator that does regression will output the correlation coefficient. Software that performs regression analysis usually does not output the correlation coefficient directly because correlation and regression are not interchangeable. In the next section you will learn a statistic that is part of regression that can be used to obtain the correlation coefficient.

In many cases, the correlation coefficient is used along with regression analysis as another measure of how good the regression model is. It is important to recognize that the correlation coefficient can be used as a statistic in its own right when prediction of one variable as a function of the other is not appropriate or not necessary.

For example, suppose that a company is interested in knowing whether there is a relationship between the score on an aptitude test and the number of months that a person remains in an entry level position. The company does not necessarily want to *predict* the number of months in the entry level position; it simply wants to know if the test score is related to that variable. In this case the company could calculate the correlation between score on the test and number of months in the entry level job.

EXAMPLE 15.18   Is TQM Working?

Calculating the Correlation Coefficient

The company looking at its TQM program wonders what the correlation coefficient is for the relationship between machine speed and productivity. It decides to calculate the correlation. The relevant numbers are

$$\Sigma x = 10,340 \qquad \Sigma y = 391.4 \qquad \Sigma xy = 162,951.9 \qquad \Sigma x^2 = 4,306,918$$
$$\Sigma y^2 = 6170.22 \qquad n = 25$$

The calculation is

$$r = \frac{162,951.9 - (10,340)(391.4)/25}{\sqrt{4,306,918 - (10,340)^2/25}\sqrt{6170.22 - (391.4)^2/25}}$$
$$= \frac{1068.86}{(174.051\ \ldots)(6.516\ \ldots)} = 0.9424$$

The correlation coefficient is very close to $+1$, which indicates a strong, positive relationship between machine speed and productivity. ∎

You may know what values indicate good correlation and what values indicate that there is not really a relationship between two variables. The answer is "it depends." When data are collected under uncontrolled circumstances, there is usually a lot of variation in the data from other, uncontrolled, unmeasured variables. When this happens, the correlation between two variables might not appear to be strong because of the additional factors. This will also be reflected in the widths of the confidence and prediction intervals. Sometimes a correlation in the range of 0.6 to 0.7 (or

-0.6 to -0.7), which might not provide *useful* information, is an indication that further investigation is appropriate.

EXAMPLE 15.19 HMO Health

Calculating the Correlation Coefficient

The analysts looking at the relationship between revenues and number of members for various HMOs are pretty certain that the relationship between the two variables will be a strong, positive one, so they decide to calculate the correlation coefficient to check their assumption. The numbers that they need to perform the calculation are

$$n = 10 \qquad \Sigma x = 21.09 \qquad \Sigma y = 31.97 \qquad \Sigma xy = 77.6371$$
$$\Sigma x^2 = 53.4063 \qquad \Sigma y^2 = 117.3013$$

The calculation is

$$\frac{77.6371 - (21.09)(31.97)/10}{\sqrt{53.4063 - (21.09)^2/10}\sqrt{117.3013 - (31.97)^2/10}} = \frac{10.2124}{11.60795 \ldots} = 0.8798$$

Looking at the value, the analysts feel that the number does indicate the strong positive correlation that they expected to find. ∎

 **TRY IT NOW!**

Increasing Capacity

Calculating the Correlation Coefficient

The relevant data to calculate the correlation coefficient for the oil company problem are

$$n = 10 \qquad \Sigma x = 78 \qquad \Sigma y = 463.64 \qquad \Sigma xy = 4121.86 \qquad \Sigma x^2 = 714$$
$$\Sigma y^2 = 25{,}359.3224$$

Find the correlation coefficient for the data.

The correlation coefficient is also related to one of the quantities that we looked at in regression analysis, the coefficient of determination, R^2. The value of r is equal to the positive square root of R^2.

$$r = \sqrt{R^2}$$

ANS. 0.791

15.6 REGRESSION ASSUMPTIONS AND RESIDUAL ANALYSIS

While regression is certainly one of the most powerful and frequently used tools in statistical analysis, it is also one that is often misused. This sometimes happens because the model is built on certain *assumptions* and people are often unaware of the assumptions or choose to ignore them. As you have learned before in this book, when you ignore the *assumptions* behind a statistical tool, the results of the analysis might be totally incorrect. In this section we will look at the assumptions of the simple linear regression model and the ramifications of violating those assumptions. We will also look at some simple techniques for determining whether the assumptions of the model are appropriate for a particular set of data.

Another reason that the simple linear model is misused is that people do not realize that the model is inappropriate for the data being analyzed. This can be true even when the results of the analysis indicate that the simple linear model is appropriate. We will also look at ways to determine when the simple linear model is not appropriate.

15.6.1 Assumptions and Problems in the Regression Model

Remember that the simple linear model is given by

$$y = \beta_0 + \beta_1 x + \varepsilon$$

and that ε represents the random error in the model. Most of the assumptions that are built into the simple linear regression model involve this error term.

The basic assumptions about the error term ε are

1. It has a mean value of zero ($\mu_\varepsilon = 0$).
2. For every value of X, the standard deviation, σ, of ε is the same.
3. The distribution of ε is normal.
4. The error terms for the different observations are not correlated with each other.

What happens if these assumptions are not valid? Obviously, the simple linear model is not correct, but what does that really mean? It may mean several different things. One possible result of violating the assumptions is that you may decide that a model is significant when it is not, or vice versa. Another result is that although you may be correct in your decision about the significance of the model, the equation you obtain is completely wrong. That is, although the model may give reasonable predictions for Y, the coefficients in the model, b_0 and b_1, cannot be interpreted correctly. When this is the case, using the model for planning or control purposes is very risky.

In addition to violating the assumptions of the regression model, there are other factors that make the linear model inappropriate. Some of these factors are

- Fitting a linear model when the true relationship is nonlinear
- Influence of outliers on the regression equation

Sometimes data that appear to have a linear relationship in a scatter plot may actually have a nonlinear relationship. It is very often hard to see curves or nonlinear patterns when looking at scatter plots. The results of the analysis might indicate that the linear model is appropriate when in fact a nonlinear model would be better. Also, sometimes certain observations in the data set have a large influence on the regression equation. This often happens when the X value for an observation is far away from the other values. When these data points are removed, the entire analysis may change.

How can we tell if we are violating the assumptions or using the linear model incorrectly? One tool that answers many of these questions is called *residual analysis*. You have already learned how to calculate residuals. Now we will look at them as a diagnostic tool.

According to the assumptions of linear regression, for all values of the independent variable, the residuals of the model should be normally distributed random variables with a mean of zero and a standard deviation of σ. That is, the ε are $N(0, \sigma)$. If we plot the residuals from the regression model, the e_i's, as a function of X, we should expect to see that they are randomly distributed around a value of 0. Furthermore, the amount of variation around 0 should be the same for all values of X. When the plot exhibits departures from this it indicates that certain of the model assumptions are not valid or that a linear model was not appropriate.

In Figure 15.15 we see several different plots of the residuals versus the values of the independent variable, X. The first plot shows what the residual plot should look like when the assumptions of the linear regression model are met and when the simple linear model is appropriate. You can see that the residuals are randomly distributed around the mean value of 0 and that there is no apparent pattern in the plot. The others plots show different patterns that can result when the assumptions are violated or when the model is not appropriate. Figure 15.15b shows a typical plot that results when the linear model is used on data that have a nonlinear relationship. The residuals have a systematic pattern to them. Figures 15.15c and d show plots that can result when the assumption of equal variances is violated.

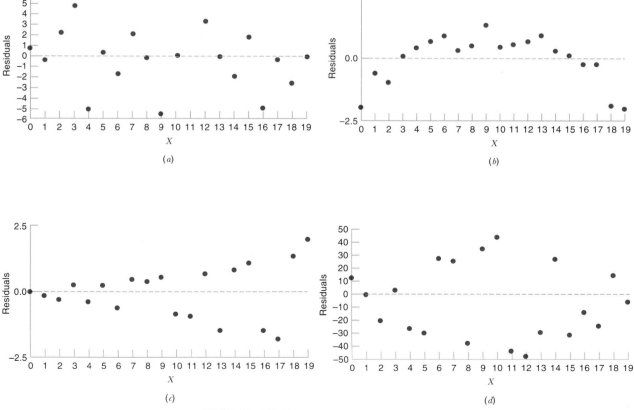

FIGURE 15.15   Examples of residual plots

Certain statistical packages plot the residuals against the fitted values ($\hat{y}$) instead of X. When the regression model is significant, X and Y are related and so the plots will not be different in shape. Also, some packages use "standardized residuals" (the residuals divided by the standard error of the residuals) instead of the residuals. Again, this does not produce a difference in the plots.

In addition to checking that the residuals have a mean of zero and that they are randomly dispersed around this value, we must check the shape of the distribution to see if it is normal. You remember that this is also an assumption for some of the hypothesis tests that you learned in Chapters 12 and 13. How can we check to see whether data come from a normal distribution? There are several ways to accomplish this. We will look at one *informal* method, making a histogram, and one *formal* method, a normal probability plot, that is available in many statistical software packages.

If you look at a data distribution to see if it might be normally distributed, you will look to see that it is approximately bell-shaped, that is, unimodal and symmetric. Figure 15.16 shows histograms of the residuals from the same plots that you looked at in Figure 15.15. In each case, a normal curve with a mean of 0 is overlaid on the histogram.

You may notice that even in the case where the residuals looked randomly distributed, it does not look like the residuals come from a normal distribution. It may be that the residuals are not normally distributed, or it may be that there is just not enough data to make a really good histogram. This is one of the problems with such an informal method.

Another, more formal method for determining whether data are normally distributed is a plot called a **normal probability plot.**

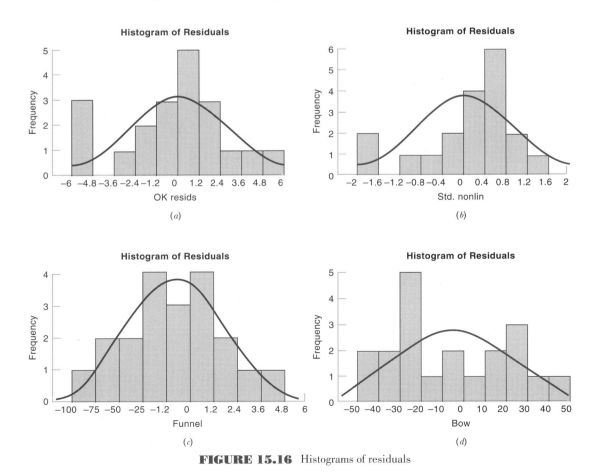

FIGURE 15.16 Histograms of residuals

A **normal probability plot** is a plot of the ordered data against their expected values under a normal distribution. When data are normally distributed, the plot will be a straight line.

The methods for creating a normal plot are well beyond the scope of this book, but they are easily created with statistical software packages. In fact, most statistical software make normal plots as part of their output from regression analysis. Figure 15.17 shows a set of residual plots, including a normal plot from Minitab, for each of the sets of residuals we have been looking at.

Another factor that can cause problems in using the linear regression model is the effects of observations that have excessive influence on the equation of the line. This can happen as a result of two things: data pairs that are outliers, and data points whose X values are very far removed from the rest of the data. Both of these factors can cause the regression model to shift toward the suspect data point and skew the resulting regression line.

There are a few ways to determine whether a data set has such an observation. One visual method, as seen in the Minitab output in Figure 15.17, is to create a control chart for the residuals and see if any of the values indicate an "out of control" situation. Another visual method is to make a boxplot of the residuals or the X data values and see if any are determined to be outliers.

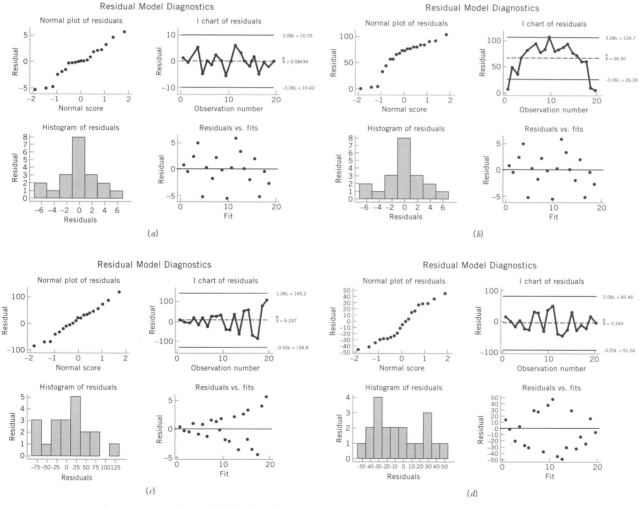

FIGURE 15.17 Regression diagnostic plots (Minitab)

Obs	Members	HMORev $	Fit	StDev Fit	Residual	St Resid
1	4.24	5.486	5.631	0.509	−0.146	−0.36 X
2	3.19	4.629	4.432	0.313	0.197	0.34
3	1.83	3.857	2.878	0.215	0.979	1.59
4	1.62	3.600	2.639	0.232	0.961	1.58
5	2.07	3.429	3.153	0.206	0.276	0.45
6	2.30	2.914	3.415	0.210	−0.501	−0.81
7	1.83	2.743	2.878	0.215	−0.136	−0.22
8	2.15	2.400	3.244	0.206	−0.844	−1.37
9	0.97	1.714	1.896	0.323	−0.182	−0.32
10	0.89	1.200	1.805	0.336	−0.605	−1.08

X denotes an observation whose X value gives it large influence.

Obs	of Sta	RadioRev	Fit	StDev Fit	Residual	St Resid
1	83	1.0500	0.4628	0.1191	0.5872	2.75P
2	57	0.3143	0.3446	0.0783	−0.0303	−0.13
3	104	0.3143	0.5583	0.1720	−0.2440	−1.40
4	35	0.3048	0.2446	0.0942	0.0602	0.27
5	21	0.2857	0.1809	0.1232	0.1048	0.50
6	63	0.2286	0.3719	0.0831	−0.1433	−0.62
7	67	0.2190	0.3901	0.0882	−0.1710	−0.75
8	41	0.2095	0.2718	0.0852	−0.0623	−0.27
9	38	0.2095	0.2582	0.0894	−0.0487	−0.21
10	20	0.1238	0.1764	0.1256	−0.0526	−0.25

FIGURE 15.18   Warning output from Minitab

It is also possible to determine whether a data point has a large influence on the model by using some statistical techniques. Many statistical software packages use these techniques and report the problem as part of the output. Figure 15.18 shows this part of the output for Minitab.

EXAMPLE 15.20   Is TQM Working?

Checking Regression Assumptions Using Residuals

The company that is using linear regression to determine whether its TQM program is working decides to make sure that what it has done so far is correct. Since the linear regression model is significant, the analysts want to make sure that it is valid and that they can use it for decision-making purposes. They use Minitab to analyze the data and looked at the output of residual plots. The output is shown below:

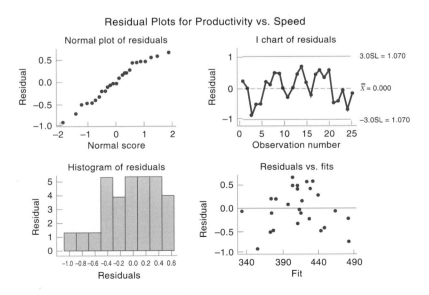

From the histogram and normal probability plots it appears that there might be a problem with the assumption of normality. The plots of the residuals versus the $\hat{y}$'s appears to be random, although there is an impression that the data actually might have a nonlinear relationship. The control chart indicates that there are no problems with extreme values.

Before the analysts decide to use this model, they will look more carefully at the data and see if another, perhaps nonlinear model, would be better. ■

15.6.2 Exercises—Learning It!

15.19 Look at the regression model you developed for the data on federal grant money and police officers hired.

Requires Exercises 15.5, 15.15

(a) Make a plot of the residuals vs. the values of the independent variable.

(b) Look at the plot you obtained. Does it appear that a linear model is really appropriate for these data? Why or why not?

(c) Looking at the same residual plot, does the assumption of equal variances appear reasonable? Why or why not?

(d) Use an appropriate graphical technique to display the distribution of the residuals. Be sure to consider how many residuals you have in selecting the technique. Does the assumption of normality appear to be reasonable?

(e) Create a normal probability plot of the residuals. What does this make you think about the normality assumption?

(f) Considering your answers to parts (a)–(e), do you think that the linear regression model was a good one for these data. Be sure to cite specifics in your answer.

15.20 Consider the regression model you found for the data on the number of postal workers in a country and the amount of mail processed.

Requires Exercises 15.1, 15.16

(a) Make a plot of the residuals versus the values of the independent variables.

(b) What does this lead you to believe about the appropriateness of the linear model?

(c) Now plot the residuals versus the predicted values of y.

(d) How do the two graphs compare? Why do you think this happened?

(e) Make a normal probability plot of the data. Do you think the assumption of normality is reasonable?

(f) Considering your answers to the parts (a)–(e), do you think that the linear regression model is appropriate for these data. Why or why not?

15.21 Look at the model you found for the data on the number of shopping centers and retail sales for the North Central states.

Requires Exercises 15.2, 15.17

(a) Make a plot of the residuals versus the independent variable.

(b) From the plot, does it appear that a linear model is appropriate?

(c) Looking at the residual plot, do you think that the assumption of equality of variances is violated?

(d) Make a graph that shows the distribution of the residuals.

(e) Does it appear that the residuals are normally distributed?

(f) Considering your answers to parts (a)–(e), do you think that the linear regression model is appropriate for these data? Why or why not?

15.22 Look at the model you found for the data on the number of automobiles in a country and the total number of kilometers traveled.

Requires Exercises 15.3, 15.18

(a) Make a residual plot of the data versus the independent variables.

(b) What does the residual plot tell you about the assumption of a linear relationship? The assumption of equal variances?

(c) Using the residual plot, the computer output, or a plot of the values of the independent variable, do you think that any observation had a lot of influence on the model? If so, which point(s) are they?

(d) Do you think that the assumption of normality is reasonable?

(e) Considering your answers to parts (a) – (d), do you think that the linear model is appropriate for these data? Why or why not?

15.7 SIMPLE LINEAR REGRESSION IN EXCEL

You can find the equation of the least squares line in Excel by creating a scatter plot and inserting a linear trend line. However, there is more to understanding the relationship between two variables than just finding the equation of the line. You can do much more to analyze the simple linear regression model in Excel by using a procedure from the data analysis tools.

15.7.1 Analyzing Quantitative Bivariate Data in Excel

Excel allows you to create a scatter plot and plot the least-squares line on the plot. Suppose that we want to plot the observations and find the least-squares line for data on the number of days that a company spends training employees at a particular job and the employees' performance as measured on a standardized aptitude test. The data are shown in Figure 15.19.

When creating a scatter plot in Excel, it is important that the independent variable be in the first column, and the dependent variable in the second column. The steps for plotting the data and finding the least-squares line are as follows:

1. Highlight the range of the data and start the Chart Wizard. Select XY (Scatter) as the chart type and Scatter (no lines) for the subtype, as shown in Figure 15.20.

2. Select **Next>** twice to get to the Chart Options dialog box; put in titles and make any other desired formatting choices.

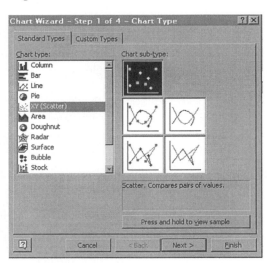

	A	B
1	**Training**	**Score**
2	1.0	41
3	1.5	60
4	2.0	72
5	2.5	91
6	3.0	99

FIGURE 15.19   Training Data   **FIGURE 15.20**   Selecting the scatter plot option

3. Select **Next>** to tell Excel where to locate the chart and **Finish** to display the chart. The chart should look like the one in Figure 15.21.

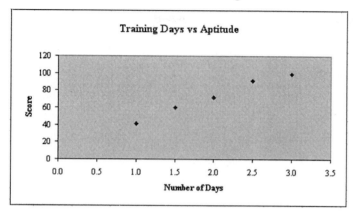

FIGURE 15.21 Finished scatter plot

Once you have the scatter plot, you can add the least-squares line as follows:

1. Click on any one of the points in the scatter plot to highlight them all.

2. From the **Chart** menu, select **Add Trendline,** as shown in Figure 15.22. The Add Trendline dialog box will open.

Make sure you single click here; double clicking will bring up a different dialog box.

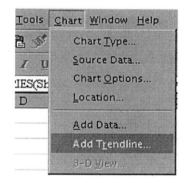

FIGURE 15.22 Chart menu

3. In the Trend/Regression section, highlight **Linear,** as shown in Figure 15.23.

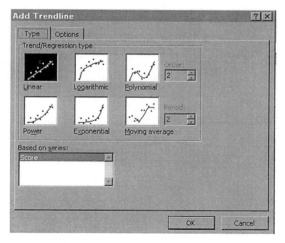

FIGURE 15.23 Selecting the type of trendline

4. Click on the **Options** tab and click the box labeled **Display equation on chart,** as shown in Figure 15.24. You can also choose to label the line with a title other than the default.

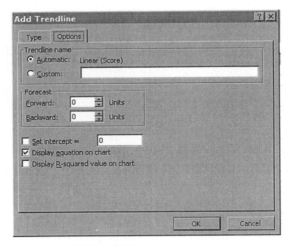

FIGURE 15.24   Trendline options

5. Click **OK;** the least-squares line and its equation will appear on the chart, as shown in Figure 15.25.

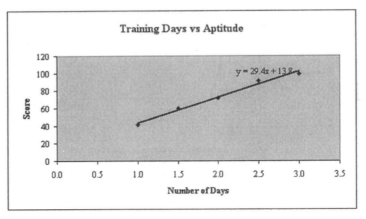

FIGURE 15.25   Scatter plot with least-squares line

Notice that Excel displays the equation with the slope term first and the intercept term last.

15.7.2 The Linear Regression Model Analysis in Excel

Let's look at the problem of predicting HMO revenues based on the number of members. The data should be located in an Excel worksheet similar to the one shown in Figure 15.26.

To find the simple linear regression model for a set of data, follow these steps:

1. From the list of data analysis tools, select **Regression.** The dialog box shown in Figure 15.27 opens.

	A	B	C
1	**HMOs**	**Members (m)**	**Rev $bn**
2	United HealthCare	4.24	5.49
3	Humana	3.19	4.63
4	FHP International	1.83	3.86
5	PacifiCare Health systems	1.62	3.60
6	U.S. Healthcare	2.07	3.43
7	WellPoint Health Network	2.3	2.91
8	Health Systems International	1.83	2.74
9	Foundation Health	2.15	2.40
10	Oxford Health Plans	0.97	1.71
11	Physician Corp. of America	0.89	1.20

FIGURE 15.26   The HMO data in Excel

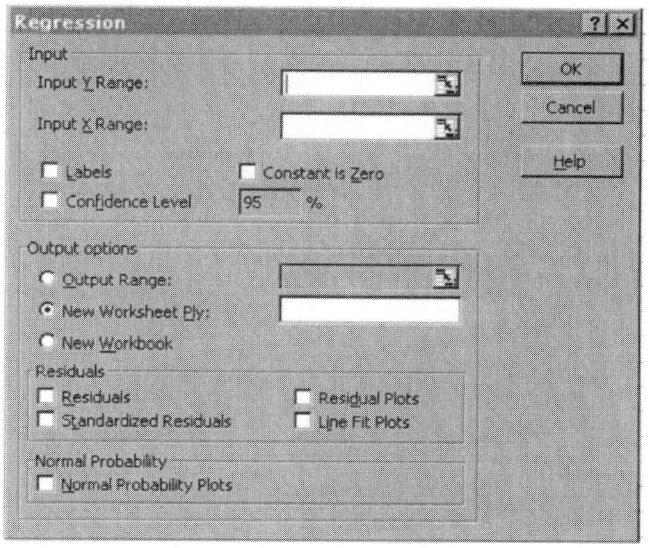

FIGURE 15.27   The regression dialog box

The dialog box has two sections, Input and Output. The input section gives Excel information about your data. The output section lets you make decisions about specific types of extra output that are available.

2. Position the cursor in the textbox labeled **Input Y Range:** and highlight the data range for the *Y* variable, in this case, Revenues.

3. Move the cursor to the textbox for **Input X Range:** and highlight the data range of the *X* variable, in this case, Members.

4. If the data ranges contain labels, click on the **Labels** checkbox. If you want confidence intervals for the regression estimates, click the checkbox for **Confidence Level.**

5. Specify the location where you want the output to appear, either in the current sheet, in a new worksheet, or in a new workbook.

In addition to standard regression output, Excel will create some of the diagnostic plots that were discussed in this chapter. You can choose to have the residuals and/or standardized residuals as part of the output. You can also have residual plots, fitted line plots, and normal probability plots. We recommend that you not create some of these plots since the values are calculated incorrectly.

6. Click the checkbox for **Residuals**. Do *not* check the **Standardized Residuals** checkbox. Excel does not calculate these values correctly.

7. Click the checkboxes for **Residual Plots** and **Line Fit Plots**. Do *not* click the checkbox for **Normal Probability Plot**. The plot is not created correctly.

The completed dialog box should resemble the one in Figure 15.28.

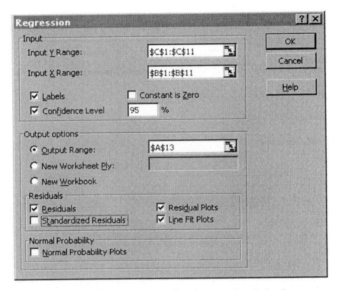

FIGURE 15.28   Completed regression dialog box

8. Click on **OK.** The output will appear in the location you specified.

The output from the regression analysis is quite extensive. While the whole output is still highlighted, you should select **F̲ormat > Column > AutoFit Selection** so that the columns adjust to fit the various output sections.

The first section of the output is the Summary Section. This section gives information about the regression model including the coefficients, the values of R^2, the standard error, and the ANOVA output. This output is shown in Figure 15.29.

The coefficients for the regression equation are at the end of the section. You see that the equation is $\hat{y} = 0.78 + 1.1439x$. In the same section, you see the information for testing whether the regression model is significant. You can perform this test

13	SUMMARY OUTPUT						
14							
15	*Regression Statistics*						
16	Multiple R	0.879773847					
17	R Square	0.774002022					
18	Adjusted R Square	0.745752275					
19	Standard Error	0.65297731					
20	Observations	10					
21							
22	ANOVA						
23		*df*	*SS*	*MS*	*F*	*Significance F*	
24	Regression	1	11.68217506	11.68217506	27.3985468	0.000788678	
25	Residual	8	3.411034941	0.426379368			
26	Total	9	15.09321				
27							
28		*Coefficients*	*Standard Error*	*t Stat*	*P-value*	*Lower 95%*	*Upper 95%*
29	Intercept	0.784464301	0.505044505	1.55325777	0.158965828	-0.380171168	1.94909977
30	Members (m)	1.143923992	0.218541241	5.23436212	0.000788678	0.639966661	1.647881323

FIGURE 15.29   Summary section of regression output

using the ANOVA output just above. In both cases, we see that the regression model is significant.

The top section of the output, Regression Statistics, includes the value for R^2 and the standard error. Excel does not include the correlation coefficient specifically, but you can get it by taking the square root of the R^2 value.

The second section of the output you created is the Residual section, as shown in Figure 15.30. This section contains the values of $\hat{y}$ and the residuals.

34	RESIDUAL OUTPUT		
35			
36	Observation	Predicted Rev $bn	Residuals
37	1	5.634702027	-0.144702027
38	2	4.433581835	0.196418165
39	3	2.877845206	0.982154794
40	4	2.637621168	0.962378832
41	5	3.152386964	0.277613036
42	6	3.415489482	-0.505489482
43	7	2.877845206	-0.137845206
44	8	3.243900884	-0.843900884
45	9	1.894070573	-0.184070573
46	10	1.802556654	-0.602556654

FIGURE 15.30   Residual output

The plots that are created as part of the output are stacked on top of each other. Figure 15.31 shows the plots after they have been moved in the worksheet.

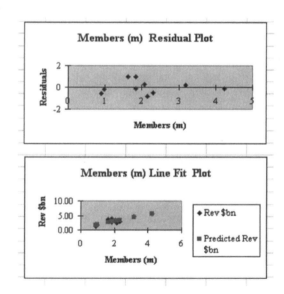

FIGURE 15.31   Plots from regression analysis

The residual plot is a plot of the residuals vs. the X variable. From this plot, you can look at whether the assumption of equal variances is violated. You can also get information about whether a linear model is appropriate, or whether any observation exerts a lot of influence on the model.

The Line Fit plot shows the predicted and fitted values of y on the same plot. This gives you information about how well the model fits the data. If the predictions were perfect, the two sets of symbols would coincide exactly.

15.7.3 Confidence and Prediction Intervals Using Excel

Although Excel does not provide an option to output confidence intervals or prediction intervals as part of the regression routine, it is not hard to calculate these intervals. All of the values that are needed are included as part of the regression output. You can use formulas to find the upper and lower values for the intervals.

Recall that the formulas for the $(1 - \alpha)100\%$ confidence interval for the mean value of Y at a particular value of X are given by

$$\hat{y}_i - t_{\alpha/2, n-2} s_{y|x} \sqrt{\frac{1}{n} + \frac{(x_i - \bar{x})^2}{\Sigma x^2 - (\Sigma x)^2/n}} \leq \mu_{y|x_i} \leq \hat{y}_i + t_{\alpha/2, n-2} s_{y|x} \sqrt{\frac{1}{n} + \frac{(x_i - \bar{x})^2}{\Sigma x^2 - (\Sigma x)^2/n}}$$

and the formulas for the corresponding prediction intervals are given by

$$\hat{y}_i - t_{\alpha/2, n-2} s_{y|x} \sqrt{1 + \frac{1}{n} + \frac{(x_i - \bar{x})^2}{\Sigma x^2 - (\Sigma x)^2/n}} \leq \mu_{y|x_i}$$

$$\leq \hat{y}_i + t_{\alpha/2, n-2} s_{y|x} \sqrt{1 + \frac{1}{n} + \frac{(x_i - \bar{x})^2}{\Sigma x^2 - (\Sigma x)^2/n}}$$

Excel allows you to obtain the predicted values $\hat{y}$ as part of the output; all other parts of the formula are either part of the output or can be calculated using other Excel functions. Unless you are adept at manipulating formulas, entering a formula this complicated in Excel is best accomplished by breaking it into smaller pieces and then combining the final results. For example, you should calculate the quantity under the square root separately and take the square root in a separate step. This minimizes possible problems with using parentheses.

We have provided a macro that will calculate 95% prediction intervals for an individual value Y at a particular value of X. To use the macro, the X and Y data should be in columns of a worksheet with no titles at the top. The X column must come first. From the list of **Tools > Macro > Macros,** choose **Prediction Interval Limits** and click **Run.** The predicted values, the residuals, and the upper and lower prediction intervals are output to the worksheet as shown in Figure 15.32.

You can use the formulas in the macro to help you set up formulas for the confidence intervals.

Note: The data are originally located in the top left of the worksheet. The macro moves it down the sheet.

Prediction Interval Limits					
	b_0	0.784464301	**xBar**	2.109	
	b_1	1.143923992	**t$_\alpha$**	2.306005626	
	α	5%	**SumX^2**	53.4063	
	n	10	**s$_{XY}$**	0.65297731	
X	**Y**	*Prediction*	*Residual*	*LPL*	*UPL*
4.24	5.49	5.6347	-0.1447	3.7249	7.5445
3.19	4.63	4.4336	0.1964	2.7630	6.1042
1.83	3.86	2.8778	0.9822	1.2923	4.4634
1.62	3.60	2.6376	0.9624	1.0392	4.2360
2.07	3.43	3.1524	0.2776	1.5730	4.7318
2.3	2.91	3.4155	-0.5055	1.8333	4.9977
1.83	2.74	2.8778	-0.1378	1.2923	4.4634
2.15	2.40	3.2439	-0.8439	1.6645	4.8233
0.97	1.71	1.8941	-0.1841	0.2137	3.5744
0.89	1.20	1.8026	-0.6026	0.1080	3.4971

FIGURE 15.32   Output from prediction interval macro

CHAPTER 15 SUMMARY

Linear regression analysis is a powerful tool for determining how two variables are related. The regression equation can be used for *description, control,* and *prediction.* Description is important when the user is simply trying to understand the way that two variables are related. Control describes when the model is used to set standards or reduce variability. Prediction is when the model is used to determine what the resulting Y value should be when X takes on certain values. Although finding the simple linear model itself is not numerically difficult, performing a complete regression analysis requires the use of computer software to completely understand the model.

We have also seen that even though the simple linear model may be significant, it might not be correct. It is necessary to test the *assumptions* of the linear model to see whether the model you obtain is appropriate.

Key Terms

Term	Definition	Page Reference
Simple linear regression model	The true relationship between the variables X and Y, the simple linear regression model, can be described by $$y = \beta_0 + \beta_1 x + \varepsilon$$	639
Deviation	The distance between the predicted value of Y, $\hat{y}$, and the actual value of Y, y, is called the deviation or error.	641
Least-squares method	The technique that finds the equation of the line that minimizes the total or sum of the squared deviations between the actual data points and the line is called the least-squares method.	642
$\hat{y}$ (y hat)	The value of $\hat{y}$ that we find is really a prediction of the mean value of Y for a given value of X.	646
Interpolation and extrapolation	Using the equation to predict values of Y within the range of the X data is called interpolation. Predicting values of Y for values of X outside the observed range is called extrapolation.	647
Residual (e_i)	The difference between the observed value of Y (y_i), and the predicted value of Y from the regression equation, ($\hat{y}_i$), for a value of $X = x_i$ is called the ith residual, e_i.	649
Standard error of the estimate ($s_{y\,|\,x}$)	The standard error of the estimate, $s_{y\,|\,x}$, is a measure of how much the data vary around the regression line.	652

Term	Definition	Page Reference
Sums of squares	**SST**—the total variation in the y values around the mean $\bar{y}$. **SST**—the variation in Y that is caused by Y's relationship with X. **SSE**—the variation in Y that remains unexplained.	661
Confidence interval	A confidence interval provides an estimate for the mean value of Y ($\mu_{y\mid x}$) at a particular value of X.	665
Prediction interval	A prediction interval gives an estimate for an individual value of Y at a particular value of X.	668
Correlation coefficient	The correlation coefficient is used as a measure of the strength of a linear relationship. A correlation of -1 corresponds to a perfect negative relationship, a correlation of 0 corresponds to no relationship, and a correlation of $+1$ corresponds to a perfect positive relationship.	671
Normal probability plot	A normal probability plot is a plot of the ordered data against their expected values under a normal distribution. When data are normally distributed, the plot will be a straight line.	677

Key Formulas

Term	Formula	Page Reference
b_1	$b_1 = \dfrac{n\sum_{i=1}^{n} x_i y_i - \sum_{i=1}^{n} x_i \sum_{i=1}^{n} y_i}{n\sum_{i=1}^{n} x_i^2 - \left(\sum_{i=1}^{n} x_i\right)^2}$ or $\dfrac{n\sum xy - \sum x \sum y}{n\sum x^2 - \left(\sum x\right)^2}$	642
b_0	$b_0 = \dfrac{\sum_{i=1}^{n} y_i}{n} - b_1 \dfrac{\sum_{i=1}^{n} x_i}{n}$ or $b_0 = \bar{y} - b_1\bar{x}$	642
Standard error of the estimate $s_{y\mid x}$	$s_{y\mid x} = \sqrt{\dfrac{\sum(y - \hat{y})^2}{n-2}} = \sqrt{\dfrac{\sum e^2}{n-2}}$	652
Standard error of the slope	$s_{b_1} = \dfrac{s_{y\mid x}}{\sqrt{\sum x^2 - (\sum x)^2/n}}$	658
Test statistic for slope	$t = \dfrac{b_1 - \beta_1}{s_{b_1}}$	657
Sums of squares	$\text{SST} = \Sigma(y - \bar{y})^2 \qquad \text{SSR} = \Sigma(\hat{y} - \bar{y})^2 \qquad \text{SSE} = \Sigma(y - \hat{y})^2$	661
Mean squares	$\text{MSR} = \dfrac{\text{SSR}}{1}$ and $\text{MSE} = \dfrac{\text{SSE}}{n-2}$	662
Coefficient of determination, R^2	$R^2 = \dfrac{\text{SSR}}{\text{SST}} 100\%$	663
Confidence interval	$\hat{y}_i - t_{\alpha/2, n-2} s_{y\mid x} \sqrt{\dfrac{1}{n} + \dfrac{(x_i - \bar{x})^2}{\Sigma x^2 - (\Sigma x)^2/n}} \leq \mu_{y\mid x_i}$ $\leq \hat{y}_i + t_{\alpha/2, n-2} s_{y\mid x} \sqrt{\dfrac{1}{n} + \dfrac{(x_i - \bar{x})^2}{\Sigma x^2 - (\Sigma x)^2/n}}$	665
Prediction interval	$\hat{y}_i - t_{\alpha/2, n-2} s_{y\mid x} \sqrt{1 + \dfrac{1}{n} + \dfrac{(x_i - \bar{x})^2}{\Sigma x^2 - (\Sigma x)^2/n}} \leq \mu_{y\mid x_i}$ $\leq \hat{y}_i + t_{\alpha/2, n-2} s_{y\mid x} \sqrt{1 + \dfrac{1}{n} + \dfrac{(x_i - \bar{x})^2}{\Sigma x^2 - (\Sigma x)^2/n}}$	669
Correlation coefficient, r	$r = \dfrac{\Sigma xy - (\Sigma x)(\Sigma y)/n}{\sqrt{\Sigma x^2 - (\Sigma x)^2/n}\,\sqrt{\Sigma y^2 - (\Sigma y)^2/n}}$	672

CHAPTER 15 EXERCISES

Learning It!

15.23 How much does advertising impact market penetration? To assess the impact of advertising in the tobacco industry, a study looked at the amount of money spent on advertising a particular brand of cigarettes and brand preference among adolescents and adults. The data are shown below:

| | | *Brand Preference* | |
Brand	Advertising ($ million)	Adolescent (%)	Adult (%)
Marlboro	75	60	23.5
Camel	43	13.3	6.7
Newport	35	12.7	4.8
Kool	21	1.2	3.9
Winston	17	1.2	3.9
Benson & Hedges	4	1.0	3.0
Salem	3	0.3	2.5

Source: *Center for Disease Control Website.*

(a) Look at the data for brand preference for adolescents and amount spent on advertising. Which variable is the dependent variable? Which is the independent variable?

(b) Create a scatter plot of advertising and adolescent brand preference. Do you think that there is a linear relationship between the two variables? Why or why not?

(c) Now create another scatter plot using adult brand preference instead. How does this plot compare to the one for adolescent brand preference? From the plots, do you think that adolescent or adult brand preference is more strongly related to advertising expenditures? Why?

(d) Find the least-squares line for adolescent brand and advertising expenditures.

(e) Interpret the meaning of the slope and intercept for the model. Do they make sense?

(f) Use the model to predict adolescent brand preference for each brand studied. How well do the predicted values agree with the actual data?

(g) Using $\alpha = 0.05$, is the model significant?

15.24 Retention of students is one of the largest problems facing colleges and universities today. Loss of students impacts revenues not only from tuition but from state and federal funding programs. If a university can understand what factors impact student retention it can form strategic plans aimed at keeping students enrolled. Academic performance is often cited as one reason that students leave. Data were collected from 20 colleges in the Midwest on the freshman retention rate (% of freshmen who stay for a second year) and the 25th percentile score on the American College Testing (ACT) examination. (The ACT test is a college entrance examination similar to the Scholastic Assessment Test, SAT.) The data are given below:

College	ACT 25th Percentile	Freshman Retention Rate
1	21	0.82
2	25	0.86
3	22	0.78
4	22	0.82
5	22	0.82
6	23	0.87
7	23	0.90
8	21	0.85
9	23	0.84
10	20	0.80
11	21	0.84

(continued)

College	ACT 25th Percentile	Freshman Retention Rate
12	23	0.82
13	21	0.82
14	23	0.83
15	26	0.78
16	21	0.79
17	22	0.83
18	20	0.78
19	22	0.78
20	21	0.61

(a) Which variable is the independent variable? Which is the dependent variable?

(b) Create a scatter plot of the data. Does it appear that freshman retention rate is related to the 25th percentile on the ACT exam?

(c) Find the equation of the regression line for the data.

(d) Plot the regression line on the same plot as the data. Do you think that the line does a good job of predicting freshman retention at a college or university? Why or why not?

(e) Calculate the standard error of the estimate, $s_{y|x}$, for the regression line.

(f) At the 0.05 level, is the model significant?

15.25 The British Bankers' Association wanted to look at the relationship between the amount of deposits made (in billions of £) and the number of customers that a bank had. Analysts collected data on six different large banks and found the following information:

Bank Name	Deposits (£ billion)	Customers (million)
Abbey National	101.7	13.6
Barclays	108.2	10.0
Lloyds	96.9	15.0
National Westminster	113.8	7.5
Woolrich	27.5	4.0
Halifax	77.1	7.6

(a) Which variable is the independent variable? Which is the dependent variable?

(b) Create a scatter plot of the data. Does it appear that the amount of deposits is related to the number of customers?

(c) Find the equation of the regression line for the data.

(d) Plot the regression line on the same plot as the data. Do you think that the line does a good job of predicting the amount of deposits? Why or why not?

(e) Calculate the standard error of the estimate, $s_{y|x}$, for the regression line.

(f) At the 0.05 level, is the model significant?

Datafile: POVERTY.XXX

15.26 An education task force looking at poverty levels in the United States has collected data for each state and the District of Columbia on the total number of people below the poverty level and the number of adults over the age of 25 who did not graduate from high school for the year 1993. The data are shown in the table below:

State	Number (1000) Below Poverty Level	Number of Adults Over 25 Not HS Graduates	State	Number (1000) Below Poverty Level	Number of Adults Over 25 Not HS Graduates
Alabama	725	843,638	District of Columbia	158	109,866
Alaska	52	43,244	Florida	2507	2,271,074
Arizona	615	491,080	Georgia	919	1,169,815
Arkansas	484	503,481	Hawaii	91	141,506
California	5803	4,450,528	Idaho	150	121,787
Colorado	354	328,056	Illinois	1600	1,735,789
Connecticut	277	457,208	Indiana	704	850,014
Delaware	73	96,472	Iowa	290	353,800

(continued)

State	Number (1000) Below Poverty Level	Number of Adults Over 25 Not HS Graduates	State	Number (1000) Below Poverty Level	Number of Adults Over 25 Not HS Graduates
Kansas	327	293,272	North Dakota	70	92,427
Kentucky	763	825,857	Ohio	1461	1,684,888
Louisiana	1119	803,872	Oklahoma	662	506,961
Maine	196	168,460	Oregon	363	343,609
Maryland	479	673,932	Pennsylvania	1598	1,994,278
Massachusetts	641	792,657	Rhode Island	108	184,344
Michigan	1475	1,356,759	South Carolina	678	687,260
Minnesota	506	488,765	South Dakota	102	98,720
Mississippi	639	549,685	Tennessee	998	1,033,914
Missouri	832	858,368	Texas	3177	2,872,559
Montana	127	96,469	Utah	203	133,315
Nebraska	169	181,072	Vermont	59	68,637
Nevada	141	167,628	Virginia	627	987,203
New Hampshire	112	127,423	Washington	634	505,783
New Jersey	866	1,205,206	West Virginia	400	398,527
New Mexico	282	229,974	Wisconsin	636	662,072
New York	2981	2,977,604	Wyoming	64	47,113
North Carolina	966	1,277,747			

(a) The task force would like to find a model that will predict the number of people living below the poverty level from the number of adults who are not high school graduates. Which is the dependent variable and which is the independent variable?

(b) Create a scatter plot of the data. Do you think that a linear relationship exists between the two variables?

(c) Find the linear regression model for the data.

(d) Interpret the meaning of the slope and the y intercept of the model. Do you think that the y intercept makes sense for these data?

(e) Use the model to predict the number of people who are below the poverty level for the states of Wyoming, Connecticut and North Carolina. What are the residuals for these three observations?

(f) At the 0.05 level of significance, is the model significant?

Thinking About It!

15.27 The Commerce Department also has data available on number of shopping centers and retail sales for the South Central states. The data are given below: *Requires Exercises 15.2, 15.17*

State	Number of Shopping Centers	Retail Sales ($ billiion)
Kentucky	593	11.7
Tennessee	1137	19.1
Alabama	601	13.2
Mississippi	418	7.2
Arkansas	339	6.4
Louisiana	676	15.6
Oklahoma	556	11.4
Texas	2824	72.7

Source: *Statistical Abstract of the United States 1995.*

(a) Find the linear regression model for the South Central states.

(b) How does the equation for the South Central states compare to the one you found for the North Central states?

(c) Would you have expected them to be exactly the same? Why or why not?

(d) What similarities would you expect them to have? What differences?

(e) Combine the data from the North Central and South Central states and find the linear regression model for the combined data.

(f) Do you think that the individual models are better or worse than the combined model? On what criteria do you base your conclusion?

Requires Exercise 15.23

15.28 Look at the data on advertising and brand preference for cigarettes again.

(a) Find the simple linear regression model that relates adult brand preference to advertising expenditures.

(b) At the 0.05 level, is the model significant?

(c) What is R^2 for this model? What does it mean? What is the correlation coefficient?

(d) Compare the model for adult brand preference to the one you found for adolescent brand preference. Which variable do you think is more strongly related to advertising expenditures? Why?

(e) Do you think that these data help to support the claims against the tobacco industry that its advertising targeted teenagers? Why or why not?

Requires Exercise 15.26

15.29 Consider the model you developed relating poverty levels and education.

(a) Find 95% confidence intervals for the mean number of people who live below the poverty level when the number of adults who are not high school graduates is (i) 500,000, (ii) 1,000,000, and (iii) 2,000,000.

(b) Find 95% prediction intervals for the number of people who live below the poverty level when the number of adults who are not high school graduates is (i) 500,000, (ii) 1,000,000, and (iii) 2,000,000.

(c) Do you think that the model is useful for predicting the number of people below the poverty level? Why or why not?

(d) The task force contends that if it can institute programs to reduce the number of adults who have not finished high school by 10% in each state, it will have a significant impact on the number of people who are living below the poverty level. Do you agree or disagree with this statement? Why?

Requires Exercise 15.25

15.30 Consider the data on the British banks.

(a) What is R^2 for this model? What does it mean? What is the correlation coefficient?

(b) Find 95% confidence intervals for the mean amount of deposits when the number of customers is (i) 12 million, (ii) 15 million, and (iii) 7 million.

(c) Find 95% prediction intervals for the amount of deposits when the number of customers is (i) 12 million, (ii) 15 million, and (iii) 7 million.

(d) Do you think that the model is useful for predicting the amount of deposits? Why or why not?

Requires Exercises 15.25, 15.30

15.31 The British Bankers' Association decided to look at another variable to predict deposits. It chose the number of branches for each bank. The data are given below:

Bank Name	Deposits (£ billion)	Branches 1996
Abbey National	101.7	867
Barclays	108.2	1997
Lloyds	96.9	2797
National Westminster	113.8	1920
Woolrich	27.5	430
Halifax	77.1	938

(a) Find the simple linear regression model that relates deposits to number of branches.

(b) At the 0.05 level, is the model significant?

(c) What is R^2 for this model? What does it mean? What is the correlation coefficient?

(d) Compare the model for predicting deposits using the number of branches to the one you found for number of customers. Which variable do you think is more strongly related to deposits? Why?

15.32 Look at the model that you developed for adolescent brand preference for cigarettes and the amount of money spent on advertising.

Requires Exercises 15.23, 15.28

(a) Make a plot of the residuals versus the values of the independent variables.

(b) What does this lead you to believe about the appropriateness of the linear model?

(c) Now plot the residuals versus the predicted values of y.

(d) How do the two graphs compare? Why do you think this happened?

(e) Make a normal probability plot of the data. Do you think the assumption of normality is reasonable?

(f) Considering your answers to parts (a)–(e), do you think that the linear regression model is appropriate for these data? Why or why not?

15.33 Look at the model you developed for shopping centers and retail sales for the South Central states.

Requires Exercises 15.17, 15.27

(a) Make a plot of the residuals versus the values of the independent variables.

(b) What does this lead you to believe about the appropriateness of the linear model?

(c) Now plot the residuals versus the predicted values of y.

(d) How do the two graphs compare? Why do you think this happened?

(e) Make a normal probability plot of the data. Do you think the assumption of normality is reasonable?

(f) Considering your answers to parts (a)–(e), do you think that the linear regression model is appropriate for these data? Why or why not?

15.34 Consider the model that you developed for poverty levels and education.

Requires Exercises 15.26, 15.29

(a) Make a plot of the residuals versus the values of the independent variables.

(b) What does this lead you to believe about the appropriateness of the linear model?

(c) Now plot the residuals versus the predicted values of y.

(d) How do the two graphs compare? Why do you think this happened?

(e) Make a normal probability plot of the data. Do you think the assumption of normality is reasonable?

(f) Considering your answers to parts (a)–(e), do you think that the linear regression model is appropriate for these data? Why or why not?

Doing It!

15.35 The provost at Aluacha Balaclava College wants to look at the faculty salary data in some other ways. She has collected data on salary, years of service, rank, school, gender, and tenure. A sample of the data is shown below:

Datafile: FACULTY.XXX

Salary ($)	Years of Service	Rank	Schools	Gender M/F	Tenure Y/N
53,316	22	ASST	BUSINESS	F	Y
64,375	11	PROF	BUSINESS	M	Y
63,501	7	ASSO	BUSINESS	M	Y
59,426	6	ASSO	BUSINESS	M	N
49,058	20	ASSO	BUSINESS	M	Y
94,969	4	PROF	BUSINESS	M	N
54,762	21	ASST	BUSINESS	M	Y
55,516	9	ASSO	BUSINESS	M	Y

(a) Make scatter plots of salary versus years of service for the entire faculty. Do you think that there is a linear relationship between the two variables? Why or why not?

(b) Find the simple linear regression model with salary as the dependent variable and years of service as the independent variable.

(c) At the 0.05 level, is there a significant linear relationship between salary and years of service?

(d) What is the value of R^2 for this model? What does it mean?

(e) Separate the data by rank and create scatter plots of salary versus years of service for each rank. How do these plots compare to the one for all of the faculty?

(f) Find a simple linear regression model for salary versus years of service for each rank. How do the regression coefficients for each rank compare to each other? How do they compare to the overall model for all of the faculty?

(g) At the 0.05 level of significance, are any of the models significant?

(h) What are the R^2 values for each of the models?

(i) Calculate the residuals for all significant models. Create a plot of residuals versus predicted values for each model. Is the simple linear model appropriate for these data? Why or why not?

(j) For each significant model, investigate the assumption of normality. Do you think the assumption is valid for these data? Why or why not?

(k) Are there any data values that might be exerting a strong influence on the model? If so, which ones? Drop any unusual values from the data and rerun the models. How does this change the results?

(l) Perform any additional analyses that you feel would be helpful to determine whether or not the simple linear model is an appropriate one for these data. Write a report with your conclusions.

CHAPTER *16*

TIME SERIES AND FORECASTING

TAKE ME OUT TO THE BALL GAME

American Pastime, Inc. is a company that provides high-quality entertainment to communities. With a clear focus on families, American Pastime plans to offer people of all ages an affordable opportunity to view exciting Double A baseball in a clean and safe environment.

To plan for this facility, American Pastime must forecast ticket sales. This information will help the company determine the number of ticket windows and determine the ordering/inventory plan for the food and souvenir items to be sold in such a way that costs are minimized.

The facility will consist of a state-of-the-art 200,000-square-foot stadium. Seating for approximately 6000 fixed seats for baseball attendees will be available: 4200 chair reserved seats and 1800 bench with back general admission seats. There will be 61 home games.

The population figures for the past 10 years for the area are available to you. The data are divided into two groups: the number of people in families with children (FWC) and the number of people in the age group from 20 to 40 with no children (NC). These two groups are your target population. Once you have the forecast for the two population groups, you must predict the number of people who will attend each game. Past experience shows that the attendance rate is 1.91 per 1000 for the FWC group and 1.75 per 1000 for the NC group.

16.1   CHAPTER OBJECTIVES

The problem of forecasting ticket sales described above is similar to many business problems that require the prediction of some variable. For example, you may wish to predict sales of a new or existing product, the closing price of a stock, the yield of an agricultural crop, the attendance at a meeting, or the donations to a nonprofit organization.

Clearly, these variables depend on many other factors. Let's consider the daily sales of loaves of bread sold at your local convenience store. The manufacturer would like to know precisely how many loaves of bread will be sold each day. Then the manufacturer could supply exactly that number and know that it will not lose any customers due to a stock out and it will not be left with loaves of bread at the end of the day that cannot be sold. But the number of loaves of bread sold per day probably depends on a number of factors, such as the number of loaves sold the previous day, the weather, the traffic, and others things. We could try to build a regression model treating the number of loaves of bread sold as the dependent variable and the others as independent variables. However, in many instances such as this one there are few data available on the independent variables. Hence, it is often impossible to build a *causal* model.

In these cases, the only alternative is to use a time series forecasting model. The main idea behind any time series forecasting model is that we can predict future values by studying the behavior of the past values. There is not a sense of cause and effect in these models but simply an attempt to *see patterns, repetitive or systematic behavior, and/or trends in the data.* The following simple example may help you understand what this means.

Suppose you observe the time that Mr. Russell goes to work every morning for 10 mornings. The data are shown below:

Monday	8:02 AM	Monday	7:59 AM
Tuesday	9:15 AM	Tuesday	9:20 AM
Wednesday	7:55 AM	Wednesday	8:10 AM
Thursday	9:08 AM	Thursday	9:09 AM
Friday	8:06 AM	Friday	8:05 AM

A quick look at these times would tell you that on Mondays, Wednesdays, and Fridays, Mr. Russell leaves somewhere around 8 AM, whereas on Tuesdays and Thursdays he leaves shortly after 9 AM. We could use this pattern to predict his departure times for the next few days. We are banking on the fact that Mr. Russell, like most of us, is a creature of habit and therefore the future looks much like the past in terms of general patterns. We do not know what is causing this pattern and we cannot know the cause by simply examining these data. In a time series method, we do not need to know what is causing the difference in his departure times on MWF versus TuTh.

Remember: The independent variable is displayed on the x axis and the dependent variable on the y axis.

Typically, when you are looking at time series data, you have several observations of the variable of interest (the dependent variable) observed over time. So, in effect, time becomes the independent variable. The graph shown in Figure 16.1 displays the number of loaves of bread sold at a typical convenience store for the last 25 days.

This scatter plot depicts the number of loaves sold along the *y* axis and the number of the day along the *x* axis. It is the same scatter plot technique that you learned in Chapter 3 except now the data points are connected. The days are arbitrarily numbered beginning with day 1. To predict the number of loaves of bread to be sold on day 26 we should exploit the information in the scatter plot. This means we should try to identify any patterns that exist in the sales over time.

This small example illustrates the nature of **time series** data.

FIGURE 16.1 Sample scatter plot of loaves sold by time

A *time series* is a set of observations of a variable at regular time intervals, such as yearly, monthly, weekly, daily, etc.

There are many different ways of extracting the information in a time series in order to predict the value of the variable in the next time period or beyond. In this chapter you will learn how to analyze time series data using four different methods: Simple Moving Average, Weighted Moving Average, Exponential Smoothing, and Regression. These are the most commonly used techniques. The moving average and exponential smoothing techniques are appropriate for stationary time series. This simply means there is no significant upward or downward long-term trend in the data. The technique of regression is appropriate for nonstationary time series, where there is some upward or downward trend over time. This is the same basic technique you learned in Chapter 15 only with the variable time as the independent variable now.

*Time series models are used to **extrapolate** past behavior in order to predict future behavior.*

This chapter covers the following material:

- Getting Started With Time Series Data
- Simple Moving Average Models
- Weighted Moving Average Models
- Exponential Smoothing Models
- Regression Models

16.2 GETTING STARTED WITH TIME SERIES DATA

16.2.1 Time Series Notation

To study time series data we must introduce some general notation. Consistent with the notation from regression, we will label the variable that we are trying to predict with the letter Y. Since each observation is taken at a particular *time*, we will subscript Y with the letter t. Thus, the data in a time series are labeled

$$y_1, y_2, \ldots, y_t$$

where

y_1 is the observation of the variable at time period 1

y_2 is the observation of the variable at time period 2

y_t is the observation of the variable at time period t

The observation that is the oldest in terms of the time that it was observed compared to the present is labeled y_1. For the bread example, the daily sales 25 days ago is the oldest observation and is therefore labeled y_1. The second oldest observation is labeled y_2 and so forth. With t observations of the variable of interest, the current time is labeled time period t and we are typically trying to predict the value for the next time period, $t + 1$. This prediction will be labeled $\hat{y}_{t+1}$. Remember that we used the notation $\hat{y}$ as the predicted value of y from the regression model. In a similar manner, the caret in $\hat{y}_{t+1}$ tells you it is an estimate and the subscript $t + 1$ tells you that it is an estimate for the next time period, $t + 1$. Sometimes you may wish to predict beyond this point. In this case you may wish to predict the value of y for time periods $t + 2$, $t + 3$, etc. These would be labeled $\hat{y}_{t+2}$, $\hat{y}_{t+3}$, etc.

Consider how the data would be labeled for the baseball problem.

EXAMPLE 16.1 Baseball Attendance

Population Data for the Past 10 Years

The population data for the community surrounding the location of the new baseball stadium are shown below. Remember that FWC is the number of people in families with children and NC is the number of people in the age group from 20 to 40 with no children.

Year	FWC	NC
1989	1,098,909	564,790
1990	1,197,185	556,734
1991	1,169,860	653,134
1992	1,350,675	632,459
1993	1,335,213	642,387
1994	1,207,658	654,890
1995	1,379,723	657,238
1996	1,321,457	657,238
1997	1,387,692	717,903
1998	1,546,920	692,340

There are really two time series here: the time series consisting of observations of the number of people in families with children and those for the families with no children. For the moment, just consider the time series labeled FWC. The value from the year 1989 is the oldest so it is labeled y_1. The most recent observation is 1998 and so it is labeled y_{10}. Thus, the current time is $t = 10$ and we wish to predict the value for $t = 11$ or the year 1999. ∎

The time period should be selected to fit the needs of the problem.

Note that the time period could be every minute, every hour, every day, every week, every month, every year, or any other period that makes sense for the particular application you are working with. For the loaves of bread example in the previous section, the time period is every day since we observed daily sales. For the baseball ticket sales example, the population must be predicted for the next year based on the last 10 years of data. So the time period is every year. If you are predicting the closing price of a stock, the time period would be every day.

16.2.2 Displaying the Time Series

Always display the data before deciding on a method of analysis.

Once you have identified the data and labeled them properly, you should display them using a scatter plot. The x axis should be time and the y axis should be the variable of interest. Let's continue working with the FWC time series.

EXAMPLE 16.2 Baseball Attendance

Scatter Plot for FWC Time Series

The scatter plot for the FWC time series is shown at the top of the next page.

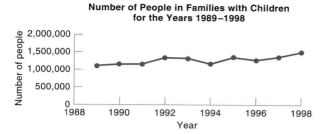

Number of People in Families with Children for the Years 1989–1998

Note that alternatively the *x* axis could have been labeled year 1 to year 10. In this case, it is probably preferable to use the actual years for the *x* axis, whereas for the bread data example, it was easiest to label the days from 1 to 25. ■

After you plot the data, you should examine the plot to see if there are any obvious patterns or trends. In this case, the time series seems to have a slight upward trend but there are no regular patterns. Now you plot the scatter plot for the NC time series and examine it for patterns or trends.

*TRY IT NOW!*

Baseball Attendance

Scatter Plot of NC Time Series

Display the NC time series as a scatter plot.

Do you see any patterns or trends? If so, what are they?

The next two sections discuss simple and weighted moving average models of time series data.

16.2.3 Exercises—Learning It!

16.1 Citing significant strides in math and science by women of all ethnic backgrounds, the College Board says that the male–female gap on Scholastic Assessment Test scores con-

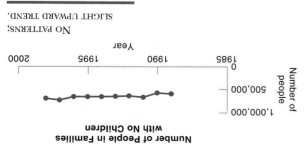

tinues to narrow among high school students. The average SAT scores for men and women from 1990 to 1994 are shown in the table below.

	Verbal		Math	
	Men	**Women**	**Men**	**Women**
1990	429	419	499	455
1991	426	418	497	453
1992	428	419	499	456
1993	428	420	502	457
1994	425	421	501	460

(a) Display each of the four time series on the same axis.

(b) Do you see any patterns?

16.2 The graph below displays the high and low weekly closing price of John Wiley and Sons stock for the year 1994. Do you see any patterns in the time series?

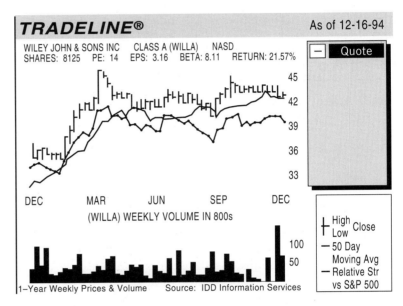

16.3 The graph below displays the dropout rate for seniors in high school from 1971 to 1994. Describe the graph and note any patterns in the time series.

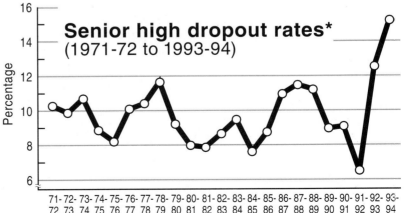

*As of 92–93, new counting methods resulted in a higher and more accurate count

16.4 The table below shows the Indian sugar production in thousands of metric tons by year.

Crop Year	Indian Sugar Production (Thousands of Metric Tons)	Crop Year	Indian Sugar Production (Thousands of Metric Tons)
74–75	4,949	85–86	7,983
75–76	5,464	86–87	9,474
76–77	6,043	87–88	10,000
77–78	8,201	88–89	10,150
78–79	7,071	89–90	10,420
79–80	5,170	90–91	9,989
80–81	6,542	91–92	9,750
81–82	9,727	92–93	10,560
82–83	9,508	93–94	11,020
83–84	7,042	94–95	11,750
84–85	7,071		

(a) Display this time series using a scatter plot.

(b) Do you notice any patterns?

16.5 The number of people living in the United States and the United Kingdom aged 15–64 is shown in the two time series below:

	1990	1991	1992	1993	1994	1995
USA	164,654,840	166,150,720	167,660,448	169,184,152	170,721,960	172,274,000
UK	37,599,962	37,668,710	37,737,588	37,806,594	37,875,732	37,945,000

(a) Use a line graph to display these time series.

(b) Do you notice any patterns in the data? If so, is the pattern the same in both countries?

Discovery Exercise 16.1
LOOKING FOR PATTERNS AND TRENDS

Examine the time series displayed in the graphs below and identify any patterns and trends that you see. Use what you know about the particular data being collected to help you identify why the data look the way they do.

(a) This chart shows the average monthly price (cents per kilowatt·hour) of electricity across the United States from 1988 to 1996.

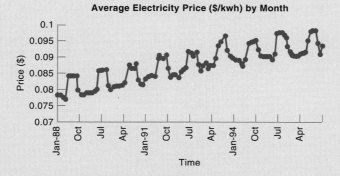

(b) The Bureau of Labor Statistics (BLS) reports data on commodity prices on a monthly basis. These data are reported on a regional and national basis. The graph at the top of page 702 displays the monthly average price (cents per pound) of white bread (pan) in the United States for the years 1980–1996.

(*continued*)

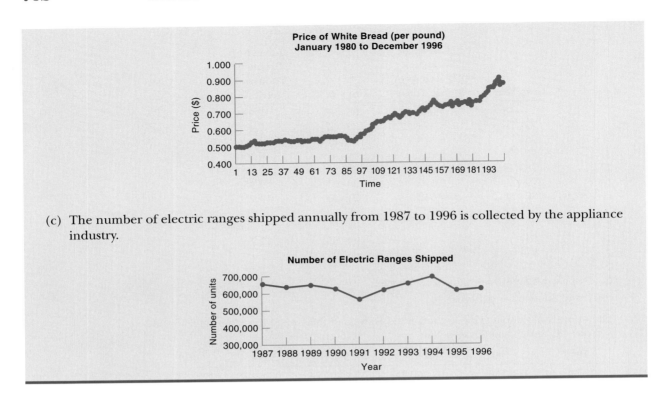

(c) The number of electric ranges shipped annually from 1987 to 1996 is collected by the appliance industry.

16.3   SIMPLE MOVING AVERAGE MODELS

Let's look again at the problem on baseball attendance. If you had to forecast the number of people in families with children for 1999 right now, how would you use the data? If it helps, think about forecasting the number of loaves of bread to be sold on day 26. Remember you have data on only the past behavior of the variable of interest. Nothing else.

If you suggested finding the average of all of the data and using that as your forecast for time period $t + 1$, then you would be thinking along the right lines. Figure 16.2 redisplays the FWC time series with an additional line drawn at the average, $\bar{y}$.

The average FWC population over the past 10 years is 1,299,529. This, then, would become our forecast for 1999 if we decided to use the overall average as our forecast. By doing so we would be saying that we believe that the time series is station-

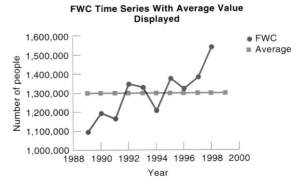

FIGURE 16.2   FWC time series with average displayed

ary and that all of the data, even the 10-year-old data, are equally valuable in constructing our forecast.

How good is this model? The graph of the straight line at the average value tells us that our forecast of the overall average is not terribly close to the data points that we have already observed. For instance, if we used the forecast of 1,299,529 to forecast for the year 1998 we would have been off by 1,546,920 − 1,299,529) = 247,391 people. Although there is not a strong trend in the data, it should be clear that the value from 1989 is not nearly as relevant in predicting the 1999 FWC population as the 1998 value. Perhaps we should not even use the 1989 value.

We are trying to find a model that comes close to as many data points as possible.

16.3.1 Calculating Simple Moving Averages

If you thought of this, then you have figured out what a moving average model is all about. Averaging the data was not a bad idea and, in fact, it is about the only thing you can do if you conclude that you have a stationary time series. But instead of averaging *all* of the data, we will average only the most recent observations. For example, we could average only the most recent 3 years as our forecast for the next year. In this case the predicted FWC population for 1999 would be calculated as follows:

$$\hat{y}_{11} = \frac{y_{10} + y_9 + y_8}{3}$$

$$= \frac{1,546,920 + 1,387,692 + 1,321,457}{3} = 1,418,690$$

Notice that this forecast is higher than the overall average forecast and ignores any data older than 3 years. Once the population for year 11 is observed then the forecast for year 12 would be calculated as follows:

Remember that t, current time, is labeled t = 10 because you have 10 observations.

$$\hat{y}_{12} = \frac{y_{11} + y_{10} + y_9}{3}$$

You can see that the observation from year 8 has been dropped and the newest observation, year 11, is included in the forecast calculation. Thus, the average "moves" along with time. This is why it is called a **moving average.**

You can use any number period MA as long as the number of periods to be averaged doesn't exceed the number of periods for which you have data, namely t.

> A *k*-period ***moving average*** is the average of the most recent *k* observations.

What we just calculated is called a 3-period moving average (MA), since we averaged the data from the most recent 3 time periods to get the forecast for the next period. You could instead use a 2-period moving average, a 4-period moving average or any number period moving average. In general, we will talk about a *k*-period moving average. The formula for a *k*-period moving average is given as

$$\hat{y}_{t+1} = \text{average of the most recent } k \text{ observations}$$

$$= \frac{y_t + y_{t-1} + \cdots + y_{t-k+1}}{k}$$

Formula for a k-period moving average

Don't let the notation confuse you; it is really just an average of *k* numbers.

16.3.2 Evaluating the Model

The next logical issue is to decide how to select the value of *k*. In other words, should we use a 2-period MA model, a 3-period MA model, or some other number period MA model? The right answer, of course, is that we should use the "best" model.

The MSE is a measure of variability—a measure of how different the predicted values are from the actual values.

Ideally, we would like the forecasting model with zero error—that is, one that predicts perfectly. Recognizing that we will never find such a model, we look for a model with the smallest possible error. This is the same approach that we took in developing the "best" regression model. In regression, we used the model to predict the value of the dependent variable, $\hat{y}$, for each observation that we had. The predicted value, $\hat{y}$, was then compared to the actual value of y. This comparison took the form of a difference, $y - \hat{y}$. These differences were then squared and averaged to yield a measure called the mean square error (MSE). The "best" regression model is the one with the smallest MSE. We will use precisely the same approach here.

Consider a 3-period moving average for the baseball attendance problem. We have 10 observations, so $t = 10$. To calculate the MSE for this 3-period MA model we must, in a sense, rewind the clock back to time $t = 4$ and use a 3-period MA to forecast for period 4, pretending for the moment that we don't know its value. The predicted value $\hat{y}_4$ can then be compared to the actual value, y_4, and the difference will become part of the mean square error calculation. Then we roll the clock forward to time $t = 5$, and use a 3-period MA to forecast for period 5 and so forth. It may seem odd to start the calculations of our predictions at time $t = 4$ when we have 3 data points prior to it. But you will quickly realize that to use a 3-period MA model, you must have 3 actual observations. So the earliest time period we can use to compare the model prediction to the actual value is for time period $t = 4$. In total we can compare the forecast to the actual values for time periods 4 through 10 or 7 time periods. We cannot use time period 11 in the calculation of the MSE because we don't yet know the actual value of the population for 1999 (time $t = 11$). The details are shown in the next example.

EXAMPLE 16.3   Baseball Attendance

Calculating the MSE for a 3-Period MA Model for FWC

The steps for calculating the mean square error (MSE) for a 3-period moving average model are shown in the spreadsheet below. The column labeled FWC shows the time series data. The column labeled Prediction shows what the 3-period MA would have been for each of the time periods from 4 to 10. The column labeled Error is calculated by subtracting the predicted value from the actual value, $y_t - \hat{y}_t$. Finally, the column labeled Error Squared is calculated by squaring each of the errors. The MSE is found by averaging all of the values in the Error Squared column.

3-Period Moving Average Model

Time	Year	FWC y_t	Prediction $\hat{y}_t$	Error $y_t - \hat{y}_t$	Error Squared
1	1989	1,098,909			
2	1990	1,197,185			
3	1991	1,169,860			
4	1992	1,350,675	1,155,318	195,357	38,164,357,449
5	1993	1,335,213	1,239,240	95,973	9,210,816,729
6	1994	1,207,658	1,285,249	−77,591	6,020,363,281
7	1995	1,379,723	1,297,849	81,874	6,703,351,876
8	1996	1,321,457	1,307,531	13,926	193,933,476
9	1997	1,387,692	1,302,946	84,746	7,181,884,516
10	1998	1,546,920	1,362,957	183,963	33,842,385,369
				MSE	14,473,870,385 ■

The MSE is a rather large number in this case. This is nothing to be concerned about. It is due to the fact that the values in the time series are large, making the errors fairly large and making the errors squared even larger. Remember that like the variance calculation, the differences (errors) are squared before averaging them so that the positive and negative errors do not cancel each other out, leaving you with an inaccurate view of the usefulness of the model.

Notice that all but one of the errors are positive. Just like in regression, a pattern in the errors should make you suspicious. It usually means that there was a pattern in the original time series that you failed to model, in other words, a pattern you missed or you ignored. In this case, the positive errors tell you that your forecast from a 3-period MA model consistently underestimates the actual population. Because of this observation you consider using only 2 periods to forecast for the next period, a 2-period MA. You also begin to suspect that the assumption that this time series is stationary may be flawed. Since the forecast is almost always lagging behind (as evidenced by the positive errors), perhaps there is a long-term trend in the population for families with children. You saw a slight upward trend in Figure 16.2 but it wasn't clear at that point whether to consider the time series stationary or nonstationary.

You should examine the errors for patterns.

The formula for calculating the mean square error (MSE) for a k-period MA model is given below:

$$\text{MSE} = \frac{\sum_{i=k+1}^{t} (y_i - \hat{y}_i)^2}{t - k}$$

Formula for MSE for k-period MA model

Notice that the first difference, $y_{k+1} - \hat{y}_{k+1}$, in the numerator corresponds to the time period numbered $k + 1$. As we have already observed, this is the earliest time period for which you can calculate a k-period forecast. The last difference in the numerator corresponds to the current time period, labeled t. To get an average (or mean) square error, the numerator is divided by the number of squared errors being added together. For the baseball example we had 7 squared errors to be averaged in the column labeled Error Squared. In general, we will have $t - k$ values in this last column.

Now try the calculations for a 2-period MA model for the FWC time series.

 **TRY IT NOW!**

Baseball Attendance
A 2-Period MA Model

Find the 2-period MA forecast for the number of people in families with children for 1999.

Complete the table below to find the MSE for the 2-period MA model for the FWC time series.

2-Period Moving Average Model

Time	Year	FWC	Prediction	Error	Error Squared
1	1989	1,098,909			
2	1990	1,197,185			
3	1991	1,169,860			
4	1992	1,350,675			
5	1993	1,335,213			
6	1994	1,207,658			
7	1995	1,379,723			
8	1996	1,321,457			
9	1997	1,387,692			
10	1998	1,546,920			

(continued)

How does the MSE for the 2-period MA model compare to the MSE for the 3-period MA?

Now we know that the 2-period MA has a smaller MSE than the 3-period MA. To see the difference in the performance of the 2-period MA model and the 3-period MA model, we can graph the original time series (FWC) and the 2 models on the same graph. This is shown in Figure 16.3.

Note that a 10-period MA model would be an overall average. We would have no way to find an MSE for this model.

But how do we know that we have the "best" model? Maybe a 4-period MA would be even better. The only way to be sure is to try all of the possible models. In this case that would mean models with $k = 2, 3, 4, 5, 6, 7, 8, 9, 10$. We have already considered the models for $k = 2, 3$, and 10. Clearly, if you have 30 observations in your time series you do not want to try all the possible values of k from 2 to 30. Even with the "power of the spreadsheet," you would not want to do this. Worse yet, imagine if you had 100 observations in your time series! Usually you can try a few models and look for a pattern in the mean square error values. The MSEs for the 2-, 3-, and 4-period MA models are summarized in the table below.

k	MSE
2	12,900,675,743
3	14,473,866,726
4	14,792,462,017

Thus it appears that the models that include more historical data (larger k values) have higher MSEs. So, if we are going to use a moving average model, we should use a 2-period model and our estimate for the FWC population for the year 1999 is 1,467,306.

There is another measure that is sometimes used instead of the MSE to evaluate the goodness of a forecasting model. It is called the mean absolute deviation or MAD. You have seen this measure in Chapter 6 when we looked at ways to calculate a number that would estimate the spread or variability of a data set. In doing inferential statistics, the MSE or variance is preferred over the MAD because of its nice mathematical properties. In forecasting models, you are simply using the MSE to evaluate

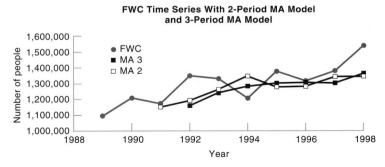

FIGURE 16.3   FWC time series with a 2-period MA model and 3-period MA model

Ans. $\hat{y}_{11} = 1,467,306$
MSE = 12,900,675,743
It is smaller.

a model. Therefore, it really doesn't matter much if you use the MSE or MAD to evaluate models as long as you are consistent. That is, if you calculate the MSE for one model you must compare it to the MSE of a second model.

To calculate the MAD, you simply use the absolute value of the difference between the observed value and the predicted value instead of squaring the difference. The absolute value function eliminates the problem of the negative errors canceling out the positive errors. The formula is given below:

$$\text{MAD} = \frac{\sum_{i=k+1}^{t} |y_i - \hat{y}_i|}{t - k}$$

Formula for mean absolute deviation (MAD) for k-period moving average

The calculation of the MAD for the 3-period MA model of the FWC time series is shown in the next example. You will notice that all columns but the last column are the same as Example 16.3.

EXAMPLE 16.4 Baseball Attendance

Finding the MAD

The table below shows the details for calculating the MAD for the 3-period MA model for the FWC time series.

3-Period Moving Average Model

Time	Year	FWC	Prediction	Error	Absolute Value of Error
1	1989	1,098,909			
2	1990	1,197,185			
3	1991	1,169,860			
4	1992	1,350,675	1,155,318	195,357	195,357
5	1993	1,335,213	1,239,240	95,973	95,973
6	1994	1,207,658	1,285,249	−77,591	77,591
7	1995	1,379,723	1,297,849	81,874	81,874
8	1996	1,321,457	1,307,531	13,926	13,926
9	1997	1,387,692	1,302,946	84,746	84,746
10	1998	1,546,920	1,362,957	183,963	183,963
				MAD	**104,776**

The MAD is simply the average of the 7 values in the last column. ∎

Like the MSE, the MAD itself is not a useful number. It is helpful as a way to help decide between different forecasting models. For the remainder of this chapter we will use the MSE to evaluate the models simply because it is more commonly used.

The models presented in this section are called **simple moving average models.** We have been referring to them as moving average models but should now note that they are *simple MA models* because the forecast is calculated as a simple average. This means that you simply add up all the values to be averaged and divide by k.

> A *simple moving average model* uses the simple average of the most recent k observations to predict for the next time period.

Each number in the average carries the same weight or importance in the averaging process. This is the type of average we have been working with throughout this book. However, it is possible to consider what is known as a weighted moving average. This is the subject of the next section.

16.3.3 Exercises—Learning It!

16.6 Reconsider the baseball attendance shown in Example 16.1.

(a) Display the people with no children (NC) time series graphically using a line graph.

(b) Find the 2-period MA forecast for the population with no children (NC) for the year 1999.

(c) Find the MSE for the 2-period MA model for the NC time series.

(d) Find the 3-period MA forecast for the population with no children (NC) for the year 1999.

(e) Find the MSE for the 3-period MA model for the NC time series.

(f) Display the NC time series and the 2-period and 3-period MA models on the same graph.

Requires Exercise 16.1

16.7 Analyze the math SAT scores.

(a) Find the 3-period MA forecast for 1995 math SAT scores for men.

(b) Find the MAD for this model.

(c) For the math SAT scores for women, find the 3-period MA forecast for 1995.

(d) Find the MAD for this model.

16.8 The price for purchasing a half-gallon of ice cream in a typical U.S. city is one of many time series collected by the Bureau of Labor Statistics. The monthly prices for 1995 and 1996 are shown below:

	1995	1996
Jan	2.659	2.665
Feb	2.55	2.673
Mar	2.686	2.752
Apr	2.629	2.728
May	2.634	2.825
Jun	2.649	2.827
Jul	2.665	2.851
Aug	2.667	2.966
Sep	2.683	3.041
Oct	2.735	3.077
Nov	2.61	2.978
Dec	2.675	2.94

(a) Display this time series as a line graph.

(b) Find the 3-period, 4-period, and 5-period MA forecast for January 1997.

(c) Find the MSE for the 3-period, 4-period, and 5-period MA models.

(d) Which model yields the smallest MSE?

(e) Find the MAD for the 3-period, 4-period, and 5-period MA models.

(f) Which model yields the smallest MAD?

(g) Based on your analysis, which model would you recommend be used and why?

16.9 The health of the construction industry is often an indication of the health of the economy. The data below show the number of bricks produced by quarter for 1995 and 1996.

Year	Quarter	Brick Production
1995	I	1,800,252
1995	II	1,849,733
1995	III	1,770,466
1995	IV	1,586,484
1996	I	1,496,798
1996	II	1,836,651
1996	III	1,689,123
1996	IV	1,654,901

(a) Display this time series as a line graph.

(b) Find the 4-period MA forecast for January 1997.

(c) Find the MSE for the 4-period MA model.

16.10 Hospitals have been paying increased attention to customer service. To determine staffing needs in the emergency room, the number of patients seen daily by the ER of a city hospital is shown below. The city has about 150,000 residents. One month of data are shown below.

Day	1	2	3	4	5	6	7	8	9	10	11	12	13	14	15	16	17	18	19	20	21	22	23	24	25	26	27	28	29	30
No. Patients	43	47	41	43	45	44	34	33	35	37	38	43	41	42	36	30	35	38	45	42	43	34	39	30	42	45	38	33	38	39

(a) Display this time series as a line graph.

(b) Find the 7-period MA forecast for the next day.

(c) Find the MSE for the 7-period MA model.

16.4 WEIGHTED MOVING AVERAGES

In the previous section you saw that for one time series there are many possible forecasting models. In fact, all of the models considered in the last section were simple moving average models. In this section we will see that by using unequal weights in the calculation of the average, there will be an infinite number of models of this type. These are called **weighted moving average models.**

> A *weighted moving average model* is a moving average model with unequal weights.

You can quickly see that, as in multiple regression analyses, building a forecasting model is partly a science and partly an art. You must use your eye to discern patterns in the data and your knowledge of the application area to decide what type of models to consider. Fortunately, with the aid of the software, we can often consider many different possible models and pick the one with the smallest MSE as our model of choice.

16.4.1 Calculating Weighted Moving Averages

Reconsider the 3-period simple moving average model for the FWC time series. The forecast for 1999 (time period 11) is calculated as follows:

$$\hat{y}_{11} = \frac{y_{10} + y_9 + y_8}{3} = 1,418,690$$

To clearly see the weight that each value has in this calculation let's rewrite the equation as

$$\hat{y}_{11} = \tfrac{1}{3}y_{10} + \tfrac{1}{3}y_9 + \tfrac{1}{3}y_8$$

Now you can easily see that each value is weighted the same, $\tfrac{1}{3}$, in the formula for a 3-period *simple* moving average.

If we consider a 3-period *weighted* moving average, we will allow the weights to be different from each other. For example, we might decide to weight the most recent observation by $\tfrac{1}{2}$ (50%), the next most recent observation by $\tfrac{1}{4}$ (25%), and the third most recent observation by $\tfrac{1}{4}$ (25%). The forecast for 1999 would then be calculated as

$$\begin{aligned} \hat{y}_{11} &= \tfrac{1}{2}y_{10} + \tfrac{1}{4}y_9 + \tfrac{1}{4}y_8 \\ &= (0.50)(1,546,920) + (0.25)(1,387,692) + (0.25)(1,321,457) \\ &= 1,450,747 \end{aligned}$$

What have we accomplished by doing this? The forecast for 1999 calculated this way is larger than the forecast from the 3-period simple MA model. By weighting the most recent observation, y_{10}, more heavily than the older data we are saying that it is more important to our forecast. In this case, this argument makes sense, since we observed a slight upward tendency in the time series. Thus, the most recent observation should carry a greater weight in the forecast for the next period.

The only rule that needs to be observed as you pick the weights is that the sum of the weights must be 1 and each weight must be a positive number between 0 and 1. We will use the term w_t to represent the weight to be used for the observation from time period t. The general formula for a 3-period weighted moving average is then

Formula for 3-period weighted moving average

$$\hat{y}_{t+1} = w_t y_t + w_{t-1} y_{t-1} + w_{t-2} y_{t-2}$$

where

$$0 \leq w_t, w_{t-1}, w_{t-2} \leq 1$$
$$w_t + w_{t-1} + w_{t-2} = 1$$

16.4.2 Finding the Best Model

Many other possible weights could be used for the 3-period weighted moving average. For instance, maybe our forecasting model would be improved if we used $w_t = 0.60$, $w_{t-1} = 0.30$, and $w_{t-2} = 0.10$. How can we tell? We can find the MSE for each model and see which one yields the lowest MSE. The MSE for two weighted MA models are shown in the next example.

EXAMPLE 16.5   Baseball Attendance

MSE Calculation for Two Different 3-Period Weighted MA Models

The table below shows the calculation of the MSE for two different 3-period weighted MA models for the FWC time series.

3-Period Weighted Moving Average Model

Time	Year	FWC	Prediction	Error	Error Squared
			Model 1: Weights: 0.50, 0.25, 0.25		
1	1989	1,098,909			
2	1990	1,197,185			
3	1991	1,169,860			
4	1992	1,350,675	1,158,954	191,721	36,756,941,841
5	1993	1,335,213	1,267,099	68,114	4,639,516,996
6	1994	1,207,658	1,297,740	$-90,082$	8,114,766,724
7	1995	1,379,723	1,275,301	104,422	10,903,954,084
8	1996	1,321,457	1,325,579	$-4,122$	16,990,884
9	1997	1,387,692	1,307,574	80,118	6,418,893,924
10	1998	1,546,920	1,369,141	177,779	31,605,372,841
			MSE		14,065,205,328

Time	Year	FWC	Prediction	Error	Error Squared
			Model 2: Weights: 0.60, 0.30, 0.10		
1	1989	1,098,909			
2	1990	1,197,185			
3	1991	1,169,860			
4	1992	1,350,675	1,170,962	179,713	32,296,762,369
5	1993	1,335,213	1,281,082	54,131	2,930,165,161
6	1994	1,207,658	1,323,316	$-115,658$	13,376,772,964

(continued)

Time	Year	FWC	Prediction	Error	Error Squared
			Model 2: Weights: 0.60, 0.30, 0.10		
7	1995	1,379,723	1,260,226	119,497	14,279,533,009
8	1996	1,321,457	1,323,653	−2,196	4,822,416
9	1997	1,387,692	1,327,557	60,135	3,616,218,225
10	1998	1,546,920	1,367,025	179,895	32,362,211,025
			MSE		14,123,783,596

From Example 16.5 we learn that model 1 using weights of 0.50, 0.25, and 0.25, has a smaller MSE than model 2, but the 2-period simple MA has a lower MSE than both of these models. Does this mean that any 3-period weighted average will have a higher MSE than the 2-period simple MA model? We can't say this for sure but it seems that by including one more observation in the model we raise the MSE by quite a bit even with the weighted models. Perhaps we should consider a weighted 2-period model. Try one and see for yourself.

 **TRY IT NOW!**

Baseball Attendance

A 2-Period Weighted MA Model

Find the 2-period weighted MA forecast for the number of people in families with children for 1999. Weight the most recent observation by 0.75 and the second most recent observation by 0.25.

Complete the table below to find the MSE for the 2-period weighted MA model for the FWC time series.

2-Period Weighted Moving Average Model

Time	Year	FWC	Prediction	Error	Error Squared
			Weights: 0.75, 0.25		
1	1989	1,098,909			
2	1990	1,197,185			
3	1991	1,169,860			
4	1992	1,350,675			
5	1993	1,335,213			
6	1994	1,207,658			
7	1995	1,379,723			
8	1996	1,321,457			
9	1997	1,387,692			
10	1998	1,546,920			

Is this model better than the simple 2-period moving average?

ANS: $\hat{y}_{11}$ = 1,507,113; MSE = 12,735,801,453; Yes.

Some software packages allow you to optimize the weights for a MA model. For example, in Excel you can use the Solver feature of the Tools menu to find the set of weights that gives you the smallest MSE for a 3-period weighted MA model.

***Remember** all the weights must be positive fractions and they must sum to one.*

The discussion in the previous paragraph is an illustration of the process by which you arrive at a "good" forecasting model. You try some models, look for patterns in the MSE, and arrive at what you feel is the "best" model. Most of the time you will not be guaranteed that there is no better model for the time series.

The formula for a *k-period weighted moving average* is shown below. It is simply a generalization of the 3-period weighted moving average formula that we have already seen.

$$\hat{y}_{t+1} = \text{weighted average of the most recent } k \text{ observations}$$
$$= w_t y_t + w_{t-1} y_{t-1} + \cdots + w_{t-k+1} y_{t-k+1}$$

where

$$0 \le w_t, w_{t-1}, \ldots, w_{t-k+1} \le 1$$
$$w_t + w_{t-1} + \cdots + w_{t-k+1} = 1$$

The formulas for calculating the MSE and the MAD are the same as for the simple moving average model.

16.4.3 Exercises—Learn It!

Requires Exercise 16.6

16.11 For the baseball attendance problem:

(a) Find the 2-period weighted MA forecast for population with no children (NC) for the year 1999. Use $w_t = 0.80$ and $w_{t-1} = 0.20$.

(b) Find the MSE for the weighted 2-period MA model. How does it compare with the MSE for the simple 2-period MA model?

(c) Find the 3-period weighted MA forecast for population with no children (NC) for the year 1999. Use $w_t = 0.60$, $w_{t-1} = 0.20$, $w_{t-2} = 0.20$.

(d) Find the MSE for the 3-period weighted MA model. How does it compare with the MSE for the simple 2-period MA model?

(e) Display the NC time series and the 2-period and 3-period weighted MA models on the same graph.

Requires Exercise 16.4

16.12 Use a 4-period weighted moving average to forecast for the 95–96 crop year of Indian sugar.

(a) Use weights of $w_t = 0.50$, $w_{t-1} = 0.20$, $w_{t-2} = 0.15$, $w_{t-3} = 0.15$.

(b) Find the MSE for this model.

(c) Try a different set of weights. Did you reduce the MSE?

16.13 A small pizza restaurant is interested in using past data to estimate the sales for the next month. The management plans to use the sales forecast to determine the amount of cheese, tomato sauce, and other ingredients to order. The sales pattern for the past 2 years is shown below:

Month	1995	1996
Jan	$5,838.90	$12,958.07
Feb	$5,226.89	$12,731.63
Mar	$6,326.66	$13,901.36
Apr	$6,815.22	$13,232.64
May	$6,665.17	$14,223.89
Jun	$10,801.43	$13,440.71
Jul	$10,171.44	$13,751.10
Aug	$6,047.26	$15,324.82
Sep	$4,840.70	$11,738.72
Oct	$5,431.87	$14,185.08
Nov	$5,536.20	$14,607.72
Dec	$5,944.26	$15,167.76

(a) Display the time series as a line graph.

(b) Use a 3-period moving average with weights of $w_t = 0.75$, $w_{t-1} = 0.15$, and $w_{t-2} = 0.10$ to forecast for the next month.

(c) Find the MSE for this model.

(d) Use a 3-period weighted moving average with $w_t = 0.89$ and $w_{t-1} = 0.04$ and $w_{t-2} = 0.07$ to forecast for the next month.

(e) Find the MSE for this model.

(f) Compare the MSE for both models. Which one is smaller? What does this tell you?

(g) Examine the line graph of this time series and explain why you should have expected this.

16.14 The cotton and wool outlook is published monthly by the Economic Research Service, U.S. Department of Agriculture, Washington, D.C. The number of 480-lb bales in the world is shown below for 1995 and 1996:

	1995	1996
Jan	29.94	28.89
Feb	30.14	28.92
Mar	30.72	28.75
Apr	29.62	28.87
May	27.54	28.91
Jun	27.42	33.34
Jul	29.45	33.78
Aug	29.84	34.33
Sep	29.68	34.95
Oct	29.50	35.59
Nov	29.37	35.63
Dec	29.18	36.16

(a) Display this time series as a line graph.

(b) Use a 6-period simple moving average to predict the number of bales for January 1997.

(c) Find the MSE for this model.

(d) Use a 6-period weighted moving average to predict the number of bales for January 1997. Use $w_t = 0.50$, $w_{t-1} = 0.30$, $w_{t-2} = 0.10$, $w_{t-3} = 0.05$, $w_{t-4} = 0.03$, and $w_{t-5} = 0.02$.

(e) Find the MSE for this model.

(f) Which model do you recommend and why?

16.15 The amount of money spent on alcohol is collected by the Bureau of Labor Statistics (BLS) using a consumer expenditure survey. The amount spent on alcohol for two different age groups is shown below:

	Under 25 Years of Age	Between 25 and 34 Years Old
1990	$318,000	$365,000
1991	$252,000	$370,000
1992	$356,000	$365,000
1993	$304,000	$307,000
1994	$247,000	$347,000
1995	$277,000	$299,000

(a) Display each of the time series as a line graph.

(b) Describe any differences you see in the two time series.

(c) Use a 3-period weighted moving average to predict the alcohol expenditure for each age group for 1996. Use $w_t = 0.60$, $w_{t-1} = 0.30$, and $w_{t-2} = 0.10$.

16.5   EXPONENTIAL SMOOTHING MODELS

In the last section we saw that it is often difficult to determine the "best" values for the weights to be used in the weighted moving average model. The **exponential smoothing model** to be discussed in this section is another averaging technique that allows you to use unequal weights. This technique specifies a formula for the assignment of these weights.

> An *exponential smoothing model* is an averaging technique that uses unequal weights. The weights applied to past observations decline in an exponential manner.

16.5.1   FORECASTING USING AN EXPONENTIAL SMOOTHING MODEL

Weighted MA models use only k periods of historical data in the calculation of the forecast for the next period.

Remember that the weights must be positive fractions and sum to one.

The exponential smoothing model is different from the weighted moving average model because *all* of the historical data in the time series are used to generate the forecast for the next period. It is similar to a weighted MA model because the forecast is a weighted average. The weights are assigned in such a way that the most recent observation, y_t, carries the largest weight. The second most recent observation, y_{t-1}, carries the second largest weight and the weights assigned to the other data points decrease *systematically*.

What does the word "systematically" mean? Generally, it means that there is some order or pattern. Let's see how we could assign the weights systematically. Use the FWC time series for the baseball attendance problem again. Suppose we decided to assign a weight of 0.70 to the most recent observation in the time series. Call this weight α. So $w_{10} = \alpha = 0.70$. We know from the above discussion that the weight for the observation at time period 9 should be less than 0.70. If we calculate the weight for this second most recent time period as $\alpha(1 - \alpha)$, then we know it would always be less than α. In this case $w_9 = 0.70(1 - 0.70) = 0.21$. The weight for the third most recent observation should be less than 0.21. It is calculated as $\alpha(1 - \alpha)^2 = (0.70)(1 - 0.70)^2 = 0.063$. If we continue with this pattern the weights to be used in forecasting for period 11 are shown in the table and in the associated graph of Figure 16.4. The table and the graph show that the weights decrease rather quickly as the age of the data increases. In fact, the graph looks like the graph of an exponential curve. Hence, the name given to this technique is exponential smoothing.

The definition of α is not the same as that used in hypothesis testing. We could use a different Greek letter to avoid any confusion with the earlier definition of α,

Time period	Weight
10	0.7000000
9	0.2100000
8	0.0630000
7	0.0189000
6	0.0056700
5	0.0017010
4	0.0005103
3	0.0001531
2	0.0000459
1	0.0000138

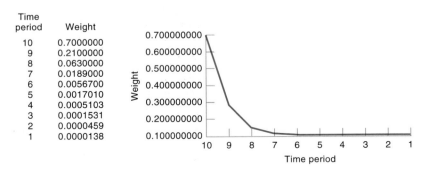

FIGURE 16.4   Weights for exponential smoothing model with $\alpha = 0.70$

but most forecasting books use α so we will do so as well. In this case, α is called the **smoothing constant.**

> The *smoothing constant, α,* is the weight assigned to the most recent observation in an exponential smoothing model.

Let's see how to calculate the forecast for the FWC time series using a smoothing constant of $\alpha = 0.7$.

EXAMPLE 16.6 Baseball Attendance

Exponential Smoothing Model

The forecast for period 11 (FWC population for the year 1999) is calculated as a weighted average of all of the historical data using the weights shown in Figure 16.4:

$$\hat{y}_{11} = (0.7)(1{,}546{,}920) + (0.21)(1{,}387{,}692) + (0.063)(1{,}321{,}457)$$
$$+ \cdots + (0.0000138)(1{,}098{,}909)$$
$$= 1{,}493{,}651 \qquad \blacksquare$$

The general formula for the forecast for the next period, $t + 1$, is shown below.

$$\hat{y}_{t+1} = \sum_{i=1}^{t} w_i y_i$$
$$= \alpha y_t + \alpha(1 - \alpha)y_{t-1} + \alpha(1 - \alpha)^2 y_{t-2} + \cdots + \alpha(1 - \alpha)^n y_{t-n} + \cdots$$

Formula for exponentially smoothed forecast for the next period

Notice that the sum starts with the most recent observation weighted by α and includes data all the way back to the beginning of the time series. By examining the weights used when $\alpha = 0.70$, you can see that any data older than 3 time periods contributes virtually nothing. Why then should we bother to include these observations in the weighted average? The answer to that rather appropriate question has to do with a practical consideration. It can be shown that the exponentially smoothed forecast calculated using the formula above is algebraically equivalent to the following equation:

$$\hat{y}_{t+1} = \hat{y}_t + \alpha(y_t - \hat{y}_t)$$

Alternative formula for computing the exponentially smoothed forecast for period $t + 1$

It is not immediately obvious when you look at the formula that all of the data in the time series are actually being used. All of the data are rolled into the forecast made for time period t, $\hat{y}_t$. The new forecast $\hat{y}_{t+1}$ is equal to the previous forecast plus some adjustment. The adjustment is based on how much error there was in the previous forecast, that is, the difference between the forecast for period t and the observation for period t, $(y_t - \hat{y}_t)$. If your previous forecast was "too high" then you need to adjust the forecast in the downward direction and the adjustment will be negative since $\hat{y}_t$ was larger than y_t. If, on the other hand, your previous forecast was "too low," then you need to make the forecast a bit larger and the adjustment will be positive since y_t was bigger than $\hat{y}_t$.

16.5.2 Evaluating the Exponential Smoothing Model

The equation shown above is the best one to use to actually calculate the forecast using exponential smoothing. This is true because you need only the most recent forecast, $\hat{y}_t$, the most recent observation, y_t, and α to complete the computation. Let's see how to use this equation and find the MSE of the exponential smoothing model for the FWC time series.

EXAMPLE 16.7   Baseball Attendance

Forecast and MSE for Exponential Smoothing Model

The first five columns of the table below display the year, the population for families with children, the forecast for each year, the error, and the adjustment to be made for the next forecast. The last column is used to find the mean square error.

Year	FWC y_t	Forecast $\hat{y}_t$	Error, $y_t - \hat{y}_t$	Alpha × Error $\alpha(y_t - \hat{y}_t)$	Error Squared $(y_t - \hat{y}_t)^2$
1989	1,098,909	1,098,909	0	0	0
1990	1,197,185	1,098,909	98,276	68,793.20	9,658,172,176
1991	1,169,860	1,167,702	2,158	1,510.46	4,656,101
1992	1,350,675	1,169,213	181,462	127,023.64	32,928,580,838
1993	1,335,213	1,296,236	38,977	27,283.69	1,519,183,299
1994	1,207,658	1,323,520	− 115,862	− 81,103.39	13,424,000,588
1995	1,379,723	1,242,417	137,306	96,114.48	18,853,048,354
1996	1,321,457	1,338,531	− 17,074	− 11,951.86	291,524,175
1997	1,387,692	1,326,579	61,113	42,778.94	3,734,771,425
1998	1,546,920	1,369,358	177,562	124,293.28	31,528,204,498
Forecast		1,493,651		**MSE**	11,194,214,145 ■

The observation for the first period becomes the forecast for the first period to get the technique started.

Notice that to create the table shown in Example 16.7, we needed a forecast for period 1. Generally, you simply set the forecast for the first time period equal to the observation for the first time period. The effect of doing this is quickly washed out. As a result, the error for period 1 is always zero, making the second forecast precisely the same as the first forecast. Notice also that you cannot really forecast beyond period $t + 1$ because you do not know what the error will be for future forecasts. Since you have assumed a stationary time series (or you wouldn't be using simple exponential smoothing), your forecast for any period beyond $t + 1$ is the same as your forecast for period $t + 1$.

To compare the usefulness of the exponential smoothing model to the moving average and weighted moving average models that we have considered, we again use the mean square error (MSE). The MSE is computed in precisely the same manner as we have done before. It turns out that to calculate the forecast for $\hat{y}_{t+1}$ you need to calculate the error and so it is a simple matter of squaring these errors and averaging them to get the MSE.

EXAMPLE 16.8   Baseball Attendance

Evaluating the Exponential Smoothing Model

In this case the MSE for the exponential smoothing model with a smoothing constant of 0.7 is smaller than for any other model we have tried for this time series. The graph shown below clearly depicts this as the time series and the forecasts are very close to each other for most periods.

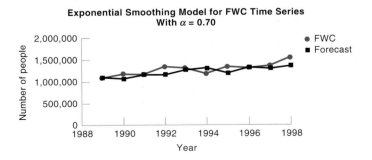

Exponential Smoothing Model for FWC Time Series
With $\alpha = 0.70$

Perhaps the MSE could be made even smaller with a different smoothing constant. Try it and see for yourself!

 *TRY IT NOW!*

Baseball Attendance

Exponential Smoothing Model

Find the forecast for the number of people in families with children for 1999 using an exponential smoothing model with a smoothing constant of 0.6.

Complete the table below to find the MSE for the exponential smoothing model ($\alpha = 0.6$) for the FWC time series.

Year	FWC	Forecast	Error	Alpha × Error	Error Squared
1989	1,098,909				
1990	1,197,185				
1991	1,169,860				
1992	1,350,675				
1993	1,335,213				
1994	1,207,658				
1995	1,379,723				
1996	1,321,457				
1997	1,387,692				
1998	1,546,920				
	Forecast			**MSE**	

Is this a better model than the model with a 0.7 smoothing constant?

As you can see, the selection of the smoothing constant affects both the forecast and the MSE for the model. Since α can vary anywhere from 0 to 1, it is impossible to try all values of α to find the model with the least MSE. Just as we did with the weighted moving average models, we can vary α and observe what happens to the MSE and attempt to zero in on a good value of α.

Some software packages allow you to optimize the value of α. For example, in Excel you can use the Solver feature of the Tools menu to find the value of α that gives you the smallest MSE.

16.5.3 Exercises—Learn It!

16.16 Use the baseball attendance for the population with no children (NC) time series.

Requires Exercises 16.6, 16.11

    (a) Forecast the attendance for 1999 using an exponential smoothing model with a value of $\alpha = 0.90$.

    (b) Find the MSE for this model.

    (c) How does this model compare with the weighted moving average models and the simple moving average models you tried in Exercises 16.6 and 16.11?

    (d) Which model do you recommend and why?

16.17 Use an exponential smoothing model with $\alpha = 0.65$ to forecast for the 95–96 crop year of Indian sugar.

Requires Exercises 16.4, 16.12

ANS. $\hat{y}_{11} = 1{,}473{,}021$; MSE = 11,378,876,337; No

(a) Find the MSE for this model.

(b) How does this model compare with the weighted moving average models and the simple moving average models you tried in Exercises 16.4 and 16.12?

(c) Which model do you recommend and why?

Requires Exercise 16.13   **16.18** Use an exponential smoothing model with $\alpha = 0.55$ to forecast the sales for the next month for the pizza restaurant.

(a) Find the MSE for this model.

(b) How does this model compare with the weighted moving average model developed in Exercise 16.13?

(c) Which model do you recommend and why?

Requires Exercise 16.10   **16.19** Use an exponential smoothing model to forecast the number of patients seen in the emergency room for the next day. The data are shown in Exercise 16.10. Use $\alpha = 0.85$.

16.20 The local liquor store has a bottle redemption center attached to it. The management is trying to determine staffing needs for the redemption center. The number of bottles brought into the redemption center varies from day to day. The data from the last 6 days are shown below:

Mon	148,454
Tues	180,000
Wed	147,053
Thurs	152,197
Fri	150,583
Sat	122,366

(a) Use an exponential smoothing model with $\alpha = 0.75$ to predict the number of bottles to be returned on the following Monday; the store is closed on Sunday.

(b) Find the MSE for this model.

16.6 REGRESSION MODELS

Remember: A stationary time series is one that exhibits no long-term trend.

All of the models that have been considered thus far in the chapter are models that could be used if you have a time series that has no trends. But what if there is a long-term trend in the data, either upward or downward? In this case you should use a technique that explicitly models the trend. You are already familiar with one such technique, regression. There are other techniques that can be used to model trends in the data. Two of these are second-order moving averages and double exponential smoothing. These are extensions of the techniques covered in the previous sections. However, the most commonly used technique when a trend is present is the technique of regression. This is the only one that will be covered in this chapter.

16.6.1 Finding the Linear Regression Model

The first step to analyzing time series data is to display the data using a scatter plot. After you construct a scatter plot and visually examine the time series data, you may observe a clear upward or downward linear trend. In this case you should try a regression model. However, it is not always crystal clear that there is a trend in the data. This is the case with the FWC time series we have been working with. There appears to be a slight upward trend, but is that enough to warrant the use of regression? You may decide, as we have done, to use a stationary technique for such data. Upon examination of the errors (or residuals as they are called in regression) you may see a pattern. If there is a pattern, there probably is some more information in the time series data that you have not explicitly modeled. In this case you may also try a regression model.

In Chapter 15, you learned all about the technique of simple linear regression. The same technique is used to model time series data. Recall that the general model for a simple linear regression model is

$$y = \beta_0 + \beta_1 x + \varepsilon$$

where

> y is the dependent variable
>
> x is the independent variable
>
> β_0 is the theoretical, unknown y intercept of the line
>
> β_1 is the theoretical, unknown slope of the line
>
> ε are independent with a normal distribution with a mean of zero and a constant standard deviation

Using observations of x and y, estimates of β_0 and β_1 can be computed. The prediction equation is given as

$$\hat{y} = b_0 + b_1 x$$

where

> b_0 is the estimate of the y intercept
>
> b_1 is the estimate of the slope

When using regression to model time series data, the independent variable is time and the dependent variable is the variable you are interested in forecasting. The prediction model thus becomes

It makes more sense to label the independent variable as t when using regression to model a time series.

$$\hat{y} = b_0 + b_1 t$$

Typically, you number the first time period as $t = 1$, the second time period as $t = 2$, and so forth. This is what we have done with the FWC time series. It would not be wrong to use the actual years for the independent variable in this case and you would get the same results for the slope (b_1) but different intercepts (b_0). You will get the same forecast as long as you use the correct number for time in the prediction equation.

Let's check to see that this is the case.

EXAMPLE 16.9   Baseball Attendance

Regression Model for FWC Time Series

The output from running regression in Excel is shown on page 720. The column of numbers containing the population for families with children (FWC) was identified as the dependent (or y) values and the column of numbers ranging from 1 to 10 was identified as the range for the independent variable (or x values).

SUMMARY OUTPUT

Regression Statistics	
Multiple R	0.8485124
R Square	0.7199733
Adjusted R Square	0.68497
Standard Error	73836.86
Observations	10

ANOVA

	df	SS	MS	F	Significance F
Regression	1	1.12138E+11	1.12E+11	20.56871	0.001911048
Residual	8	43615055164	5.45E+09		
Total	9	1.55753E+11			

	Coefficients	Standard Error	t Stat	P-value
Intercept	1096755.1	50440.17795	21.74368	2.11E−08
X Variable 1	36868.012	8129.17074	4.535273	0.001911

Extracting the values for b_0 and b_1 from this output, we can write the regression model as

$$\hat{y}_t = b_0 + b_1 t$$
$$= 1{,}096{,}755.1 + 36{,}868.012t$$

The forecast for time period 11 (year 1999) is calculated by using the regression model and substituting the value of 11 for $t =$ time. This calculation is shown below:

$$\hat{y}_{11} = 1{,}096{,}755.1 + (36{,}868.012)(11)$$
$$= 1{,}502{,}303$$

If we used the actual years as the independent variable values instead, we would get the following results:

	Coefficients
Intercept	−72196853
X Variable 1	36868.012

The model is

$$\hat{y}_t = -72{,}196{,}853 + 36{,}868.012t$$

Notice that the estimate of the slope, b_1, is the same but the intercept term, b_0, is different. The same forecast is obtained as long as we use the year 1999 as the value for time to be predicted:

$$\hat{y}_{1999} = -72{,}196{,}853 + (36{,}868.012)(1999)$$
$$= 1{,}502{,}303$$

       ■

It is clearly possible to forecast more than one period into the future. If we wished to estimate the population for the year 2000 we would simply use the year 2000 (or time $t = 12$) as the value for time in the regression model. Thus,

$$\hat{y}_{2000} = -72{,}196{,}853 + (36{,}868.012)(2000)$$
$$= 1{,}539{,}171.279$$

The farther out in time that you try to forecast, the less reliable your forecast will be since you have no additional values.

It should be noted that, by using a regression model to forecast for the next period or beyond, you are extrapolating. You were warned not to do this in the chapter on regression. However, in analyzing time series data the objective is to make a prediction for the next time period or beyond that will always be outside the range of the data. You are always extrapolating the past into the future when you work with time series data. The assumption for any time series model is that the future "looks" like the past. Without this assumption you cannot do anything with the historical data. So it is not that you were unduly cautioned in the chapter on regression but that the objective is now slightly different.

A word on extrapolation

Now that you have seen that there is no difference in the forecast resulting from different ways of coding the time values, we will always number the values of time starting with the value $t = 1$. This is often the easiest thing to do because sometimes your time variable is a calendar date, an hour of the day, or a day of the week.

Remember to label time starting with $t = 1$.

 TRY IT NOW!

Baseball Attendance
Regression Model

Find the regression model and forecast for the number of people with no children (NC) for 1999 (time $t = 11$).

The data are shown below:

Time	Year	NC
1	1989	564,790
2	1990	556,734
3	1991	653,134
4	1992	632,459
5	1993	642,387
6	1994	654,890
7	1995	657,238
8	1996	657,238
9	1997	717,903
10	1998	692,340

16.6.2 Evaluating the Regression Model

We have been using the mean square error to evaluate the moving average and exponential smoothing models. This is easily done for the regression model because any software package that you use to run the regression model will calculate the pre-

ANS. $b_0 = 563,461.53$
$b_1 = 14,445,412$
$\hat{y}_{11} = 722,361$

dicted values and the residuals for each value of y_t. These residuals can then be squared and averaged to get the MSE. These values are shown in the next example.

EXAMPLE 16.10   Baseball Attendance

Finding the MSE for the Regression Model

The first three columns of the table below were generated by Excel as part of the output of regression. The last column is simply the residuals (or errors) squared and then this column is averaged to get the MSE.

RESIDUAL OUTPUT

Observation	Predicted Y	Residuals	Residuals Squared
1	1133623.1	−34714.14545	1205071895
2	1170491.2	26693.84242	712561223.4
3	1207359.2	−37499.1697	1406187728
4	1244227.2	106447.8182	11331137996
5	1281095.2	54117.80606	2928736933
6	1317963.2	−110305.2061	12167238484
7	1354831.2	24891.78182	619600802.1
8	1391699.2	−70242.2303	4933970918
9	1428567.2	−40875.24242	1670785443
10	1465435.3	81484.74545	6639763742
MSE			**4,361,505,516**

■

A comparison of the MSE for the regression model to the MSE values for all of the other models of this time series that we have tried shows that the regression model is the best. This makes sense since we did see a slight upward trend in the data when we examined the scatter plot and we saw a pattern in the errors when we used a moving average model (all but one were positive).

In addition to using the MSE as a way to evaluate the regression model, we can also use the methods learned in Chapter 15. Specifically the value of R^2 should be examined and a hypothesis test for the significance of the slope term should be done. The next example serves as a reminder of these methods using the FWC time series.

EXAMPLE 16.11   Baseball Attendance

R^2 and Test for Significance for the Regression Model

The information needed to evaluate the regression model is shown in the output in Example 16.9. The coefficient of determination for the regression model is $R^2 = 72\%$. Thus, 72% of the variability in the y values is explained by the independent variable of time.

The hypothesis test for the slope is

$$H_0: \quad \beta_1 = 0$$
$$H_A: \quad \beta_1 \neq 0$$

The t-test statistic is shown in the coefficient table as 4.535273 with a p value of 0.0019. The low p value (less than 0.01) tells us to reject the null hypothesis and conclude that the slope is nonzero. This tells us that the model is useful.   ■

 *TRY IT NOW!*

Baseball Attendance

Evaluating the Regression Model for the NC Time Series

Evaluate the regression model for the number of people in families with no children (NC). Find the MSE, the value of R^2 and test for the significance of the slope term.

16.6.3 Exercises—Learning It!

16.21 The Bureau of Labor Statistics (BLS) reports data on commodity prices on a monthly basis. These data are reported on a regional and national basis. The following graph displays the monthly average price (cents per pound) of white bread in the United States for the years 1980–1996.

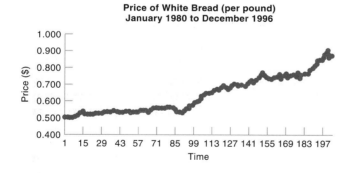

Price of White Bread (per pound)
January 1980 to December 1996

The output from running regression in Excel is shown below:

	Coefficients	Standard Error	t Stat	P-value
Intercept	0.456250652	0.004516911	101.0094	7.1E−175
Time	0.001819793	3.82101E−05	47.62591	8.7E−112

(a) Predict the average price of bread for the next 3 months (time = 205, 206, 207).

(b) Test the significance of the slope for this model.

(c) Is the graph consistent with the results of your hypothesis test? Explain why or why not.

16.22 The way companies do business is changing dramatically because of technology. The values shown below are the number of FAX machines per 1000 people for the years 1992–1995 for the United States (USA) and the United Kingdom (UK).

	1992	1993	1994	1995
USA	36	44	54	65
UK	17	22	26	31

(a) Find the slope and intercept for the regression model for the USA data.

(b) Use the slope and intercept to predict the number of FAX machines per 1000 people in the USA in 1996.

(c) Find the slope and intercept for the regression model for the UK data.

(d) Use the slope and intercept to predict the number of FAX machines per 1000 people in the UK in 1996.

(e) Comment on the rate of increase of FAX machines in the USA compared to the UK.

16.23 The appliance industry continues to grow and change rapidly. The number of camcorders shipped for the years 1987 through 1996 are shown in the table below.

Year	Number Shipped
1987	1,604,153
1988	2,044,045
1989	2,286,326
1990	2,961,691
1991	2,864,395
1992	2,814,979
1993	3,088,427
1994	3,208,651
1995	3,560,497
1996	3,634,038

(a) Display these data as a line graph.

(b) Find the slope and intercept of the linear model that best fits these data.

(c) Use the slope and intercept to predict the number of camcorders shipped in 1997.

(d) Interpret the slope in terms of this application.

(e) Who might use this model and for what purpose?

Requires Exercises 16.13, 16.18

16.24 Reconsider the sales of pizzas shown in Exercise 16.13.

(a) Find the slope and intercept of the regression model for these time series data.

(b) Compare the MSE for the regression model to the MSE for the weighted moving average model and the exponential smoothing model.

16.7   TIME SERIES ANALYSIS IN EXCEL

You learned how to do one of the forecasting methods described in this chapter when you learned about the linear regression model. Excel also has analysis tools for Moving Average and Exponential Smoothing models. In this section, we will look at two methods for the Moving Average model. For the Exponential Smoothing model, we will use the Excel data analysis tool.

Look at the problem of predicting the number of families with children that will attend Double A baseball games in a new stadium that is being built. The data collected are population data from the area from 1989 to 1998. The population figures are for two groups, families with children (FWC) and people 20–40 years old with no children (NC), both groups the stadium would like to attract. The data, in an Excel worksheet, are shown in Figure 16.5.

	A	B	C
1	Year	FWC	NC
2	1989	1,098,909	564,790
3	1990	1,197,185	556,734
4	1991	1,169,860	653,134
5	1992	1,350,675	632,459
6	1993	1,335,213	642,387
7	1994	1,207,658	654,890
8	1995	1,379,723	657,238
9	1996	1,321,457	657,238
10	1997	1,387,692	717,903
11	1998	1,546,920	692,340

FIGURE 16.5   Population data for baseball stadium

16.7.1 Moving Average Models in Excel

It is possible to analyze a moving average forecasting model in Excel using the **Add Trendline** option with a graph. To do this, you must first create a graph of the time series you want to analyze. The type of graph you want to create in Excel is a scatter plot, which you learned about in Chapter 15. Select the range that contains your data and make a scatter plot of the data. It should look similar to the one in Figure 16.6.

Once the chart is created, follow these steps:

1. Click on the chart to select it, and click on any point on the line to select the data series. When you click on the chart to select it, a new option, **Chart,** is added to the menu bar.

2. From the **Chart** menu, select **Add Trendline.** The dialog box shown in Figure 16.7 opens.

3. Click on the Trend/Regression type box for **Moving average.** You can specify the number of periods you want to use in the model by entering the value in the textbox labeled **Period:** Type in 3, since we want a three-period moving average for this example. Click **OK;** a new line for the Moving Average model is inserted into the chart, as shown in Figure 16.8 on page 726.

Although this tool is simple to use, it is limited. The trendline appears on the chart, but you do not get the actual predicted values from the model anywhere in the worksheet; thus, you will be unable to calculate the error. Also, there is a trendline model only for the Moving Average model, and not for Exponential Smoothing. For these reasons, you might want to use the data analysis tools for time series. You can do this as follows:

1. From the list of Analysis Tools, select **Moving Average** and the dialog box shown in Figure 16.9 (page 726) opens.

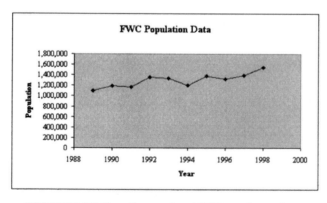

FIGURE 16.6   Scatter plot of FWC population data

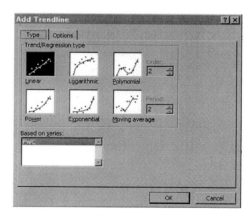

FIGURE 16.7   The Add Trendline dialog box

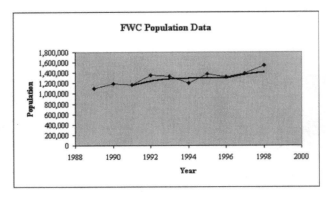

FIGURE 16.8   Chart with moving average trendline

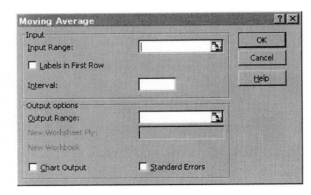

FIGURE 16.9   Moving Average dialog box

2. Position the cursor in the textbox labeled **Input Range:** and highlight the range that contains the data. Highlight only the actual data, not the column with the time periods. Put a check in the **Labels in First Row** box if the range contains a label.

3. Position the cursor in the textbox for **Interval:** and type in the number of periods you want to use for the moving average.

We do not use the standard errors for the predicted values, so there is no reason to include them.

4. Now, position the cursor in the box labeled **Output Range:** and enter the location in the worksheet where you want the predicted values from the model to be located. Excel automatically skips the first value and enters NA for any values that cannot be predicted. For example, in a three-period moving average, the first value that can be predicted is y_4. To line up the predicted values with the actual data, position the cursor in the cell adjacent to the *second* data value. You cannot have the output from this tool placed in a new worksheet or new workbook.

5. Finally put a check in the box for **Chart Output** and click **OK.** The worksheet will update to show the predicted values from the moving average model and a graph of the actual data and the predictions from the model, as shown in Figure 16.10.

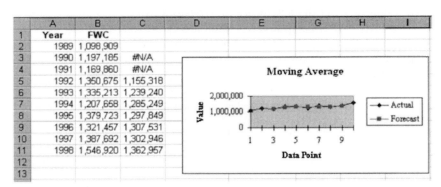

FIGURE 16.10   Output from moving average tool

If you are going to look at several models and try to choose the best model, then you will need to calculate the MSE for the model. To do this, we will use formulas in the worksheet:

Be sure to remember the equal sign to tell Excel you are entering a formula.

1. First you must calculate the error for each prediction. Just as we saw in regression analysis, the errors provide valuable information about how well the predicted values match the actual data. Position the cursor in the cell adjacent to the first predicted value and type in a formula to find the difference between the data value and the predicted value.

2. Next, you need to square the deviations. Position the cursor in the cell adjacent to the first error, and type in the formula "=E5^2".

3. Copy these formulas to the appropriate cells for the other data values and predicted values.

4. In the cell directly below the squared errors, you will need to type in the rest of the formula for the MSE. That is, you must add up all of the squared deviations and divide by the number of values, $t - k$, in this case, 7. We can use the **SUM** and **COUNT** functions to do this, as shown in Figure 16.11.

Note: Your cell references should be the ones for your own data locations.

$$=SUM(E5:E11)/COUNT(E5:E11)$$

FIGURE 16.11   Summing the squared deviations

The MSE will now appear in the cell just below the squared deviations. Figure 16.12 shows the worksheet with all of the calculated values.

	A	B	C	D	E
1	Year	FWC		Error	Squared Error
2	1989	1,098,909			
3	1990	1,197,185	#N/A		
4	1991	1,169,860	#N/A		
5	1992	1,350,675	1,155,318	195,357	38,164,357,449
6	1993	1,335,213	1,239,240	95,973	9,210,816,729
7	1994	1,207,658	1,285,249	-77,591	6,020,415,008
8	1995	1,379,723	1,297,849	81,874	6,703,406,459
9	1996	1,321,457	1,307,531	13,926	193,924,192
10	1997	1,387,692	1,302,946	84,746	7,181,884,516
11	1998	1,546,920	1,362,957	183,963	33,842,262,727
12				MSE	14473866726
13					

FIGURE 16.12   Table with errors and MSE

16.7.2 Exponential Smoothing Models in Excel

The simplest way to analyze a time series using an Exponential Smoothing model in Excel is to use the data analysis tool. This tool works almost exactly like the one for Moving Average, except that you will need to input the value of α instead of the number of periods, k. The dialog box for the Exponential Smoothing analysis tool is shown in Figure 16.13.

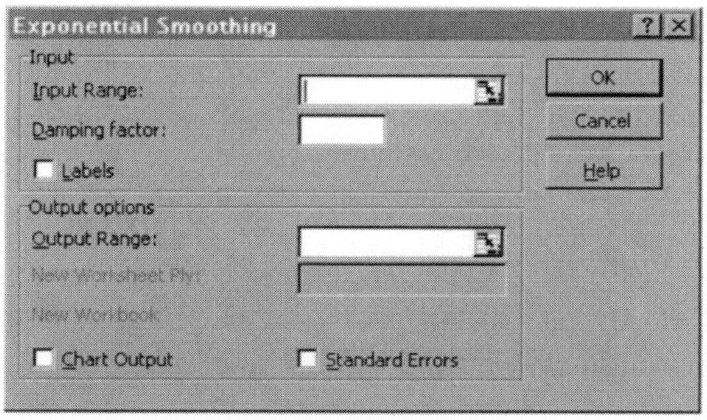

FIGURE 16.13   The Exponential Smoothing dialog box

Once you have entered the data range and the damping factor, α, and indicated what output you want and a location, the analysis is the same as the one for the Moving Average model.

CHAPTER 16 SUMMARY

A lot of time series data are collected on virtually every aspect of life. The government, businesses, and even individuals keep track of many variables over periods of time. This type of data contains information on patterns and trends that can be used to forecast the behavior of the variable for the future. According to Peter Bernstein in his book entitled *Against the Gods: The Remarkable Story of Risk,* the idea that defines the boundary between modern times and the past is the mastery of risk. This means that the future is more than a whim of the gods and that men and women are not passive. Bernstein says that the ability to define what may happen in the future and to choose among alternatives lies at the heart of contemporary societies.

This chapter has provided you with a brief introduction to the topic of forecasting. You have learned some of the basic terminology and techniques in this interesting discipline. You learned how to use moving averages and weighted moving averages to predict future values of a numeric variable. You also saw an additional application of the technique of regression. Finally, in this chapter you learned how to evaluate the model in order to choose the "best" model.

Key Terms

Term	Definition	Page Reference
Time series	A **time series** is a set of observations of a variable at regular time intervals, such as yearly, monthly, weekly, or daily.	697
A k-period moving average	A **k-period moving average** is the average of the most recent k observations.	703
Simple moving average model	A **simple moving average model** uses the simple average of the most recent k observations to predict for the next time period.	707
Weighted moving average model	A **weighted moving average model** is a moving average model with unequal weights.	709
Exponential smoothing model	An **exponential smoothing model** is an averaging technique that uses unequal weights. The weights applied to past observations decline in an exponential manner.	714
Smoothing constant	The **smoothing constant, α,** is the weight assigned to the most recent observation in an exponential smoothing model.	715

Key Formulas

Term	Formula	Page Reference
Simple moving average forecast	$\hat{y}_{t+1}$ = average of the most recent k observations $$= \frac{y_t + y_{t-1} + \cdots + y_{t-k+1}}{k}$$	703
Mean square error (MSE)	$$\text{MSE} = \frac{\sum_{i=k+1}^{t}(y_i - \hat{y}_i)^2}{t - k}$$	705
Mean absolute deviation (MAD)	$$\text{MAD} = \frac{\sum_{i=k+1}^{t}|y_i - \hat{y}_i|}{t - k}$$	707
Weighted moving average forecast	$\hat{y}_{t+1}$ = weighted average of the most recent k observations $$= w_t y_t + w_{t-1} y_{t-1} + \cdots + w_{t-k+1} y_{t-k+1}$$	712
Exponential smoothing forecast	$$\hat{y}_{t+1} = \alpha y_t + \alpha(1-\alpha)y_{t-1} + \alpha(1-\alpha)^2 y_{t-2}$$ $$+ \cdots + \alpha(1-\alpha)^n y_{t-n} + \cdots$$	715

CHAPTER 16 EXERCISES

Learning It!

16.25 A number of cities in the United States have had severe ozone problems caused by the burning of fossil fuels. The problem has been severe in cities such as Denver and Los Angeles (LA). The average monthly ozone rate in LA from 1979 to 1996 is given in the datafile. The first few lines of this datafile are shown below:

Datafile: OZONE.XXX

Jan 1979	2.7
Feb	2
Mar	3.6
Apr	5
May	6.5
Jun	6.1

(a) Find the MSE for a 12-month simple moving average model of these data.

(b) Find the MSE for an exponential smoothing model of these data with $\alpha = 0.90$.

(c) Find the MSE for the regression model of these data.

(d) Select the model with the lowest MSE and forecast the average monthly ozone rate for January 1997.

16.26 Most hospitals monitor length of stay of patients. These data help administrators to effectively use resources and staff. The average length of stay (in days) by month for a medium-sized New England city hospital are shown below:

Month	Length of Stay
Jan 1996	7.15
Feb	4.97
Mar	5.3
Apr	5.9
May	5.62
Jun	5.55
Jul	4.69
Aug	5.16
Sep	4.85
Oct	5.4
Nov	8.12
Dec	5.7

(a) Display these data as a line graph.

(b) Find the MSE for a 3-month simple moving average model of these data.

(c) Find the MSE for a 3-month weighted moving average model using $w_t = 0.70$, $w_{t-1} = 0.20$, $w_{t-2} = 0.10$.

(d) Find the MSE for an exponential smoothing model of these data with $\alpha = 0.80$.

(e) Select the model with the lowest MSE and forecast the average length of stay for January 1997.

16.27 Technology is affecting virtually every aspect of our lives. An increasing number of school districts across the United States are utilizing CD-ROM technology in educating children. The percentage of school districts using CD-ROM technology is shown below:

Year	% Using CD-ROM
1989	3.0
1990	4.5
1991	9.1
1992	11.4
1993	19.0
1994	42.6
1995	58.7

(a) Display this time series as a line graph.

(b) Which of the forecasting techniques would you recommend for these data and why?

(c) Develop a forecast for the percentage of school districts using CD-ROM technology in 1996 using the technique you identified in part (b).

16.28 Virtually all medical research involves the collection and analysis of data. There are many job opportunities in the field of biostatistics. The American Statistical Association (ASA) did a salary survey of faculty teaching in Biostatistics Departments. The data shown below are the median starting salaries for assistant professors from 1990 to 1996:

1990	$45,000
1991	$45,000
1992	$49,440
1993	$48,000
1994	$49,578
1995	$50,250
1996	$53,250

(a) Display these data as a line graph.

(b) Find the MSE for a 2-month moving average model.

(c) Find the MSE for an exponential smoothing model with $\alpha = 0.70$.

(d) Using the model with the smaller MSE, predict the starting salary for assistant professors of Biostatistics in 1997.

Datafile: RIVER.XXX   **16.29** The data presented in the file RIVER.XXX show the average monthly flow of the Elbe River in Canada from 1982 to 1996. The Department of Natural Resources monitors the water flow to detect any shifts in the ecology of the region. The first few lines of the data file are shown below:

Jan 1982	170
Feb	376
Mar	658
Apr	311
May	205
Jun	385

(a) Find the MSE for a 12-month moving average model.

(b) Find the MSE for a 12-month weighted moving average model with the following weights: 0.34, 0.10, 0.11, 0, 0, 0, 0, 0, 0.02, 0.14, 0.25, 0.04. These weights are the optimal weights for a 12-month moving average model.

(c) Find the MSE for an exponential smoothing model with $\alpha = 0.55$. This is the optimal value of α for an exponential smoothing model.

(d) Find the MSE for a regression model.

(e) Using the model with the smallest MSE, predict the average monthly river flow for January 1997.

16.30 Each year during the "dry" season, a number of western states experience forest fires. Some of these fires are ignited by natural elements such as lightning, while others are started by arson. The FIRES.XXX file shows the number of acres lost in forest fires by year from 1941 to 1996. The first few lines of this datafile are shown below: *Datafile: FIRES.XXX*

1941	1,009,000
1942	1,475,000
1943	856,000
1944	3,027,000
1945	4,271,000
1946	3,126,000
1947	1,115,000

(a) Find the MSE for a 3-year, 6-year, and 9-year moving average model.

(b) Using the model with the smallest MSE, forecast the number of acres to be lost in 1997.

Thinking About It!

16.31 The forecast for the bottle redemption center done in Exercise 16.20 is based on one week of data. You now have two more weeks of bottle returns as shown below: *Requires Exercise 16.20*

	Week 1	Week 2	Week 3
Mon	148,454	154,737	133,633
Tues	180,000	183,968	180,202
Wed	147,053	137,972	151,757
Thurs	152,197	132,140	136,850
Fri	150,583	135,596	156,593
Sat	122,366	153,452	140,539

(a) What do you notice when you look at the graph for 3 weeks of data?

(b) What suggestions do you have for forecasting the number of returns for week 4?

16.32 When you talk to the owner of the pizza restaurant you discover that in January 1996 they began a delivery service to local businesses. What do you recommend be done with the data prior to January 1996 in building a forecasting model? *Requires Exercise 16.13*

16.33 Frequently when you are asked to analyze time series data, data are missing for some of the time periods. Think about how you would handle missing data. What suggestions do you have for dealing with missing data?

16.34 When you speak to the administrators of the hospital in Exercise 16.26 you discover that the length of stay data are collected by groups of physicians. Ten physician groups are tracked at this hospital. Suggest ways in which these data could be misused to the detriment of the patients. *Requires Exercise 16.26*

16.35 In most of the exercises in this chapter we have selected the model with the smallest MSE as the "best" model.

(a) If we used the MAD as our criterion instead of the MSE, would we always get the same "best" model?

(b) What characteristics in the dataset would lead you to the same "best" model regardless of whether you used the MSE or the MAD?

(c) What characteristics in the dataset would lead you to different "best" models depending on whether you used the minimum MSE or the minimum MAD as your criterion?

Doing It!

Datafile: POWER.XXX

16.36 Due to population and economic development, the power consumption in Nigeria has steadily increased over the past several years. The data in the file POWER.XXX show the average monthly power consumption from January 1989 to December 1996 in kilowatt-hours. A portion of the datafile is shown below:

Jan 1989	56,634
Feb	56,427
Mar	65,170
Apr	63,847
May	56,864
Jun	67,288

(a) Display these data as a line graph.

(b) Find the MSE for a 12-month moving average model.

(c) Find the MSE for a 12-month optimized moving average model.

(d) Find the MSE for an exponential smoothing model with an optimized value of α.

(e) Find the MSE for a regression model.

(f) Using the model with the smallest MSE, forecast the consumption for 1997.

Datafile: HOGS.XXX

16.37 The daily opening, high, low, and closing prices of live hogs traded on the commodities market are shown in the datafile HOGS.XXX from 1 July 1996 to 31 December 1996.

LH_M97	Open	Hi	Lo	Close
1-Jul-96	77.775	77.775	77.775	77.775
2-Jul-96	78.000	78.800	78.050	78.300
3-Jul-96	78.250	79.000	78.000	79.000
5-Jul-96	79.000	79.600	78.800	79.600
8-Jul-96	79.450	79.450	79.000	79.400
9-Jul-96	79.050	79.250	78.600	79.125
10-Jul-96	78.975	79.100	78.975	79.100

(a) Display each of the 4 time series on one graph.

(b) What patterns do you see in the data? Do all 4 series "move" in the same manner?

(c) Find the model that minimizes the MSE for each of the 4 series. Consider only a 6-day moving average, a 6-day weighted moving average with optimal weights, an exponential smoothing model with optimal α, and a regression model.

(d) Find the model that minimizes the MAD for each of the 4 series. Consider only a 6-day moving average, a 6-day weighted moving average with optimal weights, an exponential smoothing model with optimal α, and a regression model.

(e) Summarize your findings and recommend a model to forecast the opening, high, low, and closing prices for 2 January 1997. Why is no forecast needed for 1 January 1997?

WORKSHOP 4

CHAPTER *17*

THE ANALYSIS OF QUALITATIVE DATA

COLLEGE DRINKING

For many students, consumption of alcohol is part of the "college experience." Some colleges and universities attract students based on their reputation as "party schools." Students do not often think about the ramifications of their drinking until they experience problems related to it, such as failing classes or flunking out entirely, unwanted pregnancies, or encounters with law enforcement. University administrations are often unaware of or in denial of the problems and do not have programs in place to help students address the problems of drinking and alcohol-related health behaviors.

The problem of excessive alcohol consumption in college is one that has been documented many times, in many different ways. A random national survey involving 140 colleges and universities, conducted by Harvard University, looked at the problems connected with *binge drinking*, which is defined as the consumption of five or more drinks during one episode of drinking for men and consumption of four or more drinks during one episode of drinking for women. The study found that as the number of binge drinking episodes for an individual increased, the number of alcohol-related problems experienced increased as well. The types of problems that the two studies examined included having a hangover, missing class, having unplanned sex, and driving while intoxicated.

A senior Public Health major at a university in the Northeast conducted a survey similar to the Harvard survey to examine the problem of binge drinking on the officially "alcohol-free" campus. The survey was administered to students registered in a Public Health course entitled "Wellness." Although the sample is not really a random sample of the student population, the course was one that is required of all students at the University. The study was limited to students 25 years of age or under and participation was voluntary.

The study resulted in 221 usable responses. Data were collected in four different areas: information about drinking habits (4 questions), problems experienced related to the students' drinking (15 questions), problems experienced because of other students' drinking (8 questions), and some demographic information (9 questions).

734

A sample of some of the data is shown below:

Five	Four	Three	Last	Binger	Hangover	Miss	Behind
0	0	0	2	1	1	1	1
0	0	0	4	1	1	1	1
0	1	1	3	2	1	1	1
0	0	0	3	1	1	1	1
0	0	0	3	1	1	1	1
0	0	0	4	1	1	1	1

(The variable names are abbreviated and are explained in the text as needed.)

17.1 CHAPTER OBJECTIVES

You have learned that while *qualitative data* are important in understanding the results of statistical analyses, there are not many statistical tools that deal specifically with these types of data. You learned some graphical techniques for displaying qualitative data in Chapter 3 and you learned how to calculate probabilities for these variables in Chapter 7. In Chapters 12 and 13 you learned some techniques for testing hypotheses about population proportions, which arise from *qualitative data.*

This chapter deals primarily with the analysis of qualitative data. Almost all of the techniques that are used to analyze qualitative data are known as *chi-square tests.* In this chapter you will learn to use chi-square tests for

- testing whether a particular probability model fits a set of data (goodness of fit test)
- testing equality of proportions for more than two populations
- testing whether two qualitative variables are dependent or independent

17.2 TEST FOR GOODNESS OF FIT

In Chapters 12 and 13 you learned several different hypothesis tests that *assumed* that the data came from a population that was normally distributed. At the time you may have wondered how you could know if the assumption was valid for a particular set of data. In Chapter 15 you saw a normal probability plot, which you used informally to determine whether the residuals were normally distributed.

There are some other, *informal,* ways to check to see if the assumption of normality is valid, such as creating a histogram, dotplot, or boxplot of the data and seeing if the shape resembles that of the normal distribution. This is not a bad idea, but whenever we can substantiate an *eyeball* test with a formal statistical technique, we are more secure in the decisions we make.

One method of testing to see if data come from a population with a certain distribution is to perform a test called a **chi-square goodness of fit test.**

> The *chi-square goodness of fit test* checks to see how well a set of data fit the model for a particular probability distribution.

17.2.1 The Chi-Square Test

You know by now that every hypothesis test has

- a set of hypotheses, H_0 and H_A
- a test procedure that defines how the data are collected and processed
- a test statistic that has a certain sampling distribution
- one or more critical values
- a decision or conclusion that can be made as a result of the test

For a chi-square goodness of fit test, the hypotheses are

The chi-square goodness of fit test is always a one-tail or one-sided test

H_0:   The data come from a population with a specific probability distribution (normal, binomial, etc.).

H_A: The data do not come from a population with the specified distribution.

The test procedure is to collect a set of sample data and to create a frequency histogram for the data. The test then compares the **observed** frequency distribution of the data to the frequency distribution that would be **expected** if the null hypothesis were true.

The *observed frequencies* are the actual number of observations that fall into each class in a frequency distribution or histogram.

The *expected frequencies* are the number of observations that should fall into each class in a frequency distribution under the hypothesized probability distribution.

EXAMPLE 17.1   College Drinking

Choosing the Hypothesized Distribution

Because the data collected for the survey on binge drinking were not a random sample, the researcher was interested in knowing how well the sample represented the actual student population. The researcher thought that the distribution of students among the four classes—freshman, sophomore, junior, and senior—was approximately uniform. That is, the probability distribution for the students he expected was

Class	Freshman	Sophomore	Junior	Senior
Probability	25%	25%	25%	25%

This type of distribution is known as a **uniform distribution.** ■

A *uniform distribution* is one in which each outcome or class of outcomes is equally likely to occur.

EXAMPLE 17.2 College Drinking

Setting Up the Goodness of Fit Test

In this situation, the hypotheses we wish to test are

H_0:   The distribution of students is uniform over the classes.
H_A: The distribution of students is not uniform.

Because of a mixup in the early administration of the survey, the question about class was left off some of the surveys. As a result there were only 171 responses to this question. The data are shown below:

Class	Observed Frequency
Freshman	86
Sophomore	36
Junior	30
Senior	19
Total (n)	**171**

From simply looking at the data it might be reasonable to wonder whether the distribution is uniform, since the number of freshmen is larger than that of the other classes combined. ■

Remember that we often test hypotheses when observed data seem unusual or inconsistent with what we expect.

EXAMPLE 17.3 Quality Problems

The Chi-Square Goodness of Fit Test

A company that manufactures CD jewel cases contracts with various computer software manufacturers to provide a product that is 10% defective. One of the software manufacturers is wondering about the quality of the product it receives and decides to check on the quality of the cases.

Since the company has just received a shipment of product it decides to sample 1000 boxes of the jewel cases and take 5 jewel cases from each. The reason for checking 5 jewel cases is that this is consistent with the company's current acceptance sampling procedures. The company will inspect and count the number of defective cases. If the vendor is conforming to the contract, then the data collected on the number of defective cases should have a binomial distribution with $\pi = 0.10$ (or 10%).

The hypotheses the company wishes to test are

H_0:   The distribution of defective cases is binomial with $n = 5$ and $\pi = 0.10$.
H_A: The distribution of defective cases is NOT binomial with $n = 5$ and $\pi = 0.10$.

It decides to perform the test at the 0.05 level of significance. ■

You might be wondering why we would want to do this, since you already know how to do hypothesis tests about proportions. This is a good question. The reason is tied to the *information* we are trying to obtain. If the software manufacturer tests the hypothesis $\pi = 0.10$ (against $\pi \neq 0.10$ or $\pi > 0.10$), then the conclusion will only be about the percent defective over the 5000 jewel cases in the sample. Depending on the length of time over which the product was manufactured, an average percent defective may not be a good indication of what is happening.

The chi-square goodness of fit test will give information about the distribution of the number of defective cases in a sample of size 5, which may enable the software manufacturer to obtain additional information about when or how the population changed. Remember that when you learned about the binomial probability distribution, one of the assumptions we made was that π, the proportion of successes in the population, remains constant over time. In a manufacturing process this is not always a safe assumption, since the process may shift over time. If the process changes but *averages* 0.10 over time, the hypothesis test about proportions might not be able to identify the problem. If we test the distribution as well as the value of π, we may be able to identify a problem. You will see what we mean as the example progresses.

 *TRY IT NOW!*

Seat-Belt Usage
Setting Up the Hypotheses for the Goodness of Fit Test

Analysts for insurance companies assume that the number of drivers who wear seat belts is a binomial random variable with $\pi = 0.70$. To test this assumption they decide to set up checkpoints and sample 10 drivers every 2 hours. Set up the hypotheses to perform an appropriate chi-square goodness of fit test.

In a chi-square goodness of fit test, we are always trying to decide whether the data we observed fit a particular probability distribution model. We can use this model to determine how many data points we would *expect* to fall in each class of the frequency distribution since we know the theoretical probabilities of observing any of the possible values.

EXAMPLE 17.4   College Drinking

Finding the Expected Frequencies

If the distribution of students over the classes is really uniform, then we would expect there to be the same number of students in each class; that is,

$$(25\%)(171) = 42.75 \text{ students in each year}$$

If we add a column to the frequency table we see that indeed the observed frequency does differ from what we expected to happen.

Class	Observed Frequency	Expected Frequency
Freshman	86	42.75
Sophomore	36	42.75
Junior	30	42.75
Senior	19	42.75
Total (n)	**171**	**171**

We can see the deviation from what is expected visually by looking at the frequency histograms for both the observed and expected distributions.

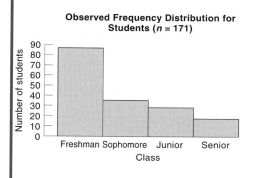

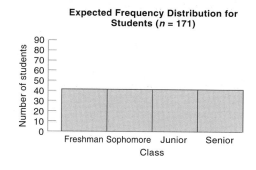

In the case of the uniform distribution, there is not much calculation involved in finding the expected frequencies of the probability distribution. For other probability distributions, you will need to do some additional calculations to find the expected frequencies.

EXAMPLE 17.5   Quality Problems

Calculating the Expected Frequencies

The software manufacturer who is wondering about the quality of the jewel cases it purchases from an outside vendor has collected the data and obtained a frequency distribution and histogram:

Number of Defective Cases	Frequency
0	485
1	340
2	127
3	48
4	0
5	0
Total	**1000**

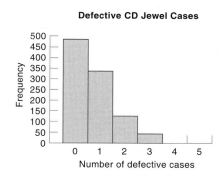

The data have the right shape for a binomial distribution with a small value of π but there seem to be too many crates that have a high number of defective cases.

The next step is to determine the expected frequencies for the binomial distribution with $n = 5$ and $\pi = 0.10$. Looking in the binomial tables, the manufacturer finds that the probability distribution is

x	0	1	2	3	4	5
$p(x)$	0.590	0.328	0.073	0.008	0.000	0.000

The *expected* frequency for each class of the frequency distribution, e_i, is calculated by multiplying the probability for that class by the sample size:

$$e_i = np_i$$

where n is the sample size of the data, and p_i is the probability that the value in the ith class will occur. For this problem, $n = 1000$, so we have

x	np_i
0	$(1000)(0.590) = 590$
1	$(1000)(0.328) = 328$
2	$(1000)(0.073) = 73$
3	$(1000)(0.008) = 8$
4	$(1000)(0.000) = 0$
5	$(1000)(0.000) = 0$

Putting it all together in one table, we see that there is definitely a difference between the observed and expected frequencies.

The reason for the discrepancy in the totals is rounding. If the binomial tables had 4 decimal places you could find the expected frequencies to 1 decimal place and avoid the problem.

Number of Defective Cases	Observed Frequency	Expected Frequency
0	485	590
1	340	328
2	127	73
3	48	8
4	0	0
5	0	0
Total	**1000**	**999**

 **TRY IT NOW!**

Seat-Belt Usage
Calculating the Expected Frequencies

The insurance analysts collect data for 1000 samples of 10 drivers and obtain the frequency distribution shown in the table. Find the expected frequency distribution for the data if the distribution is really binomial with $n = 10$ and $\pi = 0.70$.

Number Wearing Seat Belts (x)	Observed Frequency	$p(x)$	Expected Frequency
0	0		
1	0		
2	1		
3	6		
4	33		
5	116		
6	213		
7	275		
8	216		
9	119		
10	21		
Total	**1000**		

For the moment, don't worry about rounding if the expected frequency column does not add to 1.

Create frequency histograms for both the observed and the expected frequency distributions. At this point, does it appear that the observed data conform to the binomial distribution with $n = 10$ and $\pi = 0.70$? Why or why not?

Once we have collected the data, formed a hypothesis, and found the expected frequencies for the hypothesized distribution, we need to find a way to quantify the results of the test. The next section describes the way that this is done.

17.2.2 The Chi-Square Statistic

You know from previous work that often it is departure from what we expect to happen that leads us to perform a hypothesis test. The question is, how can we quantify the "deviation" from what is expected, and how will we know when what we observe deviates by more than it should or more than is likely by pure chance? To answer these questions we need to define a test statistic that measures the deviation of interest, and has a sampling distribution whose behavior we understand. This test statistic is known as the ***chi-square statistic*** and is calculated as

$$\chi^2 = \frac{\sum_{i=1}^{k}(o_i - e_i)^2}{e_i}$$

Chi-square statistic

where

o_i = observed frequency in the ith class of the frequency distribution

e_i = expected frequency in the ith class of the frequency distribution

k = the number of classes in the frequency distribution

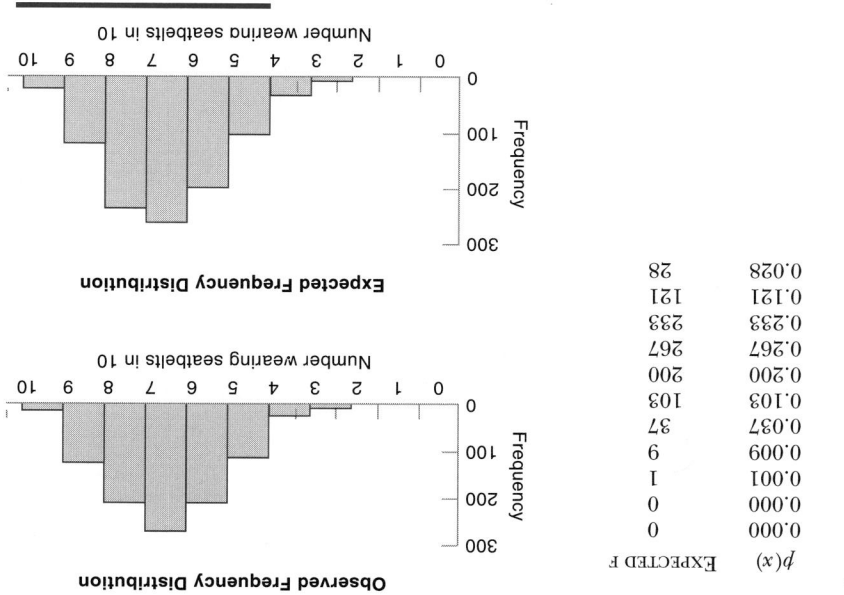

$p(x)$	EXPECTED F
0.000	0
0.000	0
0.001	1
0.009	9
0.037	37
0.103	103
0.200	200
0.267	267
0.233	233
0.121	121
0.028	28

ANS.

This test statistic has a chi-square distribution with $k - p - 1$ degrees of freedom, where p = the number of parameters of the theoretical distribution that were estimated from the data.

This statistic is a little different from any that you have encountered before. We are trying to measure the amount by which the observed frequency distribution deviates from the expected frequency distribution. Clearly, looking at the difference between the two quantities, $o_i - e_i$, is a good start. Just as clearly, though, if you simply add the deviations they will cancel each other out, since some will be positive and some will be negative. So, just as we did when we learned about the variance and the standard deviation, we square the quantities so that they are all positive. Adding the squared deviations together is a measure of *total* deviation. Dividing by e_i is similar to calculating the *percent error* instead of the *absolute error*.

EXAMPLE 17.6 College Drinking

Calculating the Chi-Square Statistic

For the example about the distribution of students, we can calculate the chi-square statistic by extending the frequency table that we have developed:

Class	Observed Frequency (o_i)	Expected Frequency (e_i)	$o_i - e_i$	$\dfrac{(o_i - e_i)^2}{e_i}$
Freshman	86	42.75	43.25	43.76
Sophomore	36	42.75	−6.75	1.07
Junior	30	42.75	−12.75	3.80
Senior	19	42.75	−23.75	13.19
Total (*n*)	171	171		61.82

You see that the value of the test statistic is 61.82. This is a measure of the **total** deviation of the observed frequencies from the expected frequencies. The question now is, what does this tell us about the distribution of students? ∎

EXAMPLE 17.7 Quality Problems

Calculating the Chi-Square Statistic

From the data that were collected it appears that there are definitely differences between the observed data and what would be expected if indeed the population is binomial with $\pi = 0.10$ and $n = 5$. To quantify the total difference we need to calculate the chi-square test statistic:

Number of Defective Cases	Observed Frequency	Expected Frequency	$o_i - e_i$	$\dfrac{(o_i - e_i)^2}{e_i}$
0	485	590.5	− 105.5	18.85
1	340	328.1	11.9	0.43
2	127	73.0	54	39.95
3	48	8.1	39.9	196.54
4	0	0.5	− 0.5	0.5
5	0	0.0	0.0	0.0
Total	1000	1000.2	− 0.2	256.27

From the calculation of the chi-square statistic we see that the measure of total deviation between the expected and observed frequencies is 256.77. We also see that the classes in the frequency distribution for values of 2 and 3 contribute the most to the total.

TRY IT NOW!

Seat-Belt Usage

Calculating the Chi-Square Statistic

Fill in the table below to calculate the value of the chi-square statistic for the data obtained by the insurance analysts.

Number Wearing Seat Belts	Observed Frequency	$p(x)$	Expected Frequency	$o - e$	$\dfrac{(o - e)^2}{e}$
0	0	0.000	0		
1	0	0.000	0		
2	1	0.001	1		
3	6	0.009	9		
4	33	0.037	37		
5	116	0.103	103		
6	213	0.200	200		
7	275	0.267	267		
8	216	0.233	233		
9	119	0.121	121		
10	21	0.028	28		
Total	**1000**	**0.999**	**999**		

Once we have quantified the difference between the observed and expected frequency distributions, we are ready to determine whether or not the difference is significant.

17.2.3 The Critical Value and the Decision Rule

We now have a way to measure the deviation of the observed data from what is expected, the chi-square statistic, and we know that this test statistic has a χ^2 distribution with $k - 1 - p$ degrees of freedom. All that remains of the test is finding the critical value and making a decision about our hypothesis.

To find the appropriate critical value we will need a level of significance for the test, α. Chi-square goodness of fit tests are always one-sided and upper-tailed tests. This is because we are looking at the *total deviation* and trying to see if that *exceeds*

7.18

$\dfrac{(o - e)^2}{e}$	$o - e$	
—	0	
—	0	
0.00	0	
1.00	−3	
0.43	−4	
1.64	13	
0.85	13	
0.24	8	
1.24	−17	
0.03	−2	
1.75	−7	

ANS.

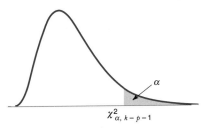

FIGURE 17.1 The upper tail of a chi-square distribution

some reasonable level. To find the critical value we look up $\chi^2_{\alpha, k-p-1}$ for the appropriate values of α and $k - p - 1$ as shown in Figure 17.1.

To make a decision about our hypothesis we compare the value of the test statistic to the critical value. Since it is an upper-tail test, if the chi-square test statistic is greater than the χ^2 value from the table we will reject H_0 and conclude that the data did not come from the hypothesized distribution. If the chi-square statistic is less than the critical value, we fail to reject H_0 and conclude that there is no evidence that the hypothesized distribution is not appropriate.

There are two things we must know to find the critical value for the test: the level of significance of the test, α, and the degrees of freedom, $k - p - 1$.

EXAMPLE 17.8   College Drinking

Performing the Chi-Square Test

For the problem with the student distribution, suppose we would like to perform the test at the 0.05 level of significance. This would mean that there is a 5% chance that the test will conclude that the data collected did not come from the uniform distribution when in fact it did.

To determine the number of degrees of freedom, we first look at the number of classes in the frequency distribution, k. In this case, $k = 4$. The value p represents the number of population parameters we estimated from the data. In this case we did not use the data to estimate any parameters. Thus, the number of degrees of freedom for the test is $4 - 0 - 1 = 3$. From the chi-square table, we look up $\chi^2_{0.05,3}$ and find that the critical value is 7.82.

To make a decision, we compare the chi-square test statistic, 61.82, to the critical value of 7.82. Since 61.82 is larger than 7.82 (the test statistic falls beyond or outside the critical value) we reject H_0. That is, the conclusion of the test is that the data do not come from a population that has a uniform distribution. This indicates that the survey was not evenly spread out over the four years.   ∎

EXAMPLE 17.9   Quality Problems

Performing the Chi-Square Test

The software company found the value of the chi-square statistic for the data on the defective jewel cases to be 256.27. To decide what this means, it needs to find the critical value for the test, compare the test statistic to the critical value, and make a decision.

To find the critical value the company analysts need to know α and the degrees of freedom for the test. They have already decided to test at the 0.05 level of significance. The number of degrees of freedom for the test is calculated to be

$$\text{Degrees of freedom} = k - p - 1 = 6 - 0 - 1 = 5$$

Remember that p is the number of parameters of the hypothesized distribution that were estimated from the data. The binomial distribution has 2 parameters, n and π, both of which were given or assumed, not estimated.

This critical value, $\chi^2_{0.05,5}$, is 11.07. Since 256.27 is clearly beyond the critical value, the software company can reject H_0 and conclude that the number of defective jewel cases in a sample of 5 does not have a binomial distribution with $\pi = 0.10$. That is, the data do not fit the assumed probability model. ∎

TRY IT NOW!

Seat-Belt Usage
Finding the Critical Value and Performing the Test

The insurance analysts decide that they want to test the goodness of fit hypotheses at the 0.01 level of significance.

How many degrees of freedom will the critical value for the test have?

Find the critical value for the test.

Based on the chi-square test statistic and the critical value, what can you conclude about the distribution of the number of people in a sample of size 10 that wear seat belts?

One of the things that happens as we learn more and more statistical techniques is that we lose sight of the fact that we are doing the analyses to help us identify problems, that is, to understand why things happen and to make informed decisions. After the chi-square goodness of fit test, what do we know and how can we use the information?

EXAMPLE 17.10   College Drinking

After the Test is Over. . .

As a result of the chi-square test, the researcher who administered the survey on binge drinking knows that the distribution of students in his sample is not uniformly

ANS. DEGREES OF FREEDOM = 10. CRITICAL VALUE IS 23.21. WE CANNOT REJECT H_0. THERE IS NO REASON TO BELIEVE THAT THE HYPOTHE-SIZED DISTRIBUTION IS INCORRECT.

distributed over the four classes. This *means* that his sample *may not be representative of the population he is trying to study* and that he will have to be careful about any conclusions or generalizations he makes.

In this case, the researcher began to think a little more about his null hypothesis. Does it really make sense that students are uniformly distributed over the four classes? After some thought he realized that if you consider the problems of student retention (loss of students from transfer or dropout) and the transfer in of students from other universities and community colleges, it is more likely that the true percentage of students in each class decreases as the class level increases. He contacted the university administrator who keeps track of such things and found out that the historical distribution of students is the one in the table below:

Class	Expected Distribution (%)	Expected Frequency
Freshman	45.2	77.3
Sophomore	18.6	31.8
Junior	18.0	30.8
Senior	18.2	31.1
Total (n)	**100**	**171**

The researcher then decided to compare the actual frequency distribution to the new distribution obtained from the university and redo the chi-square test. The comparison is shown in the table below:

Class	Observed Frequency (o_i)	Expected Frequency (e_i)	$o - e$	$\dfrac{(o - e)^2}{e}$
Freshman	86	77.3	8.7	0.98
Sophomore	36	31.8	4.2	0.55
Junior	30	30.8	−0.8	0.02
Senior	19	31.1	−12.1	4.71
Total (n)	**171**	**171.0**		**6.26**

Since nothing else changed, the critical value of the test is still $\chi^2_{0.05,3} = 7.82$. The value of the test statistic, 6.26, is not outside the critical value so there is not sufficient evidence to reject H_0. This means that there is no reason to say that the sample did not come from a population with the hypothesized distribution and so it is reasonable to assume that the sample is representative of the student population of the university *as defined by the university administration.*   ∎

One of the benefits of the chi-square test is that it lets us proceed with confidence in our analysis. Knowing that the sample does indeed represent the population of interest gives much more credibility to any conclusions drawn from the analysis.

EXAMPLE 17.11   **Quality Problems**

After the Test Is Over. . .

The company that used the outside vendor as a source of CD jewel cases knows as a result of the test that the number of defective cases in a sample of size 5 does not have a binomial distribution with $\pi = 0.10$. The company does *not* know if the distribution is binomial with a different value of π, or if some other factor makes the binomial distribution inappropriate. This is the time to use some common sense, knowledge of how the data were collected, and descriptive and graphical statistics to see if the company can identify the problem.

There are many things that the company analysts could do, depending on how they collected the data and what variables they recorded, but in this case they decided to plot the data in the order in which the cases were produced. Each sample that they took came from a different box and they recorded the lot number and production data along with the data on the number of defective cases. A time plot of the data indicated that there was an increase in the number of defectives in the sample starting at about the 750th sample. Since the first plot was very dense the analysts decided to also plot every fifth sample so that they could see the shift more clearly.

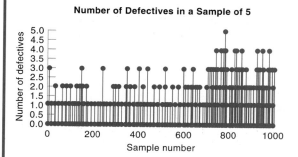

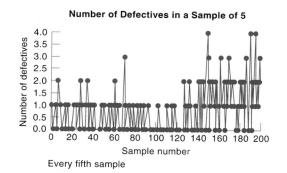

As a result of the chi-square test and the plot of the data they learned that the percent defective had apparently changed during production. This is why the binomial distribution was not correct. ∎

17.2.4 Testing for Normality and Other Considerations

The results of the chi-square goodness of fit test are similar to the test you learned for testing whether the variances of two populations are equal. The test does not result in a definitive answer that the hypothesized distribution is correct, just as the hypothesis test to see if two variances are equal does not tell you that the variances are equal. Rather, if you reject H_0, it will tell you when the hypothesized distribution is *not* correct. When you are using a chi-square test to test the assumption of normality, you will either decide that the assumption of normality is not appropriate, or that since there is no evidence to the contrary, the assumption is a reasonable one.

EXAMPLE 17.12 Jar Weights

Testing for Normality

The quality specifications for a 15.0-ounce jar of peanut butter state that the weights of the jars should be normally distributed with a mean of 15.1 oz. The Quality Focus Group for the product line selects a random sample of 500 jars to determine whether the specifications are being met. It decides to test at the 0.05 level of significance.

The hypotheses for the test are

H_0: The jar weights are normally distributed with $\mu = 15.1$ oz.

H_A: The jar weights are not normally distributed with $\mu = 15.1$ oz.

The group collects the data and uses a computer package to create a frequency distribution and histogram.

Weights (<)	Observed Frequency
14.75	3
14.80	5
14.85	34
14.90	43
14.95	79
15.00	83
15.05	92
15.10	77
15.15	56
15.20	20
15.25	5
15.30	1
15.35	1
15.40	1
Total	**500**

Observed Frequency for 500 Peanut Butter Jar Weights

The software package creates classes the same way you have learned. For example, 14.80 is the same as 14.75 < x ≤ 14.80.

The data are symmetric, although this does not mean that they are definitely normally distributed.

To calculate the expected frequency for each class the analysts need to calculate the probability that a randomly selected jar will fall into the weight range defined by each class. For the first class they need to find $P(14.70 < X \le 14.75)$.

Remember from Chapter 7 that you need to turn the two values into standard normals (Z's) and look them up. But to do that you need μ and σ! The specifications for the jars give only a value for μ, so the analysts will have to use the data to estimate σ.

Using the same software package, they find that the sample standard deviation, s, is 0.102 oz. So, to find the probability of interest they use

$$\text{For 14.70:} \quad Z = \frac{14.70 - 15.10}{0.102} = -3.92$$

$$\text{For 14.75:} \quad Z = \frac{14.75 - 15.1}{0.102} = -3.43$$

Looking up the Z values in the table and subtracting the results they find that $P(14.70 < X \le 14.75) = 0.0003 - 0.0000 = 0.0003$. When they were doing the calculations, they noticed that there was a discrepancy between the total for the expected frequency and the total of 500 observations. Looking at the table, they realized that this was probably because there were a few additional classes on the high side that had nonzero probabilities. They expanded the table to include these classes.

To find the expected number of jars out of the 500 that would weigh between 14.70 and 14.75 oz the analysts multiplied the probability by 500 and obtained $(0.0003)(500) = 0.15$. They found the probabilities and expected frequencies for all of the classes:

Weight (<)	Frequency	$P(a < X < b)$	Expected Frequency	$\frac{(o_i - e_i)^2}{e_i}$
14.75	3	0.0003	0.1	84.10
14.80	5	0.0013	0.7	28.13
14.85	34	0.0055	2.7	355.99
14.90	43	0.0179	8.9	130.33
14.95	79	0.0458	22.9	137.71
15.00	83	0.0927	46.4	28.93
15.05	92	0.1486	74.3	4.23

(*continued*)

Weight (<)	Observed Frequency	Expected $P(a < X < b)$	Frequency	$\dfrac{(o_i - e_i)^2}{e_i}$
15.10	77	0.1879	94.0	3.07
15.15	56	0.1879	94.0	15.36
15.20	20	0.1486	74.3	39.66
15.25	5	0.0927	46.4	36.91
15.30	1	0.0458	22.9	20.92
15.35	1	0.0179	8.9	7.03
15.40	1	0.0055	2.7	1.11
15.45	0	0.0013	0.7	0.67
15.50	0	0.0003	0.1	0.13
Total	**500**	**1.0000**	**500**	**894.28**

Looking at the chi-square statistic, there would appear to be some kind of problem. The data may not be *exactly* normally distributed, but they are not that far off. Why is the value of the test statistic so large? ■

Before we proceed any further with the example we need to discuss some of the problems with the chi-square goodness of fit test. The problems are related to the number of observations that are used to create the frequency distribution.

You may remember from Chapter 3 that a fairly large sample size is required to obtain a good frequency distribution. If your frequency histogram is highly variable due to lack of data, the frequencies in some classes will have greater differences from the expected frequencies.

Another related problem is that when the total sample size is small, the expected frequencies for each class will also be small. Since the chi-square statistic divides by the expected frequency, this can inflate the value of the statistic artificially and lead to rejection of H_0 when, in fact, it is true. For this reason, it is recommended that the chi-square test not be used if the expected frequency in any cell is less than 5. If this is not the case, it is possible to combine adjacent cells to get the expected frequencies above 5, but this results in a loss of degrees of freedom.

Remember, dividing by numbers close to 0 causes the result to be large.

EXAMPLE 17.13 Jar Weights

Collapsing Classes when the Expected Frequencies Are <5

Looking at the table of the observed and expected frequencies, the Quality Focus Group for the product called in a statistician to help them. The statistician explained that there are too many classes with very small expected frequencies that are causing the chi-square statistic to be inflated. The group decides to collapse cells on both ends of the distribution to get the expected frequencies above 5. The table was adjusted and the chi-square statistic recalculated:

Weight (<)	Observed Frequency	$P(a < X < b)$	Expected Frequency	$\dfrac{(o_i - e_i)^2}{e_i}$
14.75	3 ⎫	0.0003	0.1 ⎫	
14.80	5 ⎬ 85	0.0013	0.7 ⎬ 12.4	423.89
14.85	34	0.0055	2.7	
14.90	43 ⎭	0.0179	8.9 ⎭	
14.95	79	0.0458	22.9	137.71
15.00	83	0.0927	46.4	28.93
15.05	92	0.1486	74.3	4.23
15.10	77	0.1880	94.0	3.07
15.15	56	0.1880	94.0	15.36

(continued)

Weight (<)	Observed Frequency	$P(a < X < b)$	Expected Frequency	$\dfrac{(o_i - e_i)^2}{e_i}$
15.20	20	0.1486	74.3	39.66
15.25	5	0.0927	46.4	36.91
15.30	1	0.0458	22.9	20.92
15.35	1 ⎫	0.0179	8.9 ⎫	
15.40	1 ⎬ 2	0.0055	2.7 ⎬ 12.4	8.78
15.45	0 ⎪	0.0013	0.7 ⎪	
15.50	0 ⎭	0.0003	0.1 ⎭	
Total	**500**	**1.0000**	**500**	**719.46**

After collapsing classes there were only 10 classes left. Also, the normal distribution has two parameters, μ and σ, and the data were used to estimate one of them, σ.

Clearly the chi-square statistic is still large, but it has been considerably reduced.

To complete the test, the Quality Focus Group found the critical value for the test. It calculated the degrees of freedom to be

$$\text{Degrees of freedom} = k - p - 1 = 10 - 1 - 1 = 8$$

Since it wanted to test at $\alpha = 0.05$, it used $\chi^2_{0.05,8} = 15.51$.

Since the value of the test statistic, 719.46, is beyond the critical value of 15.51, the group rejects H_0 and concludes that the jar weights are not normally distributed with a mean of 15.1 oz and a standard deviation of 0.102 oz. ∎

Again, we need to look at the conclusion of the chi-square goodness of fit test and realize that, in fact, the conclusion comes as a package deal. In the previous example the conclusion is that the data do not come from a normal distribution with a mean of 15.1 oz. It is not necessarily true that the data are not normally distributed—they might be normally distributed with a *different mean* than the one hypothesized. It is also not necessarily true that the mean is not 15.1 oz; the mean could be 15.1 oz, but the distribution *might not be normal*. To really know what is going on here, the Quality Focus Group might want to perform a hypothesis test about the value of the mean to see if that is the problem, or rerun the chi-square test using the data to estimate μ.

17.2.5 Exercises—Learning It!

17.1   The administration of a university has been using the following distribution to classify the ages of their students:

Age Group	Estimated % of Student Population
Less than 18	2.7
18–19	29.9
20–24	53.4
Older than 24	14

A recent student survey provided the following data on age of students:

Age Group	Frequency
Less than 18	6
18–19	118
21–24	102
Older than 24	26

(a) Set up a table that compares the expected and observed frequencies for each group.

(b) Based on the table, do you think that the data represent the estimated distribution?

(c) Set up the hypotheses for the chi-square goodness of fit test.

(d) Perform the goodness of fit test at the 0.05 level of significance.

(e) Based on the chi-square test, is the estimated age distribution that the university is using correct?

17.2 As part of a survey on the use of Office Suites Software, the company doing the polling wanted to know whether its population was uniformly distributed over the following age distribution: under 25, 25 to 44, 45 and up. The company looked at the data it had collected so far and found the following distribution:

Age Group	Number of Respondents
Under 25	73
25 to 44	61
45 and up	66
Total	**200**

(a) Based on the data, do you think that the respondents are uniformly distributed over the age categories?

(b) Set up the hypotheses to test whether the data are uniformly distributed over the age categories.

(c) Find the expected frequency distribution and perform the chi-square goodness of fit test.

(d) At the 0.05 level of significance, would you say that the respondents were uniformly distributed over the age groups?

17.3 The Transit Authority in a large city estimates that 80% of business commuters get a seat for their entire commute. It decided to take a random sample of its subway system over the course of a 12-week period to see if its estimate is correct. It set up an exit poll at the most common destination station and asked 15 randomly selected commuters whether or not they got a seat for the entire commute. The data the Transit Authority obtained are as follows:

Number of Commuters in 15 Who Got Seats	Count
0	0
1	0
2	0
3	0
4	0
5	0
6	0
7	0
8	1
9	3
10	10
11	21
12	31
13	20
14	9
15	5
Total	**100**

(a) Why is the number of commuters in 15 who got seats a binomial random variable?

(b) Set up the hypotheses to test whether the sample data come from a binomial distribution with $n = 15$ and $\pi = 0.80$.

(c) Find the expected frequencies for the hypothesized distribution.

(d) Perform the chi-square goodness of fit test at the 0.01 level of significance.

(e) Is it reasonable to assume that the data come from a binomial distribution with $n = 15$ and $\pi = 0.80$?

17.4 One of the assumptions of the small sample (t) hypothesis test about a mean is that the underlying population is normally distributed. The accompanying data represent the diameter of the holes in washers that were purchased by a company. The specification on the washers is that the mean diameter of the hole be 0.5000 in.

0.5053	0.5098	0.4606	0.4606
0.4711	0.4627	0.4800	0.4800
0.4672	0.5642	0.5495	0.5495
0.4672	0.5346	0.5745	0.5745
0.5340	0.3767	0.3933	0.3933

(a) Make a frequency distribution and a histogram for the data. Do not use more than 5 classes.

(b) From the histogram, would you be willing to believe that the data are normally distributed?

(c) Set up the hypotheses to test that the data come from a normal distribution with a mean of 0.5000 and a standard deviation of 0.0600 in.

(d) Find the expected frequency distribution of the data.

(e) At the 0.05 level of significance, does it appear reasonable to assume that the data are normally distributed?

17.5   A large banking corporation believes that 80% of the loan applications it receives are approved within 24 hours. It decides to take a random sample of 10 loan applications every day for 3 months and record the number of the applications that are approved within 24 hours. The following data are obtained:

Number of Loan Applications in 10 Approved in 24 Hours	Frequency
4	1
5	5
6	11
7	19
8	27
9	18
10	7
Total	**88**

(a) Set up the necessary hypotheses to test whether the data come from a binomial distribution with $n = 10$ and $\pi = 0.80$.

(b) Find the expected frequency distribution for the data.

(c) At the 0.05 level of significance, is it reasonable to assume that the number of loan applications that are approved in 24 hours has a binomial distribution with $\pi = 0.80$?

17.3 TESTING PROPORTIONS FROM MORE THAN TWO POPULATIONS

In Chapter 13 you learned how to compare parameters from two different populations. In particular, you learned that you can use a Z test to determine whether the population proportions for two populations are equal. What if you are interested in comparing the population proportions for *more than two populations?*

Using a chi-square test, it is possible to compare proportions for more than two populations. Although it would be possible to compare a set of c populations by using the Z test you learned in Chapter 13 to test all of the possible pairs of populations, this is not a good thing to do. The reason for doing the single test is that each hypothesis test you do has a type I error probability associated with it. When you do multiple tests there is an increased chance that you will decide that two population proportions are different when in fact they are not. The single test controls the probability of making this mistake.

17.3.1 Testing Proportions for More Than Two Populations

In general, some characteristic of importance (a success) is defined for several different populations. We want to determine whether the proportion of successes in each population is the same. That is, the general set of hypotheses we wish to test is

$$H_0: \quad \pi_1 = \pi_2 = \cdots = \pi_c$$
$$H_A: \quad \text{At least one } \pi_i \text{ is different.}$$

To test these hypotheses for two populations you take a sample from each population and count the number of times that the characteristic of interest occurs. The data collection for more than two populations is the same.

In Chapter 7 you learned how to summarize data that involved two qualitative variables—the contingency table. For this test the two variables are the population that the sample item comes from ($i = 1, 2, \ldots, c$) and whether the item is a success (the characteristic is present) or a failure (the characteristic is not present). An example of the contingency table is shown in Table 17.1.

TABLE 17.1   Data Table for Testing Equality of Proportions

Number of	Population 1	2	. . .	c	Totals
Successes	s_1	s_2	. . .	s_c	s
Failures	f_1	f_2	. . .	f_c	f
Totals	n_1	n_2	. . .	n_c	n

The proportion of successes for each population can be calculated as

$$p_i = \frac{s_i}{n_i}$$

EXAMPLE 17.14   College Drinking

Setting Up a Chi-Square Test of Proportions

For the purposes of the study on binge drinking, students were classified as non-drinkers, non-binge drinkers, infrequent binge drinkers, and frequent binge drinkers. One question of interest to the researcher was whether there was a difference in academic responsibility among the three different categories of drinkers. He decided to look at the question of whether the proportion of people who missed class because of drinking is the same for each of the three groups. He created the contingency table shown below:

Miss Class	Type of Drinker Non-binge	Infrequent	Frequent	Total
Never	77	39	31	147
Once or More	7	19	39	65
Total	84	58	70	212

The populations in this case are the different types of drinkers and the characteristic of interest is whether the student missed class. The hypotheses are

$$H_0: \quad \pi_N = \pi_I = \pi_F$$
$$H_A: \quad \text{At least one of the } \pi_i \text{ is different.}$$

From the data you can see that $p_N = 0.083$, $p_I = 0.328$, and $p_F = 0.557$. The proportions certainly appear to be different but because of sampling error, a statistical test is necessary. ∎

The hypotheses are equivalent to asking the question, Is missing class related to the amount of drinking by a student? This is similar to the question that we asked when looking at regression models. The difference is that the variables are *qualitative*.

EXAMPLE 17.15   Comparing Machine Quality

Setting up a Chi-Square Test of Proportions

Suppose that a manufacturing facility had three different machines, all of which produced the same product. In an attempt to better understand their product, the management of the facility wants to know if the percent defective product is the same for all machines. That is, they wish to test the following hypotheses:

$$H_0: \quad \pi_1 = \pi_2 = \pi_3$$
$$H_A: \quad \text{At least one of the } \pi_i \text{ is different.}$$

The statistical tool used to test these types of hypotheses is a *chi-square test* similar to the one used to test goodness of fit.

The manufacturing facility collected finished product from the end of the production line. For each product analysts collected data on the machine and whether the product was defective. They assembled the data into a contingency table, which is shown below:

Number of	Machine 1	2	3	Totals
Defective	12	22	14	**48**
Good	188	148	196	**532**
Totals	**200**	**170**	**210**	**580**

From the data the analysts can estimate the proportion defective for each machine, $p_1 = 0.06$, $p_2 = 0.13$, and $p_3 = 0.07$. Clearly, the numbers are different, but the analysts know enough about sampling error to know that they cannot tell for sure without out a statistical test. ∎

Contingency tables like the one in the example above are tedious to do by hand for large data sets, but are easily done using computer software packages like Minitab or Excel.

 **TRY IT NOW!**

Technical Support

Setting Up the Chi-Square Test for Proportions

A company that sells computer software has three different locations set up to provide customers with technical support for their products. The support representatives keep a log for each call to technical support, and as part of that log, they record whether the problem was resolved successfully.

The company analysts are interested in knowing whether the percentage of calls that are successfully resolved is the same for each location. They randomly select logs from each location and collect data on the number of calls that result in a successful resolution of the problem. The data are summarized in the table below:

Number of	Location 1	Location 2	Location 3	Totals
Successful calls	257	264	283	804
Unsuccessful calls	43	86	97	226
Totals	300	350	380	1030

Set up the hypotheses for the software company.

Calculate the proportion of successfully resolved calls for each location. Based *solely* on these numbers, do you think that the proportion of successfully resolved calls for all three locations is the same?

17.3.2 Description of the Test

In Section 17.1 you learned that a chi-square test compares *expected and observed frequencies*. The observed frequencies are the raw data that are collected and the expected frequencies are the frequencies that would be predicted if the null hypothesis were true. The data in Table 17.1 can be analyzed using exactly this method.

If the null hypothesis, H_0, is true, then the data for all of the populations being sampled can be combined to give one overall estimate of the true proportion of successes, π. This overall estimate, p, can then be used to predict, for example, the number of defectives that the manufacturing facility should *expect* to see in the sample from each machine. In general, the estimate for π is calculated by counting all of the successes observed and dividing by the total number of objects sampled. That is, the **overall proportion of successes** is given by

$$p = \frac{s_1 + s_2 + \cdots + s_c}{n_1 + n_2 + \cdots + n_c} = \frac{s}{n}$$

Overall proportion of successes

and the **expected number of successes** in the sample from population i is

$$e_i = \frac{s}{n} n_i = \frac{sn_i}{n} = pn_i$$

Expected number of successes

These are all equivalent ways of calculating the expected frequency. For the chi-square test on proportions the last expression is probably easiest to use. For other chi-square tests, people find the second expression simplest.

ANS. $H_0: \pi_1 = \pi_2 = \pi_3$, $H_A:$ AT LEAST ONE π_i IS DIFFERENT. $p_1 = 0.86$, $p_2 = 0.75$, $p_3 = 0.74$. NO.

Since we know the sample size for each population and the expected number of successes in the sample, the easiest way to find the expected number of failures is by subtraction.

EXAMPLE 17.16 College Drinking

Calculating the Expected Frequencies

For the college drinking survey the researcher needed to estimate π, the overall proportion of students who miss class because of drinking. The value of π is estimated assuming that the null hypothesis is true. The overall proportion of students who miss class because of drinking is

$$p = \frac{7 + 19 + 39}{84 + 58 + 70} = \frac{65}{212} = 0.307$$

The expected number for each group is calculated as

$$e_N = (0.307)(84) = 25.8$$
$$e_I = (0.307)(58) = 17.8$$
$$e_F = (0.307)(70) = 21.5$$

The next step was to calculate the number of students who did not miss class for each group:

Non-binge drinkers:	$84 - 25.8 = 58.2$
Infrequent binge drinkers:	$58 - 17.8 = 40.2$
Frequent binge drinkers:	$70 - 21.5 = 48.5$

To get a better picture, the observed and expected frequencies were put into a single table as shown below. The expected frequencies are below the observed frequencies and are in parentheses.

	Type of Drinker			
Miss Class	**Non-binge**	**Infrequent**	**Frequent**	**Total**
Never	77	39	31	**147**
	(58.2)	(40.2)	(48.5)	
Once or More	7	19	39	**65**
	(25.8)	(17.8)	(21.5)	
Total	**84**	**58**	**70**	**212**

EXAMPLE 17.17 Comparing Machine Quality

Calculating the Expected Frequencies

For the manufacturing facility that wanted to compare machine quality, the estimate for π, the overall proportion defective, is found to be

$$p = \frac{12 + 22 + 14}{200 + 170 + 210} = \frac{48}{580} = 0.083$$

The expected number of defectives for each machine are

$$e_1 = (0.083)(200) = 16.6$$
$$e_2 = (0.083)(170) = 14.1$$
$$e_3 = (0.083)(210) = 17.4$$

The next table shows the observed and expected frequency of defectives for each machine. The expected frequencies are below the observed frequencies and enclosed in parentheses.

Number of	Machine			Totals
	1	2	3	
Defective	12	22	14	48
	(16.6)	(14.1)	(17.4)	
Good	188	148	196	532
	(183.4)	(155.9)	(192.6)	
Totals	200	170	210	580

■

TRY IT NOW!

Technical Support

Calculating the Expected Frequencies

The data for the computer software company interested in its technical support locations are

Number of	Location			Totals
	1	2	3	
Successful calls	257	264	283	804
Unsuccessful calls	43	86	97	226
Totals	300	350	380	1030

As with the goodness of fit test, the expected frequencies might not add up to the total because of rounding.

Estimate π, the percentage of calls that are resolved successfully, assuming that the three locations are the same.

Use the overall proportion of successful calls to find the expected frequency of successful calls for each location.

17.3.3 Performing the Test

Once we have the expected frequency for each population, all that remains to be done is to calculate the chi-square test statistic, find the critical value for the test, and make a decision.

We have already defined the chi-square test statistic in general and have shown how it is used for the goodness of fit test. The application of the chi-square statistic to testing proportions is exactly the same:

Numbering of the cells is not critical, although generally we number cells in a table row by row.

$$\chi^2 = \sum_{\text{all cells}} \frac{(o_i - e_i)^2}{e_i}$$

where

o_i is the observed frequency in the ith cell of the table

e_i is the expected frequency in the ith cell of the table

The degrees of freedom for a chi-square test involving a table are given by

$$\text{Degrees of freedom} = (r - 1)(c - 1)$$

where r is the number of rows in the table and c is the number of columns.

EXAMPLE 17.18   College Drinking

Performing a Chi-Square Test for Equality of Proportions

The data on missing class and drinking that the researcher has summarized so far are shown again below:

Miss Class	Type of Drinker Non-binge	Infrequent	Frequent	Total
Never	77 (58.2)	39 (40.2)	31 (48.5)	147
Once or more	7 (25.8)	19 (17.8)	39 (21.5)	65
Total	84	58	70	212

To perform the chi-square test at a 0.05 level of significance, the researcher needs to calculate the test statistic and find the critical value for comparison. The test statistic is calculated as shown:

$$\frac{(77-58.2)^2}{58.2} + \frac{(39-40.2)^2}{40.2} + \frac{(31-48.5)^2}{48.5} + \frac{(7-25.8)^2}{25.8} + \frac{(19-17.8)^2}{17.8}$$
$$+ \frac{(39-21.5)^2}{21.5} = 40.45$$

To find the critical value he needs to first find the degrees of freedom of the test, $(r-1)(c-1)$:

$$(2-1)(3-1) = 2$$

The critical value is then $\chi^2_{0.05,2} = 5.99$.

To finish the test he compares the test statistic of 40.45 to the critical value and finds that since 40.45 is outside the critical value the decision is to reject H_0 and conclude that the proportion of students who miss class is not the same for each population. That is, missing class *is related* to the amount that a student drinks. ∎

Software packages make it easy to perform the calculations for the expected frequencies, and some software will display these values as part of their output, as shown in Figure 17.2.

	Non-Binge	Infrequent	Frequent	All
Never	77	39	31	147
	58.25	40.22	48.54	147.0
Once or	7	19	39	65
more	25.75	17.78	21.46	65.0
All	84	58	70	212
	84.00	58.00	70.00	212.0

Chi-Square = 40.484, DF = 2, P-Value = 0.00

Cell Contents—
      Count
      Exp Freq

FIGURE 17.2   Minitab output from chi-square analysis

EXAMPLE 17.19   Comparing Machine Quality

Performing a Chi-Square Test for Equality of Proportions

The manufacturing facility has collected and summarized its data and calculated the expected number of defectives for each machine:

Number of	Machine			Totals
	1	2	3	
Defective	12	22	14	48
	(16.6)	(14.1)	(17.4)	
Good	188	148	196	532
	(183.4)	(155.9)	(192.6)	
Totals	200	170	210	580

It wants to perform the test at the 0.05 level of significance. It calculates the chi-square statistic as

$$\frac{(12-16.6)^2}{16.6} + \frac{(22-14.1)^2}{14.1} + \frac{(14-17.4)^2}{17.4} + \frac{(188-183.4)^2}{183.4} + \frac{(148-155.9)^2}{155.9}$$
$$+ \frac{(196-192.6)^2}{192.6} = 6.94$$

Since the level of significance is 0.05 and the degrees of freedom is $(2-1)(3-1) = 2$, the critical value for the test is $\chi^2_{0.05,2} = 5.99$.

Comparing the test statistic of 6.94 to the critical value of 5.99, the manufacturer rejects H_0 and concludes that the proportion defective for the three machines is not the same. ∎

 **TRY IT NOW!**

Technical Support

Performing the Chi-Square Test for Proportions

The computer software company with the different technical support locations wants to complete the test to determine whether the percentage of successfully resolved calls is the same at all three locations. It wants to test at the 0.01 level of significance.

(*continued*)

Calculate the value of the chi-square test statistic and complete the test.

Is the proportion of successfully resolved calls the same at each location?

17.3.4 Using the Results of the Chi-Square Test for Proportions

You might be wondering (and with good reason) just how you can *use* the chi-square test for equality of proportions to gain useful information or to make decisions. The results of the chi-square test are similar to the results of analysis of variance that you learned in Chapter 14. When the test does not lead to rejection of the null hypothesis, there is not really anything else to do. However, when the test leads to rejection of the null hypothesis, you conclude that at least one of the populations is different from the others. Does it mean that one of the populations is different from all of the others? Does it mean that each population is different from every other population? How useful is this kind of conclusion?

Remember that statistical analyses are not answers in themselves, but rather tools that can be used to identify problems or aid in decision making. The key words here are *identify* and *aid*. If you have done a chi-square test for proportions, then it is likely that you suspect that something is going on that is causing the populations to be different. The results of the test *verify* that your suspicions are justified. The test result tells you that further investigation is warranted.

How might you go about determining exactly which of the populations are the same and which are different? Certainly, you will know that the two samples that are farthest apart are probably different, but how about the others in between? How do they compare? For ANOVA there are formal techniques for testing which means are different. For chi-square tests, the techniques are much less formal.

Often, descriptive and graphical techniques can help. You might simply plot the sample proportions for each population on the same graph and see how they compare.

EXAMPLE 17.20   College Drinking

After the Chi-Square Test

As a result of the chi-square test, the researcher knows that at least one of the populations is different from the others with respect to missing classes. The problem is that that is *all* he knows. The chi-square test does not tell him whether all three populations are different from each other or if the two binge drinking categories are similar but different from the non-bingers.

He decides to plot the proportions for each sample on a graph. From the graph it appears that all three groups are different from each other. The chi-square test verifies his original ideas, but if he wants more definitive conclusions he might need to use a designed experiment.

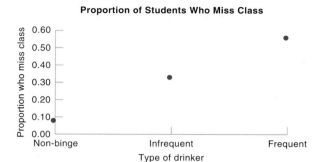

As you can see from the previous example, the answer is not often obvious from a plot. If the sample sizes are relatively the same you could find confidence intervals for each of the samples and compare them.

EXAMPLE 17.21 Comparing Machine Quality

After the Chi-Square Test

Now that the manufacturing facility knows that at least one of its machines is different from the others, it needs to try to identify which one(s) are different so that it can determine where the problem lies. The company decides to plot the proportion defective for each machine:

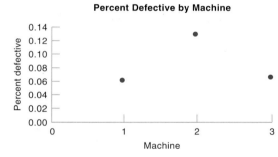

From the plot it certainly appears that machines 1 and 3 are comparable, while machine 2 has a much higher proportion defective. Just to be sure the company decides to calculate and plot 95% confidence intervals for the proportion defectives:

Machine 1: $0.06 \pm 1.96 \sqrt{\dfrac{(0.06)(0.94)}{200}} = 0.06 \pm 0.0329 = (0.0271, 0.0929)$

Machine 2: $0.13 \pm 1.96 \sqrt{\dfrac{(0.13)(0.87)}{170}} = 0.13 \pm 0.0504 = (0.0796, 0.1804)$

Machine 3: $0.07 \pm 1.96 \sqrt{\dfrac{(0.07)(0.93)}{210}} = 0.07 \pm 0.0337 = (0.0363, 0.1037)$

Note: There has been some rounding in these calculations.

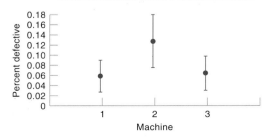

From the plot they see that the confidence intervals for machines 1 and 3 almost completely overlap, while the confidence interval for machine 2 barely overlaps the other two intervals. Thus, the company is pretty sure that machine 2 is different from the other two and that it produces a higher proportion of defective items. ■

An important thing to remember when doing the chi-square test for proportions, or any statistical test, is that the test is not the *answer* or *solution* to the problem. It is simply verification that something is *causing* the differences that you observed. Once the statistical analysis is done, it always requires knowledge of the situation to identify the causes and correct the problem if that is needed.

17.3.5 Exercises—Learning It!

17.6 In an experiment to study the attitude of voters concerning term limitations in Congress, voters in Indiana, Ohio, and Kentucky were polled with the following results:

Opinion	Indiana	Kentucky	Ohio
Support	82	107	93
Do not support	97	66	74

(a) Set up the hypotheses to test whether the proportion of voters who support congressional term limits is the same for all three states.

(b) Calculate the proportion of voters that support congressional term limits for each state individually. Based on these values, do you think there is a difference in the proportions?

(c) Calculate the overall proportion of voters who support term limits for Congress.

(d) Calculate the expected frequencies for each cell and find the value of the chi-square test statistic.

(e) At the 0.05 level of significance, is there a difference in the proportion of voters who support congressional term limits among the three states?

17.7 In a survey about satisfaction with local phone service, those respondents who rated their current service as excellent and those who rated it Poor–Very Poor were asked to classify their current local service provider. The results are given in the table below:

	Type of Company				
Current Service Source	**Long Distance**	**Local Phone**	**Power**	**Cable TV**	**Cellular Phone**
Excellent	264	444	131	215	198
Poor–Very Poor	1394	1318	485	431	572

(a) Set up the hypotheses to test whether the proportion of people who rated their company as excellent is the same for each type of company.

(b) Calculate the overall proportion of people who rate their current phone service as excellent.

(c) Find the expected frequency for each cell and calculate the chi-square test statistic.

(d) If you wanted to perform the test at the 0.05 level of significance, what would the critical value of the test be?

(e) At the 0.05 level of significance, is there a difference in the proportion of people who rate their local phone service as excellent among the different types of companies?

17.8 A computer magazine surveyed its readers to determine how likely it was that people who planned to purchase new computers in the near future would buy a portable/notebook or desktop model. The results are tabulated below:

	When Purchase Will Be Made		
Type of Computer	**0–3 months**	**3–6 Months**	**6–12 Months**
Notebook/Portable	34	156	258
Desktop	56	346	128

(a) Suppose that you were interested in determining whether there is a relationship between the type of computer that a person is planning to buy and when the person plans to make the purchase. Set up the hypotheses to test this. Which variable represents the population?

(b) Calculate the overall proportion of people who plan to buy a notebook/portable computer.

(c) Use this value to find the expected frequency for each cell.

(d) Calculate the value of the chi-square test statistic.

(e) At the 0.01 level of significance, is there a relationship between the type of computer that will be purchased and the time when the purchase will be made?

15.9 In a survey on the use of Office Suite software, 201 people who responded to the survey were asked about their experience with such software. The people who were administering the survey were interested in determining whether females had different experience with such software than males. The data they compiled are shown in the table:

<table>
<tr><td colspan="5" align="center">*Experience With Office Suite Software*</td></tr>
<tr><td>Gender</td><td>Use It Now</td><td>Don't Use It Now, But Have in Past</td><td>Know Someone Who Uses It</td><td>No Experience With It</td></tr>
<tr><td>Female</td><td>46</td><td>3</td><td>3</td><td>4</td></tr>
<tr><td>Male</td><td>118</td><td>12</td><td>5</td><td>10</td></tr>
</table>

(a) Set up the hypotheses to determine whether experience with Office Suite software is related to gender.

(b) Calculate the overall proportion of female respondents in the survey.

(c) Use the overall proportion to calculate the expected frequencies for each cell.

(d) If you want to do this test at the 0.10 level of significance, what is the critical value?

(e) At the 0.10 level of significance, are gender and experience with Office Suite software related?

17.4 THE CHI-SQUARE TEST FOR INDEPENDENCE

In the previous section you learned that it is possible for there to be a relationship between two qualitative variables. For example, suppose the Human Resources department of a large company collected data on employees' job levels and the type of medical coverage that the employees chose. Both of these variables, job level and type of medical coverage, are qualitative, but it makes sense that there might be a relationship between the two. We might expect that people with higher level, higher paying jobs would have better health plan coverage. How can we determine whether such a relationship exists? That is, how can we determine whether two qualitative variables are dependent or independent?

The chi-square test for proportions involved a contingency table that had 2 rows, successes and failures, and a column for each population. This test can be extended to look at two qualitative variables where each row corresponds to a different value of the first variable and each column corresponds to a different value of the second variable.

17.4.1 Probability and Independence

By now you are quite familiar with the idea that we rarely know the populations that we are studying, and instead we use data to *estimate* or *approximate* different characteristics of these populations. You learned in Chapter 7 that data collected over time are often used as *estimates* of the probabilities that different events will occur. These probabilities are known as *empirical* probabilities.

When we collect data on two qualitative variables we can display the data using a contingency table. The rows of the table represent the possible categories for one of the variables and the columns of the table represent the categories of the other variable. We can use the frequencies in the table to estimate different probabilities about the elements of the population. We can estimate whether an element will have a certain characteristic (event A), a particular pair of characteristics (event A AND B), or one or more of the characteristics (event A OR B).

EXAMPLE 17.22 Health-Care Plans

Classifying Events

Suppose that the Human Resources department of the company mentioned previously looked at the records of 300 employees and collected data on two characteristics: Job Classification and Health Plan. The data are summarized below:

We are assuming here that the sample is representative of the employees of the company and we are using the data to estimate the probabilities for the company as a whole.

Job Classification	Type of Coverage			Total
	Physician Network	HMO	No Coverage	
Salaried professional	35	12	3	50
Salaried clerical	21	67	12	100
Hourly	6	112	32	150
Totals	**62**	**191**	**47**	**300**

If we define A as the event that a person in the company selects the HMO for medical coverage, and B as the event that a person in the company is an hourly worker, then

$$P(A) = \frac{191}{300} \qquad P(B) = \frac{150}{300} \qquad P(A \text{ AND } B) = \frac{112}{300}$$ ■

Definition of Independence in Probability

When you looked at bivariate relationships in Chapter 15, you were trying to decide whether the value of one quantitative variable depended on another or if they were independent. We can also use the word **independent** in terms of events and probability.

> Two events are ***independent*** if the probability that one event occurs in any given trial of an experiment is not affected or changed by the occurrence of the other event.

In probability, two events, A and B, are independent *exactly when*

Formula for independent events

$$P(A \text{ AND } B) = P(A) \times P(B)$$

The phrase *exactly when* means that the statement can be used in both directions. That is, if we know two events are independent, then to find the probability that both will occur we can multiply the individual probabilities together. It also means that if we know that the probability that both occur is equal to the product of the individual probabilities, we can conclude that the events are independent. We will use this fact in the chi-square test for independence.

EXAMPLE 17.23 Health-Care Plans

Independence and Probability

In the example about the Human Resources department we can check to see if events A and B are independent by comparing the quantity P(A AND B) to the quantity P(A) × P(B):

$$P(A) \times P(B) = \left(\frac{191}{300}\right)\left(\frac{150}{300}\right) = 0.318 \quad \text{and} \quad P(A \text{ AND } B) = \frac{112}{300} = 0.373$$

Thus, events A and B are not independent *in this sample*. We know that when we use sample data the probabilities are *exactly correct* only when applied to the sample. Does it mean that the two events A and B are not independent in the population? Not really. ■

You know from your study of sampling error and sampling distributions that if the company were to take another sample of 300 employees the results would not be identical. If both samples are representative of the population of interest, then the results should be similar, but there will be some sampling error. We need a test for independence that takes into account the fact that we are dealing with *sample data* and not population values. In the next section we see how the chi-square test that we have been studying can be used to accomplish this task.

17.4.2 Description of the Test

We would like to decide whether or not two qualitative variables are related. The hypotheses that we will test are

H_0: The two variables are independent of each other.

H_A: The two variables are not independent.

When we collect data on two qualitative variables and summarize them using a contingency table, the number in each cell of the table is the *observed* frequency associated with one particular category of the first variable *and* one particular category of the second variable. A general example of the contingency table for two variables is shown in Table 17.2.

Variable 1 has r possible categories and variable 2 has c possible categories. The value o_{12} is the number of sample elements that were in category 1 of variable 1 and category 2 of variable 2.

TABLE 17.2 Contingency Table for Test for Independence

	Variable 2				
Variable 1	1	2	...	c	Totals
1	o_{11}	o_{12}	...	o_{1c}	r_1
2	o_{21}	o_{22}	...	o_{2c}	r_2
$\vdots$	$\vdots$	$\vdots$	$\vdots$	$\vdots$	
r	o_{r1}	o_{r2}	.	o_{rc}	r_r
Totals	c_1	c_2	...	c_c	n

EXAMPLE 17.24 College Drinking

Setting Up the Contingency Table

In addition to questions used to classify drinking habits, the survey on drinking asked students about where they lived: residence halls or dormitories, fraternity or sorority houses, other university housing, or off campus house or apartment. In analyzing the results of the survey, the researcher asked whether the amount of drinking by students depends on where they live. His hypotheses were

H_0: The amount that a student drinks is independent of where he or she lives.

H_A: The amount that a student drinks depends on where he or she lives.

He compiled a contingency table of the data:

| | *Type of Drinker* | | | |
Where They Live	Non-binge	Infrequent	Frequent	Total
Residence hall or dormitory	35	25	46	**106**
Fraternity or sorority	0	1	0	**1**
Other university housing	0	2	1	**3**
Off-campus house or apartment	49	30	24	**103**
Total	**84**	**58**	**71**	**213** ■

To do a chi-square test, we must find the expected frequency for each cell in the table and compare it to the observed frequency. You remember from the other chi-square tests that we have done that we find the expected frequencies by *assuming* that H_0 is true and calculating the probability for each cell under that assumption. For the test of independence, the null hypothesis is that the two variables *are independent*. To find the probabilities for each value of a variable, you use the row or column totals, the same way you did in Chapter 7 when you learned about probability. That is, for row i,

$$P(i) = \frac{r_i}{n}$$

and for column j,

$$P(j) = \frac{c_j}{n}$$

Remember that the definition of independence works two ways. If we know that two events are independent, then we can multiply their probabilities together to find the probability that they will both occur. That is, when we have independence for each cell in the contingency table, the probability that variable 1 will have category i and variable 2 will have category j is

$$P(i \text{ AND } j) = \frac{r_i}{n} \frac{c_j}{n}$$

To find the **expected frequency** for any cell we multiply the probability by the sample size, so

Formula for expected frequency in a cell

$$e_{ij} = \left(\frac{r_i}{n} \frac{c_j}{n} \right) n = \frac{r_i c_j}{n}$$

EXAMPLE 17.25   College Drinking

Finding the Expected Frequencies

The data on where a student lives and the type of drinker that he or she is are shown again below:

	Type of Drinker			
Where They Live	**Non-binge**	**Infrequent**	**Frequent**	**Total**
Residence hall or dormitory	35	25	46	**106**
Fraternity or sorority	0	1	0	**1**
Other university housing	0	2	1	**3**
Off-campus house or apartment	49	30	24	**103**
Total	**84**	**58**	**71**	**213**

The calculations for the expected frequencies for each cell are

$$e_{11} = \frac{(106)(84)}{213} = 41.80 \qquad e_{12} = \frac{(106)(58)}{213} = 28.86 \qquad e_{13} = \frac{(106)(71)}{213} = 35.33$$

$$e_{21} = \frac{(1)(84)}{213} = 0.39 \qquad e_{22} = \frac{(1)(58)}{213} = 0.27 \qquad e_{23} = \frac{(1)(71)}{213} = 0.33$$

$$e_{31} = \frac{(3)(84)}{213} = 1.18 \qquad e_{32} = \frac{(3)(58)}{213} = 0.82 \qquad e_{33} = \frac{(3)(71)}{213} = 1.00$$

$$e_{41} = \frac{(103)(84)}{213} = 40.62 \qquad e_{42} = \frac{(103)(58)}{213} = 28.05 \qquad e_{43} = \frac{(103)(71)}{213} = 34.33$$

The expected frequencies are shown with the observed frequencies in the table below:

	Type of Drinker			
Where They Live	**Non-binge**	**Infrequent**	**Frequent**	**Total**
Residence hall or dormitory	35	25	46	**106**
	(41.80)	(28.86)	(35.33)	
Fraternity or sorority	0	1	0	**1**
	(0.39)	(0.27)	(0.33)	
Other university housing	0	2	1	**3**
	(1.18)	(0.82)	(1.00)	
Off-campus house or apartment	49	30	24	**103**
	(40.62)	(28.05)	(34.33)	
Total	**84**	**58**	**71**	**213** ∎

    Remember that in any chi-square test the expected frequency for each cell or class should be at least five. When this is not the case, the categories must be collapsed or joined. For the goodness of fit test, the data are quantitative, so you always join classes that are adjacent. In the test for proportions or the test for independence this is not the case. The categories should be collapsed on a *logical* basis, that is, one that preserves the intent of the question.

EXAMPLE 17.26   College Drinking

Adjusting the Contingency Table

When he looked at the contingency table on type of drinker and residence, the researcher saw that two of the categories for residence did not have expected frequencies of at least 5. He knew that this meant he should collapse some of the categories.

His first thought was to combine the two low-frequency categories (Fraternity or sorority and Other university housing) into a single category designated Other university housing, but he realized that this would still not raise the expected values above 5.

His final decision was to collapse all of the on-campus housing into a single category so that he met the requirements of the chi-square test. The modified contingency table is shown below:

Where They Live	*Type of Drinker* Non-binge	Infrequent	Frequent	Total
On campus	35 (43.38)	28 (29.95)	47 (36.67)	**110**
Off-campus house or apartment	49 (40.62)	30 (28.05)	24 (34.33)	**103**
Total	**84**	**58**	**71**	**213** ∎

EXAMPLE 17.27   Health Care Plans

Setting up the Contingency Table

For the Human Resources department, the hypotheses to be tested are

H_0:   Choice of health-care plan and job classification are independent.

H_A: Choice of health-care plan and job classification are not independent.

For the employee/health plan data the expected frequencies are calculated as follows:

$$e_{11} = \frac{(50)(62)}{300} = 10.3 \qquad e_{12} = \frac{(50)(191)}{300} = 31.8 \qquad e_{13} = \frac{(50)(47)}{300} = 7.8$$

$$e_{21} = \frac{(100)(62)}{300} = 20.7 \qquad e_{22} = \frac{(100)(191)}{300} = 63.7 \qquad e_{23} = \frac{(100)(47)}{300} = 15.7$$

$$e_{31} = \frac{(150)(62)}{300} = 31.0 \qquad e_{32} = \frac{(150)(191)}{300} = 95.5 \qquad e_{33} = \frac{(150)(47)}{300} = 23.5$$

The expected frequencies together with the original data are shown below. There are clearly discrepancies between the observed and expected frequencies, but it is not clear if they are simply the result of random variation or if they are indicative that H_0 is false.

Job Classification	*Type of Coverage* Physician Network	HMO	No Coverage	Total
Salaried professional	35 (10.3)	12 (31.8)	3 (7.8)	**50**
Salaried clerical	21 (20.7)	67 (63.7)	12 (15.7)	**100**
Hourly	6 (31.0)	112 (95.5)	32 (23.5)	**150**
Totals	**62**	**191**	**47**	**300** ∎

As in the case with the goodness of fit test and the test for proportions, it is important to remember that the expected frequency in each cell should be at least 5. If this is not the case, then categories for one or the other or both of the variables need to be combined. Since the categories are not numerical, the categories should be grouped on similarities.

 *TRY IT NOW!*

Drinking Survey

Setting Up the Contingency Table for a Test for Independence

The Public Health student who did the study on drinking also collected data on the number of times that the student drove while intoxicated in the last two weeks (coded). The contingency table for the usable responses is given below:

Class	Number of Times Drive While Intoxicated			
	Not at All	Once	Twice or More	Total
Freshman	72	5	9	86
Sophomore	19	8	9	36
Junior	16	8	6	30
Senior	8	4	7	19
Total	115	25	31	171

The university is interested in knowing if the number of times a student drove while intoxicated is related to his or her class in school. It feels that this information will help target student audiences for programs on drinking and driving.

Set up the hypotheses that the university should test.

Calculate the expected frequencies for each cell and put them in the appropriate location in the table above.

Are any of the expected frequencies less than 5? If so, can you suggest a logical way to combine categories to avoid this problem?

17.4.3 Performing the Test for Independence

Once we have the expected and observed frequencies for each cell in the table, nothing new needs to be learned to perform the test. The chi-square test statistic is calculated in the same way as for the test of proportions:

$$\chi^2 = \sum_{\text{all cells}} \frac{(o_{ij} - e_{ij})^2}{e_{ij}}$$

and the degrees of freedom are calculated as

$$(r - 1)(c - 1)$$

EXAMPLE 17.28   College Drinking

Performing the Chi-Square Test for Independence

The researcher looking at whether where students live is related to the amount that they drink calculated the chi-square statistic for the revised contingency table:

| | *Type of Drinker* | | | |
Where They Live	Non-binge	Infrequent	Frequent	Total
On campus	35	28	47	**110**
	(43.38)	(29.95)	(36.67)	
Off-campus house or apartment	49	30	24	**103**
	(40.62)	(28.05)	(34.33)	
Total	**84**	**58**	**71**	**213**

$$\frac{(35 - 43.38)^2}{43.38} + \frac{(28 - 29.95)^2}{29.95} + \cdots + \frac{(30 - 28.05)^2}{28.05} + \frac{(24 - 34.33)^2}{34.33}$$
$$= 1.62 + 0.13 + \cdots + 0.14 + 3.11 = 9.63$$

He decided to do the test at the 0.05 level of significance and calculated the degrees of freedom of the test to be $(2 - 1)(3 - 1) = 2$. The critical value of the test is $\chi^2_{0.05,2} = 5.99$. Since the test statistic of 9.63 is outside the critical value, he rejects H_0 and concludes that the variables residence and type of drinker are not independent. That is, the amount of drinking that students do depends on where they live. ■

EXAMPLE 17.29   Health Care Plans

Performing the Chi-Square Test for Independence

The Human Resources department that was looking at the relationship between employee job classification and type of health plan chosen calculated the chi-square statistic for its data as

$$\frac{(35 - 10.3)^2}{10.3} + \frac{(12 - 31.8)^2}{31.8} + \cdots + \frac{(112 - 95.5)^2}{95.5} + \frac{(32 - 23.5)^2}{23.5}$$
$$= 58.88 + 12.36 + \cdots + 2.85 + 3.07 = 101.35$$

The analysts decide to perform the test with $\alpha = 0.05$. The degrees of freedom for the test are $(3 - 1)(3 - 1) = 4$, and they find the critical value to be 9.49. Since the chi-square test statistic is definitely outside the critical value, they reject H_0 and conclude that the type of job and health plan chosen are not independent; that is, they are related. ■

📖 *TRY IT NOW!*

Drinking Survey
Performing the Chi-Square Test for Independence

The university that is looking at the relationship between class year and drinking and driving wants to perform the test at the 0.05 level of significance.

Calculate the value of the chi-square statistic for the data.

Don't forget to collapse the categories so that there are no cells with expected values below 5.

Find the critical value and perform the test.

Are class and drinking and driving independent?

17.4.4 Exercises—Learning It!

17.10 A report by the Department of Justice on rape victims reports on interviews with 3721 victims. The attacks were classified by age of the victim and the relationship of the victim to the rapist. The results of the study are given below:

		Relationship of Rapist	
Age of Victim	Family	Acquaintance or Friend	Stranger
Under 12	153	167	13
12 to 17	230	746	172
Over 17	269	1232	739

(a) Set up the hypotheses to test whether age of victim and relationship of rapist are independent.

(b) Calculate the expected frequencies for each cell.

(c) How many degrees of freedom will the chi-square test for indpendence have? Using a level of significance of 0.01, what is the critical value for the test?

(d) Calculate the value of the chi-square test statistic.

(e) Is the age of the victim independent of the relationship of the rapist?

17.11 A company that manufactures cardboard boxes for packaging cereals wants to determine whether the type of defect that a particular box has is related to the shift on which it was produced. It compiles the following data. In each case, if a box had multiple defects the most serious defect was recorded.

	Type of Defect		
Shift	Printing	Rips/Tears	Size
1	55	60	85
2	58	63	79
3	89	63	48

(a) Set up the appropriate hypotheses for the test.

(b) Calculate the expected frequencies for each cell and calculate the value of the chi-square test statistic.

(c) How many degrees of freedom does the test have?

(d) At the 0.05 level of significance are defect type and shift related?

17.12 A company that depends heavily on advertising for selling its products wants to know if its various advertising media have different effectiveness relative to the age of the customer. The company uses its warranty return cards to collect the following information:

Type of Advertisement	Age of Customer				
	21–30	31–40	41–50	Over 50	Totals
Store display	21	28	8	6	63
Catalog	8	5	1	1	15
Magazine	1	23	8	1	33
Newspaper	18	14	2	4	38
Totals	48	70	19	12	149

(a) Set up the hypotheses to test whether type of advertisement and age of customer are independent.

(b) What population will the test apply to?

(c) Calculate the expected frequencies for each cell.

(d) How many degrees of freedom will the test have?

(e) Without collapsing any cells, calculate the value of the chi-square test statistic.

(f) At the 0.05 level of significance, are type of advertisement and age of customer independent?

(g) Do you think that categories should have been collapsed? Why or why not?

(h) What categories would you recommend collapsing to fix the problem?

17.13 The people who did the survey on Office Suite software were also interested in whether experience with such software was related to age. They created a contingency table to look at this question:

Experience	Age Group		
	Under 25	25 to 44	45 and Up
Use it now	11	51	101
Don't use it now, but have in past	2	2	11
Know someone who uses it	2	2	4
No experience	2	3	10

(a) Set up the hypotheses to test whether experience with Office Suite software is independent of age group.

(b) How many degrees of freedom will the chi-square test have?

(c) Calculate the expected frequencies for each cell.

(d) Without collapsing any categories, calculate the value of the chi-square test statistic.

(e) At the 0.05 level of significance, is experience with Office Suite software independent of age?

(f) Are the cells with frequencies less than 5 a factor in this case? Why or why not? Do you think that you should redo the test after collapsing categories?

17.5 THE CHI-SQUARE TEST FOR INDEPENDENCE IN EXCEL

Excel provides the tools needed to do a chi-square test for independence, although they are not found as a single analysis tool. You are already familiar with one of them, the pivot table. In this section, we will explain how to use Excel to perform a chi-square test for independence.

17.5.1 Creating the Contingency Table in Excel

The contingency table provides the backbone for the chi-square test for independence in Excel. We will look at the data from the binge drinking survey, and test whether where a student lives is independent of the type of binge drinker. A small portion of the data in an Excel worksheet is shown in Figure 17.3.

	A	B	C
1	BINGER	MISS	LIVE
2	1	1	4
3	1	1	4
4	2	1	1
5	1	1	4
6	1	1	4
7	1	1	1
8	1	1	4
9	3	1	4
10	1	1	4
11	1	1	4
12	2	2	1
13	3	2	4
14	0	1	4
15	3	1	1
16	3	3	1

FIGURE 17.3   Binge drinking data

The first step is to create a contingency table in Excel using the **Pivot table** tool. From the **Data** menu, select **Pivot Table Report** and follow the steps of the pivot table wizard to create the table. In this example, we will put the variable for type of binge drinker (BINGER) in the columns and the variable for where a student lives (LIVE) in the rows. At step 4 of 4, after you indicate where you want the table to be placed, do not select **Finish.** Instead, click **Options...** and the dialog box shown in Figure 17.4 opens.

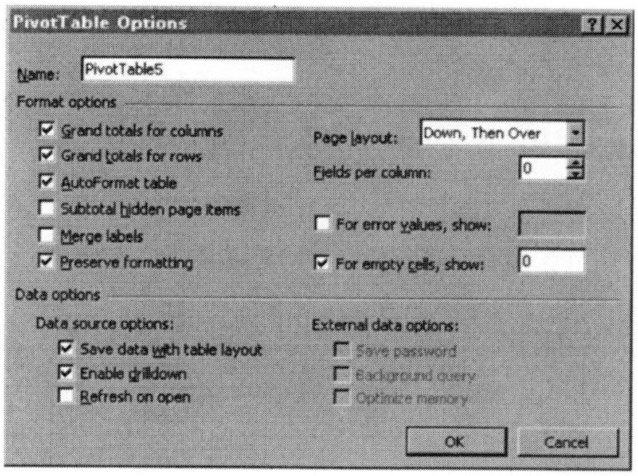

FIGURE 17.4   Pivot table options dialog box

Make sure that the checkbox next to **For empty cells, show** is checked, and in the textbox next to it, type "0". To perform a chi-square test, Excel will not accept empty cells in the contingency table. Click **OK** and then **Finish.** The contingency table for the data is shown in Figure 17.5 (page 774).

Since we are interested only in students who are drinkers, we do not want the Binger column for 0, which indicates that the student does not drink. To eliminate this column, double click on the field button BINGER and the pivot table field dialog box shown in Figure 17.6 opens.

O	P	Q	R	S	T
Count of BINGER	BINGER				
LIVE	0	1	2	3	Grand Total
1	11	35	25	46	117
2	0	0	1	0	1
3	0	0	2	1	3
4	28	49	30	24	131
Grand Total	39	84	58	71	252

FIGURE 17.5   Contingency table for binge drinking data

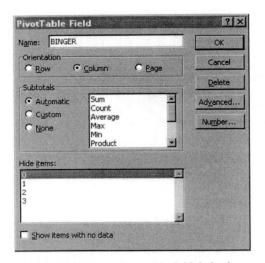

FIGURE 17.6   Pivot table field dialog box

At the bottom of the box under **Hide Items:,** highlight "0" and click **OK.** The pivot table will adjust to hide this column. To work with this table, use **Paste Special > Values** to make a copy in a different location in the worksheet. You can then enter titles that are more descriptive and change format features.

17.5.2 Creating the Table of Expected Frequencies in Excel

The contingency table provides the actual frequencies for each cell. To perform the chi-square test for independence, we also need the expected frequencies for each cell. You can use formulas in Excel to calculate these. Excel expects to find these frequencies in a separate table, directly below the table of actual frequencies.

Remember that the expected frequency for any cell is calculated by

$$e_{ij} = \left(\frac{r_i}{n} \times \frac{c_j}{n} \right) n = \frac{r_i c_j}{n}$$

where r_i is the row total, c_j is the column total, and n is the total number of observations.

1. Copy the row and column headings of the pivot table to a location just below it. This will define the table for the expected frequencies.

2. Position the cursor in the first cell of the table, for non-binge drinkers who live in dorms or residence halls, and type "=$S20*$P$24/$S$24". In this formula, S20 is the location of the row total for dorms and residence halls, P24 is the location of the column total for non-binge drinkers, and S24 is the location of the grand total. You must use relative references if you are going to copy the for-

mula to the rest of the cells and not retype them. The relative references in this formula will allow you to copy the formula to the rest of the cells in this column. To copy to the rest of the cells in a row, make the first reference $S$20 and the second P$24. The table of expected frequencies showing the formulas in each cell is found in Figure 17.7.

26		Type of Drinker		
27	LIVE	Non-Binge	Infrequent	Frequent
28	Residence Hall or Dormitory	=$S$20*P$24/$S$24	=$S$20*Q$24/$S$24	=$S$20*R$24/$S$24
29	Fraternity or Sorority	=$S$21*P$24/$S$24	=$S$21*Q$24/$S$24	=$S$21*R$24/$S$24
30	Other University Housing	=$S$22*P$24/$S$24	=$S$22*Q$24/$S$24	=$S$22*R$24/$S$24
31	Off Campus House or Apartment	=$S$23*P$24/$S$24	=$S$23*Q$24/$S$24	=$S$23*R$24/$S$24
32				

FIGURE 17.7 Table of expected frequency formulas

17.5.3 Performing the Chi-Square Test

Excel has a function, **CHITEST,** that performs the chi-square test for independence. The function accepts the data ranges for the observed and expected values as input and returns the p value of the test as output.

To perform the test for the binge drinking data:

1. Position the cursor in the cell where you want the p value of the test to be located, and start the function wizard. Choose **Statistical** as the function category and **CHITEST** as the function name. The **CHITEST** dialog box opens as shown in Figure 17.8.

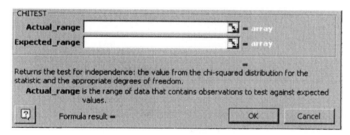

FIGURE 17.8 The CHITEST dialog box

2. Position the cursor in the textbox for **Actual_range** and highlight the range of cells that contain the observed frequencies. Be careful not to highlight the row and column totals.

3. Do the same thing for the **Expected_range,** highlighting the location of the expected frequencies.

4. Click **OK;** the result of the test will appear in the designated cell. In this case, the value returned is 0.01744.

If the test is performed at the 0.05 level of significance, we conclude that type of drinker and where a student lives are not independent.

From the examples in this chapter, you know that there was a problem with this test because of cells with expected frequencies less than 5. You can use the pivot table to combine rows and columns and then perform the test again. To combine rows or columns in a pivot table, use the **Group** command found on the pivot table toolbar. You can also access this command by highlighting the rows or columns you want to combine, right clicking the mouse, and selecting **Group and Outline > Group** from the menu.

CHAPTER 17 SUMMARY

The chi-square test involves comparing observed and expected frequencies for different classes of data. The test is quite versatile and can be used to test *goodness of fit, equality of proportions for more than two populations,* and *independence of qualitative variables.*

As is true with *any* statistical tool, the results of a chi-square test do not *solve* a problem or make a decision for the user. They simply point out when further action is indicated and when it is not. Solving the real problem requires knowledge of the situation and sometimes further data collection and analysis.

Key Terms

Term	Definition	Page Reference
Chi-square goodness of fit test	The **chi-square goodness of fit test** checks to see how well a set of data fits the model for a particular probability distribution.	735
Expected frequencies, e_i	The **expected frequencies** are the number of observations that should fall into each class in a frequency distribution under the hypothesized probability distribution.	736
Independent events	Two events are independent if the probability that one event occurs is not affected or changed by the occurrence of the other event.	764
Observed frequencies, o_i	The **observed frequencies** are the actual number of observations that fall into each class in a frequency distribution or histogram.	736
Uniform distribution	A **uniform distribution** is one in which each outcome or class of outcomes is equally likely to occur.	736

Key Formulas

Term	Formula	Page Reference
Chi-square critical value	$\chi^2_{\alpha,\, k-p-1}$	744
Chi-square statistic	$\chi^2 = \dfrac{\sum_{i=1}^{k}(o_i - e_i)^2}{e_i}$	741
Degrees of freedom for goodness of fit test	$k - p - 1$	742

Term	Formula	Page Reference
Degrees of freedom for test for independence	$(r-1)(c-1)$	770
Expected frequency for test for independence	$e_{ij} = \left(\dfrac{r_i}{n}\dfrac{c_j}{n}\right)n = \dfrac{r_i c_j}{n}$	766
Expected frequency for test for proportions	$e_i = \dfrac{s}{n}n_i = \dfrac{sn_i}{n} = pn_i$	755
Probability for independent events	$P(A\ AND\ B) = P(A) \times P(B)$	765
Overall proportion of success	$p = \dfrac{s_1 + s_2 + \cdots + s_c}{n_1 + n_2 + \cdots + n_c} = \dfrac{s}{n}$	755

CHAPTER 17 EXERCISES

Learning It!

17.14 A random sample of 69 Porsche drivers were asked how many miles they had driven their vehicle in the past calendar year. A frequency table for the data is given below:

Number of Miles	Observed Frequency
$0 < x \le 4000$	20
$4000 < x \le 8000$	22
$8000 < x \le 12{,}000$	14
$12{,}000 < x \le 16{,}000$	6
$16{,}000 < x \le 20{,}000$	1
$20{,}000 < x \le 24{,}000$	4
$24{,}000 < x \le 28{,}000$	0
$28{,}000 < x \le 32{,}000$	2

(a) Set up the hypotheses to test whether the number of miles driven by Porsche drivers is normally distributed with a mean of 7500 miles with a standard deviation of 6500 miles.

(b) Find the expected frequencies for each cell.

(c) Calculate the chi-square test statistic. If necessary, collapse cells so that the expected frequency for each category is at least 5.

(d) At the 0.05 level of significance, what can you conclude about the number of miles driven annually by Porsche drivers?

17.15 To begin a study comparing the monthly salaries of the auditors for a particular company to accountants' salaries in the industry in general, a random sample of 30 auditors was taken and their monthly salaries recorded. The data are shown here:

2832	3032	3122	3208	3325
2843	3050	3123	3224	3328
2875	3050	3128	3231	3375
2878	3087	3130	3233	3396
2995	3093	3133	3237	3429
3010	3096	3175	3239	3450

(a) Create a frequency distribution for the data using six classes and display the data graphically.

(b) Do you think that the assumption that the data are normally distributed is at least reasonable? Why or why not?

(c) Set up the hypotheses to test whether the data come from a normal population that has a mean of $3150 and a standard deviation of $165.

(d) Perform the chi-square goodness of fit test with a level of significance of 0.05.

(e) What does the test tell you about the data?

17.16 The problem of workplace violence is growing. A survey of 600 full-time American workers on workplace violence concentrated on those respondents who were victims of harassment, threat of physical violence, and actual physical violence. In follow-up interviews the victims were asked to identify the major effect that the violence had on them. The data are shown below:

Major Effect on Worker	Type of Violence		
	Harassment	**Threat**	**Physical**
Psychological	56	28	15
Disrupted work life	39	13	12
Physical injury or sickness	15	5	10
No negative effect	5	7	6

(a) Suppose that the group that conducted the study were interested in whether the type of violence that a person experienced is related to the effect of the violence on their life. Set up the hypotheses for this test.

(b) Find the expected frequencies for each cell.

(c) Without collapsing any categories, how many degrees of freedom will the test have?

(d) At the 0.05 level of significance, are type of violence and effect on a worker independent?

17.17 In a survey sponsored by the National Cancer Institute (NCI) and the American Cancer Society in 1993, adults (people 18 years of age and older) in eight states were surveyed on their attitudes toward smoking in public places. One question of interest was whether the respondent favored banning smoking in fast-food restaurants. The data collected are shown in the table below:

Ban Smoking in Fast-Food Restaurants	Louisiana	Missouri	New Jersey	Ohio	Oklahoma	South Carolina	Texas	Washington
Yes	129	124	133	130	107	210	204	272
No	146	124	128	128	145	161	201	159

(a) Set up the hypotheses to test whether the proportion of adults who favor banning smoking in fast-food restaurants is the same for all eight states.

(b) Calculate the overall proportion of adults who favor a ban on smoking in fast-food restaurants.

(c) Use the overall proportion to find the expected frequencies for each cell.

(d) Calculate the chi-square statistic.

(e) At the 0.05 level of significance, do the data indicate that the proportion of adults who favor banning smoking in fast-food restaurants is the same for all eight states?

17.18 In the survey about using Office Suite software the researchers were interested in finding whether the proportion of people who installed the software themselves was different for different age groups. They created the contingency table shown below:

Install Software	Age Group		
	Under 25	**25 to 44**	**45 and Up**
Yes	8	45	94
No	9	14	31

(a) Set up the hypotheses to test whether the proportion of people who install their own Office Suite software is the same for all three age groups.

(b) Use the data to calculate the overall proportion of people who install their own Office Suite software.

(c) Find the expected frequencies for each cell and calculate the chi-square statistic.

(d) At the 0.05 level of significance, what can you conclude about the proportion of people who install their own Office Suite software?

17.19 A software company that was looking at the time to failure of the diskettes it uses decides to look at its two suppliers of the product. The specifications for the product state that the life of the diskette should be normally distributed with a mean of 500 hours and a standard deviation of 5 hours. The data for each of the suppliers are shown below:

Supplier A						Supplier B				
474	492	498	504	511		487	492	495	497	499
486	492	500	505	511		488	492	495	497	499
489	494	501	506	512		489	492	495	497	499
490	494	501	507	513		489	492	495	497	501
490	494	501	508	514		489	493	495	498	502
490	495	502	508	515		491	493	496	498	503
491	496	502	509	517		491	494	496	498	503
491	498	504	509	519		491	494	496	498	505
491	498	504	510	528		492	494	496	499	506

(a) Create a frequency distribution and a histogram for each supplier.

(b) From the histogram, does it appear that the data for each supplier are normally distributed?

(c) Do a chi-square goodness of fit test for each supplier to see if it is reasonable to say that the diskettes meet the specifications. Use a level of significance of 0.05.

17.20 As part of an annual survey by the Department of Transportation, randomly selected households are asked, among other questions, about the number of vehicles in the household and the availability of public transportation. The results for people who lived in urban areas are tabulated below:

Public Transportation?	*Number of Vehicles*					
	0	**1**	**2**	**3**	**4**	**5+**
Yes	974	2592	2427	687	194	59
No	98	378	540	199	58	23

(a) Set up the hypotheses to determine whether the number of vehicles in a household and the availability of public transportation are independent.

(b) How many degrees of freedom will the critical value for the test have?

(c) Calculate the expected frequencies and compute the test statistic.

(d) At the 0.05 level of significance, what can you conclude about the number of vehicles in a household and the availability of public transportation for households in urban areas?

17.21 In a survey of adults who were currently smokers done in 1992 the following data were collected on gender and education level:

	Education Level			
Gender	Less Than 12 Years	12 Years	13 to 15 Years	16 Years or More
Male	359	546	293	173
Female	364	662	392	175

(a) Set up the hypotheses to test whether the proportion of female smokers is the same for each level of education.

(b) Calculate the overall percentage of women who smoke.

(c) Calculate the expected frequency for each cell.

(d) At the 0.05 level of significance, what can you say about the proportion of women who smoke for the different levels of education?

Thinking About It!

17.22 One of the big controversies during the Vietnam War era was whether the draft lottery was truly random. Many people thought that the lottery was biased against people who were born in certain months of the year. The accompanying data show the number of birth dates in each month that were chosen in the first half of the draft lottery (there are 366 possible birth dates, so these are the first 183 birth dates that were chosen).

Month	Number of Birth Dates in the First Half of the Draft	Month	Number of Birth Dates in the First Half of the Draft
January	13	July	14
February	12	August	18
March	9	September	19
April	11	October	13
May	14	November	21
June	14	December	25

(a) Just by looking at the data, do you think that the claim was worth investigating? Why or why not?

(b) At the 0.05 level of significance do the data indicate that the distribution of the first 183 draft dates was something other than random? (*Hint:* Think about what random would mean in this situation.)

Requires Exercise 17.12   **17.23** Consider the company that is looking at the relationship between type of advertising media and age of customer. The data are shown again below:

Type of Advertisement	Age of Customer				
	21–30	31–40	41–50	Over 50	Totals
Store display	21	28	8	6	**63**
Catalog	8	5	1	1	**15**
Magazine	1	23	8	1	**33**
Newspaper	18	14	2	4	**38**
Totals	**48**	**70**	**19**	**12**	**149**

(a) Create a revised contingency table that collapses categories and fixes the problem of cells with expected frequencies less than 5.

(b) Perform the chi-square test for independence again.

(c) Compare the results of this test with the results of the test where categories were not collapsed.

Requires Exercise 17.19   **17.24** Consider the data from the problem about the diskette suppliers.

(a) Why is doing a chi-square goodness of fit test for each supplier to see if they meet specifications different than doing a test to see whether the mean for each supplier is 500 hours and a test to see if the variance for each supplier is 25 hours²?

(b) Just because you reject the hypothesis that the data are normally distributed with a mean of 500 hours and a standard deviation of 5 hours, does that mean that the data are *not* normally distributed?

(c) Calculate the sample mean and standard deviation for each supplier from the data provided.

(d) Do a chi-square test for each supplier to see if the data are normally distributed using the sample means and standard deviations.

(e) How does this change the test procedure itself?

(f) How do the results of these tests compare to the results of the previous tests?

Requires Exercise 17.17   **17.25** Look at the results of the chi-square test on proportions for the data on smoking in fast-food restaurants.

(a) Prepare a plot of the proportion of adults who favor banning smoking in fast-food restaurants by state.

(b) When you reject H_0 you decide that at least one population (in this case state) is different. Use your plot to come to a more informative conclusion.

Datafile:
PORSCHE.XXX   **17.26** Look at the data on the number of miles driven by the sample of Porsche drivers.

(a) Make a boxplot of the data.

(b) Do the data look normally distributed?

(c) Delete the outliers and replot the data. Did this change your opinion about the normality of the data? Why or why not?

(d) Calculate the mean and standard deviation of the data without the outliers.

(e) Perform a chi-square goodness of fit test to determine whether the data come from a normally distributed population with the calculated mean and standard deviation.

(f) What are your conclusions?

Doing It!

17.27 The Public Health student who was looking at the issue of college drinking administered a survey that asked questions on four different areas: drinking habits, problems related to drinking, problems related to other people's drinking, and living habits. A portion of the data from the survey and an explanation of the variables are given below:

Five	Four	Three	Last	Binger	Hangover	Miss	Behind	Regret	Forget
0	0	0	2	1	1	1	1	1	1
0	0	0	4	1	1	1	1	1	1
0	1	1	3	2	1	1	1	1	1
0	0	0	3	1	1	1	1	1	1
0	0	0	3	1	1	1	1	1	1
0	0	0	4	1	1	1	1	1	1

Argue	Engage	Protect	Damage	Police	Injured	Medical	Drive	DWI	Ride
1	1	1	1	1	1	1	1	1	1
1	1	1	1	1	1	1	1	1	1
1	1	1	1	1	1	1	1	1	1
1	1	1	1	1	1	1	1	1	1
1	1	1	1	1	1	1	3	1	3
1	1	1	1	1	1	1	1	1	1

Age	Gender	Fulltime	Greek	Live	Roommate	Race	Class
18	1	Y	N	4	P	2	1
62	2	Y	N	4	S	1	
18	2	Y	N	1	R	1	1
31	1	Y	N	4	R	1	
19	2	Y	N	4	P	1	2
20	2	Y	N	1	R	1	2

For the purposes of this survey a "drink" means any of the following:

12-ounce can or bottle of beer
4-ounce glass of wine
12-ounce bottle or can of wine cooler
1 oz (shot) of liquor straight or in a mixed drink

Binge drinking is defined as

The consumption of five or more drinks in one episode of drinking for males.
The consumption of four or more drinks in one episode of drinking for females.

The variables *Five*, *Four*, and *Three* refer to the two-week period just before the survey was administered and answer the questions indicated. They are coded as follows:

0	None
1	Once
2	Twice
3	Three to five times
4	Six to nine times
5	Ten or more times

- *Five*   "How many times have you had five or more drinks in a row?"

  Four   "How many times have you had four drinks in a row (but no more than that)?"

  Three   "How many times have you had three drinks in a row (but no more than that)?"

- The variable *Last* answers the question When did you have your last drink? The answers are coded as follows:

0	I never had a drink
1	Not in the past year
2	More than 30 days ago, but less than a year ago
3	More than one week ago, but less than 30 days ago
4	Within the last week

- The variable *Binger* is defined as the type of drinker that a student is based on his or her answer to the first four questions.

0	Nondrinker
1	Non-binge drinker
2	Infrequent binge drinker
3	Frequent binge drinker

The next 15 variables answer questions that start with the statement "Since the beginning of the school year, how often has your drinking caused you to. . . " The answers are coded as follows:

0	Not at all
1	Once
2	Twice or more

- *Hangover*   ". . . have a hangover?"
- *Miss*   ". . . miss a class?"
- *Behind*   ". . . get behind in school work?"
- *Regret*   ". . . do something you later regret?"
- *Forget*   ". . . forget where you were or what you did?"
- *Argue*   ". . . argue with friends?"
- *Engage*   ". . . engage in unplanned sexual activity?"
- *Protect*   ". . . not use protection when you had sex?"
- *Damage*   ". . . damage property?"
- *Police*   ". . . get into trouble with campus or local police?"
- *Injured*   ". . . get hurt or injured?"
- *Medical*   ". . . require medical treatment for an alcohol overdose?"
- *Drive*   ". . . drive after drinking alcohol?"
- *DWI*   ". . . drive after having five or more drinks?"
- *Ride*   ". . . ride with a driver who was high or drunk?"

- The variable *Age* is the age of the student.

- The variable *Gender* is the gender of the student and is coded as

1	Male
2	Female

- The variable *Fulltime* is Y if the student is full-time and N if the student is part-time.

- The variable *Greek* is Y if the student is a member of a fraternity or sorority and N if not.

- The variable *Live* records where the student lives and is coded as follows:

    1 Residence hall or dormitory
    2 Fraternity or sorority house
    3 Other university housing
    4 Off-campus house or apartment

- The variable *Roommate* describes with whom the student lives and is coded as follows:

    A Alone
    R Roommate(s) or housemate(s)
    S Spouse
    P Parent(s) or other relative(s)
    O Significant other
    C Children

- The variable *Race* describes the race or ethnic category of the student and is coded as

    Blank Hispanic (separate question)
    1 White
    2 Black/African American
    3 Asian/Pacific Islander
    4 Native American/Native Alaskan
    5 Other

- The variable *Class* describes the student's year at the university and is coded as

    1 Freshman
    2 Sophomore
    3 Junior
    4 Senior

(a) Look at the variable *Drive,* which refers to whether a student drove after drinking. Create a contingency table for this variable and the variable *Binger.*

(b) At the 0.05 level of significance, is whether a student drove after drinking independent of the type of drinker he or she is?

(c) If you exclude nondrinkers and look at only those students who drink, does the answer change?

(d) From the variables *Live* and *Roommate* derive a way to classify students as residents/commuters and to separate commuters living at home from commuters living with nonrelatives.

(e) Create contingency tables that will allow you to look at the relationship between how often a student missed class *(Miss)* and resident/commuter students. Would you expect these two variables to be related? Why or why not?

(f) Do a chi-square test to determine whether these two variables are related.

(g) Look at the variables *Miss* and *Gender* and do the same analysis.

(h) Look at the variables *Binger* and *Age.* Is there a relationship between them?

(i) From the variable *Age* create a new variable that classifies students as above and below the legal drinking age (assume that it is age 21 in the state of interest). How is this new variable related to the type of drinker that a student is? to *Drive?* to *Miss?*

(j) Is there a relationship between *Race* and *Binger?*

(k) Consider all of the other variables relating to behavior after drinking. Determine whether these variables are related to the type of drinker that a student is.

(l) Consider any other variables and relationships that you think are important and investigate them. Prepare a report for the university administration that describes the drinking behaviors of the students at the university.

17.28 The Chamber of Commerce that is studying the credit problems of small businesses asked the businesses 3 questions to classify their business and 7 questions related to the issue of credit problems. A portion of the datafile and an explanation of each variable follow.

Size	Employees	Nature	Problem	Understd	Concern	Call	Loan	Collateral	Access
2	2	1	1	2	1	2	2	0	2
1	2	3	1	2	2	0	2	2	1
4	3	1	2	1	2	2	2	2	0
1	1	1	2	1	2	2	2	2	2
1	2	1	2	0	0	0	0	0	0
3	1	5	2	0	2	2	2	2	1
3	3	2	2	1	1	2	2	2	1
2	2	3	2	1	2	2	2	2	2
3	5	2	2	1	2	2	2	1	2

- The variable *Size* refers to the annual sales of the company and is coded as follows:

  1   Under $1 million   4 $11–20 million
  2   $1–5 million   5   Over $20 million
  3   $6–10 million

- The variable *Employees* refers to the number of employees that the company currently employs. This variable was coded as

  1   0–5 employees
  2   6–10 employees
  3   11–50 employees
  4   51–150 employees
  5   151–250 employees
  6   Over 250 employees

- The variable *Nature* refers to the type of business and is coded as

  1   Manufacturing
  2   Retail
  3   Service
  4   Real Estate
  5   Other

The next seven variables contain the response to the questions or statements indicated and are coded as follows:

  1   Yes
  2   No

- *Problem*   "Are you experiencing credit-related problems?"

- *Understd*   "The bank understands my problems."

- *Concern*   "I am concerned that my note might be recalled."

- *Call*   "The bank is planning to recall my loan."

- *Loan*   "The bank has called my loan."

- *Collateral*   "The bank has demanded more collateral."

- *Access*   "Access to credit is affecting my business."

(a) Look at the variable *Nature,* which refers to the type of business. Is the sample of businesses that were surveyed by the Chamber of Commerce uniformly distributed over the different types of companies?

(b) Answer the question posed in part (a) for *Size* and *Employees.*

(c) Create a contingency table for the variables *Problem* and *Nature.* Is the proportion of companies that are having credit-related problems different for the different types of companies?

(d) Answer the same question for the variable *Nature* and each of the variables *Understd* and *Access.*

(e) Investigate the relationship between the variables *Size* and *Access* and *Size* and *Problem.*

(f) Investigate the relationship between the variables *Employees* and *Access* and *Employees* and *Problem.*

(g) Is there a relationship between the size of the company and the type of business that the company does?

(h) Is there a relationship between the size of the company and the number of employees in the company?

(i) Look at the variables related to loan recall, *Concern, Call,* and *Loan.* Investigate whether these variables are related to *Size, Employees,* and *Nature.*

(j) Look at any other relationships that you think might be useful to the Chamber of Commerce.

(k) Prepare a report for the Chamber of Commerce about your findings.

NONPARAMETRIC STATISTICS

SURGICAL LEADERSHIP DEVELOPMENT PROGRAM

Administrators in the Surgical Department of a large hospital were dissatisfied with the way their current training program for surgical residents was working. In particular, they were not happy with the leadership skills developed and demonstrated by the residents and the chief resident. In turn, the residents were not satisfied with what the program provided them in the way of support. The administrators hired a management consultant firm to develop a new training program. The new program reorganized the resident teams and the way that the teams, chief resident, and surgical attending physicians interacted.

To assess the impact of the new program over time, the administrators developed a set of questionnaires which were given out to personnel in each department that interacted with the residents and to the residents themselves. The groups were the Surgical Residents(SR), Surgical Attending Physicians (SA), Surgical Nursing Staff (SNS), Emergency Department Physicians (EDP), and Emergency Department Nursing Staff (EDN). The survey had a series of statements and used a 5-point Likert scale (1 = Strongly disagree to 5 = Strongly agree) to assess the respondents' agreement with the statement. A portion of the data are shown below:

Category	Time	Q1	Q2	Q3	Q4	Q5	Q6
SNS	Jun-98	3	3	2	4	2	4
EDP	Jun-98	4	2	4	3	4	3
SR	Jun-98	5	4	5	1	2	5
SR	Jun-98	4	4	3	4	4	4
SR	Jun-98	4	4	5	5	2	4
SR	Jun-98	4	4	4	2	4	2
SR	Jun-98	4	4	4	4	3	4
SR	Jun-98	4	4	5	5	3	4
SR	Jun-98	3	3	4	2	1	2
SR	Jun-98	4	4	4	4	4	4
SR	Jun-98	5	4	5	4	4	4
SR	Jun-98	4	5	5	5	4	4

The survey was given to all groups at the start of the new program, to selected groups after six months, and to the entire group again at the end of the year. The administrators were interested in how perceptions compared for different groups and whether and how perceptions changed over time.

18.1 CHAPTER OBJECTIVES

In hypothesis testing you learned techniques for testing inferences about *quantitative* data. In particular, you learned that when you have large sample sizes you can usually assume that the sampling distribution of the test statistic is normally distributed. In other cases, when the sample size is small, or when you are testing some population parameters such as variances, the test procedures you learned rely on the fact that the data collected come from populations that are normally distributed. What about data for which this assumption is not true? What can we do to make inferences about these populations?

This problem arises in two ways. The first occurs when the data collected are simply not normally distributed—that is, they come from populations with skewed distributions. The second occurs when the data are not truly quantitative, as, for example, when the data represent rankings or ratings. Remember that this type of data is called ordinal data. Even though ordinal data can be numerical, they are not really quantitative.

Procedures for working with these types of data are called nonparametric, or distribution-free, statistical methods. Many of the techniques are analogous to ones that you learned in earlier chapters. The five basic steps of hypothesis testing still apply.

This chapter will cover the following material:

- Wilcoxon rank sum test for comparing the centers of two populations
- Kruskal–Wallis test—a nonparametric alternative to ANOVA

18.2 METHODS FOR COMPARING TWO POPULATIONS

18.2.1 The Wilcoxon Rank Sum Test

The **Wilcoxon rank sum test** is used to determine whether two independent populations have the same distribution or are different. It can also be used to determine whether one population has values that are significantly higher or lower than those of another population. If the populations that are sampled have the same shape (not necessarily normal) and variability, then the Wilcoxon rank sum test is equivalent to comparing the centers (i.e., means) of the two populations.

The Wilcoxon rank sum test is sometimes referred to as the Mann–Whitney test because it appeared simultaneously in the literature.

> The **Wilcoxon rank sum test** is used to compare the distributions of two independent populations when the samples are small and the data are not normally distributed, or when the data are not quantitative.

The Wilcoxon rank sum test ranks the observations from the two populations together and then looks at the sum of the rankings for each population. If the two populations are the same, then the sums should be approximately equal. If one of the populations is lower or higher than the other, then the sum of the ranks should change accordingly.

By convention, the Wilcoxon test calls the population with the smaller sample size population 1 and the one with the larger sample size population 2.

EXAMPLE 18.1   Surgical Residents

Identifying the Populations

The administrators of the Surgery Department who were about to begin the new training programs wanted to know about how the surgical residents were viewed by different departments. In particular, they want to know whether there was a difference in the way surgical residents were perceived by nurses in two different departments. Their feeling was that the nurses outside the Surgical Department were more satisfied with the surgical residents. They decided to compare the responses for the Emergency Department nurses and the surgical nursing staff to the statement:

Members of the resident teams appear to work well together.

The survey administered in the two departments had 7 responses from the ED nurses and 10 responses from the SD nurses. Thus the Emergency Department was designated population 1 and the Surgical Department was population 2. The data are as follows:

Emergency Department	Surgical Department
4	2
4	2
4	4
4	3
5	4
4	4
3	2
	2
	4
	4

The Wilcoxon rank sum test is also used when the data are quantitative, but samples are small and the normality assumption does not hold.

EXAMPLE 18.2   Employee Breaks

Identifying the Populations

The employees in the assembly department of a large manufacturing company have asked to have their current break policies changed. They would like to be able to take breaks as they feel necessary, rather than at scheduled times. The workers believe that the increased freedom will increase productivity, and so they convince management to do a pilot study to collect some data. A small group of employees is chosen and randomly assigned to two groups. One group will remain on the existing break schedule and the other will be allowed to take breaks whenever they want to. During this time, productivity (in terms of number of items produced) will be monitored for each employee. Six employees are assigned to the fixed break group and five to the free break group.

The data for the week for each group were:

Free	Fixed
351	357
316	347
480	380
446	259
470	342
	282

Since the free group has a smaller sample size, it is designated as population 1.

The hypotheses for the Wilcoxon test will differ slightly, depending on what assumptions you make. If you can assume that the two populations have similar shapes and variability, then the actual hypotheses will be about differences in the means. If you cannot make these assumptions, then the hypotheses will be about the actual population distributions.

EXAMPLE 18.3 Surgical Residents

Setting up the Hypotheses

To see whether the populations are similar in shape and variability, the hospital administrators looked at dotplots of the data.

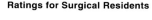

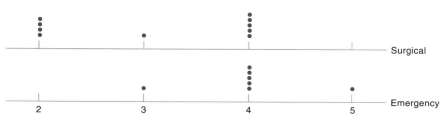

From the dotplot, it appears that the variability is similar but the shapes are very different. Since the Surgery Department thinks the Emergency Department nurses are more satisfied (higher rating numbers) than the Surgical Department nurses, for this statement, the hypotheses are:

H_0:   Responses for the ED nurses are not higher than those for the SD nurses.

H_A: Responses for the ED nurses are higher than those for the SD nurses.   ■

EXAMPLE 18.4 Employee Breaks

Setting up the Hypotheses

To set up the hypotheses for the study on employee breaks, the department management creates dotplots of the data:

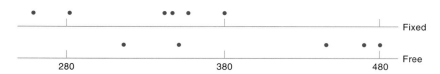

Since the variability and shapes look similar, the hypotheses are about the mean of each population:

H_0:   The mean productivity for the free break group is not higher than for the fixed break group.

H_A: The mean productivity for the free break group is higher than for the fixed break group.   ■

The Wilcoxon rank sum test takes the observations from both populations and puts them in order from lowest to highest. Each observation is assigned a rank, which is the position of that observation in the ordered list. When there is a tie, each observation is assigned the average of the ranks that would have been assigned if there had not been a tie. The populations are separated again, and the ranks for each population are added to obtain the sum.

If there are n_1 observations in one population and n_2 observations in the second, then the rankings will go from 1 to n, where $n = n_1 + n_2$.

If the two populations are not different, then the rankings should be evenly distributed across the two samples. That is, each sample should account for about half of the ranking totals. The Wilcoxon test looks at the sum of the rankings and the amount that is assigned to the smaller population. It uses the following mathematical formula for the sum of integers from 1 to n:

$$\frac{n(n + 1)}{2}$$

Formula for sum of first n integers

For example, the sum of the integers from 1 to 5 is $\dfrac{5(5 + 1)}{2} = \dfrac{(5)(6)}{2} = \dfrac{30}{2} = 15.$

This saves a lot of time in summing the rankings. Imagine adding the numbers from 1 to 100 by hand!

If the sum of the ranks for one population is too large or too small, then we would conclude that the populations are not the same. The critical values for the hypothesis test are found in Table A.7 for sample sizes up to 10. The values given are the lower and upper critical values for tail areas of 0.025 and 0.05. The one-tail 0.025 table is shown below:

One-Tail α = 0.025;   Two-Tail α = 0.05

n_2	n_1	T_L	T_U	T_L	T_U	T_L	T_U	T_L	T_U	T_L	T_U	T_L	T_U	T_L	T_U	T_L	T_U
		3		4		5		6		7		8		9		10	
4		6	18	11	25												
5		6	21	12	28	18	37										
6		7	23	12	32	19	41	26	52								
7		7	26	13	35	20	45	28	56	37	68						
8		8	28	14	38	21	49	29	61	39	73	49	87				
9		8	31	15	41	22	53	31	65	41	78	51	93	63	108		
10		9	33	16	44	24	56	32	70	43	83	54	98	66	114	79	131

From the table you see that the critical value for a lower-tail test with α = 0.025 and $n_1 = 4$ and $n_2 = 7$ is 13.

For larger sample sizes the sampling distribution that applies is the normal probability distribution, and the standard normal table is used.

EXAMPLE 18.5 Surgical Residents

Finding the Critical Value

The hospital administrators decided to perform the test at the 0.05 level of significance. With $n_1 = 7$ and $n_2 = 10$, the critical value for the upper-tail test was determined to be 80. ■

EXAMPLE 18.6 Employee Breaks

Finding the Critical Value

The management of the manufacturing company looked at the results of the analysis. For $n_1 = 5$ and $n_2 = 6$ at the 0.05 level of significance, the critical value of the test was 41. ■

The next step in the hypothesis testing procedure is to determine the value of the test statistic. The data from the two samples must be combined and then put in order from smallest to largest. Each value is assigned a rank from 1 to n. If values are repeated in the data—that is, if there are ties in the rankings—then the rank assigned is the average of the ranks for the tied data.

EXAMPLE 18.7 Surgical Residents

Determining the Rankings

To find the rankings, we must combine. Since the Emergency Department had fewer observations, it is designated population 1; thus $n_1 = 7$ and $n_2 = 10$. For the data on surgical residents, the ordered data and rankings are shown below:

Emergency	Surgical	Ranking	Position
	2	2.5	1
	2	2.5	2
	2	2.5	3
	2	2.5	4
3		5.5	5
	3	5.5	6
4		11.5	7
4		11.5	8
4		11.5	9
4		11.5	10
4		11.5	11
	4	11.5	12
	4	11.5	13
	4	11.5	14
	4	11.5	15
	4	11.5	16
5		17	17

At first glance this might seem complicated, but it really is not. The first rating, 2, occurred four times, so the average of the positions involved was assigned:

$$Rank = \frac{1 + 2 + 3 + 4}{4} = \frac{10}{4} = 2.5$$

The next data value, 3, occurred two times, so the average of those positions (5 and 6) was assigned:

$$Rank = \frac{5 + 6}{2} = \frac{11}{2} = 5.5$$

The next data value, 4, occurred 10 times, in positions 7 through 16, and so the rank is

$$Rank = \frac{7 + 8 + \cdots + 15 + 16}{10} = \frac{115}{10} = 11.5$$

Finally, the last rating, 5, has rank 17. ■

Finding the ranks is really not difficult if you carefully organize your work. It is important to keep track of which population each value is in so that you can separate them again later.

EXAMPLE 18.8   Employee Breaks

Finding the Ranks

The management of the manufacturing company looked at the data collected for the week of the pilot study. They calculated the rank for each data value. Since none of the data values repeated, each rank is the same as the position of the data value in the list.

Free	Fixed	Rank	Position
	259	1	1
	282	2	2
316		3	3
	342	4	4
	347	5	5
351		6	6
	357	7	7
	380	8	8
446		9	9
470		10	10
480		11	11

■

 **TRY IT NOW!**

Restaurant Data

Setting Up the Test

A large restaurant is interested in seeing whether offering "specials" during the week will increase business. The owners decide to run a trial for two months, offering the specials once a week. They will compare the trial results with those of the previous two months to see if they are correct. Because one of the weeks in the special menu time period included a holiday, they have only seven observations for that period. The data (in number of customers) are shown below:

Regular	Specials
44	47
45	49
46	49
48	52
42	53
47	46
48	53
44	

Which group is population 1 and which is population 2?

Create dotplots of the data. Do you think you can assume that the shape and variability for each population are the same?

Set up the hypotheses for the test.

At the 0.05 level of significance, what is the critical value for the test?

The value of the test statistic is the sum of the ranks for the smaller population. When the sample sizes are large the test statistic is approximately normally distributed with mean

$$\mu = \frac{n_1(n_1 + n_2 + 1)}{2}$$

Mean of Wilcoxon rank sum test statistic

and standard deviation

$$\sigma = \sqrt{\frac{n_1 n_2(n_1 + n_2 + 1)}{12}}$$

Standard deviation of Wilcoxon rank sum test statistic

For example, if $n_1 = 20$ and $n_2 = 25$, then the test statistic T, the sum of the ranks, would be normally distributed with mean $\mu = \dfrac{20(20 + 25 + 1)}{2} = \dfrac{(20)(46)}{2} = 460$ and standard deviation $\sigma = \sqrt{\dfrac{(20)(25)(46)}{12}} = \sqrt{\dfrac{23{,}000}{12}} = \sqrt{1916.6667} = 43.78.$

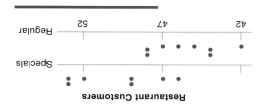

ANS. Specials is population 1. It seems reasonable to assume that the shape and variability are similar; H_0: The mean for the specials menu is not higher than for the regular menu, H_a: The mean for the specials menu is higher than for the regular menu. With $n_1 = 7$ and $n_2 = 8$, the critical value is 71. Dotplot:

To understand where these formulas come from, remember that for samples of sizes 20 and 25 you would have a total of 45 rankings. The sum of the numbers from 1 to 45 is $\frac{(45)(46)}{2} = 1035$. Since the sample from population 1 is smaller than the sample from population 2, you would expect the sum of the ranks, if the populations were the same, to be a little less than half of the total, which is 517.5. That is what the mean, 460, represents.

EXAMPLE 18.9 Surgical Residents

Summing the Rankings

The rankings for the Emergency Department nurses were:

Rating	Rank
3	5.5
4	11.5
4	11.5
4	11.5
4	11.5
4	11.5
5	17
Sum	80

The test statistic, the rank sum, was found to be 80. ∎

EXAMPLE 18.10 Employee Breaks

Finding the Ranks

From the data table, the manufacturing company found that the sum of the ranks for the free break group was $3 + 6 + 9 + 10 + 11 = 39$.

Free	Fixed	Rank	Position
	259	1	1
	282	2	2
316		3	3
	342	4	4
	347	5	5
351		6	6
	357	7	7
	380	8	8
446		9	9
470		10	10
480		11	11

∎

 **TRY IT NOW!**

Restaurant Data
Finding the Sum of Ranks

Consider the restaurant data that you looked at previously. Use the next table to find the rank for each data value.

Specials	Regular	Rank	Position
			1
			2
			3
			4
			5
			6
			7
			8
			9
			10
			11
			12
			13
			14
			15

Find the sum of the rankings for the "specials" menu sample.

The last step of the process is to compare the test statistic to the critical value and make our decision. The results of the Wilcoxon rank sum hypothesis test are the same as for any hypothesis test procedure. You decide either to reject H_0 or fail to reject H_0.

EXAMPLE 18.11 Surgical Residents

Concluding the Test

Since the rank sum was 80 and the critical value for the test was 80, the hospital administrators knew that they should reject H_0. The data indicated that the Emergency Department nurses were more favorable toward the surgical residents than the Surgical Department nurses.

From this they decide that they would like to look at other, more specific statements to try to determine in what areas of leadership the differences exist. ■

Remember that simply rejecting or not rejecting the null hypothesis is not enough. You must use the information to make an informed decision about the problem.

Specials	Regular	Rank	Position
	42	1	1
	44	2.5	2
	44	2.5	3
	45	4	4
46		5.5	5
	46	5.5	6
47		7.5	7
	47	7.5	8
	48	9.5	9
	48	9.5	10
49		11.5	11
49		11.5	12
52		13	13
53		14.5	14
53		15	15

ANS. The sum of the rankings is 78.

EXAMPLE 18.12   Employee Breaks

Concluding the Test

Since the rank sum for the data was 39 and the critical value of the test was 41, the management of the manufacturing company concluded that they could not reject H_0. The data did not provide evidence that the change in break policy would increase productivity.

At this point management told the employees that there was no justification for changing the policy, but that they would continue to study the issue and see what else could be done.   ∎

TRY IT NOW!

Restaurant Data
Concluding the Test

Compare the rank sum for the restaurant data to the critical value for the test. What is your conclusion?

If you were the manager of this restaurant, what action would you take? Do you think you would consider any other factors before making this decision? If so, what would you do and why?

18.2.2 Exercises—Learning It!

18.1   A paper company is using two different vendors to supply the packing cases used to store products in the warehouse. Recently, employees have been noticing that there are more cases being rejected for printing defects. They believe that the majority of the rejected cases come from vendor X. They decide to run some tests and collect data by looking at 10 cases from each of 5 lots for both vendors. They count the number of defective cases in each sample and record that data, as shown below:

Vendor X	Vendor Y
1	2
4	0
2	1
3	2
2	3

(a) Create dotplots for each vendor. What assumptions do you think are reasonable?

(b) Set up the hypotheses to see whether the number of defective cases from vendor X is higher than for vendor Y.

(c) Find the ranks for each data value and the rank sum for each sample.

(d) At the 0.05 level of significance, what can you conclude?

18.2 An Internet service provider is asking customers to rate its current service. Company officials are interested in knowing whether people who have been customers for more than a year are less satisfied than new customers (less than six months). They use a 5-point scale where 1 = Extremely satisfied and 5 = Extremely dissatisfied. The data (already sorted) are given below:

Old Customers	New Customers
1	1
2	1
3	1
3	1
3	1
3	2
3	2
4	3
5	4
5	4

(a) Create dotplots for each sample. What assumptions do you think are reasonable?

(b) Set up the hypotheses to see whether the old customers are less satisfied than the new customers.

(c) Find the ranks for each data value and the rank sum for each sample.

(d) At the 0.05 level of significance, what can you conclude?

18.3 A business with an Internet presence wonders about the effectiveness of banner ads. They decide to create two different Web sites for their company. One of the sites will use a banner ad, while the other will not. They decide to count the number of daily hits for each site for a one-week period. The data are given below:

Banner Ad	No Banner Ad
19	12
19	13
19	15
21	16
21	19
25	20
28	32

(a) Create dotplots for each sample. What assumptions do you think are reasonable?

(b) Set up the hypotheses to see whether banner ads produce more hits to the Web site for this company.

(c) Find the ranks for each data value and the rank sum for each sample.

(d) At the 0.05 level of significance, what can you conclude?

18.4 A golf ball manufacturer wants to compare two different ball designs. The product managers take 10 balls of each type and hit them using a controlled environment. They measure the distance that each ball travels to where it first hits the ground. The data (in yards) are shown in the next table.

Ball Design 1	Ball Design 2
212	201
231	208
232	210
233	217
240	220
241	227
266	232
311	237
313	238
325	265

(a) Make plots of the data and decide whether the shapes and variability of the data are similar.

(b) Set up the hypotheses to test whether there is a difference in the amount of travel for the two different designs.

(c) Carry out the Wilcoxon rank sum test at the 0.05 level of significance.

(d) What can you conclude about the distances for the two different designs?

18.3   METHODS FOR COMPARING MORE THAN TWO POPULATIONS

18.3.1 The Kruskal–Wallis Test

Previously you learned that analysis of variance (ANOVA) is the statistical technique used to compare the means of more than two populations. ANOVA relies on the assumptions that the populations are normally distributed and that they have equal variances. If these assumptions do not hold, then it is necessary either to transform the data or to use an analogous nonparametric statistical test.

The nonparametric technique is the **Kruskal–Wallis test.** The idea behind the test is very similar to that for the Wilcoxon rank sum test. Just as in the Wilcoxon rank sum test, if the populations have similar shapes and variation, then the hypotheses will be about the means of the populations; otherwise they will be about the general distributions of the populations.

> The *Kruskal–Wallis test* is a nonparametric method for comparing the data from more than two populations.

EXAMPLE 18.13   Surgical Residents

Setting up the Test

One of the things that the hospital administration was interested in was whether perceptions of the surgical residents changed during the new training program. They thought that this would be an indication of whether or not the program was effective. To investigate this, they administered the same questionnaire to certain groups at three different times: at the beginning of the training, six months into the program, and at the end of the program.

They decided to look at the surveys for the surgical attending physicians, the people who work most closely with the surgical residents. The data for the statement

Members of the resident teams appear to work well together.

are shown in the next table.

Jun-98	Dec-98	Jun-99
4	4	2
4	4	2
2	4	4
3	4	3
4	4	4
4	2	4
3	3	5
2	4	4
4	4	4
4	3	4
4	4	5
4	4	4
4	2	
4	4	
	2	
	3	
	4	
	4	
	4	
	4	

Plots of the data were made to see what assumptions could be made. Except for two observations in June 1999, the assumptions of equal shape and variation seem reasonable.

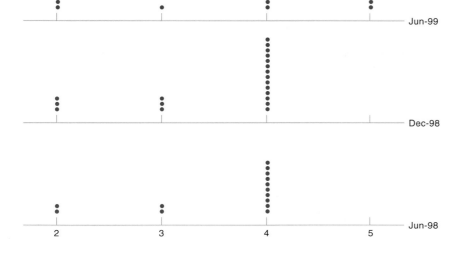

The hypotheses are:

H_0:   The mean rating for each time period is the same.

H_A: At least one mean is different.

EXAMPLE 18.14   Tax Preparation Software

Setting up the Test

A company that does tax preparation for the general public is considering changing the software that it currently uses to process returns. The tax managers wonder how the three alternatives that they are considering compare in ease of use. They decide to take the last 21 tax returns that they have done and randomly assign one return to each of the three software packages. The tax return data are input into the respective software and the time until a useable, correct return is obtained is measured (in hours). The data and plots are shown below:

Software A	Software B	Software C
2.0	2.4	3.1
1.9	2.2	2.2
2.3	1.8	2.8
2.3	2.3	2.5
1.9	1.6	2.3
1.7	1.9	2.2
2.2	1.9	2.9

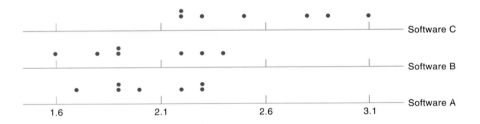

Times for Tax Return Software

Plots of the data indicate that the assumptions of equal shape and variability are not unreasonable, and so the hypotheses are:

H_0:   The mean times for the three software packages are the same.

H_A: At least one mean is different.   ■

The Kruskal–Wallis test proceeds very similarly to the Wilcoxon rank sum test. The data are joined together and ranked, and then separated back into their original samples. The idea is that if the populations are similar, then the ranks should be about equally distributed among the populations. The Kruskal–Wallis test looks at the mean rank for each group rather than the sum of the ranks.

EXAMPLE 18.15   Surgical Residents

Calculating the Mean Rank

For the surgical attending physicians, the sorted and ranked data are shown in the next table.

Jun-98	Rank	Dec-98	Rank	Jun-99	Rank
2	4	2	4	2	4
2	4	2	4	2	4
3	10.5	2	4	3	10.5
3	10.5	3	10.5	4	29
4	29	3	10.5	4	29
4	29	3	10.5	4	29
4	29	4	29	4	29
4	29	4	29	4	29
4	29	4	29	4	29
4	29	4	29	4	29
4	29	4	29	5	45.5
4	29	4	29	5	45.5
4	29	4	29		
4	29	4	29		
		4	29		
		4	29		
		4	29		
		4	29		
		4	29		
		4	29		
Average Rank	**22.8**		**22.5**		**26.0**

■

EXAMPLE 18.16 Tax Preparation Software

Finding the Average Rank

The tax return company assembled the data, ranked them, and calculated the average ranks, as shown below:

Software A	Rank	Software B	Rank	Software C	Rank
1.7	2	1.6	1	2.2	10.5
1.9	5.5	1.8	3	2.2	10.5
1.9	5.5	1.9	5.5	2.3	14.5
2.0	8	1.9	5.5	2.5	18
2.2	10.5	2.2	10.5	2.8	19
2.3	14.5	2.3	14.5	2.9	20
2.3	14.5	2.4	17	3.1	21
Average Rank	8.6		8.1		16.2

■

In ANOVA, the idea was that if the means of the populations were not different, then the variation among the means of the populations should be similar to the variation within each population. The Kruskal–Wallis test uses the same idea, but it tests the variation among the mean ranks, rather than the means of the data values themselves. In addition, the Kruskal–Wallis test compares the variation among the populations to the total variation in ranks, rather than the variation within each population.

The test statistic for the Kruskal-Wallis test is

$$H = \frac{\text{SSA}}{\text{SST}/(n-1)}$$

Kruskal–Wallis test statistic

SSA is calculated in the same way as for ANOVA; that is, for the Kruskal–Wallis test,

SSA for Kruskal–Wallis test

$$SSA = \sum_{j=1}^{c} n_j(\bar{r}_j - \bar{\bar{r}})^2$$

where

n_j is the number of observations in the sample from population j

c is the total number of populations

$\bar{r}_j$ is the average rank of the sample observations from population j

and

$\bar{\bar{r}}$ is the average of all of the ranks.

SST, the total variation in the ranks, is calculated as

SST for Kruskal–Wallis test

$$\sum_{j=1}^{c} \sum_{i=1}^{n_j} (r_{ij} - \bar{\bar{r}})^2$$

When there are no (or few) ties in the data, there is a simpler way to calculate the test statistic H:

Approximate Kruskal–Wallis test statistic

$$H = \frac{12}{n(n+1)} \sum_{j=1}^{c} \frac{R_j^2}{n_j} - 3(n+1)$$

where R_j is the sum of the ranks for the sample observations from population j, c is the number of populations, and n is the total number of observations ($n = n_1 + n_2 + \cdots + n_c$).

There are tables of exact critical values for the Kruskal–Wallis test. However, if the sample sizes are all larger than 5 and if the shapes of the underlying distributions are approximately the same, then the sampling distribution of H is the chi-square distribution with $c - 1$ degrees of freedom. Since these assumptions are not very restrictive, the chi-square distribution is almost always used.

EXAMPLE 18.17 Surgical Residents

Calculating the Test Statistic

When the data are this complicated, it is probably easiest to run an ANOVA using the ranks as the data to find H.

Since the data for the surgical resident ratings had many ties, the exact formula was used to find the value of the test statistic. In this case:

$$SSA = 14(22.8 - 23.5)^2 + 20(22.5 - 23.5)^2 + 12(26.0 - 23.5)^2 = 104.27$$

Similarly, SST was calculated to be 5581.5.

Thus the test statistic, H, was found to be

$$H = \frac{104.27}{5581.5/45} = 0.84$$

For a level of significance of 0.05, the critical value of the test is $\chi^2_{0.05,2} = 5.991$. Since the test statistic does not fall outside the critical value of the test, there is not enough evidence to say that the means for the three time periods are different.

From this analysis, the hospital administrators find that the training program did not change the attitudes of the surgical attendings toward the surgical residents. ∎

EXAMPLE 18.18 Tax Preparation Software

Finding the Average Rank

The data for the tax preparation software did not have many ties and so the approximation can be used to find the value of the test statistic. The sums of the ranks for the three samples are 60.5, 57, and 113.5, respectively. The test statistic is calculated to be

$$H = \frac{12}{(21)(22)} \left(\frac{60.5^2}{7} + \frac{57^2}{7} + \frac{113.5^2}{7} \right) - 3(22) = 7.44$$

The tax company decided to use a level of significance of 0.05 and so the critical value of the test is $\chi^2_{0.05,2} = 5.991$.

Since the test statistic is greater than the critical value of the test, the tax managers conclude that the mean time for at least one of the software packages was different. Looking at the data, it would appear that the first two packages, A and B, were similar but that software package C took considerably longer. This information would help them in deciding which package to use. ∎

 TRY IT NOW!

Travel Expenses
Performing the Test

A large company was interested in looking at the differences in the amount of money that was being charged for travel reimbursement in three different departments. The reason for this was that each department used different methods for booking travel arrangements and the company thought that there might be significant differences in savings. If so, they wanted to take advantage of the method that had the lowest cost. They looked at travel expenses turned in for the past three weeks in each department and obtained the following data:

Accounting	Marketing	Sales
529	604	567
451	633	409
546	568	504
372	457	605
426	520	682
603	509	477
	551	480

Make dotplots for the data. Do you think it is reasonable to assume similar shapes and variability?

Write down the hypotheses for the test.

(continued)

Calculate the ranks for each department, the sums of the ranks, and the average ranks.

Which formula for the test statistic is better to use for these data?

Calculate the value of the Kruskal–Wallis test statistic.

At the 0.05 level of significance, what can the company conclude about travel expenses?

Do you think this conclusion is correct? What other factors should have been considered in collecting the data?

18.3.2 Exercises—Learning It!

18.5 The Internet service provider who was interested in how customers rated its service decided to include another group in the study. The new group consisted of people who had been customers between six months and a year. Company officials conducted a new survey using the same scale (1 = Extremely satisfied and 5 = Extremely dissatisfied) and the results were:

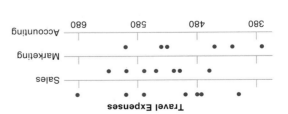

Accounting	Rank	Marketing	Rank	Sales	Rank
372	1	457	5	409	2
426	3	509	9	477	6
451	4	520	10	480	7
529	11	551	13	504	8
546	12	568	15	567	14
603	16	604	17	605	18
		633	19	689	20
Sum	47		88		75
Average	7.833333		12.6		10.71429

Since there are no ties, use the approximate test statistic. $H = 2.09$. Since H is not outside the critical value, we cannot reject H_0. There is no difference in the means for the three departments. No, they need to consider length of trips, number of trips, number of days involved, etc. $X_{0.05,2}^2 = 5.991$.

More than a year	Six months to one year	Less than six months
1	1	1
2	1	1
3	2	1
3	3	1
3	3	1
3	3	2
3	4	2
4	5	3
5	5	4
5	5	4

(a) Create dotplots for all three samples. What assumptions do you think are reasonable?

(b) Set up the hypotheses to see if there is a difference in satisfaction for the three groups.

(c) Find the rank for each data value, and find the rank sum and average rank for each sample.

(d) Perform the Kruskal–Wallis test at the 0.05 level of significance. What can you conclude?

18.6 A business that is interested in starting an on-line shopping service is interested in finding out whether or not there are differences in how women shop on-line. They are interested in capturing people who are already connected to the Internet, so they run a Web-based survey. They ask respondents how many purchases they have made on-line in the last three months. In addition, they ask demographic questions about gender, age, and level of education. The data for women respondents in three age categories are shown below:

21–30	30–45	45–60
2	4	3
2	5	3
3	6	4
4	6	5
4	8	6
5	9	6
8	9	9

(a) Create dotplots for all three samples. What assumptions do you think are reasonable?

(b) Set up the hypotheses to see if there is a difference in the number of purchases made on-line by women for the three age groups.

(c) Find the rank for each data value, and find the rank sum and average rank for each sample.

(d) Perform the Kruskal–Wallis test at the 0.05 level of significance. What can you conclude?

18.7 The Regional Planning Board for a metropolitan area is wondering about how often people use public transportation. They would like these data to use with other data they have collected on housing starts to plan for future expansion of services. They divide the area into three different geographical regions and conduct a survey of households in each region. The respondents are asked how many times the head of household used public transportation on weekdays for a two-week period in February. The results are given in the next table.

Region A	Region B	Region C
3	4	5
3	5	6
4	5	6
4	5	6
4	5	7
5	5	7
5	5	

(a) Create dotplots for all three samples. What assumptions do you think are reasonable?

(b) Set up the hypotheses to see if there is a difference in the use of public transportation for the three regions.

(c) Find the rank for each data value, and find the rank sum and average rank for each sample.

(d) Perform the Kruskal–Wallis test at the 0.05 level of significance. What can you conclude?

18.8   The food services manager at a large company is looking at the types of hotdogs that are available. He wonders if there is really a difference in the number of calories for different kinds of hotdogs. He decides to look at three different types of hotdogs and randomly selects brands from each type. The number of calories was recorded for each, as shown below:

Beef	Meat	Poultry
111	147	152
132	153	152
139	179	129
152	182	129
181	190	113
181		102

(a) Create dotplots for all three samples. Do you think it is reasonable to assume that the shapes and variability are similar?

(b) Set up the hypotheses to see if there is a difference in calories for the different types of hotdogs.

(c) Find the rank for each data value, and find the rank sum and average rank for each sample.

(d) Perform the Kruskal–Wallis test at the 0.05 level of significance. What can you conclude?

18.4 USING NONPARAMETRIC TESTS

At this point you might be wondering several things. Will I ever figure out which test to use? What happens if I use the wrong test? How good are these nonparametric tests anyway?

If you have any doubts about which test to use and whether it matters, do them both. If they agree, then the decision you make is probably correct. If they disagree, then you need to look at the data more carefully before you choose the test.

Determining which test to use is important. What is basic to this skill is knowing what assumptions each test makes and *carefully checking* those assumptions. This is why it is so important to really look at your data with descriptive methods *before* you start any analysis.

For the most part, the thing that is most affected by using a statistical test incorrectly is the probability of making a Type II error. For example, if you use a small-sample *t* test for the population mean when the underlying population is not normally distributed, then the chance that you will make a Type II error is higher than if the population had been normally distributed. That is, your test is not as *powerful* at detecting false hypotheses.

When data are normally distributed, the classical statistical tests you learned first are always more powerful than the nonparametric tests, although, in many cases, not overwhelmingly so. When the data are not normally distributed, the nonparametric tests will be the more powerful tests.

CHAPTER 18 SUMMARY

When you learned the traditional hypothesis testing techniques, you learned about the assumptions that needed to be made for the tests to apply. At the time, you did not really understand what to do if the assumptions could not be made. Nonparametric statistics is one tool that can be used when the traditional statistical tools cannot be used.

Key Terms

Term	Definition	Page Reference
Wilcoxon rank sum test	The **Wilcoxon rank sum test** is used to compare the distributions of two independent populations when the samples are small and the data are not normally distributed, or when the data are not quantitative.	787
Kruskal–Wallis test	The **Kruskal–Wallis test** is a nonparametric method for comparing the data from more than two populations.	798

Key Formulas

Term	Formula	Page Reference
Sum of first n integers	$\dfrac{n(n + 1)}{2}$	790
Large-sample mean of Wilcoxon rank sum test statistic	$\mu = \dfrac{n_1(n_1 + n_2 + 1)}{2}$	793
Large-sample standard deviation of Wilcoxon rank sum test statistic	$\sigma = \sqrt{\dfrac{n_1 n_2(n_1 + n_2 + 1)}{12}}$	793
Kruskal–Wallis test statistic	$H = \dfrac{\text{SSA}}{\text{SST}/(n - 1)}$	801
SSA for Kruskal–Wallis test statistic	$\text{SSA} = \sum_{j=1}^{c} n_j(\bar{r}_j - \bar{\bar{r}})^2$	802
SST for Kruskal–Wallis test statistic	$\text{SST} = \sum_{j=1}^{c} \sum_{i=1}^{n_j} (r_{ij} - \bar{\bar{r}})^2$	802
Approximate Kruskal–Wallis test statistic	$H = \dfrac{12}{n(n + 1)} \sum_{j=1}^{c} \dfrac{R_j^2}{n_j} - 3(n + 1)$	802

CHAPTER 18 EXERCISES

Learning It!

18.9 A large food company is considering changing the preparation directions on packets of instant oatmeal because it appears that customer complaints are rising. They run a palatability study on a test group of consumers. For one group they use a high level of liquid and for the other a low level of liquid. The data (ratings based on a questionnaire) are:

Low	High
16	21
35	24
39	39
77	60
84	64
97	65
104	86
129	94

(a) Create dotplots for each sample. What assumptions do you think are reasonable?

(b) Set up the hypotheses to see if there is a difference in palatability for level of liquid.

(c) Find the rank for each data value and the rank sum for each sample.

(d) Perform the Wilcoxon rank sum test at the 0.05 level of significance. What can you conclude about the effect of liquid level on palatability?

18.10 A major retailer of large appliances is interested in finding out whether type of advertising affects sales. Marketing executives pick two geographically similar areas and decide to use print advertising (newspapers, magazines) in one area and Internet advertising (banner ads, Web site ads) in another. The data, in number of sales per week for an eight-week period, are given below:

Internet	Print
10	18
13	20
15	21
15	22
16	25
19	25
20	26
21	26

(a) Create dotplots for each sample. What assumptions do you think are reasonable?

(b) Set up the hypotheses to see if sales with print advertising are greater than sales with Internet advertising.

(c) Find the rank for each data value and the rank sum for each sample.

(d) Perform the Wilcoxon rank sum test at the 0.05 level of significance. What can you conclude about sales and type of advertising?

18.11 A group of consumers in an urban area are given the opportunity to test three different types of high-speed Internet service: DSL (direct subscriber line), cable modem, and satellite. A service type is randomly assigned to each consumer and they are asked to use the service for one month. At the end of the month they are asked to rate the service using a 5-point scale where 1 = Extremely satisfied and 5 = Extremely dissatisfied. The data are given in the next table.

DSL	Cable	Satellite
1	1	1
1	1	1
1	2	1
1	2	2
1	2	3
1	3	3
1	3	4
2	4	4
2	4	5
3	4	5

(a) Create dotplots for all three samples. Is it reasonable to assume that the shapes and variability are the same for the three groups?

(b) Set up the hypotheses to see if there is a difference in satisfaction for the three high-speed Internet services.

(c) Find the rank for each data value, and find the rank sum and average rank for each sample.

(d) Perform the Kruskal–Wallis test at the 0.05 level of significance. What can you conclude?

18.12 The same company that is trying to start an on-line shopping service is interested in how often people use the Internet. The survey asks respondents how many times per week they access the Internet for purposes other than business or e-mail. Again the responses are for women for three different household income categories:

Under $25,000	$25,000 to $50,000	Over $50,000
1	2	8
3	5	12
3	6	13
3	6	13
5	6	17
10	10	18

(a) Create dotplots for all three samples. Is it reasonable to assume that the shapes and variability are the same for the three groups?

(b) Set up the hypotheses to see if there is a difference in the number of times women use the Internet for reasons other than e-mail or business for the three different income levels.

(c) Find the rank for each data value, and find the rank sum and average rank for each sample.

(d) Perform the Kruskal–Wallis test at the 0.05 level of significance. What can you conclude?

Thinking About It!

18.13 It has been a widely held belief that the switch to participative management would increase employees' "buy-in" to the company. One of the benefits that should be realized is a reduction in the number of sick days that employees use. A company that has made the switch in some departments wonders if this has been true. They decide to sample 25 employees from each of two manufacturing departments. The first has been using a participative management style for almost two years and the second is still using a traditional management style. The data on the number of sick days used by each employee in the past 12 months are found in the table.

Participative					Traditional				
1	3	5	5	6	0	5	6	7	9
1	4	5	6	7	3	5	7	7	9
2	4	4	6	8	4	6	7	7	10
2	4	5	6	8	4	6	7	8	11
3	4	5	6	8	5	6	7	8	11

(a) Create plots of the data. Do you think they are normally distributed?

(b) Use the appropriate nonparametric test to decide whether the data provide enough evidence to say that employees who use participative management styles use fewer sick days than those who use a traditional management style.

(c) Redo the test using a small-sample (Student t) test. What are the results of this test?

(d) Which test do you think should be used? Why?

18.14 A software company that was looking at the time to failure (hours) of the diskettes it uses decides to look at an alternative supplier of the product. The data for its current supplier and for the new supplier are shown below:

Current Supplier				Alternative Supplier			
486	494	502	508	489	492	495	498
490	496	504	510	489	492	496	499
491	498	505	514	491	493	497	502
491	498	506	515	492	493	497	503
494	498	507	527	492	494	497	505

(a) Create plots of the data. Do you think they are normally distributed?

(b) Use the appropriate nonparametric test to decide whether the data provide enough evidence to say that the new supplier is better than the old one.

(c) Redo the test using a small-sample (Student t) test. What are the results of this test?

d) Do you think that the nonparametric test is the better alternative here? Why or why not?

18.15 A diaper company is considering three different filler materials for its disposable diapers. Eight diapers were tested with each of the three filler materials and 24 toddlers were randomly given a diaper to wear. As the child played, fluid was injected into the diaper every 10 minutes until the product failed (leaked). The amount of fluid (in grams) at the time of failure was recorded for each diaper. The data are shown below:

Material 1	Material 2	Material 3
791	809	828
789	818	814
796	803	855
802	781	844
810	813	847
790	808	848
800	805	836
790	811	873

(a) Create plots of the data. Do you think that the three populations are normally distributed?

(b) Use the appropriate nonparametric test to decide whether the data provide enough evidence to say that the absorptions of the three materials are different.

(c) Redo the test as a one-way ANOVA. What are the results of this test?

(d) Do you think that the nonparametric test is the better alternative here? Why or why not?

18.16 Look again at the data from the golf ball company.   *Requires Exercise 18.4*

(a) Do you think it would have been reasonable to assume that the data were normally distributed? If so, how would this have changed your approach to solving the problem?

(b) If the normality assumption is not justified, what would be the impact of using a small-sample hypothesis test procedure?

Requires Exercise 18.11

18.17 The people who conducted the survey on high-speed Internet access want to provide the area with the type of service that received the highest ratings.

(a) Based on your conclusions, what would you recommend they do?

(b) As a consultant, you are concerned with the way the study was done. Write a memo to the company explaining how its study was faulty and suggest changes for a new study. Be sure to include what other factors you think should be considered and what other types of questions should be asked.

18.18 The appliance retailer looking at effects of advertising type on sales decides to look at one more type of advertising. The marketing executives use broadcast advertising (radio, television) in another, similar area for a different eight-week period. The new data, along with the old, are shown below:

Internet	Print	Broadcast
10	18	15
13	20	16
15	21	17
15	22	19
16	25	23
19	25	23
20	26	26
21	26	27

(a) Perform the Kruskal–Wallis test at the 0.05 level of significance to determine whether there is a difference in weekly sales due to advertising type.

(b) Do you think that what the retailer did was reasonable? Why or why not?

(c) Write a memo explaining why the study was faulty and suggest a better method to use.

CHAPTER *19*

THE RESEARCH REPORT

19.1 CHAPTER OBJECTIVES

After completing this chapter you should:

- Know what the contents of a research report are.
- Tailor the report format to meet the needs of:
  - different types of research (basic and applied)
  - different research goals that need reports of varying lengths
  - different audiences
- Be able to write a good:
  - Executive summary or synopsis
  - Introductory section
  - Methods section
  - Data analysis section
  - Interpretation of the results, using tables and pictorial representations, wherever appropriate
- Give your recommendations and suggestions for implementation, as necessary.
- Write the summary and acknowledgment.
- Provide the appropriate references.
- Include appropriate materials in the appendix.
- Critique research reports and published studies.
- Know the components of, and make, a good oral presentation.

Once the data analyses are completed and conclusions drawn from the findings, the investigator is ready to present the results of the research study and make suitable recommendations. This is usually done in the form of a written report, frequently followed up by an oral presentation.

It is important that the results of the study and the recommended solutions to the problem are effectively communicated to the sponsors, so that the suggestions made are accepted and implemented. Otherwise, all the effort hitherto expended on the investigation would be in vain. Writing the report concisely, convincingly, and with clarity is perhaps even more important than conducting a perfect research study. Hence, a well-thought-out written report and oral presentation are critical.

The contents and organization of both modes of communication—the written report and the oral presentation—depend on the purpose of the research study, and the audience at which it is targeted. The relevant aspects of the written and oral reports are discussed in this chapter.

19.2 THE WRITTEN REPORT

The written report enables the manager to weigh the facts and arguments presented therein, and implement the acceptable recommendations, with a view to closing the gap between the existing state of affairs and the desired state. To achieve its goal, the written report has to focus on the issues discussed below.

19.2.1 The Written Report and Its Purpose

Reports could be written for different purposes and hence the form of the written report would vary according to the situation. It is important to identify the *purpose* of the report so that it can be tailored accordingly. If the purpose is simply to offer *details on some specific factors* requested by a manager, the report can be very narrowly focused and provide the desired information to the manager in a brief format, as in Example 19.1. If, on the other hand, the report is intended to *"sell an idea"* to management, then it has to be more detailed and convincing as to why the proposed idea is an improvement and should be adopted. Here the emphasis would be on presenting all the relevant information backed by the necessary data, to persuade the reader to "buy into the idea." An example of the purpose of such a report and its contents can be seen in Example 19.2. A variation will be provided in some cases where a manager asks for *several alternative solutions* or *recommendations* to rectify a problem in a given situation. The researcher provides the requested information and the manager chooses from among the alternatives and makes the final decision. In this case, a more detailed report surveying past studies, the methodology used for the present study, different perspectives generated from interviews and current data analysis, and alternative solutions based on the conclusions drawn from the results of data analyses, will have to be provided. How each alternative will help to improve the problem situation must also be discussed. The advantages and disadvantages of each of the proposed solutions, together with a cost–benefit analysis in terms of dollars and/or other resources, will also have to be presented to help the manager make the decision. A situation such as in Example 19.3 would warrant this kind of a report. Such a report can also be found in Report 3 in the appendix to this chapter.

Still another type of report might require the researcher to *identify the problem and provide the final solution* as well. That is, the researcher might be called in to study a situation, determine the nature of the problem, and offer a report of the findings and recommendations. Such a report has to be very detailed, following the format of a full-fledged study, as detailed later in this chapter. A fifth kind of research report is the very scholarly publication presenting the *findings of a basic study* that one usually finds published in academic journals.

EXAMPLE 19.1   A Simple Descriptive Report

If a study is undertaken to *understand in detail* certain factors of interest in a given situation (e.g., variations in production levels, composition of employees, and the like), then a report describing the phenomena of interest, in the manner desired, is all that is called for. For instance, let us say a personnel manager wants to know how many employees have been recruited during the past 18 months in the organization, their gender composition, educational level, and the average proportion of days that these individuals had absented themselves since recruitment. A simple report giving the desired information is all that would be necessary.

In this report, a statement of the **purpose** of the study will be first given (e.g., it was desired that a profile of the employees recruited during the past 18 months in the company, and an idea of their rate of absenteeism be provided. This report offers those details.) The **methods** or **procedures** adopted to collect the data would then be given (e.g., the company payroll and the personal records of the employees were examined). Finally, a narration of the actual **results,** reinforced by visual tabular and graphical forms of representation of the data will be provided. Frequency distributions, cross tabulations, and other data will be presented in a tabular form, and pictorial illustrations will include bar charts (for gender), pie charts (to indicate the proportions of individuals at various educational levels), and such. This section will summarize the data and may look as follows.

> *A total of xx (number of) employees have been recruited during the past 18 months, of whom 45 percent are women and 55 percent are men. Twenty percent have a master's degree, 68 percent a bachelor's degree, and 12 percent a high school diploma. The average proportion of days that these employees remained absent during the past 18 months is z.*

These details provide the required information to the manager. It may, however, also be advisable to provide a further gender-wise breakdown of the mean proportion of days of absence of the employees in an appendix, even though this information might not have been specifically requested. If considered relevant, a similar breakdown can also be offered for people at different job levels. A short, simple report of the type discussed previously is provided in Report 1 in the appendix to this chapter. ■

EXAMPLE 19.2   Details of a Report to "Sell" an Idea

The purpose of a report may be to *sell an idea to top management*. For example, the Information Systems (IS) manager might want to convince the top executives that an **executive information system (EIS)** would greatly enhance the effectiveness of top executives, by virtue of the speed and timeliness of the electronic information delivery system. With up-to-the-minute information available at the fingertips of executives—something that the current paper reporting system lacks—informed decisions could be made with much confidence. When the executives realize that they can perform their information-intensive activities with ease and speed, and simultaneously enhance the quality of their decisions, they will readily buy into the idea. But then the research report for this purpose will have a different thrust and focus in greater detail on the following:

1. Explanation in clear and simple terms of what an EIS is, and how it will be a powerful executive tool for effective decision making.

2. How it would save on time (e.g., by giving immediate access to the specific information the executive needs, without the frustration of shuffling papers and not finding what is needed).

3. How it would be infinitely better than the current system (e.g., since all information is updated twice daily, the EIS will provide executives all the current data needed—something that enhances the quality of the decisions made).

4. How it would save resources in the long run (backed by a detailed cost–benefit analysis). For instance, compare the costs of training executives to use the system and updating information on a daily basis, *versus* the benefits of savings accrued through more informed and timely decisions, as in the case of the establishment of a viable "just-in-time" inventory system, which saves a lot of money for the organization.

5. Illustration of examples from past company history (within the past two months, if possible) of how an EIS system would have facilitated the executives to make more informed decisions in those instances, and how it could have saved the system money/resources.

6. A final convincing recommendation to adopt EIS as a way of organizational decision making. ■

A specimen of the type of report discussed above with respect to recommending sabbaticals for managers is provided in Report 2 in the appendix to this chapter.

EXAMPLE 19.3   **A Situation where a Comprehensive Report, Offering Alternative Solutions, Is Needed**

The president of a tire company wants several recommendations to be made on how the future growth of the company should be planned, taking into consideration manufacturing, marketing, accounting, and financial perspectives. In this case, only a broad objective is stated: corporate growth. There may currently be several impediments that retard growth. One has to carefully examine the situation to determine the obstacles to expansion and how these may be overcome through strategic planning from production, marketing, management, financial, and accounting perspectives. Identifying the problems or impediments in the situation would require intensive interviewing, literature review, industry analysis, formulation of a theoretical perspective, generation of several hypotheses to come up with different alternative solutions, data gathering, data analyses, and then exploration of alternative ways of attaining corporate growth through different strategies. To enable the president to assess the alternatives proposed, the pros and cons of implementing each of the alternative solutions, and a statement of the costs and benefits attached to each, would follow.

This report will be more elaborate than the previous two, detailing each of the steps in the study, emphasizing the results of data analysis, and providing a strong basis for the various recommendations. The alternatives generated and the pros and cons of each in a report such as this are likely to follow the format of Report 3 in the appendix. Report 4 in the appendix relates to the basic research of an issue that was examined by a researcher. ■

As we have seen, the contents and format of a report will depend on the purpose of the study and the needs of the audience to whom it is submitted.

19.2.2 The Written Report and Its Audience

The organization of a report, its length, focus on details, data presentation, and illustrations, will in part be a function of the audience for whom it is intended. The letter of transmittal of the report would clearly indicate to whom the report is being sent. An executive summary placed at the beginning would offer busy executives just the right amount of vital details—in less than three pages. The executive summary will help the busy managers to quickly grasp the essentials of the study and its findings, and turn to the pages that offer more detailed information on aspects that are of particular interest to them.

Some managers are distracted by data presented in the form of tables and feel more comfortable with graphs and charts, while others want to see "facts and figures." Whereas both tables and figures are visual forms of representation and need to be presented in reports, which ones are predominantly displayed within the body of the report and which are the ones relegated to an appendix is a function of the awareness of the idiosyncrasies of the ultimate consumer of the report. If a report were to be handled by different executives with different orientations, the report should be packaged in such a way that the individuals know where to find the information that meets their preferred mode of information processing. For example, in addition to mentioning market share in the text, it can be illustrated through a pie chart, and the raw data can also be presented in tabular form.

The length, organization, and presentation modes of the report will depend partially on the target audience. Some businesses might also prescribe their own format for report writing. In all cases, good reporting is a function of knowing who the audience is and the purpose of the report. As we have seen, some reports may have to be long and detailed, and others brief and specific.

Sometimes, the findings of a study could be unpalatable to the executive (e.g., the organizational policies are outdated and the system is too bureaucratic), or could reflect poorly on management, tending to make them react defensively (e.g., the system has an ineffective top-down approach). In such cases, tact should be exercised in presenting the conclusions without compromising on the actual findings. That is, although there is no need to suppress the unpalatable findings, they can be presented in a nonjudgmental or nonaccusatory manner, using objective data and facts that forcefully lead to and convince the managers of the correctness of the conclusions drawn. If this is not done, the report will be read defensively, the recommendations will not be accepted, and the problem will remain unsolved.

Tact and diplomacy, combined with honesty and objectivity, are essential in report writing and presentation. Although this is true for both internal and external research teams, the internal team's research report writing in such cases becomes even more difficult. Being a part of the very system on which such findings are reported, the internal team might be perceived as challenging the authority of the hierarchy. Hence, it is easy to be intimidated by power and authority, but the internal research team, while being polite, should package its findings in a professional, unbiased, and tactful manner, thereby preserving the integrity of the findings.

As an example of such a presentation, if the system has outmoded policies (or is highly bureaucratic), the report can convey the following. After presenting the data to support the facts, it might state that these policies (and the system) were perhaps appropriate at the time they were formulated, but the current goals of the present management, coupled with the passage of time, call for a change. It can also highlight the fact that the present system is receptive to changes and changing the policies (or the structure of the organization) will not, therefore, pose difficult problems. A similar appropriate strategy can be followed to change the top-down approach to a bottom-up management style.

19.2.3 Characteristics of a Well-Written Report

Despite the fact that report writing is a function of the purpose of the study and the type of audience to which it is presented, and has accordingly to be tailored to meet both, certain basic features are integral to all written reports. Clarity, conciseness, coherence, proper emphasis on important aspects, meaningful organization of paragraphs, smooth transitions from one topic to the next, apt choice of words, and specificity, are all important features of a good report. The report should be free of technical or statistical jargon to the extent possible, unless it happens to be of a technical or statistical nature. Care should also be taken to eliminate grammatical and spelling errors.

Any assumptions made by the researcher should be clearly stated in the report, and facts, rather than opinions, provided. The report should be organized in a manner that enhances the meaningful and smooth flow of materials, as the reader goes through it. The importance of the appearance of the report and its readability cannot be overemphasized.

Appropriate headings and subheadings organize the report in a logical manner and help the reader to follow the transitions easily. A double-spaced, typed report with wide margins on all sides enables the reader to make notes/comments while going through the contents.

It is obvious that the research report should bear a *title* that indicates in a succinct manner what the study is about. It should have at the beginning a *table of contents,* a copy of the *authorization* to conduct the study (in response to the original research proposal), and an *executive summary* (in the case of applied research) or a *synopsis* (in the case of basic research).

Contents of the research report

All reports should have an introductory section detailing the purpose of the study, giving some background of what it relates to, and stating the problem studied, setting the stage for what the reader could expect in the rest of the report. The body of the report would contain details regarding the framework of the study, hypotheses if any, sampling design, data collection methods, analysis of data, and the results obtained. The final part of the report would present the findings and draw conclusions. If recommendations have been called for, they would be included, with a cost–benefit analysis provided with respect to each. Such information would clarify the net advantages of implementing each of the recommendations. The details provided in the report should convince the reader of the thoroughness of the study, and offer a sense of confidence in accepting the results and the recommendations made. Every professional report would also point out the limitations of the study (e.g., in sampling, data collection, etc.).

Good descriptions and lucid explanations, smooth and easy flow of materials, recommendations that flow logically from the results of data analysis, and an explicit statement of any limitations to the study, offer a scientific authenticity to the report. The transmittal letter is best written with a personal touch, wherever appropriate.

In sum, a rigorous, well-conducted study loses its value when it is not properly presented. A report to be considered useful should provide a good rationale for the study, clearly present the problem studied, present the results of data analyses fully and adequately, and interpret the data in a manner understandable to the reader. The conclusion drawn from the findings should indicate a clear solution to the problem.

The report can be organized in parts, sections, or chapters and should be tailored to meet the needs of the situation. Good, crisp, and clear writing, figures, charts, and tables that succinctly support or highlight the salient issues, and attractive packaging, are some of the essential characteristics of a good report. The writing style should be simple, interesting, precise, and comprehensible. Unbiased and objective presentation of the findings and specific reference to the limitations of the study lend credibility to the research work. Tact and diplomacy are required in presenting unpalatable findings without distortion, and in an objective, nonthreatening, and useful manner that would not offend the sponsor. The format and style of reporting should be tailored to the audience and meet the purpose of the study.

The report would end with a summary and acknowledge the help received from various individuals and sources. A list of references cited in the report would then follow. Appendices, if any, would be attached to the report.

A report on the factors influencing the upward mobility of women in accounting firms can be found in Report 4 in the appendix to this chapter. We will now discuss the different parts of the report.

19.3 INTEGRAL PARTS OF THE REPORT

19.3.1 The Title Page of the Research Report

The title of the report should succinctly indicate what the study is all about. Examples of some good report titles are:

1. A Study of Customer Satisfaction with the Pizza Hut at Sunshine City, Illinois
2. Factors Influencing the Burnout of Nurses in Monroe Hospital
3. Antecedents and Consequences of White-Collar Employees' Resistance to Mechanization in Service Industries
4. Factors Affecting the Upward Mobility of Women in Accounting Firms
5. A Study of Portfolio Balancing and Risk Management in Investment Firms

The first two projects will be applied research, whereas the last three will be in the realm of basic research.

In addition to the title of the project, the title page will indicate the name of the sponsor of the study, the names of the researchers and their affiliations, and the date of the final report.

19.3.2 Table of Contents

The table of contents with page references usually lists the important headings and subheadings in the report. A separate list of tables and figures should also appear in the table of contents.

19.3.3 Authorization Letter

A copy of the letter of authorization from the sponsor of the study, approving the investigation and detailing its scope, will be attached at the beginning of the report. This would have been given by the sponsor in response to the research proposal submitted by the researcher soon after the initial interviews and the identification of the problem. The authorization makes clear to the reader the goals of the study.

19.3.4 The Executive Summary or Synopsis

The executive summary (or synopsis) is a brief account of the research study that provides an overview, and highlights the following important information related to the study: the problem statement, the sampling design, the data collection methods used, results of data analysis, the findings, and the recommendations, with suggestions for their implementation. The executive summary (or synopsis) will be brief—usually less than three pages in length.

An example of a synopsis of the study of customer satisfaction with the Pizza Hut in Sunshine City follows.

EXAMPLE 19.4 Synopsis of Pizza Hut Study

Introduction and Relevant Details

At the request of the manager of Pizza Hut in Sunshine City, a survey was conducted to assess customer satisfaction. The sample comprised 240 customers who were administered a short questionnaire during a two-month period from July 15 to September 14. Each day, four customers who walked into the Pizza Hut at 12:00 noon, 3:00 P.M., 6:00 P.M., and 9:00 P.M. were asked to respond to a short questionnaire on site, after they had eaten the pizza. The questionnaire, requiring fewer than three minutes for completion, asked respondents to give information on their gender and age, and to indicate on a 5-point scale, the extent of their satisfaction with (1) the flavor and texture of the pizza, (2) its taste, (3) nutritional value, (4) price, (5) the quality of service, and (6) the ambiance of the eating place. An open-ended question also asked them to offer additional comments they might wish to make. Customers dropped off their responses in a locked box with a slit at the top, kept near the exit.

Results of Data Analysis

Analysis of the data indicated that about 60 percent of the respondents were men and 40 percent women. Most of them were over 25 years of age. Customers expressed greatest satisfaction with the taste of the pizza (a mean of 4.5 on a 5-point scale), followed by its flavor and texture (mean of 4). They were neither pleased nor displeased with the price or the quality of service (3 on a 5-point scale). However, they were not particularly happy with the ambiance or the nutritional value (mean of 2.5 for each). The comments offered in the open-ended question indicated that some 25 individuals thought that the amount of cheese in the pizza might increase their cholesterol level to the detriment of their health.

Conclusions and Recommendations

These results indicate that customers do like the pizza and have no specific complaints about the price or the service. Should the manager be concerned about the displeasure of the customers with the ambiance or the nutritional value, these can be handled fairly easily. It is possible, for instance, to improve the ambiance with flowers and hanging baskets of plants. Lighted candles on the tables in the evenings would also enhance the atmosphere.

As for dissatisfaction with the nutritional value, information about the use of only low-fat cheese in the pizza as a health safeguard can be disseminated through the menu card and advertisements. The option of pizza with nonfat cheese may also be offered to the customers.

If enhancement of the level of customer satisfaction is desired, a short training program could be introduced for the waiters for the purpose, and their service thereafter supervised till the "service with a smile" motto is internalized by them. ■

19.3.5 The Introductory Section

The introductory section starts with a statement of the problem that is investigated. The research objective, together with background information of why and how the study was initiated, will also be stated. In the case of basic research the introductory section will offer an idea of the topic that is researched, and why it is important to study it. The arguments would focus on the relevancy, timeliness, and appropriateness of the research, in the context of current factors and trends in society and/or organizations.

The research objective and the problem statement to be studied are clearly set forth in this section.

19.3.6 The Body of the Report

In this part, the details of the interviews conducted, the literature survey, the theoretical framework, and the hypotheses are furnished. The design details such as sampling and data collection methods, as well as the nature and type of study, the time horizon, the field setting, and the unit of analysis, will be described.

Next, the details of the types of data analyses done to test the hypotheses, and the findings therefrom, will be provided. Tabular and pictorial depictions of the results of data analysis will find a place here. A few of the various ways in which data can be pictorially presented in written reports and oral presentations are illustrated in Figure 19.1 (pages 821–822).

19.3.7 The Final Part of the Report

The final part of the report will contain the conclusions drawn from the findings. In most cases (depending on the scope of the project), a list of recommendations for implementation will follow. Frequently, a cost–benefit analysis will also be provided. Any limitations to the study, as for example, flaws in sampling due to circumstances beyond one's control, will find a place herein. A brief summation paragraph will also be provided at the end.

19.3.8 Acknowledgment

Next, the help received from others is acknowledged. Usually, the people who assisted in the study by collecting the questionnaires, acting as liaisons, helping in data analysis, and so on, are recognized and thanked. The organization is thanked for the facilities provided, and organizational members are thanked for responding to the survey.

It should now be easy to see, given the variety of information covered in the report, why it is important to have appropriate headings and subheadings throughout. This enables the reader to progress through the report smoothly, easily, and quickly, and leaving wide margins on all sides enables the reader to jot down points or make notes, wherever considered necessary, as the report is being read.

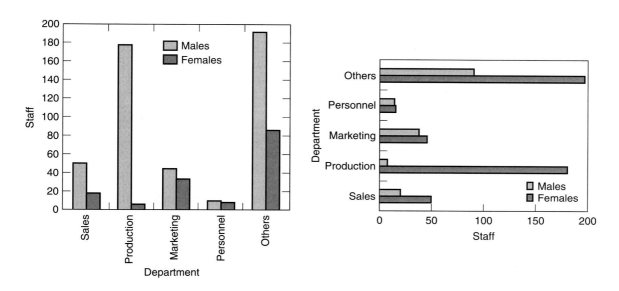

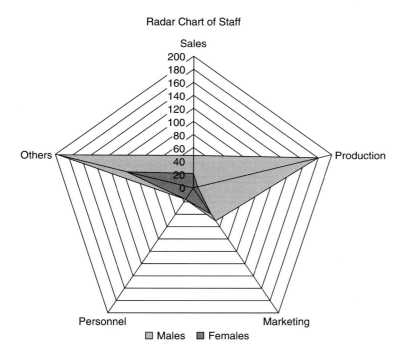

FIGURE 19.1 Pictorial representation of data

(continued)

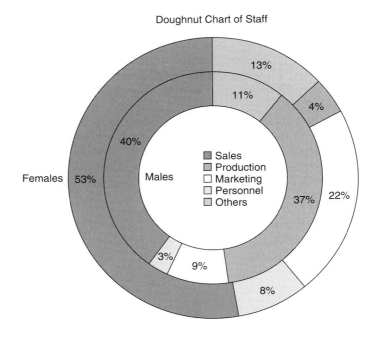

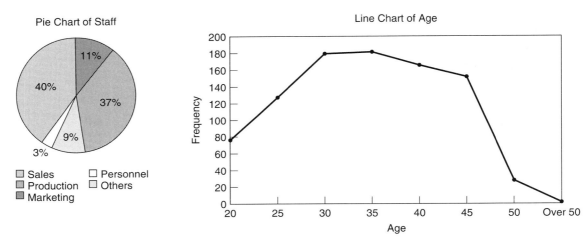

FIGURE 19.1 *(continued)*

19.3.9 References

Immediately after the acknowledgments, starting on a fresh page, a list of the references cited in the literature review and at other places in the report will be given. Footnotes, if there are any in the text, are referenced either separately at the end of the report, or at the bottom of the page where the footnote occurs.

19.3.10 Appendix

The appendix, which comes last, is the appropriate place for the organization chart, newspaper clippings or other materials that substantiate the text of the report, detailed verbatim narration of interviews with members, and anything else that would help the reader follow the text. The appendix should also contain a copy of the questionnaire administered to the respondents. If there are several appendices, they could be referenced as Appendix A, Appendix B, and so on, and appropriately labeled.

The preceding will make clear that the Table of Contents (mentioned earlier) following the title page and the letter of transmittal, would look somewhat as indicated below, with some variations.

Table of Contents

Letter of Authorization

Introduction
- Problem Studied
- Background Information
- Research Goals

Preliminary Details
- Unstructured and Structured Interviews
- Literature Survey
- Theoretical Framework
- Hypotheses Formulated

Research Design
- Type and Nature of the Study
- Sampling Design
- Data Collection Methods
- Data Analytic Techniques Used

Results of Data Analysis
- Hypothesis Substantiated / Unsubstantiated

Conclusions

Recommendations

Limitations of Study

Acknowledgments

References

Tables

Graphs

Appendices

19.4 ORAL PRESENTATION

Most organizations (and class instructors) require about a 20-minute oral presentation of the research project, followed by a question-and-answer session.

The oral presentation requires considerable planning. Imagine a study that spanned over several months has to be presented in 20 minutes to a live audience! Those who have only superficially read the report, or not read it at all, have to be convinced that the recommendations made in the study would prove to be beneficial to the organization. All this will have to be effectively accomplished in a short space of time.

The challenge is also to present the important aspects of the study in an interesting manner while still providing statistical and quantitative information, that many in the audience may find rather dull. Different stimuli (overheads, slides, charts, pictorial and tabular depiction, etc.) have to be creatively provided to the audience to sustain their interest throughout the presentation. To make all this possible, time and effort have to be expended in planning, organizing, and rehearsing the presentation.

Slides, overheads, charts, graphs, handouts—all in large, bold print, and preferably in multicolors—help the presenter to sustain the interest of the audience. They also help in talking about and explaining the research project coherently, without *reading* from prepared notes.

Factors irrelevant to the written report, such as dress, mannerisms, gestures, voice modulation, and the like, take on added importance in oral presentations. Speaking audibly, clearly, without distracting mannerisms, and at the right cadence for the audience to comprehend, is vital for holding their attention. Varying the length of the sentences, establishing eye contact, tone variations, voice modulation, and the rate of flow of information, make all the difference to audience receptivity. Use of 3 × 5 cards for orderly presentation helps smooth transitions during the presentation. Thus, both the content of the presentation and the style of delivery should be well planned.

19.4.1 Deciding on the Content

Because a lot of material has to be compressed into a 20-minute presentation, it is necessary to decide what points will be focused on and how much importance will be given to each. Remembering that the listener absorbs only a small proportion of all that he or she has heard, it is important to determine what the presenter would like the listener to walk away with, and then organize the presentation accordingly.

Obviously, the problem investigated, the results found, the conclusions drawn, the recommendations made, and the ways they can be implemented are of most interest to organizational members, and need to be emphasized during the presentation. The design aspects of the study, details of the sample, data collection methods, details of data analysis, and the like can be mentioned in passing to be picked up at the question-and-answer session by the interested members.

Depending on the type of audience, however, it may become necessary to place more stress on the data analytic aspects. For example, if the presentation is made to a group of statisticians in the company, or in a research methods class, the data analyses and results will receive more time than if the project is presented to a group of managers whose main interest lies in the solution of the problem and implementation of the recommendations. Thus, the time and attention devoted to the various components of the study will require adjustment, depending on the audience.

19.4.2 Visual Aids

Graphs, charts, and tables help to drive home the points one wishes to make much faster and more effectively, true to the adage that a picture is worth a thousand words. Visual aids provide a captivating sensory stimulus which sustains the attention of the audience. Modern technology makes it possible for color graphics to be produced on personal computers and projected onto the screen. Slides, transparencies, flipcharts, the chalkboard, and handout materials also help the audience to easily follow the points of the speaker's focus. The selection of specific visual modes of presentation will depend on such things as the size of the room, the availability of a good screen for projection, and the cost constraints of developing sophisticated visuals. All visuals should be produced with an eye on easy visibility from the far end of the presentation hall. Large, easily readable visuals that are properly labeled in big bold letters help the audience to focus on the presentation. Visuals that present side-by-side comparisons of the existing and would-be state of affairs via graphs or pie charts, drive home the points made much more forcefully than elaborate and laborious verbal explanations.

Integrated multimedia presentations using videotapes, videodiscs, CD-ROM, and the visuals described earlier, are quite common in this technological age. Digital whiteboards facilitate digital storage of intricate diagrams that can be used in conjunction with electronic projective systems to serve as electronic flipcharts. When planning a presentation using Power Point or integrated multimedia, it is important to ensure before the presentation starts that the related equipments are properly hooked up and tested so that the presentation can proceed smoothly without interruptions.

19.4.3 The Presenter

An effective presentation is also a function of how "unstressed" the presenter is. The speaker should establish eye contact with the audience, speak audibly and understandably, and be sensitive to the nonverbal reactions of the audience. Strict adherence to the time frame and concentration on the points of interest to the audience are critical aspects of presentation. Extreme nervousness throughout the presentation, stumbling for words, or fumbling with notes or audiovisuals, speaking inaudibly and/or with distracting mannerisms, straying away from the main focus of the study, and exceeding the time limit, all detract from effectiveness. One should also not minimize the importance of dress, posture, and self-confidence, on the impression created on the audience. Such simple things as covering the materials on the visuals until they need to be exhibited and voice modulation, help the audience to focus on the discussion.

19.4.4 The Presentation

The opening remarks set the stage for riveting the attention of the audience. Certain aspects such as the problem investigated, the findings, the conclusions drawn, the recommendations made, and their implementation are important aspects of the presentation. The speaker should drive home these points at least three times—once in the beginning, again when each of these areas is covered, and finally, while summarizing and concluding the presentation.

19.4.5 Handling Questions

Concentrated and continuous research on the research topic over a considerable period of time makes the presenter the person more knowledgeable about the project than anyone else in the audience. Hence, it is not difficult to handle questions from the audience with confidence and poise. It is important to be nondefensive when questions that appear to find fault with some aspect of the research are posed. Openness to suggestions also helps, since the audience might occasionally come up with some excellent ideas or recommendations the researcher might not have thought of. Such ideas must always be acknowledged graciously. If a question or a suggestion from a member in the audience happens to be flawed, it should be addressed in a nonjudgmental fashion.

The question-and-answer session, when handled well, leaves the audience with a sense of involvement and satisfaction. Questions should be encouraged and responded to with care. This interactive question-and-answer session offers an exciting experience both to the audience and to the presenter.

As may readily be seen, a 20-minute presentation and a short question-and-answer session thereafter do call for substantial planning, anticipation of audience concerns, psychological preparedness, and good management skills.

Reporting has to be done in an honest and straightforward manner. It is unethical to suppress findings that are unpalatable to the sponsors, or reflect poorly on management. As suggested earlier, it is possible to be tactful in presenting such findings, without withholding or distorting information to please the sponsors. Internal researchers, in particular, will have to find ways of presenting unpopular information in a tactful manner. It is also important to state the limitations of the study—and almost every study has some limitation—so as not to mislead the audience.

CHAPTER 19 SUMMARY

The components of various types of written research reports were discussed in this chapter. It was emphasized that the purpose of the report and the intended audience are critical factors in deciding what aspects of the study will be stressed the most. Examples of different kinds of reports were offered and additional examples can be found in the appendix to this chapter.

Ways of making effective oral presentation were also discussed, stressing both the contents of the presentation and the style of delivery.

CHAPTER 19 EXERCISES

Thinking About It!

19.1 Why is it necessary to have the letter of authorization in the report?

19.2 Discuss the purpose and contents of the Executive Summary.

19.3 What are the similarities and differences of basic and applied research reports?

19.4 How have technological advancements helped in writing and presenting research reports?

19.5 Why is it necessary to specify the limitations of the study in the research report?

19.6 What aspects of a class research project would you stress in the written report and in the oral presentation?

Doing It!

19.7 Critique Report 4 in the appendix. Discuss it in terms of good and bad research, suggesting how the study could have been improved, what aspects of it are good, and how scientific it is.

APPENDIX

SAMPLE REPORTS

REPORT 1: SAMPLE OF A REPORT INVOLVING A DESCRIPTIVE STUDY

SEKRAS COMPANY

TO: Mr. L. Raiburn, Chairman
       Strategic Planning Committee

FR: Joanne Williams
       Public Relations Officer

RE: Report requested by Mr. Raiburn

Attached is the report requested by Mr. Raiburn
If any further information or clarification is needed, please let me know.

Encl: Report

REPORT FOR THE STRATEGIC PLANNING COMMITTEE

Introduction

Vice President Raiburn, Chairman of the Strategic Planning Committee, requested two pieces of information:

1. The sales figures of the top five retailers in the country in 1997 and in 1988.
2. Customers' ideas of what improvements can be made at Sekras to enhance their satisfaction. For this purpose, he asked that a quick survey of the company's customers be done to elicit their opinions.

Method Used for Obtaining the Requisite Information

Figures of sales of the top five retailers in the country for 1997 and 1988 were obtained from *Business Week,* which periodically publishes many kinds of industry statistics.

     To obtain customers' inputs on improvements that could be made by the company, a short questionnaire (specimen in Appendix A) was mailed to 300 of our credit card customers—100 who had most frequently used the card in the last 18 months, 100 who most infrequently used it during the same period, and 100 average users. Questionnaires in three different colors were sent to the three groups. Respondents were offered a complimentary magnet for responses received within two weeks.

     The questionnaire asked for responses to three questions:

1. What are some of the things you like best about shopping at Sekras?
2. What are some of the things that you dislike and would like to see improved at Sekras? Please explain in as much detail as possible.
3. What are your specific suggestions for making improvements, so as to enhance the quality of our service to customers like you?

Findings

1. Sales Figures of the Top Five Retailers in 1997 and 1988

Information regarding sales of the top five retailers in 1997 and 1988 is provided in Table 19.1. It is evident that the top five retailers are the same across the two time periods. Wal-Mart, however, had jumped from third place in 1988 to first place in 1997, Sears had dropped from first place to the second, and Kmart slid from second position to third during the same period. JC Penney and Dayton Hudson continued to retain the fourth and fifth positions, respectively.

It is interesting that Wal-Mart had increased its sales a spectacular five and a half times during this period, while the others had increased theirs, ranging from 1.4 to 2.2 times.

TABLE 19.1   Comparative Sales Figures of the Five Top Retail Companies During 1997 and 1988

Top Retailers in 1997			Top Retailers in 1988		
Company	Sales in Billions of $	Share Among Top Five	Company	Sales in Billions of $	Share Among Top Five
Wal-Mart Stores	113.4	47%	Sears, Roebuck	30.2	29%
Sears, Roebuck	41.5	17%	K Mart	27.3	26%
Kmart	32.1	13%	Wal-Mart Stores	20.6	20%
JC Penney	29.2	12%	JC Penney	15.2	14%
Dayton Hudson	26.9	11%	Dayton Hudson	12.2	11%

Percentage Increase in Sales From 1988 to 1997

Company	% Increase
Wal-Mart Stores	5.5
Sears, Roebuck	1.4
Kmart	1.2
JC Penney	1.9
Dayton Hudson	2.2

II. Customer Suggestions for Improvements

Of the 300 surveys sent out, 225 were received—a 75 percent response rate. Of the 100 most frequent users of our credit card to whom questionnaires were sent, 80 responded; among the most infrequent users, 60 responded; and among the average users, 85 responded.

About 75 percent of the respondents were women. The majority of the customers were between the ages of 35 and 55 (62%).

The responses to the three open-ended questions were analyzed. The information needed by the Committee on the Suggested Improvements is tabulated (see Table 19.2). Responses to the other two questions on features liked by the customers, and their specific suggestions for improvement are provided in the two tables in the appendix. The following are suggestions received from one or two respondents only:

1. Have more water fountains on each floor.
2. The pushcarts could be lighter, so they will be easier to push.
3. More seats for resting after long hours of shopping would help.
4. Prices of luxury items are too high.

TABLE 19.2 **Suggested Areas for Improvement**

Features	Frequent Users No.	Medium Users No.	Infrequent Users No.	Total No.	Total %
1. Small appliances such as mixers and blenders are often not in stock. This is irritating.	30	48	22	100	44%
2. The cafeteria serves only bland, uninteresting food. How about some spicy international food?	26	14	5	45	20%
3. Often, we are unable to locate the items we want!	3	6	14	23	10%
4. It would be nice if you could have a childcare service so we can shop without distractions.	28	32	25	85	38%
5. It is often difficult to locate an assistant who can help us with answers to our questions.	29	49	22	100	44%
6. I wish it were a 24-hour store.	17	13	7	37	16%
7. Sometimes, there is a mistake in billing. We have to make some telephone calls before charges are corrected. This is a waste of our time.	4	12	14	30	13%
8. Allocate some floor space for kids to play video games.	2	—	4	6	3%
9. Import more Eastern apparel like the kimono, saris, sarongs.	—	8	4	12	5%
10. Regulate the temperature better; often, it is too cold or too hot.	15	12	17	44	20%

From looking at Table 19.2, it is seen that the most dissatisfaction stems from (a) out-of-stock small appliances, and (b) inability to locate the store assistants who could guide customers in finding what they want (44% each). The need for childcare services is expressed by 38 percent of the customers. The next two important items (20% each) indicate that the cafeteria should cater to the international spicy type of foods and that the temperature should be better regulated. Some customers (16%) also wish the store would be open 24 hours, and 13% complain about billing errors. The rest of the suggestions are offered by fewer than 10 percent of the customers, and hence can perhaps be attended to later.

A note of caution is in order at this juncture. We are not sure how representative our sample is. We thought that a mix of high, average, and infrequent users of our credit card would provide some useful insights. If a more detailed study obtaining information from a sample of *all* the customers who come to the store is considered necessary, we will initiate it quickly. In the meantime, we are also interviewing a few of the customers who shop here daily. If we find anything of significance from these interviews, we will inform you.

Improvements Indicated by These Suggestions

Based on the current sample of customers who have responded to our survey, the following improvements and actions seem called for:

1. Small appliances need to be adequately stocked (44% complained about this). An effective reorder inventory system has to be developed for this department to minimize customer dissatisfaction and avoid loss of sales for want of stock.

2. Customers seem to have trouble locating store items and would appreciate help from store assistants (44% expressed this need). If providing assistance is a primary concern, it would be a good idea to have liveried store personnel wear badges to indicate they are there to assist customers. During idle hours, if any (when there are no customers seeking help), these individuals can be deployed as shelf organizers.

3. Need for child care has been expressed by more than a third of our customers (38%). It would be a good idea to earmark a portion of the front of the building for parents to drop off their kids while shopping. The children will have to be supervised by a trained child-care professional recruited by the organization. An assistant could later be recruited if there is a need. From the cost–benefit analysis in Exhibit 7, it may be seen that these additional expenditures will pay off multifold in sales revenues, and at the same time, create a fund of goodwill for the company.

4. Adding to the variety of foods served in the cafeteria (a need expressed by 20%) is at once a simple and a complex matter. We need further ideas and details as to what types of food need to be added. This information can be obtained through a short survey, if Mr. Raiburn so desires.

5. Billing errors should not occur (16% indicated this). Our billing department should be warned that such mistakes should be avoided and should not recur. Their performance assessment should be tied to such mistakes.

6. Regulation of temperature (16% identified this) is easy. This, in fact, could be immediately attended to by our Engineering Department personnel.

I hope this report contains all the information sought by Mr. Raiburn. As stated earlier, if the noncredit card customers have also to be sampled, it can be easily arranged.

REPORT 2: SAMPLE OF A REPORT WHERE AN IDEA HAS TO BE "SOLD"

MUELLER PHARMACEUTICALS

June 15

TO: The Board of Directors

FR: Harry Wood, VP.
(Through: President Michael Osborn)

RE: Sabbatical for Managers

Enclosed is a brief report on the need for a sabbatical policy for our managers and R & D personnel, for discussion at our next board meeting. We will also plan on a more detailed presentation at that time.

WHY SABBATICALS FOR MANAGERS ARE NECESSARY

Introduction

At the company's board meeting last month, the members were concerned that no new products have been developed during the last four years and that the profits of the company are considerably down. One of the board members suggested that a sabbatical given to the managers and key staff of our company might rejuvenate them and help creativity flow again. At that time, the matter was treated casually and not given any further consideration. Sensing the need to consider this option seriously, I have since talked to a few companies that do offer this benefit to their managers. I have also obtained some data from them, which demonstrate the efficacy of sabbaticals.

Based on the available information, there is a strong case for introducing a sabbatical policy in our company. Details of my discussions with other companies and their data are presented below.

Gist of Telephone Conversations with Vice Presidents and Presidents of Companies

I talked to the presidents, vice presidents, and directors of IBM, Tandem, Apple Computers, Eli Lilly, and Time Warner Inc. All these companies have had sabbatical policies for at least the past seven years. Some presidents to whom I spoke said they initiated the policy because they found that their own productivity increased after they had had some time away from their jobs doing different kinds of things. Some said that they introduced the sabbatical because they felt that their managerial staff experienced burnout after long years of nonstop work at a hectic pace and became ineffective.

Without exception, everyone said that it makes good business sense to offer managers a chance to refurbish their lives and recharge their batteries every six years or so, so that they come back to work with renewed vigor. Among the many advantages recounted by those to whom I spoke are:

1. More enthusiasm and zest for work.
2. Better working relationships with staff.
3. A fresh approach to problem solving with less competitiveness among the different departments.
4. More creative flow of ideas, new marketing strategies, product development ideas.
6. A more dynamic workplace in terms of interpersonal interactions, interdepartmental collegiality, and joint problem solving.

Some Hard Data

The appendix, which contains the information provided by two companies, shows that the number of new products developed quadrupled in one company and increased fivefold in the other during the years since the introduction of the sabbatical. As they themselves acknowledge, the increase cannot be attributed to the sabbatical alone, but they have also documented that most of the new products developed were under the leadership of the managers after their return from a three-month sabbatical. You will note that new product development statistics for these managers, before and after their sabbatical, are indeed compelling! Reinforcing our theory is also the decline in the figures after the fourth or fifth year of their return from sabbatical and the pickup again after the next sabbatical. Noteworthy too is that the "pickup" years were no different from the others in terms of the economic environment, technology advancement, or other factors that might have a direct impact on innovation!

I have also placed in the appendix a copy of the article on Executive Life, which appeared in the *New York Times* of June 3, 1990, which you have probably already read. Is it not incredible that many of the executives who try something new during the sabbatical ultimately want to get back to their old jobs? The case cited of the law firm partner, Axinn, who missed the rigors of his old job and could just not shake off the lawyer in him when he tried to be a rabbi-in-training during his sabbatical, is particularly interesting.

Benefits of Sabbatical

The benefits of sabbatical to the managers are obvious; they refresh themselves trying their hands at new things or doing the things they have dreamed of (such as learning to play the flute or paint or write). These activities seem to offer them a new lease on their professional lives, but the benefits to the corporation seem to be even greater, as experienced by the companies that already have this scheme in place. Apple Computer's revenues are stated to have quadrupled under the leadership of Mr. John Sculley, who took nine-week sabbaticals. Again no one is attributing a cause and effect relationship, but there might be a strong correlation possible there! Mr. Lerman, partner of Wilmer, Cutler & Pickering, strongly affirms that when managers come back from sabbatical, they are more effective and invigorated.

Recommendation

Given the qualitative and quantitative evidence generated from a number of organizations that have implemented the sabbatical policy, I strongly recommend that we also establish a sabbatical policy in our company. The suggestion is to offer a paid, three-month sabbatical for all our R & D scientists, and managerial and executive staff, after every six years of service. The costs of implementing this with respect to our senior scientists, managers, and executives are worked out and shown in Exhibit 4. The likely benefits within 10 years of our initiating such a policy in terms of new product development, increased sales, and joint problem-solving endeavors due to higher energy levels of department heads, are also shown in the same exhibit.

I will ask the HRM Director to collect information from more companies having sabbatical policies and ask him to make a presentation to the board at our next meeting. In the meantime, if you need more information or clarification, feel free to give me a call.

In conclusion, our company is at the crossroads, and our scientists and managers need to be energized to enhance their performance and productivity. Constant pressure and ceaseless toil are wearing them out. Many are frustrated by the demands imposed by the jobs. "All work and no play" has banished their zest for working and drained them of their creative ideas. It is high time we inject some vitality into our system through sabbaticals.

REPORT 3: SAMPLE OF A REPORT OFFERING ALTERNATIVE SOLUTIONS AND EXPLAINING THE PROS AND CONS OF EACH ALTERNATIVE

TO: Mr. Charles Orient, CEO
Lunard Manufacturing Company

FR: Alex Ventura, Senior Researcher
Beam Research Team

RE: Suggestions on alternative ways of cutting costs in anticipation of recession.

Enclosed is the report requested by Mr. Orient. If any additional information or clarification is needed, please let me know.

Encl: Report

REPORT ON ALTERNATIVE WAYS OF HANDLING RECESSIONARY TIMES WITHOUT MASSIVE LAYOFFS

Introduction

The Beam Research Team was asked to suggest alternative ways of tiding over the anticipated recession over the next several months, when a slowdown of the economy is expected. A recent article in *Business Week* titled "Hunkering Down in a Hurry" indicated that executives in a large number of companies are slashing costs, mostly through layoffs and restructuring. Mr. Orient wanted the Beam Research Team to suggest other alternatives besides layoffs.

This report provides five alternatives citing the advantages and disadvantages of each.

Method Used for Developing the Alternatives

The team studied the economic indicators and the published industry analyses, read the Federal Reserve Board Chairman's speeches, examined the many ways in which companies cut costs during nonrecessionary periods as well as recessions, and based on these, suggested the following five alternatives.

Alternatives Suggested

1. A moratorium on all capital expenditure.
2. Hiring freeze.
3. Recovery of bad debts through sustained efforts.
4. Trimming of operating expenditures with substantial reduction in travel and entertainment expenditures.
5. Discontinuance of the manufacture of low profit margin products.

Advantages and Disadvantages of Each of the Above

Itemized details of the cost–benefit analysis for each of the preceding suggestions are furnished in the appendix, which may be referred to. We give only the net benefits for each alternative here.

1. Moratorium on all capital expenditure

It makes good sense to desist from all capital expenditure since manufacture of most of the items will slow down during recession. Except for parts for existing machines, there is no need to buy capital equipment, and all proposals in this regard should be shelved.

This strategy will cut down the expenditure to the extent of 7 to 10 percent of revenue. See appendix for full details. A reserve fund can be created to catch up with future orders when the economy returns to normal.

2. Hiring Freeze

The annual increase in the strength of our staff during the past four years has been about 15 percent. With a slowdown of the economy, a hiring freeze in all our branch offices will save more than $10 million annually.

This might initially result in some extra workload for the staff and cause some job dissatisfaction. But once they get used to it, and the impact of the actual recession hits them, employees will be thankful for the job they have. It is advisable to explain in advance the reasons for the hiring freeze to the employees so that they understand the motive behind the company's policy and appreciate having been informed.

3. Recovery of Bad Debts through Aggressive Efforts

Bad debts of the company have been on the increase over the past three years, and no intensive efforts to recover them seem to have been made hitherto.

We suggest that collection agents who have successfully recovered bad debts for other companies be hired immediately. Such agents may have to be paid more than other collection agents, but the extra cost will be well worth it. About a billion dollars can be collected within a few weeks of their being on the job, and this will help the financial cash flow of the company.

4. Trimming of Operating Expenditures

Several operating expenses can be trimmed—the travel expenses of managers in particular—as shown in Exhibit 4 of the appendix. Videoconferencing is inexpensive and should be encouraged for most of the meetings and negotiations. This alone will result in savings of more than $175,000 per month.

Another way of considerably curtailing expenditure is to restrict entertainment expenses only for such purposes and to such managers as actively promote the business of the company and is essential for public relations.

These changes will have a negative impact on morale, but managers understand the economic situation and will adjust to the new system once the initial mental resistance wears off.

5. Eliminating the Manufacture of Low-Margin Products

The team found from a detailed study of the company records of manufacturing, sales, and profits figures for the various products, that all the items listed in Exhibit 5 of the appendix have very low profit margins. It is evident from the data provided that considerable time and effort are expended in manufacturing and selling these items.

It will be useful to phase out the discontinuance of manufacture of these items, and divert the resources to the high-profit items suggested in Exhibit 6. From the cost–benefit analysis in Exhibit 7, it may be seen that several billions can be saved through this strategy.

It is possible to put into effect all of the five preceding alternatives and handle the onslaught of the recession with confidence.

REPORT 4: EXAMPLE OF AN ABRIDGED BASIC RESEARCH REPORT

FACTORS AFFECTING THE UPWARD MOBILITY OF WOMEN IN PUBLIC ACCOUNTING

Introduction

A substantial number of women have entered the public accounting profession in the last 15 years or so. However, fewer than 4 percent of the partners in the Big Eight accounting firms are women, indicating a lack of upward mobility for women in the accounting profession. Against the backdrop of the fact that the women students perform significantly better during their academic training than their male counterparts, it is unfortunate that their intellectual ability and knowledge remain underutilized during their professional careers. The recent costly litigation and discrimination suits filed make it imperative for us to study the factors that affect the upward mobility of the women and examine how the situation can be rectified.

A Brief Literature Survey

Studies of male and female accounting majors indicate that the percentage of women accounting students has increased several fold since 1977 (Kurian, 1998). Based on the analysis of longitudinal data collected over a 15-year period, Mulcher, Turner, and Williams (1997) found that women students' grades in senior accounting courses were significantly higher than those of the male students. This higher level of academic performance has been theorized as due to the higher need and desire that women have to achieve and overcome stereotypes (Messing, 1989), having higher career aspirations (Tinsley et al., 1983), or having a higher aptitude for accounting (Riley, 1984; Jones & Alexander, 1996). Empirical studies by Fraser, Lytle, and Stolle (1978), and Johnson and Meyer (1995), however, found no significant differences in personality predispositions or behavioral traits among male and female accounting majors.

Several surveys of women accountants in the country pinpoint three major factors that hinder women's career progress in the public accounting field (see, for example, Kaufman, 1986; Larson, 1994; Walkup & Fenman, 1997). They are: (1) the long hours of work demanded by the profession (a factor that conflicts with family demands), (2) failure to be entrusted with responsible assignments, and (3) discrimination. In sum, the lack of upward mobility seems to be due to factors over which the organization has some control.

Research Question

Do long work hours, failure to be handed greater responsibilities, and discrimination account for the lack of upward mobility of women in public accounting?

Theoretical Framework

The variance in the dependent variable, **upward mobility,** can be explained by the three independent variables: long hours of work, not handling greater responsibilities, and discrimination. As women are expected to, and do indeed take on responsibility for household work and child rearing, they are unable to work beyond regular work hours at the workplace. This creates a mistaken impression on higher-ups in the organization that women are less committed to their work. Because of this perception, they are not entrusted with significant responsibilities. This further hinders their progress, because they are not afforded the same exposure to the intricacies of accounting practices as men. Hence women are overlooked at the time of promotion.

Deliberate discriminatory practices due to sex-role stereotypes, as evidenced in the well-known case of *Hopkins* vs. *Price Waterhouse & Co.,* also arrest women's progress. If women are not valued for their potential and are expected to conform to sex-typed behavior (which confines them to inconspicuous roles), their chances of moving up the career ladder are significantly reduced.

Thus, the three independent variables considered here would significantly explain the variance in the upward mobility of women in public accounting. The impracticability of putting in long hours of work, lack of opportunities to handle greater responsibilities, and sex-role stereotyping, all negatively impact upward mobility.

Hypotheses

1. If women spend more hours on the job after regular work hours, they will be given greater responsibilities.

2. If women are entrusted with higher level of responsibilities, they will have more opportunities to move up in the organization.

3. If women are not expected to conform to stereotypical behavior, their chances for upward mobility will increase.

4. All three independent variables will significantly explain the variance in women CPAs' upward mobility.

METHOD SECTION

Study Design

In this cross-sectional correlational field study, data on the three independent variables and the dependent variable were collected from women CPAs in several public accounting organizations in the country, through mail questionnaires.

Population and Sample

The population for the study comprised all women CPAs in the country. A systematic sampling procedure was first used to select 30 cities from the various regions of the country from which a sample of accounting firms would be drawn. Then, through a simple random sampling procedure, five CPA firms from each of the cities were chosen for the study. Data were collected from all the women in each of the firms so chosen. The total sample size was 300 and responses were received from 264 women CPAs, for an 88 percent response rate for the mail questionnaires. The unit of analysis was the individuals who responded to the survey.

All respondents had, as expected, the CPA degree. Their ages ranged from 28 to 66. About 60 percent of the women were over 45 years of age. The average number of children in the house below the age of 13 was two. The average number of years of work in the organization was 15, and the average number of organizations worked for was two. The average number of hours spent at home on office-related matters was 2.8.

Variables and Measures

All demographic variables such as age, number of years in the organization, number of other organizations in which the individual had worked, number of hours spent at home on office-related matters, and number of children in the house and their ages, were tapped by direct single questions.

Upward Mobility. This dependent variable indicates the extent to which individuals expected to progress in their career during the succeeding 3 to 10 years. Hall (1986) developed four items to measure this variable, an example item being: *I see myself being promoted to the next level quite easily.* The measure is reported to have convergent and discriminant validity, and the Cronbach's alpha for the four items for this sample was .86.

Sex-Role Stereotyping. This independent variable was measured using Hall and Humphrey's (1972) 8-item measure. An example item is: *Men in this organization do not consider women's place to be primarily in the home.* Cronbach's alpha for the measure for this sample was .82.

Responsibilites Assigned. This was tapped by three items from Sonnenfield and Mc-Grath (1983), which asked respondents to indicate their levels of assigned responsibility to (a) make important decisions, (b) handle large accounts, and (c) account for the annual profits of the firm. Cronbach's alpha for the three items was .71 for this sample.

Data Collection Method

Questionnaires were mailed to 300 women CPAs in the United States. After two reminders, 264 completed questionnaires were received within a period of six weeks. The high return rate of 88 percent can be attributed to the shortness of the questionnaire and perhaps the motivation of the women CPAs to respond to a topic close to their heart.

Questionnaires were not electronically administered for various reasons, including the advantage it afforded to the respondents to reply without switching on the computer.

Data Analysis and Results

After determining the reliabilities (Cronbach's alpha) for the measures for this sample, frequency distributions for the demographic variables were obtained. These may be seen in Exhibit 1. Then a Pearson correlation matrix was obtained for the four independent and dependent variables. This may be seen in Exhibit 2. It should be noted that no correlation exceeded .6.

Each hypothesis was then tested. The correlation matrix provided the answer to the first three hypotheses. The first hypothesis stated that the number of hours put in beyond work hours on office-related matters will be positively correlated to the responsibilities assigned. The correlation of .56 ($p < .001$), between the number of hours spent on office work beyond regular work hours, and the entrusted responsibilities, substantiates this hypothesis.

The second hypothesis stated that if women are given higher responsibilities, their upward mobility would improve. The positive correlation of .59 ($p < .001$) between the two variables substantiates this hypothesis. That is, the greater the entrusted responsibilities, the more are the perceived chances of being promoted.

The third hypothesis indicated that sex-role stereotyping would be negatively correlated to upward mobility. The correlation of $-.54$ ($p < .001$) substantiates this hypothesis as well. That is, the greater the expected conformity to stereotyped behavior, the lower the chances of upward mobility.

To test the fourth hypothesis that the number of hours spent beyond regular work hours on job-related matters, assignment of higher responsibilities, and expectations of conformity with stereotyped behavior will significantly explain the variance in perceived upward mobility, the three independent variables were regressed against the dependent variable. The results shown in Exhibit 3 indicate that this hypothesis is also substantiated. That is, the R^2 value of .43 at a significance level of $p < .001$, with df(3, 238), confirms that 43 percent of the variance in upward mobility is significantly explained by the three independent variables.

Discussion of Results

The results of this study confirm that the variables considered in the theoretical framework are important. By focusing solely on the number of hours worked, ignoring the quality of work done, the organization is perhaps not harnessing the full potential and encouraging the development of the talents of the women CPAs adequately. It seems worthwhile to investigate into and remedy this situation.

It would be useful if the top executive were to assign progressively higher levels of responsibilities to women. This will help to utilize their abilities fully, and in turn, enhance the effectiveness of the firm. If executives are helped to modify their mental attitudes and sex-role expectations, they would tend to expect less stereotypical behavior, and encourage the upward mobility of women CPAs. Knowing that women bring a different kind of perspective to organizational matters (Smith, 1989; Vernon, 1998), it is quite possible that having them as partners of the firm will enhance the organizational effectiveness as well.

Recommendations

It is recommended that a system be set up to assess the value of the contributions of each individual in discharging the duties, and use that, rather than the number of hours of work put in, as a yardstick for promotion.

Second, women CPAs should be given progressively more responsibilities after they have served three to five years in the system. Assigning a mentor to train them will make for smooth functioning of the firm. Third, a short seminar could be organized for executives to sensitize them to the adverse effects of sex-role stereotyping in the workplace. This will help them to utilize the talents of women CPAs. If viewed as professionals with career goals and aspirations, rather than in stereotyped ways, women CPAs will be enabled to handle more responsibilities and advance in the system. The organization will also stand to benefit by their contributions.

In conclusion, it would be worthwhile for public accounting firms to modify their mental orientations toward and expectations of women CPAs. It is a national waste not to utilize their abilities.

WORKSHOP 5

Chapter 20   Grant Writing and Publishing

GRANT WRITING AND PUBLISHING

20.1 CHAPTER OBJECTIVES

Considering all of the work involved in a business research project, you might think that the final report should be the end of it. On the other hand, you might wonder what happens to your report after all your effort. Will it sit on a shelf in an office for the rest of time or, worse yet, be shoved into a file cabinet drawer and never be thought of again?

Actually, sometimes these things do happen, but often the business research project is a springboard for some other kinds of activities. In this chapter we will look at what you can do after the final report is written. In particular, we will consider

- Publication of a paper in a research or trade journal
- Writing a grant proposal to receive outside funding to continue the research

20.2 PUBLISHING YOUR RESEARCH

It is possible that your business research project solved a problem that is common in your own business or industry. It might also have addressed or solved a problem that is universal across several different businesses. In this case you might want to consider publishing the results of your research in a research or trade journal.

Research journals are usually associated with professional organizations such as the American Marketing Association (*Journal of Marketing Research, Marketing Management*) or the American Society for Quality (*Quality Management Journal, Quality Progress*).

Trade journals or industry journals are usually associated with a particular type of business or industry and might be regulated by any number of organizations. Some examples of trade journals are the *Journal of Pyrotechnics*, which relates to companies that are involved with fireworks and other pyrotechnic displays, and the *Journal of Electronic Publishing*, which is related to publishing on the Web or other electronic media.

Every journal has its own criteria and guidelines for publications that specify length of articles, style, and formatting. You can get the guidelines from the organization that administers the publication. You would never publish your entire research report in a journal. Instead, you would take the report and rewrite it according to the guidelines of the publication you are considering.

Before you decide to publish your results in a journal, you will need to think about which journal is most appropriate and most likely to publish your work. You can find some likely sources by looking at your literature search. Which journals and publications contained the information you used to find out about your research problem? These are all good leads for finding a journal to publish your results.

Another thing to consider in submitting your work for publication is the validity of the study and the reporting of the results. It is critical that you have a solid methodology and that you address any shortcomings of the research. Some of the most common reasons that journals cite when not accepting papers for publication are not establishing the validity of the data collection methods, no mention of limitations of the study, and results not stated clearly.

In addition to or instead of publishing your results, you might consider presenting them at a professional conference or meeting. There are many formats for presenting research at such meetings such as talks, panel discussions, and poster sessions.

20.3 OBTAINING RESEARCH GRANTS

Sometimes your research uncovers the tip of an iceberg and you can see many opportunities and advantages to continuing the research on a larger scale. When this happens you might want to find an appropriate source to fund the research. Such a source would be a foundation, company, or organization that would see the benefits of your research as matching its goals and objectives.

To obtain a grant you usually have to write a proposal that states the problem you want to study. You also need to state the benefits to the organization and to either society or industry that would result from the study, and set out a timetable and a budget for carrying out the research.

20.3.1 The Connection

Grantseekers are really linked to grantmakers by virtue of an interest in solving the same problem. What grantseekers do to solve the problem is termed a **project** or **program.** The way the grantseeker communicates with the grantmaker is through a proposal that usually just answers questions about the project posed by the grantmaker. That is the connection.

Those who seek grants need to study the situation and concentrate on solving the problems. Develop projects to do so. Then, and only then, go looking for funders. They will be found by virtue of their interest in the same problem you intend to solve.

20.3.2 Using the Connection to Do It Right

Most grantmakers have deadlines for submission of proposals. Rather than reviewing proposals year round, there is a certain time (or several times) a year when the grantmaker reviews all proposals submitted for that particular offering. Often, the time from when you find out that a grantmaker is letting a request to the deadline can be very brief. Many a grantseeker gets a notice from his or her employer on December 17 with a January 1 deadline. However, the difference in whether this is just a troublesome occurrence or a family crisis is in whether the organization is doing it right. If, when presented with the December request, there is a rush of holiday shopping and giving up that holiday house tour, then that is one thing. It is a problem, but not a crisis. If, however, it means facing no trimming, farming out dear old Aunt Ida to other relatives, and ordering take-out Chinese for the holiday dinner, then that is quite another. However, if it is done right, the writing of the proposal should not take very long.

That last statement bears repeating. If it is done right, the writing of the proposal should not take very long. Writing the proposal should be about 10 to 20 percent of the issue.

Why is this so often not the case? The reason is that people too frequently pick up the funder's guidelines or Request for Proposal (RFP) and start writing. Never mind that they really do not have anything concrete about which to write. Grantmakers do not want a stream of consciousness flow of prose about your good, but half-baked, idea. They want answers to the questions about the good and very well-developed project!

What do you do? Work within the organization to consistently plan for solving problems. Design projects, not in full detail, but at least on paper, with good chances for solving the problems faced every day. Keep these in a file or in notebooks on a shelf for when a grant program that matches arises.

Never start writing a grant proposal without first designing and developing a solid project about which to write. How is that done? This chapter will act as a guide. You cannot account for the subjective decisions of the reviewers, but you can be competitive each and every time you submit.

An often asked question is, "What percentage of grant proposals get funded?" The answer is a high of 20 to 25 percent for some state programs to less than one percent at many foundations. What percentage of proposals submitted by a master grantseeker get funded? The answer to that question is 30 to 40 percent.

Therefore, grantseeking is, across the board, a job at which the organization will fail more often than it succeeds. That is the way it is. How exactly does one succeed at a job when one fails more often than one succeeds?

Consider baseball. One can no doubt agree that one of the jobs of a major league baseball player is to go to the plate and get base hits. What, then, happens to a major league baseball player when he fails to do his job 7 times out of 10—70 percent of the time. Just what happens to this miserable failure?

He is called a superstar and is paid millions of dollars a year. If he does it over his entire career, the chances are strong that he will be in the Baseball Hall of Fame. He will be a legend of the game by failing to get a hit 70 percent of the time.

If a grantseeker succeeds 30 to 40 percent of the time with your grant proposals, they will be a grants superstar and will bring hundreds of thousands of dollars into their organization. If there is a secret to grantseeking, it is to follow the grantor's directions and keep submitting proposals. If the win ratio is 3 out of 10, then it can be determined exactly how many proposals to submit to receive the funds that are needed. It is really that simple. Keep doing it, and keep doing it. If three winners are needed this year, submit 10 proposals. If six winners are needed this year, submit 20 proposals. If nine winners are needed this year, submit 30 proposals. Do the math. Submit the proposals.

20.3.3 Defining the Real Problem

Grantseeking begins with a problem. It is the reason grant funds are being sought—to solve a problem. As previously stated, it is the match, the connection between the organization, as the grantseeker, and the grantmaker. Grantmakers award grants to solve problems. At the core of every grant proposal, at its heart, lies a problem statement.

What are these problems? How can a problem be defined? The dictionary says that a problem is "a question proposed for a solution." That is a start, but what exactly is a question? The same dictionary says that a question is "a problem." Hmmm—it seems we are being led in a tight little circle. Look further. Another word often used in grantseeking as a synonym for "problem" is "need." Our friend Webster says that a need is "a lack of something useful or required." A lack of something useful is an easy idea to grasp. Teen pregnancy, poverty, illiteracy, cancer, domestic violence, child abuse, unemployment—all of these clearly bespeak a lack of something useful, something required. In general, in grantseeking, a problem means a lack of something useful, be it education, motivation, food, a job, self-esteem, health, or any other such thing.

So, what is *your organization's* problem? A teacher's answer is likely to be "low test scores." A health care professional: "People are dying too young." A law enforcement officer: "We are beginning to have youth problems—vandalism is on the rise." A community planner: "Our city accident statistics show over 90 percent occurring to folks over 60." A parent: "My child is one of a big group who is in the fourth grade and still cannot read." These are tough topics all right, but they're not problems.

"Not problems?! You go tell our parents and the school board that low test scores aren't a problem! We've got our administrators, politicians, the press, and parents all over us about low test scores. Don't tell me it's not a problem!"

Low test scores certainly do cause problems. However they are really symptoms, not the problem. The real issues—the causes—of the low test scores, the causes of rising juvenile violence, the causes of high accident rates in the elderly population, and the causes of fourth graders not being able to read are something else altogether.

Back to low test scores. It takes only a few minutes with a group of educators and parents for the real causes of low test scores to emerge: lack of motivation, absenteeism, lack of parental support, poor tests, poor teachers, poor facilities, and so on. There is absolutely nothing to be done directly about low test scores. It is too broad a topic. Think of a fourth grader who has just been assigned his or her first grade school paper. With enthusiasm, the topic proudly presented to the teacher is *Animals Throughout History*. The teacher gently explains that it might have been better if one animal and one point in time had been chosen. It is the same with problem solving and with grantseeking. The more narrowed down and focused the sights are on a target, the better the project design. Low test scores will be improved by combating the cause—the real problem.

Grantmakers require that you present them with a project that solves a problem about which you have a mutual interest. The symptom is what gets the big press, the public attention. Symptoms are headline grabbers. However, the real problem, the underlying cause, is where solutions take place. As a rule of thumb, when someone asks, "What's the problem," the first answer is likely to be a symptom, not the real problem. We call these symptoms "Broad Problems" because they are very broad and general. Broad Problems are not a bad place to start project development, but they are a terrible place to end it. Following is a step-by-step process to follow, along with explanations and examples, to help you get beneath the Broad Problems to the causes, or Real Problems, which form the foundation of every fundable project.

20.3.4 What Is Your Real Problem?

Step One: List Broad Problems

Stating a problem should not be very hard. After all, they slap us in the face every day. The first exercise to do, then, is to sit quietly and write them down. Make a list, as has been done in Exhibit 20.1.

EXHIBIT 20.1   Broad Problems List

- We have a high rate of teen pregnancy in our community.
- There are too many students not finishing high school.
- Drug abuse is growing.
- The incidents of reported child abuse are increasing.
- Youth vandalism has skyrocketed.
- Domestic violence is on the rise.
- Local employers are reporting that our graduates have low reading and writing skills.
- More homeless people are gathering down at the central park.
- We actually had a few elderly people die of heat stroke (or cold, depending on where you live), shut up in their houses.
- The emergency room at the hospital is overwhelmed by all the indigent cases on the weekends.

Do any of these look familiar? They should. They are right out of your regional newspapers. Can you tackle them all? No, of course not. The problem an organization takes on is determined by the purpose and scope of your organization. There are many ways to narrow down the topic and one way is for each organization (or, better yet, group of partnering organizations) to take on one or two problems as its mission. Focus on the one that makes the most sense to you and your organization.

In hundreds of workshops across the country, the single biggest stumbling block when developing an initial problem statement is that people begin immediately to discuss the gear, the paraphernalia, of a solution.

- "We need more room."
- "We need another resident facility."
- "We've just got to get space for 10 to 15 more people."
- "We need a swimming pool."
- "Band instruments . . . we need band instruments."

All are the paraphernalia of a solution. Picture a solution as a critter—a little munchkin with a backpack. Hanging out of the backpack are all these things, all dangling down. These things will be needed somewhere along the line to help the solution along, but they are not, in and of themselves, the solution itself. They are just "things." They are tools to be used.

If the organization looks at the list and sees a group of tools, put the tools aside for later. The problem is not that you need more room, another resident facility, space for more people, a swimming pool, or band instruments. These are tools you'll need to solve the problem, but their lack is not the problem itself.

Here is an example. Some people have difficulty getting to libraries to acquire reading and other learning materials. They have no reliable transportation or they live too far away and the library is too difficult for them to access. The solution? Take the library to them. Develop a project that succeeds in bringing the library to the folks that need it but cannot get to it. To do this, several things are required. One necessity might be a van with shelving and other storage to hold library materials. A driver and other staff will be needed for the van, along with a good system for checking out, tracking, and retrieving materials. Then there is a project component devoted to spreading the word about the availability of this service and scheduling is another component. It does not take long to see how large and complex a project this will become before it is completely planned and implemented. If an organization did not know better and was asked, "What's your problem?" it might have answered, "We need a van." You see, the van is just a tool. It certainly isn't the problem, nor is it the project. It is just a tool.

Look at another solution to this same problem with another project. The problem, remember, is that folks cannot access the materials in the library because they have no reliable transportation or they live too far away and the library is too difficult for them to access. The first solution was a mobile library. However, what if we make most of the library services available using on-line communications as well as the mail and other delivery services. Our project allows people to check out materials by electronic mail and have them delivered to their doors with a return shipping slip for sending the materials back to the library. In addition, a toll-free data access capability is added for searching library databases and reading newspapers and other such electronically friendly services. Now, some of the isolated people can access the library and its materials right from their own homes. The same problem was solved, but with another project. Does one project, the mobile van, preclude the need for the other project, electronic access? No, some people will use one service and some the other.

What's your problem? If asked, would you say, "We need phone lines?" Not now, you wouldn't. You would know better!

Step Two: Choose a Broad Problem and List the Causes

Look back at your list of Broad Problems. Pick one problem on which to concentrate. List the things you think are the likely causes of the Broad Problem. One way to get at likely causes is to ask yourself, "Why?" In the case of the example in Exhibit 20.2, the question to answer might be, "Why has youth vandalism skyrocketed in our community?" Under "Likely Causes," all the answers to the question are listed.

This exercise should provide broad causes to the broad problem—think of everything possible.

EXHIBIT 20.2 Broad Problems and List of Causes

Broad Problem—Youth vandalism has skyrocketed in our community.
Likely Causes
- Too little parental supervision
- Too little community supervision
- Nothing for young people to do
- Lots of kids 13–19
- Few school or church activities
- Not many street lights or night security lights
- Many vacation homes that are empty most of time
- Not enough police officers
- No organized "watch" groups
- Some stores lax about carding young people
- Buildings very close together with lots of alleys

Step Three: Cross Out

Look at the list. There are usually several items that, though true enough, cannot be changed—at least not by the organization or not at this time. Cross through the things the organization cannot do anything about, as there is no point in worrying about things that cannot be corrected. Look at Exhibit 20.2. The fact that there are lots of teens certainly increases the likelihood of youth vandalism since there is safety in numbers, but there is not much you can do about the numbers. So cross that one off the list. You probably cannot increase the number of officers on the police force or the way the buildings are built, so those go off too. You cannot, and would not want to, keep people from having vacation homes. Draw lines through all those.

Step Four: Circle

Now look at your list. Circle any of the causes with which your organization has a direct relationship, as we did in Exhibit 20.3 (page 846). If the organization is a youth service organization, a school, a church, or a community development group, then several items on the list will be circled.

Step Five: Group

Study the causes circled in your exercise. Are they closely connected or somewhat separate? Group the things that are very closely related, as seen in Exhibit 20.4. In our example, the groupings might look like this.

In Exhibit 20.4, parent and community supervision are closely related causes, since they are both connected to observation of and attention to community young people. Activities for youth can be developed around both schools and churches, thus providing something for young people to do, so these items are associated. The

EXHIBIT 20.3   **Circling Direct Relationships**

Broad Problem—Youth vandalism has skyrocketed in our community.

Likely Causes

- Too little parental supervision
- Too little community supervision
- Nothing for young people to do
- ~~Lots of kids 13–19~~
- Few school or church activities
- Not many street lights or night security lights
- ~~Many vacation homes that are empty most of time~~
- ~~Not enough police officers~~
- No organized "watch" groups
- Some stores lax about carding young people
- ~~Buildings very close together with lots of alleys~~

EXHIBIT 20.4   **Grouping Related Causes**

Broad Problem—Youth vandalism has skyrocketed in our community.

Likely Causes

- Too little parental supervision
- Too little community supervision
- Nothing for young people to do
- Few school or church activities
- Some stores lax about carding young people

odd item is the laxness of some businesses in carding people before selling alcoholic beverages and other regulated products.

Step Six: Choose

Our example organization may, in fact, decide to work on all three groupings of causes listed. However, each grouping deserves a separate project. The more well-defined the problem, the more likely the project will be fundable. Therefore, it is important to choose a reasonably narrowly defined problem.

From the grouped list in Exhibit 20.4, choose the grouped causes: too little parental supervision and too little community supervision. Focus the project on these two related causes.

Step Seven: Specify the Target Population

Go back to an earlier point, the confusing idea of "things," the tools that are needed with the problem. One reason many organizations have a hard time letting go of the things they need so they can work directly on the real problem is that they have lost, forgotten, or misplaced their target population. It is *not* "What do I as an administrator need?" It is *not* "What do we as employees need?" It is not even "What do we as an organization need?" That's right. It *is* "What is needed to serve the target population?" It is "How can the organization do a better job for those they are serving?" In grantseeking, those the organization serves, however it serves them, are always first. Everything comes from and points to them. The organization is asking for assistance only as a means to achieve this end.

For example, an organization may want to say, "Our problem is that our staff lacks training." The way it arrived at this conclusion was by noting that staff are not providing the services or products they should and could because they do not know how. The focus is really on providing better services and products for the target population; one way to do that is to train the staff to be more effective. However, what the organization said does not reflect its thinking. It said, "We need training." "We" refers to the organization and its staff. In grant terms, "We need training" in and of itself is not fundable. In grant terms, "We're going to provide our staff with the knowledge and skills they need to adapt the physical therapy activities to accommodate children with cerebral palsy," is definitely the core of a fundable project. Note the difference. Here is another example.

"We need computers." If an organization goes to a grantmaker with that request, it will get turned down. Why? Because that statement can be responded to with, "So what." Never allow a grantmaker to think "so what" about your request.

If the request can be answered with "so what," then think through why it is made. How does it relate to the target population? "We need computers." Why? For what will they be used? Suppose it is for a class of gifted students who should be taught high-level research skills using all types of research avenues, both primary and secondary. To teach them to use all the current tools available, computers are needed. This makes sense to a grantmaker. The focus is clearly on the target population.

If an organization is typical, it serves many target populations and many combinations of populations. If you are an educator, your target populations are: the student body; the at-risk students in the fourth grade; first graders who do not have all their immunizations; gifted seniors; preschoolers from single-parent families; migrant children; children who cannot read at grade level; boys; girls; Hispanic middle-school students; Native American preschoolers . . . the list could literally go on forever.

What if you are a physical therapist? Your target populations may be: people with hand injuries; children with cerebral palsy; obese people with knee problems; construction and heavy industry workers with low back pain; women; men; children; pregnant ladies; old folks; babies; people recovering from broken bones . . . again, the list could go on and on. You also serve many target populations.

Make a list of as many of the target populations and combinations of populations that you serve as you can think of. Think of all the combinations due to a special characteristic such as ability, health problem, gift and skill, or challenge. Think of all combinations due to age, sex, or condition.

After you have made your large list, then look back at Step Six: Choose one grouping on which to concentrate for your project. Now, identify the target population for which you will design your project. Be as specific as possible.

Return to the example Broad Problem in Exhibit 20.5. Remember that our Broad Problem was "Youth vandalism has skyrocketed in our community." Several causes were listed, and the ones our organization would logically tackle were put into three groupings. Any of the three groupings could be grounds for a grants project.

For purposes of illustration, we will choose one of the groupings with two parts: (1) too little parental supervision, and (2) too little community supervision. You may choose a single stand-alone cause. We could have chosen the stand-alone cause: Some stores are lax about carding young people.

EXHIBIT 20.5 Determining the Target Population

Broad Problem — Youth vandalism has skyrocketed in our community.

Cause (Real Problem) — (1) too little parental supervision, and (2) too little community supervision.

Target Population — middle school children, average ages 12 to 14.

After you have chosen your grouping or single issue, then write down the target population for that issue. You may have several target populations listed when you are finished, or the cause may be linked clearly to just one.

In this case, we will target middle school children, average ages 12 to 14.

You have now completed the first steps in getting a grant. Here is one more review of the seven steps we have done so far before we go on.

1. List Broad Problems—we listed the Broad Problems.

2. Choose one and list causes—we chose one of our Broad Problems and listed the causes of it.

3. Cross Out—we eliminated those causes we could not do anything about.

4. Circle—we circled the causes with which our organization has a direct relationship.

5. Group—we grouped related circled causes.

6. Select—we selected one grouping of causes on which to concentrate for our project. We also call these causes the Real Problem.

7. Specify—we specified a target population for the cause or real problem we selected in step 6.

20.4 INTRODUCTION TO FINDING FUNDERS

Researching funding sources to support solutions for problems is a time-consuming activity, and it can be confusing if an organization does not know where to look. There are many avenues for finding sources but few guides to help work through the myriad types of information available.

As discussed in Section 20.3, it is very important to focus on a particular problem, then design a project to solve that problem before attempting to research funding sources. There are so many sources, all saying they support health care, education, the arts, or some other general topic, that one can be lulled into thinking that all that is necessary is to write a "give-me-letter" to each one and voilà! The money just rolls in. If it were that easy, everyone would have many programs supported with grant funds.

Grants acquisition is highly competitive. Those that succeed recognized the importance of knowing everything possible about a funder that is to be approached. However, each and every funder that purports to support education cannot be researched before submitting a proposal. It would take years to do that and a proposal would never get in the mail. The prospects can be narrowed down through savvy research.

Grants can be thought of as originating from four basic sources: the federal government, foundations, corporations or companies, and state and local sources. While many of the basics of grantseeking remain the same, each source takes a slightly different approach.

Each source funds in different ways and with different motives. Deciding which of the four grant sources to pursue is part of the matching process.

Where does this research fit in our grants acquisition process? This section provides step-by-step processes for researching and matching foundation, government, and corporate funders.

20.4.1 The Total Grants Acquisition Process

The process of grantseeking and proposal writing is complex. Most mistakes made by novice grantseekers are due to the false assumption that writing the proposal is the

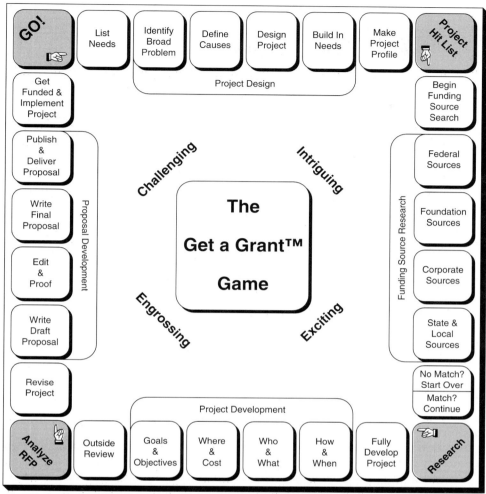

FIGURE 20.1 Grants acquisition process

focus of the process. Actually, developing something to write about is the key. In the case of a grant proposal, that "something" is a project, complete with a start, a finish, and activities in between. Writing the proposal is only 10 to 20 percent of the entire process. Figure 20.1 illustrates the grants acquisition process.

Grantseeking begins with project design—define a problem and create a project to solve it. Next comes funding source research—find a grantor with program aims that match your organization, problem, and project. Next comes project development—take your project idea and flesh it out into a fully developed project. Last comes proposal development—from your developed project, a proposal is written. Many proposals can be written from one project, but each proposal is used once for a specific grantor.

20.4.2 The Concept of Match

To have a successful, ongoing grants effort, it is important to have a number of projects in process, into the design phase, at all times. Project profiles should be kept handy for review when that request for proposal (RFP) arrives on August 1 with a deadline of August 20. With the information gathered during project design, and the knowledge of how to analyze the RFP and complete the project development, an organization should be able to meet the deadline with a competitive project and proposal.

In addition, and more important, if the entire staff is involved in project design, there is an ongoing team effort in planning and implementing changes that improve services or products for the target population. Designing the projects elicits change in and of itself as the focus changes from routine and status quo to a problem-solving atmosphere.

Finally, project profiles are needed if finding funders for projects is going to be actively pursued. The information gained in the design process is needed to find a matching funder.

What are you looking for in a match? There are many clues.

20.4.3 Match the Problem

As discussed thoroughly in Section 20.3, a project must address a problem in which the funder is intensely interested.

20.4.4 Eligibility

An organization must be eligible for funding by the grantmaker. This is both a legal and a policy issue.

Legal Issue

Both donations to and grants from a grantmaker have tax implications. Grantmakers are themselves nonprofit organizations or are for-profit corporations with a grant program that affords a tax break. The grantmaker pays no taxes on the funds granted. If the grantmaker receives donations, people making donations pay no taxes on the funds donated. As with any tax-related issue, an extensive body of law governs the process. Grantmakers must donate to organizations with the appropriate tax status or be themselves required to pay taxes on the funds they have donated.

Most organizations eligible for grant funding are classified as 501(c)(3) organizations. An organization may qualify for tax-exempt 501(c)(3) status if it is organized and operated exclusively for one or more of the following purposes: charitable, religious, educational, scientific, literary, testing for public safety, fostering national or international amateur sports competition, or the prevention of cruelty to children or animals. The organization must be a corporation, community chest, fund, or foundation to qualify.

Examples of qualifying organizations are nonprofit old-age homes, parent–teacher associations, charitable hospitals or other charitable organizations, alumni associations, schools, chapters of the Red Cross or Salvation Army, boys' and girls' clubs, and churches. A state or municipal instrumentality may qualify if it is organized as a separate entity from the governmental unit that created it and if it otherwise meets the organizational and operational tests of section 501(c)(3). Examples of qualifying instrumentalities might include state schools, universities, or hospitals.

An organization may be tax exempt, however, and not be a 501(c)(3) organization. Public schools are 509(a) organizations. They are tax exempt, but many grantmakers will not give to anything other than a 509(c)(3), thus eliminating public schools who have not set up a 501(c)(3) foundation from which to seek grant funding. If an organization is not a 501(c)(3), the advantages of becoming one should be considered.

The process has two basic steps. Step one is to create a not-for-profit corporation under the laws of the organization's state. Step two is to apply for 501(c)(3) tax-exempt status from the Internal Revenue Service.

Creating a not-for-profit corporation, in most states, is done through the state's Secretary of State's office. In some states, a lawyer must be used to create and file the appropriate paperwork. In other states, an individual may file all the paperwork without the assistance of a lawyer. If a lawyer's assistance is required, the time the lawyer spends establishing the not-for-profit corporation may become a tax-exempt donation by the lawyer to the organization once the 501(c)(3) tax-exempt status is obtained.

Application to the Internal Revenue Service for 501(c)(3) tax-exempt status is done using IRS Form 1023. The form contains instructions and checklists to help provide the information needed to process an application. Additional important information is found in IRS Publication 557, "Tax-Exempt Status for Your Organization."

Policy Issue

A grantmaker sets up policies concerning which organizations it will make awards to and which ones it will not. One grantmaker may have a policy that it will fund only religious schools, or rural health organizations, or arts programs in a certain county of a state. These policies rarely change. Some are stated clearly in the published blurb in a Foundation Center directory, and some are not so clearly stated. As a grantseeker, one must read between the lines. Look at what the grantmaker has funded. If the grantmaker says, among other things, that it funds programs in the arts, but in its five years of existence has never funded an arts program, what should that tell you? Their program should be looked at closely. If a grantmaker publishes that it funds arts programs, but on looking at the list of funded grants you notice that the only arts program listed is Picasso Bill's Art Gallery, then what chance do you have? Read between the lines.

Location. The preponderance of foundations fund programs only within a defined area of the country, usually within a city or a few counties in a state. Therefore, the closer you are to the foundation's location, the better chance you have. A number of larger foundations do fund programs nationally and internationally. The federal government funds programs nationally and internationally. Corporations are like foundations. They tend to fund programs close to their locations—programs that benefit their employees. Some corporations sponsor special nationwide competitions.

Scope and Amount. Grant funders tend to specialize in programs of a certain size and scope. The grant amounts awarded are published amounts—they are a part of the public record. One funder may tend to fund $100,000 and greater programs that are regional in scope, whereas another may fund programs from $500 to $25,000 of more highly targeted scope, serving a more specialized target population being served over a limited area of a county. An organization's project scope and the amount of money it needs should closely match the track record and published focus of the grantmaker. In other words, if the project requires a $10,000 investment, do not apply to a grantmaker who tends to fund $1,000 programs. Surprisingly, one also would not go to a grantmaker who tends to fund $100,000 programs. The organization's investment need should be closer to the average amount the funder tends to award.

20.4.5 How to Find the Match

A great deal of thought and careful study by professional grantseekers has been synthesized into steps that, if followed, should provide the information needed to effectively research a matching funder and then approach that funder to invest in your project.

Choose a Logical Starting Point

Use logical reasoning, and the information known about the project, to pick a type of funding with which to start. Here are some examples.

- If your project will serve people in a large area (a state or region), start your research will federal programs and then progress to national-level foundations.
- If your project can be a model program throughout the nation for other organizations like yours, start your research with federal programs and national-level foundations.
- If the project can only be a model program for those in your state and does not readily apply to others in the nation, start with state foundations and state and local sources.
- If the project relates specifically to improving services to your community, start with local corporations and local foundations.

Get Help

Involve your colleagues. As previously stated, finding funding sources, if done right, is time consuming, but it is one of the easiest parts of the process to pass on to a team member. Develop a research plan, and ask several people to be responsible for some part of the research. There is a place for someone who likes to surf the Net and for someone who likes to research with books. There is a place for someone who is good at follow-up and details.

School personnel can enlist the aid of parent association members, students, library–media personnel, and teachers (and their spouses), as well as interested community supporters. Health care organizations can add tasks for their volunteers, for community supporters, and for research assistants. Nonprofits can also make use of volunteer organizations and community support groups, as well as beneficiaries of their services. People are retiring young with many productive years ahead of them. Many have related expertise in management or organizational development. This wonderful community resource can be tapped into by contacting organizations of retired persons and even by running ads in the newspaper.

APPENDIX *A*

STATISTICAL TABLES

TABLE I   Random Number Table

Row #	1	2	3	4	5	6	7	8	9
1	094632795	711501513	537971597	562758635	410398128	182794408	773761503	455139927	132682754
2	033413186	653475420	289063704	485441982	460744361	328703833	289612212	569540556	620271271
3	297556368	658953044	738968017	414437050	296126017	075254187	702140315	467039889	762226273
4	472960570	785645638	574817322	817883255	976076280	843373358	118284363	445336907	327380271
5	256883707	716249997	378236162	467694224	193707682	380141891	605807481	180164558	473854769
6	179451522	878902420	602450872	987686989	686677180	242196303	517640224	691116863	275385608
7	894964682	704841116	241902107	750429362	794778197	693242123	316755091	193593484	913974355
8	738120861	744470405	873393138	758824215	394004646	496696605	006936567	163371217	727267920
9	803156944	653387115	716335974	835667154	066959782	908783760	165946696	735683921	894672507
10	187636922	953598780	481536873	055734541	493193305	566923120	435549770	007706188	839596393
11	102021077	286953643	851411058	132935798	745770831	187026467	363837178	791264282	107184709
12	734254842	133959443	113708008	443989454	786207141	772432741	682053431	048076059	617648837
13	757417865	524596578	240504889	544970942	340054233	417544234	302126745	003333205	250247568
14	227658086	233543943	487116060	577966118	524453480	934483237	367425608	431112250	536516890
15	746794759	557361146	826373105	870360802	412399571	804914923	128067420	659566961	452000520
16	235158119	336776002	728424416	086967212	040966064	335090111	985461873	832921870	461741235
17	944558037	787700710	060386364	635482046	558143223	600009181	448499754	064172342	713707601
18	658344036	271853277	275251035	744269244	877186509	130398637	367142231	846275675	485443650
19	179411521	104680475	020354893	576185422	778014690	380931445	886031872	320231466	062555147
20	865570814	699503925	628956988	683503622	276170341	744494133	081246804	523527226	198219562
21	984902072	022065717	504274676	136174524	195356906	027900159	809382340	381669407	544140648
22	139846732	496390379	582502144	768571665	177715615	830320391	105937107	329901920	618226629
23	492146502	493503513	138813631	479880385	684082619	010963692	268892703	552334849	488002392
24	739727377	916314641	263944162	861966588	286459084	491798049	760316559	837966446	371951811
25	075823838	115491339	547215506	007869049	323138362	193432798	361574944	787418390	016648846
26	915745558	104176259	349828840	546922404	266406684	490531595	336155799	242136076	641061181
27	136001350	309685268	986533618	587428568	052231831	422269870	302793461	564542482	915031158
28	691703257	926306032	988266746	716231516	519662016	986665536	993015206	066999095	731533696
29	154222042	316334873	963715901	044966315	846937935	104409586	768790545	113341348	108261519
30	983823702	641345385	203912928	219869690	208288383	861497163	149918954	160034550	759951927
31	277515731	805241329	549047139	285206828	534033122	130094940	970748730	798208639	485614463
32	566308464	543263173	711363354	339738940	051286779	714375200	698531722	072971345	762369710
33	533605194	994619567	798813607	079914804	405016946	275797011	801942743	814124918	033457635
34	374266385	237398626	014653680	107885763	848594153	093210516	751461171	121583622	388598493
35	739563239	736604709	251789737	480977217	432264262	975146204	639768152	455086460	573742841
36	539277994	536590953	293592699	279474008	525803109	281596944	199856046	646139218	124051051
37	609535944	455877404	251255783	334162937	523110770	461537085	359043224	423641232	223047648
38	254030985	962503045	747584829	988588554	631976076	901087130	961891746	149014870	557453130
39	869881662	489992047	240739861	875737562	237409010	135678267	196964820	343397286	364892121
40	935509137	382168564	659392779	628617853	533473897	569590209	333032014	937163707	460780426
41	727184789	651476235	562537081	259345598	118701307	970343244	678458590	189103174	840164700
42	076139028	276588587	329963439	184510242	904140612	761154973	175127287	520167477	242700207
43	181087637	087629199	028292894	460181436	623140518	207937371	238398056	136009368	581975565
44	514095654	401875869	095986936	976620836	391483115	713574093	679457157	184527765	553593737
45	804743163	129181317	005547120	712455031	948648814	909230622	839276576	726704039	427115217
46	492640056	038991626	749280637	430162677	414656226	603291802	983746203	756647413	333907575
47	105925664	168586200	348119547	829517480	244539107	448715108	154400559	634475954	701530403
48	688908833	510541646	706386776	935447659	798110618	127019341	889979390	455625169	283128630
49	806445894	067701203	566577304	808746117	093933115	198530698	142531634	042491555	776838859
50	249997369	047395403	102944245	987149692	239682871	971259345	193515078	797533485	459099813

TABLE 2   Binomial Probability Tables

$$P(X = x)$$

n = 5   π

x	0.05	0.10	0.20	0.25	0.30	0.40	0.50	0.60	0.70	0.75	0.80	0.90	0.95
0	0.774	0.590	0.328	0.237	0.168	0.078	0.031	0.010	0.002	0.001	0.000	0.000	0.000
1	0.204	0.328	0.410	0.396	0.360	0.259	0.156	0.077	0.028	0.015	0.006	0.000	0.000
2	0.021	0.073	0.205	0.264	0.309	0.346	0.313	0.230	0.132	0.088	0.051	0.008	0.001
3	0.001	0.008	0.051	0.088	0.132	0.230	0.313	0.346	0.309	0.264	0.205	0.073	0.021
4	0.000	0.000	0.006	0.015	0.028	0.077	0.156	0.259	0.360	0.396	0.410	0.328	0.204
5	0.000	0.000	0.000	0.001	0.002	0.010	0.031	0.078	0.168	0.237	0.328	0.590	0.774

n = 10   π

x	0.050	0.100	0.200	0.250	0.300	0.400	0.500	0.600	0.700	0.750	0.800	0.900	0.950
0	0.599	0.349	0.107	0.056	0.028	0.006	0.001	0.000	0.000	0.000	0.000	0.000	0.000
1	0.315	0.387	0.268	0.188	0.121	0.040	0.010	0.002	0.000	0.000	0.000	0.000	0.000
2	0.075	0.194	0.302	0.282	0.233	0.121	0.044	0.011	0.001	0.000	0.000	0.000	0.000
3	0.010	0.057	0.201	0.250	0.267	0.215	0.117	0.042	0.009	0.003	0.001	0.000	0.000
4	0.001	0.011	0.088	0.146	0.200	0.251	0.205	0.111	0.037	0.016	0.006	0.000	0.000
5	0.000	0.001	0.026	0.058	0.103	0.201	0.246	0.201	0.103	0.058	0.026	0.001	0.000
6	0.000	0.000	0.006	0.016	0.037	0.111	0.205	0.251	0.200	0.146	0.088	0.011	0.001
7	0.000	0.000	0.001	0.003	0.009	0.042	0.117	0.215	0.267	0.250	0.201	0.057	0.010
8	0.000	0.000	0.000	0.000	0.001	0.011	0.044	0.121	0.233	0.282	0.302	0.194	0.075
9	0.000	0.000	0.000	0.000	0.000	0.002	0.010	0.040	0.121	0.188	0.268	0.387	0.315
10	0.000	0.000	0.000	0.000	0.000	0.000	0.001	0.006	0.028	0.056	0.107	0.349	0.599

n = 15   π

x	0.050	0.100	0.200	0.250	0.300	0.400	0.500	0.600	0.700	0.750	0.800	0.900	0.950
0	0.463	0.206	0.035	0.013	0.005	0.000	0.000	0.000	0.000	0.000	0.000	0.000	0.000
1	0.366	0.343	0.132	0.067	0.031	0.005	0.000	0.000	0.000	0.000	0.000	0.000	0.000
2	0.135	0.267	0.231	0.156	0.092	0.022	0.003	0.000	0.000	0.000	0.000	0.000	0.000
3	0.031	0.129	0.250	0.225	0.170	0.063	0.014	0.002	0.000	0.000	0.000	0.000	0.000
4	0.005	0.043	0.188	0.225	0.219	0.127	0.042	0.007	0.001	0.000	0.000	0.000	0.000
5	0.001	0.010	0.103	0.165	0.206	0.186	0.092	0.024	0.003	0.001	0.000	0.000	0.000
6	0.000	0.002	0.043	0.092	0.147	0.207	0.153	0.061	0.012	0.003	0.001	0.000	0.000
7	0.000	0.000	0.014	0.039	0.081	0.177	0.196	0.118	0.035	0.013	0.003	0.000	0.000
8	0.000	0.000	0.003	0.013	0.035	0.118	0.196	0.177	0.081	0.039	0.014	0.000	0.000
9	0.000	0.000	0.001	0.003	0.012	0.061	0.153	0.207	0.147	0.092	0.043	0.002	0.000
10	0.000	0.000	0.000	0.001	0.003	0.024	0.092	0.186	0.206	0.165	0.103	0.010	0.001
11	0.000	0.000	0.000	0.000	0.001	0.007	0.042	0.127	0.219	0.225	0.188	0.043	0.005
12	0.000	0.000	0.000	0.000	0.000	0.002	0.014	0.063	0.170	0.225	0.250	0.129	0.031
13	0.000	0.000	0.000	0.000	0.000	0.000	0.003	0.022	0.092	0.156	0.231	0.267	0.135
14	0.000	0.000	0.000	0.000	0.000	0.000	0.000	0.005	0.031	0.067	0.132	0.343	0.366
15	0.000	0.000	0.000	0.000	0.000	0.000	0.000	0.000	0.005	0.013	0.035	0.206	0.463

TABLE 2 BINOMIAL PROBABILITY TABLES **A3**

n = 20

π

x	0.050	0.100	0.200	0.250	0.300	0.400	0.500	0.600	0.700	0.750	0.800	0.900	0.950
0	0.358	0.122	0.012	0.003	0.001	0.000	0.000	0.000	0.000	0.000	0.000	0.000	0.000
1	0.377	0.270	0.058	0.021	0.007	0.000	0.000	0.000	0.000	0.000	0.000	0.000	0.000
2	0.189	0.285	0.137	0.067	0.028	0.003	0.000	0.000	0.000	0.000	0.000	0.000	0.000
3	0.060	0.190	0.205	0.134	0.072	0.012	0.001	0.000	0.000	0.000	0.000	0.000	0.000
4	0.013	0.090	0.218	0.190	0.130	0.035	0.005	0.000	0.000	0.000	0.000	0.000	0.000
5	0.002	0.032	0.175	0.202	0.179	0.075	0.015	0.001	0.000	0.000	0.000	0.000	0.000
6	0.000	0.009	0.109	0.169	0.192	0.124	0.037	0.005	0.000	0.000	0.000	0.000	0.000
7	0.000	0.002	0.055	0.112	0.164	0.166	0.074	0.015	0.001	0.000	0.000	0.000	0.000
8	0.000	0.000	0.022	0.061	0.114	0.180	0.120	0.035	0.004	0.001	0.000	0.000	0.000
9	0.000	0.000	0.007	0.027	0.065	0.160	0.160	0.071	0.012	0.003	0.000	0.000	0.000
10	0.000	0.000	0.002	0.010	0.031	0.117	0.176	0.117	0.031	0.010	0.002	0.000	0.000
11	0.000	0.000	0.000	0.003	0.012	0.071	0.160	0.160	0.065	0.027	0.007	0.000	0.000
12	0.000	0.000	0.000	0.001	0.004	0.035	0.120	0.180	0.114	0.061	0.022	0.000	0.000
13	0.000	0.000	0.000	0.000	0.001	0.015	0.074	0.166	0.164	0.112	0.055	0.002	0.000
14	0.000	0.000	0.000	0.000	0.000	0.005	0.037	0.124	0.192	0.169	0.109	0.009	0.000
15	0.000	0.000	0.000	0.000	0.000	0.001	0.015	0.075	0.179	0.202	0.175	0.032	0.002
16	0.000	0.000	0.000	0.000	0.000	0.000	0.005	0.035	0.130	0.190	0.218	0.090	0.013
17	0.000	0.000	0.000	0.000	0.000	0.000	0.001	0.012	0.072	0.134	0.205	0.190	0.060
18	0.000	0.000	0.000	0.000	0.000	0.000	0.000	0.003	0.028	0.067	0.137	0.285	0.189
19	0.000	0.000	0.000	0.000	0.000	0.000	0.000	0.000	0.007	0.021	0.058	0.270	0.377
20	0.000	0.000	0.000	0.000	0.000	0.000	0.000	0.000	0.001	0.003	0.012	0.122	0.358

n = 25

π

x	0.050	0.100	0.200	0.250	0.300	0.400	0.500	0.600	0.700	0.750	0.800	0.900	0.950
0	0.277	0.072	0.004	0.001	0.000	0.000	0.000	0.000	0.000	0.000	0.000	0.000	0.000
1	0.365	0.199	0.024	0.006	0.001	0.000	0.000	0.000	0.000	0.000	0.000	0.000	0.000
2	0.231	0.266	0.071	0.025	0.007	0.000	0.000	0.000	0.000	0.000	0.000	0.000	0.000
3	0.093	0.226	0.136	0.064	0.024	0.002	0.000	0.000	0.000	0.000	0.000	0.000	0.000
4	0.027	0.138	0.187	0.118	0.057	0.007	0.000	0.000	0.000	0.000	0.000	0.000	0.000
5	0.006	0.065	0.196	0.165	0.103	0.020	0.002	0.000	0.000	0.000	0.000	0.000	0.000
6	0.001	0.024	0.163	0.183	0.147	0.044	0.005	0.000	0.000	0.000	0.000	0.000	0.000
7	0.000	0.007	0.111	0.165	0.171	0.080	0.014	0.001	0.000	0.000	0.000	0.000	0.000
8	0.000	0.002	0.062	0.124	0.165	0.120	0.032	0.003	0.000	0.000	0.000	0.000	0.000
9	0.000	0.000	0.029	0.078	0.134	0.151	0.061	0.009	0.000	0.000	0.000	0.000	0.000
10	0.000	0.000	0.012	0.042	0.092	0.161	0.097	0.021	0.001	0.000	0.000	0.000	0.000
11	0.000	0.000	0.004	0.019	0.054	0.147	0.133	0.043	0.004	0.001	0.000	0.000	0.000
12	0.000	0.000	0.001	0.007	0.027	0.114	0.155	0.076	0.011	0.002	0.000	0.000	0.000
13	0.000	0.000	0.000	0.002	0.011	0.076	0.155	0.114	0.027	0.007	0.001	0.000	0.000
14	0.000	0.000	0.000	0.001	0.004	0.043	0.133	0.147	0.054	0.019	0.004	0.000	0.000
15	0.000	0.000	0.000	0.000	0.001	0.021	0.097	0.161	0.092	0.042	0.012	0.000	0.000
16	0.000	0.000	0.000	0.000	0.000	0.009	0.061	0.151	0.134	0.078	0.029	0.000	0.000
17	0.000	0.000	0.000	0.000	0.000	0.003	0.032	0.120	0.165	0.124	0.062	0.002	0.000
18	0.000	0.000	0.000	0.000	0.000	0.001	0.014	0.080	0.171	0.165	0.111	0.007	0.000
19	0.000	0.000	0.000	0.000	0.000	0.000	0.005	0.044	0.147	0.183	0.163	0.024	0.001
20	0.000	0.000	0.000	0.000	0.000	0.000	0.002	0.020	0.103	0.165	0.196	0.065	0.006
21	0.000	0.000	0.000	0.000	0.000	0.000	0.000	0.007	0.057	0.118	0.187	0.138	0.027
22	0.000	0.000	0.000	0.000	0.000	0.000	0.000	0.002	0.024	0.064	0.136	0.226	0.093
23	0.000	0.000	0.000	0.000	0.000	0.000	0.000	0.000	0.007	0.025	0.071	0.266	0.231
24	0.000	0.000	0.000	0.000	0.000	0.000	0.000	0.000	0.001	0.006	0.024	0.199	0.365
25	0.000	0.000	0.000	0.000	0.000	0.000	0.000	0.000	0.000	0.001	0.004	0.072	0.277

TABLE 3   Standard Normal Table

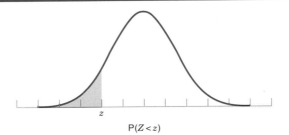

$P(Z < z)$

Second Decimal Place

z	0.00	0.01	0.02	0.03	0.04	0.05	0.06	0.07	0.08	0.09
-3.9	0.0000	0.0000	0.0000	0.0000	0.0000	0.0000	0.0000	0.0000	0.0000	0.0000
-3.8	0.0001	0.0001	0.0001	0.0001	0.0001	0.0001	0.0001	0.0001	0.0001	0.0001
-3.7	0.0001	0.0001	0.0001	0.0001	0.0001	0.0001	0.0001	0.0001	0.0001	0.0001
-3.6	0.0002	0.0002	0.0001	0.0001	0.0001	0.0001	0.0001	0.0001	0.0001	0.0001
-3.5	0.0002	0.0002	0.0002	0.0002	0.0002	0.0002	0.0002	0.0002	0.0002	0.0002
-3.4	0.0003	0.0003	0.0003	0.0003	0.0003	0.0003	0.0003	0.0003	0.0003	0.0002
-3.3	0.0005	0.0005	0.0005	0.0004	0.0004	0.0004	0.0004	0.0004	0.0004	0.0003
-3.2	0.0007	0.0007	0.0006	0.0006	0.0006	0.0006	0.0006	0.0005	0.0005	0.0005
-3.1	0.0010	0.0009	0.0009	0.0009	0.0008	0.0008	0.0008	0.0008	0.0007	0.0007
-3.0	0.0013	0.0013	0.0013	0.0012	0.0012	0.0011	0.0011	0.0011	0.0010	0.0010
-2.9	0.0019	0.0018	0.0018	0.0017	0.0016	0.0016	0.0015	0.0015	0.0014	0.0014
-2.8	0.0026	0.0025	0.0024	0.0023	0.0023	0.0022	0.0021	0.0021	0.0020	0.0019
-2.7	0.0035	0.0034	0.0033	0.0032	0.0031	0.0030	0.0029	0.0028	0.0027	0.0026
-2.6	0.0047	0.0045	0.0044	0.0043	0.0041	0.0040	0.0039	0.0038	0.0037	0.0036
-2.5	0.0062	0.0060	0.0059	0.0057	0.0055	0.0054	0.0052	0.0051	0.0049	0.0048
-2.4	0.0082	0.0080	0.0078	0.0075	0.0073	0.0071	0.0069	0.0068	0.0066	0.0064
-2.3	0.0107	0.0104	0.0102	0.0099	0.0096	0.0094	0.0091	0.0089	0.0087	0.0084
-2.2	0.0139	0.0136	0.0132	0.0129	0.0125	0.0122	0.0119	0.0116	0.0113	0.0110
-2.1	0.0179	0.0174	0.0170	0.0166	0.0162	0.0158	0.0154	0.0150	0.0146	0.0143
-2.0	0.0228	0.0222	0.0217	0.0212	0.0207	0.0202	0.0197	0.0192	0.0188	0.0183
-1.9	0.0287	0.0281	0.0274	0.0268	0.0262	0.0256	0.0250	0.0244	0.0239	0.0233
-1.8	0.0359	0.0351	0.0344	0.0336	0.0329	0.0322	0.0314	0.0307	0.0301	0.0294
-1.7	0.0446	0.0436	0.0427	0.0418	0.0409	0.0401	0.0392	0.0384	0.0375	0.0367
-1.6	0.0548	0.0537	0.0526	0.0516	0.0505	0.0495	0.0485	0.0475	0.0465	0.0455
-1.5	0.0668	0.0655	0.0643	0.0630	0.0618	0.0606	0.0594	0.0582	0.0571	0.0559
-1.4	0.0808	0.0793	0.0778	0.0764	0.0749	0.0735	0.0721	0.0708	0.0694	0.0681
-1.3	0.0968	0.0951	0.0934	0.0918	0.0901	0.0885	0.0869	0.0853	0.0838	0.0823
-1.2	0.1151	0.1131	0.1112	0.1093	0.1075	0.1056	0.1038	0.1020	0.1003	0.0985
-1.1	0.1357	0.1335	0.1314	0.1292	0.1271	0.1251	0.1230	0.1210	0.1190	0.1170
-1.0	0.1587	0.1562	0.1539	0.1515	0.1492	0.1469	0.1446	0.1423	0.1401	0.1379
-0.9	0.1841	0.1814	0.1788	0.1762	0.1736	0.1711	0.1685	0.1660	0.1635	0.1611
-0.8	0.2119	0.2090	0.2061	0.2033	0.2005	0.1977	0.1949	0.1922	0.1894	0.1867
-0.7	0.2420	0.2389	0.2358	0.2327	0.2296	0.2266	0.2236	0.2206	0.2177	0.2148
-0.6	0.2743	0.2709	0.2676	0.2643	0.2611	0.2578	0.2546	0.2514	0.2483	0.2451
-0.5	0.3085	0.3050	0.3015	0.2981	0.2946	0.2912	0.2877	0.2843	0.2810	0.2776
-0.4	0.3446	0.3409	0.3372	0.3336	0.3300	0.3264	0.3228	0.3192	0.3156	0.3121
-0.3	0.3821	0.3783	0.3745	0.3707	0.3669	0.3632	0.3594	0.3557	0.3520	0.3483
-0.2	0.4207	0.4168	0.4129	0.4090	0.4052	0.4013	0.3974	0.3936	0.3897	0.3859
-0.1	0.4602	0.4562	0.4522	0.4483	0.4443	0.4404	0.4364	0.4325	0.4286	0.4247
0.0	0.5000	0.4960	0.4920	0.4880	0.4840	0.4801	0.4761	0.4721	0.4681	0.4641

TABLE 3   STANDARD NORMAL TABLE   **A5**

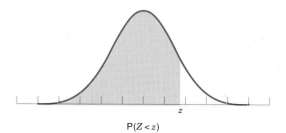

P(Z < z)

Second Decimal Place

z	0.00	0.01	0.02	0.03	0.04	0.05	0.06	0.07	0.08	0.09
0.0	0.5000	0.5040	0.5080	0.5120	0.5160	0.5199	0.5239	0.5279	0.5319	0.5359
0.1	0.5398	0.5438	0.5478	0.5517	0.5557	0.5596	0.5636	0.5675	0.5714	0.5753
0.2	0.5793	0.5832	0.5871	0.5910	0.5948	0.5987	0.6026	0.6064	0.6103	0.6141
0.3	0.6179	0.6217	0.6255	0.6293	0.6331	0.6368	0.6406	0.6443	0.6480	0.6517
0.4	0.6554	0.6591	0.6628	0.6664	0.6700	0.6736	0.6772	0.6808	0.6844	0.6879
0.5	0.6915	0.6950	0.6985	0.7019	0.7054	0.7088	0.7123	0.7157	0.7190	0.7224
0.6	0.7257	0.7291	0.7324	0.7357	0.7389	0.7422	0.7454	0.7486	0.7517	0.7549
0.7	0.7580	0.7611	0.7642	0.7673	0.7704	0.7734	0.7764	0.7794	0.7823	0.7852
0.8	0.7881	0.7910	0.7939	0.7967	0.7995	0.8023	0.8051	0.8078	0.8106	0.8133
0.9	0.8159	0.8186	0.8212	0.8238	0.8264	0.8289	0.8315	0.8340	0.8365	0.8389
1.0	0.8413	0.8438	0.8461	0.8485	0.8508	0.8531	0.8554	0.8577	0.8599	0.8621
1.1	0.8643	0.8665	0.8686	0.8708	0.8729	0.8749	0.8770	0.8790	0.8810	0.8830
1.2	0.8849	0.8869	0.8888	0.8907	0.8925	0.8944	0.8962	0.8980	0.8997	0.9015
1.3	0.9032	0.9049	0.9066	0.9082	0.9099	0.9115	0.9131	0.9147	0.9162	0.9177
1.4	0.9192	0.9207	0.9222	0.9236	0.9251	0.9265	0.9279	0.9292	0.9306	0.9319
1.5	0.9332	0.9345	0.9357	0.9370	0.9382	0.9394	0.9406	0.9418	0.9429	0.9441
1.6	0.9452	0.9463	0.9474	0.9484	0.9495	0.9505	0.9515	0.9525	0.9535	0.9545
1.7	0.9554	0.9564	0.9573	0.9582	0.9591	0.9599	0.9608	0.9616	0.9625	0.9633
1.8	0.9641	0.9649	0.9656	0.9664	0.9671	0.9678	0.9686	0.9693	0.9699	0.9706
1.9	0.9713	0.9719	0.9726	0.9732	0.9738	0.9744	0.9750	0.9756	0.9761	0.9767
2.0	0.9772	0.9778	0.9783	0.9788	0.9793	0.9798	0.9803	0.9808	0.9812	0.9817
2.1	0.9821	0.9826	0.9830	0.9834	0.9838	0.9842	0.9846	0.9850	0.9854	0.9857
2.2	0.9861	0.9864	0.9868	0.9871	0.9875	0.9878	0.9881	0.9884	0.9887	0.9890
2.3	0.9893	0.9896	0.9898	0.9901	0.9904	0.9906	0.9909	0.9911	0.9913	0.9916
2.4	0.9918	0.9920	0.9922	0.9925	0.9927	0.9929	0.9931	0.9932	0.9934	0.9936
2.5	0.9938	0.9940	0.9941	0.9943	0.9945	0.9946	0.9948	0.9949	0.9951	0.9952
2.6	0.9953	0.9955	0.9956	0.9957	0.9959	0.9960	0.9961	0.9962	0.9963	0.9964
2.7	0.9965	0.9966	0.9967	0.9968	0.9969	0.9970	0.9971	0.9972	0.9973	0.9974
2.8	0.9974	0.9975	0.9976	0.9977	0.9977	0.9978	0.9979	0.9979	0.9980	0.9981
2.9	0.9981	0.9982	0.9982	0.9983	0.9984	0.9984	0.9985	0.9985	0.9986	0.9986
3.0	0.9987	0.9987	0.9987	0.9988	0.9988	0.9989	0.9989	0.9989	0.9990	0.9990
3.1	0.9990	0.9991	0.9991	0.9991	0.9992	0.9992	0.9992	0.9992	0.9993	0.9993
3.2	0.9993	0.9993	0.9994	0.9994	0.9994	0.9994	0.9994	0.9995	0.9995	0.9995
3.3	0.9995	0.9995	0.9995	0.9996	0.9996	0.9996	0.9996	0.9996	0.9996	0.9997
3.4	0.9997	0.9997	0.9997	0.9997	0.9997	0.9997	0.9997	0.9997	0.9997	0.9998
3.5	0.9998	0.9998	0.9998	0.9998	0.9998	0.9998	0.9998	0.9998	0.9998	0.9998
3.6	0.9998	0.9998	0.9999	0.9999	0.9999	0.9999	0.9999	0.9999	0.9999	0.9999
3.7	0.9999	0.9999	0.9999	0.9999	0.9999	0.9999	0.9999	0.9999	0.9999	0.9999
3.8	0.9999	0.9999	0.9999	0.9999	0.9999	0.9999	0.9999	0.9999	0.9999	0.9999
3.9	1.0000	1.0000	1.0000	1.0000	1.0000	1.0000	1.0000	1.0000	1.0000	1.0000

TABLE 4   *t* Critical Values

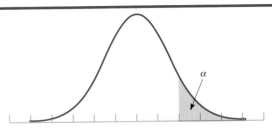

N-1

Degrees of Freedom	Upper Tail Probability (α)								
	0.15	0.10	0.05	0.025	0.015	0.01	0.005	0.001	0.0005
1	1.963	3.078	6.314	12.706	21.205	31.821	63.657	318.309	1273.155
2	1.386	1.886	2.920	4.303	5.643	6.965	9.925	22.327	44.703
3	1.250	1.638	2.353	3.182	3.896	4.541	5.841	10.215	16.326
4	1.190	1.533	2.132	2.776	3.298	3.747	4.604	7.173	10.305
5	1.156	1.476	2.015	2.571	3.003	3.365	4.032	5.893	7.976
6	1.134	1.440	1.943	2.447	2.829	3.143	3.707	5.208	6.788
7	1.119	1.415	1.895	2.365	2.715	2.998	3.499	4.785	6.082
8	1.108	1.397	1.860	2.306	2.634	2.896	3.355	4.501	5.617
9	1.100	1.383	1.833	2.262	2.574	2.821	3.250	4.297	5.291
10	1.093	1.372	1.812	2.228	2.527	2.764	3.169	4.144	5.049
11	1.088	1.363	1.796	2.201	2.491	2.718	3.106	4.025	4.863
12	1.083	1.356	1.782	2.179	2.461	2.681	3.055	3.930	4.717
13	1.079	1.350	1.771	2.160	2.436	2.650	3.012	3.852	4.597
14	1.076	1.345	1.761	2.145	2.415	2.625	2.977	3.787	4.499
15	1.074	1.341	1.753	2.131	2.397	2.602	2.947	3.733	4.417
16	1.071	1.337	1.746	2.120	2.382	2.583	2.921	3.686	4.346
17	1.069	1.333	1.740	2.110	2.368	2.567	2.898	3.646	4.286
18	1.067	1.330	1.734	2.101	2.356	2.552	2.878	3.611	4.233
19	1.066	1.328	1.729	2.093	2.346	2.539	2.861	3.579	4.187
20	1.064	1.325	1.725	2.086	2.336	2.528	2.845	3.552	4.146
21	1.063	1.323	1.721	2.080	2.328	2.518	2.831	3.527	4.109
22	1.061	1.321	1.717	2.074	2.320	2.508	2.819	3.505	4.077
23	1.060	1.319	1.714	2.069	2.313	2.500	2.807	3.485	4.047
24	1.059	1.318	1.711	2.064	2.307	2.492	2.797	3.467	4.021
25	1.058	1.316	1.708	2.060	2.301	2.485	2.787	3.450	3.997
26	1.058	1.315	1.706	2.056	2.296	2.479	2.779	3.435	3.974
27	1.057	1.314	1.703	2.052	2.291	2.473	2.771	3.421	3.954
28	1.056	1.313	1.701	2.048	2.286	2.467	2.763	3.408	3.935
29	1.055	1.311	1.699	2.045	2.282	2.462	2.756	3.396	3.918
30	1.055	1.310	1.697	2.042	2.278	2.457	2.750	3.385	3.902
40	1.050	1.303	1.684	2.021	2.250	2.423	2.704	3.307	3.788
50	1.047	1.299	1.676	2.009	2.234	2.403	2.678	3.261	3.723
60	1.045	1.296	1.671	2.000	2.223	2.390	2.660	3.232	3.681
120	1.041	1.289	1.658	1.980	2.196	2.358	2.617	3.160	3.578
Z critical value									
	1.036	1.282	1.645	1.960	2.170	2.326	2.576	3.090	3.290
Level of Significance for a one-tailed test	0.15	0.10	0.05	0.025	0.015	0.01	0.005	0.001	0.0005
Level of Significance for a two-tailed test	0.30	0.20	0.10	0.05	0.03	0.02	0.01	0.002	0.001

TABLE 5 CHI-SQUARE DISTRIBUTION TABLE A7

TABLE 5 Chi-Square Distribution Table

The entries in this table give
the critical values of χ^2 for the
specified number of degrees of freedom
and areas in the right tail.

Degrees of Freedom	Upper Tail Areas											
	0.995	0.99	0.975	0.95	0.9	0.75	0.25	0.1	0.05	0.025	0.01	0.005
1	0.000	0.000	0.001	0.004	0.016	0.102	1.323	2.706	3.841	5.024	6.635	7.879
2	0.010	0.020	0.051	0.103	0.211	0.575	2.773	4.605	5.991	7.378	9.210	10.597
3	0.072	0.115	0.216	0.352	0.584	1.213	4.108	6.251	7.815	9.348	11.345	12.838
4	0.207	0.297	0.484	0.711	1.064	1.923	5.385	7.779	9.488	11.143	13.277	14.860
5	0.412	0.554	0.831	1.145	1.610	2.675	6.626	9.236	11.070	12.832	15.086	16.750
6	0.676	0.872	1.237	1.635	2.204	3.455	7.841	10.645	12.592	14.449	16.812	18.548
7	0.989	1.239	1.690	2.167	2.833	4.255	9.037	12.017	14.067	16.013	18.475	20.278
8	1.344	1.647	2.180	2.733	3.490	5.071	10.219	13.362	15.507	17.535	20.090	21.955
9	1.735	2.088	2.700	3.325	4.168	5.899	11.389	14.684	16.919	19.023	21.666	23.589
10	2.156	2.558	3.247	3.940	4.865	6.737	12.549	15.987	18.307	20.483	23.209	25.188
11	2.603	3.053	3.816	4.575	5.578	7.584	13.701	17.275	19.675	21.920	24.725	26.757
12	3.074	3.571	4.404	5.226	6.304	8.438	14.845	18.549	21.026	23.337	26.217	28.300
13	3.565	4.107	5.009	5.892	7.041	9.299	15.984	19.812	22.362	24.736	27.688	29.819
14	4.075	4.660	5.629	6.571	7.790	10.165	17.117	21.064	23.685	26.119	29.141	31.319
15	4.601	5.229	6.262	7.261	8.547	11.037	18.245	22.307	24.996	27.488	30.578	32.801
16	5.142	5.812	6.908	7.962	9.312	11.912	19.369	23.542	26.296	28.845	32.000	34.267
17	5.697	6.408	7.564	8.672	10.085	12.792	20.489	24.769	27.587	30.191	33.409	35.718
18	6.265	7.015	8.231	9.390	10.865	13.675	21.605	25.989	28.869	31.526	34.805	37.156
19	6.844	7.633	8.907	10.117	11.651	14.562	22.718	27.204	30.144	32.852	36.191	38.582
20	7.434	8.260	9.591	10.851	12.443	15.452	23.828	28.412	31.410	34.170	37.566	39.997
21	8.034	8.897	10.283	11.591	13.240	16.344	24.935	29.615	32.671	35.479	38.932	41.401
22	8.643	9.542	10.982	12.338	14.041	17.240	26.039	30.813	33.924	36.781	40.289	42.796
23	9.260	10.196	11.689	13.091	14.848	18.137	27.141	32.007	35.172	38.076	41.638	44.181
24	9.886	10.856	12.401	13.848	15.659	19.037	28.241	33.196	36.415	39.364	42.980	45.558
25	10.520	11.524	13.120	14.611	16.473	19.939	29.339	34.382	37.652	40.646	44.314	46.928
26	11.160	12.198	13.844	15.379	17.292	20.843	30.435	35.563	38.885	41.923	45.642	48.290
27	11.808	12.878	14.573	16.151	18.114	21.749	31.528	36.741	40.113	43.195	46.963	49.645
28	12.461	13.565	15.308	16.928	18.939	22.657	32.620	37.916	41.337	44.461	48.278	50.994
29	13.121	14.256	16.047	17.708	19.768	23.567	33.711	39.087	42.557	45.722	49.588	52.335
30	13.787	14.953	16.791	18.493	20.599	24.478	34.800	40.256	43.773	46.979	50.892	53.672
31	14.458	15.655	17.539	19.281	21.434	25.390	35.887	41.422	44.985	48.232	52.191	55.002
32	15.134	16.362	18.291	20.072	22.271	26.304	36.973	42.585	46.194	49.480	53.486	56.328
33	15.815	17.073	19.047	20.867	23.110	27.219	38.058	43.745	47.400	50.725	54.775	57.648
34	16.501	17.789	19.806	21.664	23.952	28.136	39.141	44.903	48.602	51.966	56.061	58.964
35	17.192	18.509	20.569	22.465	24.797	29.054	40.223	46.059	49.802	53.203	57.342	60.275
40	20.707	22.164	24.433	26.509	29.051	33.660	45.616	51.805	55.758	59.342	63.691	66.766
60	35.534	37.485	40.482	43.188	46.459	52.294	66.981	74.397	79.082	83.298	88.379	91.952
120	83.852	86.923	91.573	95.705	100.624	109.220	130.055	140.233	146.567	152.211	158.950	163.648

TABLE 6   The *F* Distribution Table

a. Area in the Right Tail under the *F* Distribution Curve = .01

Degrees of Freedom - Numerator

Degrees of Freedom Denominator	1	2	3	4	5	6	7	8	9	10	11	12	13	14	15	16	17
1	4052.185	4999.340	5403.534	5624.257	5763.955	5858.950	5928.334	5980.954	6022.397	6055.925	6083.399	6106.682	6125.774	6143.004	6156.974	6170.012	6181.188
2	98.502	99.000	99.164	99.251	99.302	99.331	99.357	99.375	99.390	99.397	99.408	99.419	99.422	99.426	99.433	99.437	99.441
3	34.116	30.816	29.457	28.710	28.237	27.911	27.671	27.489	27.345	27.228	27.132	27.052	26.983	26.924	26.872	26.826	26.786
4	21.198	18.000	16.694	15.977	15.522	15.207	14.976	14.799	14.659	14.546	14.452	14.374	14.306	14.249	14.198	14.154	14.114
5	16.258	13.274	12.060	11.392	10.967	10.672	10.456	10.289	10.158	10.051	9.963	9.888	9.825	9.770	9.722	9.680	9.643
6	13.745	10.925	9.780	9.148	8.746	8.466	8.260	8.102	7.976	7.874	7.790	7.718	7.657	7.605	7.559	7.519	7.483
7	12.246	9.547	8.451	7.847	7.460	7.191	6.993	6.840	6.719	6.620	6.538	6.469	6.410	6.359	6.314	6.275	6.240
8	11.259	8.649	7.591	7.006	6.632	6.371	6.178	6.029	5.911	5.814	5.734	5.667	5.609	5.559	5.515	5.477	5.442
9	10.562	8.022	6.992	6.422	6.057	5.802	5.613	5.467	5.351	5.257	5.178	5.111	5.055	5.005	4.962	4.924	4.890
10	10.044	7.559	6.552	5.994	5.636	5.386	5.200	5.057	4.942	4.849	4.772	4.706	4.650	4.601	4.558	4.520	4.487
11	9.646	7.206	6.217	5.668	5.316	5.069	4.886	4.744	4.632	4.539	4.462	4.397	4.342	4.293	4.251	4.213	4.180
12	9.330	6.927	5.953	5.412	5.064	4.821	4.640	4.499	4.388	4.296	4.220	4.155	4.100	4.052	4.010	3.972	3.939
13	9.074	6.701	5.739	5.205	4.862	4.620	4.441	4.302	4.191	4.100	4.025	3.960	3.905	3.857	3.815	3.778	3.745
14	8.862	6.515	5.564	5.035	4.695	4.456	4.278	4.140	4.030	3.939	3.864	3.800	3.745	3.698	3.656	3.619	3.586
15	8.683	6.359	5.417	4.893	4.556	4.318	4.142	4.004	3.895	3.805	3.730	3.666	3.612	3.564	3.522	3.485	3.452
16	8.531	6.226	5.292	4.773	4.437	4.202	4.026	3.890	3.780	3.691	3.616	3.553	3.498	3.451	3.409	3.372	3.339
17	8.400	6.112	5.185	4.669	4.336	4.101	3.927	3.791	3.682	3.593	3.518	3.455	3.401	3.353	3.312	3.275	3.242
18	8.285	6.013	5.092	4.579	4.248	4.015	3.841	3.705	3.597	3.508	3.434	3.371	3.316	3.269	3.227	3.190	3.158
19	8.185	5.926	5.010	4.500	4.171	3.939	3.765	3.631	3.523	3.434	3.360	3.297	3.242	3.195	3.153	3.116	3.084
20	8.096	5.849	4.938	4.431	4.103	3.871	3.699	3.564	3.457	3.368	3.294	3.231	3.177	3.130	3.088	3.051	3.018
21	8.017	5.780	4.874	4.369	4.042	3.812	3.640	3.506	3.398	3.310	3.236	3.173	3.119	3.072	3.030	2.993	2.960
22	7.945	5.719	4.817	4.313	3.988	3.758	3.587	3.453	3.346	3.258	3.184	3.121	3.067	3.019	2.978	2.941	2.908
23	7.881	5.664	4.765	4.264	3.939	3.710	3.539	3.406	3.299	3.211	3.137	3.074	3.020	2.973	2.931	2.894	2.861
24	7.823	5.614	4.718	4.218	3.895	3.667	3.496	3.363	3.256	3.168	3.094	3.032	2.977	2.930	2.889	2.852	2.819
25	7.770	5.568	4.675	4.177	3.855	3.627	3.457	3.324	3.217	3.129	3.056	2.993	2.939	2.892	2.850	2.813	2.780
26	7.721	5.526	4.637	4.140	3.818	3.591	3.421	3.288	3.182	3.094	3.021	2.958	2.904	2.857	2.815	2.778	2.745
27	7.677	5.488	4.601	4.106	3.785	3.558	3.388	3.256	3.149	3.062	2.988	2.926	2.872	2.824	2.783	2.746	2.713
28	7.636	5.453	4.568	4.074	3.754	3.528	3.358	3.226	3.120	3.032	2.959	2.896	2.842	2.795	2.753	2.716	2.683
29	7.598	5.420	4.538	4.045	3.725	3.499	3.330	3.198	3.092	3.005	2.931	2.868	2.814	2.767	2.726	2.689	2.656
30	7.562	5.390	4.510	4.018	3.699	3.473	3.305	3.173	3.067	2.979	2.906	2.843	2.789	2.742	2.700	2.663	2.630
40	7.314	5.178	4.313	3.828	3.514	3.291	3.124	2.993	2.888	2.801	2.727	2.665	2.611	2.563	2.522	2.484	2.451
60	7.077	4.977	4.126	3.649	3.339	3.119	2.953	2.823	2.718	2.632	2.559	2.496	2.442	2.394	2.352	2.315	2.281
120	6.851	4.787	3.949	3.480	3.174	2.956	2.792	2.663	2.559	2.472	2.399	2.336	2.282	2.234	2.191	2.154	2.119

TABLE 6   THE *F* DISTRIBUTION TABLE   A9

TABLE 6   *(Continued)*

Degrees of Freedom - Numerator

Degrees of Freedom Denominator	18	19	20	21	22	23	24	25	26	27	28	29	30	40	60	120
1	6191.432	6200.746	6208.662	6216.113	6223.097	6228.685	6234.273	6239.861	6244.518	6249.174	6252.900	6257.091	6260.350	6286.427	6312.970	6339.513
2	99.444	99.448	99.448	99.451	99.455	99.455	99.455	99.459	99.462	99.462	99.462	99.462	99.466	99.477	99.484	99.491
3	26.751	26.719	26.690	26.664	26.639	26.617	26.597	26.579	26.562	26.546	26.531	26.517	26.504	26.411	26.316	26.221
4	14.079	14.048	14.019	13.994	13.970	13.949	13.929	13.911	13.894	13.878	13.864	13.850	13.838	13.745	13.652	13.558
5	9.609	9.580	9.553	9.528	9.506	9.485	9.466	9.449	9.433	9.418	9.404	9.391	9.379	9.291	9.202	9.112
6	7.451	7.422	7.396	7.372	7.351	7.331	7.313	7.296	7.281	7.266	7.253	7.240	7.229	7.143	7.057	6.969
7	6.209	6.181	6.155	6.132	6.111	6.092	6.074	6.058	6.043	6.029	6.016	6.003	5.992	5.908	5.824	5.737
8	5.412	5.384	5.359	5.336	5.316	5.297	5.279	5.263	5.248	5.234	5.221	5.209	5.198	5.116	5.032	4.946
9	4.860	4.833	4.808	4.786	4.765	4.746	4.729	4.713	4.698	4.684	4.672	4.660	4.649	4.567	4.483	4.398
10	4.457	4.430	4.405	4.383	4.363	4.344	4.327	4.311	4.296	4.283	4.270	4.258	4.247	4.165	4.082	3.996
11	4.150	4.123	4.099	4.077	4.057	4.038	4.021	4.005	3.990	3.977	3.964	3.952	3.941	3.860	3.776	3.690
12	3.910	3.883	3.858	3.836	3.816	3.798	3.780	3.765	3.750	3.736	3.724	3.712	3.701	3.619	3.535	3.449
13	3.716	3.689	3.665	3.643	3.622	3.604	3.587	3.571	3.556	3.543	3.530	3.518	3.507	3.425	3.341	3.255
14	3.556	3.529	3.505	3.483	3.463	3.444	3.427	3.412	3.397	3.383	3.371	3.359	3.348	3.266	3.181	3.094
15	3.423	3.396	3.372	3.350	3.330	3.311	3.294	3.278	3.264	3.250	3.237	3.225	3.214	3.132	3.047	2.959
16	3.310	3.283	3.259	3.237	3.216	3.198	3.181	3.165	3.150	3.137	3.124	3.112	3.101	3.018	2.933	2.845
17	3.212	3.186	3.162	3.139	3.119	3.101	3.083	3.068	3.053	3.039	3.026	3.014	3.003	2.920	2.835	2.746
18	3.128	3.101	3.077	3.055	3.035	3.016	2.999	2.983	2.968	2.955	2.942	2.930	2.919	2.835	2.749	2.660
19	3.054	3.027	3.003	2.981	2.961	2.942	2.925	2.909	2.894	2.880	2.868	2.855	2.844	2.761	2.674	2.584
20	2.989	2.962	2.938	2.916	2.895	2.877	2.859	2.843	2.829	2.815	2.802	2.790	2.778	2.695	2.608	2.517
21	2.931	2.904	2.880	2.857	2.837	2.818	2.801	2.785	2.770	2.756	2.743	2.731	2.720	2.636	2.548	2.457
22	2.879	2.852	2.827	2.805	2.785	2.766	2.749	2.733	2.718	2.704	2.691	2.679	2.667	2.583	2.495	2.403
23	2.832	2.805	2.780	2.758	2.738	2.719	2.702	2.686	2.671	2.657	2.644	2.632	2.620	2.536	2.447	2.354
24	2.789	2.762	2.738	2.716	2.695	2.676	2.659	2.643	2.628	2.614	2.601	2.589	2.577	2.492	2.403	2.310
25	2.751	2.724	2.699	2.677	2.657	2.638	2.620	2.604	2.589	2.575	2.562	2.550	2.538	2.453	2.364	2.270
26	2.715	2.688	2.664	2.642	2.621	2.602	2.585	2.569	2.554	2.540	2.526	2.514	2.503	2.417	2.327	2.233
27	2.683	2.656	2.632	2.609	2.589	2.570	2.552	2.536	2.521	2.507	2.494	2.481	2.470	2.384	2.294	2.198
28	2.653	2.626	2.602	2.579	2.559	2.540	2.522	2.506	2.491	2.477	2.464	2.451	2.440	2.354	2.263	2.167
29	2.626	2.599	2.574	2.552	2.531	2.512	2.495	2.478	2.463	2.449	2.436	2.423	2.412	2.325	2.234	2.138
30	2.600	2.573	2.549	2.526	2.506	2.487	2.469	2.453	2.437	2.423	2.410	2.398	2.386	2.299	2.208	2.111
40	2.421	2.394	2.369	2.346	2.325	2.306	2.288	2.271	2.256	2.241	2.228	2.215	2.203	2.114	2.019	1.917
60	2.251	2.223	2.198	2.175	2.153	2.134	2.115	2.098	2.083	2.068	2.054	2.041	2.028	1.936	1.836	1.726
120	2.089	2.060	2.035	2.011	1.989	1.969	1.950	1.932	1.916	1.901	1.886	1.873	1.860	1.763	1.656	1.533

TABLE 6   (Continued)

b. Area in the Right Tail under the F Distribution Curve = .05

Degrees of Freedom Denominator	\multicolumn{17}{c}{Degrees of Freedom - Numerator}
	1	2	3	4	5	6	7	8	9	10	11	12	13	14	15	16	17
1	161.446	199.499	215.707	224.583	230.160	233.988	236.767	238.884	240.543	241.882	242.981	243.905	244.690	245.363	245.949	246.466	246.917
2	18.513	19.000	19.164	19.247	19.296	19.329	19.353	19.371	19.385	19.396	19.405	19.412	19.419	19.424	19.429	19.433	19.437
3	10.128	9.552	9.277	9.117	9.013	8.941	8.887	8.845	8.812	8.785	8.763	8.745	8.729	8.715	8.703	8.692	8.683
4	7.709	6.944	6.591	6.388	6.256	6.163	6.094	6.041	5.999	5.964	5.936	5.912	5.891	5.873	5.858	5.844	5.832
5	6.608	5.786	5.409	5.192	5.050	4.950	4.876	4.818	4.772	4.735	4.704	4.678	4.655	4.636	4.619	4.604	4.590
6	5.987	5.143	4.757	4.534	4.387	4.284	4.207	4.147	4.099	4.060	4.027	4.000	3.976	3.956	3.938	3.922	3.908
7	5.591	4.737	4.347	4.120	3.972	3.866	3.787	3.726	3.677	3.637	3.603	3.575	3.550	3.529	3.511	3.494	3.480
8	5.318	4.459	4.066	3.838	3.688	3.581	3.500	3.438	3.388	3.347	3.313	3.284	3.259	3.237	3.218	3.202	3.187
9	5.117	4.256	3.863	3.633	3.482	3.374	3.293	3.230	3.179	3.137	3.102	3.073	3.048	3.025	3.006	2.989	2.974
10	4.965	4.103	3.708	3.478	3.326	3.217	3.135	3.072	3.020	2.978	2.943	2.913	2.887	2.865	2.845	2.828	2.812
11	4.844	3.982	3.587	3.357	3.204	3.095	3.012	2.948	2.896	2.854	2.818	2.788	2.761	2.739	2.719	2.701	2.685
12	4.747	3.885	3.490	3.259	3.106	2.996	2.913	2.849	2.796	2.753	2.717	2.687	2.660	2.637	2.617	2.599	2.583
13	4.667	3.806	3.411	3.179	3.025	2.915	2.832	2.767	2.714	2.671	2.635	2.604	2.577	2.554	2.533	2.515	2.499
14	4.600	3.739	3.344	3.112	2.958	2.848	2.764	2.699	2.646	2.602	2.565	2.534	2.507	2.484	2.463	2.445	2.428
15	4.543	3.682	3.287	3.056	2.901	2.790	2.707	2.641	2.588	2.544	2.507	2.475	2.448	2.424	2.403	2.385	2.368
16	4.494	3.634	3.239	3.007	2.852	2.741	2.657	2.591	2.538	2.494	2.456	2.425	2.397	2.373	2.352	2.333	2.317
17	4.451	3.592	3.197	2.965	2.810	2.699	2.614	2.548	2.494	2.450	2.413	2.381	2.353	2.329	2.308	2.289	2.272
18	4.414	3.555	3.160	2.928	2.773	2.661	2.577	2.510	2.456	2.412	2.374	2.342	2.314	2.290	2.269	2.250	2.233
19	4.381	3.522	3.127	2.895	2.740	2.628	2.544	2.477	2.423	2.378	2.340	2.308	2.280	2.256	2.234	2.215	2.198
20	4.351	3.493	3.098	2.866	2.711	2.599	2.514	2.447	2.393	2.348	2.310	2.278	2.250	2.225	2.203	2.184	2.167
21	4.325	3.467	3.072	2.840	2.685	2.573	2.488	2.420	2.366	2.321	2.283	2.250	2.222	2.197	2.176	2.156	2.139
22	4.301	3.443	3.049	2.817	2.661	2.549	2.464	2.397	2.342	2.297	2.259	2.226	2.198	2.173	2.151	2.131	2.114
23	4.279	3.422	3.028	2.796	2.640	2.528	2.442	2.375	2.320	2.275	2.236	2.204	2.175	2.150	2.128	2.109	2.091
24	4.260	3.403	3.009	2.776	2.621	2.508	2.423	2.355	2.300	2.255	2.216	2.183	2.155	2.130	2.108	2.088	2.070
25	4.242	3.385	2.991	2.759	2.603	2.490	2.405	2.337	2.282	2.236	2.198	2.165	2.136	2.111	2.089	2.069	2.051
26	4.225	3.369	2.975	2.743	2.587	2.474	2.388	2.321	2.265	2.220	2.181	2.148	2.119	2.094	2.072	2.052	2.034
27	4.210	3.354	2.960	2.728	2.572	2.459	2.373	2.305	2.250	2.204	2.166	2.132	2.103	2.078	2.056	2.036	2.018
28	4.196	3.340	2.947	2.714	2.558	2.445	2.359	2.291	2.236	2.190	2.151	2.118	2.089	2.064	2.041	2.021	2.003
29	4.183	3.328	2.934	2.701	2.545	2.432	2.346	2.278	2.223	2.177	2.138	2.104	2.075	2.050	2.027	2.007	1.989
30	4.171	3.316	2.922	2.690	2.534	2.421	2.334	2.266	2.211	2.165	2.126	2.092	2.063	2.037	2.015	1.995	1.976
40	4.085	3.232	2.839	2.606	2.449	2.336	2.249	2.180	2.124	2.077	2.038	2.003	1.974	1.948	1.924	1.904	1.885
60	4.001	3.150	2.758	2.525	2.368	2.254	2.167	2.097	2.040	1.993	1.952	1.917	1.887	1.860	1.836	1.815	1.796
120	3.920	3.072	2.680	2.447	2.290	2.175	2.087	2.016	1.959	1.910	1.869	1.834	1.803	1.775	1.750	1.728	1.709

TABLE 6 THE F DISTRIBUTION TABLE   A11

TABLE 6   (Continued)

Degrees of Freedom Denominator	\multicolumn{16}{c}{Degrees of Freedom - Numerator}															
	18	19	20	21	22	23	24	25	26	27	28	29	30	40	60	120
1	247.324	247.688	248.016	248.307	248.579	248.823	249.052	249.260	249.453	249.631	249.798	249.951	250.096	251.144	252.196	253.254
2	19.440	19.443	19.446	19.448	19.450	19.452	19.454	19.456	19.457	19.459	19.460	19.461	19.463	19.471	19.479	19.487
3	8.675	8.667	8.660	8.654	8.648	8.643	8.638	8.634	8.630	8.626	8.623	8.620	8.617	8.594	8.572	8.549
4	5.821	5.811	5.803	5.795	5.787	5.781	5.774	5.769	5.763	5.759	5.754	5.750	5.746	5.717	5.688	5.658
5	4.579	4.568	4.558	4.549	4.541	4.534	4.527	4.521	4.515	4.510	4.505	4.500	4.496	4.464	4.431	4.398
6	3.896	3.884	3.874	3.865	3.856	3.849	3.841	3.835	3.829	3.823	3.818	3.813	3.808	3.774	3.740	3.705
7	3.467	3.455	3.445	3.435	3.426	3.418	3.410	3.404	3.397	3.391	3.386	3.381	3.376	3.340	3.304	3.267
8	3.173	3.161	3.150	3.140	3.131	3.123	3.115	3.108	3.102	3.095	3.090	3.084	3.079	3.043	3.005	2.967
9	2.960	2.948	2.936	2.926	2.917	2.908	2.900	2.893	2.886	2.880	2.874	2.869	2.864	2.826	2.787	2.748
10	2.798	2.785	2.774	2.764	2.754	2.745	2.737	2.730	2.723	2.716	2.710	2.705	2.700	2.661	2.621	2.580
11	2.671	2.658	2.646	2.636	2.626	2.617	2.609	2.601	2.594	2.588	2.582	2.576	2.570	2.531	2.490	2.448
12	2.568	2.555	2.544	2.533	2.523	2.514	2.505	2.498	2.491	2.484	2.478	2.472	2.466	2.426	2.384	2.341
13	2.484	2.471	2.459	2.448	2.438	2.429	2.420	2.412	2.405	2.398	2.392	2.386	2.380	2.339	2.297	2.252
14	2.413	2.400	2.388	2.377	2.367	2.357	2.349	2.341	2.333	2.326	2.320	2.314	2.308	2.266	2.223	2.178
15	2.353	2.340	2.328	2.316	2.306	2.297	2.288	2.280	2.272	2.265	2.259	2.253	2.247	2.204	2.160	2.114
16	2.302	2.288	2.276	2.264	2.254	2.244	2.235	2.227	2.220	2.212	2.206	2.200	2.194	2.151	2.106	2.059
17	2.257	2.243	2.230	2.219	2.208	2.199	2.190	2.181	2.174	2.167	2.160	2.154	2.148	2.104	2.058	2.011
18	2.217	2.203	2.191	2.179	2.168	2.159	2.150	2.141	2.134	2.126	2.119	2.113	2.107	2.063	2.017	1.968
19	2.182	2.168	2.155	2.144	2.133	2.123	2.114	2.106	2.098	2.090	2.084	2.077	2.071	2.026	1.980	1.930
20	2.151	2.137	2.124	2.112	2.102	2.092	2.082	2.074	2.066	2.059	2.052	2.045	2.039	1.994	1.946	1.896
21	2.123	2.109	2.096	2.084	2.073	2.063	2.054	2.045	2.037	2.030	2.023	2.016	2.010	1.965	1.916	1.866
22	2.098	2.084	2.071	2.059	2.048	2.038	2.028	2.020	2.012	2.004	1.997	1.990	1.984	1.938	1.889	1.838
23	2.075	2.061	2.048	2.036	2.025	2.014	2.005	1.996	1.988	1.981	1.973	1.967	1.961	1.914	1.865	1.813
24	2.054	2.040	2.027	2.015	2.003	1.993	1.984	1.975	1.967	1.959	1.952	1.945	1.939	1.892	1.842	1.790
25	2.035	2.021	2.007	1.995	1.984	1.974	1.964	1.955	1.947	1.939	1.932	1.926	1.919	1.872	1.822	1.768
26	2.018	2.003	1.990	1.978	1.966	1.956	1.946	1.938	1.929	1.921	1.914	1.907	1.901	1.853	1.803	1.749
27	2.002	1.987	1.974	1.961	1.950	1.940	1.930	1.921	1.913	1.905	1.898	1.891	1.884	1.836	1.785	1.731
28	1.987	1.972	1.959	1.946	1.935	1.924	1.915	1.906	1.897	1.889	1.882	1.875	1.869	1.820	1.769	1.714
29	1.973	1.958	1.945	1.932	1.921	1.910	1.901	1.891	1.883	1.875	1.868	1.861	1.854	1.806	1.754	1.698
30	1.960	1.945	1.932	1.919	1.908	1.897	1.887	1.878	1.870	1.862	1.854	1.847	1.841	1.792	1.740	1.683
40	1.868	1.853	1.839	1.826	1.814	1.803	1.793	1.783	1.775	1.766	1.759	1.751	1.744	1.693	1.637	1.577
60	1.778	1.763	1.748	1.735	1.722	1.711	1.700	1.690	1.681	1.672	1.664	1.656	1.649	1.594	1.534	1.467
120	1.690	1.674	1.659	1.645	1.632	1.620	1.608	1.598	1.588	1.579	1.570	1.562	1.554	1.495	1.429	1.352

TABLE 7   Critical Values of the Wilcoxon Rank Sum Test

One-tail $\alpha = 0.025$;   Two-tail $\alpha = 0.05$

n_2	n_1	T_L	T_U	T_L	T_U	T_L	T_U	T_L	T_U	T_L	T_U	T_L	T_U	T_L	T_U	T_L	T_U
		3		4		5		6		7		8		9		10	
4		6	18	11	25												
5		6	21	12	28	18	37										
6		7	23	12	32	19	41	26	52								
7		7	26	13	35	20	45	28	56	37	68						
8		8	28	14	38	21	49	29	61	39	73	49	87				
9		8	31	15	41	22	53	31	65	41	78	51	93	63	108		
10		9	33	16	44	24	56	32	70	43	83	54	98	66	114	79	131

One-tail $\alpha = 0.05$;   Two-tail $\alpha = 0.10$

n_2	n_1	T_L	T_U	T_L	T_U	T_L	T_U	T_L	T_U	T_L	T_U	T_L	T_U	T_L	T_U	T_L	T_U
		3		4		5		6		7		8		9		10	
3		6	15														
4		7	17	12	24												
5		7	20	13	27	19	36										
6		8	22	14	30	20	40	28	50								
7		9	24	15	33	22	43	30	54	39	66						
8		9	27	16	36	24	46	32	58	41	71	52	84				
9		10	29	17	39	25	50	33	63	43	76	54	90	66	105		
10		11	31	18	42	26	54	35	67	46	80	57	95	69	111	83	127

APPENDIX *B*

AN INTRODUCTION TO MICROSOFT EXCEL

This appendix will introduce you to the basics of Microsoft Excel 97, if you are not familiar with the program. If you are already familiar with Excel, you can simply refer to the sections on using Excel for statistics at the end of each chapter.

B.1 STARTING MICROSOFT EXCEL

You can start Microsoft Excel 97 in several different ways:

- from the **Start > Programs** menu
- from the Microsoft Office shortcut bar

To open Excel from the Program menu, click on **Start > Programs** and then on the icon for Excel. Excel is usually located in a folder for MS Office 97 as shown in Figure B.1. If there is not a folder for Office 97, then Excel might be a separate icon on the program menu.

FIGURE B.1 Starting Excel from the Start button

If your computer has the Microsoft Office shortcut bar running, you will see an icon for Excel on the toolbar, as shown in Figure B.2. Clicking on the icon will launch Excel.

FIGURE B.2 The Microsoft Office toolbar

B.2   THE EXCEL WORKBOOK

Excel starts by opening a new workbook. A workbook in Excel is a collection of worksheets. Each worksheet is a grid with 256 columns and 65,536 rows. An Excel workbook, the menus, and toolbars are shown in Figure B.3. We will now look at the different parts of the workbook and talk about their functions.

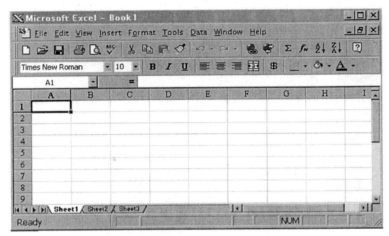

FIGURE B.3   The Excel workbook

The Worksheet

The worksheet portion of the Excel workbook is where you will store your data and do all of your calculations. A worksheet consists of 256 columns labeled with letters and 65,536 rows, labeled with numbers. An individual location in a worksheet is called a **cell.** Each cell in a worksheet is identified by its **location.** Locations are described in terms of the column letter followed by the row number. In Figure B.4, you see a typical worksheet and its formula bar. The important parts of the worksheet are labeled.

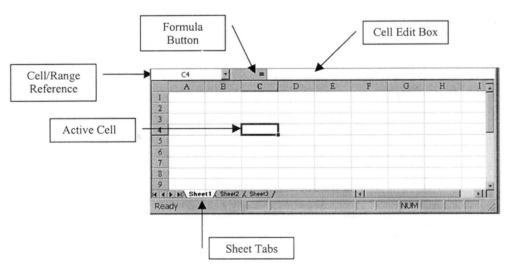

FIGURE B.4   The Excel worksheet

In this worksheet, cell C4 is the **active cell.** A cell is made active by either clicking on it or by moving to the cell using the directional keys. The active cell will have a darker border around it. The location of the active cell is shown in the Cell/Range Reference and the contents of the active cell appear in the Cell Edit box.

A workbook consists of a collection of worksheets. You can move from sheet to sheet, by clicking on the **sheet tabs** at the bottom of the workbook. It is a good idea to give different names to the worksheets to keep your worksheet organized and so that you don't waste time looking for something important. To change the name of a worksheet, double-click on the sheet tab and type in the new name.

The Excel Menu and Toolbars

The other portion of the workbook that you see when you use Excel consists of the Excel menu bar and some toolbars as shown in Figure B.5. You can specify exactly which toolbars Excel shows, but the ones shown by default are the standard and formatting toolbars.

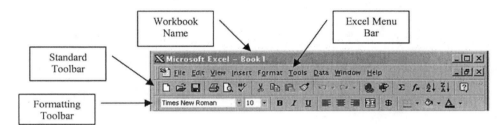

FIGURE B.5   Excel menus and toolbars

The **Standard Toolbar** contains icons for often-used tasks from the Excel menu bar. Some of these tasks are **New, Open, Print, Print Preview,** and **Save** from the File menu, **Copy, Cut,** and **Paste** from the Edit menu, and **Sort** from the Data menu. The **Formatting Toolbar** includes icons for changing the appearance of entries in the worksheet.

B.3   FILE OPERATIONS IN EXCEL

When you start Excel a new workbook automatically opens. After you have been working in Excel awhile you might want to save your work and to open your own workbooks the next time you start Excel.

Opening an Existing Workbook in Excel

There are two basic ways to open workbook files in Excel. One way is simply to double-click on the name of the file in Windows Explorer or My Computer. Excel files have the file extension "**.xls**" after the name of the file. When you open the file this way, Excel will start automatically with the file as the open workbook.

If you have already started Excel, then you can open a file using the **File** menu. Click on the **File** in the menu bar and the menu shown in Figure B.6 (page B4) will appear.

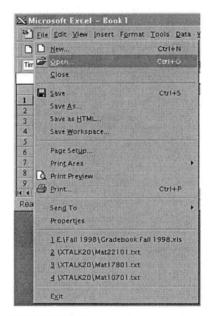

FIGURE B.6 The File menu in Excel

Choose **Open** and the dialog box shown in Figure B.7 opens.

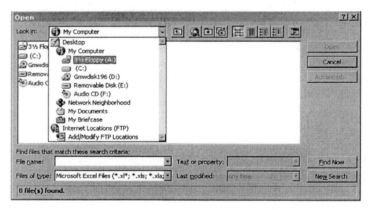

FIGURE B.7 The Open File dialog box

Clicking on the arrow next to the **Look in:** box will bring up a list of possible locations for your file. Most of the time your files will be on a floppy disk or on one of the hard drives of the computer. If the file is located on a floppy disk, choose the $3\frac{1}{2}$ Floppy option. If it is located on some other drive such as your hard drive or a network drive, you should choose the appropriate disk letter.

After you choose the drive that contains your file, a list of all Excel files on that drive will appear in the dialog box, as shown in Figure B.8. To open the file, click on the name of the file you want and choose **Open** or simply double-click on the file name.

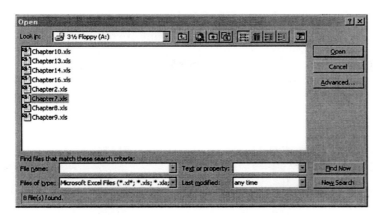

FIGURE B.8 List of Excel files on the A: drive

Saving a File in Excel

After you have done some work in Excel you will want to save your file, perhaps to work on at a later time or to hand in to your instructor. The process of saving a file in Excel is very similar to that of opening a file. Choose **File > Save** from the menu bar. If the workbook is a new one *that has not been saved before,* the **Save As** dialog box will open as shown in Figure B.9.

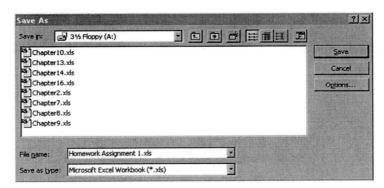

FIGURE B.9 The Save As dialog box

Use the **Save in:** list to choose the location where you want to save the file. In the **File name:** box, type in a name for the file that will help you remember what the file contains. Cute names, however funny, do not help you when you have ten files saved on the same disk! Click **Save** to save the file to the location you selected.

No computer or computer network is foolproof. Even if you are not finished with your work, you will want to save your file regularly. If you do not, and the system crashes, you will lose any work you have done. Once the file has been saved for the first time, selecting **File > Save** will automatically save the file to its previous location.

This is only one way to save a file. You can also click on the floppy disk icon in the Standard Toolbar, or hit Ctrl+S.

B.4 PRINTING IN EXCEL

You will most likely want to print out some of your work in Excel. This can be done in several different ways.

To start printing, you can click on the printer icon in the **Standard Toolbar** or select **File > Print.** This will print your current worksheet using the current printing setup. You might not like the way your output appears if you do this. You can customize the way your printout appears by using the **Print Preview** option.

To use **Print Preview** click on the icon in the **Standard Toolbar** or choose **File >
Print Preview.** When you select Print Preview your current worksheet will be shown
exactly as it will appear on the printed page as shown in Figure B.10.

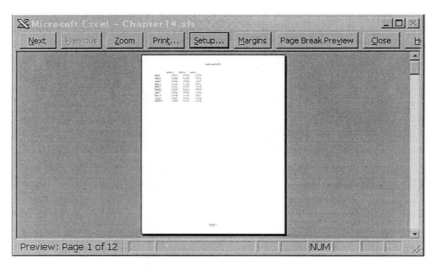

FIGURE B.10   The Print Preview dialog box

If you like the way it looks you can simply click **Print. . .** and it will be printed.
If you do not like the way it looks you can change several aspects of the printout by
selecting **Setup . . . .** The **Page Setup** dialog box opens. The dialog box has four
tabs: **Page, Margins, Header/Footer,** and **Sheet.** Each of these lets you control a different aspect of the printout.

The **Page** tab shown in Figure B.11 allows you to choose the page orientation,
portrait or landscape. Your choice will depend on how the data are located in your
spreadsheet. If you use a lot of columns and not many rows, you will want Landscape.
If you have many rows and not so many columns you will want Portrait.

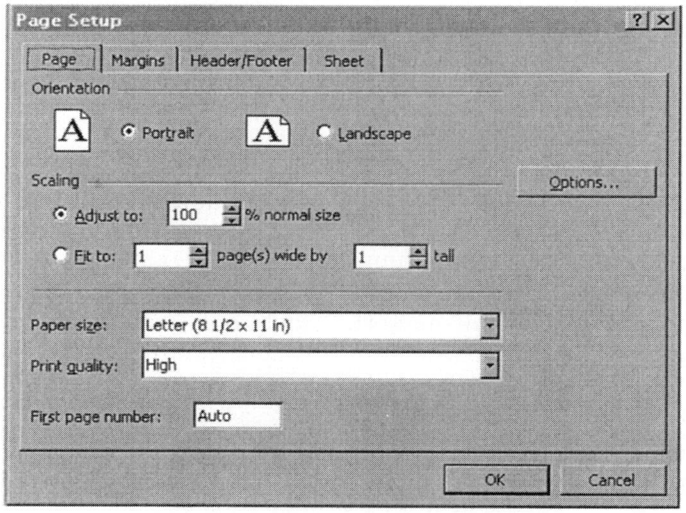

FIGURE B.11   The Page tab

The other tabs in the Page Setup dialog box let you adjust other aspects about
how your selection will print. You can adjust the margins, add headers and footers
(or delete the default headers and footers that Excel uses), and choose whether or
not you want the gridlines on the sheet to print.

B.5 OTHER USEFUL TIPS FOR EXCEL

Selecting a Range

Many activities in Excel require you to select a range of cells in the worksheet. A range of cells might be in a single column or in several columns. There are several ways to select a range:

1. Make the first cell in the range active by moving the cursor to that cell. Press the left mouse button and drag the cursor to highlight the remaining cells in the range.
2. Click the first cell in the range, and then hold down SHIFT and click the last cell in the range. You can scroll to make the last cell visible.
3. To select non-adjacent ranges, select the first range and then hold down the CTRL button while you select the other range(s).

Copying, Cutting, and Pasting

Some of the actions that you will perform most often in Excel are copying, cutting, and pasting. These commands are found in the **Edit** menu and allow you to manage the data in your worksheets. To copy a range of cells from one location in a workbook to another, you must first select the range of cells as just described. Then, from the **Edit** menu, select **Copy,** or click on the **Copy** icon on the toolbar. Move the cursor to the first cell where you want to place the copy and then select **Paste** from the edit menu or click on the **Paste** icon on the toolbar.

If you want to move a range of cells from one location to another, after selecting the range of cells, choose **Cut** from the **Edit** menu or click on the **Cut** icon on the toolbar. Then move the cursor to the destination cells and select **Paste.**

Right Clicking

Clicking the right mouse button in Excel is a useful action. When you click the right mouse button, a pop-up context-sensitive menu appears. The commands on this menu are the ones that most likely apply to the action you are performing.

Switching Among Workbooks

In Excel it is possible to keep more than one workbook open at a time. This is useful if you need to copy information or sheets from a workbook to another workbook. To switch among open workbooks in Excel, select **Window** from the menu bar. At the bottom of the menu is a list of all open workbooks, as shown in Figure B.12.

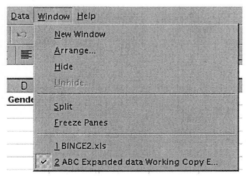

FIGURE B.12   The Window menu

To select a different workbook, click on the name of the workbook you want to use.

INDEX